International Conference on Electronic Circuits and Signaling Technologies

Journal of Physics: Conference Series Volume 2325

Online
2-3 June 2022

ISBN: 978-1-7138-6045-7
ISSN: 1742-6588

Printed from e-media with permission by:

Curran Associates, Inc.
57 Morehouse Lane
Red Hook, NY 12571

CURRAN ASSOCIATES INC. proceedings.com

Some format issues inherent in the e-media version may also appear in this print version.

Institute of Physics
Dirac House, Temple Back
Bristol BS1 6BE UK

Phone: 44 1 17 929 7481
Fax: 44 1 17 920 0979

techtracking@iop.org

Additional copies of this publication are available from:

Curran Associates, Inc.
57 Morehouse Lane
Red Hook, NY 12571 USA
Phone: 845-758-0400
Fax: 845-758-2633
Email: curran@proceedings.com
Web: www.proceedings.com

TABLE OF CONTENTS

Journal of Physics: Conference Series **2325** (2022) 011001 doi:10.1088/1742-6596/2325/1/011001

Preface

We are very pleased to introduce the Proceedings of the International Conference on Electronic Circuits and Signalling Technologies (ICECST 2022). The ICECST conference event was held at PPG Institute of Technology, from 2nd to 3rd June, 2022.

One of the significant and valuable dimensions of this first conference edition is the way to gather researchers, academicians, and engineers in the field from different countries to one place and initiates discussion on the relevant issues, challenges, opportunities, and research findings. The primary focus of ICECST 2022 is to provide an excellent platform for the conference participants to share and exchange novel and innovative ideas of original research, and to build international association. The main intend of this conference is to create a novel research landscape for the areas of electronic circuits and signalling technologies.

As more number of submissions are received from different parts of the world, Only 88 submissions were accepted as full papers for publication and presentation in ICECST 2022. These papers provide brief illustration for current research on relevant topics, covering the advanced characterization techniques in the highly complex field of signal processing and information computing and communications.

The success of the conference is due to the collective efforts of all the reviewers and advisory/review board members. We would like to express and record our gratitude and appreciation to the authors for their contributions. Many thanks to the reviewers, who helped us to maintain high quality manuscripts included in the proceedings. We also express our sincere thanks to the members of the conference technical program committee and review committee and organizing team for their continuous effort towards the success of the conference event. We wish that all the authors and delegates find ICECST 2022 proceedings interesting, exciting and inspiring.

-

Conference Chair

Dr. V. Bindhu
ICECST – 2022

Journal of Physics: Conference Series **2325** (2022) 011001 doi:10.1088/1742-6596/2325/1/011001

List of Committee Members

Chief Patron

Dr. R. Prakasam,

Principal,

PPG Institute of Technology,

Coimbatore, India.

Conference Chair

Dr. V. Bindhu,

Professor and Head, ECE,

PPG Institute of Technology,

Coimbatore, India.

Technical Program Chairs

Dr. Joy Iong-Zong Chen, Department of EEE, Dayeh University, Taiwan.

Dr. João Manuel R. S. Tavares, Professor, Faculdade de Engenharia, Universidade do Porto, Portugal.

Dr. Ram Palanisamy, Professor, Department of Business Administration, The Gerald Schwartz School of Business, StFX University, Canada.

Dr. Anshul Agarwal, Associate Professor, Electrical Engineering, National Institute of Technology, Delhi, India.

Dr. Wang Haoxiang, Go Perception Laboratory, Cornell University, USA.

Dr. Subarna Shakya, Pulchowk Campus, Institute of Engineering, Tribhuvan University, Nepal.

Technical Program Committee

Dr. R. Dhaya, Professor, Dept. of Computer Science and Engineering, King Khalid University, Kingdom of Saudi Arabia.

Dr. Hui Ming Wee, Dept. of Industrial & Systems Engineering, Chung Yuan Christian University, Taiwan.

Dr. Pavel Lafata, Faculty of Electrical Engineering, Czech Technical University, Prague.

Journal of Physics: Conference Series **2325** (2022) 011001 doi:10.1088/1742-6596/2325/1/011001

Journal of Physics: Conference Series **2325** (2022) 011001 doi:10.1088/1742-6596/2325/1/011001

Dr. Prajoy Podder, Ranada Prasad Shaha University, Bangladesh.

Dr. Nayana Shetty, NMAMIT, Nitte, Karnataka, India.

Peer Review Statement

All papers published in this volume have been reviewed through processes administered by the Editors. Reviews were conducted by expert referees to the professional and scientific standards expected of a proceedings journal published by IOP Publishing Publishing.

- **Type of peer review:** Double Anonymous
- **Conference submission management system:** Conference own submission management system
- **Number of submissions received:** 302
- **Number of submissions sent for review:** 295
- **Number of submissions accepted:** 88
- **Acceptance Rate (Submissions Accepted / Submissions Received × 100):** 29.1
- **Average number of reviews per paper:** 2
- **Total number of reviewers involved:** 150
- **Contact person for queries:**
  Name: Dr. V. Bindhu
  Email: vbindhu@ieee.org
  Affiliation: Professor, PPG Institute of Technology

Automatic Headlight Intensity Control using Light Dependent Resistor

Ambaji S. Jadhav[1], Vikram Joshi[2], Rashmi V. Pawar[1]

[1] Assistant Professor, Department of Electrical and Electronics Engineering, BLDEA's V.P. Dr. P.G. Halakatti College of Engineering & Technology, Vijayapur-586103, Karnataka, India

[2] B.E. Student 8th Sem., Department of Electrical and Electronics Engineering, BLDEA's V.P. Dr. P.G. Halakatti College of Engineering & Technology, Vijayapur-586103, Karnataka, India

Email: asjadhavec@gmail.com, vikramjoshi887@gmail.com, rashmi.ajadhav@gmail.com

Abstract: For so many years, road accidents have been a huge problem for automobile industries since they must continuously upgrade their vehicles to withstand any type of damages. In this regard, many technologies are invented to reduce the figure of accidents that happen. One of the great reasons for road accident is due to bright headlight of vehicle coming in opposite direction. High beam from the headlight causes a dangerous situation for night driving. It causes temporary blindness for the drivers that may lead to collisions or may hit people passing through the road and lives can be lost. According to the law of "Indian Roads and Transport Council", low intensity light has to be used always, however in the time of overtaking or other mandatory situations, high beam can be turned on. But due to the driver's ignorance and since they cannot see the long-range road in low intensity, they always drive with the high intensity light on. Automatic dipper, a circuit which controls the intensity of the head light of vehicle has been designed here. From this automatic dipper system, the intensity of headlight of the vehicle can be automatically controlled and can provide a better visible road using optimization techniques for light condition.

Keywords: headlight, relay, transistor, LDR (Light Dependent Resistor), variable resistor

1. Introduction

Nowadays, the number of vehicles and road transportation system are increasing rapidly. Due to this, the number of road accidents also gradually rise which has been the motivation behind this project. The "ministry of road transport and highways transport research wing" reported a project called "road accidents in India (2018)" which says road accidents in India kills 1.5lakh people annually, in which 52.02% of accidents are related to the collisions by vehicle to vehicle and more than 30% of accidents occur due to the headlight glare at night. A survey tells that 26.5% of people use dipper properly, 25.53% use it for sometime then avoid using and the remaining 48.3% drivers use high beam continuously.

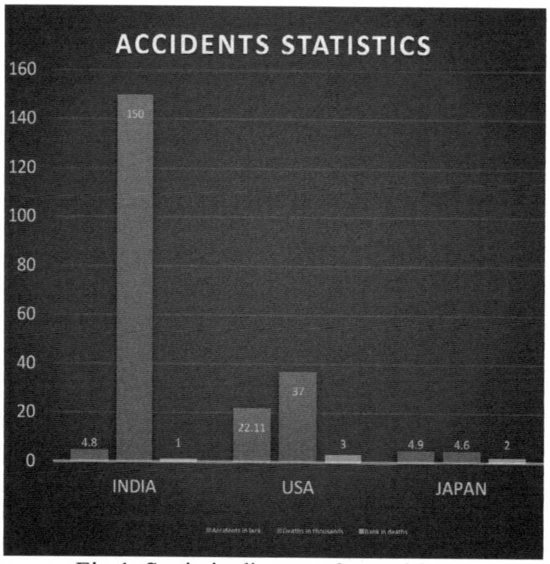

Fig.1. Statistic diagram for accidents

According to a report, in 2018: In India a total of 4.8lakh accidents have happened in which 150thousand people have died; In USA, 22.11lakh accidents have happened in which 37thousand people have died; In Japan, 4.9lakh accidents have occurred and 4.6thousand people lost their lives. Among these, India stands first in the deaths occurred due to accidents although the number of accidents is less compared to other countries, which clearly shows that the technology of the cars in India and other factors are not up to the mark.

A Swiss physician and philosopher Ignaz Paul Vital Troxler (1780-1866) proposed a theory called Troxler effect which states that, "the involuntary movements of the eye which occur even while gazing gets settled slowly i.e., if the eye could perfectly fixate on some point in our visual field by suppressing saccadic movement, a static scene would slowly fade from view after a few seconds due to the local neural adaptation of the rods, cones, and ganglion cells in the retina. In brief, any constant light stimulus will cause an individual neuron to become desensitized to that stimulus, and hence reduce the strength of its signal to the brain". Therefore, when the high beam of light directly falls over the driver, affects his sight and his concentration on the road is hindered which may lead to a crash.

Over a year, several experiments are done on vehicles in various methods to reduce the cause of accidents but failed due to some drawbacks like temperature, economic and other issues. However, from this proposed idea, those drawbacks can be compensated and can be employed in vehicles. As this system does not have any costly equipment like microprocessor or microcontroller, it's economical and can be applied to the existing and upcoming vehicles easily. Especially in a country like India, that has a greater number of lower class and middle class people, and the total number of vehicles in 2019 to be 295.8 million, there is a necessity to come up with an idea which is cost effective, easy to manufacture and easily employed to the vehicle.

2. Related Work

Paper [1] discussed various methods involving automatic headlight dipper using LDR, Microcontroller, Microprocessor etc. that were used to control the headlight intensity when the headlight of opposite vehicle falls on the light detecting device. In paper [2], some possible way to

solve the problem namely, LDR based intensity control, Fuzzy logic based intensity control, wireless sensor network, IR transmitter and receiver and camera based intensity control system .
In LDR based intensity control, a BJT or MOSFET with relay was used to detect the light. Paper [3] attempted to solve the problem using LDR, Microcontroller (arduino), and PIR sensor. The light sources here can be from the opposite vehicle, streetlights, lights from nearby shops with intensity greater than 1 candela and at a distance of 50 meters.

Paper [4] explained automatic headlight intensity control method using photo transistor, ARM controller and XBee to solve the problem. And it depicted that 33 percentage of accidents occur during the night instead of day and also tend to have higher percentage of death. As per survey done by researchers, the possible risk related to road accidents is double at night compared to daytime. In paper [5] In this paper, own circuit board using microcontroller was designed. The experimental results of this work showed that, the headlight of the vehicle deflects when the coming car light at a 230m distance. LDR was used to detect the light and transistor to amplify the signal. Paper [6] proposed a method using LDR, microcontroller, Relay, Zigbee communication to control the head light. And they have estimated that, the information that drivers gather visually is 90% and whatever the actual percentage may be, the importance of the visual system to driving cannot be doubted. Paper [7] proposed their technique using 2 LDRs, Microcontroller and Relay. LDR was used in the front of the vehicle and another LDR on roof panel, and they explained that Light Dependent Resistors are very useful especially in light/dark sensor circuits. Normally the resistance of an LDR is very high, sometimes as high as 1M ohms, but when they are illuminated with light resistance, it drops drastically.

The authors of paper [8] explained their work using LDR, PIC microcontroller and Relay, and they said that, the cause of temporary blindness is called as Troxler effect. It states that this kind of temporary blindness is otherwise known as the fading effect. The study showed that if human eyes are exposed to a very bright light source of around 10,000 lumens, a glare or temporary blindness is experienced. This glare is produced due to over exposure of the rods and cones inside our eye. Even after the source of glare is removed, an after-image remains in our eye that creates a blind spot. Paper [9] explained a technique using LDR, Transistor and Relay. This system was designed, constructed, tested and its working at a distance of 147m of the coming vehicle. In paper [10] the authors have explained a project using LDR, Doppler Radar Sensor, OFS, VIP (Video Image Processor), Ultrasonic Sensors and PIR (Passive Infrared), ILD (Inductive Loop Detectors), Magnetometers, Arduino UNO (R3) and Boar. They explained that the components proposed are also very cheap compared to the existing components in the luxury cars like BMW, Mercedes, Audi etc. Hence this can be adapted in the economy sector cars.

3. Proposed system

Initially when the headlight of the vehicle is kept in high beam on road and if another vehicle with high beam from the opposite direction hits on the light sensitive material called Light Dependent Resistor (LDR) of the current vehicle, then the LDR transmits low voltage current which cannot excite the relay. The output of LDR if fed to the transistor which is connected to the battery, which amplifies the signal supplied to the relay. 5 pin relay counter changes the current supply between two circuits. In the vehicles, the headlight system has two choices between the intensities, i.e., the relay will fluctuate the current between the high beam and low beam. When the relay is excited due to the transistor, the relay connected to high beam light at first stage changes to low beam in second stage. The intensity at which the headlight fluctuates that is detected by the LDR is controlled by the variable resistor, connected to the transistor which controls the flow of current. When the vehicles coming in opposite direction passes by, there will be no excitation to the relay and hence it suddenly changes to first stage giving low intensity light.

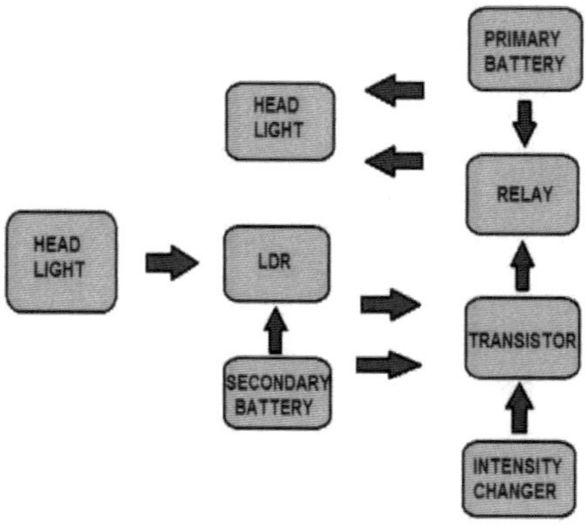

Fig.2. System architecture

a) LDR (Light Dependent Resistor):

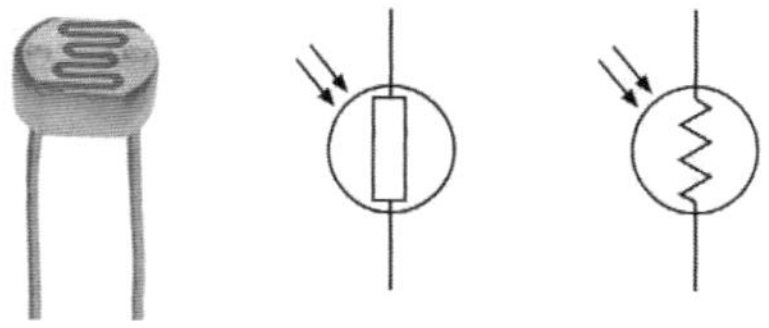

Fig.3. LDR and its symbol

LDR is a light sensitive device and it works on the principal of photo conductivity which changes its resistance value as light source falls on it. As the intensity of the light falling on it increases, the conductivity of the LDR also increases. LDR exhibits analog output. There exists a secondary option instead of LDR which is a photodiode. Both are low cost and sensitive to light, however considering the case of headlight, LDR is preferred since it is suited for rugged environment, and they are also used in streetlights. Photodiodes are used in precision equipment.

b) RELAY:

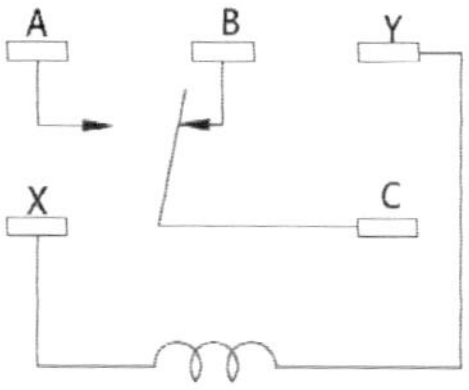

Fig.4. 5pin relay

Relay is the electrically operated switching device. As 5pin relay is used in this work, the relay consists of two circuits between which it changes the current flow and is optimized using optimization techniques [11, 12].

4. Results and Discussion

As shown in fig.4, X and Y are the terminals of electromagnetic coil. Terminal C is having metal filament which completes the circuit between the terminal A and B. In first stage, the current is flowing between the terminal C and B where the high beam light is connected. When the terminals X and Y are excited using electricity, the metal filament in C changes its direction so that the connecting between the C and B is broken and connection between the terminals C and A are made, Hence the low beam light connected to it starts glowing. As relay does not require high current and cables, it reduces the cost and weight. Relays are placed anywhere in the vehicle, and they can also be used in automobiles, motors etc.

Variable resistor:

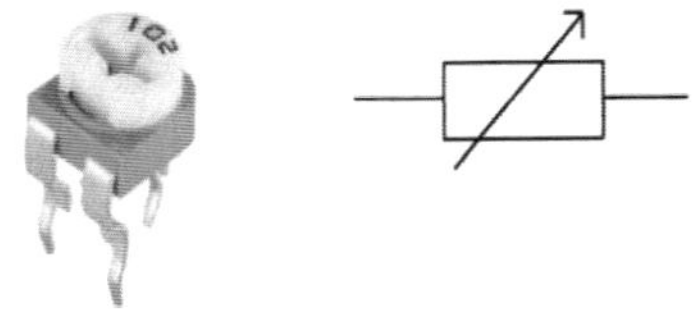

Fig.5. Variable resistor and its diagram

Variable resistor or potentiometer is a device used to vary the resistance value in the circuit depending on the position of the knob. It is commonly used in amplifiers, tuning circuits etc.

Transistor: Transistor is the basic electronic device which can be used for amplification and switching operations.

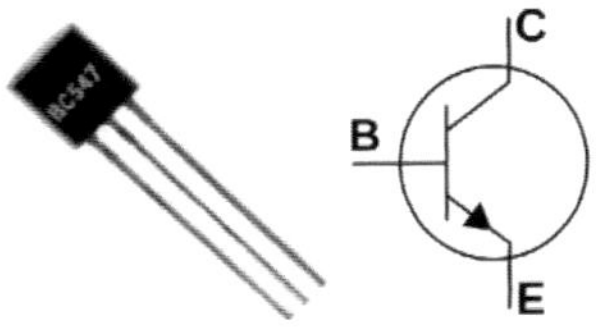

Fig.6. Transistor and its diagram

Transistor is a semiconductor device used as a switching device in electronic signals, in amplifying circuits and electrical power circuits. It is also used to control the output, however in this work it is used as an amplifier.

Every possible solution from the related work do involve the Microcontroller and other high end equipments making it complex, costly and difficult to equip in automobiles. As this proposed system solves the problem in simple way as possible it is cheaper and simple to equip and can be built with basic equipment. Moreover, it can withstand difficult conditions in automobiles.

5. Conclusion

The proposed system is simple to design and implement. As the size is compact and durable, it can be easily installed in almost all vehicles. The performance of the proposed system is good when compared with many other existing techniques.

References

[1] Lakshmi K, Nevetha R, Ilakkiya S N, Ganesan R., "Automatic Vehicle Headlight Management System to Prevent Accidents Due To Headlight Glare", *International Journal of Innovative Technology and Exploring Engineering (IJITEE)*, vol.8, no.9, pp.757-760, July 2019

[2] Devashree Chilla, Manasi Joshi , Sanjyot Kajale , Seema Deoghare, "Headlight Intensity Control Methods – A Review" , *International Journal of Innovative Research in Computer and Communication Engineering*, vol.4, no.2, pp.1140-1146, February 2016

[3] C P Sourav, Karthika K A, Athulya Saju3, Gayathri T. "Automatic Headlight Intensity Controller", *International Journal for Research in Applied Science & Engineering Technology (IJRASET)* , vol. 7, no. 5, pp.1592-1595, May 2019

[4] Karthik M.M., "Automated headlight intensity control and obstacle alerting system", *International Research Journal of Engineering and Technology (IRJET)*, vol.3, no. 6, pp.2553-2555, June-2016

[5] Mohammed Alsumady and Shadi. A. Alboon. "Intelligent Automatic High Beam Light Controller", vol.10, no. 11, *Journal of Active and Passive Electronic Devices*. pp. 1-8, June 2013

[6] Arpita K, Akhila M Jain, Avi Kumar R. "Automated Headlight Intensity Control and Obstacle Alerting System", *International Journal of Engineering Research & Technology (IJERT)*. Vol.6, no. 13, pp. 1-6, May 2018

[7] Mr. Sandip S. Jadhav, Prof. Ansar A. Mulla. "A Multi Featured Automatic Head Light Systems Prototype for Automotive Safety", *International Journal of Engineering Research and Technology*, vol.8, no.5, pp. 777-781, Number 2017

[8] Aslam Musthafa R, Bala Krishnan T, Seetha Raman N, Shankar M, Asst.prof Ms. Swathi R. "Automatic Headlight Beam Controller", *International Journal of Trend in Research and Development (IJTRD)*. pp. 15-18, Mar 2017

[9] O. Akinsanmi, A.D. Ganjang, H. U. Ezea. "Design and Development of an Automatic Automobile Headlight Switching System", *International Journal of Engineering and Applied Sciences (IJEAS)*, vol. 2, no.8, pp. 107-112, August 2015.

[10] R.D. Balaji. "A Case study on Automatic Smart Headlight System for Accident Avoidance", *International Journal of computer communication and informatics*, vol. 2, no. 2, pp. 70-77, March 2020

[11] A. S. Jadhav and Pushpa B. Patil and Dr. Sunil Biradar "Optimal Feature Selection-based Diabetic Retinopathy Detection using Improved Rider Optimization Algorithm Enabled with Deep Learning," *Evolutionary Intelligence,* vol. 14, no.4, pp. 1431-1448 April 2020.

[12] A. S. Jadhav, Pushpa B. Patil and Dr. Sunil Biradar, "Analysis of Diagnosing Diabetic Retinopathy by segmenting Blood vessels, Optic disc and retinal abnormalities" *Journal of Medical engineering and Technology*, vol.44, no.6, pp. 299-316, July 2020.

Advanced Foot step power generation system for mobile charging

K. Sai Bhargav [1], L. Guru Kiran Reddy [1], Patel Rushik P. B [1], T. Mounish Reddy [1], Bagade Shilpa [1, 2]

[1] Department of ECE, GRIET, Hyderabad, Telangana, India.
[2] Assistant Professor, Department of ECE, GRIET, Hyderabad, Telangana, India.

gurukiran0100@gmail.com, saibhargavkatta@gmail.com, rushikpatel5959@gmail.com,

Abstract: The major focus of this study is the generation of electric power from people's footsteps and the pressure applied when walking. "Advanced Foot Step Power Generation System" refers to the mechanical power transformation into electrical power as a result of the pressure generated by the footstep and the use of transducers. The power-producing floor generates power, which is essentially the conversion of kinetic energy into electrical energy. Today's electricity demand is growing, and existing power generation sources are unable to meet this worldwide challenge. The fundamental issue of the energy crisis is the demand and supply imbalance. Although it will not be enough to meet the world's overwhelming need for electrical energy, it will be able to transform and reduce dependency on traditional methods of generating electricity. It can be implemented on roadside footpaths, parks and jogging tracks, as well as many other public places, such as airports, and will have a significant impact on electrical power generation.

Keywords: Piezoelectric sensor, non- conventional renewable energy and pressure.

1. INTRODUCTION

The demand for energy has been expanding at an alarming rate in recent years, and the availability of energy supplies has decreased. To fulfil the ever- increasing needs for sustainable development, more efficient, pollution-free, and renewable energy supplies must be developed. The depletion of fossil fuels has had an impact on global economics.

Electrical energy has become a basic necessity in our day-to-day activities. The advent of the industrial revolution has drastically changed the way of life. In the past few decades, power consumption has drastically increased as electricity has been used in new industries. Conventional sources of energy include power generation from coal, petroleum, LPG, etc which involves emission of green gasses.

Furthermore, the use of fossil fuels as a source of electrical energy has contributed to a serious pollution problem in the environment. As a result, an alternative technique of generating electricity must be implemented.

[1] The usage of the "Energy Harvesting" technology, which is utilised to take any form of clean, free, and renewable energy created in the environment, could be a feasible solution to these challenges. Usually squandered and transformed into electrical energy for use in low-power applications.

[2] Energy can be collected from a variety of sources, including heat, movement, vibration, and mechanical energy. Among the various energies available in our environment, the provided research recommends energy harvesting based on the collection of energy from mechanical sources of vibration since it is a convenient choice that is independent of the amount of sunshine or other variables. Energy is collected via piezoelectric transducers, which are responsible for converting mechanical energy to electrical energy.

1.1. Assistive Technology

Electronics sensors which can detect pressure and generate electricity from the pressure are being used for piezoelectric power generation. An assistive solution is needed to be used for a reliable source of energy which can decrease emission of greenhouse gasses. It's been a long time coming, but researchers have been working on this kind of solution for a long time now. In this section, only a few of these devices have been discussed in greater detail

1.2 Issues and Challenges

Energy generation from footprints can be subjected to the frequency and the longevity of the footprints. Usage of supercapacitors in reusable batteries can play a vital role in handling fluctuations in voltage generated by the piezoelectric tiles.

Environmental Challenges

A key issue arises in production of rechargeable batteries and piezoelectric tiles as it involves mining of rare-earth materials like lithium, cobalt, and many more which eventually lead to environmental imbalance.

Technological Challenges

Longevity of piezoelectric tiles in places with high crowd movement is a challenge on which research is ongoing. Technology involving electricity transmission, rechargeable batteries and supercapacitors are still in the testing stage and will be improved over time.

2. ARCHITECTURE

BLOCK DIAGRAM:

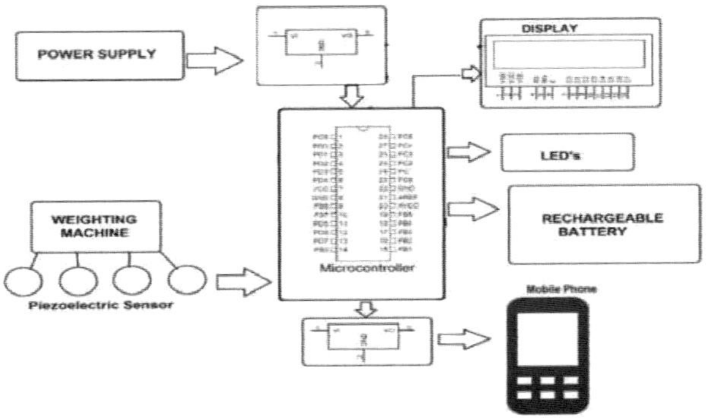

Figure.1: Block diagram

- Footstep board consists of a 4 piezoelectric sensors which are connected in parallel.
- When the pressure is applied on the sensors, the sensors will convert mechanical energy into electrical energy.
- This electrical energy will be stored in the 12v rechargeable battery connected to an inverter.
- This inverter is used to convert the 12 volt DC to the 230 volt AC is used to activate the loads.

Objectives:

- To convert Kinetic Energy into Electrical Energy
- To use the renewable source to deal with energy crisis
- To overcome the pollution in environment

3. LITERATURE SURVEY

It was described by (Anirudh chavan, 2017), [1], that work is all about optimizing and identifying the piezoelectric effect, his work also stated to design a ceramic tile with maximum piezoelectric effect. Piezoelectric material produces energy when one steps a foot on the ceramic tile which in turn produces pressure that acts a source of vibration, triggering whole process. Density fracture strength test, resistive material and shrinking measurement are the characteristics of the prototype (ceramic tile).

(N.Subba Lakshmi, 2017), [2], had presented that, in order to meet the power demand of various practical systems lot of technological advancements were done which leads to numerous ways to harvest power. In order to generate electricity through vibrations and pressure, piezoelectric transducers were used by the author. Human foot pressor, using piezoelectric tranducer, at the input acted as a source for generating power.

(Mitsuhiro Okayasu a, 2016), [3].has discussed implementation of the street light control system which combines various technologies, a statistics of traffic flow magnitude, IR sensors are used to detect the movement & density of vehicles on the road, and also usage of LDR sensor can control the switch to ON & OFF the street light can reduce the power utilization and also the mam power. PWM (pulse width modulation) can be used to control the intensity of the light with respect to the traffic conditions.

It was explained by (Gauri sahoo, 2016), [4]. that devices that are used for energy harvesting, which are based on velocity controlled harmonized switching harvesting on indicator techniques can be used for automatic unmanned energy harvesting for piezoelectric energy. SSHI techniques can be used for harvest energy. There is no need of using external power supply for controlling the switching time in SSHI techniques as they are fully powered. Further this technique provides more accurate results when compared to state of art technique.

(H.Xiong)[5] also described in piezoelectric energy harvester for roadways about the system can generate energy about 0.36kWh in an year when it is installed in Troutville weigh station, VT , USA

Hitting method and vibration method can be two methods for harvesting energy. In hitting method piezoelectric materials are deformed by striking in order to harvest energy{ (S. K. Song, 2012), (J. H. Kim, 2013) [6,7]}. In vibration method, piezoelectric material is deformed by vibration of the structure which in turn results in generating electricity { (D. Song, 2013), (C. H. Song, 2012) [8,9]}.

PROPOSED SYSTEM

Piezoelectric sensor:

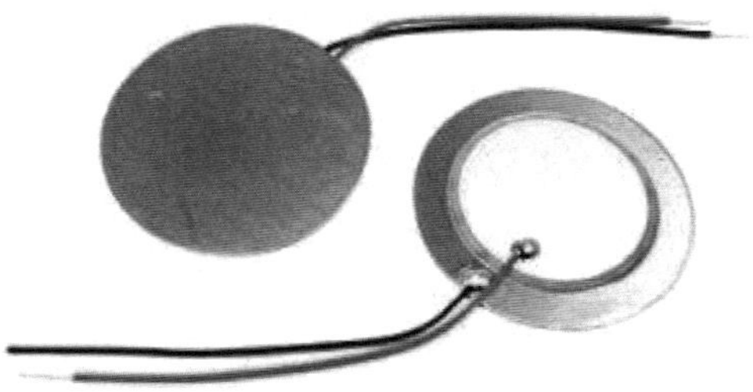

Figure. 2: Piezoelectric sensor.

A piezoelectric sensor is an electric device that measures acceleration, pressure, or force and converts that information into an electric 0signal. In numerous sectors, these sensors are mostly utilised for process control, quality assurance, and research and development. This sensor's uses include aerospace, medical, and nuclear equipment, as well as use as a pressure sensor in mobile phone touch pads. These sensors are used in the automobile sector to monitor ignition when designing internal combustion engines.

Arduino Board:

Figure. 3: Arduino Board

For artists, hackers, amateurs, and many professionals prototyping, experimenting and electronic design are made easier using Aurdino microcontroller boards. Aurdinos are also used in digital instruments and home appliances. Arduinos (we utilise the standard Arduino Uno) are based on the ATmega microcontroller, which is a single-chip computer that has a CPU, RAM, Flash memory, and input/output connections.

It's meant to connect various kinds of sensors, LEDs, small motors and speakers, servos, and other devices directly to these pins, which can read in or output digital or analogue voltages between 0 and 5 volts, unlike, say, a Raspberry Pi. The Arduino connects to your computer through USB, and you programme it in a basic language (C/C++, which is akin to Java) by uploading your compiled code to the board using the free Arduino IDE. Once configured, the Arduino may function with or without a USB connection to your computer, requiring only electricity and no keyboard or screen.

Battery:
Because of its low cost and widespread availability, lead batteries are the most widely utilised in PV systems. There are sealed and wet cell versions of these batteries. Because of their capacity to tolerate overcharge, over discharge, and shock, lead acid batteries have a high level of reliability. The batteries have a large electrolyte capacity, excellent charge acceptance, and low self-discharge. Computer Aided Design is used to test lead acid batteries. These batteries are utilised in UPS systems and inverters, and they have the ability to perform in hazardous settings.

Relay:

Figure. 4: Relay

A relay is a device that controls the operation of other devices in an electric circuit by changing its electrical or physical state. A protective relay is a relay whose primary role is to prevent or restrict damage to apparatus or to safeguard service from interruption. Circuit breaker is tripped in presence of malfunction and the device used for this purpose in electrical engineering is called protective relay. Reverse power flow, over current, over- and under- frequency and over voltage are few abnormal activities that can be detected using original protective relays - which use electromagnetic devices containing coils.

• Low voltage signals can be used to control circuits that have high voltage.

• Low current signals can be used to control circuits that have high current. Example starter solenoids.

• Detection and isolation of transmission line and distribution line failures by opening and closing circuit breakers (protection relays)

• If the two potentials are different, separate the control circuit from the control circuit. When controlling the main power supply via a low voltage switch. The latter is often used to control office lighting because low voltage cables can be easily installed in partitions. Partitions can often be moved as needed. They can also be controlled by room occupancy sensors to save energy,

• Logical function. For example, the Boolean AND function is implemented by connecting normally open relay contacts in series, and the OR function is implemented by connecting normally open contacts in parallel. The switch or Form C contact performs the XOR (Exclusive OR) function. Similar functionality for NAND and NOR is achieved using normally closed contacts. The ladder programming language is commonly used in the design of relay logic networks. o Early arithmetic. Before the advent of vacuum tubes and transistors, relays were used as a logical element in digital computers.Because relays are much more resistant than semiconductors to nuclear radiation, they are widely used in safetycritical logic, such as the control panels of radioactive wastehandling machinery.

• Time delay functions. Relays can be modified to delay opening or delay closing a set of contacts. A very short (a fraction of a second) delay would use a copper disk between the armature and moving blade assembly. The current flowing through the disk maintains the magnetic field for a short period of time and prolongs the emission time. For slightly longer delays (up to 1 minute), the dashpot is used.

• A damper is a liquid-filled piston that can escape slowly. The length of time can be changed by increasing or decreasing the flow rate. A mechanical clockwork has been installed for a long time.

Advantages of relays:

• Relays can switch between AC and DC, and transistors can switch only DC.
• Relays can switch high voltage, but transistors cannot.
• Relays are suitable for switching high currents (> 5A).
• The relay can switch many contacts at the same time.

Lcd:

Figure.5: LCD

This is an illustration of a Parallel Port. Because it does not make use of the bi-directional capability available on later ports, it should operate with the vast majority, if not all, Parallel Ports. It does not, however, illustrate how to use the Status Port as an input. Seven-segment and other multi- segment LEDs are favored over these modules. The reasons for this are as follows: Since LCDs are capable of showing unique and even characters, animations, are inexpensive, easy to program etc, they are favoured for such applications. Command registers are used store command instructions of LCD.

Various activities of the LCD such as cleaning its screen, setting cursor location, initializing it etc can be done using the commands that are stored in the command register. Data registers are used to store the data that is to be displayed by the LCD. Data is basically the ASCII value of the characters that is to be displayed on the LCD.

4. WORKING

The piezoelectric effect, which converts pressure into electrical energy, lies at the heart of this project's operation. We've connected a series of piezoelectric sensors here. The input is pressure, which is then transformed to electrical form and stored in a rechargeable battery. The Microcontroller is the system's brain, converting variable voltage to constant voltage. A diode is used to block the current passage in the wrong direction. A metal detector is installed to verify that the user is permitted, and 10 seconds of charging time is allowed. The amount of voltage stored in the battery is displayed on a Liquid Crystal Display.

Journal of Physics: Conference Series **2325** (2022) 012002 doi:10.1088/1742-6596/2325/1/012002

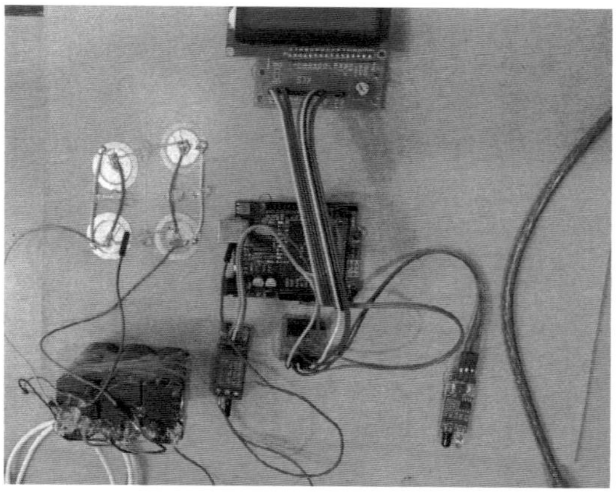

Figure.6: Picture of our hardware setup

All piezoelectric sensors are connected in series combination. The sensors are placed under a tile. Out of the other all sources of energy from human motion or movement, footstep have the highest potential power(67W,estimated in [8]), but the lowest frequency and the limited number of verticallness of shoes will make energy/power harvesting from footsteps a challenging task[9].Pressure is generated on the sensor when we walk on the tiles which are placed in crowded places. Due to this pressure AC voltage will be generated. Voltage produced by piezoelectric sensor depends on pressure applied. From the conducted investigations, the vertical pressure can be upto 130% of the person's total body weight.[12].This voltage is supplied as input to the full adder circuit which performs AC to DC conversion.

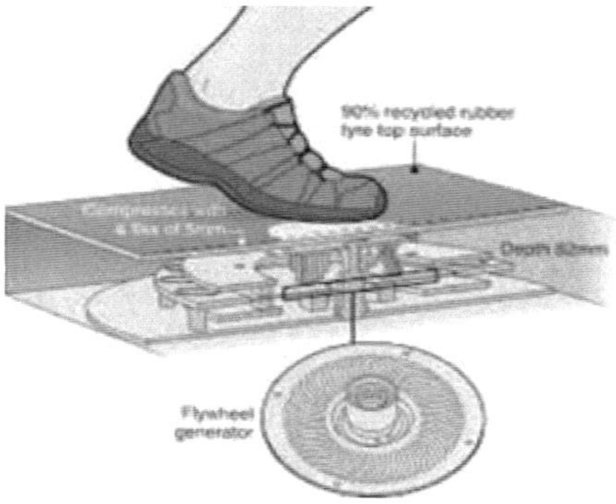

Figure.7: Energy generation through footsteps

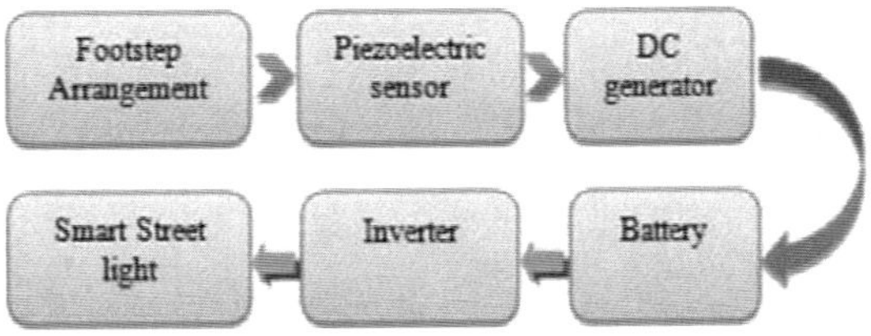

Figure.8: Steps of energy conservation

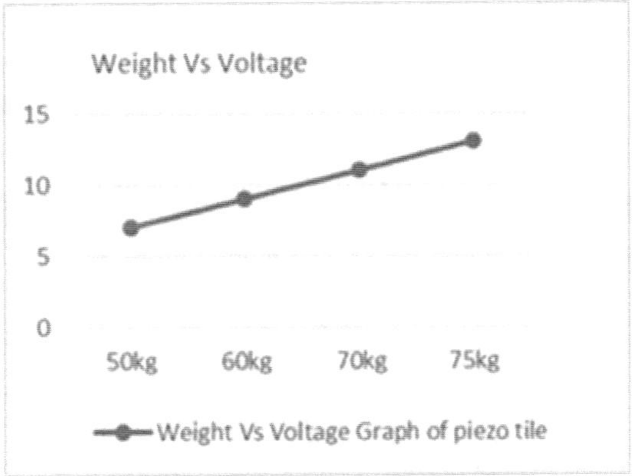

Figure.9: Graph of Weight Vs Voltage

Table 1: Weight vs Voltage Generated

S. No	Approximate weight on sensors (in kgs)	Approximate voltage generated (in Volts)
1	50	7
2	60	9
3	70	11
4	75	13

Piezoelectric sensor utilizes the piezoelectric effect, measures the pressure exerted by pedestrians and converts this pressure into electrical charge. This charge is directly related to the applied pressure.

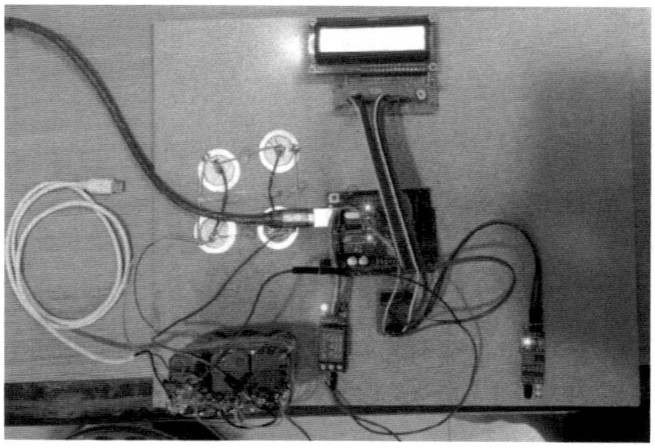

Figure.10 :Working of hardware

Figure. 11: Working of mobile charging

5. FUTURE SCOPE

A. Staircases:

Piezo electric tiles can be places under the staircases to generate power that could be stored in rechargeable batteries and this could be used to charge phones in the buses. These sensors can be placed in those regions where lots of footsteps are involved such as airports, railways stations, cricket and football grounds, large public gathering meet ups etc, and the energy generated by them can be used to power various electronic gadgets.

Above mentioned were the few ways to say energy using piezo electric concept.

B. Side walks

In population dense countries such as China, India etc, where sidewalks are always crowded can be potential place to use piezoelectric tiles as lot of footsteps would produce vibration that could generate power/energy.

This generated energy can be used to power up various appliances for public use such as street lights, traffic signals etc

. This can turn out be the most efficient way to generate huge amount of renewable energy with minimal cost and least human efforts.

C. Escalators

The concept of placing piezoelectric sensors can also be applied to escalators of various crowded places such as malls, metro stations, theaters, airports etc.

D. Dance floors:

This is the place where many people continuously produce vibrations by dancing that could generate power which can be used to power that particular dance studio at bare minimum.

6. RESULTS AND CONCLUSIONS

The primary source of energy is generated by human footsteps. There is no requirement for energy from conventional sources, and this sort of power generating produces no pollutants. It is crucial to the places, all tracks where footsteps are used to generate non-conventional energy such as electricity, that there is no need for any form of power from the mains. It is a well-known fact that non-conventional energy contributes 11% of our primary energy. If this project is implemented, it will not only exacerbate and obstruct the energy shortfall, but it will also result in positive worldwide environmental change.

The project is being tried and updated, which is the most conservative, rational solution to prosperity for the people of our world. The concept of piezoelectric energy can be used in areas with frequent power-cuts and unstable power usage. In rural areas where agriculture works involving high power usage can be benefited by piezoelectric technology and also places where high crowd movements are involved can be benefited. We need to take steps which increases power generation from renewable sources to decrease carbon footprint at global levels. In modern times the contribution of renewable energy is just 11 percent for sustainable growth this number should be increased. Various developed economies such as France, Germany are taking steps towards the growth of renewable energy and related technology.

References:

[1] Anirudh chavan, Shyam lakhadive, Vaibhav pondhe, vineeta Philip," Advanced foot step power generation using piezoelectric sensor", International journal of advance research, issue in 2017. (N.Subba Lakshmi, 2017).

[2] N.Subba Lakshmi, Jayalakshmi.A, Janani.N, Induveni.S, Divya Darshini.P," Automatic street light lightened by piezoelectric roads", International journal of innovative research in science, engineering & technolog (Mitsuhiro Okayasu a, 2016)y, mar 2017.

[3] Mitsuhiro Okayasu a,n, Keisuke Watanabe, "A study of the electric power generation properties of a lead zirconate titanate piezoelectric ceramic", Ceramics International, ELSEVIER, June,2016.

[4] Gauri sahoo,Nikhil devekar,Ruchita Rao “Smart lightning using piezoelectricity”,international journal of advanced research in electronics and communication Vol 5,july 2016.

[5] H.Xiong, “Piezoelectric Energy Harvesting for Public Roadway”, PHD Other – Dissertation, Civil Engineering, Virginia Polytechnic Institute and State University.

[6] S. K. Hong, M. S. Woo, D. Song, C. H. Yang, K. H. Baek, and T. H. Sung, "Study on reinforcement and repair of cracked piezoelectric materials," in 2012 21st IEEE Int. Symp. on Applications of Ferroelectrics Held with 11th IEEE European Conf. on Applications of Polar Dielectrics and 4th IEEE Int. Symp on Piezoresponse Force Microscopy and Nanoscale Phenomena in Polar Materials, ISAF/ECAPD/PFM, Aveiro, 2012.

[7] J. H. Kim, S. Hidaka, D. Song, K. H. Baek, S. B. Kim, and T. H. Sung, "Stress distribution design of additional substrate for piezoelectricity," Ferroelectrics, vol. 449, pp. 72-82, 2013.

[8] Song, C. H. Yang, S. K. Hong, S. B. Kim, M. S. Woo, and T. H. Sung, "Study on application of piezoelectricity to korea train express (KTX)," Ferroelectrics, vol. 449, pp. 11-23, 2013.

[9] C. H. Yang, D. Song, M. S. Woo, S. K. Hong, K. H. Baek, and T. H. Sung, "Design of vibration exciter by using permanent magnets for application to piezoelectric energy harvesting," in 2012 21st IEEE Int. Symp. on Applications of Ferroelectrics Held with 11th IEEE European Conf. on Applications of Polar Dielectrics and 4th IEEE Int. Symp on Piezoresponse Force Microscopy and Nanoscale Phenomena in Polar Materials, ISAF/ECAPD/PFM, Aveiro, 2012.

[10] T. Starner and J. A. Paradiso, “Human generated power for mobile electronics”, Low-power Electronics Design, C. Piguet,ed.,CRC Press,chapter 45,pp.1-35,2004.

[11] H. Fu, R. Xu, K. Seto, E. M. Yeatman, and S. G. Kim, “Energy Harvesting from Human Motion Using Footstep-Induced Airflow,”Journal of Physics:Conference Series,vol.660,p.012060,2015

[12] P. Niu, P. Chapman, R. Riemer, and X. Zhang, “Evaluation of motions and actuation methods for biomechanical energy harvesting,” in Power Electronics Specialists Conference, 2004. PESC 04. 2004 IEEE 35th Annual,2004,pp.2100-2106 Vol.3

Smart Home Automation using Augmented Reality and Internet of Things

R. Jaivignesh [1], R. D. Janarthanan [2], V. Gnanalakshmi [3]

[1,2] B.E, Department of ECE, Mepco Schlenk Engineering College, Sivakasi, Tamil Nadu, India.

[3] Assistant Professor, Department of ECE, Mepco Schlenk Engineering College, Sivakasi, Tamil Nadu, India.

jaivikki2000@gmail.com[1], janarthanan18bec049@gmail.com[2], v.gnanalakshmi@mepcoeng.ac.in[3]

Abstract. This paper demonstrates the concept of Augmented Reality and the Internet of Things controlling home appliances through the internet. Nowadays, Home Automation using IoT became an integral part but it can be extended to Augmented Reality. Augmented Reality is used to superimpose interactive content into real-world objects. It helps in placing the 3D objects that help in controlling different IoT devices. The main objective of this paper is to demonstrate and analyse the control of home appliances using augmented reality with the internet of things by a Wi-Fi-based microcontroller. This paper also demonstrates its approach toward multiple devices and proves that augmented reality and IoT are complementary to each other that visual interactive objects can control IoT devices. Performance measures like transmission time, RSSI were taken which plays a major role in terms of Internet of Things. The Transmission time determines the time taken by the Wi-Fi of Esp32 to connect to the blynk server. The Comparative analysis for the Transmission ON time and Transmission OFF time at different ranges is measured. The Transmission ON time is of maximum of 2.36 seconds at a distance of 5m and of 2.65 seconds at a distance of 10m without any hindrance. The Transmission time with a wall hindrance is increases than without the wall hindrance. The Transmission ON time is of 2.9 seconds and the Transmission OFF time is of 2.8 seconds. The Received Signal Strength Indicator indicates the strength of the wifi network is also measured and it indicates the network is good. The Measured value is between -23 dbm to -25 dbm which indicates a better network performance.

Keywords: Esp32; Wi-Fi-based microcontroller; IoT; Unity hub; Smart Switches; Augmented Reality;

1. Introduction

In recent times, innovation in emerging technologies has made human life easy in all ways of communication and control. The emergence of Smartphones had made a revolution in the communication field but the interactive environment is not still provided. Here in this paper, the demonstration of the work with augmented reality and IoT gives an interactive feature. Augmented

Reality is the placing of digital objects in the real world and computer-generated graphic content that is shown as real-time objects. IoT is an interactive environment where all devices are interconnected to each other and can transfer data without any human intervention [1]. In this digital world augmented reality plays a major role in terms of interaction [2]. The IoT, Augmented Reality, and voice command robots are in existence and these technologies can be also extended to realities like AR, VR, and also mixed reality [3]. Using a Wi-Fi inbuilt processor in terms of IoT, automation can be done. This also can be done with Augmented reality by extending it with IoT [4]. The combination of Augmented Reality and IoT leads the way as innovative technology. Augmented reality has rendered its applications in education, entertainment and also provides a scalable amount of devices automation [5]. Several ubiquitous technologies ensurethe interaction of humans and the world. The usage of Unity3D facilitates the mobile app for augmented reality and also in providing digital 3D objects that can be augmented in the real-world [6].Blynk API is a Cloud-Platform that is used asios and android apps to control the microcontroller over the internet. It's a digital dashboard where a graphic interface for the project is built. It can control hardware remotely and it provides better secure communication. The Wi-Fi controller Esp32 provides a better range of communication and it is the successor of esp8266 with faster Wi-Fi and Bluetooth connection that runs at a typical frequency of 160 MHz. The IoT devices are generally responsive to augmented reality apps that are controlled through the internet and the status of these devices is updated through data exchange protocol. In this paper, the Augmented Reality app is built that controls the home appliances and it is also extended to multiple appliances like bulbs, fans, laptops etc and the monitoring of the room's temperature is done.

2. Literature Survey

Augmented Reality and IoT provides a rich combination of 3D objects controlling things that make smart home automation. A web server designed to control the appliance's user friendly that makes automation through MQTT. MQTT is an application layer protocol used to show the sensor's data and also the switch controls and gives the idea of Home automation using IoT [7]. This paper generally gives the basic idea about home automation using a user-friendly web interface. A Product named IoTHome was created for the usage of smart home automation. They have used NodeMCU for control purposes and proved the feasibility of Wi-Fi control and it can be used with Google assistant and other monitoring activities. Augmented Reality makes the placing of unreal objects that seems like real objects in the real world. This is demonstrated by unity3D, and also the vuforia SDK focuses on the image target features [8]. Sung Lae Kim proposed an idea about how the unity3D objects can be placed into the real environment [9]. Augmented Reality for IoT applications is possible [10]. This paper develops AR applications and also smart glasses. The AR application makes the communication between the local server where the data is exchanged and also smart glasses to render the IoT application. This paper also gives proof of IoT with AR. The Brain-Computer Interface that used to control the home appliances using augmented reality based on brain signals through middleware and the visual stimulus controls the appliances using eye blinks [11]. Virtual reality with brain waves can be used to control IoT devices and gives proof of IoT with Virtual Reality. The sensor meter rhythms control the user application takes place through the operator's IP address. Nowadays, there are more AR Mobile apps. The method of controlling the Mobile AR appliances that works on the web in a 5G network. It supports mobility and the internet provided is sufficient and gives a lightweight, cross-platform and low-cost solution for end to end delivery. It also gives ideas about reality with virtual objects and provides an immersive experience to the end users[13]. Augmented Reality has also extended to the Healthcare side. The methods of measuring the subjective scale scores, skin level, the anxiety level are measured and shown virtually [14]. The virtual hand and visual ray pointers have been used to control appliances and serve many applications [15]. In this paper, the use of virtual buttons is introduced to control the home appliances through the Blynk server to communicate and it is extended to multiple targets also the room temperature and humidity are monitored in real-time.

3. Proposed system

The demonstration of a virtual button controlled home appliance using Blynk API and Unity3D is proposed and also to observe the real-time temperature sensor data. This paper mainly aims to give proof for the concept of Augmented Reality with IoT.

3.1 IoT Implementation

The IoT part is controlling the home appliance using esp32 that is connected to relay and Blynk API through the internet.

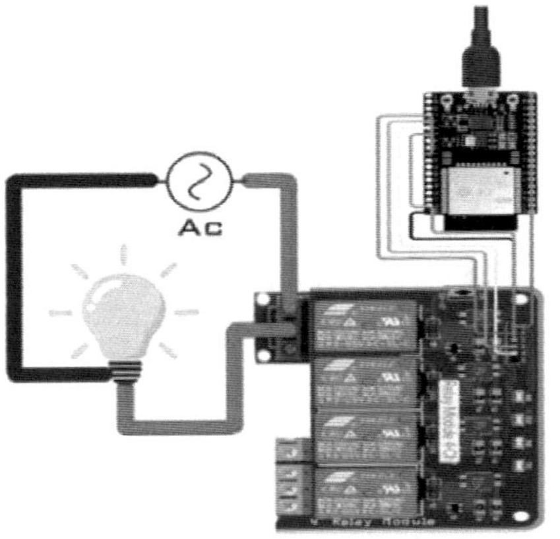

Figure 1: Connection diagram of Home Appliance with ESP32 and Relay

The above circuit represents the IoT connection of controlling an AC appliance. The esp32 is connected to the input of the relay using one of the GPIO pins. The relay module is an electronically operated switch used to turn on and off the circuit. The main usage of the relay module is to handle the voltage/current higher than the microcontroller could handle. The 5V relay module is used. It consists of three connections called terminators that have NC (Normally Open), NO (Normally Closed), COM ports [17]. The Other side has VCC, GND, IN pins that are to be connected to the board. The relay module uses an electric current to open and close the switches with the help of a coil that attracts the contacts of the switch and peels them together and a spring pushes them apart when the coil is not energized The Home appliance wire is cut into two and these are connected to the COM and NC ports and the VCC pin is connected to the Vin pin of esp32. In the Blynk android app, a new project is created and a button widget is placed in it. The Blynk app generally sends the API key token to the mail that is used to control the appliance. The C++ code is written in Arduino IDE and the code is uploaded to the esp32 board and the home appliances are controlled. To control the virtual pin of the Blynk device. The auth_token field is replaced by the BlynkAPI key token and the pin field is changed to V1(Virtual pin). And the value will be turned to 0 or 1 on and off. The Block Diagram of the IoT part is shown in fig. 2 which represents how the process of communication takes place for controlling home appliances. Fig.3 is an interfacing diagram of Home appliances with esp32.

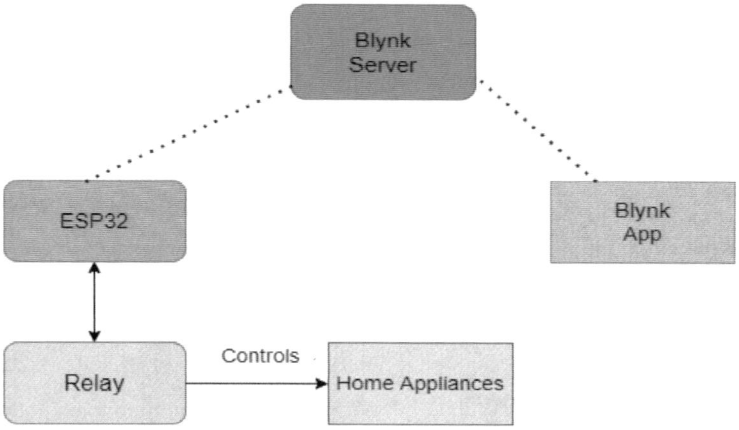

Figure 2: IoT Block Diagram of controlling Home appliances

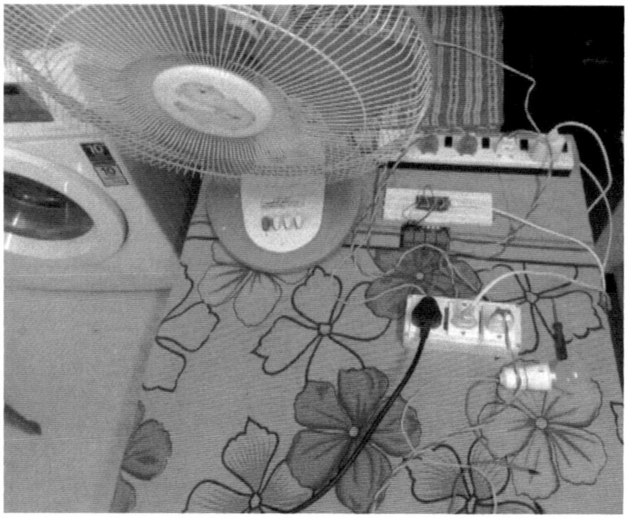

Figure 3: Interfacing Image of Home appliances with ESP32

The DHT11 is used to monitor the room's temperature and humidity. The temperature and humidity are measured and the data is sent to the ESP32 using the data pin of the DHT11 sensor. Figure 3 represents the connection diagram of DHT11 with ESP32.

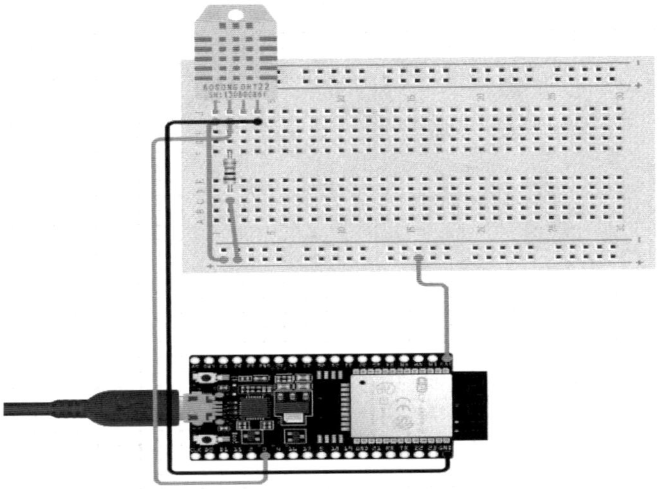

Figure 4: Connection Diagram of DHT11 with ESP32

Fig. 5 shows the image of the acquired temperature and humidity sensor data in the Arduino IDE serial monitor. The data is also sent to the Blynk server through the internet using the Blynk authentication token generated by the Blynk platform. Two measuring gauge widgets are placed in the project space of the Blynk app that shows the value of the temperature and humidity data.

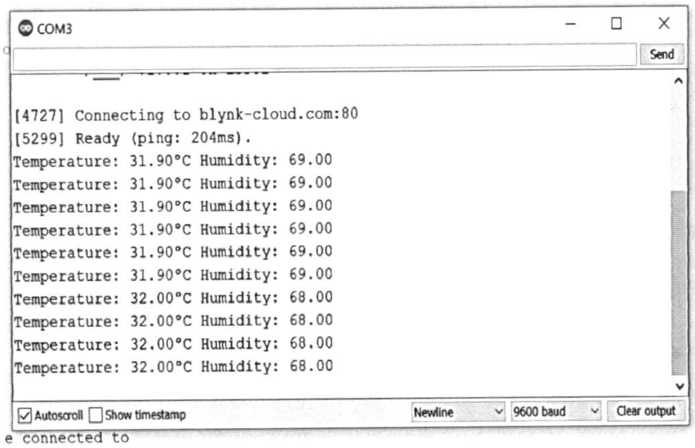

Figure 5: Temperature and Humidity value in Arduino IDE serial window

3.2 Augmented Reality Implementation

The Augmented Reality part of placing 3D objects is implemented through Unity3D and Vuforia SDK. The C# script is used to control the virtual button behaviour and update the pin status in the Blynk cloud that is uploaded to the unity hub. The image target is important to view the virtual buttons. The UI option in unity 3d already provides a button option but instead of that virtual buttons are used that can virtually control the appliance. Fig. 5 shows the AR part of the paper.

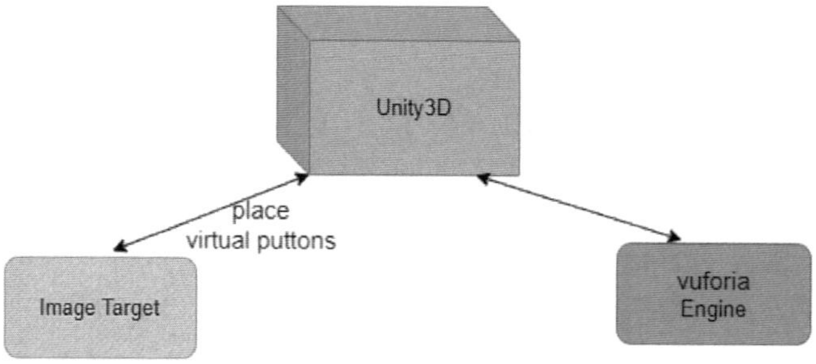

Figure 6: Augmented Reality Block for placing virtual buttons on the image target

Vuforia Engine process is for focusing the image target. First, the image target chosen is to be uploaded to the vuforia developer site and the width of 1 is set and it is downloaded as a 3D database. This is then imported to the Unity hub for placing the image target in the 3D workspace. Vuforia engine is also used to augment the virtual buttons on the image target [16-17]. Its process is shown in fig. 7 in which the image is taken into it and it updates the logic state and renders the graphics on it and it is shown on the device.

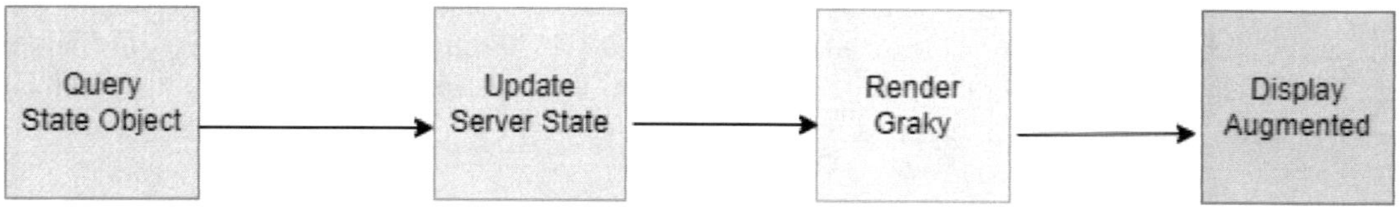

Figure 7: Vuforia Engine Process

The process that is involved in the augmented reality part is shown in fig. 6; first an image target is chosen and it is uploaded to the vuforia developer site and the database is created for that image and the license of that image is also created in vuforia. The database is then downloaded and imported into unity3d. The unity2020 version is used and the image is now imported into unity. The Vuforia package is also imported. The main camera is present default is removed and the ar camera is chosen as an augmented reality-based project is developed and then it is connected to unity3d and vuforia the license is used that is copied and pasted into the license column. The ar camera will be shown on the left side of the screen by clicking it, an image target that is imported from vuforia engine is added and so the image will be displayed in the 3D workspace. Now two virtual buttons are placed on the image target by going to the advanced options present on the right side of the workspace and clicking the 'Add Virtual Button' option also it is invisible while creating an android application. So for that, a plane is brought near to these buttons and the button images shown in fig. 8 are placed. The C# script is written particularly for the virtual buttons. This C# script is uploaded to the hub and then these Blynk addresses are pasted into the input field that is mentioned in the C# script. Before developing it can be tested using the webcam of the laptop by clicking the play button at the top. This is the proposed work of this paper that proves the IoT can be combined with Augmented Reality. The Proposed Block diagram is shown in fig. 8. A class is created in the script for these virtual buttons and if the virtual button behaviour is triggered.

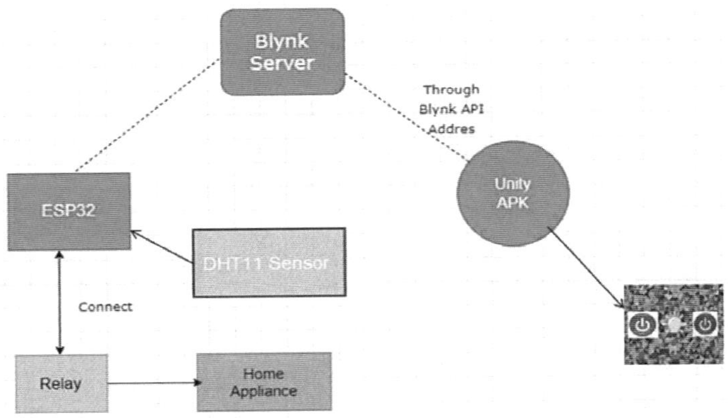

Figure 8: Proposed Block Diagram for Home appliance control

Figure 9: Virtual Buttons View

The Monitoring of Temperature and humidity is also done in the same way. First, the image target is uploaded into the vuforia engine site and the 3D database is downloaded and imported to the unity hub. It is then placed in the top view and the canvas is added to keep the input field that shows the value of temperature and humidity that is received through Blynk API using the 'GET' command and virtual pin V0 is used.The input field from the vuforia engine package are kept in it and then they are dimensionally modified and kept perpendicular to the image target. A virtual button is kept on the image target to update the temperature and humidity value. The C# script is written for it using the basic unity library functions and to get the temperature and humidity data from the Blynk cloud the GET_DATA function is used and the URL for that Blynk API is written in it hence while checking it with the webcam of the laptop the canvas with that image is shown and when the virtual button is pressed the temperature and humidity data is shown on the canvas. The virtual button is kept with the image to view the button. The image target is detected by using its colour, shape by the app. The sensitivity of the virtual buttons is kept high so that the transmission time is reduced.

4. Results and Discussions

The IoT part of the model developed controls the home appliances through the Blynk server. In the Blynk app, the virtual pin is used to control the appliance through the Blynk authentication token of the project. The control through the Blynk rest API was also demonstrated and the range of the Esp32 is

Journal of Physics: Conference Series **2325** (2022) 012003 doi:10.1088/1742-6596/2325/1/012003

checked. The status of the switches is also updated in the Blynk app. The unity hub is used to build the app. The Virtual buttons are shown while the app camera is placed in front of the image target. The Virtual buttons view is shown in fig. 10. The interruption between the image switches makes the Blynk API update the status of the home appliance ON or OFF. The App is also built for other appliances. While the image target is shown the respective home appliance is switched on and off.

The Performance measures for this paper were also measured. After implementation, the performance analysis taken like RSSI (Received Signal Strength Indicator) and the Transmission time across the different targets taken and also the analysis based on range and the across the window and free space and with a hindrance of the wall the Wi-Fi strength and the transmission time is also checked.

4.1 Transmission Time

The Transmission time is defined as the time taken to transmit data from point to point. It is affected by packet size and also some hindrances like walls and other environmental substances. The transmission time for ON and OFF buttons for different ranges is taken. The different ranges chosen are 3m, 5m, 7m, 10m for multiple targets in free space and have analyzed Wi-Fi strength also transmission time is taken if a hindrance like a wall is present and across a window. The following tables show the performance analysis on time

Table 1: Range: 3m (without any obstacle)

Target	Transmission time for ON (secs)	Transmission time for OFF (secs)
1	2.04	2.3
2	2.19	2.12
3	2.4	2.63

The Table 1 depicts the Transmission ON and OFF time for different targets at range of 3m without any obstacle. Target 3 takes maximum Transmission ON time of 2.4 seconds and Transmission OFF time of 2.63 seconds.

Table 2: Range: 5m (without any obstacle)

Target	Transmission time for ON (secs)	Transmission time for OFF (secs)
1	2.12	**2.5**
2	2.32	2.42
3	2.7	2.74

The Table 2 depicts the Transmission ON and OFF time for different targets at range of 5m without any obstacle. Target 3 takes maximum Transmission ON time of 2.7 seconds and Transmission OFF time of 2.73 seconds.

Table 3: Range: 7m (without any obstacle)

Target	Transmission time for ON (secs)	Transmission time for OFF (secs)
1	**2.5**	**2.47**
2	2.5	2.57
3	2.8	2.83

The Table 3 depicts the Transmission ON and OFF time for different targets at range of 7m without any obstacle. Target 3 takes maximum Transmission ON time of 2.8 seconds and Transmission OFF time of 2.83 seconds.

Table 4: Range: 10m (without any obstacle)

Target	Transmission time for ON (secs)	Transmission time for OFF (secs)
1	**2.75**	**2.65**
2	2.72	3.48
3	2.85	2.7

The Table 4 depicts the Transmission ON and OFF time for different targets at range of 10m without any obstacle. Target 3 takes maximum Transmission ON time of 2.85 seconds and but Target 2 takes Transmission OFF time of 3.48 seconds infers that the texture of the Target 2 image is low.

The Comparative analysis made for the targets without hindrance and the transmission time is shown in the above tables for different ranges. The Graph plot is taken for Transmission on and off time across different targets that are shown in fig. 10

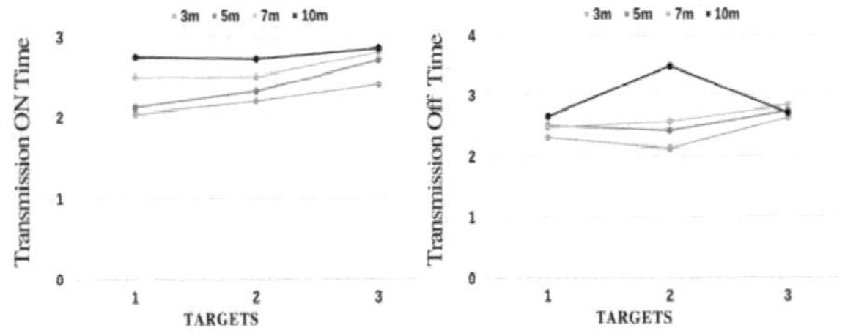

Figure 10: Comparative analysis of Transmission time without wall

From the above graph, it is preferred that the transmission on time for different targets is increased along with the increasing range. But the transmission off time is comparatively lower than the transmission off time and also for target 3 the transmission off time is approximately the same

The analysis is also taken for transmission time with wall hindrance and the Wi-Fi strength is checked. The wall is of width 30cm. The Wi-Fi strength across the wall is checked. Fig. 11 represents the analysis in different ranges across multiple targets with wall hindrance.

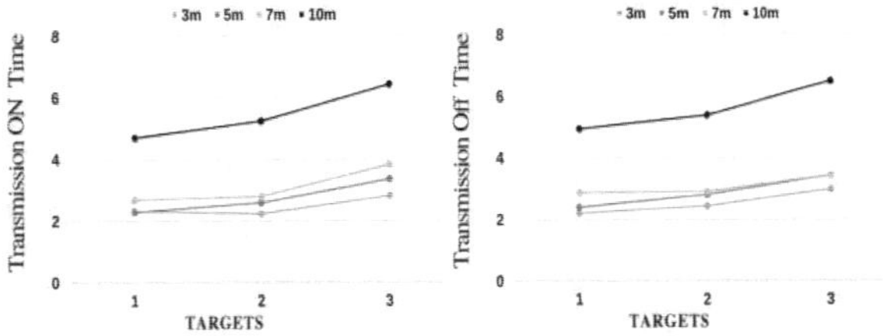

Figure 11: Comparative analysis of Transmission time across the wall

Fig.11 depicts the transmission on and off time of different targets. Both the transmission on and off time is nearly the same but the transmission off time for 5m range and 7m range takes the same transmission time. The Comparative analysis made for the targets across the window and the plot for the transmission time is shown in fig.12.

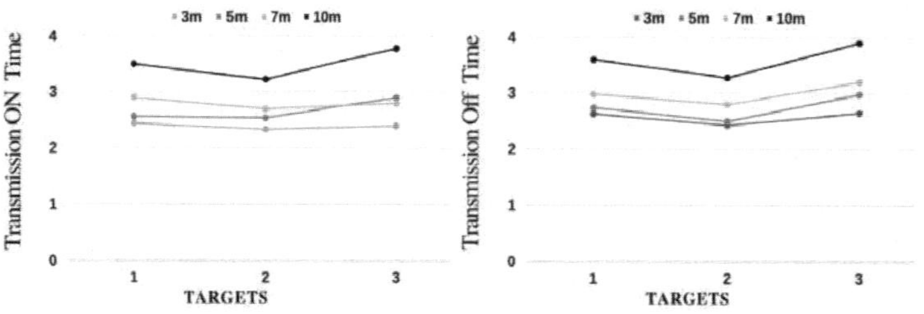

Figure 12: Comparative analysis of Transmission time across the window

4.2 Received Signal Strength Indicator(RSSI)

The RSSI is an important parameter that gives the value of the received signal strength. If the value is less than and equal to 50 defined that the wifi is great and can be used for IoT appliances. If the value is greater than 50 defines a bad network. The RSSI value is checked using Arduino ide. The RSSI value produced while controlling the home appliances is shown in fig. 13. It also changes because of the obstacles present in the environment. The esp32 is used to control the home appliances through an app developed using unity3d. The Nodemcu(ESP8266) is used to calculate the RSSI value that is shown below.

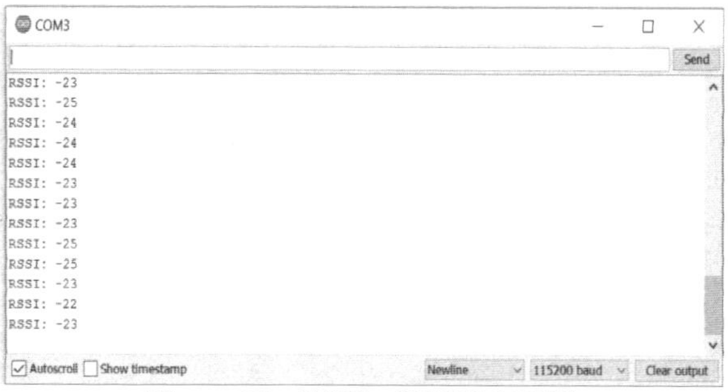

Figure 13: RSSI of ESP32 Wifi

4.3 Working Part

The apk file developed using unity hub is saved as a Multi-Target apk file. This apk file is then installed on mobile phones and when it is opened across the image target it shows virtual switches across the image target. These are used for ON (fig. 14) and OFF (fig.15) which is shown for target 1

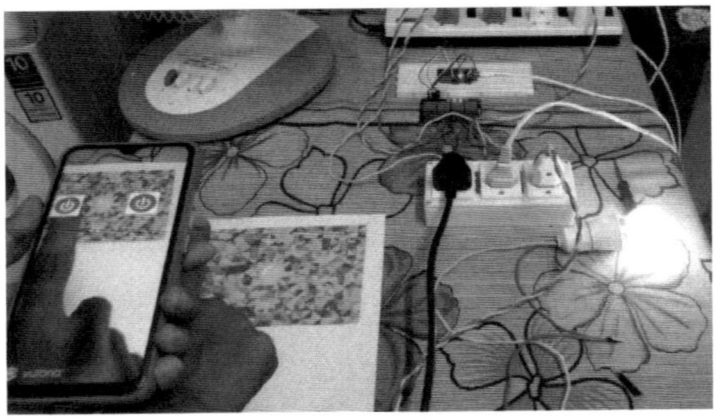

Figure 14: Target 1 Switch ON

Figure 15: Target 1 Switch OFF

The Multiple target app show virtual buttons ON (fig. 17) and OFF (fig.18) that is viewed only when target 2 image is detected by the app AR camera. The target image used is shown in fig.16.

Figure 16: Target 2 Image

Fig. 17 depicts the view of the home appliance when the green image button is triggered the relay pulls out the magnetic switch and turns the home appliance ON. The digital pin 33 is used in esp32.

Figure 17: Target 2 Switch ON

Fig. 18 depicts the view of the home appliance off when the red image button is triggered the relay pulls out the magnetic switch and turns the home appliance OFF.

Figure 18: Target 2 Switch OFF

The Multiple target app show virtual buttons ON (fig. 20) and OFF (fig.21) that is viewed only when target 3 is detected by the app AR camera. The target image used is shown in fig. 19.

Figure 19: Target 3 Image

Fig. 20 depicts the view of the home appliance ON when the green image button is triggered the relay pulls out the magnetic switch and turns the home appliance OFF. It is the view of the buttons shown while the app AR camera is shown across the image target 3. The digital pin 26 is used in esp32.

Figure 20: Target 3 Switch ON

Figure 21: Target 3 Switch OFF

The fig. 21 depicts the view of the home appliance off when the red image button is triggered the relay pulls out the magnetic switch and turns the home appliance OFF when the image target 3 is shown across the app AR camera. These are the results obtained while showing the app across each target. In these image targets, the working of the developed AR app is shown and across various situations, it is tested. The esp32 digital pins can be used for controlling more home appliances. There are 34 Gpio pins of which mostly 10 pins among them can be used for controlling home appliances. The relay can hold up to a voltage level of 240 AC voltages. Further plugs are developed and it is controlled through wifi control.

The monitoring of temperature and humidity is also shown in fig 23 which shows the value of temperature and humidity when the developed app is opened across the particular image target that is shown in fig 22. The temperature and humidity value remains updates with the help of Wi-Fi network and it keeps the same value when Wi-Fi is not connected.

Figure 22: Target Image for Temperature monitoring

Figure 23: Temperature and Humidity Monitoring

These are the results obtained while showing the app across each target. In these images, the working of the developed AR app is shown and across various situations, it is tested.

5. Conclusion

This paper proposed a smart home automation model and tested it with different devices and also different analyses taken depending upon the Wi-Fi strength providing a friendly interactive environment. The Blynk API is tested in various conditions and also the app is tested across different ranges that assure the mobility of the device. The NodeMCU is also instead of ESP32 and tested across various ranges that take more time than ESP32 and thus, sensitivity is low. Finally, it is checked with different Wi-Fi networks in different places with low Wi-Fi strength and it needed less consumption of data providing an optimized performance. The model developed controls different home appliances across different ranges.

Plans to enhance the number of devices and test the prototype apply to a large number of devices. The success of this paper depends on the transmission time, the sensitivity of the buttons, the Wi-Fi strength and the data exchange protocol.

References

[1] Tiago Andrade, Daniel Bastos, "Extended Reality in IoT scenarios: Concepts, Applications and Future Trends" in 2019 5th Experiment@ International Conference (exp.at'19) June 12th – 14th, 2019, University of Madeira, Funchal, Madeira, Portugal.

[2] Dongsik Jo and Gerard Jonghyun Kim," ARIoT: Scalable Augmented Reality Framework for Interacting with the Internet of Things Appliances Everywhere" IEEE Trans. Consumer Electronics, Vol. 62, No. 3, August 2016.

[3] Unity3D, Unity Hub definition, https://unity.com/.

[4] Harsh Kumar Singh1, Saurabh Verma2, Shashank Pal3, Kavita Pandey4, "A step towards Home Automation using IOT", doi:978-1-7281-3591-5/19/$31.00 ©2019 IEEE

[5] Waheb A. Jabbar1,2, (Senior Member, IEEE), Tee Kok Kian1, Roshahliza M. Ramli1, (Member, IEEE), Siti Nabila Zubir1, Nurthaqifah S. M. Zamrizaman1, Mohammed Balfaqih3,4, (Member, IEEE), Vladimir Shepelev3 and SoltanAlharbi 4, "Design and Fabrication of Smart Home with the Internet of Things Enabled Automation System", Digital Object Identifier 10.1109/ACCESS.2017.Doi Number

[6] Sung Lae Kim, Hae Jung Suk, JeongHwa Kang, Jun Mo Jung, "Using Unity 3D to Facilitate Mobile Augmented Reality Game Development", 2014 IEEE World Forum on Internet of Things (WF-IoT)

[7] Monica AiswaryaAnkireddy, Rajath AV, Ruthwik Ganesh M "Augmented Reality Rendered for IoT applications", Doi: 978-1-7281-2327-1/19/$31.00 ©2019 IEEE
[8] SEONGHUN PARK, HO-SEUNG CHA, AND CHANG-HWAN IM, "Development of an Online Home Appliance Control System Using Augmented Reality and an SSVEP-Based Brain-Computer Interface", date of publication November 11, 2019, date of current version November 20, 2019. Digital Object Identifier 10.1109/ACCESS.2019.2952613.
[9] C. G. Coogan and B. He, "Brain-Computer Interface Control in a Virtual Reality Environment and Applications for the Internet of Things," in IEEE Access, vol. 6, pp. 10840-10849, 2018.
[10] Xiuquan Qiao1,*, Pei Ren1, Guoshun Nan1, Ling Liu2, Schahram Dustdar3, Junliang Chen1, "Mobile Web Augmented Reality in 5G and Beyond: Challenges, Opportunities, and Future Directions", 1 State Key Laboratory of Networking and Switching Technology, Beijing University of Posts and Telecommunications, Beijing 100876, China, China Communications September 2019.
[11] Shih-ChingYeh; Yuan-Yuan Li; Chu Zhou, Effects of Virtual Reality and Augmented Reality on Induced Anxiety, IEEE Access, vol. 26, pp. 10840-10849
[12] Jorge Wagner, Student Member, IEEE, Wolfgang Stuerzlinger, "Comparing and Combining Virtual Hand and Virtual Ray Pointer Interactions for Data Manipulation in Immersive Analytics", IEEE TRANSACTIONS ON VISUALIZATION AND COMPUTER GRAPHICS, VOL. 27, NO. 5, MAY 2021.
[13] Vuforia Engine: https://developer.vuforia.com/
[14] Karuppusamy, Dr P. "A Sensor based IoT Monitoring System for Electrical Devices using Blynk framework." Journal of Electronics and Informatics 2, no. 3: 182-187.
[15] Chen, Dr Joy IongZong, and Kong-Long Lai. "Internet of Things (IoT) Authentication and Access Control by Hybrid Deep Learning Method-A Study." Journal of Soft Computing Paradigm (JSCP) 2, no. 04 (2020): 236-245.
[16] IEEE, IoT definition, https://iot.ieee.org/definition.html, PUBLISHED 27 MAY 2015.
[17] IEEE, Augmented Reality definition, https://digitalreality.ieee.org/standards.
[18] Intel, AR vs VR,https://www.intel.in/content/www/in/en/tech-tips-and-tricks/virtual-reality-vs-augmented-reality.html

BATTERY MANAGEMENT SYSTEM TO ESTIMATE BATTERY AGING USING DEEP LEARNING AND MACHINE LEARNING ALGORITHMS

S. Harippriya[1], E. Esakki Vigneswaran[2], S. Jayanthy[3]

Sri Ramakrishna Engineering College, NGGO Colony, Vattamalaipalayam, Coimbatore, Tamil Nadu 641022, India.

haripriya.2057002@srec.ac.in, esakkivignesh@srec.ac.in, jayanthy.s@srec.ac.in

Abstract: One of the greatest challenges faced by Electric Vehicle (EV) manufactures is insufficient charging stations. Estimating the aging of the battery in the electric vehicle helps the driver to predict the driving range of the vehicle. This paper proposes a battery management system that is developed to predict remaining battery charge of the Electric Vehicle. The aging of the lithium-ion (Li-Ion) battery present in the electric vehicle is predicted using different machine learning and deep learning algorithms. The parameters such as voltage, current and temperature are taken from the sensors connected to the LPC2148 ARM board and the values are given as dataset to the Long Short-Term Memory (LSTM), Decision Tree (DT), K-Nearest Neighbors (KNN), Naïve Bayes (NB) and Support Vector Machine (SVM) Algorithms. The experimental results indicate that for real-time data Naïve Bayes algorithm gave the best results in terms of metrics such as Accuracy, Precision, Recall and F1-score. Naïve Bayes produced results with the accuracy rate of 88% and used to calculate the Remaining Battery Capacity which helps predicting the aging of the lithium- ion battery.

Keywords: Electric Vehicle (EV), Lithium-Ion (Li-Ion), State Of Charge (SOC), Battery Management System (BMS), Long Short-Term Memory (LSTM), Decision Tree (DT), K-Nearest Neighbors (KNN), Naïve Bayes (NB) and Support Vector Machine (SVM).

1. INTRODUCTION

The impacts of air pollution and global warming that is caused due to greenhouse gas emissions from fuel based vehicles have increasingly attracted great attention towards the eco-friendly electric vehicles (EVs). The transport industry accounts for the large amount of greenhouse gas emission and pollution to the environment. The transport sector can be improved by the introduction of the environmentally friendly Hybrid Electric Vehicle (HEV), Battery Electric Vehicle (BEV), Fuel Cell Electric Vehicle (FCEV) and Plug-in Hybrid Electric Vehicle (PHEV). Among these Battery Electric vehicles (BEV) are currently the best choice in terms of public and personal transportation for the environment. Because of high cell voltage, low self-discharge rate, high energy density, light weight, long lifetime and low maintenance Li-ion batteries have gained high popularity when compared to other batteries. Range, cost, and battery life cycle are the main challenges for the development of Li-Ion battery systems for electric vehicles [4]. A battery can support only a finite, limited number of charge and discharge cycles. A sufficiently long battery life is necessary to avoid battery replacements during the vehicle's lifetime. To achieve this, a study on lithium-ion battery aging behavior in electric vehicles are essential. Every battery at any instant has a capacity value known as state of charge (SoC) which reduces with energy usage. Range of an EV depends on the SoC of the battery. Therefore, battery management system (BMS) is important to make sure that the battery is operated in its specified safety limits [3]. One of the most important tasks of the BMS is to predict the state of charge. SOC provides percentage of the remaining battery charge. SOC can be calculated based on three battery parameters such as voltage, current and temperature. Through SOC we can identify when the battery needs to be recharged and how long the vehicle can be driven without recharging the battery [7].

2. RELATED WORKS

Prakash et al (2021) has proposed a work that establishes the effective method for identifying the battery aging. Battery Aging provides the accurate estimation of battery health through real time simulation. This paper estimates remaining useful life and state of health by calculating the SOC, internal resistance and capacitance of Lithium-Ion battery along with the voltage transfer characteristics. Various statistical model on battery and machine learning approaches are introduced to predict SOH and RUL which are implemented on a hardware. The overall results produced by the neural network approaches in predicting the SOH are within the error rate of ±5%. The RUL is predicted using long short time memory neural network through which ±10 cycles of accuracy is obtained [2].

Kailong et al (2021) proposed a system which is used to predict remaining useful life (RUL) and future capacity using uncertainty quantification. Advanced machine-learning techniques which uses reliable uncertainty management technique to predict battery capacities and RUL for lithium-ion (Li-ion) batteries is developed. Battery capacity is then completely decomposed into intrinsic mode function (IMF) and residual value using the empirical mode decomposition (EMD) method. IMF and residual are calculated using Long short-term memory (LSTM) and Gaussian process regression (GPR) with the uncertainty level. The aging is calculated by implementing LSTM+GPR model which is then compared with GPR, LSTM, GPR+EMD, and LSTM+EMD models. The accurate results are obtained in LSTM+GPR. The proposed system shows better results in battery health diagnosis [3].

Rasool et al (2020) proposed a better management system for Lithium-Ion batteries to achieve safe use and to increase the battery lifetime. SOC estimation technique is established. Here model-based estimation method is used to find the accuracy of the model. The estimation of accuracy can be predicted considering parameters such as aging, self-discharge and temperature. An extended Kalman filter is performed to estimate better SOC. The degradation of battery capacity is evaluated using this approach. In this paper noise covariance is optimized to improve the state estimation by particle swarm optimization. The results produced are effective in both accuracy and computational simplicity [8].

Zhang et al(2020) proposed a novel sparse machine learning for real time state of charge estimation. Least squares support vector machine (LS-SVM) is developed and the process dynamics is executed in real-time for Li-Ion batteries. The computation and mapping functions are calculated using fast recursive method. A weighted regularization scheme and the differential evolution (DE) method are implemented that helps in optimizing the parameters which enhances the accuracy of the proposed system. State of charge estimation in real-time is estimated using unscented Kalman filter (UKF) based on the proposed model. The results produced by Federal Urban Drive Schedule (FUDS) test data creates an enhanced absolute error which is less when compared with conventional LS-SVMs. The proposed system produces 10 times less error when compared with conventional LS-SVMs [11].

Fen et al (2020) had proposed a model for the efficient use of energy in EV's by frequently monitoring the Li-Ion batteries. To determine the accurate lifetime of the battery a deep learning based SOC prediction model was implemented and in addition to that recursive neural network (RNN) based method was used to improve the data representation. Lithium-Ion Batteries SOC was accurately predicted using the Convolution neural network (CNN) fed with the well trained vector representation. Feature Extraction and Vector representation was used to improve the proposed model for SOC prediction. The performance produced by the system is improved by 4.3% in prediction when compared with RNN algorithm [12].

Carlos et all(2020) proposed the survey on battery state estimation using the machine learning approaches like feedforward neural network(FNN) ,recurrent neural network(RNN) ,support vector machine (SVM) ,radial basis function(RBF) and Hamming network. The comparisons are based in terms of input, outputs, data quality, stated accuracy and battery types of ML landscapes for SOC and SOH estimation. The RNN, FNN algorithms are trained 50 times while in each time 3000 epochs are obtained to improve the efficiency of the system[10]. The errors appeared to be different because of the initial values of the parameters, when the comparison must be made among the estimation techniques, the same no of initial parameters must be present, and it has to be trained and tested with similar data within proposed model [13].

Matteo et al (2019) proposes the capability to control the battery capacity degradation in terms of, recharge time, driving range and drivability for battery aging management model whose vehicle performance was guaranteed. In battery aging management, the parameters such as depth of discharge and maximum battery current are used in construction of the framework. This paper proposes a solution for a multi-input optimization problem, multi-objective which is an issue in battery management. The algorithm used in this paper gives the present battery residual capacity and forecasting the behavior of driver, also selects the best control variables over a suitable control discretization step. The Driving style gives the best aging strategy. A Markov-chain-based driving style, self-learnt system is created for producing accurate results for battery aging. The simulation of the proposed strategy produces the battery life of 200 000 (km) [14].

Yohwan et al (2019) proposed a health management system to know about the risk of the failures in advance and it has been implemented in battery systems. Depending on the charging profiles and capacity of the battery the model is analysed and characterized using the neural network algorithms. To estimate the state of health, a familiar algorithm such as Feedforward Neural Network (FNN), Convolution Neural Network (CNN) and Long -Short Term Memory (LSTM) is used. The results obtained in this of the experiment is efficient than the conventional method. The proposed multi-channel technique is based on current, voltage and temperature profiles in which the mean absolute error is reduced up to 25 to 58% [15].

3. METHODOLOGY

Fig.1 depicts the flow diagram of the proposed system where the data is split into training and testing dataset.

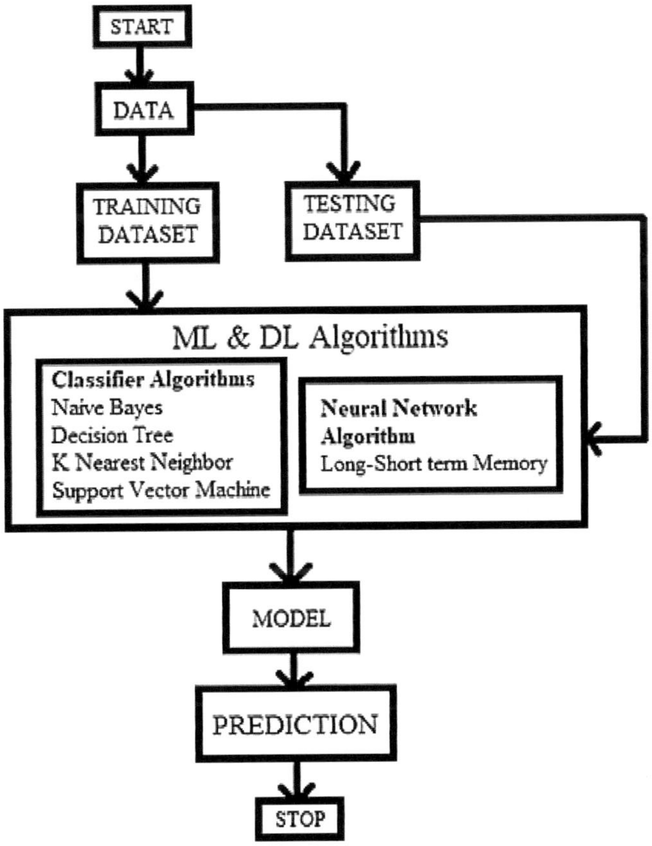

Fig.1. Flow Diagram of the proposed model

The training data is fed into ML and DL algorithms. The data is split as 80% and 20% of training and test data. The training is carried out with the machine learning and deep learning algorithm. The test data is then feedback into the algorithms to obtain the model. Once the model is obtained the prediction is performed based on the prediction formula by retraining the selected model.

3.1. BLOCK DIAGRAM

Fig 3.2.1 depicts the block diagram of the proposed system. The LPC2148 microcontroller is the development board used. The LM395 is the temperature sensor used to measure the temperature of the battery. The current sensor ACS712 measures the current discharge from the battery and the voltage sensor senses the voltage of the lithium-Ion battery. All the sensor values are stored in a spreadsheet with the help of LPC2148 which is connected to all the sensors through GPIO pins. A CPC2102 USB to TTL module is connected to ARM MC which collects the data from MC and to feds to excel in PC. Various machine learning algorithms are implemented in python IDLE to predict the aging of battery. The LPC2148 controllers main purpose is to acquire Realtime data and fed the data to the machine learning model.

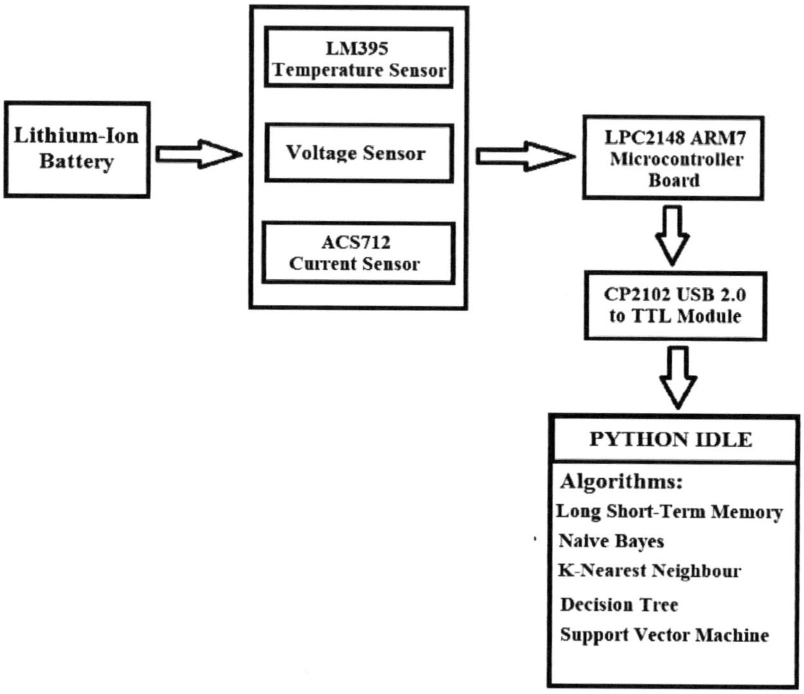

Fig.2. Block Diagram of the proposed model

4. MACHINE LEARNING ALGORITHMS

Machine learning algorithms are meant to learn from experience, it trains the computer to provide the output from the set of given input data. Machine learning algorithm are trained by datasets and provides the output for test data [9]. Computational methods are using by these algorithms for training the model. It learns the information and provided by existing data. As the number of samples increases, algorithms improve their performance. Two techniques are used in machine learning. Supervised learning predicts the future outputs by training the model on known input and output data. Unsupervised learning finds the hidden patterns in input file [1]. The classification algorithms used in this paper helps in identifying the class of data and provides metric analysis such as Accuracy, Recall, F1score and Precision.

4.1 GAUSSIAN NAÏVE BAYES

The naive Bayes classifier are designed using Bayes theorem. It considers that the existence of the feature in a class would be varying to other features. it can provide better performance when it comes to multi class predictions. this algorithm constructs the machine learning models quickly which helps in making fast predictions. this classifier is also known as probabilistic classifier. It makes prediction based on the probability of the object.

4.2 K-NEAREST NEIGHBOUR(KNN)

The K-nearest neighbours has data points that belongs to the class it shares the greatest number of common points with respect to its distance. K numbered nearest points that are present around the data point which must be predicted are taken into consideration. These K points then it belongs to a particular class. The data points belong to the class with the greatest number of points. Euclidean distance method is used to identify the distance between the points in KNN algorithm.

4.3 DECISION TREE(DT)

The Decision tree algorithm forms a tree structure to build classification or regression. The datasets are broken down into smaller combined to develop an decision tree algorithm. Finally, a tree is obtained comprising of decision nodes and leaf nodes. A decision tree algorithm uses branching method to demonstrate all the possible outcomes from a decision that forms a flow-chart-like tree structure. The nodes in the tree represents a test and the branch is the resultant of that respective test.

4.4 SUPPORT VECTOR MACHINE (SVM)

Support vector is used in both classification and regression models. It uses decision boundaries to predict the decision planes. The decision planes can be otherwise called as hyperplane. It separates two different a set of objects that are having different class of memberships in the plane. The classification is performed by finding out the hyperplane which maximizes the exact margin in a plane. By transforming the problem with the help of linear algebra the hyperplanes can be easily calculated.

4.5 DEEP LEARNING

Deep Learning is providing a framework for designing and implementing deep neural networks with algorithms and pretrained models. The neural network algorithms present in deep learning algorithms such as RNN, CNN, ANN and FNN are used to perform regression and classification on time-series, image and text data. Deep learning models can achieve accuracy in a model that can be greater than the conventional calculation or estimation methods. Large set of datasets is required to train the models in deep learning. In this project the battery aging model is trained using LSTM algorithm where the actual and predicted data's can be compared to obtain the accurate prediction results[5].

4.6 LONG- SHORT TERM MEMORY ALGORITHM(LSTM)

An LSTM network is a type of recurrent neural network (RNN) which is capable of long-term dependencies between time steps of sequence data. LSTMs are designed especially in avoiding the long-term dependency problem. The perform a chain like structure but they have their repeating modules with different structures. Long period information can be easily remembered by LSTM algorithm. The LSTM algorithm is mainly used for applications that requires long-term dependencies.

5. EXPERIMENTATIONS

The experimental setup of the Battery Management System is shown in Fig.3. The LPC2148 is connected to sensors through GPIO pins. AC5712 current sensor, Voltage sensor and a Current sensor are used to measure the Battery parameters. Parameters such as current, voltage and temperature are monitored, and the data is collected from the Battery that is connected to the setup. A CPC2102 USB to TTL module is connected between ARM MC and PC which collects the data from MC and to feds to excel in PC. The dataset is collected by measuring the parameters at different conditions. The acquired datasets are then fed into the machine learning and deep learning algorithms.

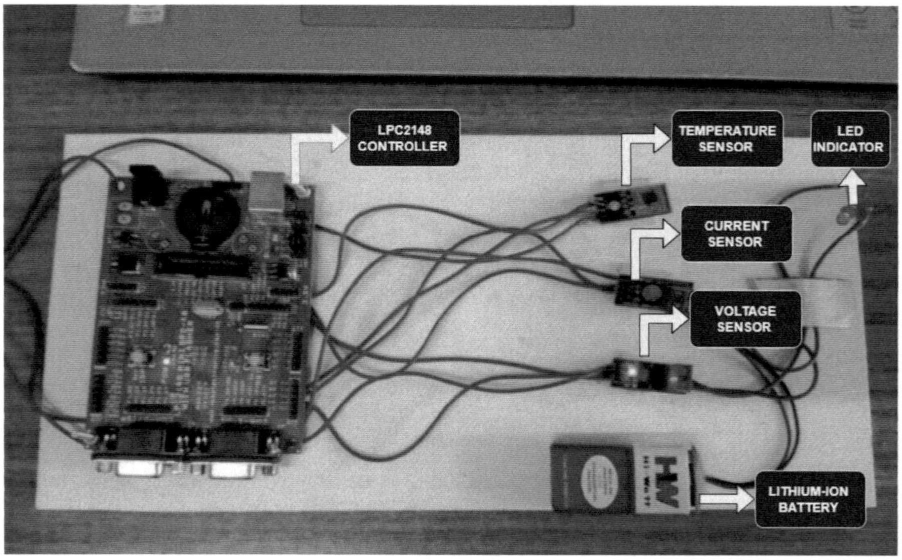

Fig.3. Experimental setup of the battery management system

6. RESULT AND ANALYSIS

The real time data sets that are acquired from the experimental setup are fed and trained in the machine learning algorithm and the various metrics like accuracy, F1-Score, Recall and Precision are analysed for the Machine Learning algorithms and the following results are obtained.

The result part also shows the metrics of the Naïve Bayes algorithm which is helpful in determining the efficiency of the algorithm. From the metrics analysed it is clear that the NB algorithm has the high accuracy of 88%

Fig.4 shows the data's such as Initial Temperature, Voltage, Current and State Of Charge in the form of graph obtained at a particular time when the battery is set to run. The real time dataset values acquired from the Li-Ion Battery is shown in the form of graph.

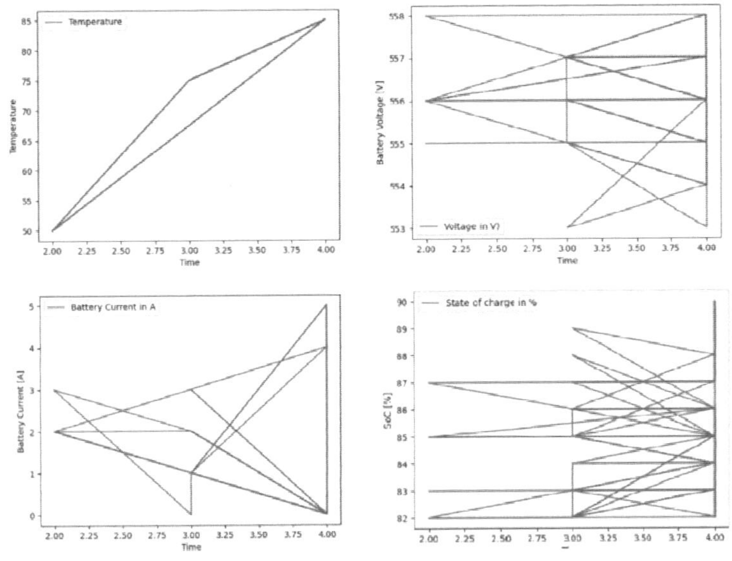

Fig.4. Data collected from Li-Ion battery

Fig.5 shows the losses obtained through number of passes. val_loss is the value obtained for cost function by cross validating the data. The cost function obtained for the training data is the loss. The loss doesn't improve from epoch 16 and thus the training of the dataset completes and early stopping occurs.

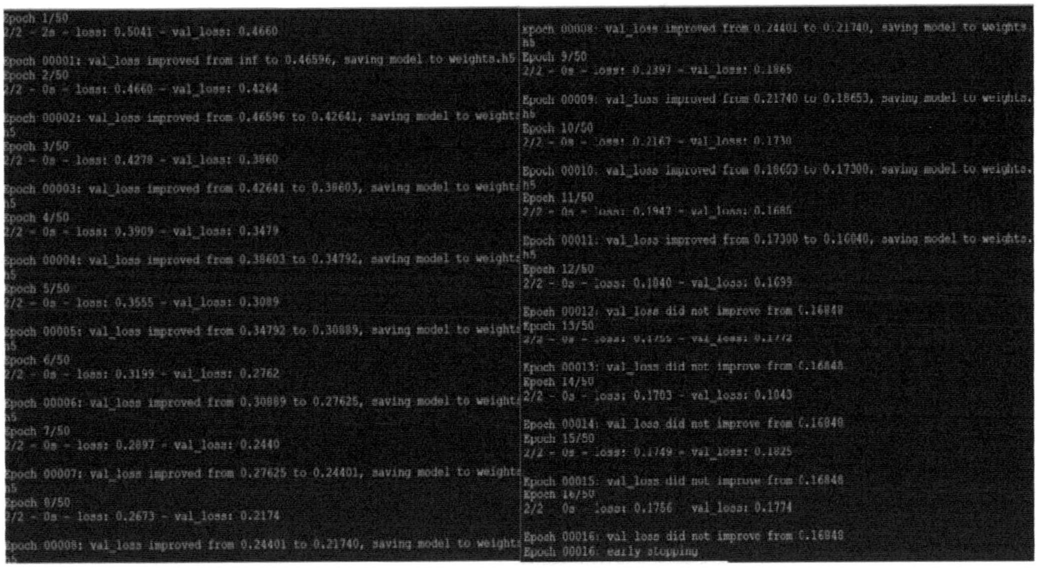

```
Epoch 1/50
2/2 - 2s - loss: 0.5041 - val_loss: 0.4660

Epoch 00001: val_loss improved from inf to 0.46596, saving model to weights.h5
Epoch 2/50
2/2 - 0s - loss: 0.4660 - val_loss: 0.4264

Epoch 00002: val_loss improved from 0.46596 to 0.42641, saving model to weights.h5
Epoch 3/50
2/2 - 0s - loss: 0.4278 - val_loss: 0.3860

Epoch 00003: val_loss improved from 0.42641 to 0.38603, saving model to weights.h5
Epoch 4/50
2/2 - 0s - loss: 0.3909 - val_loss: 0.3479

Epoch 00004: val_loss improved from 0.38603 to 0.34792, saving model to weights.h5
Epoch 5/50
2/2 - 0s - loss: 0.3555 - val_loss: 0.3089

Epoch 00005: val_loss improved from 0.34792 to 0.30889, saving model to weights.h5
Epoch 6/50
2/2 - 0s - loss: 0.3199 - val_loss: 0.2762

Epoch 00006: val_loss improved from 0.30889 to 0.27625, saving model to weights.h5
Epoch 7/50
2/2 - 0s - loss: 0.2897 - val_loss: 0.2440

Epoch 00007: val_loss improved from 0.27625 to 0.24401, saving model to weights.h5
Epoch 8/50
2/2 - 0s - loss: 0.2673 - val_loss: 0.2174

Epoch 00008: val_loss improved from 0.24401 to 0.21740, saving model to weights.h5
Epoch 9/50
2/2 - 0s - loss: 0.2397 - val_loss: 0.1865

Epoch 00009: val_loss improved from 0.21740 to 0.18653, saving model to weights.h5
Epoch 10/50
2/2 - 0s - loss: 0.2167 - val_loss: 0.1730

Epoch 00010: val_loss improved from 0.18653 to 0.17300, saving model to weights.h5
Epoch 11/50
2/2 - 0s - loss: 0.1947 - val_loss: 0.1686

Epoch 00011: val_loss improved from 0.17300 to 0.16848, saving model to weights.h5
Epoch 12/50
2/2 - 0s - loss: 0.1840 - val_loss: 0.1699

Epoch 00012: val_loss did not improve from 0.16848
Epoch 13/50
2/2 - 0s - loss: 0.1755 - val_loss: 0.1772

Epoch 00013: val_loss did not improve from 0.16848
Epoch 14/50
2/2 - 0s - loss: 0.1703 - val_loss: 0.1843

Epoch 00014: val_loss did not improve from 0.16848
Epoch 15/50
2/2 - 0s - loss: 0.1749 - val_loss: 0.1825

Epoch 00015: val_loss did not improve from 0.16848
Epoch 16/50
2/2 - 0s - loss: 0.1756 - val_loss: 0.1774

Epoch 00016: val_loss did not improve from 0.16848
Epoch 00016: early stopping
```

Fig.5. The epoch calculated during training of dataset in LSTM algorithm

Journal of Physics: Conference Series **2325** (2022) 012004 doi:10.1088/1742-6596/2325/1/012004

Fig 6 shows the battery voltage, current, state of charge and temperature plotted against predicted values and actual values trained for about 26 epochs. The actual and predicted values are same which indicates that the LSTM has produced high accurate results.

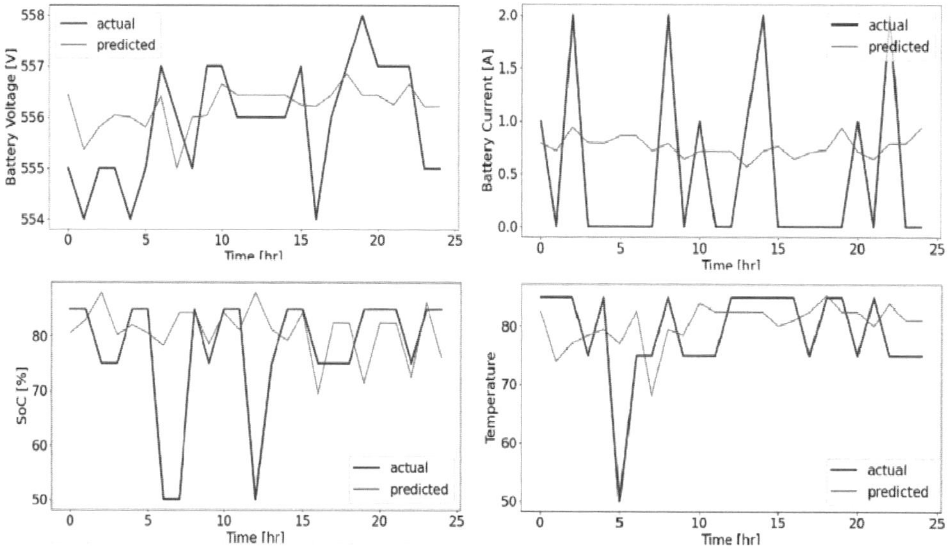

Fig.6. Predicted and actual battery voltage, current, state of charge and temperature through LSTM algorithm

Fig.7 shows for the Naïve Bayes algorithm the rate of accuracy achieved is around 88 %. Among 25 total test labels 22 labels are predicted correctly indicating true positive and negative values.

```
              precision    recall  f1-score   support

           0       0.95      0.91      0.93        22
           1       0.50      0.67      0.57         3

    accuracy                           0.88        25
   macro avg       0.73      0.79      0.75        25
weighted avg       0.90      0.88      0.89        25

[[20  2]
 [ 1  2]] cm
```

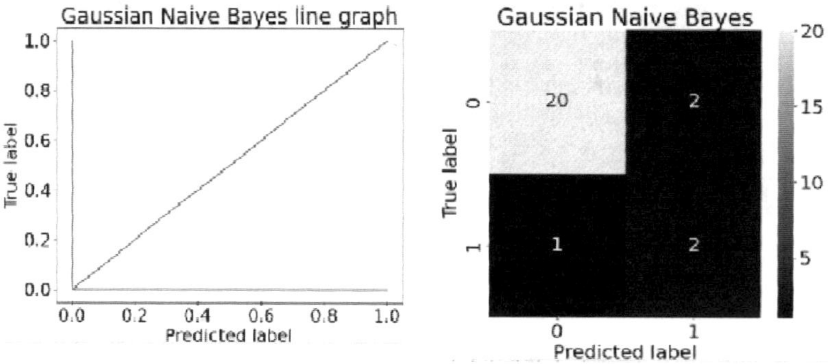

Fig.7. Results of naïve bayes algorithm

Fig.8 depicts for Decision Tree algorithm the rate of accuracy achieved is around 72 %. Among 25 total test labels 18 labels are predicted correctly indicating true positive and negative values.

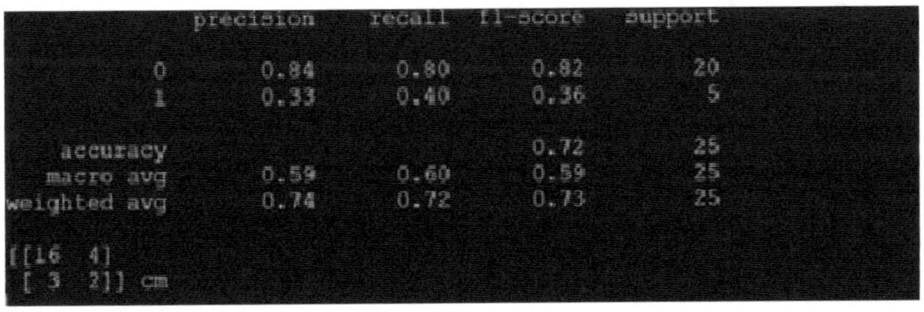

	precision	recall	f1-score	support
0	0.84	0.80	0.82	20
1	0.33	0.40	0.36	5
accuracy			0.72	25
macro avg	0.59	0.60	0.59	25
weighted avg	0.74	0.72	0.73	25

[[16 4]
[3 2]] cm

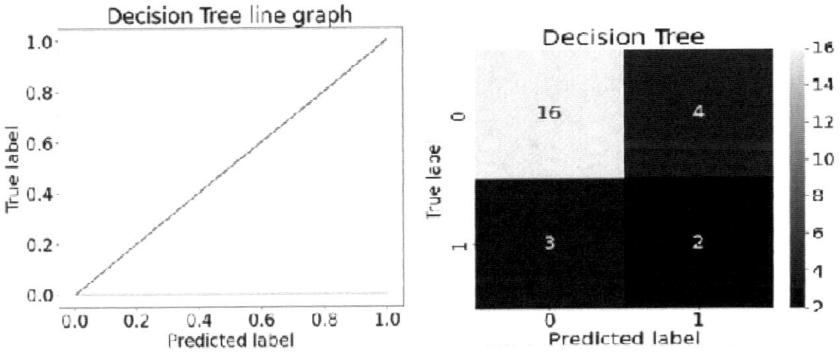

Fig.8. Results of decision tree algorithm

Fig.9 depicts for KNN algorithm the rate of accuracy achieved is around 80 %. Among 25 total test labels 20 labels are predicted correctly indicating true positive and negative values.

	precision	recall	f1-score	support
0	0.90	0.86	0.88	22
1	0.25	0.33	0.29	3
accuracy			0.80	25
macro avg	0.58	0.60	0.58	25
weighted avg	0.83	0.80	0.81	25

[[19 3]
[2 1]] cm

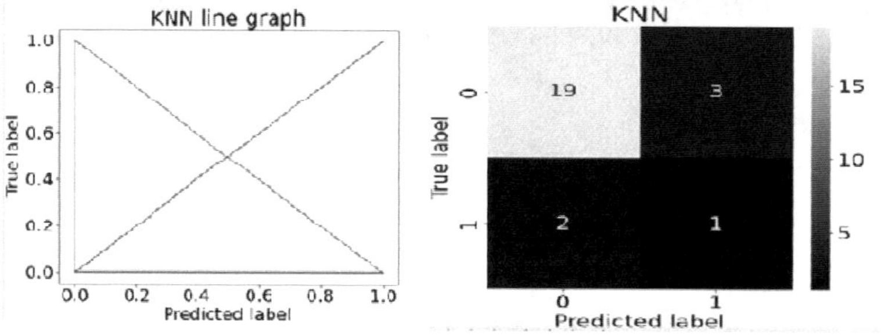

Fig.9. Results of KNN algorithm

Fig.10 depicts for SVM algorithm the rate of accuracy achieved is around 76 %. Among 25 total test labels 19 labels are predicted correctly indicating true positive and negative values.

	precision	recall	f1-score	support
0	0.79	0.95	0.86	20
1	0.00	0.00	0.00	5
accuracy			0.76	25
macro avg	0.40	0.47	0.43	25
weighted avg	0.63	0.76	0.69	25

[[19 1]
[5 0]] cm

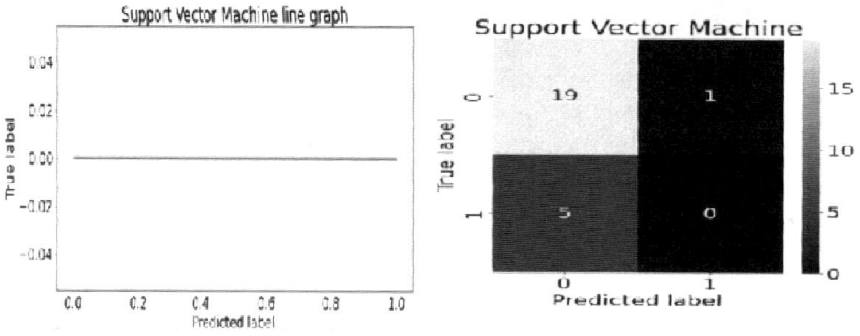

Fig.10. Results of SVM algorithm

Coulomb Counting Method

The Coulomb counting method measures battery's discharging current and integrates the discharging current over time in order to estimate SOC. To estimate the SOC(t), Coulomb counting method requires discharging I(t) and previously estimated SOC value SOC(t-1). The SOC equation is given as [6]

$$SOC(t) = SOC(t\text{-}1) + \frac{I(t)}{Qn} \Delta t \quad (1)$$

ANALYSIS REPORT

Table.1 infers that by comparing results of Classification report we can observe Naïve Bayes produces high Accuracy with high precision, Recall and F1 score values. Thus, the Naïve Bayes algorithm is best suitable to estimate remaining charge of the lithium-ion battery.

ALGORITHM	ACCURACY	PRECISION	RECALL	F1-SCORE
Naïve Bayes	**88%**	**95%**	**91%**	**93%**
Decision Tree	72%	84%	80%	82%
KNN	80%	90%	86%	88%
SVC	76%	79%	95%	86%

Table.1. comparison of classification metrices

Fig.11 depicts the battery aging prediction obtained after testing in Naïve Bayes algorithm. The Naïve Bayes algorithm produces high accuracy of 88% in Analysis2 of the project. So, the prediction results are taken for analysis from KNN algorithm. Here 1 indicates for the obtained battery voltage, current and temperature, the time for which the remaining battery capacity will be true and 0 indicates false. By this we can conclude that the battery aging can be predicted accurately through Naïve Bayes algorithm.

```
[[0.4        0.         1.        ]
 [0.6        0.4        0.71428571]
 [0.         0.4        0.71428571]
 [0.4        0.         1.        ]
 [0.4        0.4        0.71428571]
 [0.4        0.4        0.71428571]
 [0.6        0.6        1.        ]
 [0.2        0.         1.        ]
 [0.6        0.         1.        ]
 [0.8        0.         1.        ]
 [0.6        0.2        0.71428571]
 [0.2        0.         1.        ]
 [0.6        0.8        1.        ]
 [0.6        0.2        0.71428571]
 [0.2        1.         1.        ]
 [0.6        0.         1.        ]
 [0.6        0.2        0.71428571]
 [0.4        0.         1.        ]
 [1.         0.4        0.        ]
 [0.4        0.2        0.71428571]
 [0.4        0.4        0.        ]
 [0.2        0.         1.        ]
 [0.8        0.2        0.71428571]
 [0.6        0.         1.        ]
 [0.6        0.         0.71428571]]
[0. 0. 0. 0. 0. 0. 0. 0. 0. 1. 0. 0. 1. 0. 0. 0. 0. 0. 1. 0. 0. 0. 1. 0.
 0.]
```

Fig.11. Prediction result of Naïve Bayes algorithm

7. CONCLUSION

In this paper the battery management system to predict the battery capacity was developed and the method was implemented using different machine learning and deep learning algorithms. The four-classification metrics such as accuracy, precision, recall and F1-score are calculated and the naïve bayes algorithm is found to be best to predict aging of the battery by producing high accuracy of 88% while analysing real-time data. This method will allow us to predict the battery life and helps the driver to estimate the driving range of the vehicle. The real-time implementation of this technique will be highly applicable for the estimation of lithium-ion battery capacity in electric vehicles.

REFERENCES

[1] Yichun Li, Mina Maleki, Shadi Banitaan, Mingzuoyang Chen, *"Data-Driven State of Charge Estimation of Li-ion Batteries using Supervised Machine Learning Methods"*, IEEE International Conference on Machine Learning and Applications, January, 2022.

[2] Prakash Venugopal, S. Siva Shankar , C. Phillip Jebakumar, Rishab Agarwal, Hassan Haes Alhelou, S. Sofana Reka, And Mohamad Esmail Hamedani Golshan, *"Analysis of Optimal Machine Learning Approach for Battery Life Estimation of Li-Ion Cell"*, IEEE Access, November, 2021.

[3] Thiruvonasundari Duraisamy, Deepa Kaliyaperumal, *"Machine Learning-Based Optimal Cell Balancing Mechanism for Electric Vehicle Battery Management System"*, IEEE Access, September, 2021.

[4] Kailong Li, Yunlong Shang, Quan Ouyang, and Widanalage Dhammika Widanage, *"A Data-Driven Approach With Uncertainty Quantification for Predicting Future Capacities and Remaining Useful Life of Lithium-ion Battery"*, IEEE Transactions On Industrial Electronics, Vol. 68, No. 4, April 2021.

[5] Tsung-Wen Sun and Tsung-Heng Tsai, "*A Battery Management System with Charge Balancing and Aging Detection Based on ANN*", IEEE International Symposium on Circuits and Systems, April, 2021.

[6] Yizhao Gao;Kailong Liu;Chong Zhu;Xi Zhang;Dong Zhang, *"Co-Estimation of State-of-Charge and State-of- Health for Lithium-Ion Batteries Using an Enhanced Electrochemical Model"*, IEEE Transactions on Industrial Electronics, March, 2021.

[7] Yuanliang Fan; Jing Wu; Zitao Chen; Han Wu; Jianye Huang, *"Data-driven state-of-charge estimation of lithium-ion batteries"*, International Conference on Power Electronics Systems and Application, February, 2021.

[8] Rasool M. Imran, Qiang Li, And Firas M. F. Flaih, *"An Enhanced Lithium-Ion Battery Model for Estimating the State of Charge and Degraded Capacity Using an Optimized Extended Kalman Filter"*, IEEE Access, November, 2020.

[9] Liang Zhao, Wei Yao, Yu Wang, Jie Hu, *"Machine Learning-Based Method for Remaining Range Prediction of Electric Vehicles"*, IEEE Access, November, 2020.

[10] Nitikorn Junhuathon, Guntinan Sakunphaisal, Keerati Chayakulkheeree, *"Li-ion Battery Aging Estimation Using Particle Swarm Optimization Based Feedforward Neural Network"*, International Conference on Power, Energy and Innovations, October, 2020.

[11] Li Zhang, Kang Li, Dajun Du, Yuanjun Guo, Minrui Fei, And Zhile Yang, *"A Sparse Learning Machine for Real-Time SOC Estimation of Li-ion Batteries"* ,IEEE Access, August, 2020.

[12] Fen Zhao, Yinguo Li, Xinheng Wang, Ling Ba, And Tailin Liu, *"Lithium-Ion Batteries State of Charge Prediction of Electric Vehicles Using RNNs-CNNs Neural Networks"*, IEEE Access, May, 2020.

[13] Carlos Vidal, Pawel Malysz, Phillip Kollmeyer, And Ali Emadi, *"Machine Learning Applied to Electrified Vehicle Battery State of Charge and State of Health Estimation: State-of-the-Art"*, IEEE Access, March, 2020.

[14] Matteo Corno and Gabriele Pozzato, *"Active Adaptive Battery Aging Management for Electric Vehicles"*, IEEE Transactions On Vehicular Technology, Vol. 69, No. 1, January 2020.

[15] Yohwan Choi, Seunghyoung Ryu, Kyungnam Park,and Hongseok Kim , *"Machine Learning-Based Lithium-Ion Battery Capacity Estimation Exploiting Multi-Channel Charging Profiles"*, IEEE Access, June, 2019.

Low Power Analog to Digital Converter for Neuro Stimulator using Cadence Tool

Sufia Banu[1] and **Shweta Gupta**[2]

[1]Research Scholar, Department of Electronics Engineering, Jain (Deemed-to-be University), Bengaluru, India.
Assistant Professor, Department of ECE, HKBK College of Engineering, Bangalore, India.
[2] Associate Professor, Department of Electronics Engineering, Jain (Deemed-to-be University), Bengaluru, India.
sufiabanu.ec@hkbk.edu.in

Abstract. A Digital-to-Analog Converter (ADC) converts various forms of communication in the real world to portable digital integers. ADCs play a vital role as it interfaces between the analog world and digital world. Wireless sensor networks and biomedical interfaces are common examples of applications that demand ultra-low power consumption in the ADC. This research paper shows reduction in the energy consumption of an ADC used in neurostimulator which stimulates the brain for the people suffering from various disorders which includes Tumours, Parkinson's disease, Obsessive Compulsive Disorder (OCD), Epilepsy, and many more. The simulation is carried out by Cadence Virtuoso tool using various technologies.

Keywords: Analog-to-Digital Converter, Cadence, CMOS, Digital-to-Analog Converter, Low Power, Neuro-Stimulator

1. Introduction

Biomedical signals are low-frequency with millivolt-level noise. Bioelectric signals have a dynamic range of tens of microvolts to hundreds of millivolts and encompass a frequency spectrum with a maximum frequency of less than 10 kHz. Nanosensor signals are all physical in nature. Low-frequency biomedical signals have millivolt-level noise. Detectors are used to turn these data into electrical signals. The instrumentation is done to monitor biomedical signals in the human body, such as electroencephalography (EEG), electrocardiography (ECG), and electromyography (EMG). Figure 1 depicts the major block diagram of an electrocardiogram (ECG). One of the most often used domains for collecting biosignals in order for physicians to identify heart issues is the electrocardiogram ECG [2]. The electronical impulses from the human body must be transcribed to digital signals after perceiving. Signal guidelines are laid out in the digital realm. In all biomedical applications, an Analogue to Digital Converter (ADC) is essential for this transformation. The primary goal of the ADC is to convert analogue signals into digital representation. This digital signal is further processed before being sent out. Low power and extended battery life ADCs are desired for these attributes. The task at hand is to design a low-power ADC with a low-power architecture. ADCs should also have a low duty cycle, a medium granularity, and an intermediate frequency response.

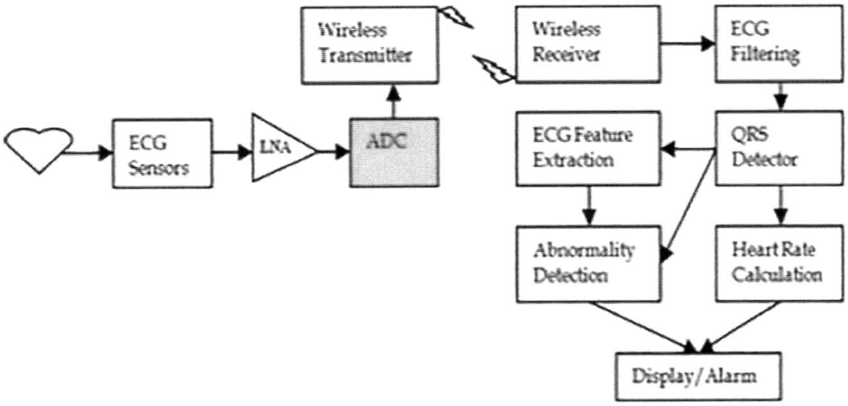

Figure 1. Block diagram of ECG [8]

These signals are transmitted using sensors. The sensor gear is intended to recognize bio-instrumentation signals generated by subject body, such as (EEG), (ECG), and (ECG) (EMG). To evaluate these bioelectric signals, moderate low-speed ADCs are being created. The focus of this dissertation is on the planning and construction of hyper ADCs. Power consumption is one of the major challenges in building tiny electronic and communication systems such as sensor network, injectable devices for diagnostics and therapeutics (such as cardiovascular illnesses), and ubiquitous surgical devices. These devices must be able to operate for a longer period of time without needing to be replaced. As a result, extensive efforts have been made to build ultra-low consumption electronic components (ICs) that use microwatts or even nanowatts of electricity. The discoveries of these investigations have resulted in increased battery life, lossless compression of batteries, and the usage of photovoltaic systems. The bulk of these processes includes wireless access through a following provider. Additional concern is data transfer rate, which is controlled by the broadcaster and receiver's power consumption boundaries. The data transfer rate is directly proportional to the sampling rate, which is due to the sampling type and ADC specification [4]. One of the most significant blocks in wireless portable instrument ICs is the ADC, which consumes a lot of power. The conventional synchronous sampling technique, in which samples are obtained evenly in the time domain and at ultra-low power, is implemented using a variety of architectures. This sampling approach, however, does not provide enough samples for sparse and laser pulse signals. Speech, ECG, and brain spikes are examples of passive signals that have become active in a short amount of time [5].

2. Related Work

S. Sirimasakul et al., discussed the design and implementation of a 0.35 m CMOS working-mode logarithmic pipeline continuous to discrete device. The ADC was used to convert the analogue output current of an ISFET-based pH sensor that is sub-microampere. Switching frequency MOSFETs and the group consisting of ideas were utilized to create affective and low-energy ADC. The ADC was demonstrated using a 1-V value of voltage and squandered 3.3 W constant power at a 1-kHz pulse amplitude [6].

Lei Zouet et al., described industrial electronics. Petroleum & energy prospecting, well drill, and self-propelling applications where the working heat exceeding 200 °C are all immensely popular. Circuit designs based on conventional uniform beam semiconductor (CMOS) technology have a maximum temperature rating of 125 degrees Celsius, which is adequate for dire circumstances. The paper described a silicon-on-insulation CMOS-based slightly elevated (HT) 9-bit serial communication register digital audio converter (SAR ADC) with a recording rate of 50 kS/s. The HT SAR ADC's design considerations were explored at the perspectives of operation selection, degree of heat aware circuit design, and implementation. The design procedures for the HT SAR ADC were examined from the angles of process selection, thermal signal conditioning, and sensing. The ADC had an effectual amount of memory (ENOB) of 8.35 bits and a figure of merit of 93 pJ/step at normal temperature. In the HT test, ENOBs of 7.3 values at 225°C and 6.9 values at 300°C were achieved. The power required from a 5-V supply is 1.52mW at normal temperature was only 2.17mW at 300 °C [7].

H Pham et al., have established a method for designing and developing organic integrated circuits employing suitable model of nanocomposite P-channel and fullerene N-channel thin-film transistors (TFTs). The transistors were first constructed, and their key physical and electrical properties were measured. These organic TFTs (OTFTs) were then constructed in Cadence with the help of an organic process engineering tool (OPDK). The modeled elements' main specifications were collected from measured data, and the suitable ones were chosen to duplicate probing curves. The computational methodology shows that the TFTs' amplitude retaliation covered all semi frequency ranges, confirming that they can be used to build integrated circuits for biomedical applications. Organic circuits that follow complementary laws worked well, such as hardware implementation, flip flops, compensators, and mixed - signal converters (ADCs). The suggested model progressively (SAR) ADC draws 883.7 W and obtains an ENOB of 5.05 bits, an SNR of 32.17 dB, and an ENOB of 5.05 bits at a dc potential of 10V and a sampling frequency of roughly 2Kilo Hz [8].

S. Kesav et al., illustrated the outline of a low-power Successive-Approximation-Register Analog voltage (SAR ADC) for biopotential image acquisition in 45nm CMOS technology. To reduce power dissipation, it was engineered with a high load resistance (Vt) cell. In terms of strength and delay, a 10-bit SAR ADC was devised and contrasted to a coarse contrast SAR ADC and a regular threshold voltage (Vt) ADC. The results demonstrated that using a high Vt SAR ADC saves up to nearly half of the power used by a reduced Vt SAR ADC not delayed. Other performance metrics investigated have included the Effective Number of Bits (ENOB) and Signal-to Noise-Ratio (SNR), Signal to Chaos along with Decline Ratio, and Fallacious Free Active Ratio [9].

X. Tong et al., demonstrated an implantable multichannel brain recording system. The assessment of a 10-bit 120 kS/s was succeedingly continuous to discrete translator (SAR ADC). To minimize the energy consumption and area occupation, more resources of VCM-based switching paradigm was given. Unlike the monotonic switching approach, the suggested scheme's switching procedure for each bit phase was nearly symmetrical, rendering comparator construction easier. Importantly, because many of these capacitors were linked to VCM during the sampling phase, this switching scheme's reset energy was zero. For improved linearity, built company sampling switches were used. The envisioned ADC had a 0.13 mm2 surface region and was produced in 0.18-m CMOS. The energy reduction was 2.97 W at a sampling rate of 120 kS/s, that included the I/O and two 4-to-1 demultiplexer. This suggested SAR ADC's meritocracy estimate was around 36.9 Volts [10].

Banu, S., & Gupta, S designed a low noise amplifier and digital to analog converter used for deep brain neurostimulator. The total power consumption was comprised of dynamic power and static power. The power consumed by the device when in ON state was the dynamic power which is due to the charging and discharging of the device while static power results in when the device was in OFF state. Paper described the reduction in power consumption while in OFF state by using high Vth transistors [11].

3. Types of ADC

The following are the six main categories of ADCs: ADCs using Successive Approximation (SAR), Delta-Sigma, Dual Slope ADC, ADC pipelined, Flash ADC and ADC for level crossing.

3.1 Successive estimation (SAR) ADC: Successive Estimation type ADC is the most generally used and popular ADC method. The basic idea of this form of A/D inverters is that it analyzes the random analogue input strength to an n-bit discrete value one section at a time, commencing with the MSB. This circuit consists of an analog to digital (SAR), DAC, and comparator. An n-bit DAC receives the SAR output. The non-inverting input of the comparator, receives the DAC's comparable analogue dc output, VD. The comparator's second input is an undisclosed analog signal voltage VA. SAR's progressive reduction logic is edged by the comparator's output. The SAR sets the least significant bits to logic high and then the other bits to logic low, culminating at a logic 1 MSB and logic 0 other bits, developed in a stream code of 1000 when the start of set of specifications is issued.

3.2 Delta – Sigma ADC: The analogue reference voltage wave is connected to the integrator's input in a converter, culminating in a potential constant change in rate, or slope, at the output terminal that equates to the input reference altitude. A comparator compares the ramp voltage to the ground potential (0 volts).

Based on nature of the compensator result is high or low, the comparator serves as a one-bit ADC, yielding 1 bit of output ("high" or "low"). This value of comparator is then routed through a high-frequency D-type flip-flop and further given back to another integrator input passage, driving the aggregator in the path of a 0 volt output. The crucial circuit is as follows: The (summing) integrator is the op-amp on the far left. The comparator, or one-bit ADC, is the next op-amp the integrator feeds into. The D-type flip-flop follows, storing the output of the comparator throughout for every clock-time and passing each of the two levels, a "high" or "low" wave to the circuit's next comparator [12].

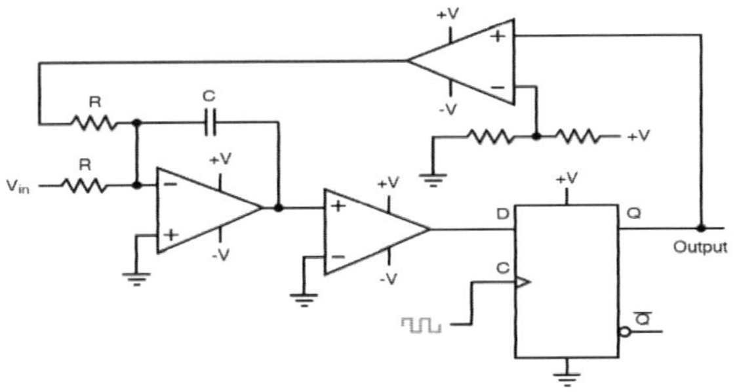

Figure 2. Schematic diagram of delta sigma ADC

3.3 Dual Slope ADC: The 0 Volts / 5Volts logic level output voltage of the flip-single-polarity flop must be modified to a +V / -V voltage wave before getting fed to the integral by this last comparator. If the integrated response is affirmative, the first comparator will send a "high" signal to the D input of the flip-flop. The gulf stream principle is exemplified here, with the first comparator identifying a difference () between the integrated output and inconsequential volts. The comparator's output is appended to the analogue signal by the integrator. The dual-slope converter is a design improvement that solves the calibration drift problem. Instead of just being reset to 0 volts, a dual-slope converter uses an integrator circuitry that is carried optimistic and pessimist in overlapping cycles to reduce power and then increase. The integrator is being in charge of by a positive analogue input signal for a predetermined amount of time in one direction of ramping (creating a cynical, change in the value, or output slope), as calculated by a counter with a precise number of clock. Then, with the same counter, in the opposite side, with a constant beginning voltage (creating a fixed rate of output DC change). The counter rests calculating when the integrator's operation is smoother with the same potential as when the rectifications of the process began. The time taken for the integrator's battery to collapse and return to the starting terminal voltage, as assessed by the amplitude accumulated by the counter, is the discrete value of the ADC circuitry.

3.4 Pipelined ADC: The pipelined mixed - signal converter (ADC) architecture has been the most typical one for sampling scales ranging from several speech sound per second (Msps) to 100Msps+. The resolution fluctuates from eight bits at speedier sample rates to 16 bits at sluggish sample rates. CCD diagnostics, acoustic medical imaging, multimedia receivers, core networks, digital video (for example, HDTV), xDSL, broadband connections, and fast Lan are all covered by these resolutions and sample rates.

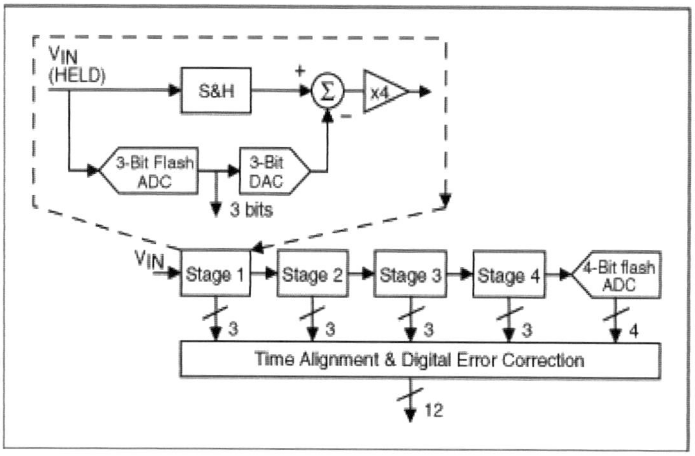

Figure 3. Block diagram of pipeline ADC

3.5 Flash ADC: A flash ADC (also known as a straightforward ADC) is a multitrack processor that utilizes a constant voltage ladder with assessments at each "tier" to ensure that the input voltage fits the succeeding working voltage. These reference ladders hire a bunch of resistors, but their own widespread use suggests that battery pack splitting is also a possibility. The outputs of these comparators are frequently passed to a multitrack encoder, which encodes the inputs to numeric representation (the extracted outputs from the comparators can be viewed as a unit value).

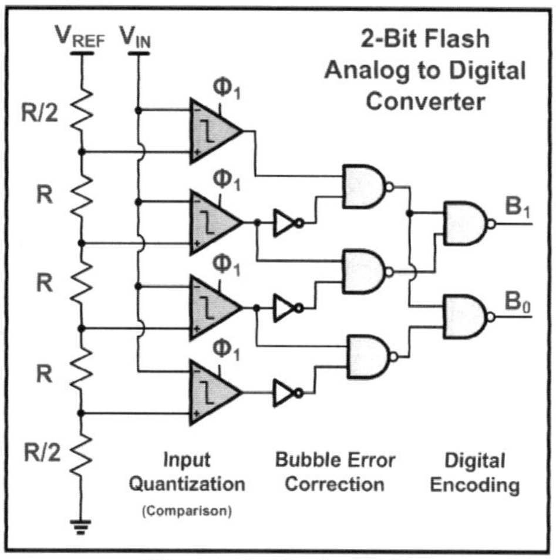

Figure 4. Flash ADC

4. Proposed work

The bioelectrical Retrieval System ASIC (Application Specific Integrated Circuit) requires a Mixed - signal Converter to digitize the analogue waves of the readout beginning end mediums. A Serial

Communication Register ADC (SAR ADC) design is chosen over alternative architectures because of its adequate resolution, quicker conversion time, and enhanced power absorption statistics. Figure 5 depicts the 10-bit SAR ADC's block diagram.

The biomedical device such as EEG, ECG and EMG picks the very weak signal which is analogous in nature which then be amplified and given to analog-to-digital converter, processed, filtered and further converted back to digital form by digital-to-analog converter. Each block must be designed with minimum power requirements, higher performance, and aggressive voltage scaling. This ADC is designed with all these constraints. If a biomedical device that approximately consumes 10 mW of power is considered, then that battery will last for 3 days. On the other hand, if the device consumes 8 μW of power, then the life of the battery will be increased. The core idea of the ADC which is used in this paper is to lower the power consumption which can handle high power biomedical applications. Specialized techniques with high threshold voltage transistors are used which achieves reduction of power consumption in bio medical applications.

Binary search algorithm is used as the basis in determining the ADC(SAR), which converts analogue to digital by using previous output bits to determine the input condition for the next intense peak. As seen in Figure 5, it is made up of four internal components. The boost converter from the ECG system is acquired by a hold and sample circuit, and the sampled DC is evaluated to the internal DAC actual output by a comparator. The comparator in the vicious circle receives the required benchmark from the DAC. The analogue output value is converted into a digital code using a successive approximation register. DAC changes this digital code to an analogue value, which operates as the reference value [6].

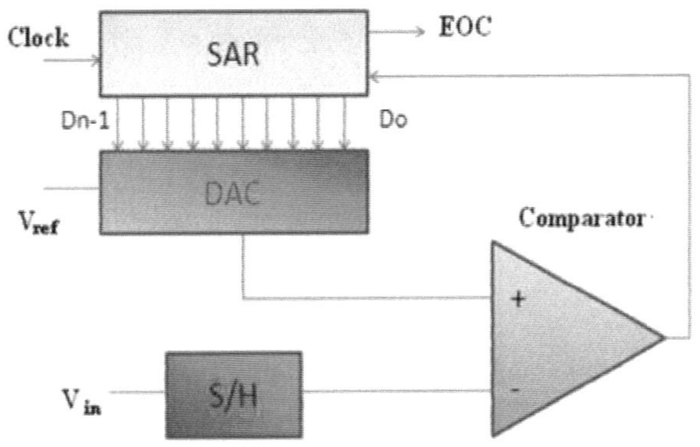

Figure 5. Block Diagram of SAR ADC

4.1 Sample and Hold Circuit
A basic analog switch is built for the sample and follows function in the proposed work, as shown in Figure 4. The supply voltage VDD is used to set the switch gate-to-source voltage. As a result, the sampling device's ON-resistance is lessened, and the switch's repeatability is strengthened. Figure 5 shows the result of the sampling and holding circuitry with sinusoidal reference wave. In 45 nm CMOS technology, the energy and delay numbers at the specified sample and follow circuit are listed in Table-I.

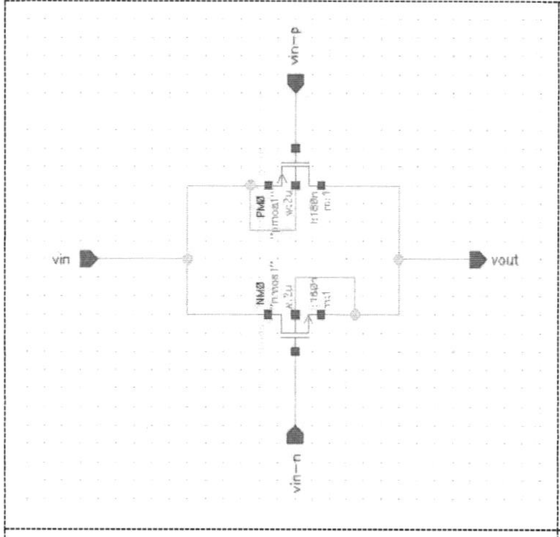

Figure 6. Sample and Hold Circuit Schematic

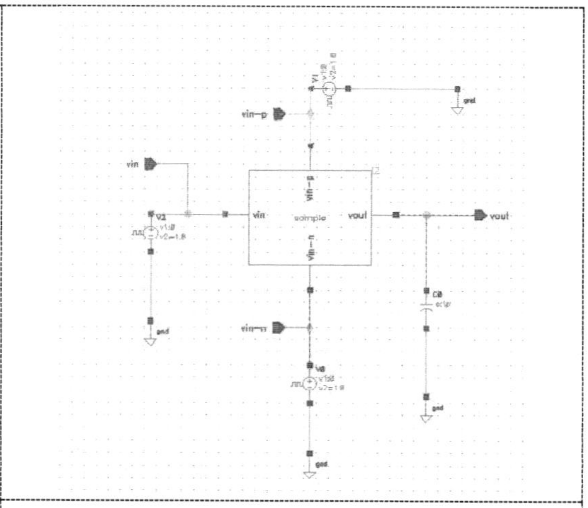

Figure 7. Test Schematic of Sample and Hold Circuit

4.2 Comparator

The audio input sample signal is significant to the result of the embedded DAC by the comparator. In comparison, the result will be either on or off logic levels. The SAR receives the value of output. The comparator's accuracy and speed are two critical factors. The comparator would have to be able to resolve minor voltage changes. The drift amplitude of the differentiator exploited in SAR ADC is transferred into the ADC's transfer characteristic. The ADC's linearity can't be changed. In this dissertation, a feedback linearization comparing is constructed using 45 nm CMOS technology. The practical circuit of an open loop comparator is shown in Figure 8.

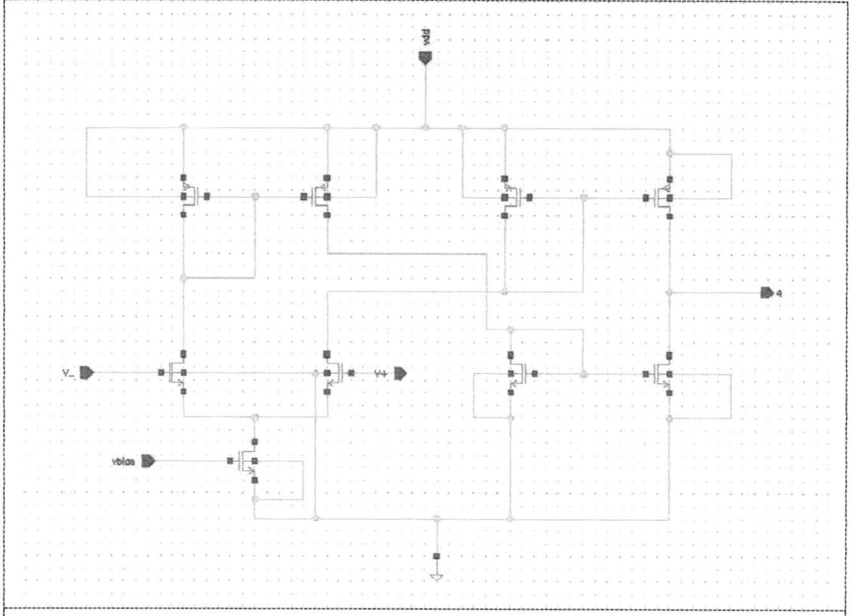

Figure 8. Schematic of a Comparator

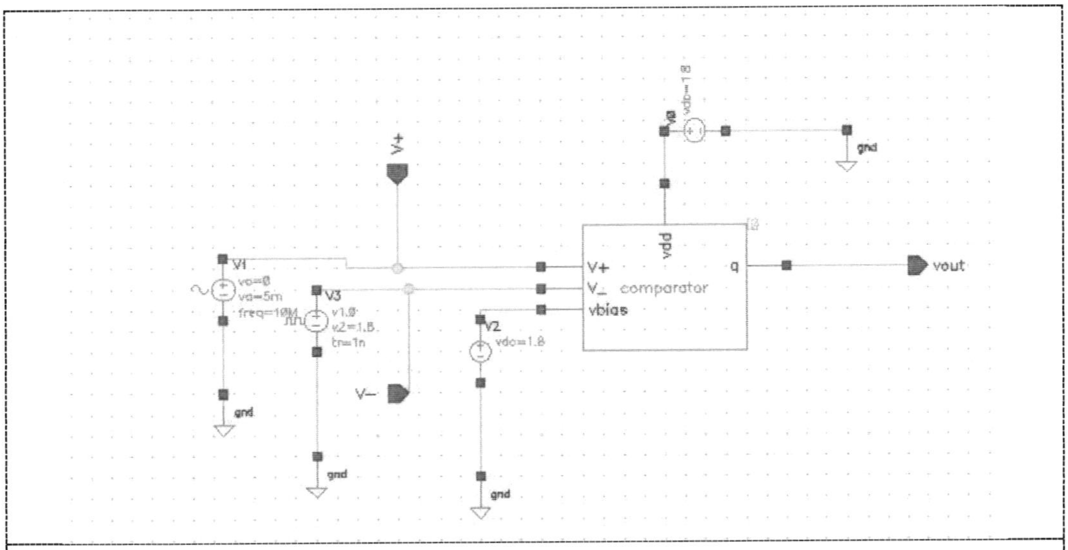

Figure 9. Test Schematic of Comparator

5. Results and Discussions

The transient response of hold and sample circuit is as shown in figure 10. When PMOS is at logic 1 and NMOS is also at logic 1, only NMOS turns ON whereas PMOS is OFF. When input Vin is high, then NMOS tries to charge up to high voltage, but it fails as NMOS is the weak passer of logic 1. Here, the output voltage is degraded by threshold voltage. It will not reach to complete 1.8 Volts, and it stops at 1.25 volts. Second condition is when NMOS is at ON state and PMOS is at OFF state, while Vin is at logic LOW causes the output to be at logic LOW. Third condition is when both NMOS and PMOS are ON, whatever input applied, the exact output is obtained at Vout terminal. Fourth condition is when both PMOS and NMOS are OFF, thus whatever input is applied, it will not be available at the Vout terminal. The Power and delay comparisons of sample and follow circuit at 180 nm and 45 nm technologies are as shown in table 1 and table 2.

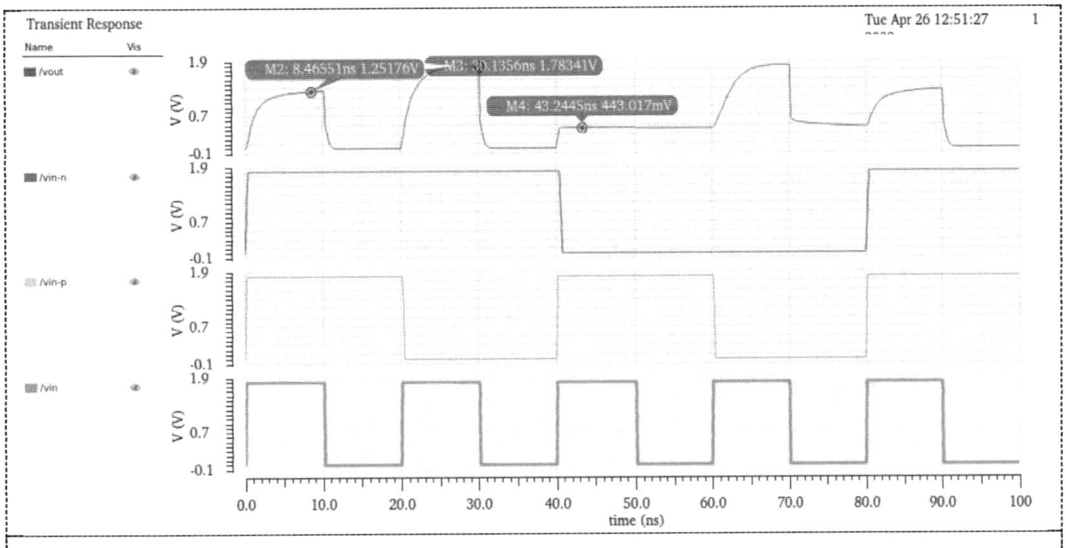

Figure 10. Transient Response of Sample and Hold Circuit.

Table 1. Power and Delay comparison for 180nm Technology

	Average Power	Delay
Normal Vth	87.52 μW	10.14 ns
High Vth	93.39 μW	14.36 ns

Table 2. Power and Delay comparison for 45nm Technology

	Average Power	Delay
Normal Vth	101.1 nW	4.785 ns
High Vth	98.43 nW	7.979 ns

6. Conclusion

Low-speed, moderate ADCs with ultra-low-power performance are required for biomedical implants. SAR ADCs are the most used ADC architecture due to their various elevated energy efficiencies. When compared to a low Vt Sample and Hold circuit, the data transfer gate circuitry of hold and sample is implemented with high threshold voltage transistors. Using 180 nm and 45 nm CMOS technology, an analogue circuit with a basic DC coupled high gain voltage is constructed with high threshold voltage Vt chambers to minimise the energy consumption and delay.

Technology scaling and integration of large number of devices on a single small chip has enabled reductions in energy consumptions which is a key factor for miniaturised, implantable high power bio medical devices. Circuit level approach is employed by using high threshold voltage transistors which reduces the leakage power. Further, a system level approach can be adopted to lower the power consumption in biomedical device. Partitioning of analog and digital domains can also be achieved as analog including higher design complexity and noise sensitivity. In future, the battery requirements of a neurostimulator can be replaced by external source by electromagnetic energy transfer. Advances in miniaturization of integrated circuits can enable the fully implanted battery-less biomedical devices.

7. References

[1] Zou, X., Xu, X., Yao, L., & Lian, Y. (2009). A 1-V 450-nW Fully Integrated Programmable Biomedical Sensor Interface Chip. *IEEE Journal Of Solid-State Circuits*, *44*(4), 1067-1077. doi: 10.1109/jssc.2009.2014707

[2] Yin, M., & Ghovanloo, M. (2010). A low-noise clockless simultaneous 32-channel wireless neural recording system with adjustable resolution. *Analog Integrated Circuits And Signal Processing*, *66*(3), 417-431. doi: 10.1007/s10470-010-9557-6

[3] Verma, N., & Chandrakasan, A. (2007). An Ultra Low Energy 12-bit Rate-Resolution Scalable SAR ADC for Wireless Sensor Nodes. *IEEE Journal Of Solid-State Circuits*, *42*(6), 1196-1205. doi: 10.1109/jssc.2007.897157

[4] Scott, M., Boser, B., & Pister, K. (2003). An ultralow-energy adc for smart dust. *IEEE Journal Of Solid-State Circuits*, *38*(7), 1123-1129. doi: 10.1109/jssc.2003.813296

[5] Gupta, S., & Kumar Singh, S. (2012). Neurostimulators Used in Brain for Parkinson's disease and Epilepsy Using Bionics. *International Journal Of Scientific Research Engineering & Technology (IJSRET)*, *1*(5), 049-052. Retrieved from http://www.ijsret.org

[6] Sirimasakul, S., Thanachayanont, A., Jeamsaksiri, W.,(2014). Low-Power Current-Mode Logarithmic Pipeline Analog-to-Digital Converter for ISFET based pH Sensor.

[7] Zou, L., Pathrose, J., & Je, M. (2015). A 9-bit successive approximation ADC in SOI CMOS operating up to 300 °C. *International Journal Of Circuit Theory And Applications*, *44*(2), 418-427. doi: 10.1002/cta.2084

[8] Pham, H., Nguyen, T., Pham-Nguyen, L., Sakai, H., & Dao, T. (2016). Design and Simulation of a 6-Bit Successive-Approximation ADC Using Modeled Organic Thin-Film Transistors. *Active And Passive Electronic Components*, *2016*, 1-11. doi: 10.1155/2016/7201760

[9] Kesav, S., K.S, N., & B.K, M. (2017). Design of Low Power SAR ADC for ECG Using 45nm CMOS Technology. *International Journal Of VLSI Design &Amp; Communication Systems*, *8*(1), 19-29. doi: 10.5121/vlsic.2017.8102

[10] Tong, X., Song, M., Chen, Y., & Dong, S. (2019). A 10-Bit 120 kS/s SAR ADC Without Reset Energy for Biomedical Electronics. *Circuits, Systems, And Signal Processing*, *38*(12), 5411-5425. doi: 10.1007/s00034-019-01138-6

[11] Banu, S., & Gupta, S. (2022). Power optimization of Low Noise Amplifier (LNA) and DAC used in Closed Loop Deep Brain Neuro-Stimulator (CDBS) at 45nm using Cadence Virtuoso. *International Journal Of Health Sciences*, 5491-5502. doi: 10.53730/ijhs.v6ns3.7153

[12] Bansal, M., Singh, H., & Sharma, G. (2021). A Taxonomical Review of Multiplexer Designs for Electronic Circuits & Devices. June 2021, 3(2), 77-88. doi: 10.36548/jei.2021.2.001

COST EFFICIENT VIRTUAL TRIAL ROOMS

M Malathi [1], R Induja [2], S Mubarak [3], T Lalitha [4]

Sri Krishna College of Technology, Coimbatore, India.

{m.malathi, 18tuit037, 18tuit057, 18tuit052}@skct.edu.in

Abstract: The recent covid-19 restrictions like social distancing has limited physical human interaction. There were days when people wish to visit the physical shopping outlets and try the best-suited outfits in the trial rooms. In the post-pandemic era, people are now preferring the E-shopping zone to select the outfits. This type of virtual shopping has its own disadvantages like the suitability of outfits to the customer's preferences. Regardless of some applications that offer outfit suitability, there are some entrust issues with online stores. This research study analyzes many general customer related concerns like fit of the outfit; appearance; color suitability with the customer skin tone and proposes the possibility of implementing virtual trial rooms. The proposed model empowers the clients to evaluate their outfits based on their stature, shape, size, and wellness estimations by standing in front of the gadget.

Keywords: Machine learning, convolutional neural network algorithm, trail rooms.

1. INTRODUCTION

Machine learning (ML) is an artificial intelligence technique (AI) that automates the process of enhancing software application accuracy in forecasting the outcomes without being specifically designed. ML algorithms predict new output values by analyzing historical data. By utilizing machine learning algorithms, users can supply a computer algorithm with a large amount of data and the computer will process that data to come up with data-driven recommendations and decisions [15]. In artificial intelligence (AI), machine learning provides the ability to learn from experience and improve without involving any explicit programming. As a part of machine learning, programs are developed by providing access to data and utilizing it to advance their education [16].

Data is either observed or obtained. For instance: explicit or implicit commands to identify the design of the data and obtain better judgments based on the provided examples. Basically, computers are supposed to work without human intervention or instruction and alter their activity appropriately. In traditional machine learning algorithms, a text will be usually treated as a sequence of keywords. However, the computer system mimics the human ability to understand the meaning of a text [17].

Supervised learning algorithms can use labeled examples of previous data to predict future events based on the information learnt from the past. In order to make predictions about the output values, the learning algorithm analyzes a known training dataset. If the system is sufficiently trained, every new input may deliver objectives [18]. As a result, the learning algorithm can also compare its output with the intended outcome to find and correct errors. A machine learning algorithm using unsupervised learning, on the other hand, is used when training is not accompanied by any classifications or labels. Learning unsupervised describes how systems can infer a function to describe a hidden structure. The method doesn't figure out the right output, but it uses datasets to infer hidden structures from unlabeled data and can draw inferences from those datasets [19]. These techniques cover both supervised and unsupervised learning because they utilize both labeled data and unlabeled data during training. They can use a small quantity of labeled data and a large amount of unlabelled data. Systems that apply this method are able to considerably improve learning accuracy. Semi-supervised learning is often used when the acquired labeled data requires skilled and relevant resources in order to be trained. Otherwise, data that is unlabeled doesn't require additional resources [20].

2. RELATED WORK

Virtual changing area applications pulled in numerous investigations. A strategy and framework were delivered to work with an acknowledgment of the body based on signals that address orders to start activities to begin activities inside an electronic commercial center for the client. A model of the client's body is created by utilizing the primary arrangement of visual information. The activity machine then builds a new model based on the second set of spatial data. The activity machine, which is addressed by a signal, controls the contrast between the primary and following models. [1]

A virtual changing area application incorporating Kinect sensors was demonstrated. The methods proposed were designed to remove the client from video transfers, detect skin color, and organizing models. The 3D areas of the joints were used to position, scale, and pivot the 2D fabric models with the client. [2]

In an arrangement of programming programs, an advanced pictures data set was introduced. The picture data set incorporates articles of clothing pictures and an essential model self-perception. This permits the customer to independently choose and take a stab at the distinctive information base article of clothing pictures. The framework delivers a picture of the customer's body in the contentions, with the client's explicit lumps, bends, and tallness which mirror the customer's body estimations. [4]

A garment depicting the use of common estimates to create simulated bodies was revealed. Clothing designs are created by using splines, then emerge around a virtual human figure to give the underlying contour. The recreation was made by applying actual boundaries dependent on genuine texture properties to get the appeared piece of clothing. Later the article on clothing creation, a web browser embedded continuously stage was utilized as an interface to the web.[5]

The application empowers online customers to perceive how an item of clothing will appear on the customer before purchasing it. The consumer installs the application on their mobile device. The customer will be asked a series of questions about his or her physical size, shape, and complexion later on.[6]

Y. LeClerc's method is tailored to realistic reconstruction challenges, such as (1) regaining form from photos with erroneous data on the measurement and (2) creating crude scene models from a variety of angles.[7].

A straightforward approach for recreating a 3D surface from stereo pictures was suggested by L. Zhang. They estimate the total depth of a mesh's vertices constructed of a patchwork of trilateral marks on the reference (model) figure to recreate a 3D surface.[8].

R. Cipolla advocated reconstructing surfaces from their 'apparent contours,' or the outlines that we see when we look at them from afar. These apparent outlines do not appear to contain enough information.[9]

Information is needed to recreate an unmarked smooth surface. There is much less evidence that they carry this information without precise knowledge of the observer's motion; in fact, we do not know whether this is true in general at the time of writing. They have, however, effectively achieved the reconstruction when only a portion of the observer's motion is known - that is when the observer's motion is restricted to a specific type of circular motion.[10].

The employment of VR and AR technology in retail applications and research is contrasted and studied. The terms "implementation," "customer acceptability," "applications," "problems," and other terms linked to them are compared [21].

Seyed Omid, is the first to use this unique scheme to categorize fashion try-on and synthesis studies. After introducing a three-level hierarchy of linked tasks, all relevant studies are identified and categorized using a multi-label method [22].

3. PROPOSED METHOD

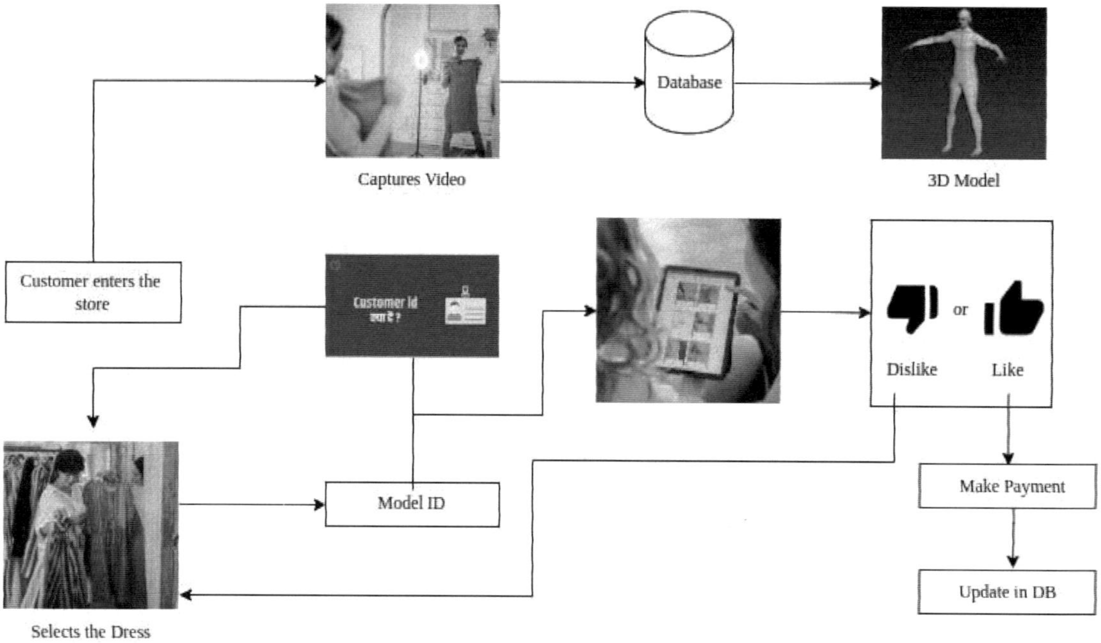

Figure 3.1 Overall architectural diagram of our model.

At whatever point the client enters the store he will be diverted to where the camera is fixed to catch the video of the client to get 2D pictures from all points of the client and a client Id is given. The picture is sent to the fundamental arrangement of the store where estimations are anticipated. The two-dimensional image will be converted into a three-dimensional model. The coordinates will be set apart on the 3D picture and the distances between the coordinates are determined utilizing the Euclidean distance and put away in the information base. With the digital replica, the customer can try on items without having to pass and switch the items. With virtual replication, the consumer never needs to worry about queue statuses again.

However, it has its personal set of disadvantages. Due to the fact that customers cannot try on clothing in the same duplicate while using virtual mirrors, the line grows. Also, digital mirrors could cost a fortune for the retail proprietor to set up and hold because they typically refrain from the usage of them. It would be amazing if every purchaser may want to attempt their garments clearly much like the digital replicas however with the aid of the usage of their cell phones.

With the use of their Ids provided in the store, customers can select clothes. On consummation of shopping, clients can envision themselves in the tab. The client moves towards the tab and says their Customer Id and Model No. of the dress to the working staff.

Behind the scenes, the estimations of the client are coordinated with the outfit estimations to really look at the accessibility which is explained in detail in 3.2. On the off chance that, accessible they can see themselves in the outfit or something bad might happen to assume the size isn't accessible the following bigger size will be thought of and imagined with a message as "accessible at ale size than required".

Assuming the client was content with their look and feel in the outfits then the outfits are added to the truck and refresh the information base. In the wake of taking care of the bill by utilizing his ID, the client information will be put away for later use on the client's demand (part clients) or probably erased from the log and the data set.

3.1. TECHNOLOGICAL ASPECT OF THE METHOD OF IMPLEMENTATION

Our profound learning model acknowledges the RGB video of the clients as information. This video is then portioned into various pictures of various stances of the client. The pictures acquired are pre-handled and resized to a decent size (128*128 pixel) utilizing TensorFlow. Our technique comprises 3 stages: [10]

- Present reproduction.
- Agreement shape assessment.
- Outline refinement and surface guide age.

The resized images are presented as a contribution to CNN's secret layers (Convolutional Neural Network). In the secret layers (Con2D layers) the body shape boundaries on each edge are anticipated. Then, at that point, we apply normal pooling (Max pooling) to get a solitary shape across the entire info grouping. Its capacities as a component extractor result in a vector for each casing. These are shipped off a Gated Recurrent Unit (GRU) layer that yields an inactive component vector for each casing dependent on the past outlines. Presently we consolidate all the data into an unposed accepted edge, where refinement is computationally more straightforward [11].

At each edge an outline puts another imperative based on body type explicitly, the arrangement of beams transitioning from the camera to the silhouette characterizes a limitation cone. The posture of the individual changes as he or she moves. Our key thought is to unpose the cone characterized by the projection beams utilizing the assessed presents. Viably, we rearrange the SMPL work for each beam. In SMPL, each vertex v disfigures as indicated by the accompanying condition:

$$v_i' = \sum_{k=1}^{K} w_{k,i}\, G_k(\theta|J(\beta))(v_i + b_{s,i}(\beta) + b_{P,i}(\theta))$$

where Gk is the worldwide change of joint k and bs, i(β) ∈ R and bP, i(θ) are components of Bs(β) and Bp(θ) compared to I – the vertex. For each beam r, we track down its nearest 3D model point. Doing this for each beam viably unposed the outline cone and places requirements on an authoritative T-present. [12] To make the surface, we twist our assessed sanctioned model back to each edge, back-project the picture tone to all apparent vertices, and lastly create a surface picture by ascertaining the middle of the most symmetrical Texel from all perspectives. At last, the result layer gives us the 3D model as its result. The person first logs onto the gadget through the use of his/her Google signal. The person is requested for his/her height and weight that's used to decide the blouse size for the person. Once logged inside the person is supplied with a catalog of garments to look from. He/ She can look for a specific blouse for an event through the use of the seek bar. The person can upload and consider the garments he/she has decided on inside the cart present. The cart affords the overall sum of money the person has to pay. Finally, the person can attempt the garments virtually. They can click pictures and the image receives saved inside the gallery of/her phone. They can then test the image and determine whether or not they need to shop for the garments or not.[13]

Board characteristics include access to the Kinect sensors that are connected to the PC, access to image and depth information streams from the Kinect picture sensors, and conveyance of a controlled form of picture and depth information that facilitates skeleton following. With a typical range of 1.2 to 3.5 meters, Kinect sensors primarily produce three kinds of data: image data, depth data, and sound data. We will use the first two data streams to improve the human model, reenactment of fabric, and GUI. In the middle camera, a 640x480-pixel RGB camera operates at 30 Hz, creating a picture stream that is communicated to the software as a progression of still images. The quality of the shading picture governs how quickly information can be transferred from the Kinect sensor cluster to the PC, which makes streamlining this program easy at various stages. The shading schemes determine whether the image data is encoded as RGB for the application code. 640480 pixels @ 30 Hz IR profundity are used to discover the IR light source, while a standard CMOS sensor on the right is used to provide depth information. According to this stream, the top 13 bits of every pixel convey in millimeters the distance to the closest article at that, particularly the x and y points in the depth camera's field of view.[14]

Journal of Physics: Conference Series **2325** (2022) 012006 doi:10.1088/1742-6596/2325/1/012006

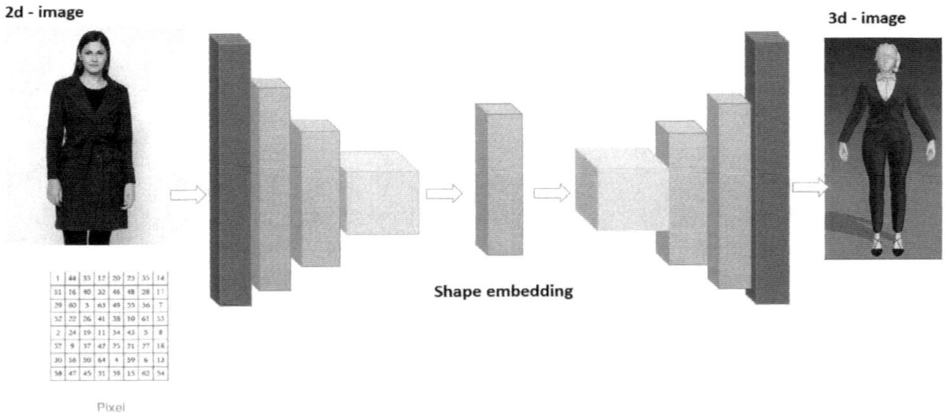

Figure 3.1.1 Conversion of 2D images to 3D model.

Contrary to 2D images, which have both a length and a width, 3D images have yet another dimension: depth. Even though 2D images can carry out this task if we take various factors into account, the result still won't be accurate. This occurs because people are not all the same size, which constitutes an anomaly in the system, thereby causing errors in measurements. But our human vision, on the other hand, sees things only in 2D. The human eye retina, which serves as a screening medium, captures images in two dimensions just like a real-world camera would. Nonetheless, humans are capable of making a good estimate regarding the right dress for different body types. This is because our human brain does some post-processing on the images captured by our retina, by taking into consideration the colors, saturation, patterns, and other features to try and interpret the object we see. In simple words, our brain converts 2D images from the retina to a comprehensive 3D image, which helps in good decision making and estimation. Following that, if we are aiming to improve the accuracy of the model, we would need to convert the 2D images to 3D images and take the offsets of the features we see into consideration.

3.2. ACCURATE MEASUREMENTS

The client was diverted towards the camera where the camera was fixed and where the client would remain to take a 2D picture was additionally fixed so the distance between the client and the camera was fixed. The 2D picture was diverted to the primary framework where body joints are utilized for accepting the estimations as follows. The Euclidean distance was estimated between the body joints and was changed over to genuine body size estimations.

Euclidean distance (E) =((a1-a2) ^2+(b1-b2) ^2) ^1/2

where E is the distance between two points (a1,b1) and (a2,b2).

Dresses of the same type of pattern will have a unique model ID, each model ID will have the available sizes stored in DB and that is matched with the size of the customer which we have calculated. In that way, customers would be able to select the outfits that would suit them perfectly without having to spend much time searching for the outfit or fretting if they would fit.

3.3. GARMENT FITTING

The fundamental thought of our dress fitting calculation is that we right off the bat build the neighborhood fitting patch to appraise the bend and twist of every vertex on body and apparel models. The element focuses are also obtained using the element extraction approach. Finally, we apply human pose estimation algorithms to the arch and twist to avoid edge defects and masking errors. These algorithms make it easier to coordinate with comparable component focuses and intelligent gain.

A few key elements focus on the two models. At last, we compute the change lattice as indicated by sets of key element focuses.

3.4. IMPACT

Disposition and aim of utilizing Virtual Fitting Room innovation will influence the goal of buying decidedly. The idea of virtual preliminary rooms is new to retailers in India, but it has been effectively attempted by e-posteriors in India just as across Europe and the US in the course of recent years. Online retailers like Barbour (UK), Ermenegildo Zenga (Italy) and Pretty Green (UK) have been utilizing the virtual preliminary rooms controlled by what fits me and what fits me.

While the shoppers' essential anxieties like fitting, look of a specific attire piece and embellishment disseminate with such innovations, retailers can likewise diminish their time utilization, decline expensive request returns, further develop transformation rate and upgrade purchaser fulfillment.

With more retailers taking on virtual preliminary rooms, it can possibly reform the manner in which purchasers look for attire and embellishments in India. [11]

4. IDENTIFIED SCOPE OF MINIMUM VIABLE PRODUCT (MVP) PERFORMANCE

Our Deep Learning models will be prepared with a small-scale cluster size of 64 for 1,00,000 cycles altogether. The underlying learning rate is set to 0.002 and lessened by an element of 10 after each 20, 000 cycles. No weight rot is utilized and dropout with a rate of 0.5 is just applied after the completely associated layers. Preparing takes right around 2 to 3 minutes on one NVidia GTX Titan X GPU. When prepared, our model can reproduce 3D models for in excess of 100 casings each second. Our model can uphold in excess of 10000 synchronous clients.

4.1. FEASIBILITY

The arrangement is especially doable contrasted with the current virtual preliminary rooms. This is on the grounds that the current innovations utilize Augmented Reality while our answer utilizes open-source programming and doesn't spend on the product part. Our costs will be just for cameras and tablets.

4.2. SECURITY

Each one of the clients will have the exceptional id to see their 3D models. So, one individual's 3D model can't be seen by someone else. The subtleties of their models will be erased once their buy is finished. This will expand the general security level of our profound learning model.

4.3. COMPATIBILITY

The proposed procedure is intended to be generally viable as long as the framework and the camera are in great condition. Windows, Mac OS X, and Linux are all supported by our model.

4.4. INSTALLATION

4.4.1. SOFTWARE INSTALLATION

- The primary framework control by the manager in the store is introduced with the Anaconda

  pilot,

  Marvelous architect programming according to prerequisites for running our model.
- Profound learning calculations carried out utilizing python language were utilized for changing over 2D pictures to the 3D item.
- Math in Computer illustrations was utilized for denoting the coordinates. The distance between the coordinates is calculated and saved in the data set for the estimations. Further similar coordinates are utilized for planning the pieces of clothing to the 3D item.

4.4.2. EQUIPMENT INSTALLATION

- Pivoting stand which moves within a specific speed and point. The client on entering the store was diverted to the turning table.

4.5. UPGRADE

- The client entering the store was diverted toward the pivoting stand.
- The camera naturally takes the client's picture and advances to the principal framework.
- The client's portable Id was considered their Id and can keep shopping.
- In the interim, the 2D picture was changed over to the 3D picture and put away in the data set with the client's ID.
- In the 3D image, the coordinates are separated. The distance between the directions is utilized for computing the estimations of the client and furthermore for fitting the piece of clothing to the model.
- After the culmination of the shopping of the client he can move towards the TAB and by saying the IDs to the administrator he can imagine and can likewise give the material depiction.
- Assuming that he was happy with the perception of looks and feel the client can add the dress to the truck.
- At last, the client can go to the counter and make a bill. The client information was put away for future solicitation or, more than likely, erased from the data set.

4.6. USER EXPERIENCE

Virtual path room gives the 3D graphical perception of the client picture with his/her ideal garments accessible in the store. Hence assists with beating the sterilization issues and keeping social separation. It helps in keeping away from the immediate contact of the client with the garments which assists with settling the spread of the Coronavirus.

5. CONCLUSION

Subsequent to applying the fabric model with the improved execution joint position, this application has turned into an adequate application to give a virtual fitting space to the client to use: Human estimation produced by the client body remains before the Kinect. A malleable and realistic-looking material model can be worn by the client. The user interface is so simple and easy to comprehend that even a beginner can understand it; also, there are many exciting and useful features in the program.

6. REFERENCES

[1] Higgins, K. R., Farraro, E. J., Tapley, J., Manickavelu, K., & Mukherjee, S. (2018). U.S. Patent No. 9,898,742. Washington, DC: U.S. Patent and Trademark Office.
[2] Isıkdogan, F., & Kara, G. (2012). A real time virtual dressing room application using kinect. CMPE537 Computer Vision Course Project.
[3] Korszun, H. A. (1997). U.S. Patent No. 5,680,528. Washington, DC: U.S. Patent and Trademark Office.
[4] Protopsaltou, D., Luible, C., Arevalo, M., & Magnenat-Thalmann, N. (2002). A body and garment creation method for an Internet based virtual fitting room. In Advances in Modelling, Animation and Rendering (pp. 105-122). Springer, London.
[5] Steermann, M. C. (2014). U.S. Patent Application No. 14/215,649.
[6] Kjærside, K., Kortbek, K. J., Hedegaard, H., & Grønbæk, K. (2005). ARDressCode: augmented dressing room with tag-based motion tracking and real-time clothes simulation. In Proceedings of the central European multimedia and virtual reality conference.
[7] S. Sullivan and J. Ponce, "Automatic Model Construction and Pose Estimation from Photographs using Triangular Splines," IEEE Transactions on Pattern Analysis and Machine Intelligence, vol. 20(10), ppl091-1097, 1998.
[8] P. Fua and Y. LeClerc, "Object-Centered Surface Reconstruction: Combining MultiImage Stereo and Shading," International Journal o f Computer Vision, vol. 16, pp35-56, 1995.
[9] L. Zhang and S. Seitz, "Image-Based Multiresolution Shape Recovery by Surface Deformation," In Proceedings o f SPIE: Videometrics and Optical Methods for 3D Shape Measurement, pp 51-61, 2001.
[10] R. Cipolla and P. Giblin, "Visual Motion o f Curves and Surfaces" Cambridge University Press, 2000.
[11] https://deepai.org/machine-learning-glossary-and-terms/convolutional-neural-network.
[12] https://www.ibm.com/in-en/cloud/learn/machine-learning
[13] Aladdin Masri, Muhannad Al-Jabi April (2019)
[14] D. Protopsaltou, C. Luible, M. Arevalo and N. Magnenat-Thalmann, "A body and garment creation method for an Internet based virtual fitting room", In Advances in Modeling Animation and Rendering, pp. 105-122, 2002.

[15] https://www.techtarget.com/searchenterpriseai/definition/machine-learningML#:~:text=Machine%20learning%20(ML)%20is%20a,to%20predict%20new%20output%20va lues.
[16] C.M. Bishop, Pattern Recognition and Machine Learning, Springer, 2006. https://codinghero.ai/7-basic-steps-in-machine-learning/
[17] S. Ray, "A Quick Review of Machine Learning Algorithms," 2019 International Conference on Machine Learning, Big Data, Cloud and Parallel Computing (COMITCon), 2019, pp. 35-39, doi: 10.1109/COMITCon.2019.8862451.
[18] Y. Kumar, K. Kaur and G. Singh, "Machine Learning Aspects and its Applications Towards Different Research Areas," 2020 International Conference on Computation, Automation and Knowledge Management (ICCAKM), 2020, pp. 150-156, doi:10.1109/ICCAKM46823.2020.9051502.
[19] N. N. Pise and P. Kulkarni, "A Survey of Semi-Supervised Learning Methods," 2008 International Conference on Computational Intelligence and Security, 2008, pp. 30-34, doi: 1109/CIS.2008.204.
[20] Kumar, T. Senthil. "Study of Retail Applications with Virtual and Augmented Reality Technologies." Journal of Innovative Image Processing (JIIP) 3, no. 02 (2021): 144-156.
[21] Mohammadi, Seyed Omid, and Ahmad Kalhor. "Smart Fashion: A Review of AI Applications in Virtual Try-On & Fashion Synthesis." Journal of Artificial Intelligence 3, no. 4 (2021): 284.

Health monitoring smart glove using esp32 microcontroller

T S Jahnavi[1], **Naveen Thumma**[2], **Kolla Abhishek Reddy**[3], **Upadarsta V R S S H Teja**[4], **M Gunasree**[5], **M Kiran**[6]

[1-5] Department of ECE, Gokaraju Rangaraju Institute of Engineering and Technology,Hyderabad, Telangana, India.

[6] Associate Professor, Department of ECE, Gokaraju Rangaraju Institute of Engineering andTechnology, Hyderabad, Telangana, India.

[*]Corresponding author's e-mail: jahnavis000@gmail.com.

Abstract. Even after voice assistants, virtual assistants, and other new technologies have come into existence, many are still facing problems to use them conveniently and effectively. Our project is mainly applicable to old people, physically challenged, and bedridden people. Our project application areas are hospitals and old age homes. During emergency situations where the user is not able to access his/her mobile phone, our prototype can handle such situations effectively by detecting the user's hand gestures and sending alerts to caretakers and hospitals. Our model also has a medical alert system. According to the prior initialization of the time for the medicines to be taken, this prototype alerts the user at that exact time. The alert message is displayed on the OLED and also through a buzzer. This glove has a temperature sensor, flex sensors, and pulse-oxy sensors embedded which give the continuous readings of user-health parameters and are displayed on the OLED placed on the glove. The main feature of the glove is to communicate the needs of the user which can be accomplished by a flex sensor.

Keywords. Hand gestures, Emergency Alert System, Medicine Alerts, Temperature, Heart beat

1. Introduction

There are a huge number of people all over the world, who are losing their lives due to delayed access to health care. There are many cases where people died only due to late access to medication or delay in reaching the hospital. So, continuous monitoring of health parameters is not only important for patients in hospitals, but also the people who are suffering from long-term health complications and who are taking health care at home by caretakers.

The second problem is that though many new technologies in the medical field came into existence not many people are aware of how to use those technologies. People who are deaf, dumb, paralyzed affected, and taken care of by caretakers need to have a system with which they can effectively communicate with caretakers or doctors.

As a solution to the above-mentioned problems, we have designed a glove that can continuously monitor patient health parameters, store the data in the cloud, and also displays the in the mobile app. The second feature of this glove is to help the user effectively communicate with the caretaker or doctor using the hand gestures of the user.

2. Basic concepts

2.1. Microcontroller

ESP32 is a powerful on chip microcontroller with inbuilt wi-fi and Bluetooth module. It is a dual core 32-bit low cost and low power microcontroller. The core of this module consists of a ESP32S chip, which is extensible and adaptive.

2.2. Flex Sensor

A flex sensor is a type of sensor that is used to determine the degree of defection or bending. This is used to employ hand gestures to communicate the user's needs. In comparison to other sensors, this sensor has a very accurate response.

2.3. Temperature Sensor

The DHT11 is a temperature and humidity sensor. The temperature and humidity sensor technology, as well as dedicated digital module collection technology, are used to assure the product's high dependability and long-term stability. The sensor has a 8-bit microcontroller and consists of a resistive wet component sense and an NTC temperature measurement device.

2.4. Touch Sensor

Touch sensor is a type of sensor which is used to detect the physical touch. It is used as a switch for activating the flex sensors in this project.

2.5. Pulse-Oxy Sensor

The pulse-oxy sensor used in this project is MAX30100. It is a sensor module, which measures the heart rate and oxygen levels in the blood.

2.6. OLED

Organic Light Emitting Diode. This 0.96-inch OLED display is based on a popular SSD 1306 display controller. It works on both I2C (TWI). The display is small in size and very crisp. It has a resolution of128x 64 pixels.

3. Literature survey

Priyanka Lokhande, Riya Prajapati, and Sandeep Pansar[1] proposed a model for sign language recognition using flex sensors. The project's heart, the Glove, is where the sign language translator begins. The black glove has nine flex sensors, four touch sensors, one two-dimensional x-y-axis accelerometer, and one one-dimensional z-axis accelerometer. The flex sensors are the most significant sensors since most letters can be recognized based on finger flexes. All fingers, with the exception of the thumb, have two flex sensors, one at the knuckle and the other at the lower joint. As a result, these fingers can flex to two different degrees.

Amandeep Kaur and Ashish Jasuja [2] proposed a system that uses specific sensors, Raspberry Pi, and IoT to monitor a person's heart rate and body temperature (vital body parameters). A wearable system that also allows for remote health monitoring is available. Bluemix uses the MQTT (Message Queuing Telemetry Transport) protocol. Accuracy and cost must be evaluated in addition to remote monitoring and system wearability. A perfect compromise between accuracy and system cost can be accomplished by selecting appropriate sensors, such as the DS18B20 (temperature sensor) and KG011 (pressure sensor) (heart rate sensor). The sensors take multiple readings of a person's heart rate and temperature. On the IBM Watson IoT platform, the data is shown as graphs.

Aniket Pramanik[3] Their project presents a digital notice board and a home automation system using a GSM SIM900 module. The purpose of this project is to provide users with a simple, rapid, and dependable way to display important notifications on an LCD by allowing them to send a message to be displayed on the LCD. An android application developed for this project may convey the message to the GSM SIM900 module, which includes a SIM card. A home automation system was also designed, allowing domestic appliances like lights and fans to be turned on and off using the same android application developed for this project.

The Android application can be used to control home appliances and display notices on an LCD display from anywhere in the world. It employs a microcontroller to control the system, GSM technology for communication, and an Android app to send the message via SMS. The project includes a 32-bit ARM-based microcontroller LPC2148, a GSM SIM900 module, an LCD, a motor, and an Android application to connect with the hardware. As long as there is mobile network connectivity, the device can be utilized anywhere, regardless of deployment location.

4. Proposed system model

The systems having the same type of application were proposed in the past. These systems are developed with only a few and less accurate sensors. There are separate systems for measuring health parameters and separate systems were proposed for gesture control.

In our model, we have combined both the applications of measuring health parameters and gesture control. In addition to this, we have added a medical alert system, which alerts the user to take the medicine on time.

This project consists of a receiver section, which consists of an OLED display, GSM module, buzzer, and Arduino which receives data from the connected glove and displays the data on the OLED display. In case of an emergency, when the health parameters go beyond the proposed values, the system alerts the caretakers by activating the buzzer and also sending emergency alerts to the user. This type of feature is highly useful in hospitals to monitor a large number of patients continuously at atime.

The patient's data and medicine status can be managed in the mobile app which retrieves the data fromthe cloud. In case of no internet access, the alerts will be sent using the GSM module which consists ofan integrated SIM for this purpose.

We have used the ESP32 microcontroller which is more accurate, efficient, and rich in features when compared to Arduino using which old systems were proposed.

All the updated sensors are used for monitoring health. For detecting the hand gestures of the user we have used flex sensors, which are highly used in many projects for measuring the amount of deflectionor bending. When the flex sensor is bent the resistance value of the sensor changes, that change in resistance is used to detect the needs of the user.

5. Implementation

This glove consists of a pulse-oxy sensor, temperature sensor, flex sensor, and touch sensor shown in Fig.6. When the finger of the user touches the touch sensor the touch sensor activates the other sensors. The other sensors start reading input values from the user. The taken values are analyzed by the ESP32microcontroller. Esp32 starts sending data to the cloud as shown in Fig.4, Fig.5 & Fig.6, andthe mobile app starts retrieving data from the cloud. If the values read by the sensors are more than the given threshold values then the caretaker receives alerts in the mobile app.

Pulse oximetry is a frequently used medical sensing method for measuring heart rate and arterial bloodoxygen levels. Pulse oximeters that employ red and infrared light to monitor Spo2 levels. A pulse oximeter measures absorption with a constant stream of light and a complex calculation, filtering out muscle, tissue, and fingernail interference.

The DHT11 is a temperature and humidity Sensor. A capacitive humidity detecting element and a thermistor for temperature detection make up the DHT11 sensor. A negative temperature coefficient thermistor is used in this sensor, which reduces resistance as the temperature rises.

A flex sensor is used to communicate the needs using hand gestures. When the user bends their finger, the strip is twisted the resistance changed, we measure those resistance values using ESP 32 microcontroller. There are three flex bending conditions. The resistance will be normal when the sensor is flat, the 2nd condition is when we bend the sensor to 45° the resistance increased when further bend the sensor to90° the resistance even managed the is the 3rdcase.

Flex sensor	Need
FS-01	Food
FS-02	Water
FS-03	Emergency

Table 1. Flex sensor application

ESP32 is a dual-core 32-bit microcontroller with inbuilt WIFI and Bluetooth modules. Here we connect the esp32 microcontroller to max30100, DHT11, and flex sensors. The SDL and SCL pins MAX30100 are connected to GPIO 21 and GPIO 22 of Esp32 respectively. VCC and GND pins of MAX30100 have connected 3V3 and GND pins of Esp32. The Serial clock pin is the clock signal which synchronizes the data transfer between the devices on the I2C bus and it's generated by the master device. The SDA pin carries the data.

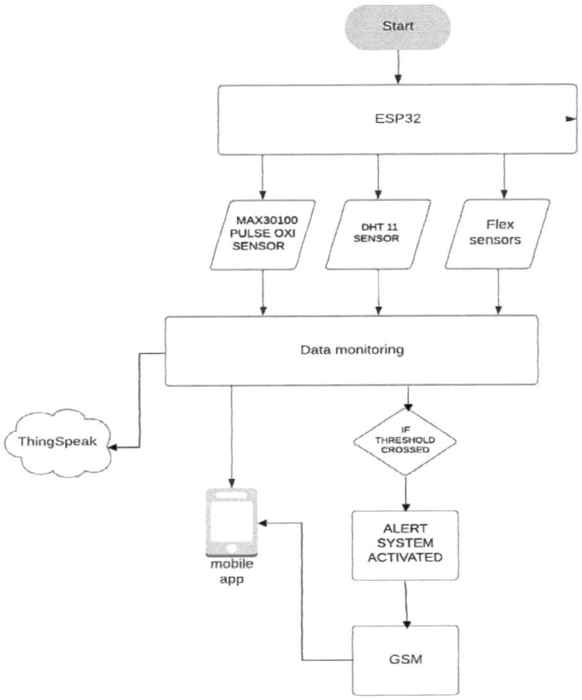

Figure 1. Block Diagram

DHT 11 sensor is used to measure Temperature. The Analog to digital converter will convert the Analog Temperature signals to digital temperature. The VCC and GND pins of DHT11 are connected to the 3V3 and GND pins of the ESP32 microcontroller. The Data pin of DHT11 is connected to the GPIO16 pin of ESP32.

Flex sensor is used to measure the amount of defection or bending.

The resistance of the Flex sensor increases when there is deflection. The Flex sensor consists of two pins, pin 1 is connected to the positive terminal of the Esp32 and pin2 is connected to the Data Pin of Esp32 and pin2 is connected to the 10KOhm resistor which is connected to the GND pin of the ESP 32.

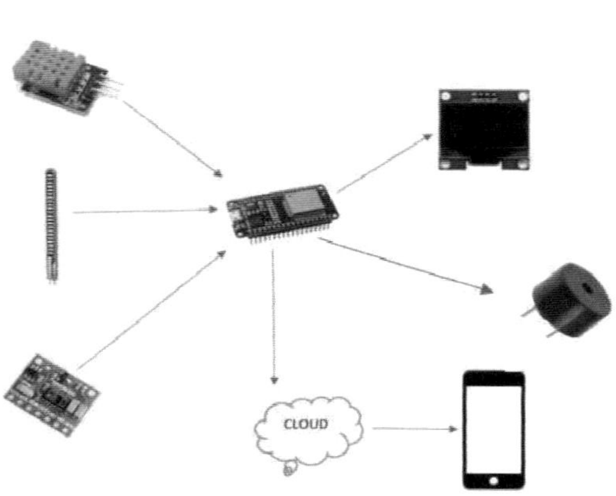

Figure 2. Transmitter module

Figure 3. Receiver module

6. Result

6.1. Description

Our results include the values of the flex sensor, temperature sensor, pulse-oxy sensor, and the emergency alerts on the OLED display of the receiver section. The emergency alerts are sent through the GSM module. These alerts are very accurate and useful for the caretakers to know the status of the user even without an internet connection. These alerts are accurate and effective to track the status of the user. The values will be recorded on the ThingSpeak cloud platform.

6.2. Experimental results

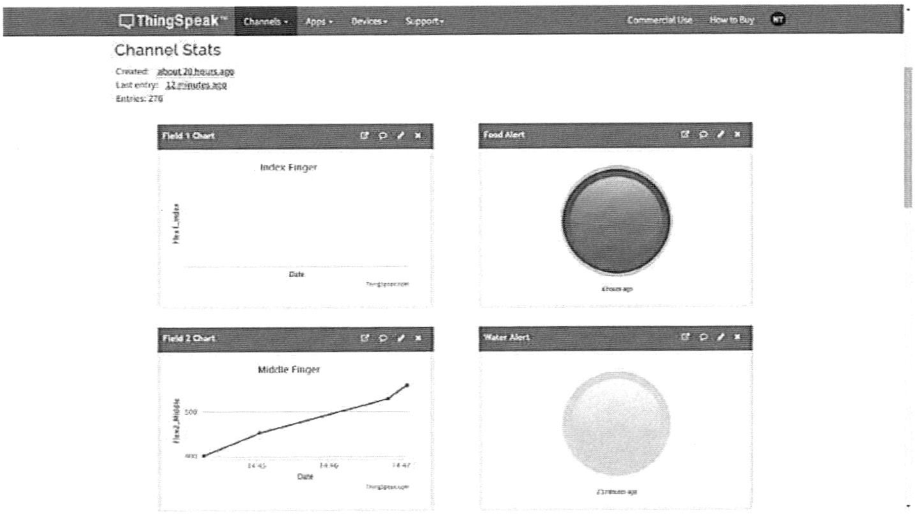

Figure 4. Flex sensor-01, 02

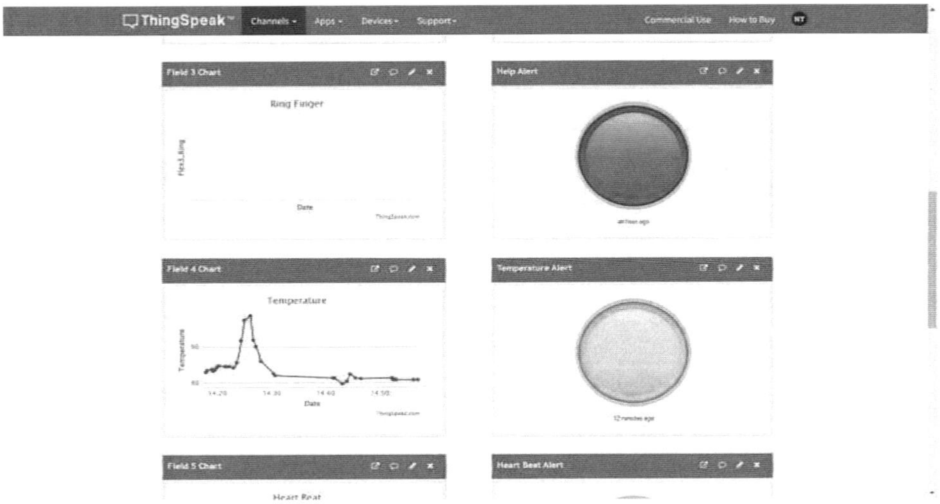

Figure 5. Flex sensor-03, temperature values

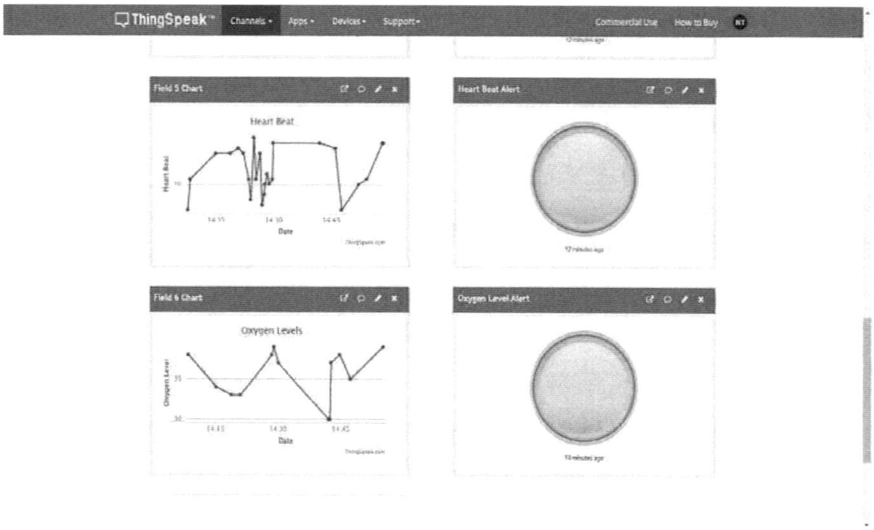

Figure 6. Heart beat and oxygen levels

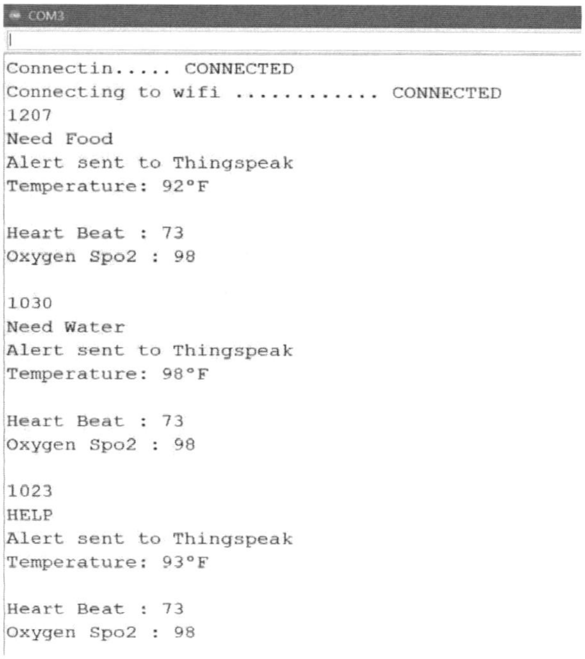

Figure 7. Output values on serial monitor

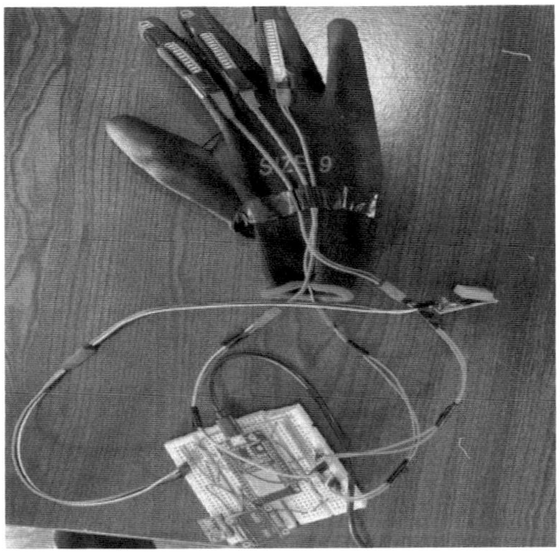

Figure 8. Glove prototype

Figure 9. Receiver prototype

7. Conclusion

Our prototype solves real-time problems faced by bed-ridden, paralyzed, deaf, and dumb people. This model develops a user-friendly interface between user and caretaker. No training is required for the user to handle this model. As the alert systems are automated, in case of emergency the hospitals are alerted and necessary actions can be taken to save the lives of the users.

This model can be developed using Machine Learning and Artificial Intelligence to enlarge the scope of communication between users and caretakers and to increase the efficiency of the current model.

This model can be further improved by developing a smart band using the EMG sensors which are used to detect the motion between the muscles using Electric Signals and Machine Learning.

8. References

[1] Priyanka Lokhande, Riya Prajapati and Sandeep Pansar *Data Gloves for Sign Language Recognition system* 2015 International Journal of Computer Applications (0975 – 8887), National Conference on Emerging Trends in Advanced Communication Technologies.

[2] Amandeep kaur and Ashish Jasuja *Health Monitoring based on IoT using Raspberry Pi* IEEE, 10.1109/ccaa.2017.8230004, December.

[3] Aniket Pramanik *Gsm Based Smart Home and Digital Notice Board* IEEE Conference on 10.1109/icctict.2016.7514549, July.

[4] R. Teymourzadeh, S. Ahmed, K. W. Chan and M. V. Hoong 2013 *Smart GSM Based Home Automation System* International Conference on Systems Process Control, IEEE Conference on, pp. 306- 309.

Journal of Physics: Conference Series **2325** (2022) 012007 doi:10.1088/1742-6596/2325/1/012007

[5] H. Elkamchouchi and A. El Shafee 2012 *Design and Prototype Implementation of SMS based Home Automation System* International Conference on Electronics Design, Systems and Applications, IEEE International Conference on, pp. 162-167.

[6] B. Ghazal, M. Kherfan, K. Chahine and K. Elkhatib 2015 *Multi Control Chandelier Operations using Xbee for Home Automation* Technological Advances in Electrical Electronics and Computer Engineering, 3 r d International Conference on, pp. 107-111.

[7] X. Liu and O. Baiocchi 2016 *A Comparison of the Definitions for Smart Sensors Smart Objects and Things in IoT* 7th IEEE Conference in Information Technology Electronics and Mobile Communication pp. 1-4.

[8] Mirjana Maksimović, Vladimir Vujović and Branko Perišić 2016 *Do it Yourself Solution of Internet of Things Healthcare System*: measuring body parameters and environmental parameters affecting health.

[9] R. Kumar and M. P. Rajasekaran 2016 *An IoT Based Patient Monitoring System using Raspberry Pi* IEEE International Conference in Computing Technologies and Intelligent Data Engineering, pp. 1-4.

[10] H. Mansor, M. H. A. Shukor, S. S. Meskam, N. Q. A. M. Rusli and N. S. Zamery, 2013 *Body Temperature Measurement for Remote Health Monitoring System*, IEEE International Conference, pp. 1-5.

[11] N. P. Jain, P. N. Jain and T. P. Agarkar 2012 *An Embedded GSM based Multiparameter Realtime Patient Monitoring System and Control-an Implementation for ICU Patients* IEEE World Congress in Information and Communication Technologies, pp. 987-992.

[12] P. A. Reddy and J. Damodhar 2012 *A Real Time Monitoring System for Physiological Signals using Wireless Sensor Network* International Journal of Engineering Trends and Technology, vol 3 (4), pp. 502-506.

[13] J. Arnil, Y. Punsawad and Y. Wongsawat 2011 *Wireless Sensor Network based Smart System for Healthcare Monitoring* international conference on robotics and biomimetics, pp. 2073-2076.

Optimized 6x6 Optical Router for Three Dimensional Optical Networks-on-Chip

Nathrao B. Jadhav [1], Bharat S. Chaudhari [1]

[1]Department of Electronics and Telecommunication Engineering, Maharashtra Institute of Technology, Paud Road, Kothrud, Pune, India – 411038.

[a]nathrao.jadhav@mitpune.edu.in

Abstract The interconnection network is very influential in Multiprocessor system-on-chip (MPSoC) performance. Due to low power efficiency and high latency, conventional interconnects are replaced by electrical networks on-chip (ENoC). However, ENoCs cannot gratify the bandwidth, latency and loss requirements within the available power budget of optical integrated circuits. The reasons being restricted bandwidth, extended delay, and reasonably high power consumption. So, ENoC are being replaced by optical networks-on-chip (ONoC). ONoC has emerged as a promising substitute to provide greater bandwidth, low latency and low power ingestion in MPSoCs. Number of cores required on-chip is increasing which leads to demand of multilayer multi-core processors. Three dimensional (3D) ONoC has become essential for such multilayer multi-core processors. For 3D ONoC, in this paper we have proposed 6x6 non-blocking optical router using micro-ring resonator (MRR) with less number of waveguide crossings and waveguide bendings. Phoenix software is used to simulate proposed optical router. Insertion loss of proposed optical router is lowest among existing non-blocking optical routers of 3D ONoC.

Keywords: *MPSoC, ENoC, 3D ONoC, MRR, Optical router*

1. Introduction

The complexity in integrating circuits is increasing [1]. The multiprocessor system-on-chip (MPSoCs) contains hundreds of cores. The performance of MPSoCs is decided by the interconnection and communication infrastructures. High performance interconnects can transmit data efficiently among cores on-chip. Vital challenge for MPSoCs is to realize a scalable on-chip network that meets the low latency and large bandwidth capacities with high power efficiency. The traditional bus based interconnects are bottleneck in the performance of on-chip multi-core processor system. The bus-based on-chip communication has been replaced by electrical networks-on-chip (ENoC) as a substitute which has improved the communication bandwidth of MPSoCs [2]. However, ENoCs cannot gratify the bandwidth, latency and loss requirements within the available power budget of optical integrated circuits because of its restricted bandwidth, extended delay, and reasonably high power consumption [3-4]. On-chip optical interconnects have appeared as a concrete alternative to the ENoCs. The optical interconnects are compatible with complementary metal oxide semiconductor (CMOS) technology that makes them highly appropriate to deploy as on-chip interconnects [5-6].

The optical networks-on-chip (ONoC) has emerged as a promising approach to provide higher bandwidth, low latency and low power consumption in MPSoCs. There is a need of optical device like optical cross connect that will make routing between processors [7]. An optical router is the key element of ONoC and it is responsible for data transfer among processors. The performance of optical router affects the overall performance of the network.

For multi-core processors, two dimensional (2D) ONoCs are designed. The optical routers for 2D ONoC architecture using MRR have been proposed in [8-15]. The number of cores required on-chip are increasing, and so complexity of the on-chip interconnects are increasing. This leads to demand of multilayer multi-core processors. For such multilayer multi-core processors, three dimensional (3D) interconnect are required. Therefore the researchers in this field have shifted their focus from 2D ONoC to 3D ONoC. The recent work has reported design of architectures, topology and optical routers for 3D ONoC. 7x7 Optical routers used for 3D ONoC are reported in [16-18]. The number of MRRs, waveguide crossings and waveguide bendings required in 7x7 optical routers are more which has resulted in higher insertion loss. 6x6 optical router requires less number of MRRs, waveguide crossings and waveguide bendings, reduced insertion loss as compared to 7x7 optical router. Therefore designing 6x6 optical router for 3D ONoC is more efficient in terms of number of optical elements required and insertion loss. In the view of these benefits, [19-21] have proposed 6x6 optical routers. The optical router in [16] has 7 ports and used 22 MRRs. In [17], 56 waveguide crossings and 36 MRRs are required to design an optical router. A mesh topology is used for ONoC design and XYZ dimension order routing algorithm. The router in [18] designed with 30 MRRs and 14 waveguide terminators. Waveguide terminators increases cost as well as energy needs. ONoC design in [19] used 7x7 and 6x6 optical routers. 7x7 router used 24 MRRs and 31 waveguide crossings and 6x6 router used 16 MRRs and 20 waveguide crossings. Topology used for this ONoC is mesh. The router in [20] has reduced waveguide crossings to 18 and waveguide bendings to 24 by keeping 16 MRRs. The 6 port optical router in [21] has 20 MRRs, 17 waveguide crossings and 22 waveguide bendings.

The work reported is summarized as follows:

i. The optical router for 3D ONoC can be 6x6 or 7x7 optical router.
ii. The 6x6 optical router requires less number of MRRs, waveguide bendings and waveguide crossings than 7x7 optical routers.
iii. Reducing either number of MRRs, the number of waveguide bendings or the number of waveguide crossings can improve the performance of optical router.

On the basis of above mentioned observations, a new 6x6 non-blocking optical router for 3D ONoC is proposed in this paper. The MRRs used for the proposed optical router are 16. The optical router is designed in such a way that the number of waveguide bendings and the number of waveguide crossings are minimized. The minimum waveguide bendings used in enhanced existing optical router are 24 which we have reduced it to 18 in our proposed router. Similarly the minimum waveguide crossings used in enhanced existing router are 18 which we have reduced it to 15 in our router. The performance parameter used to analyse the proposed router is insertion loss.

The following are the major contributions of the proposed work:

i. 6x6 optical router using MRR is designed and simulated.
ii. The proposed router has lowest waveguide bendings and waveguide crossings as compared to available non-blocking optical routers of 3D ONoC.
iii. The Waveguide crossing elements are reduced by 4 and waveguide bending elements are reduced by 6 in the proposed router as compared to existing router.
iv. The insertion loss of proposed optical router is less than existing optical routers of 3D ONoC. Average insertion loss is improved by 3% than existing optical routers of 3D ONoC.

Journal of Physics: Conference Series **2325** (2022) 012008 doi:10.1088/1742-6596/2325/1/012008

Paper is organized in four sections. Section 2 explains design and working of proposed router. The performance analysis of this design is done in section 3. Section 4 is conclusion of work.

2. Proposed Optical Router Design

A MRR is used as a switching element for the design of optical router. A MRR can be used as cross switching element (CSE) or parallel switching element (PSE). Micro-ring and two waveguides are used to use MRR as CSE or PSE. The proposed 6x6 router using MRR is as shown in fig. 1. A design of 6x6 optical router is optimized by avoiding some redundant routing paths like 'South' and 'North' port to 'East' and 'West' port. We have used XYZ dimension order routing algorithm and it does not require the mentioned routing paths. This approach has minimized the number of MRRs required for the proposed optical router to 16. Reduced number of MRRs reduces overall insertion loss and power consumption. It also makes design compact. The number of waveguide crossings required is 15 which is the minimum in existing 6x6 non-blocking optical routers. Reduced crossing results in less crosstalk noise in optical router. The number of waveguide bendings required in the proposed router is reduced to 18. Waveguide bendings and waveguide crossings are one of the major sources of losses in optical routers. Hence, the insertion loss is affected significantly by waveguide bendings and waveguide crossing s.

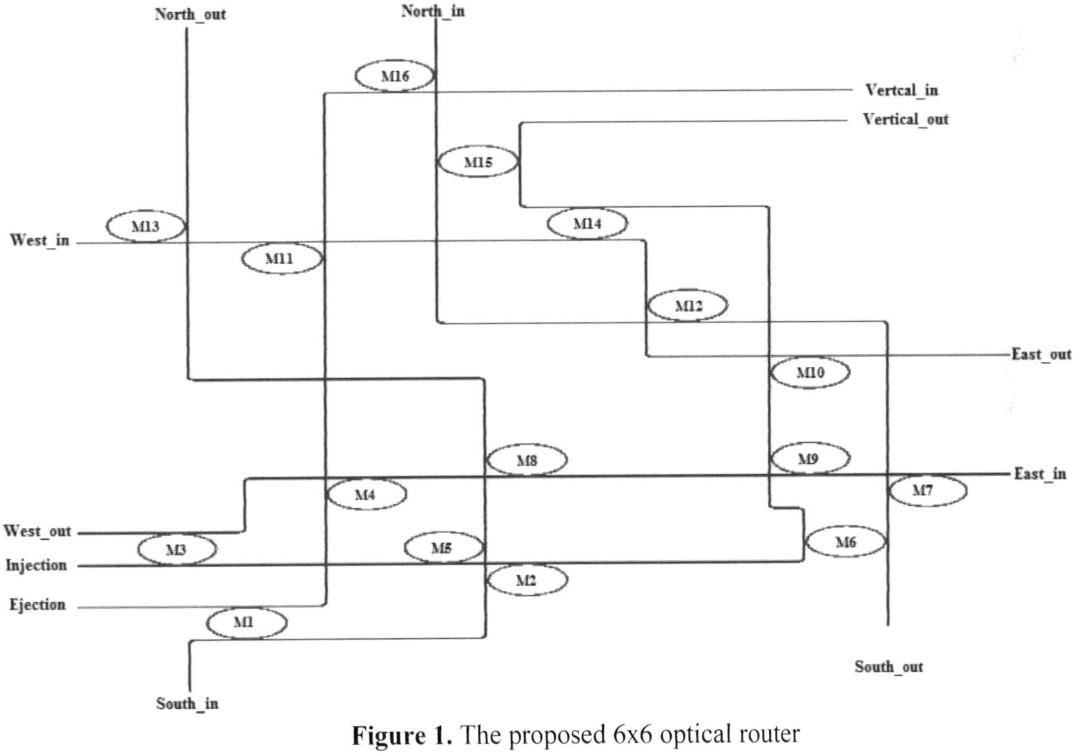

Figure 1. The proposed 6x6 optical router

The MRR matrix for routing paths of the proposed optical router is shown in Table 1. Maximum one MRR is need to be activated for data transmission from one port to other. Some of the paths even do not require activation of any MRR. These are represented by 'NONE'. The paths like 'North' port to 'East' port and 'North' port to 'West' port are not allowed and represented as '-'. The paths which are required the activation of MRR is shown by M1 to M16.

Table 1. MRR matrix for routing paths

Input	Output					
	East	West	South	North	Ejection	Vertical
East	-	None	M7	M8	M4	M9
West	None	-	M12	M13	M11	M14
South	-	-	-	None	M1	M2
North	-	-	none	-	M16	M15
Injection	M10	M3	M6	M5	-	none
Vertical	-	-	-	-	none	-

3. Performance Analysis

The performance of optical router is verified with insertion loss analysis. The power loss of optical signal in routing path due to undesired coupling is represented as insertion loss. Waveguide bendings, waveguide crossings and MRR are responsible for insertion loss. The power loss of any path can be calculated by considering all these losses.
Output power at waveguide crossing is given by

$$P_{outcr} = L_{cr} P_{in}$$

where L_{cr} is power loss occurred at crossing and P_{in} is input power and P_{outcr} is output power at crossing.

Power loss occurred even in straight waveguide. The output power at straight waveguide is given by

$$P_{outsw} = L_{w} P_{in}$$

where L_w is power loss occurred in straight waveguide.

Output power at waveguide bending is given by

$$P_{outbw} = L_{bw} P_{in}$$

where L_{bw} is power loss occurred in waveguide bending.

Output signal power at through port and drop port in MRR is given by

$$P_{Through} = L_{off} P_{in}$$

$$P_{Drop} = L_{on} P_{in}$$

where L_{off} is power loss in MRR when it is OFF and Lon is power loss in MRR when it is ON.

The optical router is designed and simulated using Phoenix software. Fig. 2 (a) shows the path of light transmitted from 'West' to 'North' port and the lost optical signal. Similarly fig. 2(b) and fig. 2(c) shows light path for 'West' to 'Vertical' port and 'Injection' to 'South' port. The waveguide crossing and waveguide bending are not required in this path and so insertion loss is lowest in this path. Fig. 3 shows the 3D simulation result for the same paths

.

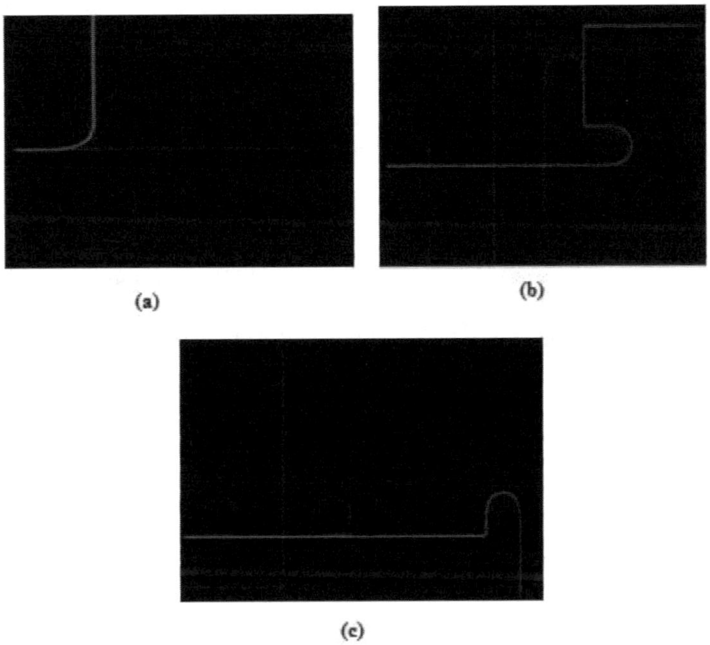

Figure 2. Simulation result of optical signal flow (a) 'West' to 'North' port (b) 'West' to 'Vertical' Port (c) 'Injection' to 'South' Port

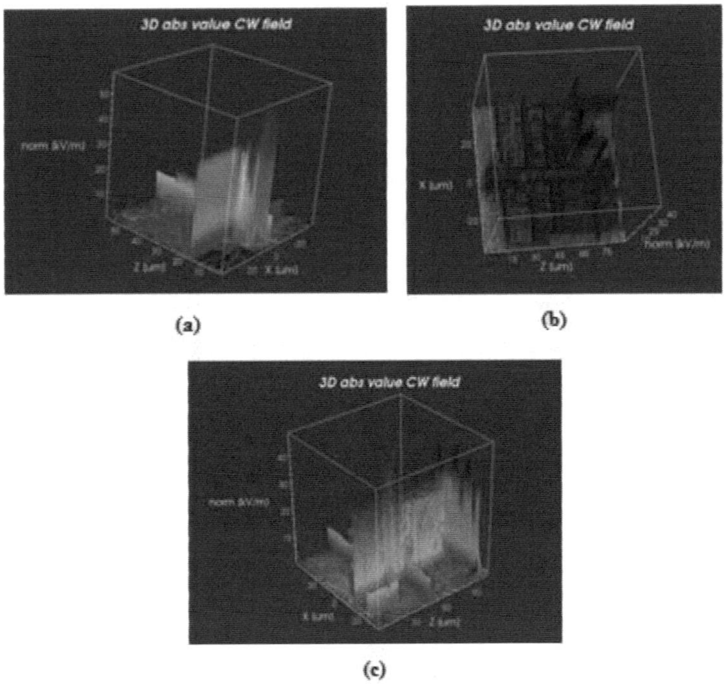

Figure 3. 3D simulation results (a) 'West' to 'North' port (b) 'West' to 'Vertical' Port (c) 'Injection' to 'South' Port

The insertion losses are calculated for non-blocking optical routers reported in the literature as well as the proposed router. Insertion loss in particular path is obtained by considering insertion loss due to MRRs, waveguide crossings and waveguide bendings in that path. If either of number of MRRs, waveguide crossings or waveguide bendings are reduced, insertion loss will be reduced. The insertion loss values used for these calculations are as per the table 2.

Table 2. Insertion loss values

Loss	Value	Reference
Loss at waveguide crossing	0.15dB	[22]
Loss at waveguide bending	0.005dB/900	[22]
Loss of micro-ring in drop	0.5 dB	[22]
Loss of micro-ring in pass	0.005dB	[22]

Fig. 4 shows the insertion losses of all the paths in optical routers in [16-18]. Optical router in [18] has the lowest loss as it has lowest number of MRRs in existing 7x7 routers.

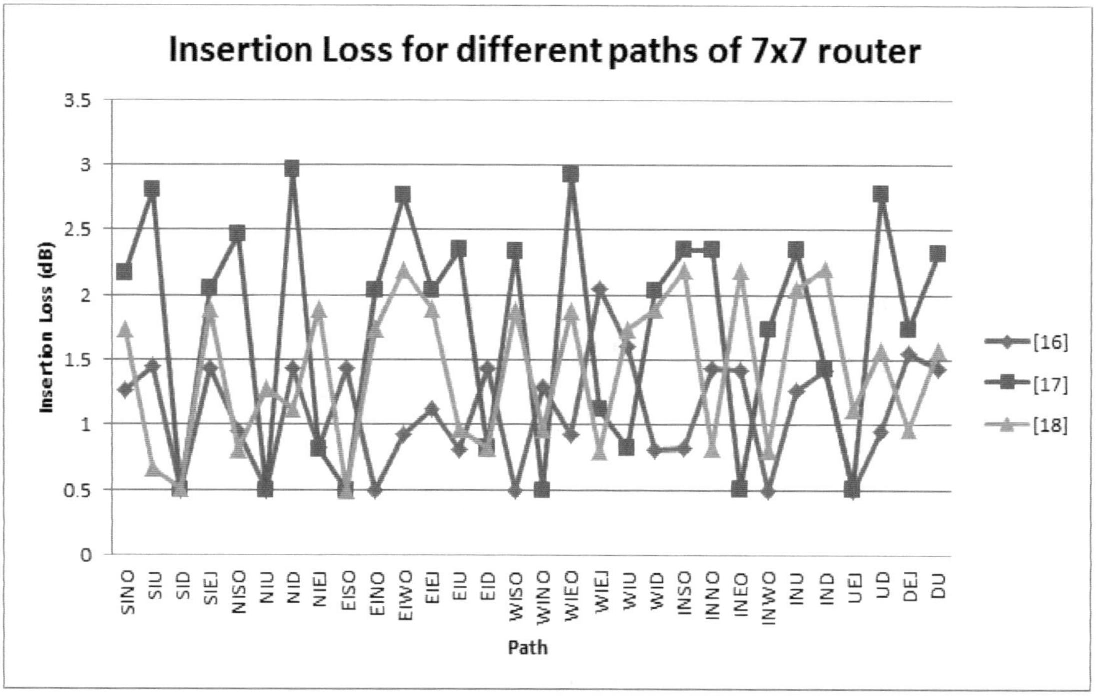

Figure 4. Insertion losses for 7x7 optical routers for different paths

Fig. 5 shows the insertion loss of different paths of 6x6 optical routers along with the proposed router. It can be seen that for most of the paths in proposed router has lower insertion loss compared to the optical routers in [19-21].

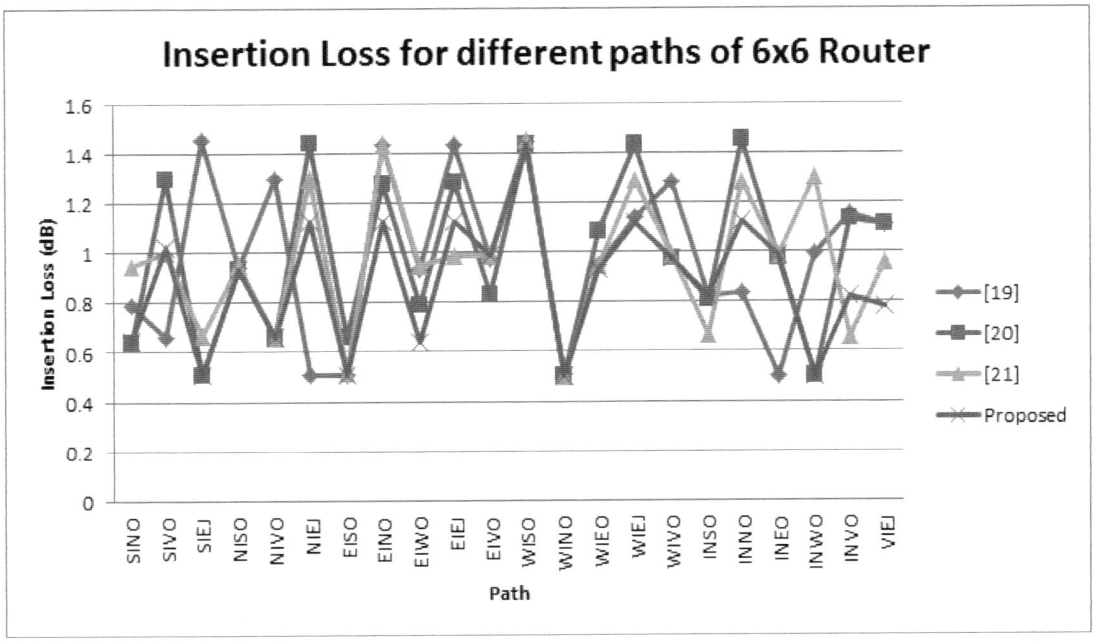

Figure 5. The insertion losses for 6x6 optical routers for different paths

Fig. 6 shows the comparison of 6x6 and 7x7 optical routers in the literature with proposed router for the average insertion loss. All 6x6 optical routers have lower average insertion loss compared to any of 7x7 optical routers. The highest average insertion loss is 1.83dB and is observed for optical router in [17]. The lowest average insertion loss is 0.88dB and it is for proposed optical router. We have reduced insertion loss significantly by designing optical router with minimum waveguide crossings and waveguide bendings.

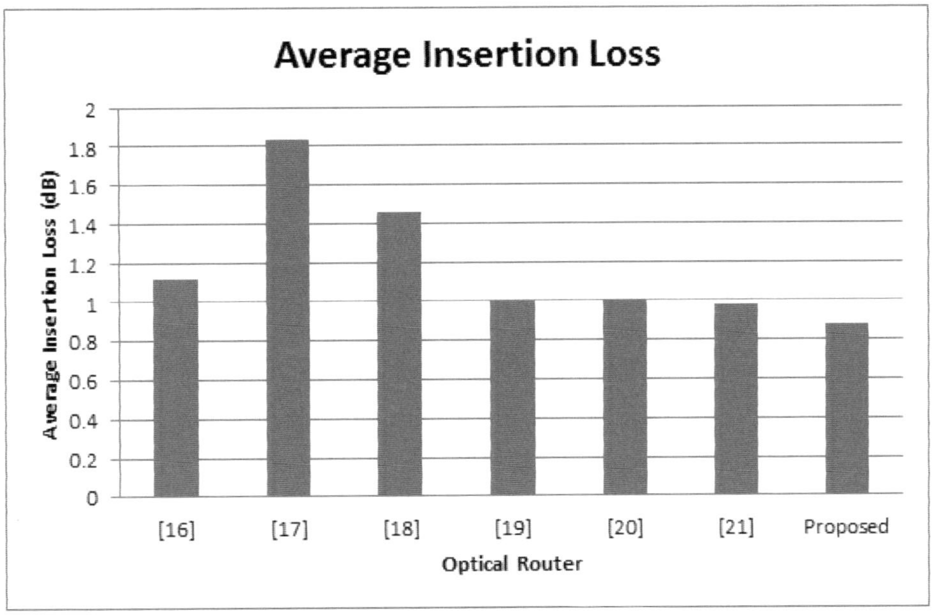

Figure 6. Average insertion loss for optical routers

4. Conclusion

In the field of integrated circuits ONoCs are now evident alternative of ENoCs due to its high bandwidth, low latency and lower power requirements. Specifically for multulayer multi-core processors, 3D ONoCs are used. Optimized design of optical router is the key requirement of 3D ONoC. In this paper an optical router for 3D ONoC is designed and simulated. The router design aims at reduced number of MRRs, waveguide bendings and waveguide crossings. The proposed design is optimized by reducing the number of waveguide crossings and waveguide bendings up to 15 and 18 respectively with 16 MRRs. As compared to existing non-blocking optical router of 3D ONoC with same number of MRRs, the waveguide crossing elements are reduced by 4 and waveguide bending elements are reduced by 6. Insertion loss is one of the optical router performance parameters. Average insertion loss is calculated and it is observed that it is improved by improved by 3% of existing optical router of 3D ONoC. The reduced insertion loss of the proposed optical router has improved the performance of ONoC significantly. This work can be extended to analyse insertion loss of proposed router on network level for different network sizes. Future work includes analysis of crosstalk noise and signal to noise ratio at network level.

References

[1] Kumar N , 2019, Energy and power efficient system on chip with nanosheet FET, Journal of Electronics, 1, 52-59.
[2] O'Connor I, and Nicolescu G, 2012, Integrated optical interconnect architectures for embedded systems, New York, Springer Science & Business Media.
[3] Ji R, Yang L, Zhang L, Tian Y, Ding J, Chen H, Lu Y, Zhou P, and Zhu W, 2011, Microring-resonator-based four-port optical router for photonic networks-on-chip, Opt. express, **19**,18945-18955.
[4] Jadhav N, Bhagat R, Paranjpe S, Dahitule S, Madke S, and Jadhav S, 2021, Micro-ring resonator based all-optical Arithmetic and Logical Unit, Optik, **244**, 167622.
[5] Das A., Chaudhari B, and Ghosh S, 1998, Characteristics of polymeric optical passive single-mode waveguiding devices fabricated by an argon-ion laser, Appl. Opt., **37**, 6779-6786.
[6] Yahya M, Wu N, Yan G, Ge F, and Ahmed T, 2019, RoR: A Low Insertion Loss Design of Rearrangeable Hybrid Photonic-Plasmonic 6x6 Non-Blocking Router for ONoCs, IEICE Electron. Express, **16**, 20190346.
[7] Li B, Soo-Jin C, Fitzgerald E, Chaudhari B, Jiang S, and Cai Z, 2004, Intelligent integration of optical power splitter with optically switchable cross-connect based on multimode interference principle in Si Ge/Si, Appl. Phys. Lett., **85**, 1119-1121.
[8] Chaudhari B, and Patil S, 2020, Optimized designs of low loss non-blocking optical router for ONoC applications, Int. J. Inf. Technol., **12**, 91-96.
[9] Shi X, Wu N, Ge F, Yan G, Xing Y, and Ma X, 2019, Srax: a low crosstalk and insertion loss 5× 5 optical router for optical network-on-chip, 45th Annual Conf. of the IEEE Industrial Electronics Society.
[10] Kalange O, Ladniya B, Kothari R, Jadhav N, and Chaudhari B, 2018, Design and Analysis of Five Port Optical Router for Optical NoC, IEEE Int. Conf. on Inventive Research in Computing Applications (ICIRCA.
[11] Yu Z, Zhang Q, Jin X, Zhao J, Baghsiahi H, and Selviah D, 2016, Microring resonator-based optical router for photonic networks-on-chip, Quantum Elec, **46**, 655.
[12] Jia H, Zhao Y, Zhang L, Chen Q, Ding J, Fu X, and Yang L, 2016, Five-port optical router based on silicon microring optical switches for photonic networks-on-chip, IEEE Photonics Technol. Lett., **28**, 947-950.
[13] Xie Y, Xu J, Zhang J, Wu Z, and Xia G, 2011, Crosstalk Noise Analysis and Optimization in Hitless Silicon-Based Optical Router for Optical Networks-on-Chip (ONoC), J.Lightw. Technol., **30**, 198-203.

[14] Ji R, Yang L, Zhang L, Tian Y, Ding J, Chen H, Lu Y, Zhou P, and Zhu W, 2011, Five-port optical router for photonic networks-on-chip, Opt. express, **19**, 20258-20268.
[15] Ji R, Xu J, and Yang L, 2013, Five-port optical router based on microring switches for photonic networks-on-chip, IEEE Photonics Technol. Lett., **25**, 492-495.
[16] Fadhel M, Gu H, and Wei W, 2021, DORR: A DOR-Based Non-Blocking Optical Router for 3D Photonic Network-on-Chips, IEICE Trans. Inf. Syst., **104**, 688-696.
[17] Zhu K, Gu H, Yang Y, Tan W, and Zhang B, 2016, A 3D multilayer optical network on chip based on mesh topology, Photonic Netw. Commun., **32**, 293-299.
[18] Gu H, and Xu J, 2009, Design of 3D optical network on chip, IEEE Symp. on Photonics and Optoelectronics.
[19] Ye Y, Xu J, Huang B, Wu X, Zhang W, Wang X, Nikdast M, Wang Z, Liu W, and Wang Z, 2013, 3-D mesh-based optical network-on-chip for multiprocessor system-on-chip, IEEE Trans. on Comput.-Aided Des. Integr. Circuits Syst., **32**, 584-596.
[20] Guo P, Hou W, Guo L, Yang Q, Ge Y, and Liang H, 2018, Low insertion loss and non-blocking microring-based optical router for 3D optical network-on-chip, IEEE Photonics J., **10**, 1-10.
[21] Guo P, Hou W, and Gu L, 2016, A Low Insertion Loss and Non-Blocking Optical Router for 3D Optical Network-on-Chip, IEEE Asia Communications and Photonics Conf. (ACP).
[22] Guo P, Hou W, Guo L, Sun W, Liu C, Bao H, Duong L, and Liu W., 2019, Fault-tolerant routing mechanism in 3D optical network-on-chip based on node reuse, IEEE Trans. Parallel Distrib. Syst., **31**, 547-564.

IOT Based Moisture Measurement and Conveyor Belt Monitoring in Yarn Mill

T. Kalavathi Devi[1], M.Madhan Mohan[1], K.N.Baluprithviraj[1], V. Poojashri[1], A. Swetha[1], P. Vasuki[1]

[1]Faculty and Scholar, Department of Electronics and Instrumentation Engineering, Kongu Engineering College, Perundurai, Erode – 638060, Tamil Nadu, India.

Corresponding author's e-mail: wisdomvasuki@gmail.com, kalavathidevi@gmail.com

Abstract: Yarn manufacturing in spinning mills is done when the twisted length of fiber goes to cheese, doubling, and winding machines. The humidity and temperature must be maintained to ensure the yarn quality. The conveyor belts have to be monitored continuously to avoid the interruption of the belts. This paper provides a solution to maintain the precise range of humidity and temperature and also to surveil the conveyor belts. The system contains a DHT sensor and an IR sensor. If the DHT sensor detects the increase in the moisture level or temperature, the signal is sent to the controller, and using a relay, either the heater or the cooler is activated respectively. The humidity level and the temperature of the yarn are displayed on the LCD. Simultaneously, the values can also be viewed on the user's mobile application Blynk using an IoT. Another feature of this application is the continuous monitoring of the conveyor belt. If the belt stops, the IR sensor detects it, and the user is alarmed by a buzzer indicating that action is required to restart. This method reduces yarn breakage and improves the quality of the yarn. It provides continuous monitoring; the IoT application provides real-time values.

Keywords - moisture monitoring, conveyor, continuous monitoring, humidity, temperature, health hazards

1. Introduction
Clothing has been one of the essential needs of humans for years together. Today, the textile industries account for 14% of the overall industrial production and add 4% to India's GDP and also provide the second-largest employment after agriculture. The primary material needed for manufacturing clothes is yarn. Yarn is a continuous length of fiber twisted together. It is employed in the production of textiles, weaving, embroidery, and rope making. India's share in yarn manufacturing is 21% which is the second leading manufacturing country in the world. There are more than 1220 cotton mills in India, out of which 283 mills are composite mills, and the others are spinning mills. The Indian textile industry is likely to grow at a rate of 20 percent annually to reach 30 billion dollars over the next five years.

1.1. Industrial visit to the yarn mill
Spinning is the initial process for any manufacturing process. The spinning mill consists of three machines, a cheese machine, a doubling machine, and a winding machine. The process of winding the

yarn into a paper or wooden cone is called cheesing and it is done in a cheese machine. It has 28 drums on the left and right sides and each is driven by 2hp motors. The cost of one drum is Rs 3600 which leads to a total cost of Rs 2, 01,600. The doubling machine is also called yarn twisting which performs the process of twisting two or more strands of yarn together. There are three machines that perform the spinning, in which two machines have 474 spindles and the other spindle has 440 spindles and is driven by a 10hp motor. One spindle costs Rs 1000 so, it totally costs Rs 13, 88,000. Winding is the process of packing a yarn or thread onto a core. It has a winding drum or also called a cone holder. It has 60 needles on the right and left sides each. It is driven by a 3 hp motor. The cost of one needle is Rs 2750, so for 120 needles, totally it costs Rs 3, 30,000.

1.2. Yarn manufacturing process

The detailed process in manufacturing consists of nine steps which are performed sequentially. The first step is bale opening in which the cotton tufts are opened to loosen the matted fibers using rotating spiked rollers. The next process is the mixing of the fibers to get the essential properties of the fiber. After this, the fiber is sent to the blow room where it is cleaned and opened. In the next process of carding, the minute impurities are removed from the fiber and the yarn is straightened and aligned. Next, the fiber is paralleled by a draw frame and it is stretched or straightened and made into a single thread called roving. Here, the fiber is fed into the simplex where it is made much thinner by drafting and mild twisting. The roving is sent to a ring frame which is made into yarn by further twisting and drafting until fineness is achieved. Lastly, the fiber is fed into the auto cone winding machine where large yarn packages are created. Faults are also removed with the help of the yarn clearer. There are three sections in the yarn industry that use three major machines such as cheese machine, doubling machine, and winding machine. Figure 1 indicates the overview of the yarn manufacturing process in Cheese, Doubling, and Winding machines.

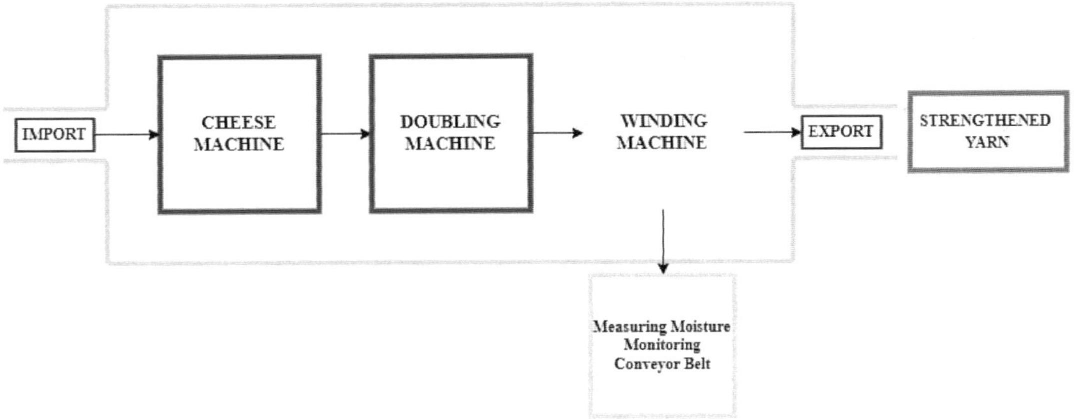

Figure 1. Overview of yarn manufacturing in Cheese, Doubling and Winding machines

1.2.1. Cheese machine. The process of winding the yarn into a paper or wooden cone is called cheesing in the textile industry, and the cone is called cheese. The cone resembles a wheel of cheese; the yarn is wound on a paper tube called a paper bobbin in a cotton mill. Since a roll of yarn built up on a paper tube resembles a bulk cheese which is a package of a large amount of cheese, the cone is said to resemble a wheel of cheese. A bobbin is a cylindrical frame that allows the yarn to be wrapped around it. The type of bobbin used in cotton yarn mills is the paper cheese bobbin. So, the yarn is wind into the bobbin made of paper. The cheese of yarn thread is a larger tube with a total diameter of more than 30 cm and the ability to hold over 10kg of yarn thread. It is used for large-scale production as it does not need to be replaced frequently. Figure 2 shows the cheese machine

1.2.2. Doubling machine. During the cheesing process, the yarn is transferred to a cone. In the doubling process, the two ends of the yarn are combined and wound on a cone without twisting. The main goal of the doubling process is to produce a stronger yarn by improving the yarn regularity. In the doubling process, the thread is cleaned from the waste. The weak and overly thick sections are removed. Figure 3 shows the doubling machine also called as yarn twisting machine.

1.2.3. Winding machine. Winding is the process of creating large yarn packages that can easily be unwounded. It is used to get the continuous length of yarn on cones for the weaving process. It is also used to remove yarn faults and obtain a clean yarn. Some examples of yarn faults are oil stains, hairiness, and foreign matters. These faults lead to some negative effects on the fabric such as damaged appearance and shade variation. These faults can be removed by careful material handling, maintaining prescribed roller settings, maintaining optimal relative humidity in the spinning room and ensuring that no foreign matters enter while winding. Figure 4 shows the winding machine.

Figure 2. Cheese machine

Figure 3. Doubling machine

Figure 4. Winding machine

1.3. Importance of moisture measurement of yarn

All types of yarn are hygroscopic, which means they absorb or release moisture depending on the relative humidity of the environment. If the atmosphere is drier than the textile's equilibrium relative humidity, then the thread will give up its moisture to the air. If the air is more humidified, then the textile's moisture content will increase. The moisture of the thread contributes to the properties of the thread such as rigidity, tensile strength, dimensions, etc. For instance, cotton thread absorbs moisture more rapidly when exposed to high atmospheric dampness. As a result, the weight and strength of the material will increase, and also the other properties will be changed. But if we consider the man-made fibers like acetate and viscose, the strength of the thread reduces when exposed to atmospheric dampness. But synthetic fibers like terylene and nylon are not affected much as they are not fully hygroscopic.

The moisture of the thread is examined by measuring the humidity level and the temperature of the yarn. Humidity is the amount of water vapor present in a given amount of air. It is measured as relative humidity. The yarn spinning mill has to maintain relative humidity. Low relative humidity in the yarn produces fly dust and as a result, production decreases. High relative humidity also affects production so, maintaining an optimum relative humidity is essential. The relative humidity is expressed in a form of 0 % to 100 %. The cotton and linen yarn has to be processed at very high levels of humidity, around 60-80% RH as they are very brittle. In this project, the set point is given as 65 % that is if the humidity increases more than 65 %, the heating mechanism is switched on. By humidifying every process in the manufacturing of yarn, the manufacturer can ensure that the product remains flexible and is prevented from breaking.

Temperature is the amount of warmth or coldness measured in an object. Maintaining temperature is very important to delivering a good product of yarn. It plays a vital role in the moisture absorption of cotton fibers. The temperature and relative humidity collectively contribute to the elongation, softness, and strength of the cotton fibers. When the temperature rises, the electrical conductivity increases and affects the moisture absorption of the fibers. So maintaining the correct temperature is essential for maintaining the quality of the thread. The optimum temperature for yarn development is 25 – 35°C. Cotton is subject to self-heating/spontaneous combustion. In this project, the set point is made at 37 degrees Celsius that is if the temperature increases more than 37 degrees Celsius, the cooling mechanism is turned on. In some cases, the textile machinery develops excess heat, which may cause an undesired ambient temperature rise which can be prevented by maintaining the temperature.

1.4. Importance of conveyor in yarn machine

Belting conveyor belts are used in the textile industry for the transfer of packaged goods as well as general cargo. In the yarn industry, the yarn is wound onto a slightly tapered tube called corners. The automation in this section is linking corners directly to the ring frame. These trays are transferred to the winding division through several conveyor belt lines. Finished packages are also expelled by a mechanical ejection system onto a conveyor belt. The cones are placed into hollow tubes for packaging. After being packed, these empty tubes are transported into empty cardboard boxes using conveyor belts. Since these processes are being accomplished using conveyor belts, they have to be maintained without the belts stopping.

1.5. IoT based measurement and monitoring

The implementation of IoT-based measurement helps us to work in the comfort of the users. The change in the humidity and temperature of the yarn in the mill can be studied using the IoT application installed on our mobile phones. Integration of Industrial IoT Solutions with Conveyor Belts will reduce downtime of conveyor systems with the help of predictive maintenance. The IoT enables more information from the sensor systems to be stored and accessed at any time. The industry can be monitored all day, and any fault can be identified and corrected at any time.

2. Observation in Literature review

A solution for the detection of defects of yarn in the textile industry is by capturing the image of the yarn by a camera and later developed using filters. The defects in the geometric contour and surface area are identified by the image processing technique. The results showed better reliability in number evaluation and the defect types [1]. The processed fabrics after knitting undergo several testing parameters such as drying time, air permeability, immersion time, bursting strength, and so on. Using ANOVA on a single factor, the results of the test were analyzed statistically using Scanning Electron Microscope(SEM) analysis [2]. A wireless moisture sensor based on passive ultra-high-frequency (UHF) radio frequency identification (RFID) technology was developed where sensors are flexible, lightweight, cheap, and easy to make and made using PVA material. As it provides, permanently wireless reachable evidence of moisture coverage, it can be utilized in the frozen food deliver chain, or for health monitoring in wearable functions [3]. The size of cotton thread for weaving at minor RH for which studies were conducted at 28°C and different relative humidity levels on the moisture recover tensile force and abrasion resistance of cotton yarns with diverse size formulations. According to the results of the study, the collection of appropriate size formulations allowed equivalent sized yarns to be woven at minimum relative humidity [4]. A new technique based on the appearance of Yarn is evaluated through Machine Vision and Image Processing methods. In order to improve the efficiency, flexibility, and robustness of the yarn defect system, this process is proposed to test yarn diameter, defects, and hairiness. Statistical methods were used to analyze the yarn defects in order to further assess the quality of yarn [5, 8]. The assessment of the effect of heat-setting conditions on yarn quality was performed with the use of a uniformity tester, yarn strength tester, moisture regain tester, and electronics balance. A number of quality parameters such as unevenness, imperfections, yarn hairiness, yarn strength, moisture, and yarn cone weight were measured before and after heat setting. Due to better yarn quality, it is recommended that the finished yarn is transported to the knitting and weaving industries subsequent to 8 hours of heat setting [6, 9]. The mechanical properties of conveyor belts in terms of their usage at the three main stages of manufacturing were analyzed. The fibers of polyester warp and polyamide weft were used to produce nine different fabrics of 1 m width. The results of the experiment indicate that the mechanical properties of products at different steps of conveyor belt model are affected by the number of intersections of yarns in woven fabrics [7, 10]. The comfort characteristics of denim fabrics with different weft yarns of cotton, polyester, and core-spun Lycra have been discussed. [11] Research shows that cotton weft yarn provides higher air and water vapor permeabilities than polyester and Lycra cotton weft yarn for unwashed denim fabrics. Every step of the process should be verified according to the customer's requirements based on the combined information of the various fault detection techniques. During the testing [12], it should be possible to measure each parameter of the yarn according to the technologies developed, such as the artificial neural network model and modeling and simulation in the appropriate software. The soft wastes from spinning were used to create fiber blends from yarn waste and fabric clippers reclaimed from yarn waste. ANOVA statistical procedures were used to assess fiber and yarn properties. A Tukey HSD (Honestly Significant Difference) analysis confirmed the degree of significance [13] of the relationship between the waste types and fiber. By systematically varying yarn counts [14, 16] and by systematically varying the fabric cover factor, the mechanical, comfort, and hand properties of fine cotton fabrics were examined. The fabric test results for fabrics that underwent the same pre-treatment processes revealed an improvement in fabric wrinkle recovery, smoothness, and resilience, as well as an increase in cover factor values. During production, the textile industry [15] generates two types of waste, hard waste, and soft waste. A limited volume of these materials is mixed with virgin cotton in Blow Rooms. A variety of machines are used to transform these fabrics into reclaimed or regenerated fibers. Depending on the desired end product, reclaimed fibers and virgin cotton are mixed in a preset volume.

2.1 Summary of literature review

The defects in the yarn of the textile industry have been analyzed using several technologies like image processing; Ultra-High-Frequency (UHF), Radio Frequency Identification (RFID), and machine vision techniques have been used to ensure the quality of the yarn in the various sections of the yarn mill. The breakage of the yarn occurs due to the inadequate amount of moisture in the thread. To maintain the required moisture level, the yarn has to be maintained at the prescribed range of temperature and humidity level. The temperature and relative humidity collectively contribute to the elongation, softness, and strength of the cotton fibers. The conveyor belt in the mill also has to be monitored continuously to avoid them from stopping due to unavoidable reasons. These methods have drawbacks such as additional setups like cameras and filters, using several testers like evenness tester, and yarn strength tester to ensure the quality of the thread. Our method reduces the use of more setups and a single sensor can be used to measure both humidity and temperature also the conveyor belt can be monitored using an IR sensor. Another advantage is that the real-time values can be viewed through an IoT application.

3. Proposed Methodology

The continuous monitoring in the mill is made easy by developing a method that maintains the quality of the yarn by monitoring the temperature and humidity of the thread. The system consists of a humidity (DHT) sensor and IR sensor. The Node MCU acts as a controller. The main purpose of this method is to continuously observe the thread and the conveyor belt. When the humidity sensor detects the increase in moisture content in the thread, it sends the signal to the controller, and the Node MCU processes the signal and activates the heater, and heat is supplied to the thread to reduce the moisture content. Similarly, when the sensor detects the increase in the temperature of the thread, it sends the signal to the controller, and the Node MCU processes the signal and activates the cooler, to reduce the temperature of the thread.

Figure 5 shows the block diagram of IoT-based moisture measurement and conveyor belt monitoring with a DHT11 sensor, an IR sensor, a couple of relays, a buzzer, an I2C converter, a cooling fan, a heater, an LCD and ESP32 microcontroller. The cooling fan is ON when the temperature rises beyond the set value. Similarly, the heater is ON when the humidity level is beyond the set point. A buzzer is turned ON automatically when the conveyor belt stops its rotation.

The moisture content and temperature are monitored continuously by the DHT sensor as shown in figure 6. If the sensor detects the increase of moisture or temperature, it sends the signal to the controller, and using a relay, either the heater or cooler is activated respectively. The humidity level and temperature are displayed on the LCD display which utilizes an I2C converter to make the display easier. Simultaneously, they can also be viewed on the user's mobile application Blynk using IoT. IoT allows to measure all kinds of data remotely and provide this information to the user in real-time. The heater and pump are activated through the Blynk app via ON and OFF mode.

The low-cost and high-performance ESP boards are ideal for various applications. The module allows microcontrollers to connect to 2.4 GHz Wi-Fi. It can be used with the optional ESP-AT firmware to provide connectivity to external host devices. The high-performance processing and storage capabilities of the module allow it to be utilized with various sensors and devices. Its integrated connectivity allows it to operate under various operating conditions.

Another feature in this application is continuous monitoring of the conveyor girdle as exposed in figure 7. If the belt stops due to unavoidable conditions, the IR sensor which detects continuously sends the signal to the controller. The user is notified by a buzzer that the belt is stopped and immediate action is performed.

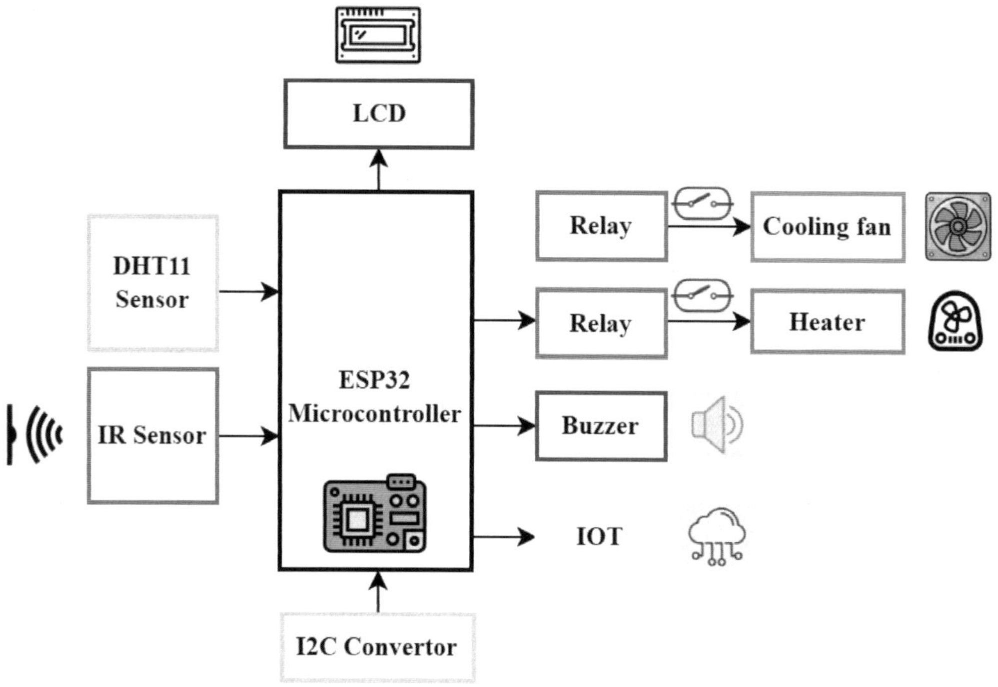

Figure 5. Block diagram for IoT based moisture measurement and conveyor belt monitoring

Figure 6. Prototype of IoT based moisture measurement in yarn mill

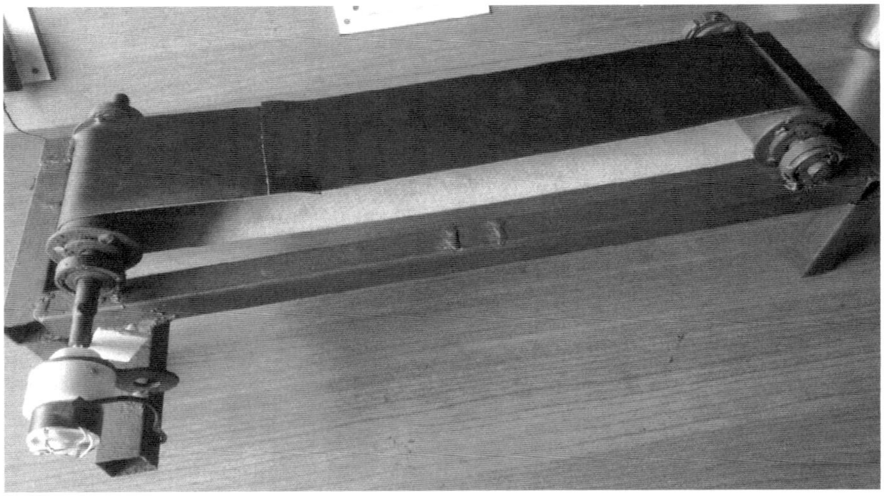

Figure 7. Prototype of Conveyor belt monitoring in Yarn mill

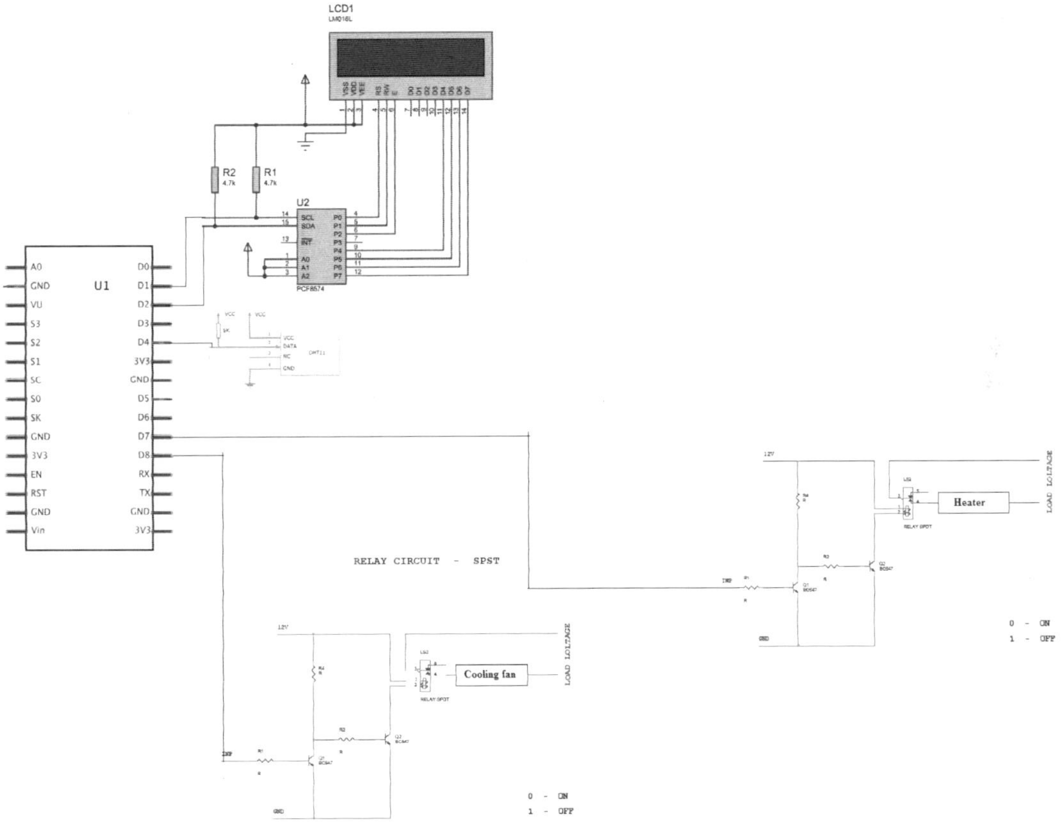

Figure 8. Circuit Diagram of IoT based moisture measurement and conveyor belt monitoring

In NODMCU U1 pin is connected to the PCF8574 14th pin, the D2 pin is connected to the PCF8574 15th pin, the D4 pin is connected to the DHT11 sensor, the D7 pin is connected to the resistor, and that resistor is connected to the one resistor followed by BC547 transistor and that is connected to relay and that relay is linked to the heater, D8 pin connected to the one resister followed by BC547 transistor and that is coupled to relay and that relay is associated to the cooling fan. In PCF8574 P0 pin is connected to the LCD pin 4, the P1 pin is connected to the LCD pin 5, the P2 pin is tied to the LCD pin 6, the P4 pin is connected to the LCD pin 11, P5 pin is connected to the LCD pin 12, P6 pin connected to the LCD pin 13, P7 pin connected to the LCD pin 14. In LCD, 2 and 3 pins are connected to the resistor 4.7k and that resistor is connected to PCF8574 pins 14 and 15, LCD 1st pin is connected to the ground as shown in Figure 8.

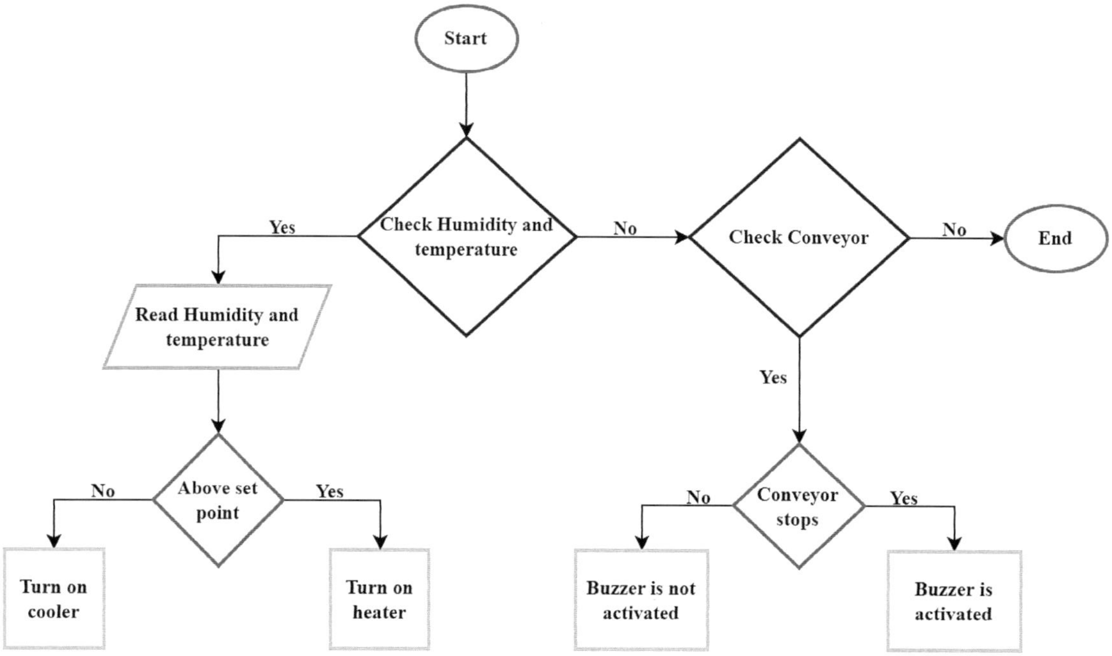

Figure 9. Flow chart of IoT based moisture measurement and conveyor belt monitoring in yarn mill

Algorithm for the same figure 9 is shown below,

Step 1: Humidity and temperature is measured by an DTH11 sensor.

Step 2: The data is fed into the microcontroller and based on the set point, either the heating or cooling effect is provided with the help of a relay.

Step 3: The continuous motion of the conveyor is being sensed.

Step 4: In case of sudden stop of the conveyor, the user is alerted through a buzzer and also IoT application.

A circuit can be developed as shown in Figure 9with DHT11 and IR sensor to provide the variation of the equation. The equation is that of a straight line

Journal of Physics: Conference Series **2325** (2022) 012009 doi:10.1088/1742-6596/2325/1/012009

$$V_{out} = mV_{in} + V_0 \quad (1)$$

Where,

m is slope of the line and represents the gain (m>10) or attenuation (m<1)

V_0 is the intercept, the value $V_{out} = V_0$ (when $V_{in} = 0$)

For humidity sensor,

$$3.5 = m(6.5) + V_O \quad (2)$$

For IR sensor,

$$0.4 = m(100) + V_O \quad (3)$$

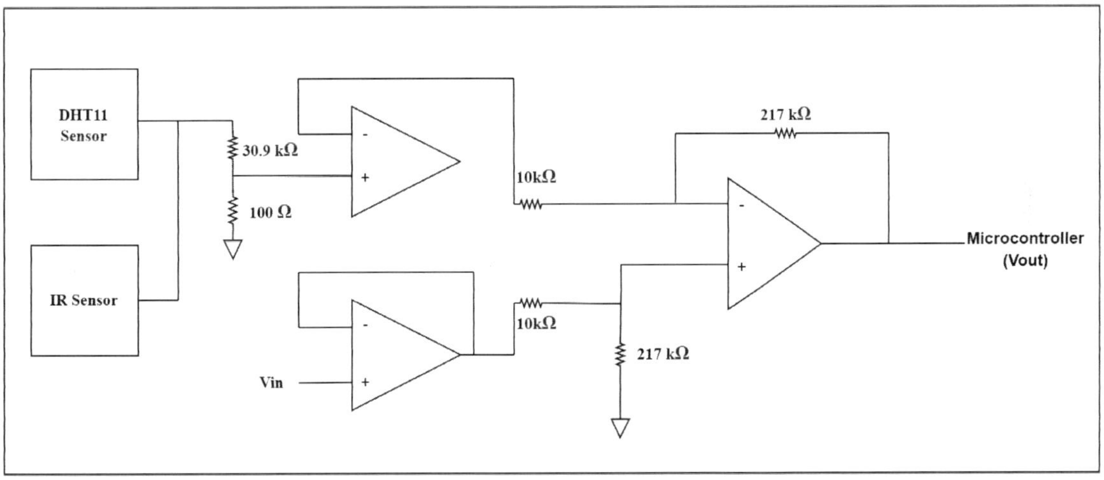

Figure 10. Signal Conditioning Circuit for DHT11 and IR Sensor

From equation 2 and 3,

$$V_{out} = (0.032)\, V_{in} - 3.7155 \quad (4)$$

From equation 4, the differential amplifier with a gain of 0.032 and a fixed input of 3.71 volts to the inverting side is shown in Figure 10 with instrumentation amplifier as shown in table 1.

Table 1. Sensor ranges and corresponding output

Sensors		**Input range**	**Output range**
DHT11	Humidity	65-80 RH	3.5 - 5.5 v
	Temperature	25-35 °C	
Infrared (IR)		100-500 cm	0.4 - 3.25 v

3.1. Importance of IoT based moisture measurement and conveyor belt monitoring

This method is employed to identify the level of moisture content in the yarn, achieved by using an IoT platform that controls the switching of the heater and cooling arrangement based on the moisture

Journal of Physics: Conference Series **2325** (2022) 012009 doi:10.1088/1742-6596/2325/1/012009

content of the cotton thread and gives real-time surveillance to the owners. That reduces the breakage in the yarn and also maintains the quality. The conveyor belt is continuously observed to avoid automatic stopping of the belt due to unavoidable reasons. It promotes predictive maintenance and accident prevention as this system constantly examines any unusual behaviour of the conveyor belt.

3.2. Internet of Things (IoT)

The term Internet of things refers to the interconnected devices and systems that can be accessed through the various networks and communications platforms that exist today. The network-connected sensors and other monitoring device improve the quality and manufacturing of a product.
Figure 11 explains the process of using an IoT. The input is the data from the DHT sensor and IR sensor. These values are stored in an IoT cloud using a Wi-Fi module. The output is the display of the values on a mobile application. Using IoT, the data can be viewed from anywhere on the application.

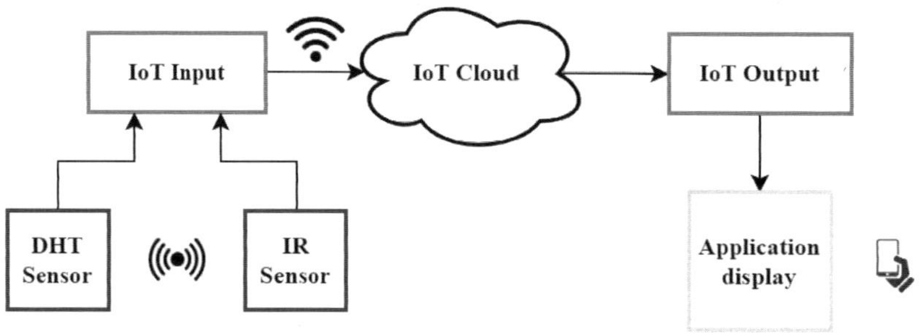

Figure 11. Block Diagram of IoT

4. Results and Discussion

The outputs are displayed under three conditions:
a) At room temperature
b) When humidity is above set point (65 RH)
c) When temperature is above set point (37 °C)

4.1 Output at room temperature

Figure 12. LCD output at room temperature

Figure 13. IoT application output at room temperature

At room temperature, the temperature and humidity are below-set points as shown in Figures 12 and 13. During this condition, both the heater and cooler are in OFF condition.

4.2 Output when humidity is above set point (65 RH)

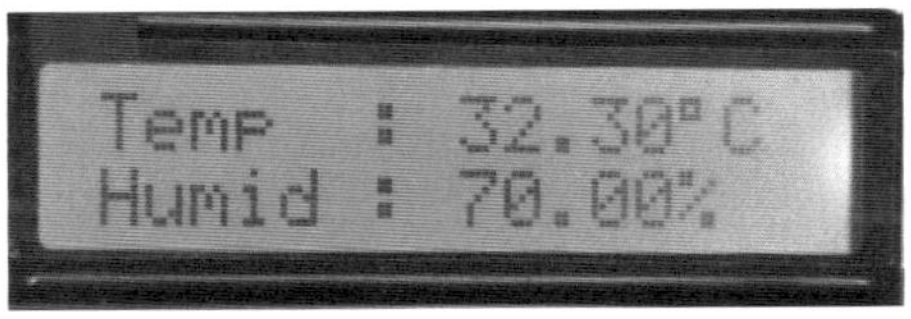

Figure 14. LCD output when humidity is above set point

Figure 15. IoT application output when humidity is above set point

In figures 14 and 15, the output humidity is beyond the set point (i.e., 65 RH) which results in turning ON a heater.

4.3 Output when temperature is above set point (37 °C)

Figure 16. LCD output when temperature is above set point

Figure 17. IoT application output when temperature is above set point

In figures 16 and 17, the output temperature is beyond the set point (i.e., 37 ℃) which results in turning ON a cooler(cooling fan).

4.4 Corresponding outputs of temperature and humidity

Table 2. Temperature VS Cooler table

TEMPERATURE ℃	COOLER
31.7	OFF
32	OFF
33	OFF
34	OFF
37	ON
40	ON

Table 3. Humidity VS Heater table

TEMPERATURE ℃	COOLER
31.7	OFF
32	OFF
33	OFF
34	OFF
37	ON
40	ON

Tables 2 and 3 represent the outputs of temperature and humidity and their corresponding cooler and heater condition. It is seen that the cooler is ON when the temperature rises beyond the set point (37 °C). Similarly, the cooler is ON beyond the set point (65 RH).

5. Conclusion

IoT-based moisture measurement and conveyor belt monitoring are built. By adopting this method of controlling the moisture, breakage in the yarn can be reduced and minimizes health-related issues because the workers suffer from various respiratory problems and other health conditions due to the presence of dust in their workplace. They were not taking any precautionary measures to avoid inhaling the dust. On the other hand, monitoring the conveyor belt conserves time while working, helps in transportation, and enhances the product quality by maintaining the desired parameters. It provides continuous monitoring where users are alerted through a buzzer and an IoT application which increases the quality of the yarn.

References

[1] Laa Arafa Badr and Ashraf El-Nahrawy 2016 Moisture properties of raised 3-thread fleece fabric knitted with different face and fleecy yarns *Alexandria Engineering Journal* **55** 2881-92
[2] Shams Nateri A, Ebrahimi F and Sadeghzade N 2014 Evaluation of yarn defects by image processing technique *Optik* **125** 5998-6002
[3] Xiaochen Chen, Han He, Meinan Gou, Yunshan Yang, Lauri Sydänheimo, Leena Ukkonen and 2020 Johanna Virkki Passive Moisture Sensor Based on Conductive and Water-Soluble Yarns *IEEE Sensors Journal* **20** 10989-995
[4] Wu, H., Shen, Y., Yao, Y., and Zhang B 2021 Size formulations for cotton yarn weaving at lower relative humidity *Textile Research Journal* **91** 168–174
[5] Z. Li, P. Zhong, X. Tang, Y. Chen, S. Su and T. Zhai 2020 A New Method to Evaluate Yarn Appearance Qualities Based on Machine Vision and Image Processing *IEEE Access* **8** 30928-937
[6] Rokonuzzaman, Md, M. A. A. Mamun, and A. K. M. A. H. Asif 2017 Effect of Heat Setting Conditions on the Quality of Yarn *Frontiers in Science* **7** 46-49
[7] Barburski M 2016 Analysis of the mechanical properties of conveyor belts on the three main stages of production *Journal of Industrial Textiles* **45** 1322-34
[8] Kalavathi Devi T, Mouleeshuwarapprabhu R, Uma Devi S, Sakthivel P, Poojashri V, Swetha A and Vasuki P 2021 Sensor Technology and Regulation method for sustaining the pH value in

sugar Industry *IEEE Fourth International Conference on Electrical, Computer and Communication Technologies (ICECCT)* 1-5.

[9] Kalavathi Devi T, Priyanka EB, Sakthivel P, Stephen Sagayaraj A 2022 Low Complexity Modified Viterbi Decoder with Convolution Codes for Power Efficient Wireless Communication *Wireless Personal Communications* **122** 685-700.

[10] Kalavathi Devi Thangavel, Sakthivel Palaniappan 2020 Low power sleepy keeper technique based VLSI architecture of Viterbi decoder in WLANs *Australian Journal of Electrical and Electronics Engineering* **17** 263-268

[11] Vinay Midha, Suresh Kumar S and Nivas Kumar M 2017 Investigation on permeability and moisture management properties of different denim fabrics after repeated laundering *The Journal of the Textile Institute* 71-77.

[12] Kulkarni K M, Sahu A R 2022 Brief Overview on Study of Various Parameters Affecting the Productivity of Cotton Yarn *Smart Technologies for Energy, Environment and Sustainable Development, Springer* **1** 955-963.

[13] Hafsa Jamshaid, Uzair Hussain, Rajesh Mishra, Martin Tichy, Miroslav Muller 2021 Turning textile waste into valuable yarn *Cleaner Engineering and Technology* **5** 2666-7908.

[14] Nilay Ork Efendioglu, Mehmet Mete Mutlu, Oktay Pamuk 2021 An investigation on usability of 3D visualization and simulation programs in leather apparel *Journal of the Textile Institut*e 1-16.

[15] Wagaye B T, Adamu B F, Jhatial A K 2020 Recycled Cotton Fibers for Melange Yarn Manufacturing *Cotton Science and Processing Technology, Springer* 529–546.

[16] Karrupusamy P 2021 Advanced metering infrastructure with secure chord lookup protocol for IoT Systems *Journal of Electrical Engineering and Automation* **3** 112-117

Comparative Analysis of Phasor Estimation Techniques for PMU Applications

Hari Krishna Achuthan Parthasarathy [1], Madhusudan Saranathan [1*], Adhitya Ravi [1], M.C. Lavanya [2], Dr. V. Rajini [3]

[1] Department of Electrical and Electronics Engineering, Sri Sivasubramaniya Nadar College of Engineering, Chennai, India.

[2] Deputy Director, R&D Division, National Institute of Wind Energy, Chennai, India

[3] Professor & Head, Department of Electrical and Electronics Engineering, Sri Sivasubramaniya Nadar College of Engineering, Chennai, India.

*madhusudan183001048@eee.ssn.edu.in (Corresponding author)

Abstract. Phasor Measurement Units (PMU) are valuable sources of data which increase the visibility into several fields of application such as power production, transmission and distribution. With the initial goal of obtaining phasor values at fixed points in the network and syncing it to a standard time frame, the technology has morphed into an all-encompassing unit, which has its uses in the field of protection, control and automation. With the application of these units becoming a standard benchmark, the improvement in the current available units is a huge requirement. This study presents a comprehensive analysis of various techniques that is used for the estimation of phasors.

Keywords: Phasor Measurement Unit, standard time frame, estimation of phasors, Discrete Fourier Transform, Fast Fourier Transform, Sliding Discrete Fourier Transform, Least square error.

1. Introduction

The characteristics of the distribution network have profoundly changed as a consequence of the integration of renewable energy resources. As a result, grid supervision and control have become more complex. The distribution network has become exceedingly complicated and there are a lot of uncertainties in the system to consider because of multiple nodes and dynamic nature of the grid. SCADA (Supervisory Control and Data Acquisition) is an existing architecture that takes data from remote terminal units located in different sections of the power system and is being implemented in practically every substation now.

Voltage, current and reactive power flow can all be measured with this acquisition technique but at a poor resolution (number of samples per unit time) and the measurements are not synchronised. SCADA, on the other hand, can only monitor the system's steady state. In order to facilitate dynamic observability of the grid and to mitigate the complexities caused by SCADA, a distribution level phasor measurement unit or the micro - phasor measurement unit was designed and developed. In contrast to SCADA, it may collect 120 samples per second and monitor the network's dynamic state. Aside from that, it can also measure the phase angle and time stamp each measurement.

In addition, PMU allows for wide-area monitoring and control, while SCADA only allows for local monitoring. Phasor Measurement Units (PMU) are used to determine the voltage / current signal values at different locations within a power network. The phasor data acquired via the phasor evaluation approach are used by the PMUs in their computations. Therefore, Phasor estimation techniques are essential for accurate data interpretation and analysis, thereby having a significant impact on the design and implementation of control mechanisms and protection systems [5].

P. Karuppusamy in his research evaluated the amount of variation in the estimated phasor which was caused by external disturbances using adaptive estimators. The performance of the adaptive estimator was demonstrated successfully using simulations [9].

This study analyzes various phasor estimation algorithms like Discrete Fourier Transform (DFT), Fast Fourier Transform (FFT), Sliding Discrete Fourier Transform (SDFT), Least Square Error method (LSE) to calculate the magnitude and phasor values of the input signal. These techniques are implemented in MATLAB script and are compared on the basis of execution time.

2. Phasor Estimation Techniques

The estimation techniques utilized within a PMU and the length of cycles needed to compute a phasor are not standardized. Algorithms used till now can be classified into DFT based algorithms (DFT, SDFT, FFT) and non-DFT based algorithms (LSE) [6-8].

Four phasor estimation methods namely, Sliding Discrete Fourier Transform (SDFT), Discrete Fourier Transform (DFT), the Least Squares Technique and Fast Fourier Transform (FFT) were explored using MATLAB software for three phase voltage signals and their phasors were estimated. Additionally, the time taken to compute the phasors by each algorithm was recorded and compared.

2.1. Discrete Fourier Transform

Discrete Fourier transform is used to transform an incoming time dependent signal, or a signal based in time domain, to the frequency domain. This transformation retains all the information in both the domains. The frequency domain yields a complex value of the sinusoid which can be resolved into the magnitude and phase of the respective signal [10,12].

The DFT of a signal is given by:

$$\mathbf{X}(\mathrm{k}) = \frac{1}{\mathrm{M}}\sum_{\mathrm{m}=0}^{\mathrm{M}-1}\mathrm{x}(\mathrm{m})\mathrm{e}^{-\mathrm{j}2\pi\mathrm{m}\frac{\mathrm{k}}{\mathrm{M}}} \qquad (1)$$

where x is the sampled input signal, M is the total number of samples. Once the DFT is found, magnitude and phase can be found using:

$$\mathrm{Magnitude}[\mathrm{X}(\mathrm{k})] = \sqrt{(\mathrm{Real}[\mathrm{X}(\mathrm{k})])^2 + (\mathrm{Img}[\mathrm{X}(\mathrm{k})])^2}$$

$$\mathrm{Phase\ Angle} = \tan^{-1}\frac{\mathrm{Im}[\mathrm{X}(\mathrm{k})]}{\mathrm{Re}[\mathrm{X}(\mathrm{k})]} \qquad (2)$$

The fundamental frequency from the spectrum of the transform is given by:

$$X^{m}=\frac{\sqrt{2}}{M}\sum_{m=0}^{M-1} Xm\cos\left(\frac{2\pi m}{M}\right)+j\frac{\sqrt{2}}{M}\sum_{m=0}^{M-1} Xm\sin\frac{2\pi m}{M} \quad (3)$$

A non-recursive DFT estimation uses the above equation to find the magnitude and phase. Recursive DFT makes use of the previous set of data from previous iterations and creates relative computation scheme, and yields the next fundamental frequency component as:

$$X^{m+1} = X^{m} + \frac{\sqrt{2}}{M}(xm + M - xm)e^{-jm\theta} \quad (4)$$

Figure 1 summarizes the process of computing phasors using DFT in the form of a flow chart.

2.2. *Fast Fourier Transform*

The Fast Fourier Transform is a numerical algorithm that computes the DFT of an n-dimensional signal. Equation 5 depicts the equation form of the measured signal.

$$x(t) = \sum_{h=1}^{H} V_h\, Sin(2\pi hft + \theta_h) \quad (5)$$

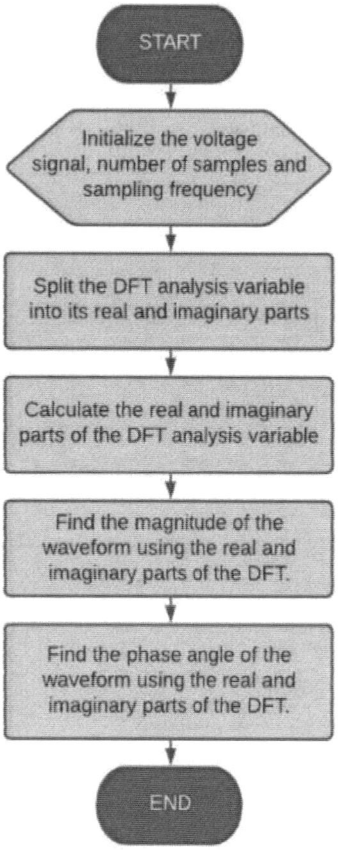

Figure 1. Flow chart for assessing phasors using DFT

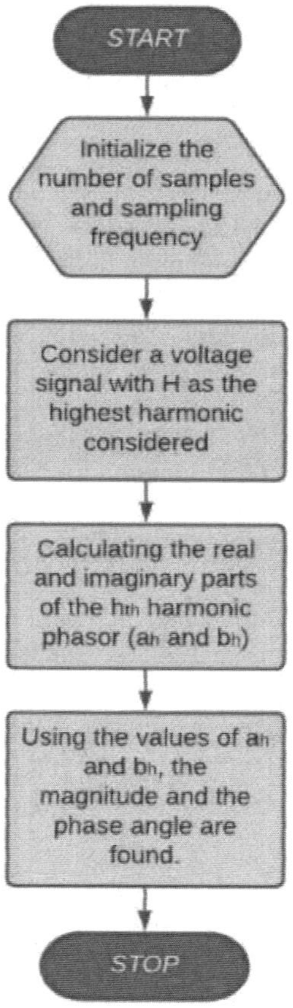

Figure 2. Flow chart for assessing phasors using FFT

A where H is the highest harmonic order being judged, f is the actual fundamental frequency which is given by $f_0 + \Delta f$, where f_0 is the fundamental frequency and Δf is the frequency deviation.

V_h and θ_h represent the magnitude and the primary phase angle of the phasor with h^th harmonics. When computing harmonic phasors, the traditional FFT ignores the frequency departure of the real signal and uses f_0 as the input. FFT technique involves the estimation of two parameters, that is, a_h and b_h as given in equations 6 and 7.

$$a_h = \frac{2}{T_0}\int_{t-T_0}^{t} x(t)\, Cos(2\pi\, hf_0 t)\, dt \quad (6)$$

$$b_h = \frac{2}{T_0}\int_{t-T_0}^{t} x(t)\, Sin(2\pi\, hf_0 t)\, dt \quad (7)$$

Where b_h represents the imaginary part of hth harmonic phasor, a_h is the real part of the hth harmonic phasor and $T_0$is the time taken for a single cycle.

Using the values of the imaginary and real part of the computed phasor, the magnitude Ah and $\emptyset_h$ can be calculated using equations 8 and 9. The process flow is depicted in the form of a flowchart in Figure 2.

$$|A_h| = \sqrt{a_h^2 + b_h^2} \quad (8)$$

$$\emptyset_h = \text{atan}(\frac{a_h}{b_h}) \quad (9)$$

It can be observed that the real and the imaginary parts of the computed h^{th} harmonic phasor (a_h and b_h) does not take into consideration the frequency deviation and thus, FFT technique does not provide exact values of magnitude and phase angle. It will always have errors, because in practical cases the rated frequency and the actual frequency is not always the same[1].

2.3. Sliding Discrete Fourier Transform

On a per-sample basis, the Sliding DFT technique gives frequency domain updates with much less computations compared to Discrete Fourier Transform or Fast Fourier Transform. The flowchart summarizing the process of estimating phasors using Sliding DFT has been presented in Figure 3. Consider a discrete time signal represented as a sequence at the time instant n. The windowed sequence x(n) comprising of N-Samples is represented as:

$$\mathrm{x(n)} = \{\mathrm{x(n-N+1), x(n-N+2), \dots\dots\dots\dots x(n)}\} \quad (10)$$

The N-point DFT on the above signal will result in:

$$\mathrm{X(n)} = \{\mathrm{X_0, X_1 \dots\dots, X_{N-1}}\} \quad (11)$$

The sliding DFT algorithm states that the DFT of the windowed sequence x(n) , that is at the instance n , consisting of N samples can be computed using the N point DFT of the previous instance (n-1)

$$X_K(n) = [X_K(n-1) - x(n-N) + x(n)]e^{-j2\pi\frac{k}{N}} \quad (12)$$

The SDFT algorithm is based on the conclusion that for two consecutive time instants, say at n and n-1, the windowed sequences x(n) and x(n-1), contain principally the same elements. Consider an example where the window length is 16, at time instant 15.

$$x(15) = \{x(0), x(1), \dots \dots \dots \dots x(15)\} \quad (13)$$

This sequence is fed to the 16-DFT and at the time instant n=16, $x(16) = \{x(0), x(1), \dots \dots \dots \dots x(16)\}$ is used in the 16-DFT computation [2], [3].

The SDFT is calculated in 2 steps:

a) Replacement of the initial element of the sequence $x(n-1)$ with final element of the sequence $x(n)$, that is with respect to the above example, replace X_0 with X_{16} .

The N-DFT formula is given by,

$$X_k = \sum_{n=0}^{N-1} x(n)W_N^{nk} = x(0) + x(1)W_N^k + \dots + x(n-1)W_N^{(N-1)k} \quad (14)$$

Where, $W_N^{nk} = e^{\frac{j2\pi nk}{N}}$.

It is essential to observe that the first term of the DFT remains unmodified.

Thus, the intermediate sequence is

$\mathbf{x_i}(15) = \{x(16), x(1), x(2), \dots \dots, x(15)\}$.

Now, Denoting the DFT of $x_i(15)$as $X_I(15)$,

we obtain, $X_I(15) = X(15) - x(0) + x(16)$.

b) To perform on the sequence, circular shifting, that got replaced, by a single sample.
$X(16) = X_I(15) * e^{\frac{j2\pi k}{N}}$

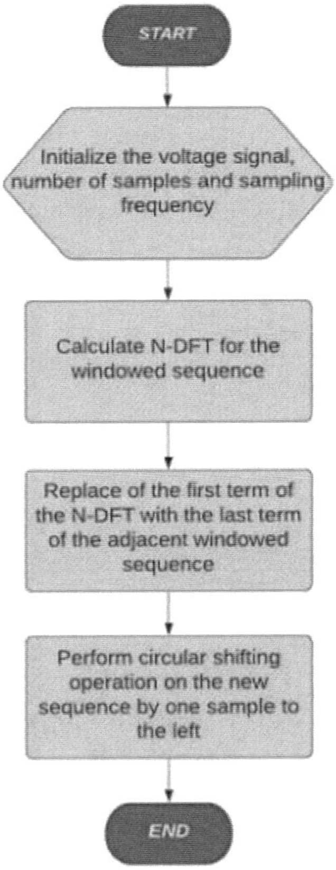

Figure 3. Flowchart for estimating phasors using SDFT

2.4. Least Square Method

Least Squares method is statistical which aims at providing the best value by minimizing the sum of the squares of the residues [11]. Figure 4 depicts a flowchart which presents the process for computing phasors using Least square error method.

Let us consider the following equation,

$a + bt = m$

where a, b are unknowns,

t is the time instant,

m is the measured value

The motive of this method is to minimize the error, where error is

$$\in = \text{Estimated Value} - \text{Measured Value}$$

In this method, we will have to minimize the error to compute the unknown values.

Now, let's say we have 'n' measurements.

They are,

$$a + bt_1 = m_1$$
$$a + bt_2 = m_2$$
$$.$$
$$.$$
$$a + bt_n = m_n$$

So, totally there are 'n' equations involved.

Now let's consider a' and b' to be the estimated values. Thus, it can be inferred that,

$$a' + b' t_1 - m_1 = \epsilon_1$$

$$\vdots$$

$$a' + b' t_n - m_n = \epsilon_n$$

Representing the above equations in a matrix,

$$\begin{bmatrix} 1 & t_1 \\ 1 & t_2 \\ . & . \\ . & . \\ 1 & t_n \end{bmatrix} \begin{bmatrix} \hat{a} \\ \hat{b} \end{bmatrix} - \begin{bmatrix} m_1 \\ m_2 \\ . \\ . \\ m_n \end{bmatrix} = \begin{bmatrix} \epsilon_1 \\ \epsilon_2 \\ . \\ . \\ \epsilon_n \end{bmatrix}$$

$$=> [A][x'] - [m] = [\epsilon]$$

So, $[\epsilon]^T = ([A][x'] - [m])^T$.
Thus,

$$[\epsilon]^T[\epsilon] = ([A][x'] - [m])^T ([A][x'] - [m]) \quad (15)$$

The aim of the algorithm is to minimize error.
Thus, the error equation can be differentiated and equated to zero to minimize.

As a result,
$$[x'] = \{([A]^T[A])^{-1}[A]^T\}^{-1} [m]$$

This equation can be written as,

$$[x'] = [A_p]^{-1} [m] \quad (16)$$

Where, $[x']$ is the estimated phasor
$[A_p]^{-1}$ is the psuedo – inverse matrix
$[m]$ is a matrix containing measured values

Knowing the general formula, the phasors for a three phase system can be found.

Current in analog form can be discretized by using an analog to digital converter.

The discretized version of current is:

$$i_k = I_1 (\mathrm{Sin}(kw_oT_S + \emptyset) + I_S \,\mathrm{Sin}(k(w_0 - w_m)T_S + \emptyset_s) + k_o e^{-\frac{T_S k}{\tau}} \quad (17)$$

where the first term relates to fundamental frequency, the second term relates to the sub synchronous frequency components and the last term refers to the decaying component.

w_o is the fundamental frequency and is expressed as $w_o = 2\pi f_o$. Fs is the sampling frequency. K is the gain of the DC component and τ is the decaying time constant.

To find the phasor it is essential to estimate the magnitude, phase angle and the frequency. So, using trigonometry formulas, it can be inferred that,

$$i_k = I_1(\mathrm{Sin}\,(kw_oT_s)\mathrm{Cos}\emptyset + \mathrm{Cos}\,(kw_oT_s)\mathrm{Sin}\,\emptyset) + I_S\left(\mathrm{Sin}\,\left(k(w_o-w_m)T_s\right)\mathrm{Cos}\emptyset_s + \mathrm{Cos}\,\left(k(w_o-w_m)T_s\right)\mathrm{Sin}\,\emptyset_s\right) + k_o e^{-\frac{T_S k}{\tau}}$$

The above equation can be rewritten as,

$$i_k = a_{11}x_1 + a_{12}x_2 + a_{13}x_3 + a_{14}x_4 + a_{15}x_5 + a_{16}x_6 \quad (18)$$

Where the known values are,
$a_{11} = \mathrm{Sin}(kw_0T_S)$, $a_{12} = \mathrm{Cos}(kw_0T_S)$
$a_{13} = \mathrm{Sin}(k(w_o-w_m)T_s)$,$a_{14} = \mathrm{Cos}(k(w_o-w_m)T_s)$
$a_{15} = 1$, $a_{16} = -kT_S$.

And the unknown values are,
$x_1 = I_1\mathrm{Cos}\emptyset$
$x_1 = I_1\mathrm{Sin}\emptyset$
$x_1 = I_S\mathrm{Cos}\emptyset_S$
$x_1 = I_S\mathrm{Sin}\emptyset_S$
$x_1 = k_o$
$x_1 = \frac{k_o}{\tau}$

Writing in Matrix format,

[A] [X] = [B]

Where, [X] is the matrix containing unknown values and [B] is similar to [m] matrix in (16)

$$[X] = [x_1x_2x_3x_4x_5x_6]^T = \{([A]^T[A])^{-1}[A]^T\}^{-1}\ [B] \quad (19)$$

From the above equation (19), x_1, x_2, x_3, x_4, x_5 and x_6 can be found from which I, $\emptyset, I_S, \emptyset_S, k_o$ and τ can be computed.

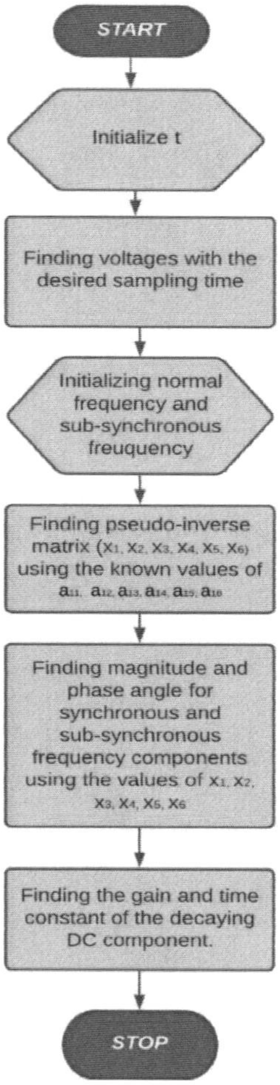

Figure 4. Flow chart for phasor estimation using Least Square Error Technique

3. Comparison of phasor estimation techniques

A simple MATLAB script was developed to understand and compare the time complexity of the three phasor estimation techniques viz., Least Square Error (LSE), Sliding Discrete Fourier Transform (SDFT) and Discrete Fourier Transform (DFT) using an ideal three phase sine wave for 1000 samples.

Table 1. Formatting sections, subsections and subsubsections.

S. No.	Phasor Estimation Technique	Average CPU Time (seconds)
1.	Discrete Fourier Transform	0.153
2.	Fast Fourier Transform	0.012
3.	Sliding Discrete Fourier Transform	0.029
4.	Least Square Error Technique	0.109

Fourier Transform was the slowest among the four. However, it is the most commonly used technique because of its simplicity. Least Square Error had the second slowest response. Nonetheless, it allows the computation of phasor corresponding to the sub synchronous frequency components and decaying DC components which is not possible with DFT, SDFT or FFT. Sliding Discrete Fourier Transform was found to be the second fastest. It exploited the similarity of consecutive sample windows to accomplish computational efficiency. Fast Fourier transform was the fastest of the four.

4. Conclusion

A comprehensive analysis of four phasor estimation techniques namely, discrete Fourier transform, fast Fourier transform, sliding DFT and least square error technique were studied in MATLAB. Upon comparing the profile time of each technique, it was observed that FFT estimated the phasors in less time compared to other techniques.

With increasing developments in the field of communications and electronics, a phasor measurement units could be implemented so as to not only increase the efficiency of current practices but also increase the level of automation in all aspects of power generation, transmission and distribution.

5. Acknowledgement

We thank National Institute of Wind Energy for giving us the opportunity to engage on this project. Further, we'd want to offer our deepest appreciation. to Dr. Rajesh Katyal, Deputy Director General, National Institute of Wind Energy, for providing us with this opportunity.

References

[1] S. Li et al., "PMU-based harmonic phasor calculation and harmonic source identification," in 2018 2nd IEEE Conference on Energy Internet and Energy System Integration (EI2), 2018, pp. 1–6.Another reference

[2] E. Jacobsen and R. Lyons, "The sliding DFT," IEEE Signal Process. Mag., vol. 20, no. 2, pp. 74–80, 2003.

[3] E. Jacobsen and R. Lyons, "An update to the sliding DFT," IEEE Signal Process. Mag., vol. 21, no. 1, pp. 110–111, 2004.

[4] Prabhash Nanda, C.K. Panigrahi, Abhijit Dasgupta, Phasor Estimation and Modelling Techniques of PMU- A Review,Energy Procedia,Volume 109,2017,Pages 64-77, ISSN1876-6102,https://doi.org/10.1016/j.egypro.2017.03.052.

[5] Meriem MA, Hamid BE, Abderrahmane OU. An Investigation of Different PMU phasor estimation techniques Based on DFT Using MATLAB. In2020 International Conference on Electrical Engineering (ICEE) 2020 Sep 25 (pp. 1-6). IEEE.

[6] Prabhash Nanda, C.K.Panigrahi, and Abhijit Dasgupta, "PMU Implementation in a Modern Power System," International Journal of Innovative Research in Science, Engineering and Technology. Vol. 6,Issue 8, August 2017.

[7] Prajakta R. Bedse, and Prof. Nayana N. Jangle,"Review on PMU using Recursive DFT Algorithm", International Conference on Computing, Power and Communication Technologies (GUCON), Galgotias University, Greater Noida, UP, India. Sep 28-29, 2018.

[8] Subhashree Priyadarshini, and Chinmoy Kumar Panigrahi, "Non Recursive DFT Based Model of Synchrophasor Unit", International Journal of Innovative Technology and Exploring Engineering (IJITEE), Volume-9 Issue-2, December 2019.

[9] Karuppusamy, P. (2021). Design of Adaptive Estimator for Nonlinear control system in Noisy Domain. Journal of Electronics and Informatics, 3(3), 150-166. doi:10.36548/jei.2021.3.001.

[10] S. V. Hareesh, P. Raja and M. P. Selvan, "An effective implementation of Phasor measurement unit (PMU) by using non-recursive DFT algorithm," 2015 International Conference on Condition Assessment Techniques in Electrical Systems (CATCON), 2015, pp. 195-199, doi: 10.1109/CATCON.2015.7449533.

[11] W. Xiong, L. Wang, R. Wu, X. Ren, H. Liu and T. Bi, "A Phasor Measurement Method Considering Decaying DC Component," 2020 IEEE/IAS Industrial and Commercial Power System Asia (I&CPS Asia), 2020, pp. 1-6, doi: 10.1109/ICPSAsia48933.2020.9208439.

[12] D. Petri, D. Fontanelli, D. Macii and D. Belega, "A DFT-based synchrophasor, frequency and ROCOF estimation algorithm," 2013 IEEE International Workshop on Applied Measurements for Power Systems (AMPS), 2013, pp. 85-90, doi: 10.1109/AMPS.2013.665623.

River Water Quality Prediction and index classification using Machine Learning

Jitha P Nair [1], Vijaya M S [2]

[1] Research Scholar, Department of Computer Science, PSGR Krishnammal College for Women, Peelamedu, Coimbatore, India.

[2] Associate Professor, Department of Computer Science, PSGR Krishnammal College for Women, Peelamedu, Coimbatore, India.

jithapnair20@gmail.com

Abstract: Various pollutants have had a substantial impact on the quality of water in recent years. The quality of water directly impacts human health and the environment. The water quality index (WQI) is an indicator of effective water management. Water quality modelling and prediction have become essential in the fight against water pollution. The research aims to build an efficient prediction model for river water quality and to categorize the index value according to the water quality standards. The data has been collected from eleven sampling stations located in various locations across the Bhavani River, which flows through Kerala and Tamilnadu. The water quality index is determined by 27different parameters affecting water quality like dissolved oxygen, temperature, pH, alkalinity, hardness, chloride, coliform, etc. Data normalization and feature selection are done to construct the dataset to develop machine learning models. Machine learning algorithms such as linear regression, MLP regressor, support vector regressor and random forest has been employed to build a water quality prediction model. Support vector machines (SVM), naïve bayes, decision trees, MLP classifiers, have been used to develop a classification model for classifying water quality index. The experimental results revealed that the MLP regressor efficiently predicts the water Quality index with root mean squared error as 2.432, MLP classifier classifies the water quality index with 81% accuracy. The developed models show promising output concerning water quality index prediction and classification.

Keywords: River water quality, Prediction model, Classification model, Exploratory data analysis, Machine learning algorithms.

1. Introduction

Water is the most vital resource for life, as it is required for the survival of all living things, including humans. Better quality and quantity of water is essential for life on earth. Some pollution levels are satisfactory to aquatic species, but when the level increases the oxygen content in the water decreases and lead to disasters. A large percentage of environmental water sources, such as lakes, rivers, and streams, have quality standards that demonstrate their worth. Guidelines are applied to all types of water bodies for all applications and uses.

Irrigation water does not need to be either too saline or harmful to the plant or soil, thus ruining the ecosystem. Water quality also requires different qualities based on certain various processes for industrial applications. Natural water resources are among the cheapest options for freshwater, such as ground and surface water. Human and industrial activity and other natural processes can pollute natural resources. So, rapid industrial growth has led to a significant decline in water quality. The quality of drinking water is significantly affected by the infrastructure, lacking public awareness, and poor hygiene standards. The effects of contaminated drinking water are quite severe health issues, the environment, and infrastructure. According to UN research, around 1.5 million people die every year from water-borne diseases. 80% of health problems have been reported to be caused by polluted water in developing countries. Annual reports include 2.5 billion humans affected by water-borne diseases and five million deaths. The death rate of humans mainly focuses on crimes, accidents and terrorist attacks.

Novel approaches to analyzing and forecasting water quality (WQ) are critical. It is recommended that the temporal dimension of predicting water quality patterns be studied to monitor the seasonal shift of the WQ. However, using a specific model variation to forecast water quality outcomes performs better than using a single model. Several approaches to predicting and simulating water quality are being proposed. Statistical techniques, visual modelling, algorithm analysis, and predictive algorithms are commonly used. Multivariate statistical techniques were used to determine the correlation and relationship between different water quality parameters. For transitional probability, regression analysis, multivariate interpolation, and geostatistical approaches were used.

Massive population growth, the use of fertilizers and pesticides, the industrial revolution, seem to have serious consequences for water quality environments. The models for predicting water quality are extremely useful for monitoring water contamination. Modelling and predicting water quality are employed with mechanism oriented and no-mechanism-oriented models. The mechanism model is sophisticated and it simulates the water quality using advanced system structure data, it is regarded as a multifunctional model that can be applied to any water body.

This paper aims to develop an accurate model for forecasting river water quality using a developed framework. River water quality data has been collected from eleven sampling stations across Bhavani River which flows through Kerala and Tamilnadu and analyzed to understand feature distribution and correlation. Machine learning algorithms are used to predict water quality index values and to classify water quality indexes. Mean squared error, mean absolute error and root mean squared error is used for performance evaluation of prediction algorithms. Water quality index classification models are evaluated using the metrics such as accuracy, precision, recall, and F1-score.

2. Literature review

This study analyzes the approaches that were used to effectively address water quality challenges. In most studies, traditional statistical analysis and lab analysis is implemented to determine water quality, but other studies apply machine learning approaches to find an optimal solution to the water quality problem.

Sillberg et al. [1] have developed a machine learning-based technique integrating attribute-realization (AR) and support vector machine (SVM) to classify the Chao Phraya River water quality. Using the linear function, the AR has identified the most important elements for improving river quality. NH3-N, TCB, FCB, BOD, DO, and Sal was the most contributing characteristics in the categorization, with contributed values in the range of 0.80–0.98, compared to 0.25–0.64 for TDS, Turb, TN, SS, NO3-N, and conductivity. The best classification results were achieved using the SVM linear approach, which had an accuracy of 0.94, a precision average of 0.84, a recall average of 0.84, and an F1-score average of 0.84. When applied to three to six parameters, the validation revealed that AR-SVM was a powerful method for identifying river water quality with 0.86–0.95 accuracy.

Yilma et al. [2] used an artificial neural network to simulate the Akaki river water quality index. The index was calculated using twelve water quality indicators from 27 dry and wet season sample locations. All projected findings, except one upstream location, have revealed low water quality. Through 12 inputs and one output, the neural network model was trained and verified using the number of hidden layers (2–20), hidden layer neurons (5, 10, 15, 20, and 25), transfer training, and learning functions. According to their research, artificial neural network with eight hidden layers and 15 hidden neurons accurately predicted the WQI with an accuracy of 0.93.

Ding et al. [3] have designed a hybrid intelligent method that incorporates Principal Component Analysis (PCA), Genetic Algorithm (GA), and Back Propagation Neural Network (BPNN) techniques for predicting river water quality. This research included 23 different water quality indicator variables, each of which has a complex non-linear relationship with water quality. PCA boosted the training pace of follow-up algorithms substantially, while GA optimized the BPNN parameters. According to the findings, the average prediction rates for non-polluted and polluted water quality were 88.9% and 93.1%, respectively, while the overall prediction rate was about 91%.

Ahmed et al. [4] assessed the water quality index (WQI) using supervised machine learning algorithms, where an individual index was used to summarise the overall quality of water and water quality class. The proposed techniques, as well as gradient boosting with a learning rate of 0.1 and polynomial regression with a degree of 2, most effectively predicted the WQI, and that WQI was later evaluated with a mean absolute error (MAE) of 1.9642 and 2.7273. In this case, the MLP with the configuration (3, 7) has the highest accuracy for classification with 85.07%.

Zhang, Zhu, Yue, and Wong [5] presented an anomaly detection algorithm for water quality data that employs dual time-moving windows to identify anomaly data from historical patterns in real time. The algorithm was built on statistical models, specifically the auto-regressive linear combination model. The authors tested the algorithm with three month PH data from a river basin and analysed using real water quality monitoring station data. The experimental results depicted that the algorithms have a lower rate of false positives and better anomaly detection performance than the AD and ADAM algorithms.

Sakizadeh [6] used 16 water quality metrics and Bayesian regularization to predict the WQI. The research work found correlation coefficients of 0.94 and 0.77 between observed and predicted values.

The majority of the studies used manual lab analysis, failed to calculate the water quality index standard, and even included so many parameters. As seen, machine learning can produce good results for detecting anomalies in water quality, and the existing work is inspired by related work. Machine learning algorithms have the potential to significantly reduce the number of incorrect predictions.

3. Overview of Study Area and Dataset

Bhavani river flows through Tamilnadu and Kerala, India. The river originates from Nilgiri hills, then enter silent valley national park, Kerala and flows through Tamilnadu. Bhavani river is a long perennial long with 217 km and is fed by both southwest and northeast monsoons. Its watershed drains 0.62 million hectares spread across Tamil Nadu (87%), Kerala (9%), and Karnataka (4%). Fig. 1. depicts the flow of the Bhavani River, which flows primarily through the Attappady Plateau of Palakkad district and then travels to the Tamil Nadu districts of Coimbatore and Erode. Approximately 90% of the river water is used for agricultural irrigation. Data collected from different stations of the Bhavani River includes Kottathara, Thavalam, Chalayur, Karathur, Cheerakuzhi, Elachivazhi, Badrakaliamman kovil, Sirumugai, Bhavanisagar, Bhavani, Sathyamangalam.

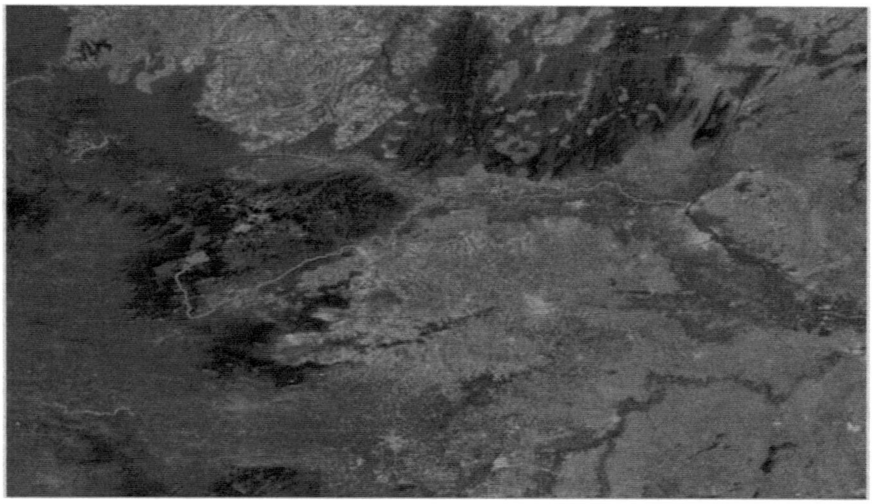

Fig.1. Map of Bhavani River

3.1 Data Collection

The main stations in the Bhavani River basin include Kottathara, Thavalam, Chalayur, Karathur, Cheerakuzhi, Elachivazhi, Badrakaliamman kovil, Sirumugai, Bhavanisagar, Bhavani, Sathyamangalam. Temperature, pH, dissolved oxygen, turbidity, chloride, and other parameters which determine the water quality index are collected from the eleven-sampling station of the Bhavani River. The 10560 data samples with 31 attributes used in this research work were collected from the water quality monitoring stations for the period from January 1, 2016, to December 31, 2020. The time fluctuation trend of water quality indicators of the Bhavani River's water source was examined for these samples. The values of each water quality parameter over sometime were represented as time-series data. Annual trends in temperature, pH, conductivity, BOD, COD, Nitrate-N, fluoride, potassium, TC, FDS, chloride, sodium, TDS, hardness and sulphate are shown in Fig.2. The sample data of river water quality monitored from eleven sampling stations is shown in Table 1.

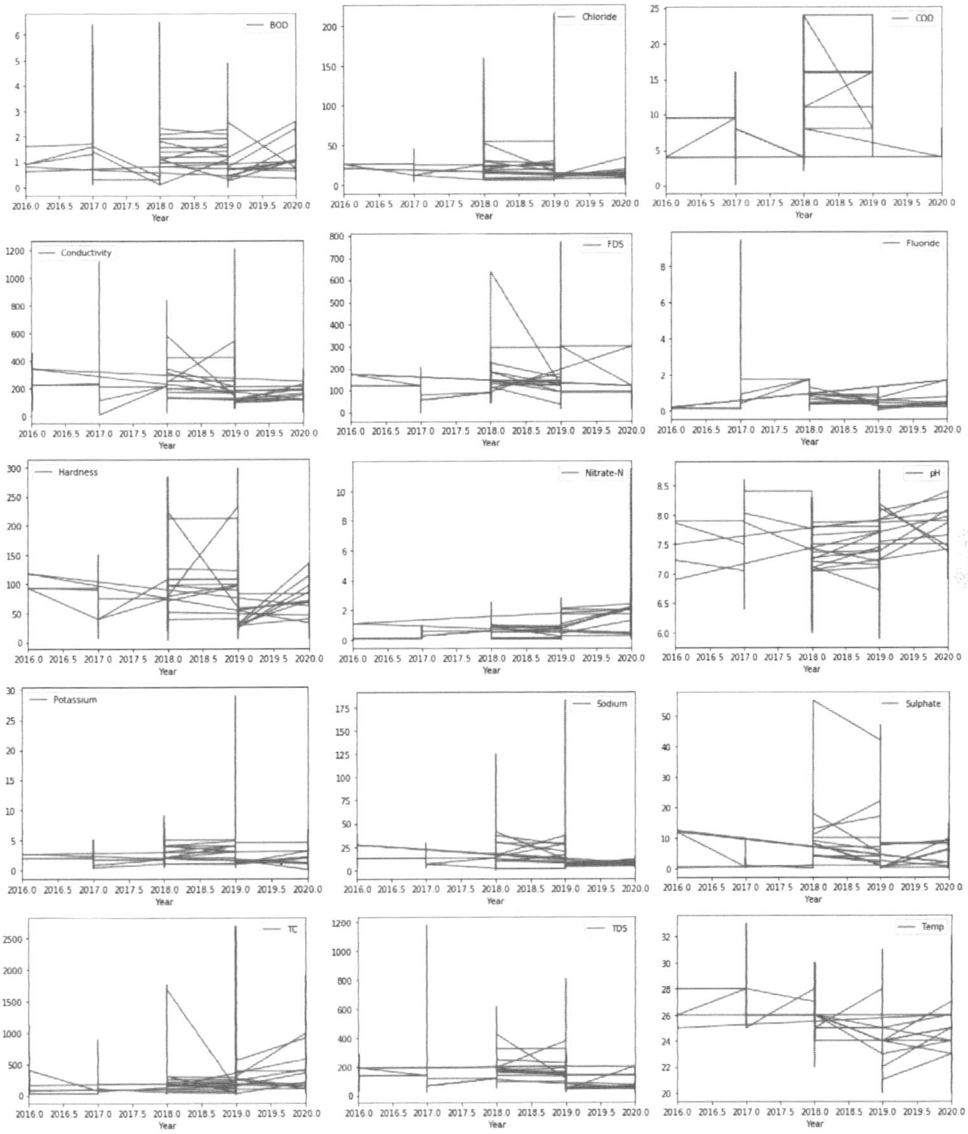

Fig. 2. Variation trend of Bhavani River from 2016 to 2020 of some parameters

Table1: Data Collected from Sampling Stations

Date	02/06/2016	05/02/2016	05/09/2016	03/03/2019	04/06/2019	04/07/2019	05/04/2018	08/03/2018	10/02/2018	06/05/2018`	12/03/2018
Temp	25	26	25	24	23	24	26	26	24	24	26
pH	7.15	7.5	6.9	7.87	8.12	8.07	7.06	7.82	7.27	7.2	7.24
Conductivity	340	340	340	116	95	102	132	316	189	583	240
Chloride	21	21	21	11	13	9	14	33	22	52	19
COD	4	3.9	4	4	4	4	8	8	16	24	11
Sodium	27.1	27.1	27.1	5.26	4.21	5.13	10	38	9	41	16
TSS	300	300	300	69	56	61	4	6	18	4	1
TDS	190	190	190	41	47	52	108	213	177	424	193
FDS	174	174	174	300	300	300	94	144	140	638	115
DO	6.99	7.51	7.4	7.24	7.97	7.58	6.8	6.93	5.43	5.4	7.47
Nitrate	1.1	1.1	1.1	0.243	0.734	1.062	0.9	0.87	0.9	1	0.912
TC	88	79.414	160	147	163	295	130	192	155	1700	191
FC	80	80	80	39	51	18	27	109	56	430	150
Station	1	2	3	4	5	6	7	8	9	10	11
Latitude	76.591123E	76.72055E	76.657335E	76.657335E	76.72055E	76.683688E	77.408333E	77.001587E	77.114998E	77.682777E	77.23003E
Longitude	11.093055N	11.188917N	11.147534N	11.147534N	11.188917N	11.163695N	11.4975N	11.329323N	11.474167N	11.433611N	11.506945N
Year	2016	2016	2016	2019	2019	2019	2018	2018	2018	2018	2018

3.2 Calculation of the WQI: The water quality index is measured using the weighted arithmetic water quality index method. The most commonly measured water quality parameters like pH, BOD, COD, DO, Nitrate, sodium, sulphate, chloride, faecal coliform, etc. are calculated according to the following formula (1):

$$WQI = \frac{\Sigma wiqi}{\Sigma wi} \quad (1)$$

qi is a relative value of water quality that is specific to each parameter and i rep the number of parameters taken into consideration. Wi is a factor that measures recent the importance of a parameter in the calculation of the WQI index is the relative weight. Qi is calculated by applying formula 2 below.

$$qi = 100 * \frac{vi-vo}{si-vo} \quad (2)$$

where: v_i represents the value experimentally and determined the analyzed parameter, v_o represents the ideal value of that parameter whereas the ideal value is zero for all other parameters except DO = 14.6 mg/l and pH = 7.0. Si represents the standard, legally accepted, value for the water category in which the analyzed water sample was included. The w_i factor is calculated by using formula 3.

$$wi = \frac{K}{si} \quad (3)$$

where K is a constant, which can result from applying the formula 4,

$$K = \frac{1}{\Sigma\left(\frac{1}{si}\right)} \tag{4}$$

Tables 2 represent the unit weight of each parameter and their permissible limits for finding WQI. Based on the value obtained for the weighted arithmetic WQI method, the water ecological status can be determined, as illustrated by Table 3.

Table 2: Parameters with Permissible limits and Unit Weights

Parameters	Permissible Limits	Weights
Temp(oC)	28	0.035714286
pH	8.5	0.117647059
Conductivity	150	0.006666667
Turbidity	5	0.2
Phenolpth Alkalinity	20	0.05
Total Alkalinity	200	0.005
Chloride	250	0.004
COD	10	0.1
TKN	100	0.01
Ammonia	50	0.02
Hardness	100	0.01
Ca. Hardness	75	0.013333333
Mg. Hardness	30	0.033333333
Sulphate	200	0.005
Sodium	200	0.005
TSS	300	0.003333333
TDS	1000	0.001
FDS	200	0.005
Phosphate	0.3	3.333333333
Boron	1	1
Potassium	2.5	0.4
BOD	3	0.333333333
Fluoride	1.5	0.666666667
DO	7.5	0.133333333
Nitrate-N	0.503	1.988071571
TC	100	0.01
FC	60	0.016666667

Table 3: Weight Arithmetic Water Quality Index standards for Water Quality

WQI	**Water Quality Class**	**Water Quality**
0-30	A	Excellent
31-60	B	Good
61-90	C	Poor
91-120	D	Very Poor
>121	E	Unsuitable

Calculation of the water quality index has been done using the above formulas and follows the standards of weighted arithmetic water quality index mentioned in Table 3. Water quality index calculation requires permissible values and unit weight of each parameter as shown in Table 2.

After calculating the water quality index value for each sample, it was added to the respective instance. Water quality index class for each instance is identified using the existing standards and added as class labels to the respective instance. Finally, the river water quality dataset is developed with 33 attributes and 10560 labelled instances.

3.3 Exploratory Data Analysis
Exploratory data analysis was applied to the Bhavani River water quality dataset obtained from eleven sampling stations with 27 water quality parameters from 2016 to 2020. Unsupervised pattern recognition and display algorithms were used to extract correlations and similarities between variables and categorise river water quality samples data. A water quality dataset with 33 attributes and 10560 instances is used for analyzing and visualizing the importance of each attribute in the work. Univariate and multivariate analysis was performed using various interesting statistical graphs such as heatmaps, histograms, pair plots, and box plots. The heatmap analysis shows the correlation of the parameters with the water quality index, some parameters such as boron and TSS were negatively correlated and had duplicate instances. The boxplot and histogram analysis shows that the parameters such as conductivity, total coliform have a wide range of values, and hence data normalization is required for these parameters to set a smaller range. Thus, exploratory data analysis provided insights that led to data preprocessing.

3.4 Data Preprocessing
Preprocessing data helps to increase data quality and efficiency. The quality of raw data is harmed by its inconsistency and noise. The exploratory data analysis performed in this research work provided a proper understanding of the data that the dataset contains duplicates, negatively correlations among the parameters and distribution of values.

Data Cleaning: The practice of rectifying duplicate, incorrect, or incomplete data from a dataset is known as data cleaning. The river water quality dataset developed in this work contains 33 attributes and 10560 instances. The two attributes such as boron and TSS are negatively correlated and these two attributes are not contributive to water quality index prediction. Hence these two attributes are removed from the dataset for building an efficient model.

Normalization Method: Normalization is a technique used for standardizing the attribute values in the dataset. The result of EDA shows that the two parameters of the river quality dataset such as conductivity and total coliform n to be normalized to a specific range as the range of values of the parameters is wide. Total coliform has a minimum value of 10 and a maximum value of 2500 but most of its values lie in the range of 10 to 300. Similarly, conductivity has a minimum value of 1 and a maximum value of 1200, most of the values lie between 60 and 210. Here Z-score normalisation is applied to these two parameters for standardizing the parameter values by using the mean and standard deviation. It is done using the following formula:

$$v' = \frac{v - \bar{A}}{\sigma A} \quad (5)$$

4. Methodology

The proposed framework of the WQI prediction model consists of various building blocks such as data collection and data exploration, data preprocessing, construction of WQI prediction and classification models, performance evaluation. Machine learning approaches have been used to build the prediction of water quality index value and classification. Various metrics such as mean squared error, root mean squared error, R^2 is used for evaluating the performance of the prediction model, and the WQI classifier is evaluated using accuracy, precision, recall, F1 score. The architecture of the proposed WQI prediction model is shown in Fig.3. and described below.

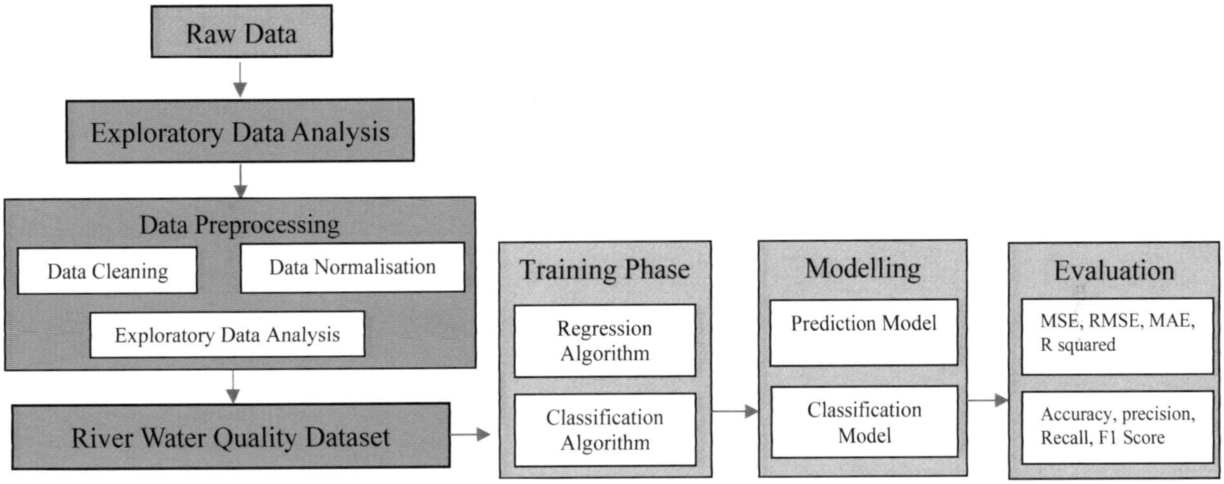

Fig 3. A framework of the proposed work

4.1 Feature Selection

The goal of the feature selection method is to exclude features that do not make a significant contribution to defining the water quality index. Feature selection has an impact on the measurement of data dimensions, whether for training data or testing data. Select K best, is applied to the developed dataset to find the best contributive attributes in determining the water quality index. Parameters such as phenolpth alkalinity, ammonia, TKN, and phosphate have less importance in predicting the water quality index. The parameters which are having high importance in developing the water quality index are mainly conductivity, alkalinity, chloride, hardness, sulphate, sodium, phosphate, potassium, BOD, fluoride, DO, nitrate, coliform and so on. This feature selection process yielded a better river water quality dataset which contains 10560 parameters and 27 attributes for building the prediction and classification model.

4.2 Building WQI Prediction model

The water quality prediction model is developed by learning the trends in the river water quality dataset using linear regression, random forest, support vector regression, and multilayer perceptron regression. Here water quality index is the target variable and all other parameters in the dataset are independent variables in modelling the regression process. Regression can be used to forecast a response using a new set of predictors.

Linear regression is used to find the best-predicted weights, that is, the weights with the smallest residuals. They usually minimise the sum of squared residuals (SSR) for all observations to get the best weights. This method is known as the method of ordinary least squares. The regression model is evaluated to ensure that it matches the data required to forecast and predict true or false occurrences.

Support Vector Regression is a supervised algorithm that is derived from wide margin kernel methods for analysis and is used to predict discrete values. It has all of the properties of support vector machine maximal margin algorithms, such as duality, sparseness, kernel, and convexity. It has evolved into a powerful predictive data analysis technique with a wide range of applications. The basic concept of SVR is to find the best fit line and the best fit line in SVR is the hyperplane with the maximum number of points.

Random forest is a supervised machine learning algorithm used for regression problems by handling continuous variables containing datasets. Random forests are built from subsets of data, and the final output is based on average or majority ranking, which eliminates the problem of overfitting. The random forest chooses observations at random, builds a decision tree, and uses the average result.

An MLP is a type of feed-forward artificial neural network that has at least three layers of nodes (neurons), including an input layer, one or more hidden layers, and an output layer. From the input to the output, the nodes are fully coupled in the form of a directed graph. Except for the input nodes, all nodes have an associated activation function that is used to compute the node output using weighted inputs from other nodes. The weights between the nodes are updated iteratively for reducing the error function in an MLP model, which is trained to utilize a backpropagation mechanism with gradient-descent as an optimization algorithm. For all hidden layer nodes in regression, a relu activation function is employed, and for the output layer nodes, a linear activation function is used. The result is a predicted real-valued quantity based on the input sample $x \epsilon X$.

4.3 Building WQI classification model.

The water quality index classification model is developed by learning the patterns in the river water quality dataset. The water quality index is used as a class label to categorise the water quality into five standards-based water quality parameters. Machine learning algorithms such as Support vector machine (SVM), Decision Tree, Naive Bayes, and Neural Networks, are used to construct classification models.

SVM is a supervised algorithm that can be used to classify and predict data. Each data point in n-dimensional space is displayed independently, allowing the two classes to be identified easily. In the fields of technology, pattern recognition, and learning classification, SVMs are gaining traction. A linear or non-linear separation surface can be used to classify the input region. In support vector classification, the separation function is a linear combination of kernels related to the support vector.

A decision tree classifier has a simple structure that can be kept in a small amount of space and efficiently classifies fresh data. Decision tree classifiers can automatically choose features and reduce complexity, and their tree structure provides easily understood and interpretable information about the classification's predictive or generalization capacity. The main goal of the decision tree growth algorithm is to determine which characteristic to test at each node in the tree.

The Naive Bayes algorithm is a form of classification method based on Bayes' theorem when the target value is set then the other attributes are independent variables. The Bayesian technique employs probability and statistical knowledge to predict and classify datasets. Using the Bayesian approach with prior and posterior probability, the bias and overfitting concepts of applying sample information can be avoided.
MLP Classifier is a machine-learning-based classification system. For all hidden layer nodes, a relu activation function is utilized, and for the output layer nodes, a softmax activation function is employed. The result is a vector that contains the odds that sample $x \epsilon X$ belongs to each class, which is the same as a categorical probability distribution. The class with the highest probability is the final result. For each class label per input, a distinct loss is calculated, and the outcome is the total of all those losses.

4.4 Performance Evaluation

The performance of the developed models in predicting the water quality index and classifying the water quality index is evaluated to find the best algorithm. The efficient prediction algorithms are found when the RMSE value is very less and the best classification model is evaluated by checking the accuracy. The following are the statistical parameters that were used:

1. Mean Squared Error (MSE) :

$$MSE = \frac{1}{n}\sum_{i=1}^{n}(Y_a - Y_b)^2 \tag{6}$$

2. Mean Absolute Error (MAE)

$$MAE = abs(Y_a - Y_b) \tag{7}$$

3. Root Mean Squared Error (RMSE)

$$RMSE = \sqrt{(Y_a - Y_b)^2}/_n \tag{8}$$

where Y_a and Y_b are the actual responses and the predicted value, respectively, and n is the total number of variables.

4. Accuracy

$$Accuracy = \frac{TP+TN}{TP+FP+TN+FN} \tag{9}$$

5. Precision

$$Precision = \frac{TP}{Tp+FP} \tag{10}$$

6. Recall

$$Recall = \frac{TP}{TP+FN} \tag{11}$$

7. F1-Score

$$F1 - Score = \frac{2*precision*recall}{Precision+recall} * 100 \tag{12}$$

where TP, TN, FP, and FN are the true positive, true negative, false positive, and false negative, respectively. Evaluation of machine learning algorithms for developing water quality index in prediction and classification are evaluated using the above equations to check the efficiency of algorithm using river water data.

5. Experiment

Here the experiments have been carried out to develop an accurate water quality index prediction model and the corresponding classification model using the Bhavani River water quality dataset. A dataset with 10560 instances and 27 attributes has been split into training and testing sets with 80%of instances for training and 20% of instances for testing. The support vector regressor, linear regression, random forest, and MLP Regressor algorithms were implemented to build water quality index prediction model and were implemented to build water quality index classification model using support vector machine, Naive Bayes, decision tree, MLP classifier. Machine learning algorithms developed a relationship with the independent and dependent parameters while modelling and then the test data is used to determine whether the models are effective with respect to various performance evaluation metrics.

5.1 Water Quality Prediction

In this section regression models are developed by training river water quality dataset through implementing MLP regressor, linear regression, support vector regressor, and random forest using python libraries. The performance of the prediction models is evaluated for its efficiency to forecast the WQ using mean squared error, mean absolute error, root mean squared error, and R-squared values.

From the prediction results, it was observed that mean absolute error value of support vector regressor based prediction model is 3.623, whereas prediction models based on linear regression, random forest and MLP regressor gives 2.85911, 2.015, and 1.9143 respectively. Thus, high error rate was produced by support vector regressor and less error by MLP regressor as illustrated in Fig.4a.

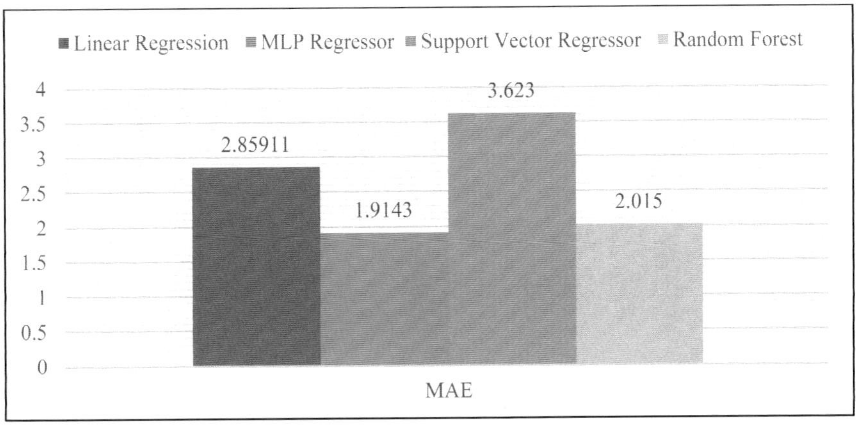

Fig.4a. Mean Absolute Error of Prediction Models

Similarly, it was found that the root mean squared error value of support vector regressor based prediction model is 4.3281, whereas prediction models based on linear regression, random forest and MLP regressor gives 3.8428, 3.79783, and 2.432 respectively. Thus, high error rate was produced by support vector regressor and less error by MLP regressor as shown in Fig.4b.

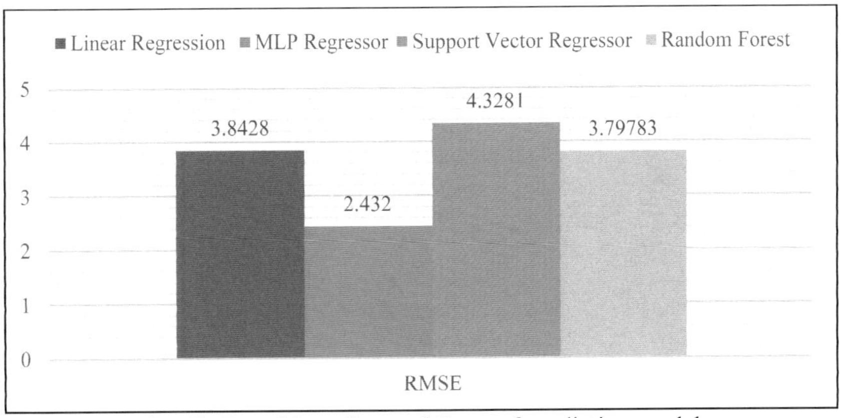

Fig. 4b. Root Mean Squared Error of prediction models

In the next case, the mean squared error value of support vector regressor based prediction model is 20.237, whereas linear regression, random forest and MLP regressor shows the respective metric values for prediction as 11.3318, 9.8278, and 7.1032. Thus, high error rate was produced by support vector regressor and less error by MLP regressor as depicted in Fig.4c.

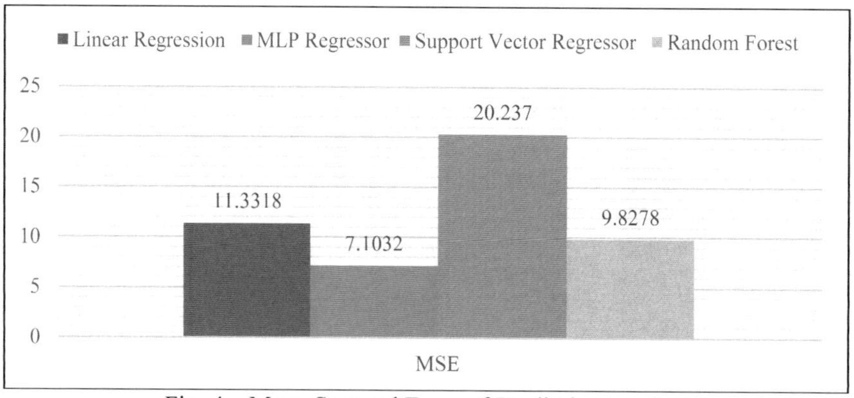

Fig. 4c. Mean Squared Error of Prediction Models

It was found that the R-squared value of support vector regressor based prediction model is -2.7132, whereas prediction models based on linear regression, random forest and MLP regressor yields 0.6375, 0.6923, and 0.7342 correspondingly. Thus, R-squared value is negatively produced by support vector regressor which means it has less accuracy and have high accuracy for MLP regressor as shown in Fig.4d.

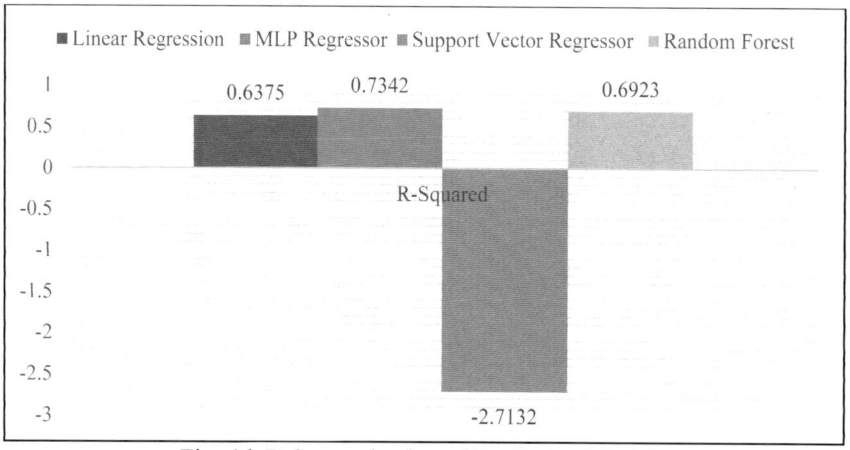

Fig. 4d. R-Squared value of Prediction Models

The comparative performance results of prediction algorithms such as linear regression, support vector regressor, random forest, MLP regressor with respect to mean absolute error, root mean squared error, mean squared error and R squared value is depicted in Table 4. It is confirmed that the MLP regressor based water quality index prediction model produced high accuracy with less error rate whereas support vector regressor based model show less accuracy and high error rate as illustrated in Fig.4e.

Table 4: Comparitive Performance results of water quality index prediction models

Models	MAE	RMSE	MSE	R-Squared
Linear Regression	2.85911	3.8428	11.3318	0.6375
MLP Regressor	1.9143	2.432	7.1032	0.7342
Support Vector Regressor	3.623	4.3281	20.237	-2.7132
Random Forest	2.015	3.79783	9.8278	0.6923

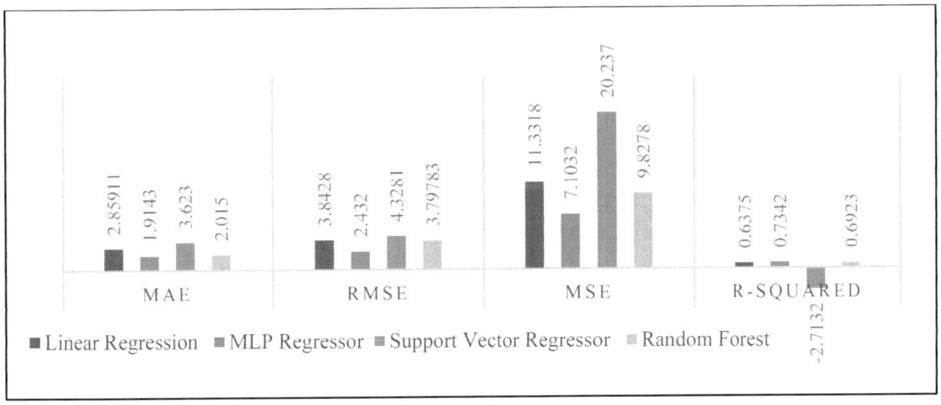

Fig. 4e. Comparitive Performance analysis of water quality index prediction models

5.2 Classification of water quality index

In this section classification models are constructed using the river water quality dataset by applying MLP classifier, support vector machine, naïve bayes, and decision tree using python libraries. The performance of the classification models is evaluated for checking the efficiency in classifying the WQI using metrics such as accuracy, precision, recall and F1 score.

From the experimental results it was found that the accuracy of MLP classifier based classification model is 0.8132, whereas classification models based on naïve bayes, decision tree and support vector machine yield 0.7738, 0.74, and 0.61 correspondingly. Thus, accuracy of MLP classifier is higher as compared to other classifiers whereas support vector machine has low accuracy as shown in Fig.5a.

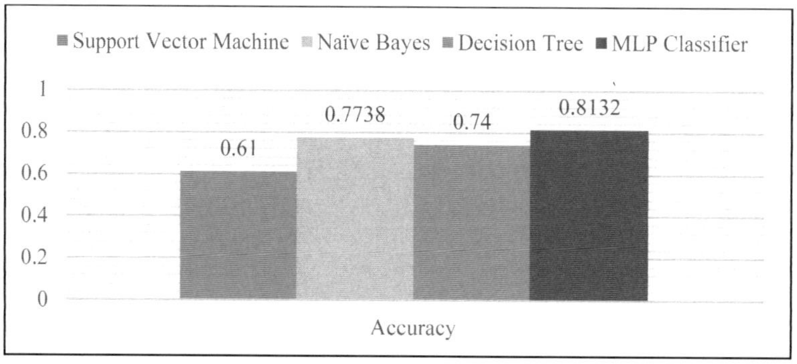

Fig. 5a. Accuracy of Classification Models

Similarly, it is observed that the results of classification model show the precision of MLP classifier based classification model is 0.632, whereas classification models based on naïve bayes, decision tree and support vector machine yield 0.5421, 0.512, and 0.429 correspondingly. Thus, precision of MLP classifier is higher as compared to other classifiers whereas support vector machine has low precision as depicted in Fig.5b.

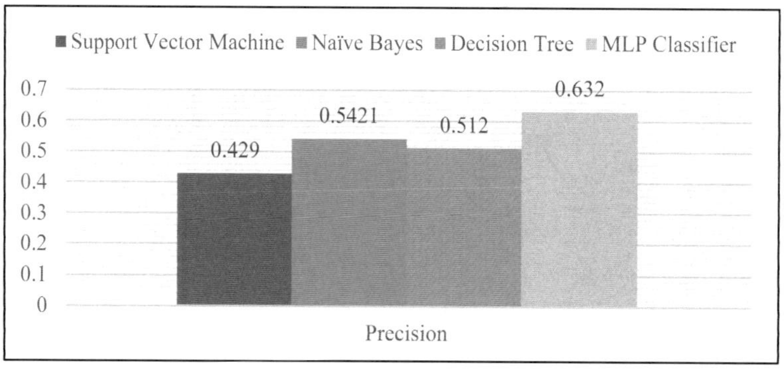

Fig. 5b. Precision analysis of Classification Models

In next case, the results of various classifiers shows that the recall value of MLP classifier based classification model is 0.6123, whereas classification models based on naïve bayes, decision tree and support vector machine gives 0.5523, 0.512, and 0.429 correspondingly. Thus, recall value of MLP classifier is higher as compared to other classifiers whereas support vector machine has low precision as illustrated in Fig.5c.

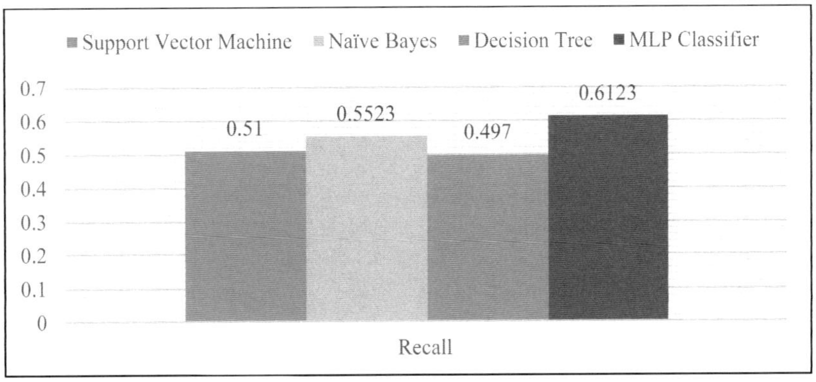

Fig. 5c. Recall evaluation of Classification Models

It was found that the results of classifiers with respect to F1 score confirms that MLP classifier based classification model is 0.5913, whereas classification models based on naïve bayes, decision tree and support vector machine yields 0.5031,0.507, and 0.42 respectively. Thus, F1score value of MLP classifier is higher as compared to other classifiers whereas support vector machine has low precision as illustrated in Fig.5d.

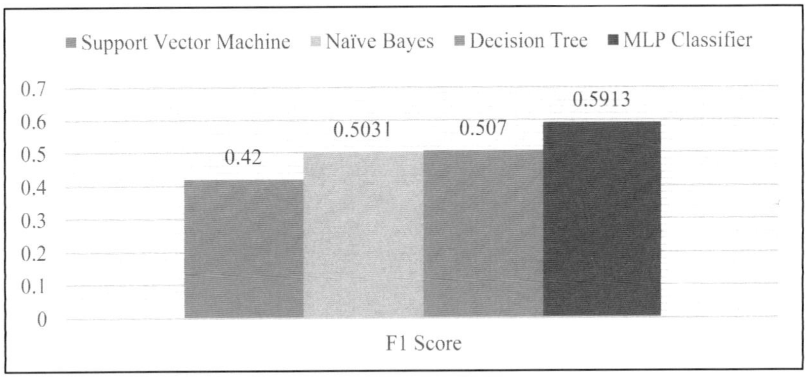

Fig. 5d. F1 score evaluation of Classification Models

The comparative performance results of classification algorithms such as support vector machine, naïve bayes, decision tree, MLP classifier with respect to accuracy, precision, recall and F1score, is illustrated in Table 5. It is proved that the MLP classifier employed in water quality index classification yield high accuracy whereas support vector classifier based model show less accuracy as depicted in Fig.5e.

Table 5. Performance results of water quality index classification models

Algorithms	Accuracy	Precision	Recall	F1 Score
Support Vector Machine	0.61	0.429	0.51	0.42
Naïve Bayes	0.7738	0.5421	0.5523	0.5031
Decision Tree	0.74	0.512	0.497	0.507
MLP Classifier	0.8132	0.632	0.6123	0.5913

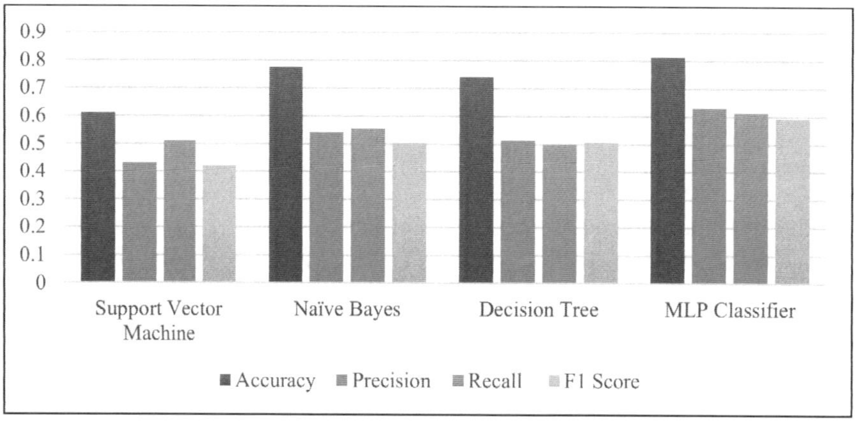

Fig.5e. Performance analysis of water quality index classification models

6. Conclusion

This study investigated the efficiency of machine learning algorithms in predicting river water quality and to classify the water quality index. Machine learning techniques like the random forest, MLP regressor, linear regression, and support vector regressor, were implemented to build WQI prediction models and decision tree, MLP classifier, naive Bayes, and support vector machine algorithms were implemented to build WQI classifiers. The Bhavani River water data with parameters like BOD, DO, TC, nitrate, pH, temperature, and so on, were collected, modelled and employed in building models. The performance of the models in forecasting river water quality index was evaluated using performance metrics. The MLP regressor shows less RMSE value while predicting the water quality index, and the MLP classifier gives high accuracy for classifying the water quality class. It is concluded that, MLP regressor and MLP classifier out performs than the other models in forecasting water quality index. In future, hybrid models with deep learning algorithm can be built to improve the efficiency of the water quality prediction.

References

[1] C.V. Sillberg, P. Kullavanijaya, O. Chavalparit, Water quality classification by integration of attribute-realization and support vector machine for the chao phraya river, Journal of Ecological Engineering 22 (2021), 70–86.

[2] M. Yilma, Z. Kiflie, A. Windsperger, N. Gessese, Application of arti- ficial neural network in water quality index prediction: a case study in little Akaki River, Addis Ababa, Ethiopia, Modeling Earth Systems and Environment 4 (2018), 175–187.

[3] Y.R. Ding, Y.J. Cai, P.D. Sun, B. Chen, The use of combined neural networks and genetic algorithms for prediction of river water quality, Journal of Applied Research and Technology 12 (2014), 493–499.

[4] U. Ahmed, R. Mumtaz, H. Anwar, A.A. Shah, R. Irfan, J. García-Nieto, Efficient water quality prediction using supervised machine learning, Water 11 (2019), 2210.

[5] Zhang, J., Zhu, X., Yue, Y., & Wong, P. W. (2017). A real-time anomaly detection algorithm/or water quality data using dual time-moving windows. 2017 Seventh international conference on innovative computing technology (INTECH) (pp. 36–41). IEEE.

[6] Sakizadeh, M. Artificial intelligence for the prediction of water quality index in groundwater systems. Model. Earth Syst. Environ. 2016, 2, 8.

[7] Fitore Muharemi, Doina Logofătu & Florin Leon (2019) Machine learning approaches for anomaly detection of water quality on a real-world data set, Journal of Information and Telecommunication, 3:3, 294-307

[8] Tejas Subramanya, Davit Harutyunyan, Roberto Riggio, Machine learning-driven service function chain placement and scaling in MEC-enabled 5G networks, Computer Networks, Volume 166, 2020,106980, ISSN 1389-1286

[9] J. P. Nair and M. S. Vijaya, "Predictive Models for River Water Quality using Machine Learning and Big Data Techniques - A Survey," 2021 International Conference on Artificial Intelligence and Smart Systems (ICAIS), 2021, pp. 1747-1753

[10] Kalimur Rahman, Saurav Barua, H.M. Imran, Assessment of water quality and apportionment of pollution sources of an urban lake using multivariate statistical analysis, Cleaner Engineering and Technology, Volume 5, 2021, 100309, ISSN 2666-7908

[11] Arunkumar, R & Thambusamy, Velmurugan. (2021). An Exploratory Data Analysis Process on Groundwater Quality Data. 54. 41-48

[12] Marisol Vega, Rafael Pardo, Enrique Barrado, Luis Debán, Assessment of seasonal and polluting effects on the quality of river water by exploratory data analysis, Water Research, Volume 32, Issue 12, 1998, Pages 3581-3592, ISSN 0043-1354

[13] Stroomberg G. J., Freriks I. L., Smedes F. and Co®no W. P. (1995) In Quality Assurance in Environmental Monitoring, ed. P. Quevauviller. VCH, Weinheim.

[14] G Tan, J Yan, C Gao, and S Yang, Prediction of water quality time series data based on least squares support vector machine, Procedia Engineering, Vol. 31, 2012, pp. 1194-1199.

[15] WC Leong, A Bahadori, J Zhang, and Z Ahmad, Prediction of water quality index (WQI) using support vector machine (SVM) and least square-support vector machine (LS-SVM), International Journal of River Basin Management, Vol. 19, 2021, pp. 149-156.

ROBOT ARM WITH GESTURE CONTROL

N Nissi Kumar, P Sai Tharun, K. Prakash, L.M.N Mythresh, Naga Jyoshna

Department of ECE, GRIET, Hyderabad, Telangana, India.

nissikumar86@gmail.com, tharun.pilli@gmail.com, kethavatprakash1@gmail.com, Mythresh@gmail.com, jyoshnajetty@gmail.com

Abstract: The efficiency of work flow is constrained in human-inaccessible regions, such as high radiation exposure un power plants and defenses. The notion for a low-cost microcontroller-based robotic arm is presented in this research. The gesture is controlled by an accelerometer and a microprocessor in this robotic arm. In the field, where there are higher radiation concerns, this arm will assist in activities such as gripping things and wirelessly remote-controlled working of gadgets. The use of 3-D printing of an arm for prosthetic replacements can enhance this process.

Keywords: Gesture robot, Accelerometer

1. INTRODUCTION

The usage of robots in situations where conditions are uncertain, such as rescue missions, robots can be programmed to follow the orders of a human operator and complete the task as instructed. Gestures are very significant in people's daily lives. The gestures control device is only confined to space- electronics and partially in from any physical movements such as the face, hands, and so on. By making it easier to communicate with computers, The Human Machine Interface is improved via gesture recognition. We've used gesture controls in prostheses and power plant operations. This study describes how gestures may be used to operate a robotic arm wirelessly. HMI stands for human-machine interface.

2. LITERATURE SURVEY

Gesture recognition has been developed to replace the traditional way of controlling robots via buttons and other means with hand gesture-based control [1]. Robots have worked to develop a "Human Machine Interfacing Device." It is feasible to man oeuvre a robot using the gesture recognition idea [2]. The accelerometers' low-moderate cost and relative compact size make them an ideal instrument for detecting and recognizing human motions [3].

3. METHODOLOGY

Controlling Your Hands Hand gestures, rather than buttons, are used to operate robots. The robot is divided into two sections: one for transmitting and one for receiving. The Accelerometer is installed on the user's hand at the Transmitting part, recording the gesture and moving the robot accordingly. The accelerometer's input voltages are assigned to appropriate levels using a comparator IC. The encoder received 3-D gesture locations from the accelerometer as input signals. Before being broadcast via an RF Transmitter module, the four-bit data is encoded by an Encoder IC. In the receiving part, the RF receiver module receives encoded data.. Using a decoder IC, the data is then decrypted. A microprocessor processes the decoded data and sends it to a motor driver, which spins in a certain way to move the robot in the same direction as the hand.

4. MODEL DESIGN

Digital model data from a mechanical model may be used to create objects of any shape or geometry. which layers the model in thermos satiating plastic or any other substance of interest layer by layer. The prosthetic arm's STL file may be obtained from the DIY website. Because of its resistance to heat and flexibility, ABS (acrylonitrile butadiene styrene) is utilized as a printing medium.

ARDUINO

Arduino is a company that creates microprocessor and microcontroller kits that are based on open-source computer hardware and software. We picked the Arduino 328 microcontroller since it has a lot of capabilities, such as a 32Kb ISP flash memory with read/write functionality, this makes resetting and adjusting the device's sensitivity a breeze. Arduino has a variety of places where you may learn how to programmed the gadget.

Figure:1: Arduino 328

SERVO MOTORS

Servo motors are little motors that provide a lot of power. An electronic motor, a feedback variable resistor, and a controller are the three main components of a servo motor. The accelerometer, which is used for gesture detection, gives out signals that control it. Pulse width modulation is the ideal transmission technology for radio-controlled modes of operation. Five servo motors with high speeds

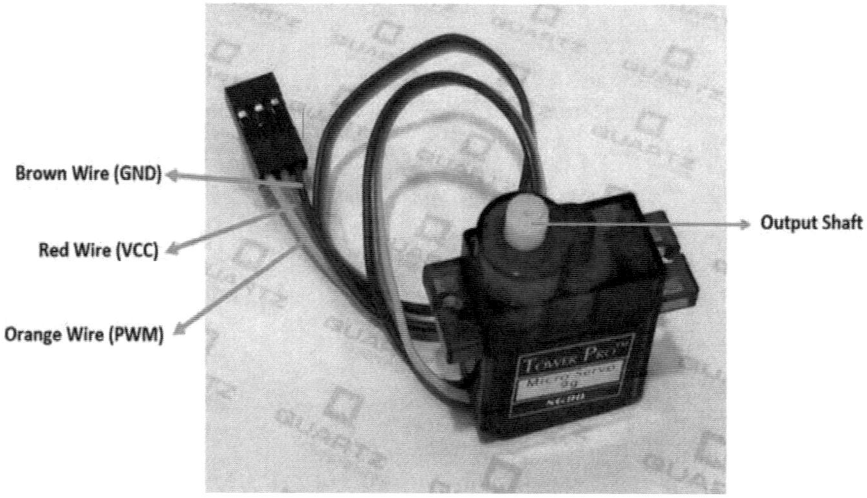

Figure:2: Servo Motors

RF LINK

The accelerometer's data is sent to the CPU through RF link 343.Because of its inexpensive cost and extended operating range of up to 10 meters, this item was chosen. The encoder and decoder integrated into the rf link transform parallel data to serial data and back to serial data, respectively. There will be no discrepancy between other signals of higher frequencies since we employ a specific frequency for conveying the information.

ACCELEROMETER

The accelerometer converts the acceleration induced by arm movement into electrical impulses by measuring it. The accelerometer utilized is extremely sensitive and can precisely detect all three dimensions.

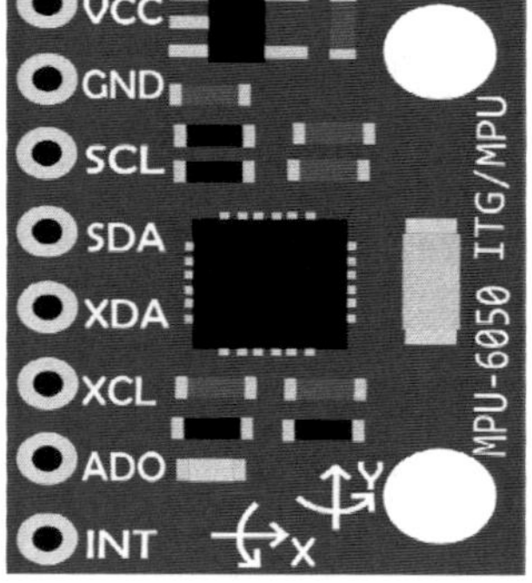

Figure:3: Accelerometer (MPU 6050)

HAND ASSEMBLY

The inmove robot's prototype hand was a mechanical model version. The STL file was uploaded to the site and printed immediately; the model took eleven hours to print and was pipette at the end. Each finger was strung with fishing lines that ran in a zigzag pattern across the finger's hinges. The elastic rubber bands on the top of the fingers were stretched. The fishing lines were securely attached to the output shafts of the servo motors.

ARM ASSEMBLY

The arm was installed on an arm stand that resembled a tripod stand, and the stand's clips were adjusted using the tripod stand's clips, and the power cables were extended from the rear side of the arm setup.

FLEX SENSOR

Flex sensor is a particular component that is used to calculate the motion or the bending of the motion. Most of the material is done with carbon and plastic, Due to the carbon's connection to the flex sensor, any motion of the sensor will cause a change in resistance. The moments of the robotic arms have become precise and meticulous, this also helps in acuteness. There are enormous types and uses of flex sensors for rehabilitation, security systems, intensity control and music interface. This is one of the most important aspects of the experiment

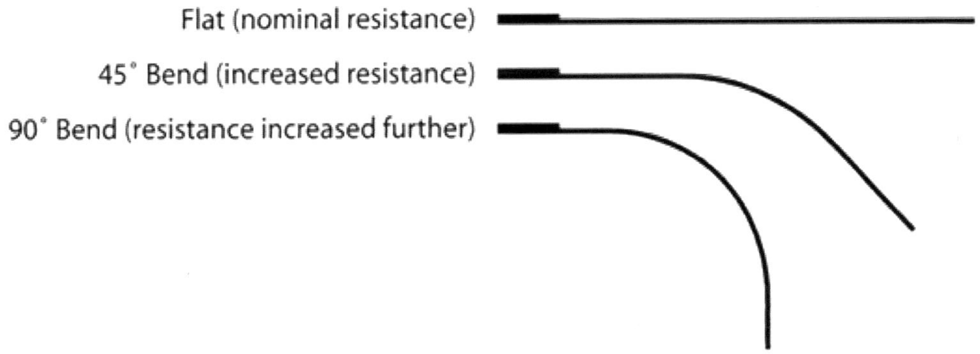

Figure:4: Flex Sensor

BREAD BOARD:

A breadboard is a tool for quickly constructing and testing circuits. The breadboard's many holes enable the insertion of circuit components like as ICs and resistors. A visual example of a bread board is shown below.

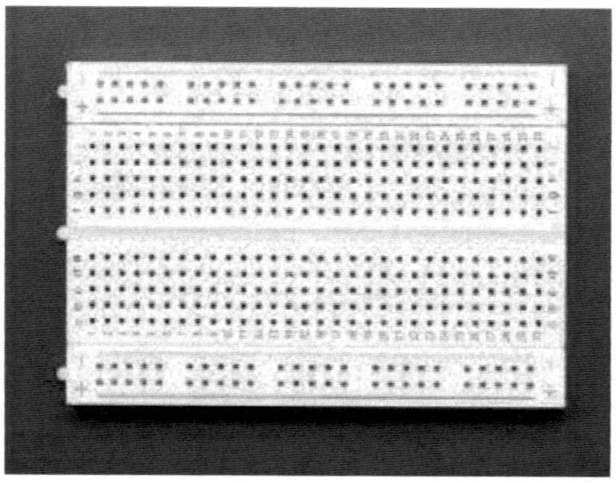

Figure:5: Bread Board

5. CIRCUITRY

Because the system is wirelessly operated, just a few cables were needed, and no main current supply was used; Instead, the modules were powered by two 9-volt battery.

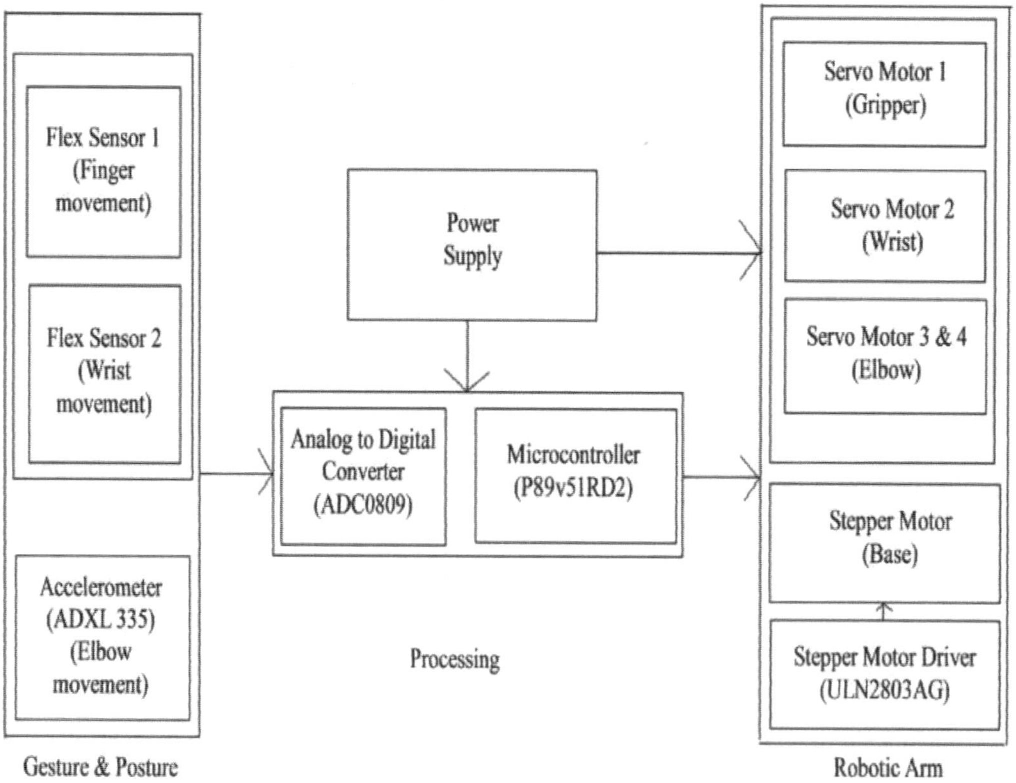

Figure:6: Circuitry

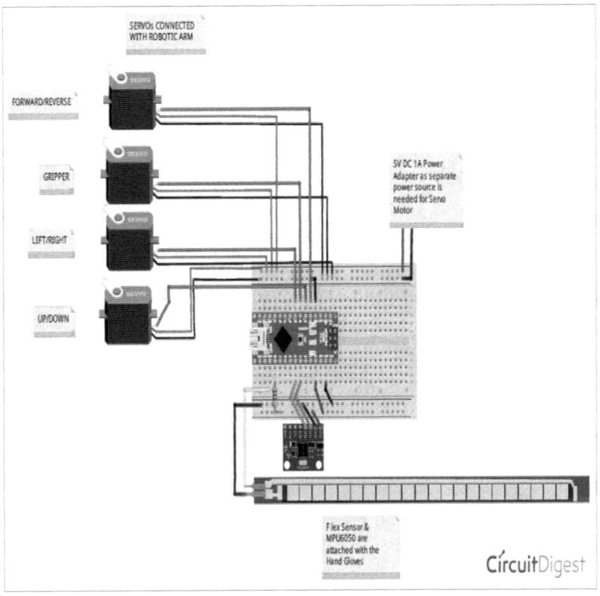

Figure:6.1: Circuitry

6. PROGRAMMING:

The application was downloaded from the Arduino website. The most significant part of configuring the gadget was the servo motor's operation time. We used 5 output pins on the Arduino board to supplement the 32 I/O pins. The servo motor's location dictates how far the figure it controls will move. The Arduino was designed to take signals from the accelerometer and convert them into pulses for the servo motor.

7. RESULTS AND DISCUSSIONS:

The motion or movement of the arm was recognized and imitated outright by the mechanical replica. This prosthetic robotic arm can help or assist any human being with their day-to-day tasks. However, the degree of rotation and functions pertinent to flexibility we're not quite same as the human's hand's function. In spite this, the grasp or grip of the robotic arm diligent concocted by us was strong enough to handle a pen ball, grip other tiny things and perform other functions like waving hand, fist pump, making a signal using fingers and etc.

8. CONCLUSION

Generally, prosthetics with simple configuration and structure have the potency to substantially ameliorate or improve the quality of life of other amputees. There is no need to carry cumbersome batteries because it itself is battery operated that can run as long as 4 hours. The aesthetic painting engraved on the prosthetic arm resembles natural flesh, which helps the amputee to gain confidence. It was our utmost endeavor to assist and help the mankind by concocting a low-cost prosthetic device capable of performing a bevy of functions.

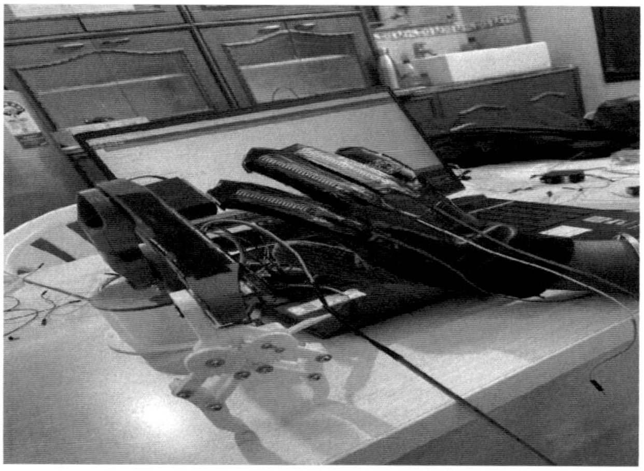

Figure:7: Conclusion

9. REFRENCES

[1] Sci-Hub | [IEEE 2018 41st International Conference on Telecommunications and Signal Processing (TSP) - Athens, Greece (2018.7.4-2018.7.6)] 2018 41st International Conference on Telecommunications and Signal Processing (TSP) - Robotic Arm Control with Hand Movement Gestures | 10.1109/TSP.2018.8441341 **https://sci-hub.hkvisa.net/10.1109/TSP.2018.8441341**

[2] Accelerometer Based Gesture Controlled Robot Using Arduino Swarnaprabha Jena, Sworaj Kumar Nayak, Saroj Kumar Sahoo, Sibu Ranjan Sahoo, Saraswata Dash, Sunil Kumar Sahoo Electronics And Communication Engineering, Centurion University Of Technology And Management, IndiaSci-Hub | [IEEE 2012 International Symposium on Innovations in Intelligent Systems and Applications (INISTA) - Trabzon, Turkey (2012.07.2-2012.07.4)] 2012 International Symposium on Innovations in Intelligent Systems and Applications - Development of a robotic-arm controller by using hand gesture recognition | 10.1109/inista.2012.6246985

[3] S. Waldherr, R. Romero And S. Thrun, 2000, "A Gesture Based Interface For Human-Robot Interaction", In Autonomous Robots In Springer, Vol. 9, Issue 2, Pp. 151-173Available At **Http://Www.Cs.Cmu.Edu/~Thrun/Papers/Waldherr.GesturesJournal.Pdf**

[4] Wang, B., And Yuan, T., "Traffic Police Gesture Recognition Using Accelerometer", IEEE SENSORS Conference, Lecce-Italy, pp. 1080- 1083,Oct. 2008.

[5] Song, M., Kim, B., Ryu, Y., Kim, Y., And Kim, S., "A Design Of Real Time Control Robot System Using Android Smartphone" The 7th International Conference On Ubiquitous Robots And Ambient Intelligence (URAI), Busan- Korea, Nov. 2010.

[6] **Https://Diyhacking.Com/Hand-Gestures-Robot/**

[7] Shruthi B. N, Shivraj, Sumathi S, "Hand Gesture Based Direction Control Of Robocar Using Arduino Microcontroller", International Journal Of Recent Technology And

[8] Http://Www.Eleinmec.Com/Datasheets/Ds_Holtek_ Ht12d.Pdf [11] Http://Oap.Sourceforge.Net/Datasheets/Rf.Pdf [12] Https://Www.Engineersgarage.Com/Electronic-Components/L293dMotor-Driver-Ic

Reliability Study of MEMS Resonator: A Review

Poorvi K. Joshi

Department of Electronics Engineering,
Shri Ramdeobaba College of Engineering and Management, Nagpur (MS), India.

joshipk@rknec.edu

Abstract. Reliability study is required for all the industrial applications to understand the behavior and functionality of device with respect to technological, environmental and operational issues. This paper aim to study and analyzed the exiting contribution related to MEMS resonator in order to explore the reliability issues. It also explored the reliability related to packaging, designing and functioning of MEMS resonator. It summarized the analysis of Environmental effect, design parameters, intermolecular forces, varying electrical load and vibrations etc. An approach and methodology for testing and simulation of MEMS resonator in view of reliability analysis is presented.

Keywords: MEMS; Resonator; Electrical Load and Vibrations.

1. Introduction

Over past few years the Micro Electro Mechanical System [MEMS] technology based on chip resonators have shown significant potential for sensing, detecting and high frequency signal controlling and processing applications. MEMS resonators are being explored to fulfil the demands or needs of these speedy wireless communication circuits which may be utilized as oscillators and/or filters part of radio transceiver. Micro resonators of radio frequency range is involved into a variety of commercial applications as a replacement of traditional bulky devices due to their small size, high frequency, high quality factor and mostly importantly CMOS compatibility, which is unlike in the quartz resonator.
Commercialization of MEMS resonator depends on the reliability and sustainability under various operating conditions and environment. Various types of MEMS resonators are clamped-clamped resonator, cantilever resonator, fixed-fixed beam resonator, square plate resonator, disc resonator, ring resonator and comb drive resonator etc.

A MEMS resonator produced a vibration at high frequencies upon actuation by resonance frequency which activates the attached electronic circuits. Resonance occurs when device is capable of storing and transforming energy from one mode to another. At resonant frequency system oscillates with higher amplitude. MEMS resonators operates for fixed resonant frequency and is most important characteristics, any shift in it, affects the performance of the device and suitability for specific application. Materials of resonators plays important role in characterization of the device, common material used to produce resonators are silicon and piezoelectric. Piezoelectric material produced electric charge due to mechanical stresses, and reverse effect i.e. strain is produced by application of electric field which creates vibrations of quartz crystal.

MEMS resonators characterize on the basis of quality factor (Q-factor), frequency tuning, motional impedance, thermal stability and collapsing voltage. With advancement in micromachining technology, on chip micro resonators can produced high-Q if operated in vacuum to reduce losses due to fluidic damping. Mechanical resonators are preferred over electrical resonators due to high Q-factor for various frequency reference applications [1].Quality factor (Q) is a ratio of maximum energy (total energy) stored to energy dissipated (lost) in one cycle. So by minimizing the energy losses, Q-factor of resonator can be improved. It is a measure of the selectivity of a tuned circuit and is dimensionless number. Quality factor is important in transceivers for synchronizing, for reliable operation Q should be high.

Transduction methods commonly used for actuations and detection of resonance of resonators are Electrostatic, piezoelectric, thermal and magnetic. In case of electrostatic actuation input and output is electrical signals.

This review focuses on various research carried out with respect to MEMS resonator. Reliability and reproducibility of device is most critical issue in design of MEMS resonator under various dynamic inputs. Sensitivity, robustness, stability and resilience are key performance indicator of MEMS resonator under variable loads. Aim of this review is to understand importance of various parameters and characteristics of MEMS resonator for reliability study. To developed an approach to carry out detail reliability study of MEMS resonator failures during design, manufacturing, packaging and functional levels. Also to understand the key and open issues in reliability analysis also suggest an approach for testing and reliability assessment of MEMS resonator. To realized the scope of reliability study of MEMS resonator this review is presented in four sections: First section elaborated work related to design issues in MEMS resonator. Manufacturing and packaging related issues are presented in second section. Third section gives insight to functional level issues and related work. Finally, in fourth section approach for further study on reliability of MEMS resonator is proposed with conclusion.

1.1. Design Related Work.

MEMS device design is significant due to their reliability, survivability, impact during fabrication, shipping, performance under electric, mechanical and thermal loads. Repeatability and reproducibility are key performance indicators of MEMS devices to understand its dynamic behaviour [2]. C. Arft [3] developed MEMS oscillator for harsh automotive environment with better reliability and resilience compared with quartz crystal-based oscillator. Test methods are presented to evaluate resilience during performance under shock and vibrations. M.S. Ahmed [4] carried out simulation of nonlinear model of single and dual microbeam resonator to analyzed dynamic response under varying mechanical shock load and electric actuation. Static response of both microbeam resonator is done to determine pull-in voltage. Dual microbeam system shows better stability for shock load than single microbeam. Tripathi A. and Bajaj A [5] discussed the reliable and reproducible performance of nonlinear resonator in view of uncertainties in design during manufacturing. Uncertainty analysis is done using simulation. Lucas V et al [6] modelled and simulate the propagation of surface uncertainties for MEMS micro beam resonator with respect to resonance frequency.

Chorsi H. T. [7] developed mathematical model for evaluating the dynamic behaviour of MEMS disk resonator. They concluded that Polydiamond disk resonator exhibits high quality factor at ultra-high frequency, low actuation voltage and a weak nonlinearity makes it more suitable for RF filters and channelizer. Chorsi, M.T. [8] presented accurate process to model MEMS disk resonator and investigated the influence of intermolecular forces. Governing equation of motion is derived and its stability is determined. Dynamic behaviour of the system is accurately predicted by the model. Chorsi H. T [9] investigated dynamic behaviour of MEMS disk resonator analytically. Frequency response of the system was studied under the influence of intermolecular forces. Dynamic response with respect to design parameters was discussed. Chorsi, M. T. et al [10] studied the dynamics of electrostatically actuated radial-contour mode microring resonator and various damping effects. Also dynamic response is discussed for various parameters like ring radius, quality factor and actuation voltage. M. Agarwal et al [11] investigated the effect of vibrations and bias voltage noise on phase noise of MEMS resonator in order to design better vibration sensitive MEMS resonator based oscillator.

1.2. Packaging Related Work

Packaging of MEMS devices very important issue as it is responsible for providing stable environment inside. Due to small size, MEMS devices are sensitive to external environment, contamination. For commercialization and reliability packaging is important issue [12]. S. Yoneoka et al [12], presented characterization of MEMS resonator which is fabricated and packed in a poly SiC with thin film encapsulation. They observed that pressure inside the package can be controlled and frequency drifts of resonator are less than frequency noise level from experimentation. Temperature coefficient of frequency (TCf) is to be to achieve stable resonant frequency over range of temperature. Stresses induced due to change in temperature and various factors affecting the TC are resonator material and dimensions. Melamud et al [13] evaluated the temperature coefficient of frequency of MEMS resonator and studied the effects of compressive and tensile stresses induced by die and its packaging. B. Kim et al [14] elaborated the importance and various challenges in packaging for MEMS devices. E. M. Ferrer et al [15] carried out pressure cooker, high temperature storage and high temperature cyclic test to study reliability for high quality factor electrostatic MEMS resonator with standard plastic packaging. C. Manier et al[16] presented an approach for wafer packaging of MEMS resonator and integration of interposer.

1.3. Functional Related Work

MEMS Resonator consists of two electrodes one is actuating and other one is sensing and vibrating mechanical structure and requires air gap between structure and electrode. Time varying voltage signal is applied to actuation electrode and when frequency of signal reaches natural frequency of resonator beam its start vibrating. This creates DC bias capacitor between sensing electrode and resonator beam which produce motional current at output electrode. Electrostatic actuation provides fast response, ease in design and low power consumption. For mechanical MEMS resonator driving force is Coulomb force of capacitor formed by oscillating beam and fixed electrode. For ultra-high frequencies piezoelectric transduction is useful. Externally transduced capacitive resonators display high impedance than piezoelectric resonators. G. Bahl et al [17] presented theoretical electrostatic-based model of charge trapped in dielectric inside transduction gap of resonator and compared with experimental data.R. Melamud et al [18] fabricated silicon and silicon dioxide composite micromechanical resonator and characterized for temperature dependence of frequency. Iqbal F. and Lee B [19] carried out frequency domain analysis of electrostatically actuated MEMS resonator to measure uncertainties in resonant characteristics. Analysis is carried out using analysis of variance (ANOVA) technique. Chorsi, M.T [20] investigated the static and dynamics analysis of electrostatically actuated MEMS resonator with and without controller. Effect of various controller and changing coefficients of quadratic controller is presented using frequency analysis curves. Prasanth C. [21] did the quantitative analysis of pull in voltage for electrostatically actuated non-uniform microcantilever MEMS resonator. Also the results are compared through simulation. They concluded that the taper ratio and length of microcantilever influences pull in voltage.

2. Reliability Issues

Reliability study is required for all the industrial applications to understand the behaviour and functionality of device with respect to technological, environmental and operational issues. Reliability is important aspect to enhance the application of MEMS devices. Resonant frequency depends on physical and material properties and is also depends on various factors viz. environment, vibrations, shocks, damping, temperature, pressure etc. Various technical issues due to small size of MEMS resonator are fatigue which causes aging and induced stresses, shift in frequency due to temperature fluctuation, surface contamination. MEMS resonators operates for fixed resonant frequency which is most important characteristics, any shift in it, affects the performance of the device and suitability for specific application. Failures and its different mechanisms in MEMS resonator are key issue in assessing the reliability and failure of device [22]. Material properties, their degradation processes are important issues in reliability study of MEMS resonator. Reliability analysis is helpful in enhancing the operational performance of the device and also in designing high quality factor MEMS resonators.

There are various types of operation failures viz. wear, fatigue, delamination, stiction, creep, corrosion, dielectric charging, electromigration, micro welding, radiation and humidity etc. [23]. Evaluation of mechanisms of occurrences for these failures and sources needs to be elaborated in order to evaluate overall reliability of the MEMS resonator.

There are various factors affecting the response due to variation in resonant frequency like; technological aspects, principle of operation, geometry, material properties, ambient conditions, environmental factors (temperature, pressure and damping) etc. A micromechanical resonator has potential of integration, throughout the low to ultra-high frequency band. Mechanical stress is one of the important factors in shifting the resonance frequency of resonators. Thermal Stability is drift in resonance frequency with change in temperature. In order to develop reliable product with its expected in-use lifetime every step in manufacturing is to be recorded systematically and categorized for the root cause of failures, also to anticipate the failure mode. This will help in achieving better MEMS resonator reliability and disciplined approach during designing, manufacturing and quality checks minimizes the risk of failures.

MEMS resonator reliability consists of lot many physics of failure and failure mechanisms. Device can fail during the design stage if design rules are not followed, it will not perform its intended functions or not at all operable. All the modelling and simulation software are having a limitation in terms of accuracy due to mesh density [22]. Parameters like surface finish, tolerances of part, profile and the required fit also influence the reliability of device. Robustness and reliability of MEMS resonators are influenced due to variation in final manufactured device from design specification, due to manufacturing process uncertainties and tolerances [24]. Structural failure in MEMS occurs due to the compression stress at the sharp edges at coupling region [25].

3. Reliability Testing

There are various testing techniques to evaluate reliability, classified as functional testing, where performance of MEMS resonator is compared with standard specification as per the operation [26]. In this type of testing, observation and measurement of surface structures, images of various elements, cracks, and stresses developed etc. by application of electrical and mechanical loading reliability is assessed. Next is failure testing, where performance deviation during its operation and failures are studied. This testing is also referred as reliability testing. Failures can be permanent or degradation type. It is difficult to differentiate between the functional testing and failure testing [26]. Various destructive and nondestructive testing methods for reliability assessment are accelerated life test, electrostatic testing, and impact loading etc.

Reliability assessment with experimentation, modelling techniques and analytical analysis opens up new challenges in reliability study. Identification of failure mode, statistical analysis of data and defining the model of the systems are key issues in reliability analysis. Reliability analysis is helpful for the manufacturer to optimize the design and manufacturing process of the device. Presently various researchers carried out reliability study of MEMS resonator with respect to electrical, mechanical loading, environmental effect, and packaging techniques etc. Lots of literature is available on study of various failures like creep, stiction, wear, fatigue and fracture along with the causes of it with detailed analysis. Various studies are based on the mathematical formulation, physical modelling of system, simulation and analytical model, along with reliability improvement approaches for MEMS resonator. MEMS devices reliability depends on distribution of failures over it lifespan. Bathtub curve is important and widely accepted engineering tool in reliability. Bathtub profile is characteristics of failure curve with respect to time of products. Sustainability of product throughout its life cycle is deciding factor for success in market and it is a major challenge for manufacturer in this growing and competitive scenario. Reliability engineering involves assessing and analyzing the failures in product. Failure rate is defined as percentage of products fails per unit time. It varies throughout the life cycle of product and it is plotted against the time, characteristic bathtub curve obtained as shown in 'figure 1' for most of the product. Bathtub curve consists of three region Infant mortality, Useful life and Wear out stage.

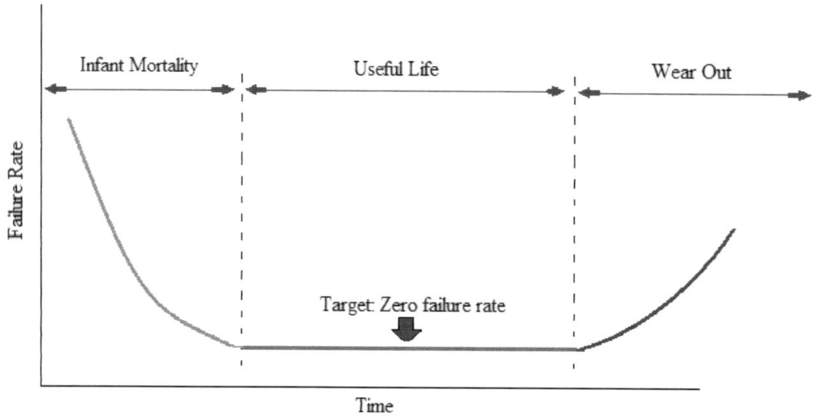

Fig. 1. Bathtub Curve

In first region, failures occur during the design and manufacturing process due to which product fails during start of operation. Improvement in design and manufacturing process needs to be done before product reaches to market or customer. Infant mortality region follows a Weibull distribution model. Failures occur at constant rate in useful life region and follows exponential distribution model. In wear out or aging region failure rises with time due to wear and tear rather than the random failure. Wear out region follows Weibull distribution model.

Reliability of any device and failure rate is estimated by life testing carried out under specific conditions and varying the factors affecting the performance of product. Physics of failure or reliability physics is the technique to evaluate the effect of mechanical, chemical and environmental factors affecting the performance of the system over its lifetime. With advancement in MEMS technology and exposure to new application areas, reliability is important for wide acceptance of MEMS devices. MEMS reliability analysis is extremely important to identify and understand the different failure mechanisms that can be implicit such as mechanical, thermal, and chemical aspects related to the diversity of materials that can be used.

Analysis of MEMS resonators behavior due to variation in AC excitation voltage, DC biasing voltage and temperature is proposed in reliability testing. Dielectric failure or any signature observed during the testing is to be analysed for study. Failures during the testing trials will be observed and analyzed to established reliability of the device. Purpose of this study is to formulate the detailed procedure (test plan) for conducting reliability test of MEMS resonators. FMEA will be used to for assessment of reliability and formulation of reliability model.

This paper aims to suggest an approach to study the different mechanisms of failure and different tests for analysis of reliability of MEMS resonator.

- Selection of high frequency MEMS resonator.

The global rollout of 5G, and the rapid expansion of the Internet of Things, creates significant new RF filtering challenges. RF filter technology is advancing to overcome these challenges. Among the chief developments: RF filters are evolving to support higher frequencies and increased bandwidth for the new 5G and Wi-Fi band expansion.

- Designing of testing circuit for reliability analysis and characterization of MEMS resonator.

A testing circuit is prepared for testing Resonator as shown in 'figure 2' and 'figure 3' The PCB is made up of Roger RO4350B material. On the PCB, the SMA connectors are connected to the Resonator through transmission line. The Transmission lines are the coplanar waveguides who's Thickness, width, and dielectric Constant values are calculated using 50-ohm calculator.

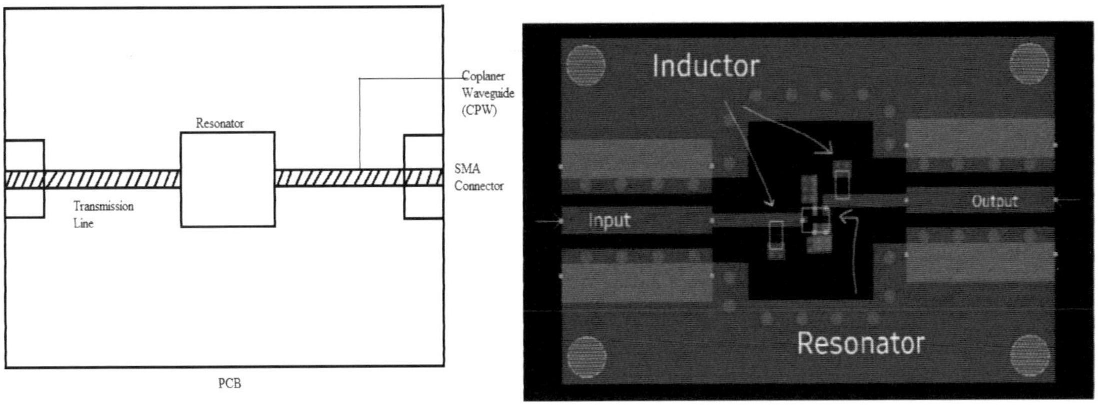

Figure 2. PCB of Testing Circuit

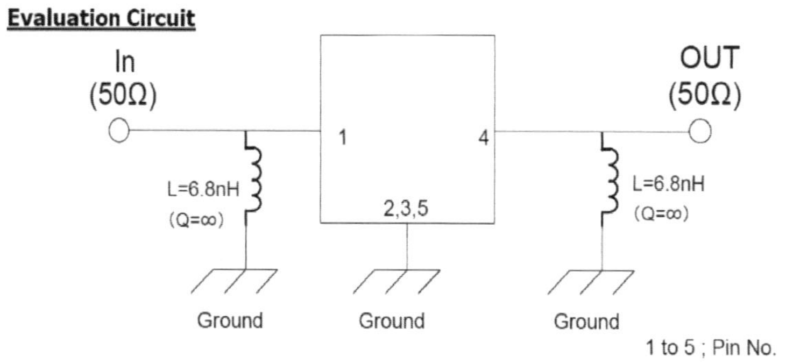

Figure 3. Evaluation Circuit of Resonator

- Different accelerated testing to be performed on MEMS resonator.

Thermal Analysis for FBAR: The temperature of Resonator is increased beyond its operating temperature and then it is cooled down to normal room temperature and the measurements are taken on VNA. The temperature of Resonator is measured using Temperature Measuring Gun as shown in 'figure 4'. The resonator was kept in High Temperature Oven .The measurements are done above the operating range of the Resonator. We have started the measurement by keeping the device for 48 Hours at 100^{0}C and measuring its S parameters. The same procedure is repeated by increasing the Temperature by 10 degrees and keeping the device by next 48 Hrs. And so on.

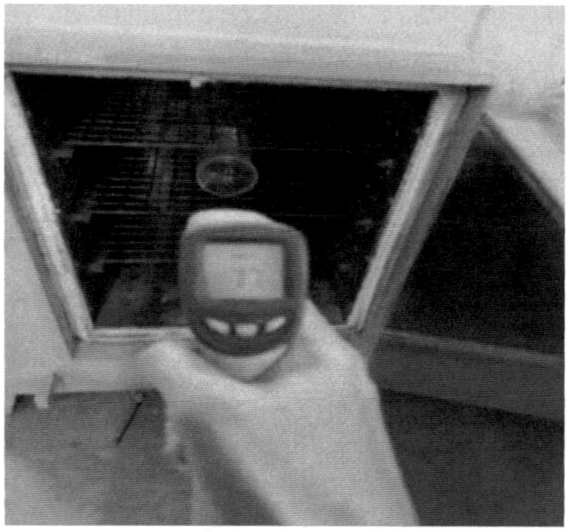

Figure 4. Temperature Measuring Gun

- Aim of testing is to prolong the useful life of MEMS resonator and prediction of start of wear out phase.
- After understanding the mechanism with which resonator fails, will use simulation for analyzing the device functioning and accordingly changes will be made in the fabrication process so as to make the resonator more reliable.

4. Discussion and conclusion

This paper presents the analysis of existing contribution in the area of packaging, design and functional related issues in MEMS resonator. It particularly focuses on important issues related with reliability of resonators. It also discussed the suggested approach based on the findings of exiting literature as summarized in last section. Accelerated testing based approach is suggested to understand the failure mechanisms. Based on this, through simulation, device functioning can be analyzed. Understanding of various designs, structural, operational, environmental and material parameters are important before start of reliability study. Identification of various failure modes and mechanisms during experimentation, modeling and simulation useful to measure reliability to predict operational life of MEMS resonator. Reliability of MEMS resonators cannot be described scientifically due to complex phenomenon and diversity as it involves multiple domains. So, a scope is there to explore it effectively. This paper tries to explore opportunity to study the complex phenomenon of failures and try to prolong the useful life region as mentioned in bath tub curve. It also tries to predict the start of aging phase and suggest any replacement of elements instead of scraping whole device.

References

[1] Herrera-May, A.L.; Soler-Balcazar, J.C.; Vázquez-Leal, H.; Martínez-Castillo, J.; Vigueras-Zuñiga, M.O.; Aguilera-Cortés, L.A., "Recent Advances of MEMS Resonators for Lorentz Force Based Magnetic Field Sensors: Design, Applications and Challenges", Sensors 2016, 16 (9), 1359, pp: 1-25, doi:10.3390/s16091359

[2] Lakrad F. and Belhaq M., "Quasi-periodically Actuated Capacitive MEMS", Structural Nonlinear Dynamics and Diagnosis. Springer Proceedings in Physics, vol 168, pp: 183-200, 2015, Springer, Cham, doi: 10.1007/978-3-319-19851-4_10

[3] Arft C., Lu Y., and Parvereshi J., "MEMS Oscillators with Improved Resilience for Harsh Automotive Environments," SAE International Journal of Passenger Cars - Electronic and Electrical Systems-V125-7EJ, Vol.: 9, issue: 1, pp: 212-218, 2016, doi: 10.4271/2016-01-0101.

[4] Ahmed, M.S., Ghommem, M. and Abdelkefi, A., "Nonlinear analysis and characteristics of electrically-coupled microbeams under mechanical shock", Microsystem Technologies, Volume 25, Issue 3, pp 829–843,2019,doi: 10.1007/s00542-018-4056-2

[5] Tripathi A. and Bajaj A., "On Regularization in Design for Reliability for Nonlinear Planar Beam-Type Resonators", Journal of Vibration and Acoustics, 2017, Vol: 139, issue: 4, pp: 1-33, Doi: 10.1115/1.4036624.

[6] Lucas V., Golinval J.C. , Voicu R., Danila M.,Gavrila R., Müller R.,Dinescu A., Noels L. & Wu L., "Propogation of material and surface profile uncertainties onMEMS micro-resonators using a stochastic second-order computational multi-scale approach", International Journal for Numerical Methods in Engineering, pp:1-44, 2017, Doi:10.1002/nme.5452.

[7] Chorsi H. T., Chorsi, M. T.andGedney, S. D, "A Conceptual Study of Microelectromechanical Disk Resonators," in IEEE Journal on Multiscale and Multiphysics Computational Techniques, vol. 2, pp. 29-37, 2017,doi: 10.1109/JMMCT.2017.2670621

[8] Chorsi, M.T. and Chorsi, H.T., "Modeling and analysis of MEMS disk resonators", Microsystem Technologies, June 2018, Volume 24, Issue 6, pp 2517–2528, doi: 10.1007/s00542-017-3645-9

[9] ChorsiH. T., Chorsi, M. T., Gedney, S. D., "Numerical Modeling of MEMS Resonat*ors",* arXiv prints, 2016, doi: 10.48550/arXiv.1611.05310

[10] Chorsi, M. T., Chorsi H. T., Gedney, S. D., "Radial-contour mode microring resonators: Nonlinear dynamics", International Journal of Mechanical Sciences, Volume 130, September 2017, Pages 258-266, doi: 10.1016/j.ijmecsci.2017.05.051

[11] M. Agarwal et al., "Effects of Mechanical Vibrations and Bias Voltage Noise on Phase Noise of MEMS Resonator Based Oscillators," 19th IEEE International Conference on Micro Electro Mechanical Systems, Istanbul, Turkey, 2006, pp. 154-157, doi: 10.1109/MEMSYS.2006.1627759

[12] S. Yoneokaet al., "Characterization of Encapsulated Micromechanical Resonators Sealed and Coated With Polycrystalline SiC," in Journal of Microelectromechanical Systems, vol. 19, no. 2, pp. 357-366, April 2010, doi: 10.1109/JMEMS.2010.2040460

[13] R. Melamud et al., "Effects of stress on the temperature coefficient of frequency in double clamped resonators," The 13th International Conference on Solid-State Sensors, Actuators and Microsystems, 2005. Digest of Technical Papers. TRANSDUCERS 05, Seoul, South Korea, 2005, pp. 392-395 Vol. 1, doi: 10.1109/SENSOR.2005.1496438

[14] B. Kim et al., "Using MEMS to Build the Device and the Package," TRANSDUCERS 2007 - 2007 International Solid-State Sensors, Actuators and Microsystems Conference, Lyon, 2007, pp. 331-334, doi: 10.1109/SENSOR.2007.4300135

[15] E. M. Ferrer, M. S. Pandian, A. Uranga, N. Barniol, W. S. Tay and C. B. Fei, "Reliability study on thin film capped monolithic CMOS-MEMS resonator with standard plastic packaging," 2018 IEEE Micro Electro Mechanical Systems (MEMS), Belfast, 2018, pp. 555-558, doi: 10.1109/MEMSYS.2018.8346613

[16] C. Manieret al., "Wafer level packaging for hermetical encapsulation of MEMS resonators", 2015, Symposium on Design, Test, Integration and Packaging of MEMS/MOEMS (DTIP), Montpellier, 2015, pp. 1-6, doi: 10.1109/DTIP.2015.7161027

[17] G. Bahlet al., "Model and Observations of Dielectric Charge in Thermally Oxidized Silicon Resonators," in Journal of Microelectromechanical Systems, vol. 19, no. 1, pp. 162-174, Feb. 2010, doi: 10.1109/JMEMS.2009.2036274

[18] R. Melamudet al., "Temperature-Insensitive Composite Micromechanical Resonators," in Journal of Microelectromechanical Systems, vol. 18, no. 6, pp. 1409-1419, Dec. 2009,doi: 10.1109/JMEMS.2009.2030074

[19] Iqbal F., and LeeB., "A Study on Measurement Variations in Resonant Characteristics of Electrostatically Actuated MEMS Resonators",Micromachines,vol: 9, pp: 1-13, 2018, doi: 10.3390/mi9040173.

[20] Chorsi, M.T., Azizi, S. andBakhtiari-Nejad F., "Application of quadratic controller to control the pull-in instability of a micro-resonator",International Journal of Mechanics and Materials in Design, March 2015, Volume 11, Issue 1, pp 111–123, doi: 10.1007/s10999-014-9281-y

[21] Prasanth C., Harsha C. S., andPratiher B., "Electrostatic pull-in analysis of a nonuniform micro-resonator undergoing large elastic deflection", Proceedings of the Institution of Mechanical Engineers, Part C: Journal of Mechanical Engineering Science,Volume: 232, issue: 18, pp: 3337–3350, 2017,doi: 10.1177/0954406217736079

[22] J. Iannacci, "Reliability of MEMS: A perspective on failure mechanisms, improvement solutions and best practices at development level", Displays, Volume 37, April 2015, pp: 62-71, DOI: 10.1016/j.displa.2014.08.003

[23] Daniel J. Fonseca and Miguel Sequera, "On MEMS Reliability and Failure Mechanisms", International Journal of Quality, Statistics, and Reliability, Volume 2011, Article ID 820243, pp: 1-7, doi:10.1155/2011/820243

[24] A. Tripathi, A. K. Bajaj, "On Regularization in Design for Reliability forNonlinear Planar Beam-Type Resonators", Journal of Vibration and Acoustics, 2017, pp: 1-33, doi:10.1115/1.4036624

[25] W. Jin, R. G. Polcawich, P. A. Morton and J. E. Bowers, "Piezoelectrically tuned silicon nitride ring resonator", Optics Express, Vol. 26, Issue 3, 2018, pp. 3174-3187, doi: 10.1364/OE.26.003174

[26] In book: System-on-Chip Test Architectures (Systems on Silicon), Xingguo Xiong, Chapter: 13, Publisher: Morgan Kaufmann Publishers, Editors: Laung-Terng Wang, Charles Stroud, and Nur A. Touba

Reduction of Electromagnetic Interference Using EMI Filter in a Buck Converter

D Ruth Anita Shirley[1], Dr S Sasipriya[2], T Sibi[3], B Ragul[4], S Praveen[5]

[1] Princeton University, Princeton NJ 08544, USA [1]Assistant Professor, Department of Electronics and Communication Engineering, Sri Krishna College of Engineering and Technology, Coimbatore, India.
ruthshirley06@gmail.com

[2]Professor, Department of Electronics and Communication Engineering, Sri Krishna College of Engineering and Technology, Coimbatore, India.

[3,4,5] Department of Electronics and Communication Engineering, Sri Krishna College of Engineering and Technology, Coimbatore, India.

Abstract: The presence of non-isolated DC-DC buck converter plays a crucial role in electrification of aircraft, bio-medical and automotive devices. To obtain output devoid of interference from external sources, maximizing specific power or power density of the EMI filter is necessary. This requires a filter that is capable of eliminating or filtering these unwanted surges or spikes in the power converters. In this paper an EMI filter is placed in the buck converter operating with common mode/differential mode. An EMI filter is first designed taking into account the impact of interference in the power rails and the ground rails. The model is implemented and the output of the buck converter in the presence and absence of the proposed EMI filter is analysed.

Keywords: Buck Converter, Electromagnetic interference, spikes, EMI filters, power density

1 First Section

The shipboard, aircraft and automotive system electrification makes use of the non-isolated DC-DC converters widely. A common DC bus with several variety of loads that can exchange powers along with generators, and energy storage is essential for such systems typically [1], [2]. The DC-DC converters ensure the dynamic resiliency and system stability thereby enabling dynamic voltage control at load interfaces and bidirectional current control between the battery storage and common bus. A buffer is formed between the DC busses by the non-isolated DC-DC converter and shipboard DC zonal architecture, covering the length of the ship as well as the distribution systems (AC and DC) within the zones through inverters. When compared to the high frequency transformer isolated converters, the non-isolated DC-DC converters are preferable due to the loss, weight and space constraint of the transportation systems.

Typically, small voltage ratios are observed between the load/battery and the common bus [3]. For example, in ships with 650 V load, 1000 V common bus is used and in aircraft with 250 V battery, 270 V common bus is used. In transportation systems, for the power electronic converters, one of the primary consideration is to achieve high power density in [4], [5]. However, at all common interfaces, the power quality standards and conducted electromagnetic interference (EMI) must be enforced for ensuring system compatibility. For DC-DC converters, the differential-mode (DM) and common-mode (CM) filters have to be designed systematically and holistically, which can be achieved by the compatibility and power density requirements simultaneously. In highly constrained system environment, the objectives are best met by designs that provide a wide range of solutions while ensuring coverage through key design variables such as wide-bandgap (WBG) devices like power semiconductors and power converter topologies. While considering the electromagnetic compatibility (EMC), several challenges are faced with the electric traction systems that make use of static converter.

Massive high frequency interferences are generated by each power converter leading to onboard electronic system malfunction. EMI filters [6] – [7] can be used to overcome this issue and considerably reduce these conducted emissions. By acting on the converter controls on the power semi-conductor components, the dv/dt component can be slowed down during transition [8]. Such solutions can be combined along with the EMI filters to achieve optimal functionality. Capacitors and coupled inductors are combined to form the EMI filters. Load impedances and networks are considered while choosing the filter topology. Power converters commonly make use of the DM and CM filters. The filter efficiency is strongly affected by the passive components [7]. The EMI filter performance is negatively influenced by the coupled inductors' equivalent parallel capacitances (EPC) and the capacitors' equivalent series inductor (ESL) which are the parasitic elements of these components [8]. The use of passive components and their parasitic elements in high frequency models helps in optimizing the filter characteristics during simulation and the design process.

The non-isolated synchronous buck converter in this paper makes use of the equivalent circuit modeling approach and CM and DM decomposition technique discussed in [9], [10]. The non-isolated DC-DC converter topology is evaluated using a methodological approach based on the results depending on the CM and DM producing behavior and its perspective. This leads to an optimal filter design considering the design objectives at the system level. The buck converter system analysis reveals two significant insights. First, a balanced arrangement is reached by the EMI filter by reduction of CM current through CM voltage cancellation that is caused by the buck converter and its positive rail achieved by the concentrated filter inductor. Second, the parasitic capacitor effects dominate the design standards of the EMI filter and LISN measurements for the canonical buck converter.

2 Literature Survey

Synchronous buck topology based voltage regulator design is commonly used in several applications. The power supply voltage at the input is stepped down to match the operational voltage required by the microprocessor. Operation of DC-DC converters occurs at a few MHz of switching frequency. Levelling up of around 1 GHz of FCC failing is observed due to faster ringing and switching frequency. It can be a serious concern to use this kind of power supply considering the electromagnetic interference [10]. Several researchers have covered the potential EMI issue in [11-13]. In the frequency spectrum, the overall peak amplitude can be reduced by eliminating the prevalent harmonics observed in the input current. This is achieved without disturbing the average duty cycle of the DC-DC converter by operating it with soft switching scheme. In DC-DC buck converters, the conducted harmonic interference is mitigated successfully with the application of the spreading spectrum concept [14]. Several researchers have turned from such devices towards the electromagnetic radiation over the last few years. Along with minimization of magnetic components, the interaction of active components creates an EMI spectrum that is enhanced by the DC-DC converters due to their high switching frequencies.

Based on the transmission line (TL) approach for the typical power converter, a circuit model is introduced by Antonini et al. [15]. This technique is used for estimating the distribution of current along the traces of printed circuit board (PCB). The near electromagnetic fields are predicted using these current distributions. When the differential mode (DM) current effect is observed in a dominant manner in low frequency range, the near-field radiated emission model is effective [15]. This is because, the common-mode (CM) currents are not estimated by the TL model. In inverter-fed motor drive system with pulse width modulation (PWM), the EMI emission radiated and its experimental measurements are presented in [16]. For different feed and ground wire configurations, comparison of the radiated EMI emissions is performed between the motor terminal and the PWM inverter. Rather than reducing the loop area created by the feeding and grounding wires, reduction of radiated EMI is more effective when shielded three-core cable is used as observed from the experimental results. In the motor drive system, on the feeding wires, damping the CM current oscillations enable further reduction of radiated EMI.

In a PWM motor drive system, the radiated electromagnetic near fields are simulated using a two-step technique in [17]. In the motor drive system, high-frequency model devices and components are used for simulating the CM current through ground paths in the first step. The radiated fields are supplied with a source from the CM ground current simulated in the first step and 3D finite element analysis is performed for evaluating the EMI that is radiated. Time harmonic electromagnetic wave is used for modeling this radiated EMI. In power conversion systems, the radiated EMI based issues are addressed by several researchers as observed in [18]. However, in the power converters with far-field radiated EMI, the emission mechanism has not been discussed in these literature. In high-speed electronic systems and PCBs, the radiation issue has been addressed in [18]. These solutions can also be extended to power electronic devices to address the radiation issues.

In DC-DC converters, for reducing the impact of EMI, various posteriori solutions are introduced by several researchers. In order to reduce EMI, a PCB layout optimization scheme is proposed in [19] for reducing the effect of EMI issues in PCBs. A passive lossless snubber cell is introduced for reduction of ringing EMI noise in the research presented in [16]. The EMI issue is most commonly addressed using the EMI filter [18]. Additional components are required for such approaches which makes them unsuitable for applications such as compact SiP/SoC as the PCB area and manufacturing cost increases considerably [12, 17]. In a DC-DC converter, the major source of EMI is its switching activity. Hence, the switching activity has to be addressed while reducing the EMI. When compared to the previous solutions, an improved low-frequency EMI performance is produced and the need for additional passive off-chip components must be eliminated by such a solution [19]. At the design stage, for EMI reduction, the controller of the DC-DC converter is widely applied with spread spectrum techniques in existing literature. At the switching signal, primarily, EMI is directly prevented using this priori approach. Harmonic peaks are reduced by altering the power spectrum shape in the spread-spectrum model. Carrier frequency modulation (CFM) was used for EMI reduction using the first spread-spectrum technique [20]. Recent research work carried out by [21] and [22] show the presence of minute spikes in electronic appliances that ultimately lead to malfunctioning of the device.

In this proposed work, DM and CM decomposition and equivalent circuit modeling is presented to analyse and determine the output circuit. An optimally designed EMI filter is used to determine the reduction of EMC. In this methodology, it is identified that the effects of parasitic capacitors have a higher impact on the design of EMI Filterwhen compared with the CM voltage measured using the converter.

3 Proposed Work

3.1. CM and DM Analysis

With 'N' conduction paths, CM and DM voltage can be measured using the expressions:

$$v_{mn} = v_{mQ} - v_{nQ} \tag{1}$$

$$i_{mn} = \frac{1(i_m - i_n)}{2} \tag{2}$$

$$v_{cm} \triangleq \frac{1}{N} \sum_{n-1}^{N} v_{nQ} \tag{3}$$

$$i_{cm} \triangleq \sum_{n-1}^{N} i_n \tag{4}$$

where i_{cm} and v_{cm} represents the common-mode current and voltage.

Here 'Q' denotes an arbitrary reference point. Based on these expressions, the following conclusions are drawn:

- The sum of current flowing through the N parallel lines is represented as 'CM current'.
- The average voltage of N lines with respect to 'Q' is denoted as 'CM voltage'.
- The current that travels through the first line pair and returns via the second is the DM current and
- The voltage difference between any two lines is the DM voltage.

In general, the output will hold both the CM and DM components and decomposition will result in a spatial transformation, resulting in currents and voltages of N-1 DM along with a CM solution. Accordingly, it is possible to define the transformation matrices with the help of the following equations:

$$i_{dcm} \triangleq \prod_{N}^{i} i_n \tag{5}$$

$$v_{dcm} \triangleq \prod_{N}^{i} v_{nQ} \tag{6}$$

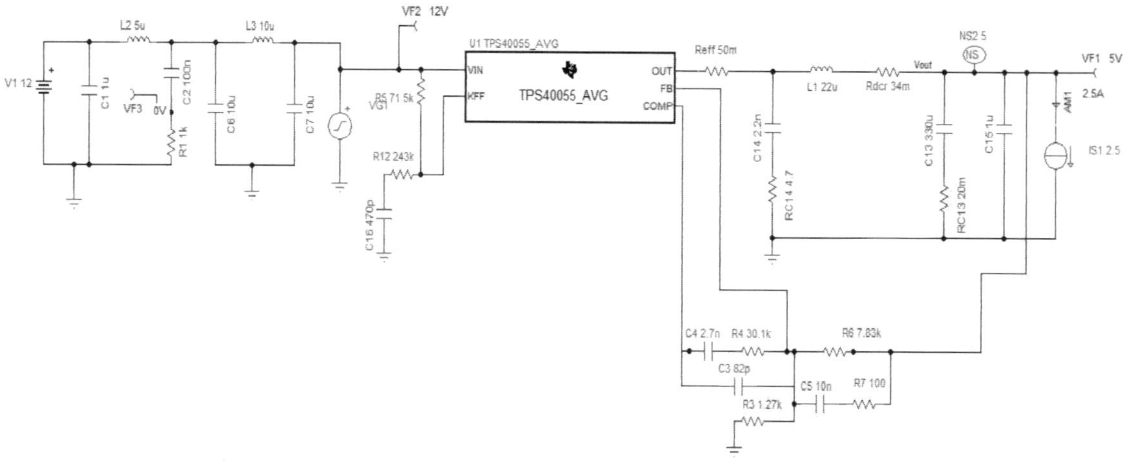

Figure 1. Buck Converter with EMI Filter

This technique will be useful when the total lines constituting CM path is more than one and also in asymmetric conduction paths. Fig.1. shows the EMI filter placed in a buck convertor. Taking into consideration the electromagnetic compatibility with other devices, asymmetry should be avoided as it paves way to coupling between CM and DM. Hence without prior knowledge on the working of DM-CM couplings and CM sources, the optimal methodology is decoupling.

The voltage drop across the filters can be represented by using the expressions:

$$\begin{bmatrix} v_{L+} \\ v_{L-} \end{bmatrix} = QL_f \begin{bmatrix} 1-k_f & 0 \\ 0 & k_f \end{bmatrix} \begin{bmatrix} i_+ \\ i_- \end{bmatrix} \quad (7)$$

Similarly, the decomposition matrices for DM/CM can be described as

$$T_2' = \begin{bmatrix} 1 & -1 \\ \frac{1}{2} & \frac{1}{2} \end{bmatrix} \quad (8)$$

Using this methodology of DM/CM decomposition and equivalent modelling, the complete description of the synchronous buck convertor is modeled as shown in Figure 1.

3.2. Filter Design Process

Electric vehicles, electric aircrafts, shipboard integrated power and energy systems are some of the transportation mechanisms that require low conducted and radiated EMI along with high power quality. These requirements have to strictly adhere to the EMI standards enforced on a global scale. Similarly, the attainment of specific power and maximum power density remains an important objective due to limitations in weight/space. Since most of these applications will have the power converter system playing a crucial part, optimization takes place at the system level. This is possible by maximizing specific cost, minimizing failure rate and maximizing efficiency.

In case of WBG devices like GaN or SiC MOSFETS have numerous opportunities that require power density increase in order to meet their requirements. Hence there is a subsequent increase in specific cost, failure rate, efficiency, specific power and power density in a virtual prototyping or multi-objective optimization process.

The optimization process takes place in the following order:

- Identification of filter structure for CM and DM
- Development of transfer function for insertion loss because of the noise sources (EMI)
- Design and identification of filter, capacitor, winding techniques, geometrics, core materials, choice of resistance etc.
- Formulate necessary constraints and metrics
- Use optimization and ensure fitness function

4 Results

In the absence of the EMI filter, the DC-DC buck converter designed is simulated in an LISN environment using LT Spice. Figure 2 represents the signal to noise ratio (SNR) is measured in dB and is found to be decreasing in a linear fashion.

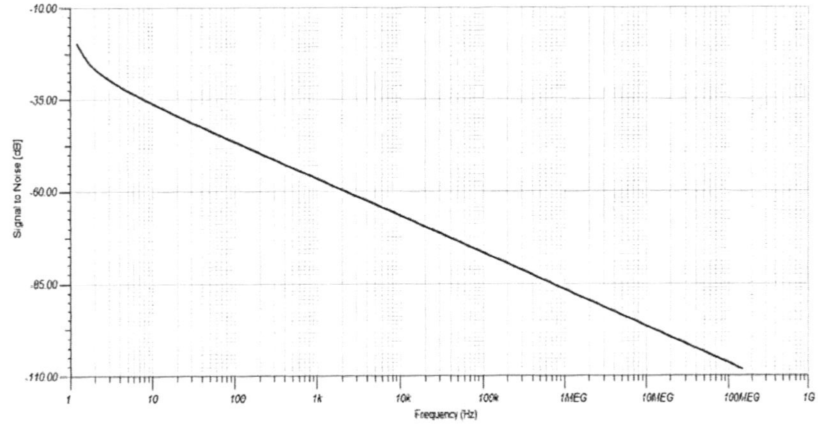

Figure 2. Signal to Noise Ratio (SNR) in absence of EMI filter

On the other hand, the same circuit is modified with the inclusion of the EMI filter and the SNR thus measured is found to be decreasing in the exponential fashion as shown in Figure 3.

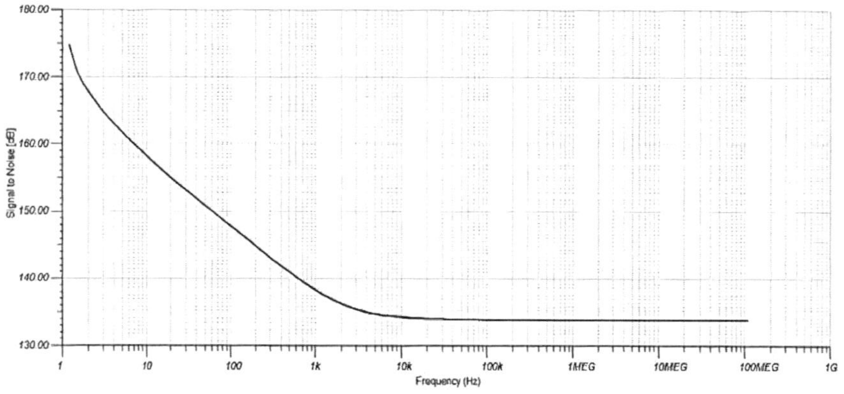

Figure 3. SNR in the presence of an EMI Filter

Figure 4.shows the phase and gain of the proposed filter within the proposed frequency range of 10-100kHz.

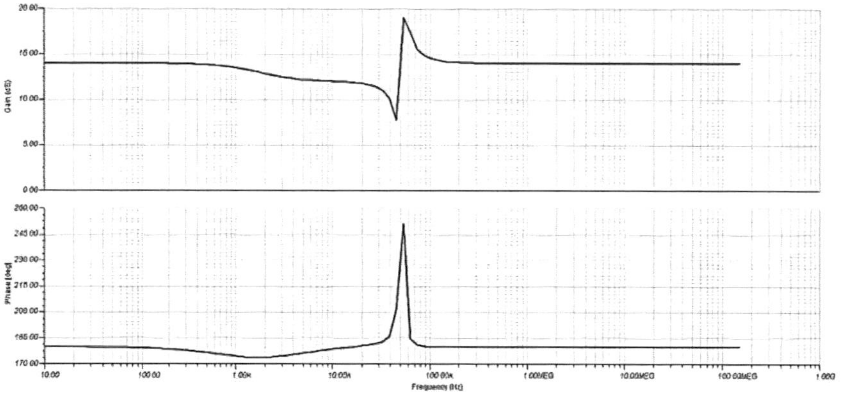

Figure 4. Phase and Gain of the Buck Converter

5 Conclusion

The proposed DC-DC buck converters are an important part in the electrification of aircraft, bio-medical and automotive devices. In this proposed methodology, an EMI filter is designed and an equivalent modeling approach is followed to eliminate unwanted noise or distortion. The model is implemented and the output of the buck converter in the presence and absence of the proposed EMI filter is analysed. It is found that the in the presence of the filter, there is noticeable decrease in the interference due to external EMI field and objects.

References

[1] Marinakis, F.L. Luo, Hong Ye2003Power Electronics and Drive Systems*Investigation of EMI, EMS and EMC in power DC/DC Converters*vol 1.

[2] D. R. Anita Shirley, "Systematic diagnosis of power switches," *2014 International Conference on Embedded Systems (ICES)*, 2014, pp. 32-34, doi: 10.1109/EmbeddedSys.2014.6953045.

[3] M.H. Nagrial, A. Hellany 1999Electromagnetic Compatibility*EMI/EMC issues in switch mode power supplies (SMPS)* p 180 – 185.

[4] C. Kraft 2000Telecommunications Energy Special*Conducted noise from 48 volt DC-DC*

[5] *converters used in telecommunications systems and its mitigation for EMC*p327 - 331 .

[6] Bhargava A, Pommerenke D, Kam K.W, Centola F, Lam C.W 2011IEEE Trans. Electromagn. Compat *DC-DC buck converter EMI reduction using PCB layout modification* p 53, 806–813.

[7] Tseng C.J, Chen C.L 1998 IEEE Trans. Ind. Electron*A passive lossless snubber cell for nonisolated PWM DC/DC converters* p 45, 593–601.

[8] Chu Y, Wang S, Wang Q 2016IEEE Trans. Power Electron*Modeling and stability analysis of active/hybrid common-mode EMI filters for DC/DC power converters* p 31, 6254–6263.

[9] Pareschi F, Rovatti R, Setti G 2015IEEE Access*EMI reduction via spread-spectrum in DC/DC converters: State of the art, optimization, and tradeoffs* p 3, 2857–2874.

[10] Lin F, Chen D.Y 1994IEEE Trans. Power Electron*Reduction of power supply EMI emission by switching frequency modulation* p 9, 132–137.

[11] Hardin K.B, Fessler J.T, Bush R August 1994In Proceedings of the IEEE Symposium on Electromagnetic Compatibility(Chicago, IL, USA)*Spread spectrum clock generation for the reduction of radiated emissionsI* p 22–26.

[12] A. Rosales, A. Sarikhani, and O. A. Mohammed May 2011IEEE Trans.Magn*Evaluation of radiated electromagnetic field interference due to frequency switching in PWM motor drives by 3D finite elements* vol 47 no 5 p 1474–1477.

[13] T. H. Hubing, J. F. KaufmanFeb. 1989IEEE Trans. Electromagn. Compat*Modeling the electromagnetic radiation from electrically small table-top products* vol 31 no 1 p 74–84.

[14] Shirley, D.R.A., Sundari, V.K., Sheeba, T.B., Rani, S.S. (2021). Analysis of IoT-Enabled Intelligent Detection and Prevention System for Drunken and Juvenile Drive Classification. In: Kathiresh, M., Neelaveni, R. (eds) Automotive Embedded Systems. EAI/Springer Innovations in Communication and Computing. Springer, Cham. https://doi.org/10.1007/978-3-030-59897-6_10

[15] D. M. Hockanson, J. L. Drewniak, T. H. Hubing, T. P. Van Doren, F. Sha,M. J. Wilhelm Nov 1996IEEE Trans. Electromagn. Compat*Investigation of fundamental EMI source mechanisms driving common-mode radiation from printed circuit boards with attached cables*vol 38 no 4 p 557–566.

[16] H. Shim T. H. Hubing Nov 2005IEEE Trans. Electromagn. Compat*Model for estimating radiated emissions from a printed circuit board with attached cables due to voltage-driven sources* vol 47 no 4 p 899–907.

[17] M. Senthilkumar, D.Ruth Anita Shirley *et al* 2021 *IOP Conf. Ser.: Mater. Sci. Eng.* **1084** 012051

[18] H. Ke, K. Morishita, T. Hubing, N. Kobayashi, T. *Harada*May 2009IEEE Trans. Electromagn. Compat*Modeling radiated emissions due to power bus noise from circuit boards with attached cables* vol 51 no 2 p 412–416.

[19] S. Deng, T. H. Hubing, D. G. Beetner May 2008IEEE Trans. Electromagn. Compat*Using TEM cell measurements to estimate the maximum radiation from PCBs with cables due to magnetic field coupling* vol 50 no 2 p 419–423.

[20] X. Tong, D. W. P. Thomas, A. Nothofer, P. Sewell, C. Christopoulos May 2010IEEE Trans. Electromagn. Compat*Modeling electromagnetic emissions from printed circuit boards in closed environments using equivalent dipoles* vol 52 no 2 p. 462–470.

[21] H. Weng, D. G. Beetner, R. E. DuBroff , Nov 2011IEEE Trans. Electromagn. Compat Prediction of radiated emissions using near-field measurementsvol 53 no 4 p 891–899.

[22] Tenginakai, Pruthvi, V. Keerthana, Sowmini Srinath, Fauzan Syed, and P. Parimala. "Meanderline Pattern Wearable Textile Antenna for Position Identification in Military

Applications." In Intelligent Data Communication Technologies and Internet of Things, pp. 847-860. Springer, Singapore, 2021.

[23] Vivekanandam, B. "Nakagami-m Fading Detection with Eigen Value Spectrum Algorithms." Journal of Electronics and Informatics 3, no. 2 (2021): 138-149.

Comparative Analysis of Single-Phase-Clocked Low Power Flip-Flops with Transmission-Gate Flip-Flop

Akhlesh Gupta [1], Ashwani K. Rana [2]

[1,2] Department of Electronics and Communication Engineering, National Institute of Technology, Hamirpur, 177005, India.

Corresponding author's e-mail: akgolugupta@gmail.com

Abstract. Flip-flops are fundamental building block in designing the digital circuits. They contribute significant amount of power consumption in the digital systems. In the modern era of VLSI design, power consumption has become crucial factor. As the technology is scaled down below 65nm, leakage power contributes significantly to overall power consumption of digital circuits. Since, Flip-flop is a fundamental building block in designing digital systems, reducing the power consumption of Flip-flop is required to achieve low power consumption of digital systems. Recently single-phase clocking technique has been employed in designing FFs to achieve low power consumption. This technique reduces excess loading on the clock signal. In this article performance of recent single-phase clocked FFs is compared with the conventional transmission-gate Flip-flop. Also, drawbacks of these FFs are discussed. For the comparison of performance of single-phase-clocked FFs with conventional transmission-gate FF, all the simulations are performed at 45-nm predictive technology on Tanner EDA tool.

Keyword. Flip-Flop (FF), Near threshold voltage operation (NTV), Subthreshold, Contention, Low power.

1. Introduction

The continuous growth of process technology has pushed the performance of digital systems extensively and enhanced the integration of digital systems enormously. However, with the increased integration of digital circuits, there arises a critical issue of power consumption. In the recent years, the area of Internet of Things (IoT) has been growing progressively and IoT devices such as smart watch, smart health monitoring devices, smoke alarm systems etc. have become an integral part of daily lives [1,2]. These IoT devices require low power consumption [3–6]. FFs, being the essential building block of digital systems consume a large portion of total power [7]. Hence designing FFs which consume low power can greatly reduce the overall power of digital systems.

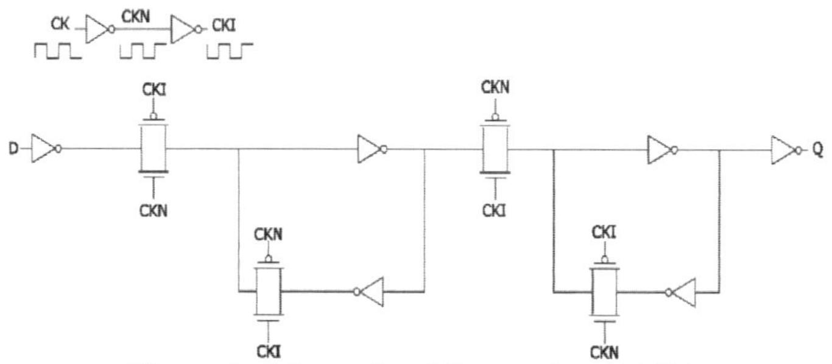

Figure 1: Schematic of Conventional TGFF

QPSK modulation technique can be used to transmit digital information. In this technique pseudo-noise sequence generator, bit splitter and frequency divider are employed. FF is the key component in designing QPSK modulator [8].

As, power follows the square law proportionality with respect to voltage, one effective method to reduce the power consumption of the large integrated devices is lowering the supply voltage [9–13]. But this technique degrades the performance of digital systems. As we scale down the supply voltage, threshold voltage should also be scaled accordingly to maintain the performance of the system. But lowering the threshold voltage causes increase in the leakage power consumption. Also, to avail the power reduction advantage of this technique, FF should be designed as such that it is operable under super-threshold as well as near/subthreshold conditions.

Currently most favoured FF used in the modern digital systems is transmission-gate FF (TGFF) as it is completely static, without any contention and the design works fairly at low voltages. The main issue with this design is excess loading on the clock network. As it can be seen from fig. 1, nodes CKN and CKI will toggle irrespective of the applied data, and 8 transistors are driven by these nodes. Because of toggling of these two nodes, excessive power is wasted even if the data activity drops to 0%.

In the recent years, single-phase clocking technique has been employed to design the low power FFs to reduce the excess loading on the clock signal. In single-phase clocking scheme every clock signal is routed on a single wire. Background study on FFs is given in section 2 [14–16]. Many designs of low power FFs has been presented [17–20]. In this paper, evaluation of these FFs is done and compared with conventional transmission-gate FF. Comparison of power consumption of these FFs is evaluated with respect to supply voltage as well as at different switching activity of data. Also, Clock-to-Q delay variation with respect to supply voltage is done to evaluate the performance in comparison of transmission-gate FF.

Section 2 gives the background study on FFs. Section 3 gives detailed description of recent single-phase clocked low power FFs with their drawbacks. Section 4 describes the simulation results of these FFs and comparison of various performance parameters is done with respect transmission-gate FF. In the end, a conclusion with application of FF is drawn in section 5.

2. Background study

Shandilya R. and Sharma R.K. [14] have proposed a dual-edge triggered FF. In the presented design, to reduce the switching activity of data, explicit pulse triggering scheme is employed. Since most of the power dissipated in the FF is caused by switching activity of data. By reducing the switching activity of data, overall power consumption is reduced in the proposed design. When input data is in ON for a long period of time, excess amount of switching activity is removed by managing the discharge path. This scheme is called as conditional discharge scheme. The proposed design reduces significant amount of power consumption.

Moreau L., Dekimpe R. & Bol D. [15] have proposed true single-phase clocked retentive flip-flop employing clock gating scheme. In this article a feedback loop is introduce for the retention of data. The feedback loop is managed by clock gating scheme. Clock gating scheme enables the retention mode. The proposed design achieves 62 % power reduction in comparison of conventional transmission-gate FF (TGFF). Main disadvantage in this design is area overhead which is caused by clock gating scheme.

Lin J., Sheu J., Hwang Y., Wong C. and Tsai M. [16] have presented a true single-phase clocked flip-flop. The presented design in this article is obtained by utilizing the logic structure reduction technique. 19 transistors are used to implement the proposed design. During the operation of FF, no floating nodes are present which reduces the leakage power consumption. High power reduction is obtained in this design. Delay of the proposed FF is also improved.

3. Evaluation of various single-phase-clocked FFs

In this segment different prior-art single-phase clocked FFs are discussed. Also, their drawbacks are explained. Also, their suitability at low voltages is described. All of the FFs discussed in this section use single-phase clocking scheme to optimize the number of clock dependent transistor in comparison of TGFF.

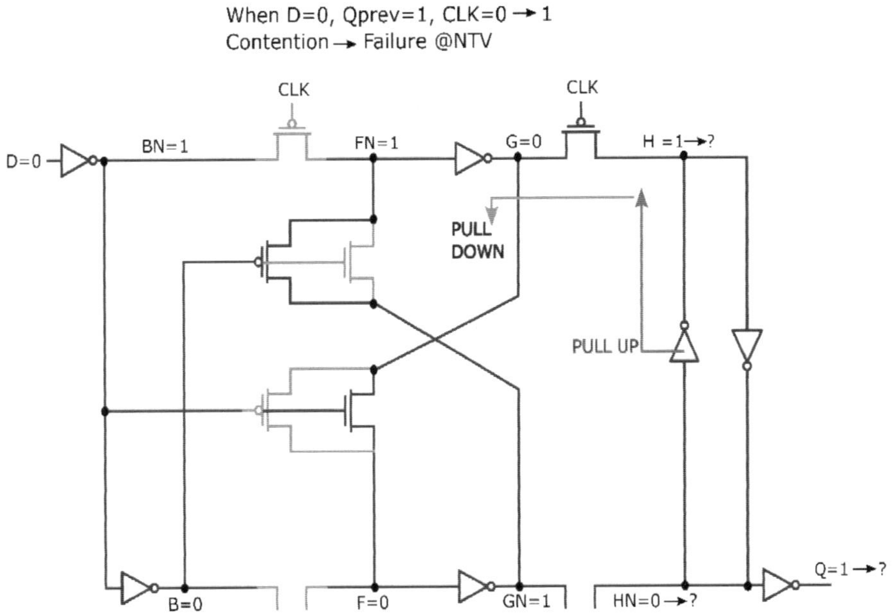

Figure 2: Schematic of ACFF showing contention issue

3.1 Adaptive-coupling Flip-flop (ACFF)

In adaptive-coupling FF [17], a new scheme known as adaptive-coupling is used to minimize the number of clock dependent transistors in comparison of TGFF. In this FF the coupling element is designed by the parallel combination of a PMOS and NMOS. Gates of both PMOS and NMOS are operated by the same signal. This technique makes it suitable to achieve better tolerance with respect to process variation. However, ACFF suffers from strong current contention in case of slave latch operation [fig. 2]. This problem restricts its use at NTV. To overcome the contention issue in the ACFF and for proper operation at NTV, size of transistors in the slave latch needs to be increased and also extra load is required.

3.2 Topologically-compressed Flip-flop

Most of the power in an FF is consumed by those transistors which are dependent on the clock. To minimize the power consumption while maintaining the performance, number of clock dependent transistors are reduced in this FF. As depicted in fig. 1, there are 8 clock dependent transistors. To eliminate the clock operated transistors directly from this circuit is quite tedious task. One of the reasons is that transmission gates require 2-phase clock, that's why the transistor driving the clock signal can not be removed. Hence, a combinational design is taken as the base circuit and topological compression scheme is employed to obtain the TCFF. To obtain the TCFF as depicted in fig. 3, logically equivalent transistors are merged and redundant transistors are removed [18]. Merging of transistors is shown in from fig. 4 to fig. 6. It has only three clock dependent transistors. But contention problem exists in this design as well, restricting its use for low supply voltage operation. To eliminate the contention issue in TCFF, size of transistors is increased. Also, due to sharing of many transistors, voltage level of internal nodes decreases greatly i.e., some internal nodes do charge completely up to V_{dd} or discharge to ground, which in turn affect the robustness of the device.

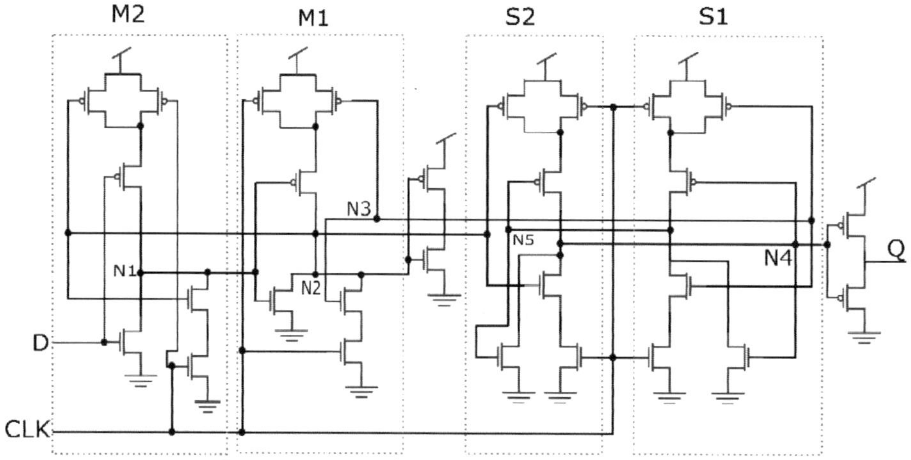

Figure 3: Transistor level schematic TCFF

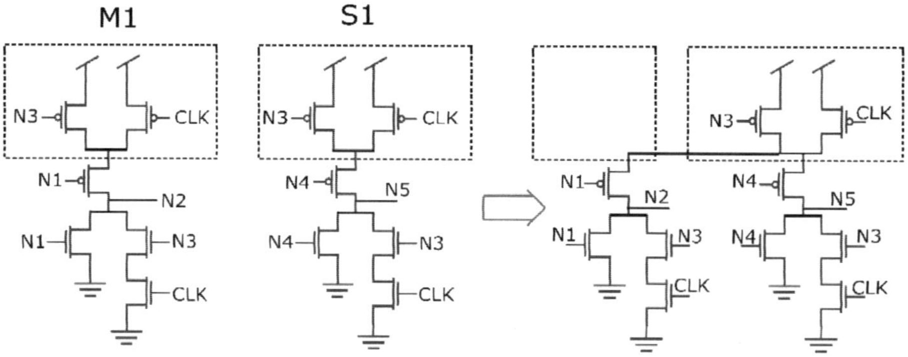

Figure 4: Merging of Transistor in PMOS side in fig. 3

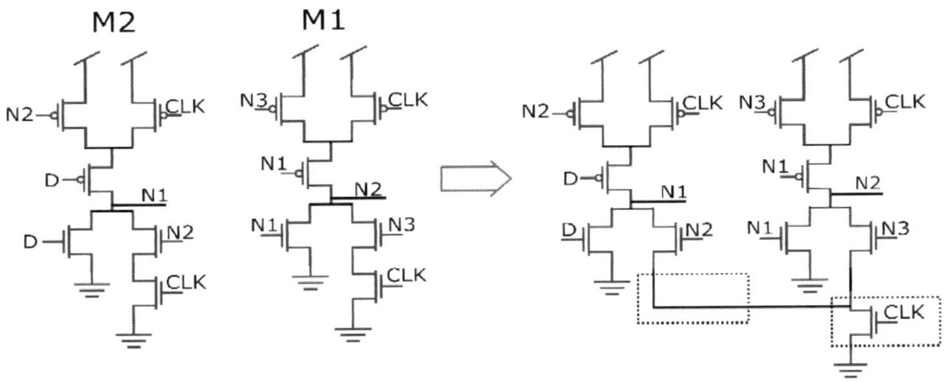

Figure 5: Merging of Transistor in NMOS side in fig. 3

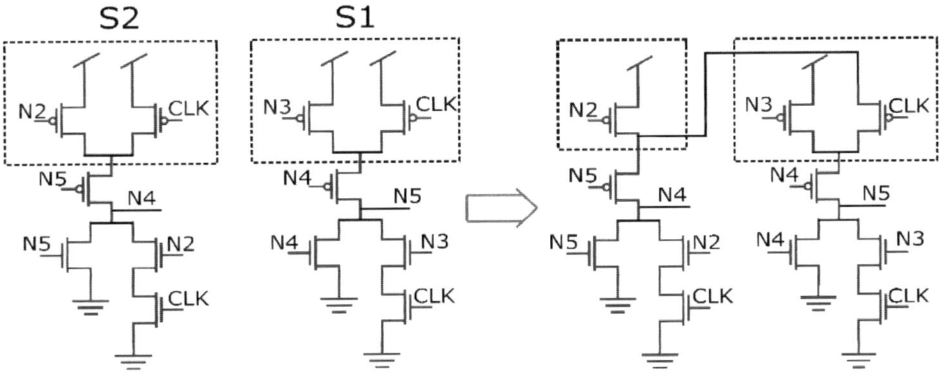

Figure 6: Further Merging of Transistor in PMOS side in fig. 3

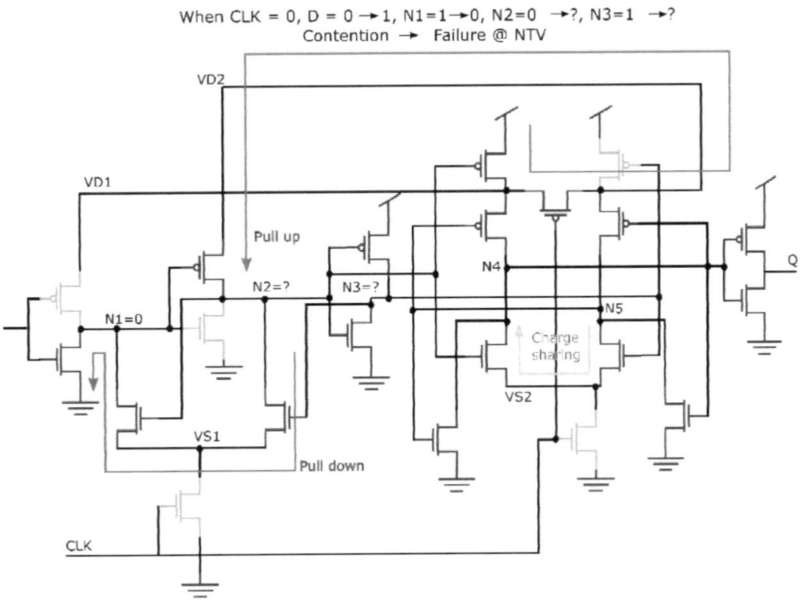

Figure 7: Optimized Transistor level schematic of TCFF

3.3 Static single-phase contention-free flip-flop

Although, significant power reduction is achieved in both ACFF and TCFF but their major shortcoming is that they fail to function at NTV (near-threshold voltage) because of the existence of contention issue in both FFs. For an FF to function at wide range of supply voltage and to achieve less power consumption following criteria should be taken care of while designing the FFs.

- Circuit design should be fully static as dynamic elements degrades the performance because of their susceptibility towards PVT changes.
- There should be no contention in the design because contention causes the FFs to fail to function fairly at low voltages.
- Single-phase clocking technique should be employed which reduces the clock dependent transistors as clock dependent transistors contribute the significant power consumption.
- While designing the FF, area overhead should be taken care. It should be least in comparison of TGFF.
- To overcome the contention issue occurred in ACFF and TCFF a static single-phase contention-free flip-flop (S^2CFF) is presented in [19]. This FF is completely contention free and reliable at NTV. However, main shortcoming of this design is, unnecessary power is consumed due to precharging and discharging of internal node N2. As shown in fig. 4, When input signal is 0 for a long period, internal node N2 is precharged to VDD by T7 when clock is 0 and is discharged to GND by T8 and T10 when clock signal is rising to 1 [fig. 4]. However, output state is not affected by this precharging and discharging, that's why this operation is unnecessary and wastes excess amount of energy.

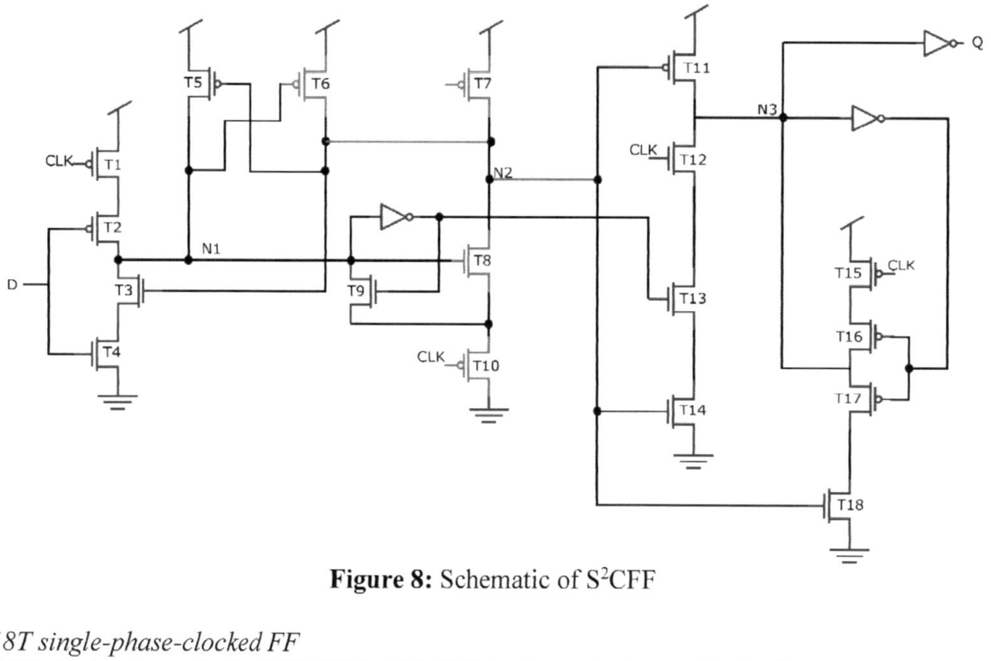

Figure 8: Schematic of S^2CFF

3.4 18T single-phase-clocked FF

The 18T single-phase-clocked FF (SPC-18T) [20] is shown in fig 5. This is also a combinational-type FF similar to TCFF. In this FF, shared transistors count is reduced in comparison of TCFF. But this FF is also not completely redundant free. In the fig. 5, it is observed that node N1 precharges and discharges unnecessarily which in turn increase the power consumption of SPC-18T FF. Also, there is contention problem in this FF. Considering Q = 1 previously, when D = 0 and CLK = 0, node N1 and N2 is charged to VDD. Now when CLK changes from 0 to 1, transistors T5, T2 and T14 may pull down the voltage of node N2. This causes contention issue for output Q. Contention issue in SPC-18T restricts its use at NTV. Although power consumption of SPC-18T is less in comparison of S^2CFF but still SPC-18T suffers from contention issues.

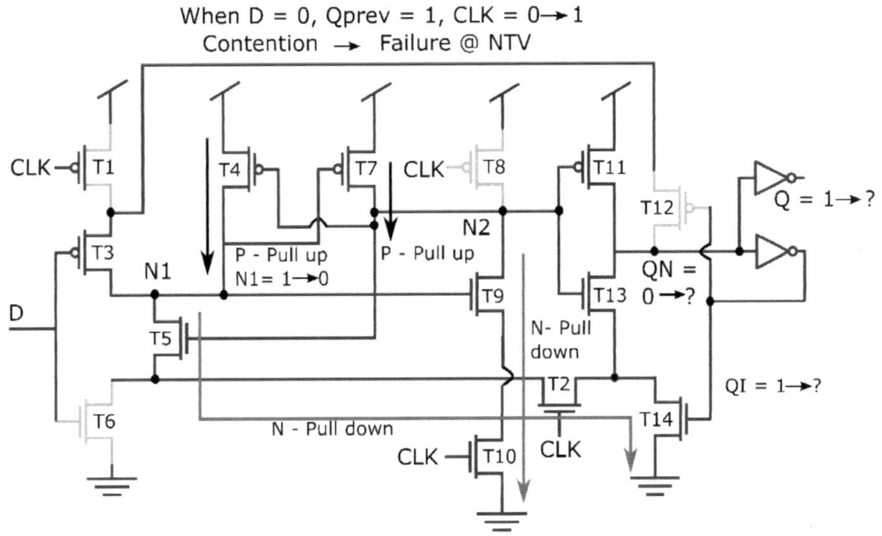

Figure 9: Schematic of SPC-18T

4. Simulation results and comparative analysis

For the evaluation and comparative analysis of single-phase clocked low power FFs (ACFF, TCFF, S^2CFF and SPC-18T) with TGFF, all the simulations are performed on 45nm predictive technology model using Tanner EDA tool. For the fair comparison width of NMOS and PMOS are taken same in the simulation of each FF described in section 3. The comparative analysis and performance evaluation of these FFs are as follows.

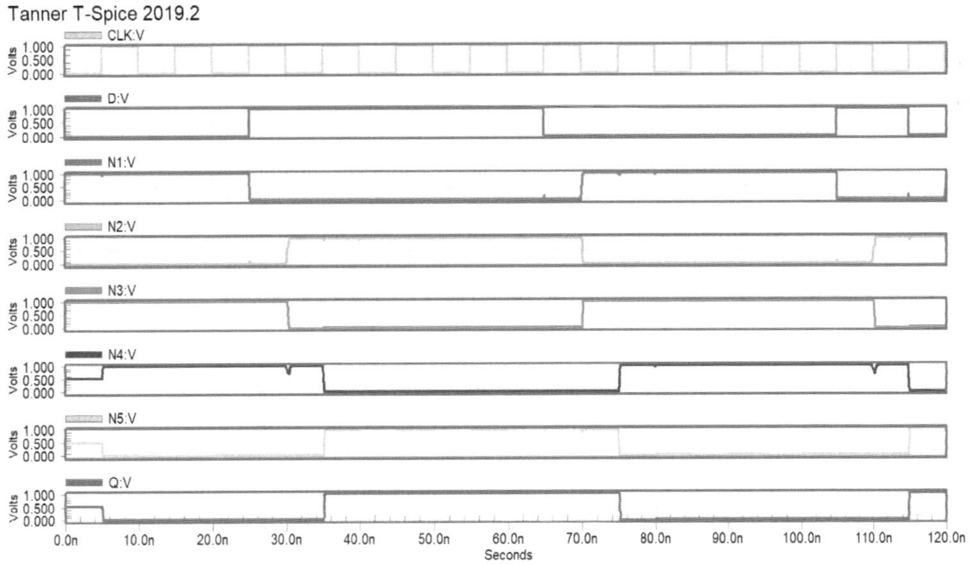

Figure 10: Transient waveform of TCFF at 1V

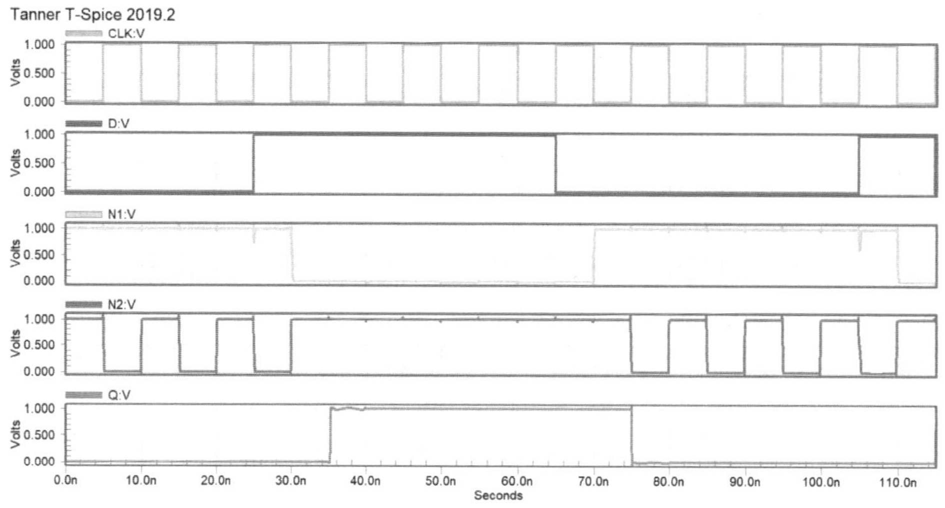

Figure 11: Transient waveform of S²CFF at 1V

Fig. 10 and fig. 11 shows the transient waveform of TCFF and S²CFF at 1V simulate on Tanner EDA tool. Frequency of clock signal (CLK) is taken as 100 MHz. Similarly, simulation of TGFF, ACFF and SPC-18T is done at 1V with clock frequency of 100 MHz on Tanner EDA tool for the comparative analysis.

In the fig. 12, comparison of power of various single-phase clocked low power FFs along with the power consumption of TGFF is demonstrated at different supply voltages ranging from 0.6 to 1 V. Frequency of clock signal is taken as 100 MHz. Switching activity of data is taken 10%. It is evident from the fig. 12 that power consumption of the ACFF is least among all low power FFs in comparison of TGFF. But major issue in ACFF is contention in the slave part which restricts its use for NTV operations. TCFF and SPC-18T also consists of contention problem. Hence, they are also not suitable at low voltages. S²CFF consume more power than other three single-phase low power FFs due to redundant transition of node N2 (fig. 8) but the advantage of S²CFF is that it works fairly at low voltages. Hence, S²CFF can be used in place of conventional transmission-gate FF due to its proper working at NTV to achieve significant power reduction.

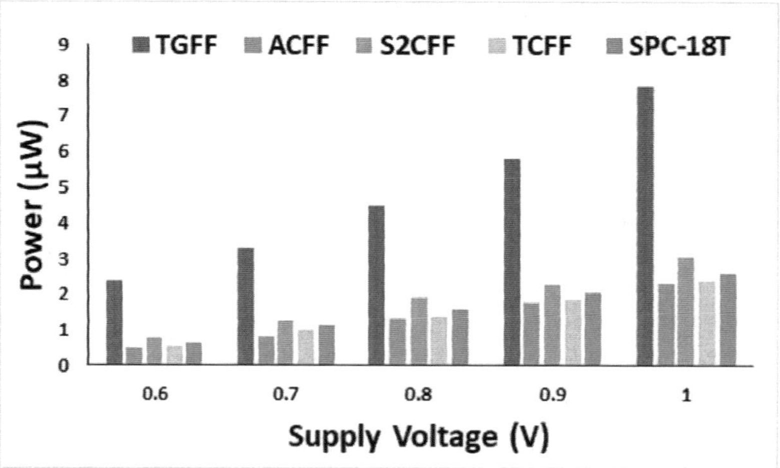

Figure 12 Simulation result of Power at 1 V

In fig. 13, comparison of Clock-to-Q delay is demonstrated at different supply voltages ranging from 0.6 to 1 V. Clock-to-Q delay of ACFF is least among all the single-phase clocked low power FFs. But, again due to contention problem it's use is restricted at lower voltages. Clock-to-Q delay of S^2CFF slightly larger than ACFF but less than TCFF and SPC-18T. As S^2CFF is static and there is no contention, if this FF is replaced with TGFF, 15.11% reduction in Clock-to-Q delay is also achieved.

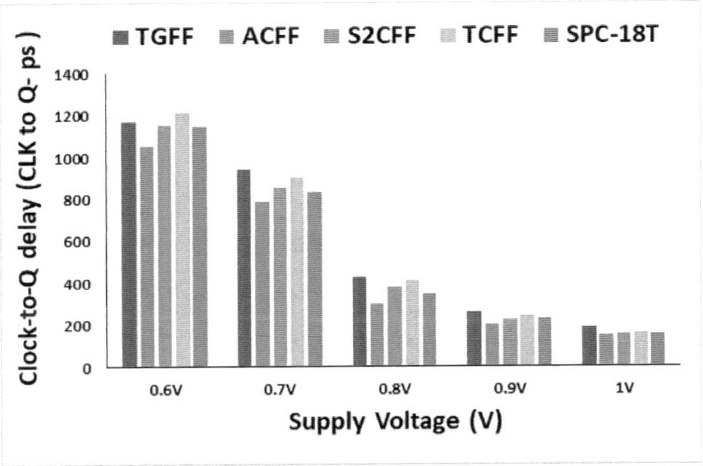

Figure 13 Variation of CLK-to-Q delay of transmission-gate FF and single-phase clocked low power FFS with varying voltage

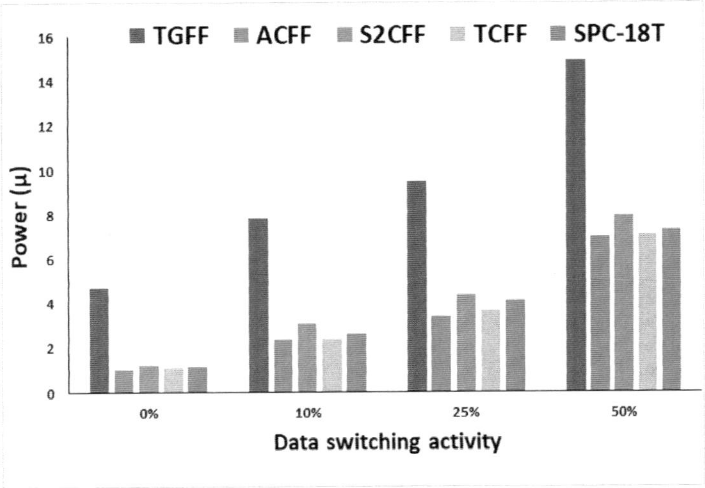

Figure 14 Variation of CLK-to-Q delay of transmission-gate FF and single-phase clocked low power FFS with varying voltage

In Fig. 14, power of TGFF and various low power FFs is compared with varying switching activity of data at supply voltage of 1 V. Frequency of clock signal is taken as 100 MHz Generally, in integrated devices, switching activity varies between 10% to 15%. It is clear from fig. 14 that ACFF consumes least amount of power among all low power FFs analysed with respect to TGFF. But the performance of ACFF is degraded at lower voltages due to contention issue. Power consumption of S^2CFF is highest among all four low-power FFs but main advantage is, there is no contention in S^2CFF. S^2CFF reduces the power consumption up to 60.61% in comparison of TGFF when switching activity of data drops to 0%. However, in TGFF, internal nodes CKN and CKI still toggles even if switching activity of data is 0% which leads to wastage of power consumption. For TCFF, no internal node has unnecessary precharge operations and its energy efficiency is high when switching activity of data is 0. But main issue in these FFs (ACFF, TCFF and SPC-18T) is existence of current contention. Also, in TCFF many transistors are shared. Because of that power consumption of these three FFs increases rapidly with switching activity of data.

Comparative analysis of TGFF FF with various single-phase low power FFs is shown in Table 1. ACFF exhibit least power consumption but it faces contention problem in slave part because of that it does not function well at low voltage. Contention issue is faced in TCFF and SPC-18T. Also, SPC-18T is not completely redundant free. Because extra power is wasted in the form unnecessary transition of internal node. In TCFF another issue is that many transistors are shared due to that voltage level is dropped at low voltage operation. This sharing of transistors degrades the performance and robustness of TCFF at low voltage operation. S^2CFF consume a little extra amount of power as compared to ACFF but its major advantage is that S^2CFF function fairly at low voltages. Hence, S^2CFF is best alternative among all four single-phase low power FFs analysed to replace TGFF in large integrated devices.

$PDP_{CLK\text{-}to\text{-}Q}$ of all the FFs is evaluated and compared. Table 1 shows the power-delay-product of TGFF and various single-phase low power FFs. Thus, from the above discussion it can be concluded that S^2CFF is most favourable single-phase low power FF to achieve energy efficient, contention-free operation among all four FFs (ACFF, TCFF, S^2CFF, SPC-18T) and can be used in integrated devices as an alternative of TGFF.

Table 1. Performance Comparison between various single-phase clocked FFs and TGFF

	TGFF conventional	ACFF ISSC	TCFF JSSC 2014	S^2CFF ISSCC 2014	SPC-18T JSSC 2019
Technology-Reported	--	40nm	40nm	45nm	65nm
Single-Phase clock	No	Yes	Yes	Yes	Yes
Operation @NTV	Yes	No	No	Yes	No
Contention	No	Yes	Yes	No	Yes
Power (μW) @ 1V, 100 MHz, 10% data Switching activity	7.82	2.32	2.4	3.08	2.59
Power (μW) @ 0.6V, 100 MHz, 10% data Switching activity	2.38	0.54	0.94	0.48	0.62
$Delay_{CK\text{-}to\text{-}Q}$ (ps) @ 1V	187.05	151.2	165.65	158.77	160.05
$PDP_{CK\text{-}to\text{-}Q}$ (ps) @ 1V	1.46	0.35	0.39	0.48	0.41

5. Conclusion

This article presents comparative analysis of TGFF and various single-phase clocked low power FFs. In the presented work, various parameters such as power and Clock-to-Q delay are evaluated and compared. Also, the limitations of low power FFs and transmission gate FF are analysed. By performing simulations on 45nm predictive technology, it was found that power consumption in ACFF is least but it does not function properly at low voltage due to contention issue. In TCFF and SPC-18T have also contention issue. Hence, they are also not good design for low voltage operation. In among all the four FFs (ACFF, TCFF, S^2CFF, SPC-18T) analysed, S^2CFF is most favourable design at both superthreshold as well as near/subthreshold condition. Also, S^2CFF is completely static. Hence, among all four FFs, S^2CFF can be used as an alternate design in integrated devices in place of TGFF. Only shortcoming of S^2CFF is, unnecessary precharging and discharging of internal node N2. Because of this unnecessary transition extra power is wasted. This drawback can be removed by employing the data-aware scheme in the precharging path of node N2. By employing this technique, power consumption in the S^2CFF can be further reduced. Some of the most important applications of FFs are counters, storage registers, shift registers, frequency dividers, data synchronizers.

References

[1] Atzori L, Iera A and Morabito G 2010 The Internet of Things: A survey *Comput. Networks* **54** 2787–805

[2] Chang K-L, Chang J S, Gwee B-H and Chong K-S 2013 Synchronous-logic and asynchronous-logic 8051 microcontroller cores for realizing the internet of things: A comparative study on dynamic voltage scaling and variation effects *IEEE J. Emerg. Sel. Top. circuits Syst.* **3** 23–34

[3] Tekeste T, Saleh H, Mohammad B, Khandoker A and Ismail M 2017 A nano-watt ECG feature extraction engine in 65-nm technology *IEEE Trans. Circuits Syst. II Express Briefs* **65** 1099–103

[4] Tekeste T, Saleh H, Mohammad B and Ismail M 2018 Ultra-low power QRS detection and ECG compression architecture for IoT healthcare devices *IEEE Trans. Circuits Syst. I Regul. Pap.* **66** 669–79

[5] Pullini A, Rossi D, Loi I, Tagliavini G and Benini L 2019 Mr. Wolf: An energy-precision scalable parallel ultra low power SoC for IoT edge processing *IEEE J. Solid-State Circuits* **54** 1970–81

[6] Cerqueira J P, Repetti T J, Pu Y, Priyadarshi S, Kim M A and Seok M 2020 Catena: A near-Threshold, sub-0.4-mw, 16-core programmable spatial array accelerator for the ultralow-power mobile and embedded internet of things *IEEE J. Solid-State Circuits* **55** 2270–84

[7] Shin J L, Golla R, Li H, Dash S, Choi Y, Smith A, Sathianathan H, Joshi M, Park H, Elgebaly M and others 2012 The next generation 64b SPARC core in a T4 SoC processor *IEEE J. Solid-State Circuits* **48** 82–90

[8] Rameshkumar R and Swaminathan J N 2019 A New Hardware Design and Implementation of QPSK Modulator in Normal and 4-QAM Mode for 5G Communication Networks *International Conference on Inventive Computation Technologies* pp 255–62

[9] Andrei A, Eles P, Jovanovic O, Schmitz M, Ogniewski J and Peng Z 2009 Quasi-static voltage scaling for energy minimization with time constraints *IEEE Trans. very large scale Integr. Syst.* **19** 10–23

[10] Zhang X, Boussaid F and Bermak A 2014 32bit x 32bit Multiprecision Razor-Based Dynamic Voltage Scaling Multiplier With Operands Scheduler *IEEE Trans. Very Large Scale Integr. Syst.* **22** 759–70

[11] Tsou W-J, Yang W-H, Lin J-H, Chen H, Chen K-H, Wey C-L, Lin Y-H, Lin S-R and Tsai T-Y 2017 Digital low-dropout regulator with anti PVT-variation technique for dynamic voltage scaling and adaptive voltage scaling multicore processor *2017 IEEE International Solid-State*

Circuits Conference (ISSCC) pp 338–9

[12] Jain S, Lin L and Alioto M 2017 Dynamically adaptable pipeline for energy-efficient microarchitectures under wide voltage scaling *IEEE J. Solid-State Circuits* **53** 632–41

[13] Lee J, Member S, Zhang Y, Dong Q, Lim W, Saligane M, Kim Y, Jeong S, Member S, Lim J, Member S, Yasuda M, Miyoshi S, Kawaminami M, Blaauw D and Sylvester D 2020 A Self-Tuning IoT Processor Using Leakage-Ratio Measurement for Energy-Optimal Operation **55** 87–97

[14] Shandilya R and Sharma R K 2018 High Speed Low Power Dual-Edge Triggered D flip-flop *Proc. 2017 Int. Conf. Intell. Comput. Control. I2C2 2017* **2018-Janua** 1–5

[15] Cells C 2019 A 0 . 4V 0 . 5fJ / cycle TSPC Flip-Flop in 65nm LP CMOS with Retention Mode Controlled by 6–9

[16] Lin J F, Sheu M H, Hwang Y T, Wong C S and Tsai M Y 2017 Low-Power 19-Transistor True Single-Phase Clocking Flip-Flop Design Based on Logic Structure Reduction Schemes *IEEE Trans. Very Large Scale Integr. Syst.* **25** 3033–44

[17] Teh C K, Fujita T, Hara H and Hamada M 2011 A 77% energy-saving 22-transistor single-phase-clocking D-flip-flop with adaptive-coupling configuration in 40nm CMOS *2011 IEEE Int. Solid-State Circuits Conf.* **39** 338–40

[18] Kawai N, Takayama S, Masumi J, Kikuchi N, Itoh Y, Ogawa K, Ugawa A, Suzuki H and Tanaka Y 2014 A fully static topologically-compressed 21-transistor flip-flop with 75% power saving *IEEE J. Solid-State Circuits* **49** 2526–33

[19] Kim Y, Jung W, Lee I, Dong Q, Henry M, Sylvester D and Blaauw D 2014 A static contention-free single-phase-clocked 24T flip-flop in 45nm for low-power applications *Digest of Technical Papers - IEEE International Solid-State Circuits Conference* vol 57 (IEEE) pp 466–7

[20] Cai Y, Savanth A, Prabhat P, Myers J, Weddell A S and Kazmierski T J 2019 Ultra-low power 18-transistor fully static contention-free single-phase clocked flip-flop in 65-nm CMOS *IEEE J. Solid-State Circuits* **54** 550–9

Fe_3O_4/TiO_2/GRAPHENE HYBRID NANOCOMPOSITE TO IMPROVE THE LIFESPAN OF DISTRIBUTION TRANSFORMERS

Saravanakumar K[1], J.Samson Issac[2] , J.Dhanaselvam[3] , R.Rajesh[4], Ajith.B.Singh[5], K.Geetha[6]

[1, 3, 5]Assistant Professor, Department of Instrumentation and Control Engineering, Karunya Institute of Technology and Sciences, Coimbatore, India.

[2]Assistant Professor, Department of Biomedical Engineering, Karunya Institute of Technology and Sciences, Coimbatore, India.

[4]Research Scholar, IIT Madras, Chennai, India.

[6]Undergraduate Student, Department of Instrumentation and Control Engineering Department of Instrumentation and Control Engineering, Sri Krishna College of Technology, Coimbatore, India.

saravanakumar.k@skct.edu.in

Abstract. The actual lifetime of a transformer is between 30-50 years, while a foundation can be in service for 100 years. The aim of this research is to get the exact remaining lifetime of the transformers and to suggest a method to improve it. Hybrid nanoparticles are used to improve the thermal properties and dielectric strength of transformer oil. Titanium oxide (TiO_2), Iron oxide (Fe_3O_4), and Graphene is used in specific concentration to prepare the hybrid nanoparticles. Energy Dispersive X-Ray Analysis (EDX), X-Ray Diffraction Analysis (XRD), Fourier Transform Infrared Spectroscopy (FTIR) are carried out to know whether the nanoparticles are properly synthesized. Initially Infrared thermography analysis using Fluke TiX-580 is being carried out to measure the surface temperature of the transformers and transformers under analysis are grouped. Nanofluid Transformer oil is prepared and Break Down Voltage (BDV) Test is taken to analyse the strength. Transformers are filled with nanofluid transformer oil and Infrared Thermography analysis is performed again to analyse the surface temperature. It is showing that lifetime of transformers is increased from 6 to 8% after using nanofluid transformer oil.

Keywords: Transformer, Nanomaterials, Break Down Voltage, Degree of Polymerisation, and Thermography.

1. Introduction

Transformers are one of the most important devices in a distribution network. Sudden failure of transformers during working may cause irreparable costs and even network failures [1]. Understanding the importance of transformers in power distribution, estimating their remaining life can be a good strategy for extending the life of transformers. Estimating transformer life can be very important for transformer loading and the determination of the appropriate points to operate the network [2].

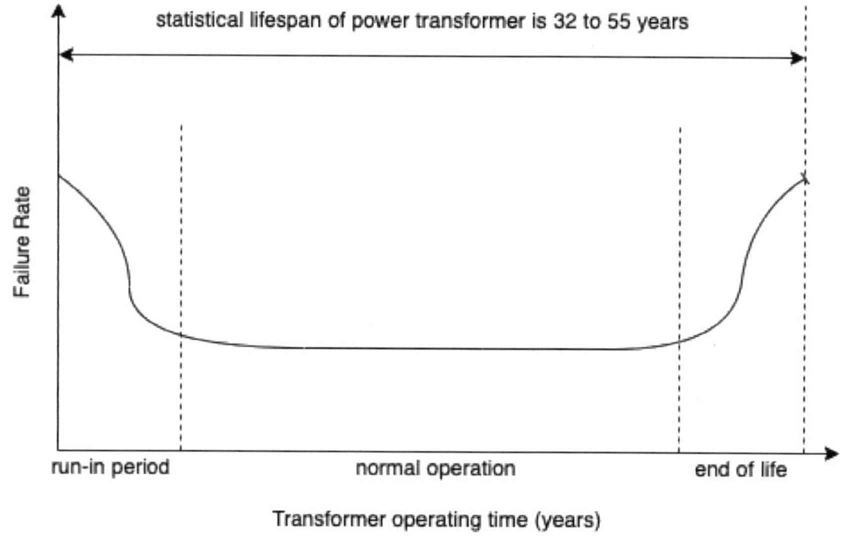

Figure 1. Failure rate of power transformers.

Figure 1, that shows the failure rate of transformers with respect to transformer operating time [3]. Transformer mineral oil is obtained from the crude oil refining. Three percentage of crude oil is suitable for the production of transformer oil. Due to the importance of power transformers in the power grid, constant maintenance of oil quality is essential [4]. In service, insulating oil is exposed to heat, oxygen and electrical discharges which can cause deterioration. The Lifetime of a power Transformer, Annual losses due to strength deliver interruptions attain into tens of thousands and thousands or maybe billions of dollars. The trouble of strength generation, transmission and distribution is of strategic importance. One of the crucial factors of the strength grid is the transformer. Knowing the situation of a transformer and its predicted lifetime lets in to successfully manipulate the prices and decrease feasible losses [5]. Transformer screw ups are due to many elements, which include broken windings, tank leaks, faucet changer screw ups, infection of insulation oil, excessive temperature and overloads etc. how can all elements be taken into consideration to increase transformer lifetime First, transformer kind have to be taken into consideration: whether or not the transformer is sealed or has a breather valve, if the transformer has nitrogen protection. Another crucial piece of data is the yearly load cycle and operation temperatures [6]. This is crucial to set the weight limits for transformers of assure class. Regulatory documentation units the common transformer lifetime duration. It varies from USA to USA, however in popular remains round 25- 30 years.

Various research works are going on in studying the life time of transformers. The remaining life of the transformer is studied, estimated by analysis of insoluble gas in petroleum and gas chromatography [7]. Another work performs thermal studies through the application of Computational Fluid Dynamics (CFD) for prototype transformers, which is considering six different cooling configurations [8]. Life estimation for oil immersion and distribution transformers is performed by considering weather-dependent intelligent load models and unbalanced loads [9]. Dielectric strength of transformer oil is improved by making a nanofluid transformer oil by considering hybrid $TiO_2/Al_2O_3/MoS_2$ nanoparticles [10]. AR sedighi et.al used Infrared thermography to identify the surface temperature of the transformer to identify the remaining life time [11]. The very important semiconductor photocatalyst selected here is TiO_2. Oxidation property, chemical characteristics, less cost and non-toxic nature of TiO_2 is very useful here in the hybrid nanocomposite prepared [12-13].

In this work the transformers under lifetime testing are grouped according to their age. Grouping of transformers is carried out by taking the surface temperature of the transformers using Infrared Thermography, which is a direct measure of transformer's age. Normal lifetime of the distribution transformers are 40 years from date of manufacture [14]. Hybrid nanofluid transformer oil is prepared by synthesizing Titanium oxide (TiO_2), Iron oxide (Fe_3O_4), and Graphene. The prepared nanofluid transformer oil is used in transformers which are being grouped and thermography test is taken again to identify the surface temperature of the transformers.

2. Experimental Section

2.1. Synthesis of Fe_3O_4/ TiO_2 /GRAPHENE nanomaterials

TiO_2, Fe_3O_4, and graphene are commercially available nanomaterials. The dielectric properties of transformer oil affect its service life as well as the distribution network's safety grade. Nanomaterials have the ability to enhance and improve the properties of transformer oil [15]. The effects of nanomaterial and surfactant concentrations on breakdown strength were studied. The results shows that adding Fe_3O_4 nanomaterials to transformer oil can improve its dielectric strength. Failure of a transformer has the potential to disrupt operations and result in financial losses. The obtained results indicating that the addition of TiO_2 nanomaterials to electrical device oil can improve its dielectric strength and also the properties [16-17].

The dielectric retention times of the transformer oil determine its service life and the degree of protection of the distribution network [18]. Nanomaterials have attracted a great deal of interest in recent years because they could grace the dielectric residences of transformer oil, and they now have a variety of abilities [19]. In this article, a solid nanofluid of Fe_3O_4 nanomaterials with exclusive focus is placed inside the transformer oil using hexdecyltrimethyl ammonium bromide (CTAB) as a surfactant. Stresses are examined. The results found showed that adding Fe_3O_4 nanomaterials to transformer oil can improve the dielectric performance. The influence of surfactant care on the switching capacity of transformer oil is examined.

The effect of nanomaterials' surface modification on the breakdown strength of colloidal systems has been investigated. The natural concentration of graphite is then dispersed in 60ml of concentrated HNO_3 for 8 hours with magnetic stirring. The graphene is then treated with HNO_3 and collected by washing 5 times with water and drying at 55 °C under vacuum. 500 mg expanded graphite is dispersed in 50 ml methanol and appealed in the methanol solution for 10 minutes. The graphite solution was placed in an 80 cm^3 Teflon autoclave and microwaved for 30 minutes at 800 W and 200°C. The reaction is then stopped by immersing the Teflon in a cool ice bath. The graphene sheets are vacuum-dried overnight at 70°C. SEM is then used to characterize the structure and morphology of the final product graphene sheet.

The uniform and solid suspension of nanomaterials within side the base liquid is very hard because of the appealing Wonder Walls appealing forces among them that result in the aggregation effect. In order to acquire a nanofluid with a homogeneous composition, nanomaterials have to be delivered to the bottom liquid under precise environmental situations and with the aid of using unique techniques [20]. Based at the quantity of nanofluid coaching steps, there are numerous techniques for getting ready nanofluids that are divided into fundamental techniques: One-step and Two-step techniques. The step technique is the maximum broadly and economically technique to supply nanofluids on a large scale [21]. For the coaching of nanofluids (primarily based totally on the two-step technique) and with inside the first step, nanomaterials utilized in this technique are produced as dry powders with the aid of using chemical or bodily techniques. Nanomaterials inclusive of TiO_2 (99.9%purity, density:3.9g.cm^{-3}) is purchased. TiO_2 having the average length 18nm. In the second one step, the Nano-sized powder can be dispersed right into a fluid with the assist of extensive magnetic pressure and ultrasonic agitation. In these paintings and for nanofluids coaching, the nanomaterials had been weighed with the aid of using an excessive accuracy scale with the precision of one mg. Then all nanomaterials dried in a drying oven for 8h at 200 °C to erase their moisture. After the nanomaterials have cooled down slowly, they may be blended with transformer oil. To make a homogenous dispersion of nanomaterials with inside the oil, the combos had been stirred for 30 min the use of a magnetic stirrer. Finally, with inside the very last stage, the nanofluids had been solicited for two hours to make sure that the nanomaterials had been nicely dispersed with inside the oil section and all the organized samples had been degassed at much less than 1kPa for forty-eight h earlier than test. In order to maximize the breakdown voltage for every nanofluid, most appropriate contents for TiO_2 nanofluid is 0.2 half wt. %, respectively.

In addition, for the foremost attention of nanomaterials, the electrical conduction of nanofluid found to be 3. The addition of nanomaterials with the transformer oil decreases the distinctive resistance and improves the electric conductivity thereby up the breakdown voltage. Moreover, the thermo-physics responsible forth development in thermo-physical and electric residences are mentioned clearly, on the thanks to be terribly useful for the layout of energy transmission/distribution systems. A positive quantity of the chosen graphene, Fe_3O_4, and TiO_2 nanomaterials became weighed in step with the favored attention [22-23]. The nanofluids have been organized through dispersing nanomaterials into transformer oil through the software of ultrasonic route. All the organized samples with distinctive concentrations have been placed into vacuum drying oven at much less than a kPa for nearly 24 hours to push aside the impact of micro bubbles fashioned throughout the sonication process.

2.2. Break Down Voltage Test:

The Break Down Voltage (BDV) of the transformer oil or Oil Dielectric Energy is the maximum dielectric energy of the oil without breakdown. The BDV rating of the transformer oil can be very important for the trouble-free operation of the transformer. BDV transformer oil testing is mandatory to ensure transformer oil health [24]. According to IEC, the minimum BDV of transformer oil should be 30 kV in accordance with the BDV Transformer Oil Test Standard. Figure 2 is a kit for performing the breakdown voltage test. Transformer Oil Body deals with transformer oil used in transformers for insulation and cooling. Different insulating materials have dielectric strengths of one type. The insulating fabric must withstand both the tension and its dielectric energy. As the voltage rises above the dielectric strength of the insulating fabric, the current tag will begin to creep through, and eventually the fabric will likely be fully punctured. Transformer oil is undoubtedly hygroscopic, which means it quickly absorbs moisture. The dielectric energy of oil degrades as it absorbs moisture. For this reason, a valve filled with silica gel (moisture-absorbing fabric) is fitted to transformers so that external moisture can be trapped in the silica gel even when breathing. The core and winding of the power transformer are immersed in transformer oil. The hot temperature generated by iron losses and copper losses in the transformer must be dissipated outside the transformer to limit the temperature to its stable limit. The hot temperature generated in the transformer is transferred to the oil and the oil transfers the hot temperature to the outer shell of the tank and the temperature is released to the atmospheric air. The first-class transformer oil is of utmost importance for the reliable operation of the oil transformer.

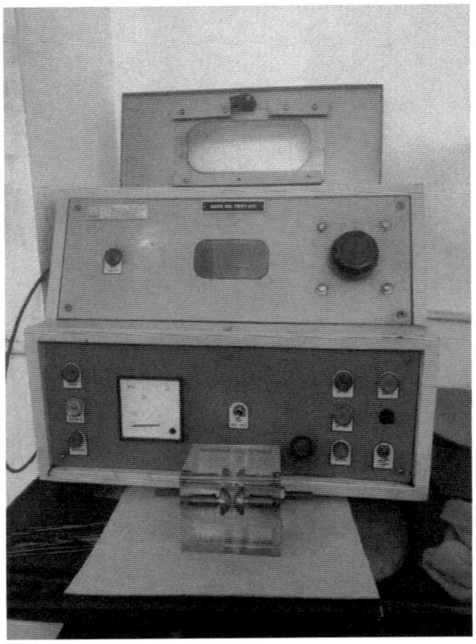

Figure 2. BDV Test kit.

2.3. Infrared Thermography:

Thermography is a non-adverse check technique that can be used to hit upon terrible connections, unbalanced loads, deteriorated insulation, or different capability issues in energized electric components. These issues can also additionally result in extra electricity use, accelerated preservation costs, or catastrophic gadget failure ensuing in unscheduled provider interruptions, gadget damage, and/ or different issues. Thermography, additionally known as infrared inspection, is primarily based totally upon the sensing of warmth emitted from the floor of an item with inside the shape of Infra-red radiation. Infrared thermography analysis [25]. Fluke Thermal Imager Tix-580 is used for the analysis of transformer surface temperature. Temperature difference between transformer hot spot (maximum temperature) and cold spot (minimum temperature) is obtained from the thermal image and temperature difference is calculated.

3. Result and Discussion

3.1. Scanning Electron Microscopy (SEM) of nanomaterials:

SEM research was conducted to investigate microstructure and crystallization. From Figure 3 it is being observed that size of nanoparticles present in the synthesized hybrid composition is below 100 nm. It is also clearly visible that large amount of TiO_2 nanoparticles, Iron Oxide nanoparticles are located at graphene oxide flakes in an area of 100 nm. Crystal size of TiO_2 is 9 nm. Size of Graphene is reduced to less than 100 nm with the help of NaOH treatment. Large graphene oxide fragments with TiO_2 and Iron Oxide particles are also observed. These wider graphene oxide layers are because of graphene oxide flakes during reaction with C = O during hydrolysis process.

Figure 3. SEM image of Fe_3O_4/ TiO_2 /GRAPHENE.

3.2. X-Ray Diffraction (XRD) Analysis of Nanomaterial:

XRD patterns of Fe_3O_4/ TiO_2 /GRAPHENE hybrid nanocomposite is shown in Figure.4. Diffraction peaks at 2q = 30.1, 35.4, 43.2 and 62.5 corresponds to planes (220), (311), (400) and (511) respectively. Diffraction peak of Fe_3O_4 is strengthened in Fe_3O_4/ TiO_2 /GRAPHENE hybrid nanocomposite by making the weight % of Fe_3O_4 upto 4.1% and the crystalline nature of Fe_3O_4/ TiO_2 /GRAPHENE hybrid nanocomposite is maintained similar to TiO_2.

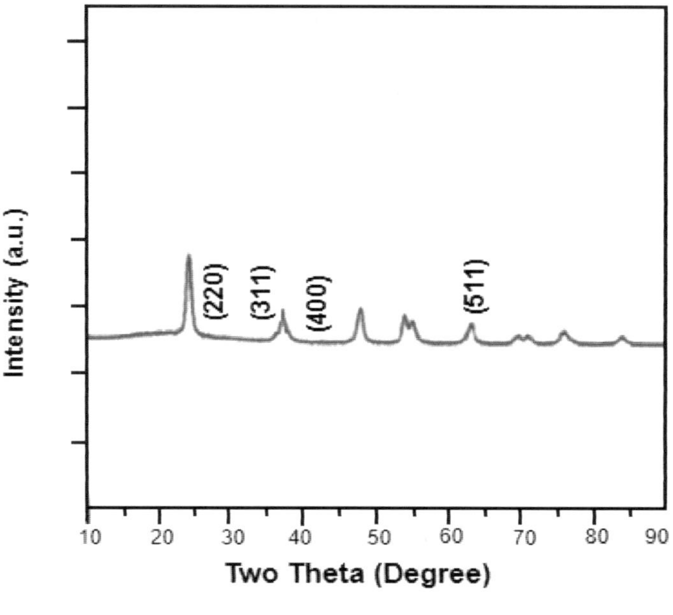

Figure 4. SEM image of Fe_3O_4/ TiO_2 /GRAPHENE.

3.3. Energy Dispersive X-Ray (EDX) analysis of nanomaterial:

Figure 5 shows the EDX analysis of hybrid nanoparticles. EDX (Energy Dispersive X-ray Spectrometer) with a solid angle of 0.28 sr performs highly sensitive analysis with nanometer resolution. It is clearly visible that Titanium oxide, Iron and Graphene are present in the hybrid composition. Ti is concentrated more than Fe. Ti covers in the outer structure of Fe particles.

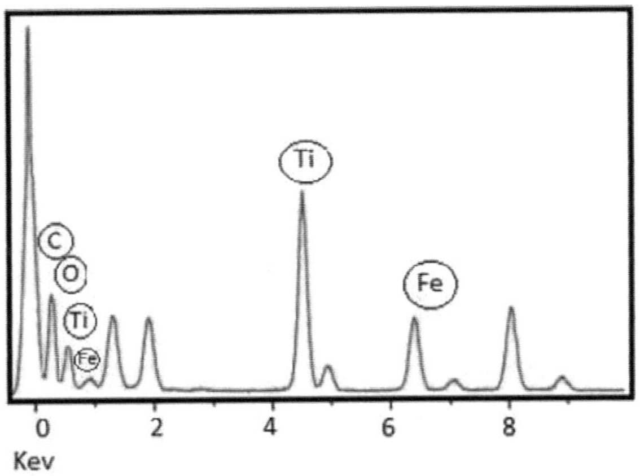

Figure 5. EDX analysis of Fe_3O_4/ TiO_2 /GRAPHENE.

3.4. Fourier Transform Infrared (FTIR) analysis:
In this work, functional groups were analyzed using FTIR spectroscopy (System ReactIR 702L, TEMCT). This showed good dipole variation and low symmetry. FTIR analysis to identify unknown materials and to conform production materials after synthesis to obtain graphical results on nanomaterials Fe_3O_4, TiO_2 and graphene. FTIR spectra of Fe_3O_4/Graphene/TiO_2 composite are shown in Figure 6. The peaks of 3419 and 1550 cm^{-1} are due to H-O-H expansion and contraction vibration and bending vibration of free water. New absorption band appeared in FTIR spectrum of 1616 cm^{-1}. This is an indicator of C = O oscillations and confirms successful packaging of GO into Fe_3O_4 nanoparticles. The absorption peaks of the obtained Fe_3O_4/Graphene/TiO_2 composite located at around 1627, 1400 and 1065 cm^{-1}.

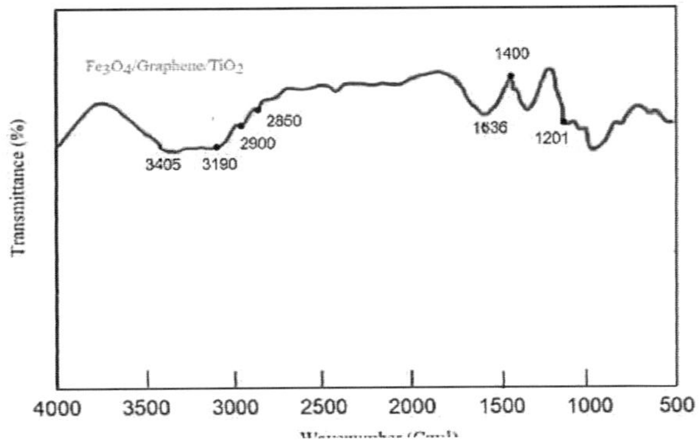

Figure 6. FTIR analysis of Fe_3O_4/ TiO_2 /GRAPHENE.

3.5. Thermography test result on transformer:

Table 1. Thermography test results to group the transformers which are under testing.

Average Surface Temperature (℃)	Age of Transformers (Years)	Remaining Life Time (Years)
6.4	1-3	37-40
7.4	4-6	34-36
7.8	7-9	31-33

8.1	10-12	28-30
8.5	13-15	25-27
8.7	16-18	22-24
9.3	19-21	19-11
10.2	22-24	16-18
10.7	25-27	13-15
11.5	28-30	10-12
11.9	31-33	7-9
12.4	34-36	4-6

Table 1. shows that the maximum and minimum temperature difference of the transformer body. It is between 6 and 13 degrees Celsius approximately during the 40 years of their service life.

Table 2. Thermography test results of normal transformer oil.

Age of Transformers (Years)	Hot Spot temperature (̊c)	Cold Spot temperature (̊c)	Surface Temperature difference (̊c)
2	41.5	35.2	6.3
11	43.16	35.06	8.1
27	44.1	33.4	10.7
30	45.1	33.6	11.5
35	47.2	34.8	12.4

Table 2. shows the thermography test results of the transformers which are being taken for analysis. Hot spot temperature is the maximum temperature measured in the transformer surface and cold spot is the place where surface temperature is minimum. Among the group of transformers made with the help of thermography results, transformers with the age of 2,11,27,30 and 35 are taken for test.

Table 3. Thermography test results of nanofluid transformer oil.

Age of Transformers (Years)	Hot Spot temperature (̊c)	Cold Spot temperature (̊c)	Surface Temperature difference (̊c)
2	40.8	35.6	5.2
8	43.1	35.3	7.8
23	44	33.7	10.3
27	44.9	34	10.9
31	46.8	35	11.8

Table 3. shows the thermography test results of the transformers being tested after nanofluid transformer oil is filled in the transformers. It is evident that surface temperature difference is reduced. The temperature distribution at the transformer tank turned into analyzed. In all gadgets below analysis, irrespective of the season, the freshest regions have been recorded at the top surfaces of the transformers. Higher temperatures have been detected for the transformers operating below better voltage 400/110 kV as in comparison to the transformers operating below 220/110 kV voltages. For transformers operating in parallel below voltage 220/110 kV it turned into found that the kingdom of the transformer (the circumstance of the electric insulation) had a substantial influence at the temperature of its floor ($P \leq 0.01$). Also, the season had a substantial effect at the floor temperature of the transformer ($P \leq 0.01$), irrespective of its type (circumstance) [26]. The coolest regions (but additionally with appreciate to warm spots) have been recorded at the decrease element of the tank. With a well-insulated transformer, the lower areas on the device surface were cooler. The coldest areas

(also related to hotspots) were recorded at the bottom of the tank. With a well-insulated transformer, the largest temperature differences were found between the top and bottom areas of the tank (in terms of hotspots) un affected by the season, resulting in cooling being very efficient is. In summer, intense sunlight during the day and high humidity has a negative effect on the cooling of transformers. The hot oil floating in the transformer tank is mainly concentrated in the upper part of the device, where the hottest points (the maximum temperature) were recorded.

Figure 7. Transformer body thermography

Grouping of transformers will be carried out according to their lifetime and to know the exact lifetime of transformer thermal analysis will be performed on transformers. Figure 7 shows way of performing thermography using thermal camera which shows the exact temperature in upper bound and lower bound, with lower bound and upper bound temperatures the difference between the temperatures can be calculated and the difference shows the exact lifetime of transformer. The 2012 model transformer is located in arivozhi Nagar at Coimbatore has been used for thermography test. In table 1, it shows the exact lifetime time of transformer based on its temperature difference in upper bound and lower bound. Upper bound temperature is a maximum temperature of the body of the transformer and lower bound is the minimum temperature of the body of transformer, and the difference between these will give a transformer's remaining age.

Cooling with the use of a forced oil circuit in combination with a forced air circulation system, with high air temperatures (heat waves) and high humidity may prove in effective and cause overheating of a transformer. As a consequence, the insulation conditions deteriorate even further (oil and paper aging occurs). Also, in the course of analysis of the thermograms for the poorly insulated transformers, the biggest differences were detected between the hot spots at the middle surface and at the bottom of the enclosure during the summer [27-28]. Depending on external conditions, the transformer surface was subjected to cooling, with cooler parts of the oil flowing down to the bottom of the device. In the case of bad weather conditions (high temperatures during the day, intensive sunshine on the surface, high humidity, no wind), the oil, despite the forced circuit and airflow to the tank, would almost completely not cool down, and, as a result, overheated areas were detected both in the upper, middle and lower parts of the tank.

Figure 8. The measurement of temperature difference after considering the parameters required for thermal imaging camera software ℃

Figure 8, the hot spot temperature is 43.16 degrees and the cold spot temperature is 35.06 degrees, so the difference between these comes under 10-12 year aged one and the temperature difference is about 8.14 degrees. Normal transformer oil is replaced with nanofluid transformer oil and the test has been performed with thermography. Figure 8. Indicates the thermal image of nanofluid transformer oil filled transformer. It has been observed for the changes in upper bound and lower bound temperature and here the temperature has been decreased and figure 8 shows it. Hotspot temperature is 43.1 degrees and cold spot is 35.3 degrees. This gives a reduction in temperature difference to 7.8 degrees. That means remaining life time of the transformer is increased to 32 years from 29 years.

Figure 9. Thermal image of transformer after mixing nanofluid

Further trends were detected for a transformer with degraded insulation. The hot spots in the upper part of the tank as well as the hot spots in the middle part of the tank sometimes appeared in a similar temperature range as the lower part of the tank. For the autumn and winter measurements there is no difference between the middle and lower of the transformer tank. In many thermograms, the hotspot switches the highest temperature was detected in the lower parts of the tank and not in the middle parts. These cases corresponded to thermograms taken over summer nights, which may be due to the fact that during many hours of the day the temperature and humidity of the air were high, making cooling much more difficult. The device was anything but satisfactory

4. Conclusion

In this present study, a new hybrid nanomaterial based nanofluid transformer oil is prepared and tested in distribution transformers which are under working condition. Transformers are undergoing thermal imaging study to group them according to their age. One transformer from each group is selected and transformer oil which is already there is replaced with nanofluid transformer oil. Thermal study is carried out again to measure the hot spot and cold spot temperatures in the transformer surface. Temperature difference between hot spot and cold spot is the direct measure of age of the transformer. Ater the replacement of normal oil with nanofluid transformer oil, surface temperature difference is reduced significantly. In this work, the transformer temperature is reduced to 1 to 1.5 degree Celsius which means life time of the transformers are increased from 3 to 4 years. Break Down Voltage of the transformer oil also not affected at all after the oil replacement. From the received outcomes, it's far observed that the nanomaterials complement the thermal residence of the transformer oil through approximately 28.3% however, the dielectric residences have been barely growth through approximately 2%. The addition of Fe3O4, TiO2 and graphene nanomaterials which increase the polymerization factor of the transformer oil by which it increases the lifetime of transformer.

References

[1] Gutten, M., Korenciak, D., Kucera, M., Janura, R., Glowacz, A. and Kantoch, E., 2018. Frequency and time fault diagnosis methods of power transformers. Measurement Science Review, 18(4), pp.162-167.

[2] Eini, B.J., Saghaei, A., Hosseini, S.H. and Hamedanian, A., 2013. Defect detection and preventive maintenance prioritization of distribution cubicles by infrared statistical image processing.

[3] Pathak, J., Jiang, Y., Honavar, V. and McCalley, J., 2006, January. Condition data aggregation with application to failure rate calculation of power transformers. In Proceedings of the 39th annual Hawaii international conference on system sciences (HICSS'06) (Vol. 10, pp. 241a-241a). IEEE.

[4] Fallah-Shojaie, A., et al. "Experimental evaluation on the dielectric breakdown voltage of fresh andused transformer oil mixed with titanium dioxide nanomaterials in the Galan electrical distribution company." 2013 21stIranianConference on Electrical Engineering(ICEE). IEEE, 2013

[5] Liu, Dangling, et al. "Characterization of high-performance AIN nanomaterial-based transformer oil nanofluids."IEEE Transactions on Dielectrics and Electrical Insulation23.5(2016): 2757-2767.

[6] Rafi, Muhammad, et al. "The impacts of nanotechnology on the improvement of liquid insulation of transformers: Emerging trends and challenges. "Journal of MolecularLiquids302(2020): 112482.

[7] Dervos, C.T., Paraskevas, C.D., Skafidas, P.D. and Stefanou, N., 2006. Dielectric spectroscopy and gas chromatography methods applied on high-voltage transformer oils. IEEE transactions on dielectrics and electrical insulation, 13(3), pp.586-592.

[8] Nogueira, G.C., Medeiros, L.H., Oliveira, M.M., Barth, N.D., Bender, V.C., Marchesan, T.B. and Falcão, C.E., 2022. Thermal Analysis of Power Transformers with Different Cooling Systems Using Computational Fluid Dynamics. Journal of Control, Automation and Electrical Systems, 33(1), pp.359-368.

[9] Gorginpour, H., Ghimatgar, H. and Toulabi, M.S., 2022. Lifetime Estimation and Optimal

Maintenance Scheduling of Urban Oil-Immersed Distribution-Transformers Considering Weather-Dependent Intelligent Load Model and Unbalanced Loading. IEEE Transactions on Power Delivery.
[10] Sumathi, S., Rajesh, R. and Subburaj, P., 2021. Investigation of dielectric strength of transformer oil based on hybrid TiO2/Al2O3/MoS2 Nanofluid using taguchi and response surface methodology. IETE Journal of Research, 67(6), pp.817-825.
[11] Sedighi,A.R.,etal."Life estimation of distribution transformers using thermography: A case study."Measurement149(2020):106994.
[12] Nasr, M., Balme, S., Eid, C., Habchi, R., Miele, P. and Bechelany, M., 2017. Enhanced visible-light photocatalytic performance of electrospun rGO/TiO2 composite nanofibers. The Journal of Physical Chemistry C, 121(1), pp.261-269.
[13] Nada, A.A., Nasr, M., Viter, R., Miele, P., Roualdes, S. and Bechelany, M., 2017. Mesoporous ZnFe2O4@ TiO2 nanofibers prepared by electrospinning coupled to PECVD as highly performing photocatalytic materials. The Journal of Physical Chemistry C, 121(44), pp.24669-24677.
[14] Jung, Jong-Wook, and Jin-Soo Jung. "The effect of thermal ageing on the electrical characteristics of insulating oil for pole transformers." 2008 International Conference on Condition Monitoring and Diagnosis. IEEE, 2008.
[15] One pot synthesis of graphene based on microwave assisted solvothermal technique F.S. Al-Hazmia,*, Ghada H. Al-Harbi b, Gary W. Beall a,c, A.A. Al-Ghamdia, A.Y. Obaid d,Waleed E. Mahmouda,e.
[16] Siva Kumar, S., teal. “Property enhancement of transformer oil with suspension of TiO2 nanomaterials."JournalofNanofluids3.4(2014):336-340
[17] Live, Yu-Zhen, et al. "Preparation and breakdown strength of TiO2 fluids based on transformer oil."2010Annual Report Conference on Electrical Insulation and Dielectric Phenomena. IEEE, 2010.
[18] Qing, Soo Hui, et al. "Thermal conductivity and electrical properties of hybrid SiO2-graphene naphthenic mineral oil nanofluid as potential transformer oil." Materials Research Express 4.1 (2017):015504.
[19] Synthesis of Fe3O4/graphene/TiO2 Composites for the Highly Selective Enrichment of Phosphopeptides from Biological Samples Jin Lu, Chunhui Deng, * Xiangmin Zhang, and Pengyuan Yang Department of Chemistry& Institutes of Biomedical Science, Fudan University, Shanghai200433, China.
[20] Almeida, Charishma, et al. "Experimental Studies on Thermophysical and Electrical Properties of Graphene–Transformer Oil Nanofluid" Fluids 5.4(2020):172.
[21] Kadim, Emran Jawad, et al. "Nanomaterials Application in High Voltage Insulation Systems."IEEE Transactions on Dielectrics and Electrical Insulation28.4(2021):1380-1399.
[22] Bahn, Hyokyung. "Efficient Management of Probe-Based Nano Storage Devices." Journal of Electronics 1, no. 01 (2019): 1-11.
[23] Nirmal, D. "Design and Efficiency Analysis of Nanocarbon Interconnect Structures." Journal of Electronics 1, no. 01 (2019): 12-23.
[24] Wang, Qi, et al. "Preparation of three types of transformer oil-based Nano fluids and comparative study on the effect of nanomaterial concentrations on insulating property of transformer oil." Journal of Nanotechnology 2016.
[25] Sun, Potao, et al. "Influence of thermal aging on the breakdown characteristics of transformer oil impregnated paper." IEEE Transactions on Dielectrics and Electrical Insulation 23.6 (2016): 3373- 3381.
[26] Fallah-Shojaie, A., et al. "Experimentalevaluation on the dielectric breakdown voltage of fresh and used transformer oil mixed with titanium dioxide nanomaterials in the Gi-lan electrical distribution company." 201321stIranianConferenceonElectrical Engineering(ICEE). IEEE, 2013.

[27] Pugazhendhi, Sugumaran C. "Experimental evaluation on dielectric and thermal characteristics of nano filler added transformer oil." 2012 International Conference on High Voltage Engineering andApplication.IEEE, 2012.

[28] Sokolov,Victor,etal."Transformer fluid: a powerful tool for the life management of an ageing transformer population" Proceedings oftheTechCon.2003.

Automatic Speed Control System in Speed Zone Areas Using OpenCV

Tejo Sumanth Pavuluri[1], **T. Vamsi Krishna**[1], **Naredla Sandeep**[1], **A. Jaya Abhitej Reddy**[1] and **Dr. Mamatha Samson**[1]

[1]Dept. of ECE, Gokaraju Rangaraju Institute of Engineering and Technology, Hyderabad, Telangana, India.

pavuluritejo@gmail.com, vamsikrishna29.tamma@gmail.com, naredlasandeep143143@gmail.com, abhithejreddy42@gmail.com, mamata2001@gmail.com

Abstract. Road accidents near school or hospital zones are all too common in today's environment. We see a lot of people driving at high speeds under the influence of alcohol or for thrill causing road accidents. So, vehicle speed control must be taken into consideration to avoid casualties due to over speeding. The purpose of this system is to control the speed of the vehicle by recognizing speed sign boards on the roadside with the help of Rasberry Pi board and image processing. The camera is positioned at the front of the vehicle to capture the images of sign boards to process them. The captured images are then compared with the stored images using Python open CV to control the vehicle's speed in the speed restricted areas. The objective of this project is to reduce the accidents caused due to over speeding.

Keywords: Control Systems; OpenCV; Speed control; Accident Reduction.

1. Introduction

Nowadays, we witness people speeding while inebriated or for the pleasure of it, resulting in traffic accidents [1][2]. Due to excessive speeding, the driver gets less amount of time to react to dangerous situations trying to avoid accidents. Speeding endangers not only the speeder's life, but the lives of everyone else on the road as well. It is a major problem that needs to be considered to avoid road accidents. So, our project helps in maintaining the speed of the vehicle by capturing the images of the speed sign boards and comparing them with the stored dataset to maintain right amount of speed.

The advancements in the processor technologies have opened the door to new systems aimed to prevent road accidents caused by driver's failure to to detect speed sign boards alongside the roads [3]. Many methods have been proposed to limit speed of the vehicle from over speeding with the help of various technologies like Zigbee transmitter, RF technology and color detecting as shown in figure 1.

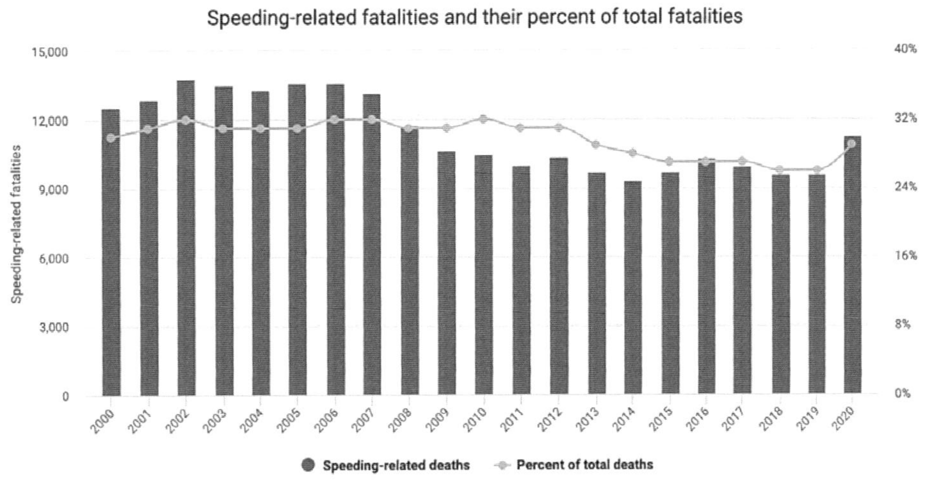

Fig. 1. Fatalities caused due to Over Speeding

For instance, the Zigbee detects the speed limit of a particular lane in the order of acoustic messages which is received by a receiver and is then sent to a microcontroller to adjust the speed of a vehicle [1]. Other such common methods include the use RF technology, where both RF transmitter and RF receiver are installed in the speed sign board and the vehicle respectively. These RF tags are used to relay the data provided by the signals to adjust the speed of the automobile [6]. These tags are configured to transmit signals to the receiver when the vehicle in the proximity [3][5].

Using vehicle to vehicle communication systems, these automatic speed control systems can also be used to reduce the speeding of the vehicle to avoid accidents. This can be accomplished with the help of sensors installed in the vehicles which provide relative distances and speeds between them [4].
In this paper, we have come up with a new idea where our system works on the principle of image processing technique with camera installed in the vehicle which captures the images of the speed sign boards and are sent to the Rasberry pi4 for further processing to limit the speed of the vehicle. Image processing is the process of converting a physical image to a digital representation and then conducting operations on it to extract valuable information.

2. Literature Review

2.1. Over speed violation management of a vehicle through Zigbee. International Journal of Engineering and Technology[1]

The proposed method was successful in establishing a mechanism for reducing traffic violations. To encourage careful and conscientious driving, the driver is made aware of his driving behavior and violations are recorded. The Zigbee technology is used for wireless transmission

2.2. Automatic Speed Control System by the Color Sensor for Automobiles -An Innovative Model Based Approach.[2]

Various color strips are designated on highway roads or common roads where the speed control within the limits is required, and the car will be equipped with a color sensor that will recognize the color marked and keep the vehicle speed within that limit.

2.3. Automatic Vehicle Speed Control With Wireless In-Vehicle Road Sign Delivery System Using ARM 7 [3]

A prototype has been developed to control the speed of automobile with the use of RF technology using ARM 7 microcontroller. This system consumes less power and is easy to implement.

2.4. Throttle And Brake Control System For Automatic Vehicle Following [4]

Vehicles are installed with sensors that can calculate relative distances and speed between the vehicle and in in front of it to avoid collision between them. Even when the leading vehicle exhibits irregular speed behavior, this proposed system ensures smooth vehicle following.

3. Proposed Method and Its Working

We have developed a system to control the speed of the vehicle which also avoids road accidents due to over speeding with the help of Speed Sign Board Detection and Recognition technique. The captured image is processed and analyzed to limit the speed of the vehicle as per the speed sign board.

To recognize and detect the desired traffic sign, the recorded video must be converted into frames. Approximately 30 frames are generated in 1 second while transforming video signal into frames. However, 30 frames per second contains redundant data, i.e., frames that take longer to identify and detect the necessary traffic sign. As a result, the system becomes slow and is unable to respond quickly enough to manage actions. As a result, redundant data is eliminated, and only one frame per second is used for comparison. As a result, the time required for data processing is significantly reduced, resulting in improved overall system performance as shown in figure 2.

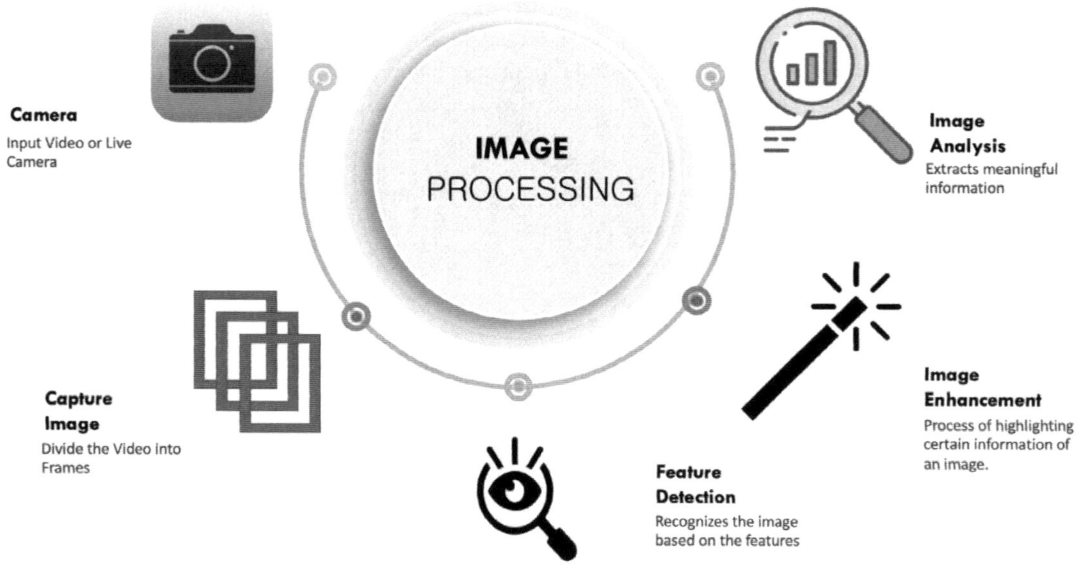

Fig. 2. Flowchart for Image Processing

The Rasberry pi4 which uses OpenCV python codes for processing the captured image and compare them with the stored dataset. The image processing technique consists of the following steps:

3.1 Capturing of image:

An image is captured from an input hardware device like camera. Our system coverts the video signal into multiple frames which are further processed. Redundant frames which are useless are removed.

3.2 Feature Detection:

The machine recognizes the Image based on the features. All the images of the same object are not the identical. For example, consider different images of a dog shot at different angles, with different focus, etc. All these images depict the same object. The machine recognizes the image by identifying important features such as the ears, nose, mouth, etc. But these features are not always in the same part of the image so; the machine shouldn't care if the features are tilted, if they are of different texture, if they are bit closer or bit farther. Even if our feature is a bit distorted the machine should be flexible enough to detect them. Pooling helps in achieving this. Maximum pooling or average pooling can be considered. The output after pooling the feature map is known as pooled feature map.

3.3 Image Enhancement:

Image Enhancement is a process of modifying digital images such that the results are more suited for display or further analysis. We can, for example remove noise, making it easier to recognize targets.

3.4 Image Analysis:

Image Analysis is a method of extracting useful information from the images as shown in figure 3.

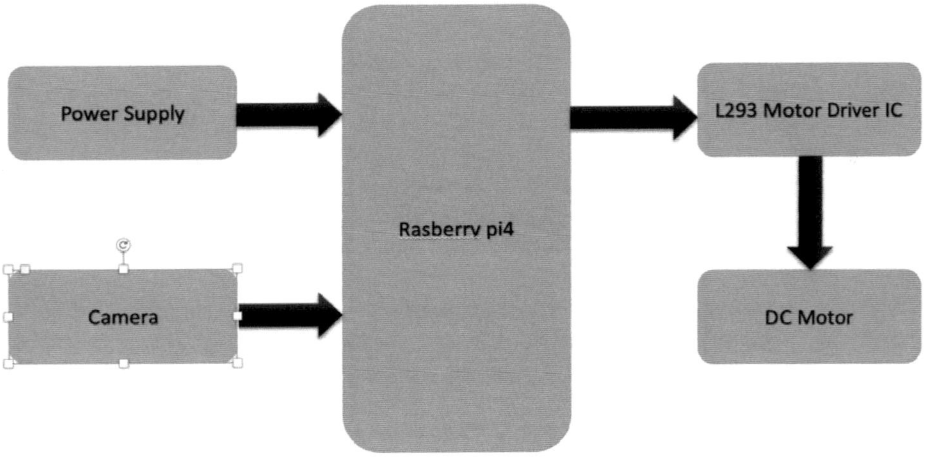

Fig. 3. Block Diagram of Proposed System

4. System Components and Software

4.1 Rasberry pi4: Rasberry pi4 is a single board computer fromRasberry pi foundation with a micro usb as a power source.
4.2 Camera: The camera is used to get the live feed of real world traffic signs which can be sent to the raspberry pi to process the images and limit the speed of the vehicle.
4.3 DC Motor: A Dc Motor is used in our system to mimic the vehicles. A Dc Motor is an electric machine that converts electrical energy into mechanical energy.

4.4 L283d Motor Driver IC: A 16 pin motor driver IC which is used to drive the dc motor in either direction.

4.5 OpenCV: Open Source Computer Vision has a programming feature library that is primarily for real-time machine visions. It has over 2500 optimize algorithms for a set of classical algorithms and state-of-the-art algorithms in machine visions. Open CV is a computer vision library that we used for face detection, object detection, image recognition, and tracking.

5. Results and Conclusion

This speed control system ensures that the number of incidents in the vicinity of the school zones is kept to a minimum and makes silence near to the hospital zones. This system is low-cost, long-lasting, low-power, and provides maximum performance. This study has been discovered that using a vehicle speed control device helps to reduce the number of accidents that occur as a result of a driver's failure to follow roadside speed signboards in special zones. It illustrates the value of automatically reducing vehicle speed and its contributions to pedestrian and road user safety. This vehicle speed control are effective, and they contribute significantly to improving safety by ensuring the safety of both passengers and pedestrians on the roads. It enables the vehicle to act independently to slow down the vehicle as it approaches at a higher speed, which actively and efficiently reduce accidents caused by the driver's fault as shown in figure 4.

Figure 4. Outputs

6. Acknowledgements

It is a great sense of relief and accomplishment whenever we complete a task. My group has taken a lot of efforts for the completion of this project. However without the kind support and assistance of many individuals, it would not have been feasible. We are deeply grateful to our supervisor, **Dr. G. Mamatha (Professor, ECE)** for her invaluable guidance and encouragement who helped us reach milestones of our project.

We are grateful to **Dr. Hima Bindu Valiveti, HOD, ECE dept.** for providing us the resources and equipment necessary for completing our project. We would like to thank **DR. J PRAVEEN (Principal)** and **DR. JANDHYALA N MURTHY (Director)** for providing us with great infrastructure that helped us achieving our project goal. Finally, we thank all our friends, faculty, and seniors for supporting us throughout the whole journey.

7. References

[1] Rubini, R. & Makeswari, A.. (2013). Over speed violation management of a vehicle through Zigbee. International Journal of Engineering and Technology. 5. 340-344.

[2] Kewate, Sunil & Karmare, S.V. & Sayankar, Nehal & Gavhale, Siddharth. (2016). Automatic Speed Control System by the Color Sensor for Automobiles -An Innovative Model Based Approach. 10.13140/RG.2.1.3791.4643.

[3] Sattibabu, Gummarekula et al. “Automatic Vehicle Speed Control With Wireless In-Vehicle Road Sign Delivery System Using ARM 7.” International Journal of Technology Enhancements and Emerging Engineering Research 2 (2014): 32-34.

[4] P. Ioannou & Z. Xu (1994) THROTTLE AND BRAKE CONTROL SYSTEMS FOR AUTOMATIC VEHICLE FOLLOWING, I V H S Journal, 1:4, 345-377.

[5] R.SenthilKumar, E.Priyanghaa, S.Nandhini, A.Priyanka, "AUTOMATIC VEHICLE SPEED REDUCTION SYSTEM IN VARIOUS ZONES", International Journal Of Advanced Research in Basic Engineering Sciences and Technology (IJARBEST), VOLUME 2,SPECIAL ISSUE 10 - MARCH 2016, pp.1819-1823.

[6] J. Tsuchiya, M. Hayashi and R. Kohno, "Platoon characteristics of automatic vehicle speed control for vehicles equipped with DS-UWB radar," 2010 IEEE 11th International Symposium on Spread Spectrum Techniques and Applications, 2010, pp. 125-130, doi: 10.1109/ISSSTA.2010.5653334.

[7] S. A. Prakash, A. Mohan R, R. M. Warrier, R. Arun Krishna, S. Bhaskar A and A. K. Nair, "Real Time Automatic Speed Control Unit for Vehicles," 2018 2nd International Conference on I-SMAC (IoT in Social, Mobile, Analytics and Cloud) (I-SMAC)I-SMAC (IoT in Social, Mobile, Analytics and Cloud) (I-SMAC), 2018 2nd International Conference on, 2018, pp. 157-161, doi: 10.1109/I-SMAC.2018.8653721.

[8] K. Yogheedha, A. S. A. Nasir, H. Jaafar and S. M. Mamduh, "Automatic Vehicle License Plate Recognition System Based on Image Processing and Template Matching Approach," 2018 International Conference on Computational Approach in Smart Systems Design and Applications (ICASSDA), 2018, pp. 1-8, doi: 10.1109/ICASSDA.2018.8477639.

[9] Sathyapriya, A. & Sathiya, K. & Sneha, T. & Raja, D. & Thiyagarajan, Manikandan. (2020). Automatic Speed Control System in Vehicles Using VANET. 10.1007/978-981-15-5029-4_60.

[10] J. Wang, J. Tong, Q. Gao, Z. Wu, S. Bi and H. Wang, "Device-Free Vehicle Speed Estimation With WiFi," in IEEE Transactions on Vehicular Technology, vol. 67, no. 9, pp. 8205-8214, Sept. 2018, doi: 10.1109/TVT.2018.2840052.

[11] Lee, YH. "Automatic Speed Control System for Vehicle Approaching and Driving on a Curve."

Proceedings of the ASME 2008 International Mechanical Engineering Congress and Exposition. Volume 17: Transportation Systems. Boston, Massachusetts, USA. October 31–November 6, 2008. pp. 345-353. ASME.

[12] D. Gruyer, S. Glaser, B. Vanholme and B. Monnier, "Simulation of automatic vehicle speed control by transponder-equipped infrastructure," 2009 9th International Conference on Intelligent Transport Systems Telecommunications, (ITST), 2009, pp. 628-633, doi: 10.1109/ITST.2009.5399281.

[13] H. Takahashi, "Automatic speed control device using self-tuning fuzzy logic," 1988., IEEE Workshop on Automotive Applications of Electronics, 1988, pp. 65-71, doi: 10.1109/AAE.1988.47596.

[14] M. K. Liubakka, D. S. Rhode, J. R. Winkelman and P. V. Kokotovic, "Adaptive automotive speed control," in IEEE Transactions on Automatic Control, vol. 38, no. 7, pp. 1011-1020, July 1993, doi: 10.1109/9.231457.

[15] Dhaya, R. "CCTV Surveillance for Unprecedented Violence and Traffic Monitoring." Journal of Innovative Image Processing (JIIP) 2, no. 01 (2020): 25-34

Early Detection of Seizure Using Electrical Brain Activity

Sangeetha V [1], Shanthini E [1], Aravindhan T [2], Dineshkumar S [2], Ashwin Shivaa D[2]

[1] Faculty of Electronics and Communication Engineering, Sri Ramakrishna Engineering College, Coimbatore, Tamil Nadu, India.
sangeetha.v@srec.ac.in, shanthini.e@srec.ac.in

[2] Student of Electronics and Communication Engineering, Sri Ramakrishna Engineering College, Coimbatore, Tamil Nadu, India.
aravindhan.1802019@srec.ac.in, dineshkumar.1802035@srec.ac.in, ashwinshiva.1802065@srec.ac.in

Abstract: Epilepsy is a chronic brain condition characterized by an uncontrollable electric blast in the brain, manifested as Epilepsy seizures. More than one percent of the world population is affected by Epilepsy seizures. This framework employs a few corporal boundaries, such as temperature, pulse, and movement limits. The device measures the electrical activity of the brain by using an EEG sensor and alerts the patient according to the electrical brain activity. This paper focuses on the development of early detection of seizures by EEG to alert the patient to changes in his/her body.

Keywords: EEG (Electroencephalogram), ICT (Information and Communication Technologies)

1. INTRODUCTION

Health is nowadays defined as the use of information andcommunication technologies (ICT) for health. Health can also be considered a tool used in the healthcare environmentfor prevention, diagnosis, treatment, and monitoring tasks. Its incorporation implies several improvements such as better management, saving costs, and efficiency enhancements of the health system. These new technologies include different products and health services, such as mobile applications, telemedicine systems, wearables devices combined withsensors, big data (large amounts of data), and the Internet of things.

Customarily, recognizable proof of EEG indicators has been carried out by using grasp visual translation. As it may, this interaction is tedious and can sometimes be inclined to human blunder. Consequently, computerizing the most frequent way of distinguishing Epilepsy seizures using Electrical Brain signals has seriously changed into a large issue. Subsequently, the mechanized attention of Epilepsy seizures works with the utilization of programmed equipped frameworks and logs of seizures, which are useful to epilepsy patients. In this specific situation, signs and symptoms created by the EEG are a substantial scientific symptomatic instrument and be deciphered precisely by way of the physician for the proper conclusion of nervous device science. Consequently, draw close to the ability of the cerebrum and as a result work with the investigation and explain the conclusion to advise adequate cure for the epileptic patient. In this particular situation, our universal intention cautions the affected person and his company so they can keep away from potential risks.

It is assessed that 70% of patients answer emphatically to treatment with an anticonvulsant prescription. In any case, the excess 30% don't have a technique to stay away from or diminish Epilepsy seizures. In the main case, it is important to change the drug for every patient; the subsequent case requires a framework to recognize and alarm a parental figure. A gadget ready to identify these seizures and save an authentic record could be exceptionally helpful. In any case, the most serious effects of Epilepsy seizures such as brain damage or death can be avoided.

2. RELATED WORKS

The early detection of seizures is extremely useful for keeping away the patient from death and unsafe treatment. Numerous techniques for distinguishing seizures have been discussed by the analyst.

T R Burchfield discussed a biomedical application related to wireless sensor networks (WSN). They used an accelerometer to determine if the movement of a person's arm is similar to a seizure. The system used some smart algorithms to decide whether the detected movement corresponded to a seizure movement or not. In addition, this work tries to monitor patients in a non-invasive, non-intrusive manner.[13]

M Z Poh et al developed a technique for monitoring sympathetic worried device endeavor at some stage in Epilepsy seizures with the aid of the potential of a wearable sensor. This study showed that EDA arises during seizures and can be used for their detection.The authors presented the results of a pilot study and their results showed that seizures can induce EDA.[2]

B Huige et al. showed the procedure of sensor hubs to follow human stances. Each of the hubs is geared up with a magnetometer and an accelerometer for movement location. To get a greater battery cycle, additional battery-saving techniques are added.[3]

S. Patel treated the transition from fixed to wearable sensors that are used to detect Epilepsy seizures. In this paper, the authors detected epilepsy seizures through motion detectionand electroencephalogram (EEG). Results suggest that wearable devices can detect these seizures with the ability to differentiate between seizures and normal movements in a patient's life. In our case, we pretend to make this differentiation by monitoring three different parameters.[4]

Bouaziz, Bassem and Chaari, Lotfi, and Batavia proposed a method to look at the presentation of recurrence and time-space signals. This technique stayed away from the complicated extraction interaction of the elements and utilized an essential design of CNN. The utilization of the recurrence area signals gives a typical accuracy of 96.7, 95.4, and 92.3% for the three examinations while the signs for the time-space give a normal exactness were 91.1, 83.8, and 85.1% in the three analyses.[10]

Journal of Physics: Conference Series **2325** (2022) 012018 doi:10.1088/1742-6596/2325/1/012018

C.L. Liu proposed a method to predict the incidence of epilepsy seizures, a multi-view CNN framework was used. Contemplating the space domain and recurrence region in two inputs and the use of a basic layer to research the representation of entities is the core issue of the structure. As the ultimate classifier, an Support vector machine is used. Experimental findings exhibit that all cited procedures are outperformed by using the proposed multi-view CNN private ranking, as it achieves a commonplace below the curve of 0.83 and 0.90 on respective subjects in the records collection.[12]

Karuppusamy, Dr P proposes a method for the detection of Brain tumors. This research work provides a hybrid manta ray foraging optimization for feature selection from brain tumor MRI images. A convolution neural network is applied to detect early-stage brain tumors using MRI images. Experimental results show that the proposed method outperforms other methods.[6]

There are many papers with a technological-medical purpose. In past years, this research area has grown up. Specifically, wearable sensors are treated in papers where energy saving, accuracy, and motion detection are protagonists.In the epileptic detection area, we can find works that try to detect seizures but most of them have into account motion and EEG or EDA, both of them obtained from sensors that are expensive or outsized for wearable devices. Our prototype is focused on monitoring EEG Signals, Electrical Brain Activities, heart rate, temperature, and vibration. These parameters are affected by most epilepsy types.

3. EXISTING METHOD

The proposed approach is shown in Figure 1. In this underlying advance, the procedure incorporates the normalization of the Electrical Brain signals to eliminate the characteristics fundamental to action.

Figure 1. Overview of Existing Method

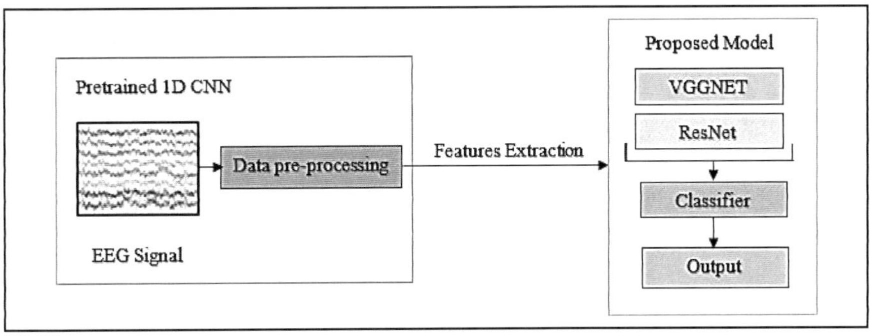

3.1 Dataset Information

To play out these examinations, the dataset gathered at Children's Hospital located in Boston was used. The recorded data are taken from Twenty-Three epilepsy patients including eighteen women and five men. This informational index incorporates 686 EEG scalps multi-channel accounts from the patients. Almost all of the patient's improvement takes place around 25 s for each account of EEG. For recording the information index, the EEG signal mounting framework is gathered worldwide and is utilized.

3.2 Pre-processing

In a 1D Convolutional neural network, the center strikes in a solitary course. It requires 2D

Journal of Physics: Conference Series **2325** (2022) 012018 doi:10.1088/1742-6596/2325/1/012018

wellsprings of records that are two-layered. Regardless, there is proof that the Electrical Brain signal is related in series. We have concerned with this hard EEG information as shown as a three-layered crew in which we have the dedication to width, adventures every so often, and in level, there are the terminals. The information is translated as a 3-layered show where all the time steps are in the width and all of the cathodes in the level.

3.3 The proposed framework

3.3.1 Seizure Detection Using VGGNET

This block diagram explains the engineering utilized for the VGG network. And this design was discussed by **Ka-ren Simonyan** in 2014, it was discussed as an extremely profound organization for picture acknowledgment. Its principal commitment has shown that network profundity is an urgent component to acquiring better acknowledgment or better arrangement exactness in CNNs.

Figure 2. Proposed VGGNET model

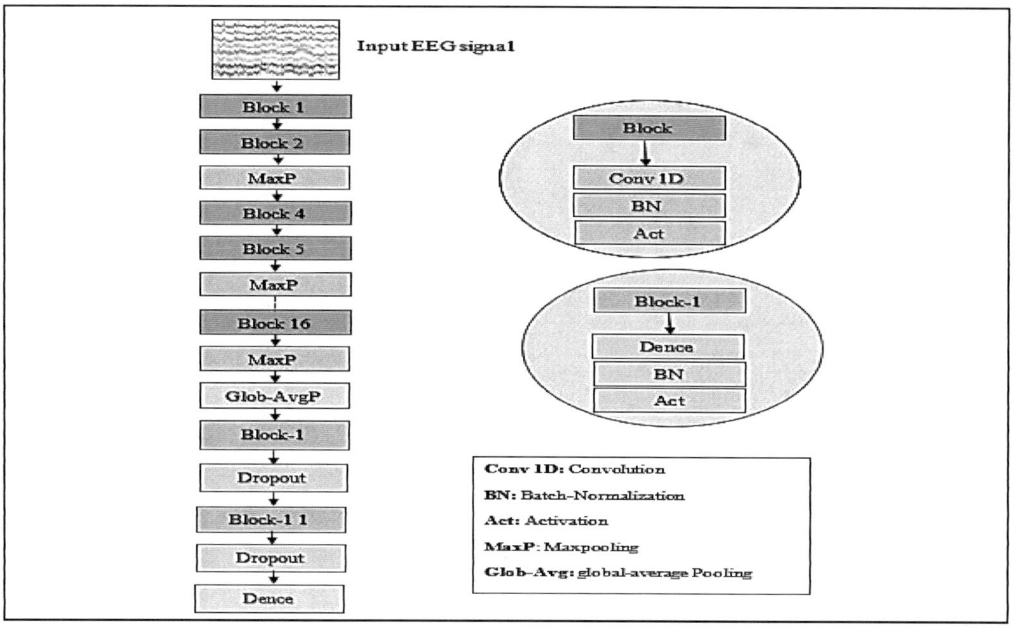

3.3.2 Seizure Detection Using ResNet

Figure 3 explains the design utilized for Residual Network, or ResNet, a quite certain brain network presented in 2015 **by Kaiming and Jian Sun** in their article" **Deep Learning for Image Recognition"**. Because of the utilization of worldwide normal grouping rather than a completely associated platform. The model size is a lot more modest, which diminishes the ResNet size model. The extraordinary piece of ResNet is the lingering learning block. This implies that all layers ought to take care of in the following layer and around 2 to 3 distance jumps straightforwardly into the layers.

Figure 3. Proposed ResNet Model

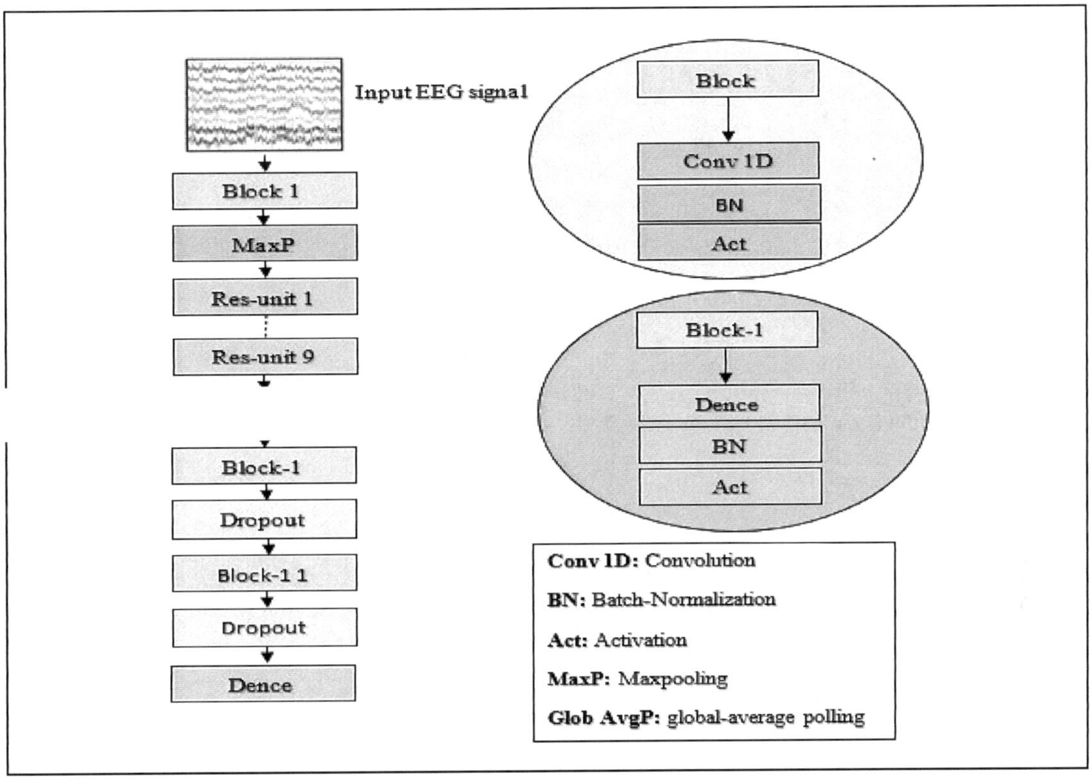

4. PROPOSED METHOD

The people who are diagnosed with seizure is made to wear a cap, inside the cap 3 EEG sensors are attached in a position such as T8, P8& FP2 This sensor monitors the electrical activities of the brain. Even though the normal EEG Test contains 10-20 electrodes placed, Seizure activities are mainly monitored using the ppositionsT8, P8 & FP2. Normal person electrical activities range below 50 microvolts per second during the time of the seizure the person may experience the electrical disturbance up to 70 microvolts per second and sometimes about 70.

If the person's electrical activities reach the abnormal level i.e., during the pre-ictal stage the person may experience symptoms like vision difficulties, oddness in taste or smell, etc ... The electrodes in the cap report this abnormal activity, sending the emergency warning message to the patient's mobile number and family or loved ones. The patient by seeing this text message can make himself prepared for the seizure which may attack at any minute.

Every time an electrical disturbance or abnormal happens it is sent to the IoT analytics platform for future research & medicine recommended for the patient.

In the pre-ictal stage (Aura), some of the time an individual might have an emanation that doesn't advance to turn into a more extreme or delayed seizure. Certain individuals don't encounter an air by any means, and their seizure has no prior advance signs.

The main advantage of this method is that using EEG signals, by using EEG signals the accuracy to detect seizures is improved.

Figure 4. A person with Electrodes attached to the sensor

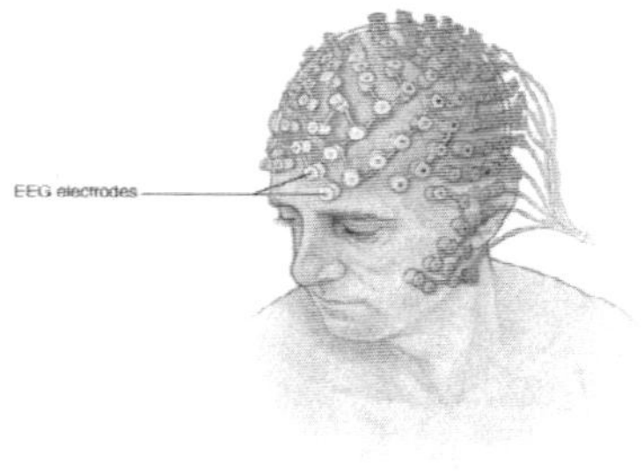

Several types of motions can potentially produce "**artifacts**" on an EEG recording of mimic brain waves. The person responsible for interpreting your EEG will take these movements into account.

Figure 5. A picture Description of normal and abnormal wave

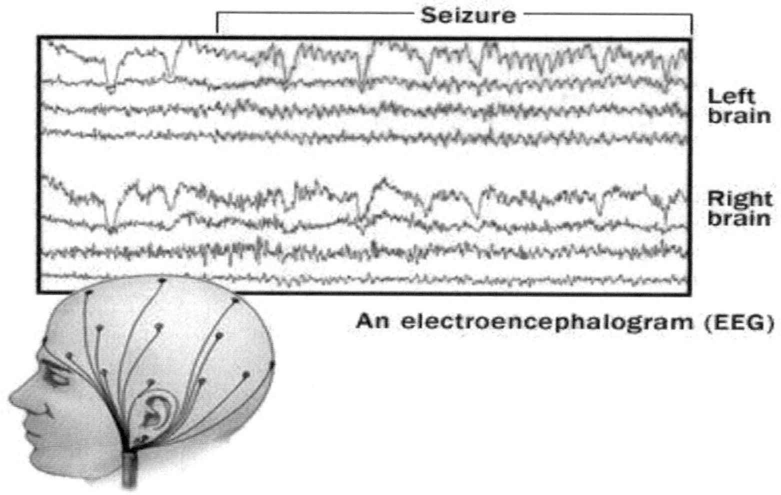

5. EXPERIMENTAL RESULTS

As we see in this Figure, there is a sudden abnormal electrical activity in the brain. When the electrical value nears 5000 there is a possibility of a seizure attack. So, we are taking that value as the parameter and if the graph reaches that value, it alerts the user with the message "Seizure Detected" and holds the recent medical record of a user.

Figure 6. A Cap which is used to measure Electrical Brain signals

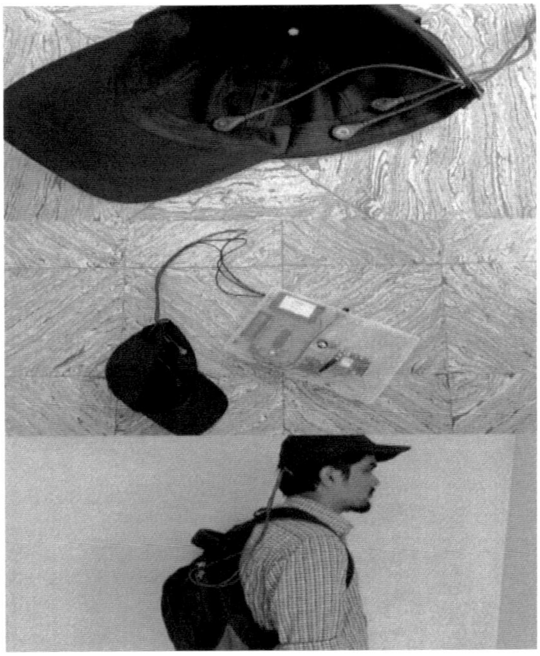

Figure 7. Seizure Detected

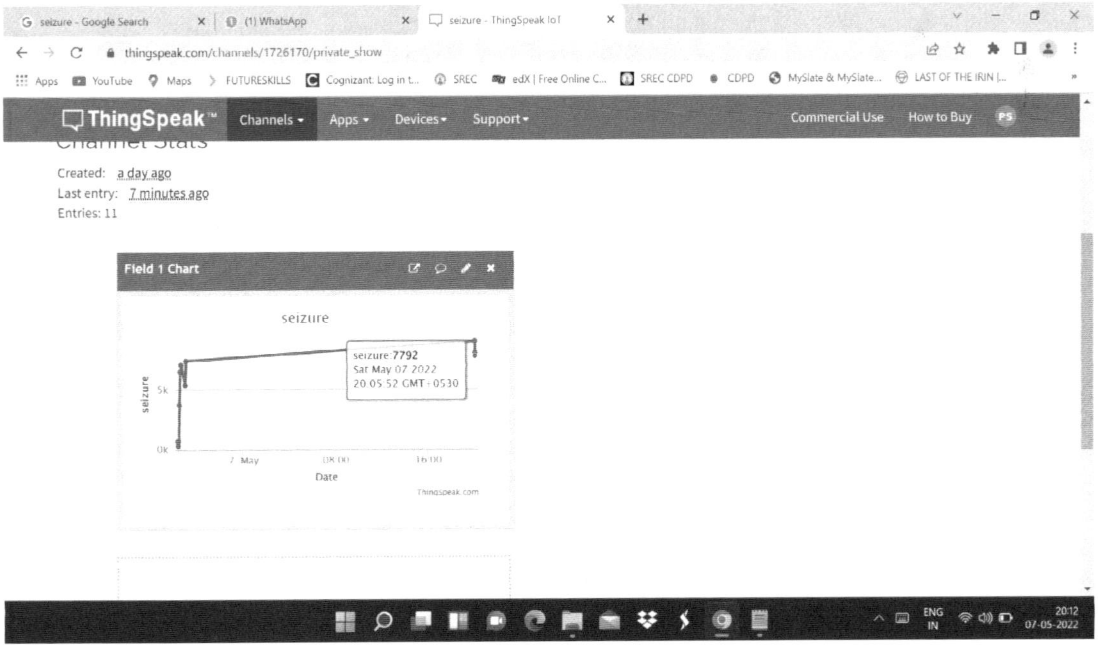

Figure 8. Seizure Detection Alerting Message

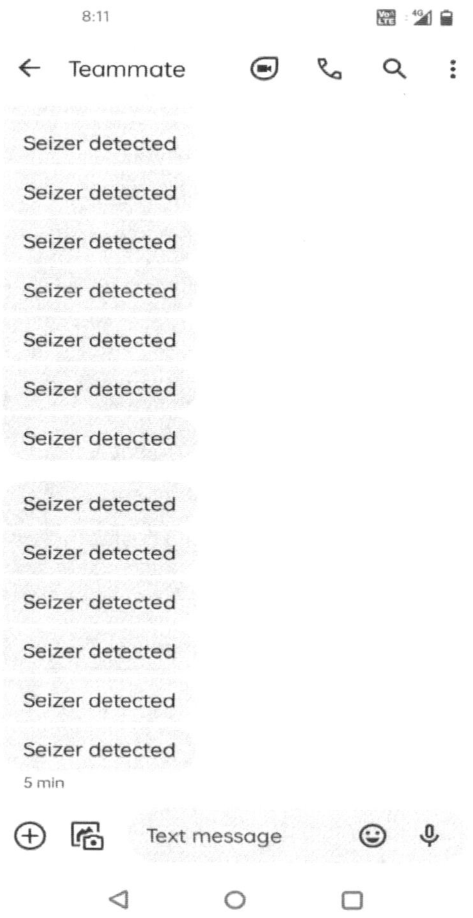

6. CONCLUSION

Early identification of seizures can keep the patient from death. In this work, we proposed a strategy given grouping the crude Electrical Brain signals from the scalp. The goal of this study is to assist the patient with early identification of seizures before they happen to lessen the gamble of seizure-related issues turning out to be more convoluted. The main aim of this paper is to improve the accuracy of the detection of epilepsy seizures. The results obtained for this proposed model are promising with precise and high accuracy rates.

7. REFERENCES

[1] A Wearable System Design for Epileptic Seizure detection, E. Shanthini, N. Sai Prasad, C. Keerthana & L. Sowmiya, Lecture notes on Data engineering and communication technology Volume 55, Pages 83-91,2021.

[2] M. Z. Poh, T. Loddenkemper, N.C.Swenson, S. Goyal, J. R. Madsen, and R.W. Picard, "Continuous monitoring of electrodermal activity during Epilepsy seizures using a wearable sensor". Annual International Conference of the IEEE Engineering in Medicine and Biology 2010, Buenos Aires, Argentina, Aug.31 – Sep. 4, 2010. pp. 4415-4418.

[3] B. Huyghe, J. Vanfleteren, and J. Doutreloigne, "Design of flexible, low-power and wireless sensor nodes for human posture tracking aiding Epilepsy seizure detection". In Proc. of the 2009 IEEE Sensors, Christchurch, New Zealand.Oct.25-28, 2009. pp. 1963-1966.

[4] S. Patel et al., "Detecting Epilepsy seizures using wearable sensors," In proc. of the 2009 IEEE 35th Annual Northeast Bioengineering Conference, Boston, MA, Apr.3-5, 2009, pp. 1-2.

[5] S. Sendra, A. Laborda, J. R. Díaz, and Jaime Lloret, "A Smart Bluetooth-based Ad Hoc Management System for Appliances in Home Environments", *Lecture Notes in Computer Science. Springer-Verlag Berlin Heidelberg, 2014.*Vol. 8487, pp. 128–141.

[6] Karuppusamy, Dr P. "Hybrid Manta Ray Foraging Optimization for Novel Brain Tumor Detection." Journal of Soft Computing Paradigm (JSCP) 2, no. 03 (2020): 175-185. Tumor Detection." Journal of Soft Computing Paradigm (JSCP) 2, no. 03 (2020):

[7] 175-185. R.B. Yaffe et al, "Physiology of functional and effective networks in epilepsy. Clin Neurophysiol 126(2):227–236, 2015.

[8] A.S, Zandi, M. Javidan, G.A Dumont, R. Tafreshi, "Automated real-time Epilepsy seizure detection in scalp EEG recordings using an algorithm based on wavelet packet transform", IEEE Trans Biomed Eng 57(7):1639–1651, 2010.

[9] M. Le Van Quyen, J. Martinerie, V. Navarro, P. Boon, M. D'Havé, C. Adam, B. Renault, F. Varela, and M. Baulac, "Anticipation of Epilepsy seizures from standard EEG recordings," The Lancet, vol. 357, no. 9251, pp. 183–188, 2001.

[10] Bouaziz, Bassem and Chaari, Lotfi and Batavia, Hadj and Quintero Rincon, Antonio Epilepsy seizure Detection Using a Convolutional Neural

[11] T. Yan, S. Zhao, S. Uono, X. Bi, A. Tian, S. Yoshimura, et al. (2016b). Target object moderation of attentional orienting by gazes or arrows. Attend. Percep. Psychophys. 78:2373. DOI: 10.3758/s13414- 016-1182-8

[12] C.L. Liu, "Epilepsy seizure Prediction With Multi-View Convolutional Neural Network", Ministry of Science and Technology, Taiwan, under Grant MOST 107-2221-E-009-109-MY2 and Grant MOST 107-2218-E-009-005.

[13] Accelerometer-Based Human Abnormal Movement Detection in Wireless Sensor Networks T. Ryan Burchfield and S. Venkatesan Computer Engineering Program University of Texas at Dallas

Power Quality Enhancement by minimizing the effect of Voltage Sag in Non-linear Load Using D-STATCOM

Vivek Sharma [a], **Vinod Kumar Chandrakar** [b]

Department of Electrical Engineering, G H Raisoni College of Engineering, Nagpur, India

[a]vivek.sharma.mtechps@ghrce.raisoni.net, vivek.sharma6000@gmail.com
[b]vinod.chandrakar@raisoni.net

Abstract - This research article discusses and examines the voltage sag and power quality issues in distribution networks with D-STATCOM. Power quality issues are examined under various system conditions. Load voltage v/s current characteristics of a nonlinear load is widely used in metal processing industries i.e. ARC Furnace, which is having Non-Linearity in the Load characteristics, is developed in MATLAB environment & recorded using Scopes. The reactive power control with D-STATCOM is applied to enhance the voltage profile and thus the power quality at the load point bus. The significant improvement in voltage and THD calculated by FFT analysis are improved as per IEEE standard.

Keywords - D-STATCOM, Arc Furnace, Voltage Sag, FACTS, MATLAB

1. Introduction

The large and complex Power System Network consist of various types of Generating Stations, large Transmission System along with complex Distribution Networks; which are connected with various types of Linear & Non-Linear loads. The Non-Linear loads are like rectifiers, IGBTs, SCRs, Arc Furnaces etc. [1-2]. These loads require flow of sinusoidal voltage and current for their efficient operations [3, 6].

The Power Distribution System used for industrial equipment, is subjected to regular power quality issues as per IEEE standards i.e. frequency deviations, Wave shape of Voltage and Current, Voltage level, Symmetry of three phase system etc. This deviation may lead to interruptions, flickers, Noise, Sag & Swell [5, 8].

Voltage Sag is a Power Quality problem in which RMS voltage magnitude decrease for short duration; generally because of different faults in power system [4, 7]. Due to voltage sag or voltage dip, the equipment capacity is scaled down or shut down abruptly. The machine failures and reduction of quality of work; affects the efficiency of mechanical output of the industrial machines driven by electrical energy due to deterioration in power quality. The failure of industrial & house hold appliances cause huge monetary & time losses [9-11].

In the manufacturing & processing industries various types of nonlinear loads viz. large capacitor rectifiers, electric arc furnaces, variable frequency drives are being used. The input voltage required for efficient working of plant & machinery, which involve electrical components, shall be purely sine wave & at component name plate input voltage frequency [15, 20]. However due to nonlinear nature of the load characteristics; and shape of the current & voltage wave form changes along with significant voltage dips at the interruption points [12-13].
The Voltage Source Converter (VSC) based Flexible

Alternating Current Transmission System (FACTS) devices are used in Distribution Network for reactive power compensation as well as to enhance the merits of system power by mitigating or minimizing the issues viz. voltage sag & Swell, non linearity of load characteristics etc [16-17].

The PI based D-STATCOM which is capable to enhance the voltage waveform profile through compensation of reactive power by firing angle control strategy. Many reviewed research papers presented the control of firing angle based on measured voltage at device location. With load having nonlinear characteristics & under varying system conditions; the execution of PI Based controller devolves. The VSC based compensating devices are having smooth and fast controlling ability. These devices are proved for power quality improvement issues [18-19].

Total Harmonic Distortion (THD) signifies a value that denotes Harmonic Distortion present in the wave form. Voltage with fundamental frequency is responsible to deliver useful power in all Alternating Current loads. The undesirable frequencies which are superimposed on fundamental are referred as Harmonics. Ideally the supplied Voltage shall have no harmonics. But due to some factors viz. fault, Non Linear Loads, Noise, AC to DC convertor & vice versa etc are responsible to insert harmonics in the power system network. The total Harmonic Distortion (THD) [5,20].

General expression of alternating voltage of power system network; with harmonics can represents as

$$v(t) = V_{dc} + \sum_{n=1}^{\infty} V_{mn} \sin(n\omega t \pm \phi_n) \qquad \text{...... (1)}$$

$$where, (n = 1, 2, 3, \dots.)$$

$v\ (t)$ – Alternating Voltage of Power System Network
V_{dc} – DC component of $v(t)$
V_{mn} – Maximum Voltage of nth Harmonic
$n\omega$ – Angular Frequency of nth harmonic voltage
ϕ_n – Phase difference of nth harmonic voltage

Root Mean Square (RMS) value of line voltage $v(t)$ may be written as equation (2)

$$V_{rms} = \sqrt{[(V_{dc})^2 + (V_{1.rms})^2 + (V_{2.rms})^2 + (V_{3.rms})^2 + \cdots]} \qquad \text{...... (2)}$$

Method to calculate THD value of $v(t)$ is given as below in equation (3)

$$THD = \frac{\sqrt{(V_{rms})^2 - (V_{1.rms})^2}}{(V_{1.rms})} \qquad \text{..... (3)}$$

Where,

V_{dc} – DC component of $v(t)$
V_{rms} – rms value of voltage $v(t)$
$V_{1.rms}$ – rms value of voltage with fundamental frequency
$V_{2.rms}$ – rms value of voltage with Second Harmonics
$V_{3.rms}$ – rms value of voltage with Third Harmonics

The measured THD value represents Wave forms which don't have harmonics contain, have nil Total Harmonic Distortion (THD) value. Harmonics in the system make THD value non zero and the requirement arises to eliminate harmonics from the power system network. This article focuses on investigation of the Voltage Sag considering the connected load, a major nonlinear load, used in manufacturing industries i.e. Electric Arc Furnace with the help of MATLAB Simulation Software. The Simulation Model of Arc Furnace is prepared. The variable considered for analysis is 3000A & 0.9pf lag as it gives improved results by minimizing the non linearity for arc furnace type of non linear load [1]. The Sag conditions are created by applying the L-G fault on the Transmission Line and further, their effects are recorded using Scopes. The PI based firing angle control strategy for D-STATCOM is developed to achieve improved Voltage profile and to mitigate the impact of Sag on the power system. The Voltage v/s Current Characteristics of the Arc Furnace load and the THD analysis with and without D-STATCOM are plotted & tested, respectively, in MATLAB environment.

In this proposed work, initially Voltage Profile is observed with line fault which create Sag & later improvement in voltage waveform with the help of D-STATCOM is also presented for Arc Furnace type of Variable load. The firing of D-STATCOM is controlled by difference of two signals i.e. Measured Voltage (Vm) and Reference Voltage Vr). The Proportional Gain (Kp) and Integral Gain (Ki) are tuned with the help of trial and error method.

2. Sample Test System

The simulation of the example considered for analysis of Power System Network is shown in Figure 1. D-STATCOM, in the network at the effective location, improves the Voltage Profile by minimizing the Sag. Arc furnace type of nonlinear load is used for test system & its Simulink model is shown in Figure 2. V-I characteristics of the Nonlinear loads is also developed. It also excludes THD from the Electrical power system.

An example of the sample power network constructed with Transformer 1 and Transformer 2. Transmission Line 1 and Transmission Line: 2 and the D-STATCOM ratings used in simulation are given in Table 1. The L-G fault at Bus: 1 for duration of 0.1 Sec is applied on the power network for the investigation the Sag condition.

Table 1: Rating Of Simulated Network's Components

Components		**Ratings & Descriptions**
Transformer	**T1**	120kV/25kv, 47 MVA, Y-Δ
	T2	25 kV/415V, 6 MVA, Δ-Y
Transmission Line	**TL1**	21km, π-Network
	TL2	2km, Reactance Model
D-STATCOM	**D1**	± 3MVAR
Variable Load	**Arc Furnace**	3000A, 0.9pf

The PI based Control strategy of shunt compensator is shown in figure 2. The AC voltage regulator change Iq(ref) and compared with Iq signal for developing the alpha signal in current regulator circuit. The alpha signal is useful for generating the firing pulse for the six (06) pulses VSC. The Vdc is kept constant during normal and abnormal conditions.

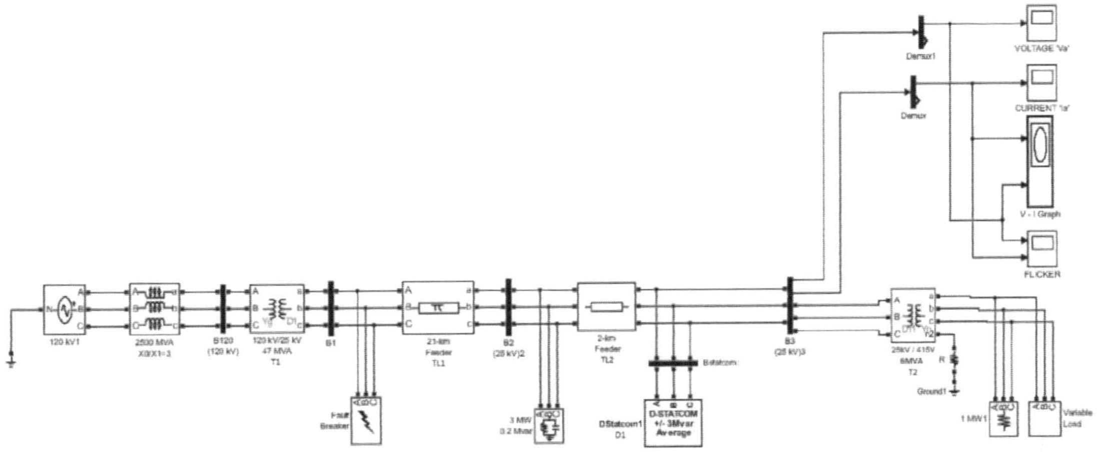

Figure 1: Simulated Power System Network

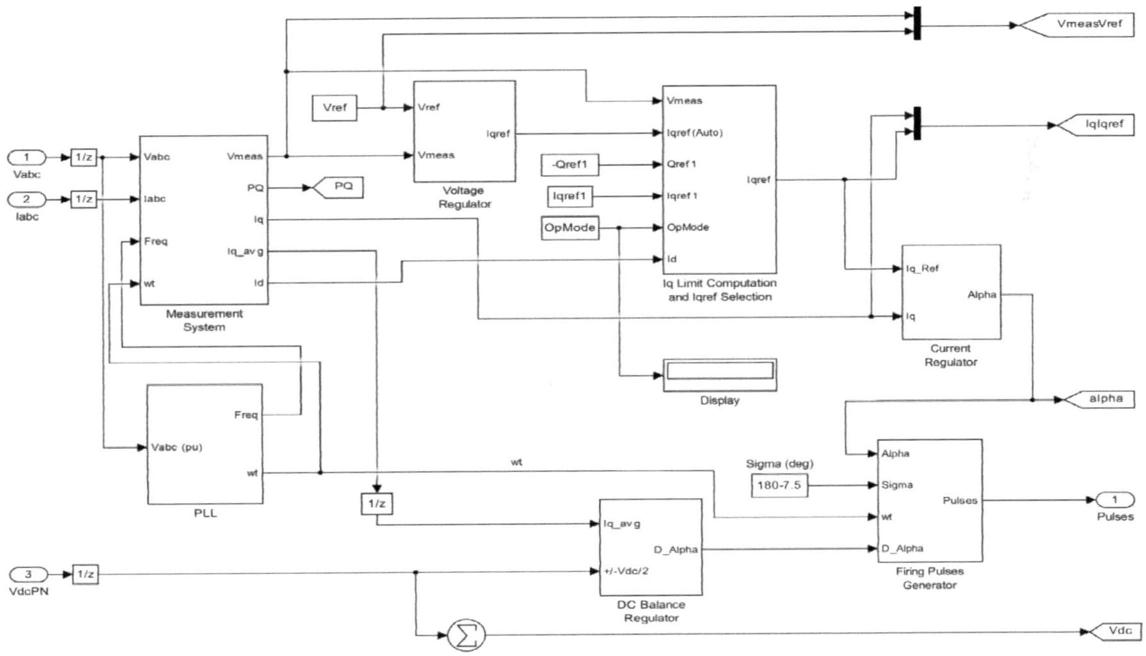

Figure 2: Control strategy of D-STATCOM

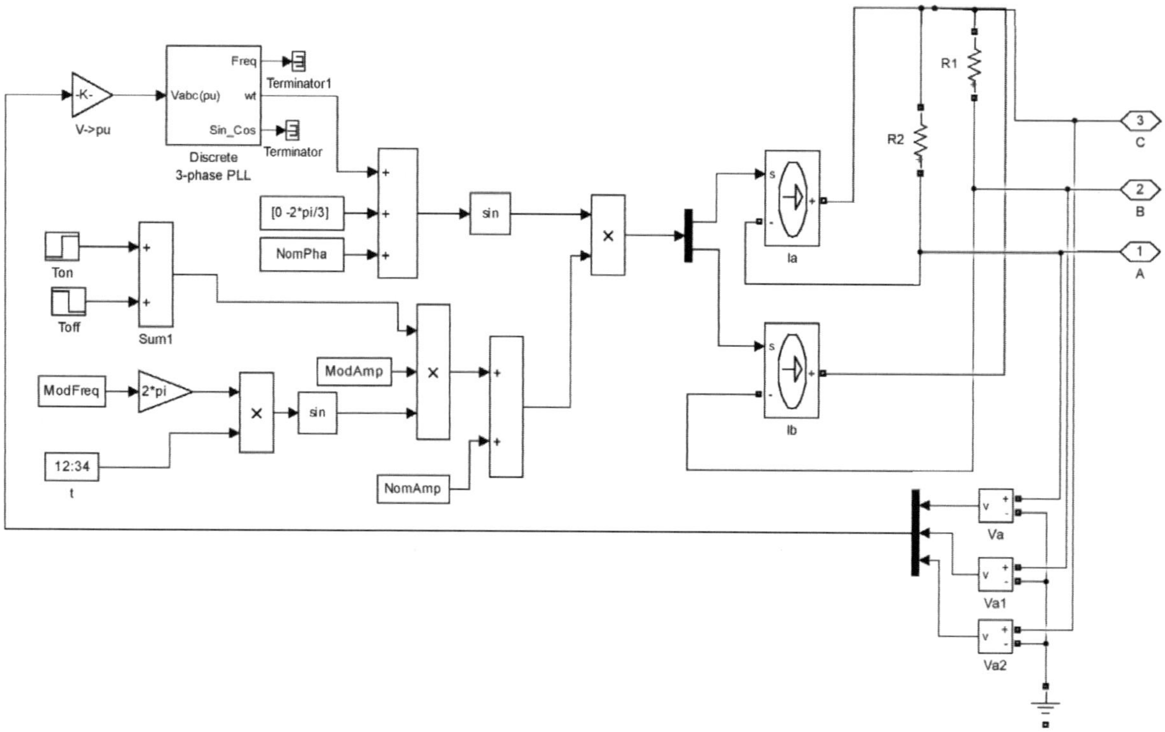

Figure 3: Simulink Model of Arc Furnace Load

3. System Response & Analysis

Non- Linear Load i.e. Arc Furnace, is connected to the equivalent Simulink model at the distribution end. The V-I characteristics & Voltage profile of Load under 0.9pf lagging condition are shown in figure 4(a) and figure 4(b); & 4(c) respectively. Figure 4(a) shows the non-linearity in the Voltage v/s current characteristics of the considered variable load. As depicted by Figure 4(b) & 4(c), the Y axis parameter i.e. voltage is significantly changing by applying the reactive power compensation using D-STATCOM.

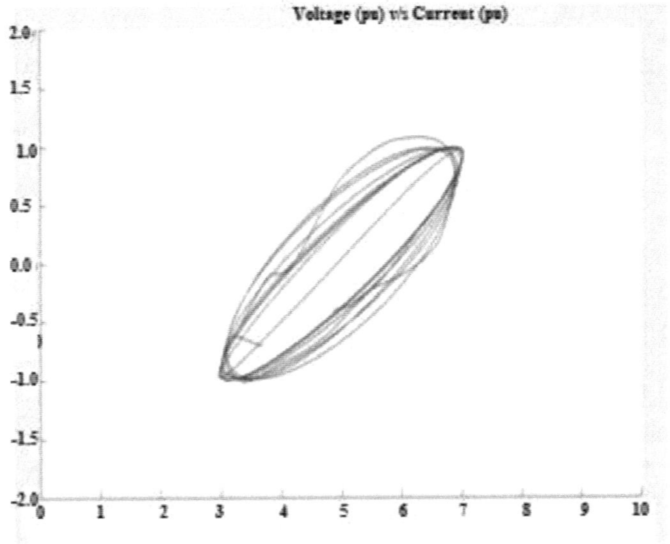

Figure 4(a): Load characteristic under 0.9pf lagging with D-STATCOM

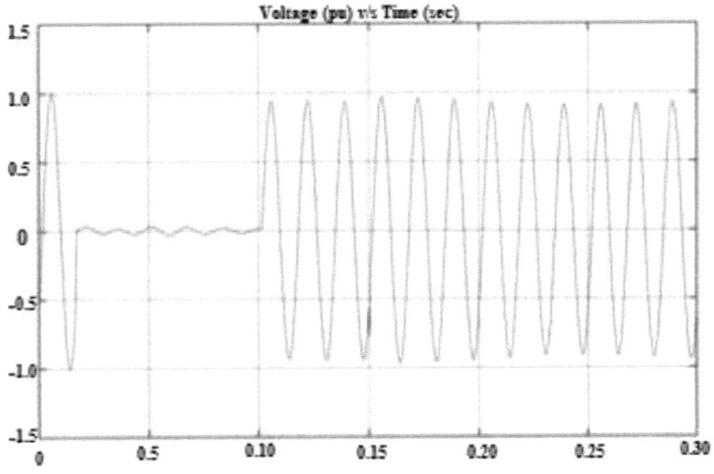

Figure 4(b): Voltage response under 0.9 pf LoadCondition with voltage sag

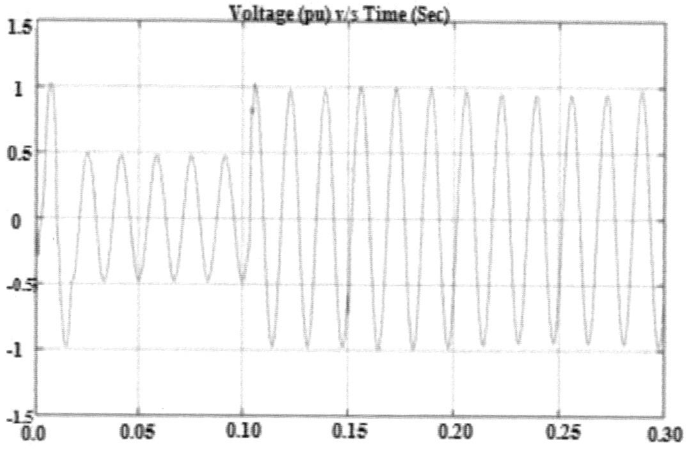

Figure4(c): Voltage response with D-STATCOM

Figures 4(b) and Figure 4(c) depict that Voltage Profile is getting improved from nearly 0.1pu to 0.5pu approx. with the help of D-STATCOM.

4. Thd Analysis

The FFT analysis is applied with and without Sag condition in the voltage wave form. The results are shown in table 2 below-

Table 2: VOLTAGE (pu) & (%) THD VALUES

Conditions		**V (pu)**	**THD (%)**
Sag absent	**Without D-STATCOM**	~ 1pu	5.96
	With D-STATCOM	1pu	2.61
Sag Present (for 0.1Sec duration)	**Without D-STATCOM**	~ 0.1pu	34.04
	With D-STATCOM	~ 0.5pu	27.92

It can be seen from Table 2 that the voltage THD (%) values decrease with D-STATCOM.

5. Conclusion

This research work discusses the enhancement of voltage profile with the participation of D-STATCOM at Bus: B3. Arc Furnace is connected at the distribution end; which also have Non-Linear characteristics. The elaborated MATLAB / Simulink model has been enforced and examined for D-STATCOM. The behavior of the Non-Linear load is observed under 0.9 pf loading conditions. Load characteristic is developed which also have elliptical nature which signals about the Non Linearity of the connected load i.e. Arc Furnace. The Voltage is having Sag/dip of nearly 90% (Vrms = 0.1pu). The voltage waveform is improved with the help of D-STATCOM to approximately to Sag/dip of 50% (Vrms = 0.5pu) and thus improve the Power Quality. This enhances the effective RMS value of voltage wave form & further power quality by improving the flow of power to load center. The voltage profile can be improved by adding D-STATCOM at the Distribution end i.e. near to Load Centre. The voltage waveform & FFT analysis under different conditions show that D-STATCOM substantially enhances the Power Quality by reducing the Voltage .Dip. The % THD is reduced from 34.04% to 27.92%. In future work may be done for further reduction of %THD value& thus power factor improvement for Distribution Arc furnace Load.

References

[1] V. Sharma and V. K. Chandrakar, "Power Quality Enhancement of Grid Connected Non Linear Industrial Load Using Distributed Static Synchronous Compensator," 2022 International Conference on Electronics and Renewable Systems (ICEARS), 2022, pp. 271-277.

[2] C. Kumar, M. K. Mishra and S. Mekhilef, "A new voltage control strategy to improve performance of DSTATCOM in electric grid," in CES Transactions on Electrical Machines and Systems, vol. 4, no. 4, pp. 295-302, Dec. 2020.

[3] L. Luo, W. Chen, Y. Tian, Z. Tan and Y. Hu, "Compensation Strategy for Multiple Series Centralized Voltage Sag in Medium Voltage Distribution Network," 2019 IEEE PES Asia-Pacific Power and Energy Engineering Conference (APPEEC), 2019, pp. 1-6.

[4] L. Luo, W. Chen, Y. Tian, Z. Tan and Y. Hu, "Compensation Strategy for Multiple Series Centralized Voltage Sag in Medium Voltage Distribution Network," 2019 IEEE PES Asia-Pacific Power and Energy Engineering Conference (APPEEC), 2019, pp. 1-6.

[5] V. T. Kullarkar and V. K. Chandrakar, "Power quality improvement in power system by using static synchronous series compensator," 20172nd International Conference for Convergence in Technology (I2CT), 2017, pp. 1035-1040.

[6] T. S. Saggu, L. Singh and B. Gill, "Harmonics Mitigation in a Steel Industry Using 11-Level Cascaded Multilevel Inverter-Based DSTATCOM," in Canadian Journal of Electrical and Computer Engineering, vol. 40, no. 2, pp. 110-115, Spring 2017.

[7] A. Singh, R. K. Singh and A. K. Singh, "Voltage sag correction in distribution system containing a real three phase electric arc furnace," 2017 4th IEEE Uttar Pradesh Section International Conference on Electrical, Computer and Electronics (UPCON), 2017, pp. 21-26.

[8] V. T. Kullarkar and V. K. Chandrakar, "Power quality improvement in power system by using static synchronous series compensator," 2017 2nd International Conference for Convergence in Technology (I2CT), 2017, pp. 1035-1040.

[9] Priyanka K. Gaurkhede and Vinod Kumar Chandrakar, "SVPWM Based STATCOM for Power Quality Improvement," International Journal of Electrical Engineering, ISSN 0974-2158 Volume 10, Number 1 (2017), pp. 1-12.

[10] A. Singh, R. K. Singh and A. K. Singh, "Voltage sag correction in distribution system containing a real three phase electric arc furnace," 2017 4th IEEE Uttar Pradesh Section International Conference on Electrical, Computer and Electronics (UPCON), 2017.

[11] C. Kumar and M. K. Mishra, "A Voltage-Controlled DSTATCOM for Power-Quality Improvement," in IEEE Transactions on Power Delivery, vol. 29, no. 3, pp. 1499-1507, June 2014, doi: 10.1109/TPWRD.2014.2310234.

[12] "IEEE Recommended Practice for Voltage Sag and Short Interruption Ride-Through Testing for End-Use Electrical Equipment Rated Less than 1000 V," in IEEE Std 1668-2017 (Revision of IEEE Std 1668-2014) , vol., no., pp.1-85, 27 Nov. 2017.

[13] Madhyama V. Wankhede and V. K. Chandrakar, "Static Synchronous Compensator (STATCOM) for the improvement of the Electrical System performance with Non Linear Load," International Journal of Power System, 2016, pp. 27-32 Vol.1.

[14] Singh, B., Dube, S.K. and Arya, S.R., 2015. An improved control algorithm of D-

STATCOM for power quality improvement. *International Journal of Electrical Power & Energy Systems*, *64*, pp.493-504.

[15] C. Kumar and M. K. Mishra, "A Voltage-Controlled D-STATCOM for Power-Quality Improvement," in IEEE Transactions on Power Delivery, vol. 29, no. 3, pp. 1499-1507, June 2014.

[16] B. Singh, P. Jayaprakash, D. P. Kothari, A. Chandra and K. A. Haddad, "Comprehensive Study of D-STATCOM Configurations," in IEEE Transactions on Industrial Informatics, vol. 10, no. 2, pp. 854-870, May 2014.

[17] V. F. Corasaniti, M. B. Barbieri, P. L. Arnera and M. I. Valla, "Hybrid Power Filter to Enhance Power Quality in a Medium-Voltage Distribution Network," in IEEE Transactions on Industrial Electronics, vol. 56, no. 8, pp. 2885-2893, Aug. 2009, doi: 10.1109/TIE.2009.2014369.

[18] Chaoui, A., Gaubert, J.P., Krim, F. and Champenois, G., 2007. PI controlled three-phase shunt active power filter for power quality improvement. *Electric Power Components and Systems*, *35*(12), pp.1331-1344.

[19] M. G. Molina and P. E. Mercado, "Control Design and Simulation of D- STATCOM with Energy Storage for Power Quality Improvements," 2006 IEEE/PES Transmission & Distribution Conference and Exposition: Latin America, 2006, pp. 1-7.

[20] Chandrakar, V. K., and A. G. Kothari. "Optimal location for line compensation by shunt connected FACTS controller." In *The Fifth International Conference on Power Electronics and Drive Systems, 2003. PEDS 2003.*, vol. 1, pp. 151-156. IEEE, 2003.

[21] V. T. Kullarkar and V. K. Chandrakar, "Power quality analysis in power system with non-linear load," Int. J. Electr. Eng, vol. 10, pp. 33-45.

Simulation Study of Hetero-Junction Single Gate Extended Source TFET

Suryavansh Singh, Saurav Guha Roy, Anjali Kumari, Sweta Chander, Sanjeet Kumar Sinha, and Rekha Chaudhary

School of Electronic and Electrical Engineering, Lovely Professional University, Phagwara, Punjab, India.

suryavanshsingh3@gmail.com, sauravguharoy2012@gmail.com, anjaliraj192001@gmail.com, sanjeetksinha@gmail.com, sweta.chander@gmail.com, rekha.ceeri@gmail.com

Abstract—The paper provides the information regarding the examination of Uni-Gate-Extended Source TFET w.r.t the doping concentration of source, drain and channel and device's dimensions. The moto of extending the source into channel is to enhance the point to point tunneling. 2D simulations in TCAD tool are carried out to analyse and investigate the performance of proposed device. Because of the transposed SS down the thermionic limit of 60 mV/decade, TFETs can be considered as the most trustworthy devices that have lower power consumption and that's why they are among the deserving devices for upcoming future CMOS logic applications. Although the TFET marks have been examined in fewer material systems, they are burdened by low ON-currents. We discussed the impact of sound on Tunnel FET devices and circuits. The dialogue is designed for the sound of a random noise effects. The paper also deals with configuration of gm and gd and their obtained significant values are rise up in the graph till 1x10-4 V-1 and 2x10-4V-1.

Keywords—Doping Concentration; Source; Hetero-Junction; TFET; Band-to-band-Tunneling

1. INTRODUCTION

As the size of MOSFETs reduces, the power supply voltage also decreases, and hence reduces power density [1][2][3]. However, Tunnel Field Effect Transistor (TFET) is used for low power electronics, and it is based on Band-to-Band Tunneling (BTBT) mechanism. TFET has excellent switching properties at room temperature like Subthreshold Swing (SS) is below than 60 mV/decade which are far better than SS of conventional MOS transistor [4]. In the study introduces use of high-gate oxide in semiconductor technology that reduces the gate leakage and enhance the stability of the devices [5][6]. Nonetheless, TFET exhibits high ON current but also shows ambipolar behavior that further increases the leakage current [7]. Ambipolar conduction is also one of the biggest concern in TFET technology as higher ambipolar current leads to decrease in power consumption [8][9]. These issues can be resolved by various techniques that have been already introduced like low-bandgap material as source [10][11], hetero-structure mechanism using III-V compound semiconductor [12][13], using the high-k spacers [14] and the pocket dropped engineering [15][16] to enhance the performance of ON-state current.

Journal of Physics: Conference Series **2325** (2022) 012020 doi:10.1088/1742-6596/2325/1/012020

In TFET, ON current can be increased by using source-pocket doping technique without increasing leakage current. Yang *et al* analysed the TFET gate capacitance and derived the analytical model of capacitance by presenting the comparative analysis of conventional SOI (Silicon on Insulator) TFET and conventional SOI MOSFET [17]. A device based on an extended source TFET (ES-TFET) to enhance the performance of AC-DC analysis. It has shown an excellent performance improvement over standard TFET.

The present work reports the electrical noise analysis of extended-source TFET (ES_TFET) of 28 nm gate length. The detailed study of device characteristics like DC analysis, AC analysis and electrical noise analysis has been presented thoroughly. The paper is organized as: Section II elaborates the proposed device schematic, device dimensions and the simulation model used. Section III presents the optimization of doping concentration. While section IV elaborates the results and discussions. At the end, the paper is settled in section V.

In this paper, numbers of noise parameters are considered such as flicker and diffusion noise having Gaussian profile at the region of Oxide and extended interface is studied by altering different structures and parameters of the conventional device. Noise analysis might be considered to achieve the charge effect so results decadency effects in the transistor's behaviour make the reliability of circuits and mess structure. The material used for the gate and gate oxide and Hafnium oxide in a order. The process lithography process source and Drain can be implemented. The part of Source and Drain is 30nm *40nm without considering the part of Extended source. In semiconductors, reducing the bandgap of model is enabled.

2. DEVICE SCHEMATIC AND DIMENSIONS

The simulation device's structure is shown in Fig.1. The device examination for the simulation is taken with 2×10^{19} cm^{-3} (Source), 1×10^{-15} cm^{-3} (Channel) , 1×10^{-18} cm^{-3} (Drain) and 1×10^{-15} cm^{-3} (Back). The device in the Fig.1 has a source length of 40 nm and height of 30 nm followed by source extension of 12 nm and height of 5 nm. The source is followed by a channel length of 28 nm and with a height of 30 nm. The channel works as a base for the settlement of oxide on it i.e., HfO_2. The dimensions of the oxide is 28×1 nm which is again overheaded by a metallic gate of length 28 nm and height of 3 nm. The material used for metallic gate is aluminium. A SiO_2 box is attached beneath the device of dimension 110×30 nm which is again based by a back of dimension 110×5 nm. The material used for source formation is Germanium, Silicon is for Channel, Drain and Back, HfO_2 as oxide, Aluminium for metallic gate and SiO_2 for Box. The doping scenario of the device follows the ***p-i-n*** configuration. The source is heavily-doped by Boron with a concentration of 2×10^{19} cm^{-3}, channel and back is doped same as that of channel i.e., with phosphorus with a concentration of 1×10^{15} cm^{-3} and drain with a concentration of 1×10^{18} cm^{-3} .In this simulation, Sentaurus TCAD simulator has been used [18].

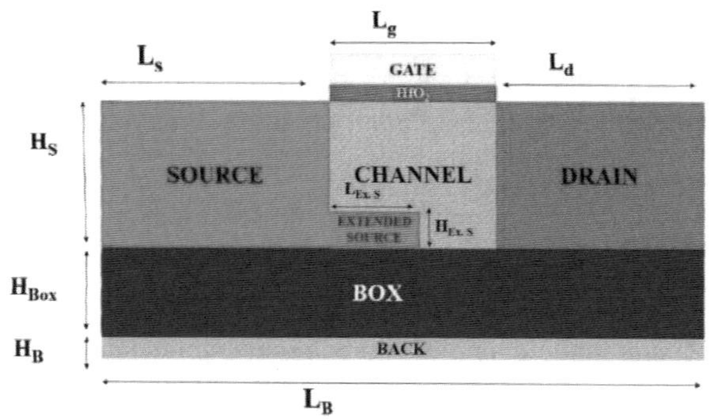

Figure 1: Schematic diagram of proposed device

Table 1: Device Dimensions

Device Parameters	Symbol	Value	Unit	Material
Length of Source	L_s	40	nm	Germanium
Length of Extended Source	$L_{Ex. S}$	12	nm	Germanium
Height of Source	H_s	30	nm	Germanium
Height of Extended Source	$H_{Ex. S}$	05	nm	Germanium
Length of Channel	L_c	28	nm	Silicon
Length of Drain	L_d	42	nm	Silicon
Height of Box	H_{Box}	30	nm	Silicon Oxide (SiO_2)
Length of Box	L_{Box}	110	nm	Silicon Oxide (SiO_2)
Height of Back	H_B	05	nm	Silicon
Length of Back	L_B	110	nm	Silicon
Length of Oxide	L_{ox}	28	nm	Hafnium Oxide (HfO_2)
Height of Oxide	H_{ox}	01	nm	Hafnium Oxide (HfO_2)
Length of Gate (Aluminium)	L_g	28	nm	Aluminium
Height of Gate (Aluminium)	H_g	03	nm	Aluminium

3. DOPING OPTIMIZATION

The source was optimised on the four different concentrations i.e., $1\times10^{18}\,cm^{-3}$, $1\times10^{19}\,cm^{-3}$ $2\times10^{19}\,cm^{-3}$ and $3\times10^{19}\,cm^{-3}$.The final optimized doping concentration was $2\times10^{19}\,cm^{-3}$. This value was kept as a final value because it was examined that there was a finest source doping that increases the TFET current at its max. A definite degree of the source depletion also helps because it keeps the low barrier hole tunneling in the source that makes up part of the tunneling way. The change in the current in the OFF region (I_{OFF}) is significant because the change in the source doping of $2\times10^{19}\,cm^{-3}$ has a relatively high effect on the barrier height of the electrons and holes formed at the virtual junction than the doping concentrations of $1\times10^{18}\,cm^{-3}$, $1\times10^{19}\,cm^{-3}$ and $3\times10^{19}\,cm^{-3}$. A current difference is also observed in the ON (I_{ON}) region. A dependence of asymmetric doping concentration among the parameters between I_{ON} and I_{OFF} takes place. While the change in the I_{ON} and I_{OFF} is negligible when the doping concentration is of source is set to a value other than $2\times10^{19}\,cm^{-3}$. There was a low effect on the barrier height of electrons and holes formed at the virtual junction.

While optimizing the doping concentrations of drain and channel there was negligible effect and sometimes no effect on I_{ON} and $I_{OFF.}$ The final optimized value of channel and drain are $1\times10^{-15}\,cm^{-3}$ and $1\times10^{-18}\,cm^{-3}$ respectively. While doping the drain with concentrations of $1\times10^{-17}\,cm^{-3}$, $1\times10^{-18}\,cm^{-3}$ and $1\times10^{-19}\,cm^{-3}$, the best I_{ON}/I_{OFF} characteristics obtained at $1\times10^{-18}\,cm^{-3}$ and in rest two concentration values the ON region diminished as of low I_{ON} and OFF region stimulates more leakage as of high I_{OFF}. While doping the channel with concentrations of $1\times10^{-15}\,cm^{-3}$, $1\times10^{-16}\,cm^{-3}$ and $1\times10^{-17}\,cm^{-3}$, the I_{ON}/I_{OFF} characteristics remains the same, as while plotting the I_d/V_g graph taking all the three concentration values the graphs get over laped over each other exactly which shows that the optimization of doping concentrations of channel doesn't put any feasible change in the characteristics of the device.

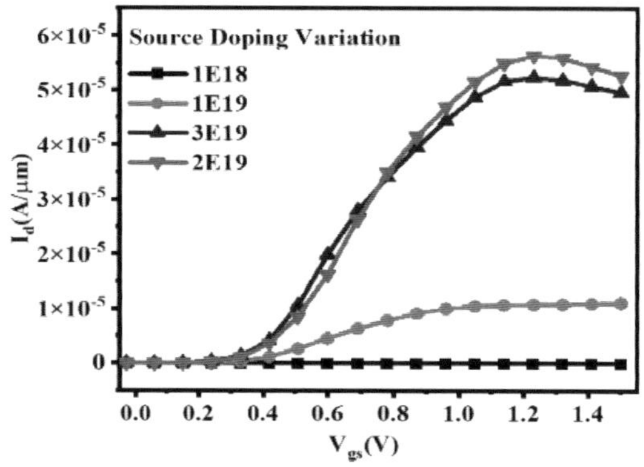

Fig.2(a)

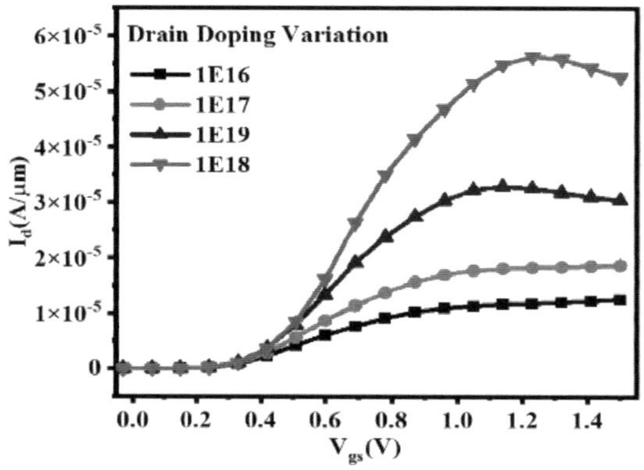

Fig.2(b)

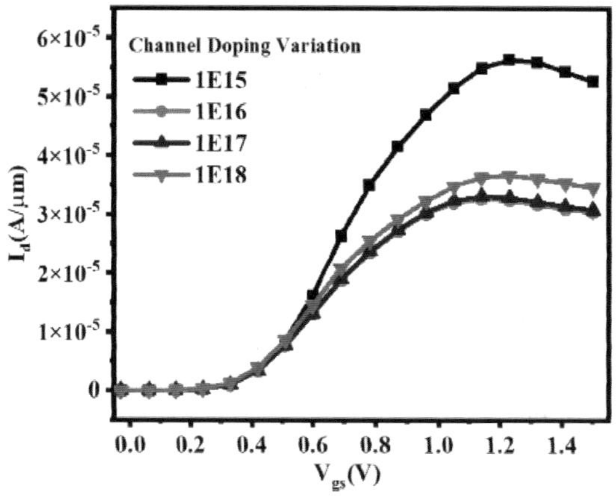

Fig.2(c)

Figure 2: Optimisation of Source, Drain and Channel (a) Source Doping Optimisation (b) Drain Doping Optimisation (c) Channel Doping Optimisation

4. RESULTS AND DISCUSSIONS

This section presents the final and detailed analysis of noise along with the different optimizations in source, drain and channel doping with different concentrations. DC analysis, AC analysis and noise analysis are also examined along with different parameters.

A. DC Analysis:

The DC analysis of the device is considered with the Energy band variations, Electric Field density along with band-to-band tunnelling in consideration with electrons and holes.

The Fig.3(a) shows the electric field variations along the whole device. The addition of extended gate in the conventional device makes it more efficient as the electric field area was plotted from 3.312e+01 to the range of 4.665e+06 (max. is 9.331e+06) ,which is a required output for a better TFET.

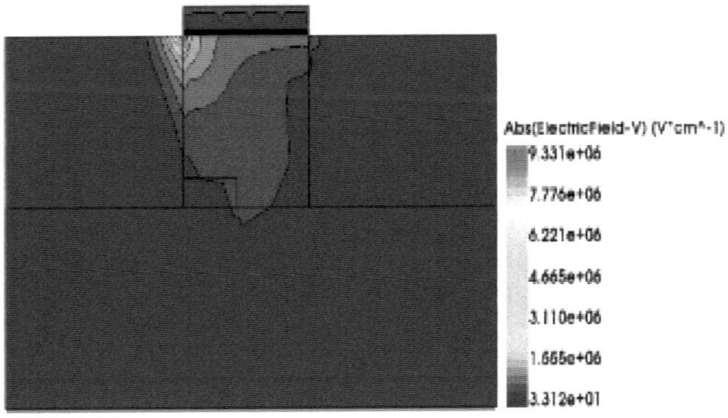

Fig.3(a)

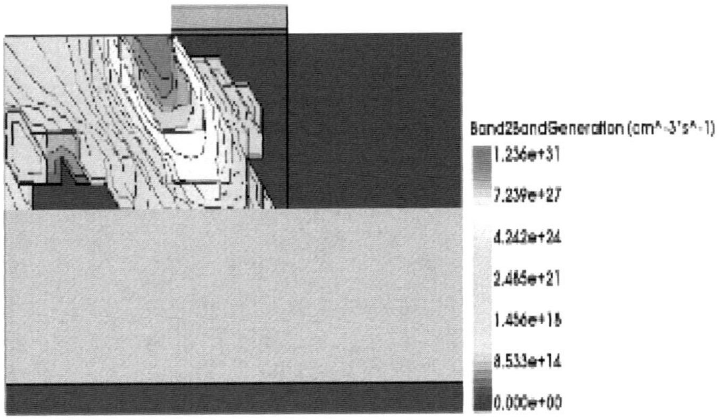

Fig.3(b)

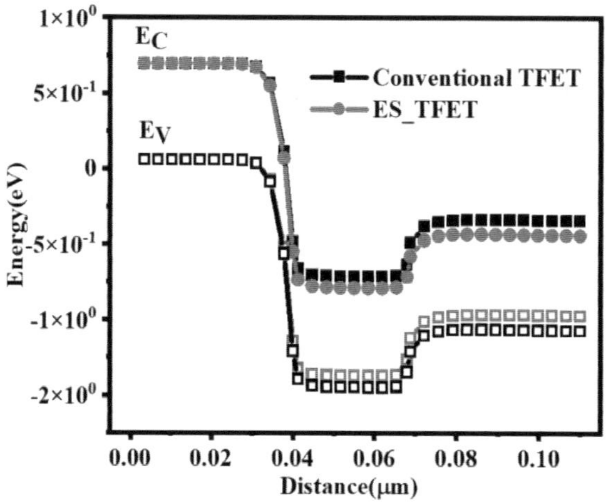

Fig.3(c)

Fig.3 Image of the different energy parameters of device (a) Electric Field (b) BTBT (c) Energy Band

The Fig.3(b) shows the band to band tunneling along the holes and electrons through the device itself. This allows TFET to achieve a higher I_{ON}-I_{OFF} rating over a particular gate power supply, thus enabling a powerful power measurement at reasonable low power performance without interfering with its current ON-OFF rate. This BTBT is much efficient than that is there in conventional TFET as the I_{ON}-I_{OFF} characteristics is poor and the concept of low power doesn't carry out over there.

The Fig.3(c) shows the energy band graphs of the final optimised device and the device without the extended source (conventional). ID_VG_n16 resembles the band of conventional device and ID_VG_f resembles the band of final optimised device. The energy band variations between both of the devices can be shown easily as the the increase in power of the tunnel fills the portable charging carrier and reduces the resistance and causes an increase in the conduction of transmission beyond the threshold voltage.

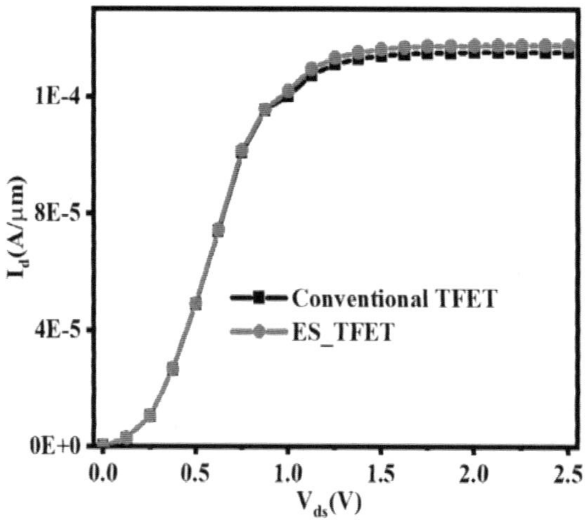

Fig.4(a)

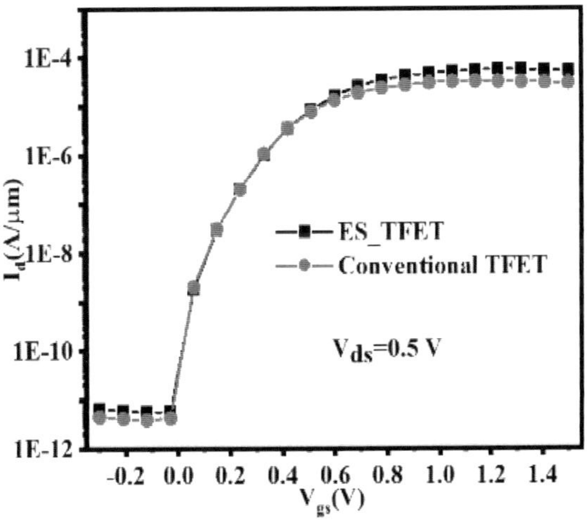

Fig.4(b)

Fig.4 Drain current- Drain voltage and Drain current- Gate voltage (a) I_dV_d characteristics (b) I_dV_g characteristics

The Fig.4(a) shows the I_dV_d charcteristics of the conventional and the final optimised device (ES TFET). The device displays the expected features due to the tunnel during fine V_{ds} (bias reversal conditions) while I_{ds} grows significantly. In two cases where V_{ds} is negative (bias). When V_{ds} is ~0V, there is an important I_{ds} regardless of V_{gs} value.Significant current performance is also evident when V_{ds} is low and V_{gs} is positive. . This is due to electrons tunneling from the conduction band of intrinsic region to the valence band of source region.

In such a way the Fig.4(b) represents the I_dV_g characteristics of the conventional device and the final device. In both of the devices we can find that the rise I_d is more in the final device than the conventional device which happens only because of the concept of extended source; the gate leakage becomes negligible which gives a proper path for the current to flow.

B. DC Analysis:

The AC analysis comprise of the comparison between conventional device and the final optimised device with respect to capacitance, trans-conductance (g_m) and output conductance (g_d).

The Fig.5(b) shows the graph plotted between output conductance and gate voltage. As the source in the final device extended towards the channel with a length of 12 nm and height of 5 nm even then the output conductance doesn't vary much, even gives the same result. The slope of both the devices is steep at the same point and get to the bottom and then constant at the very same time

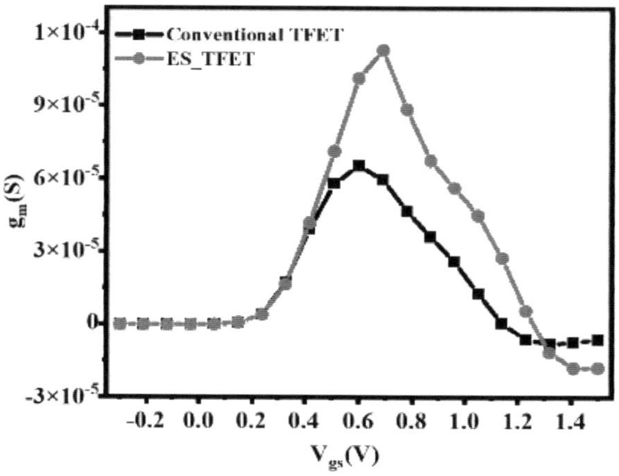

Fig.5(a)

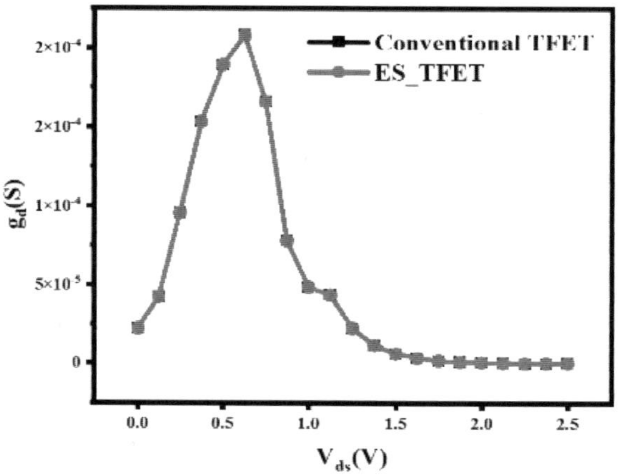

Fig.5(b)

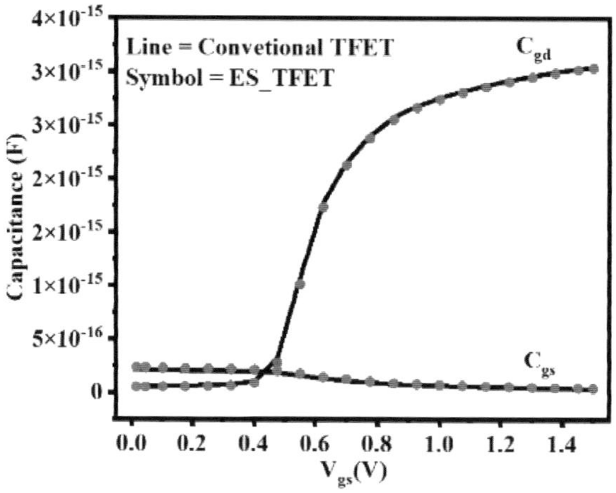

Fig.5(c)

Fig.5 Graph showing the trans-conductane, output conductance and capacitance of device w.r.t device's output voltage and current (a) $g_m V_g$ (b) $g_d V_d$ (c) FV_g

The Fig.5(c) shows the graph plotted between the Capacitance (F) and gate voltage (V_g). To improve the TFET switching speed, capacitances should be reduced. The capacitance of the final device becomes a bit for efficient by adding the external source in the conventional device. The switching speed is also get effected only a bit as the capacitance along the device is carried out by the gradual drain junction ,that is why the optimization in of the source doesn't play a crucial role in the capacitance essentiality

A. Noise Analysis:

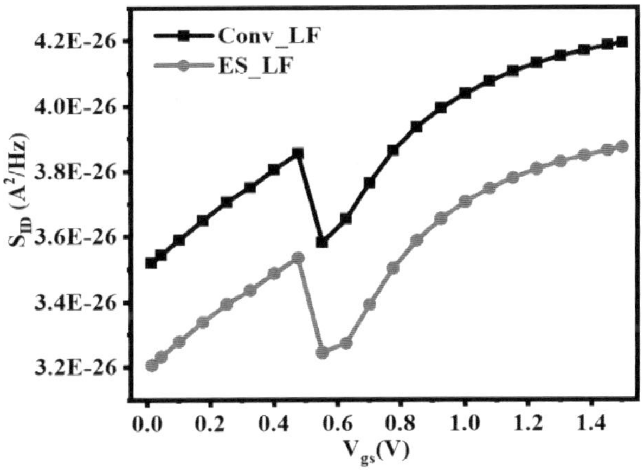

Fig.6(a)

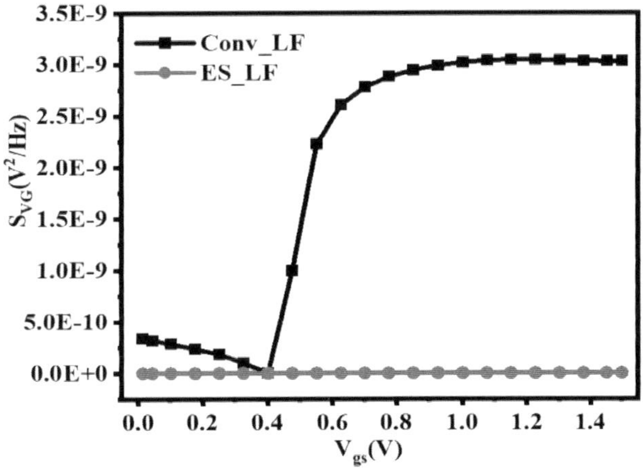

Fig.6(b)

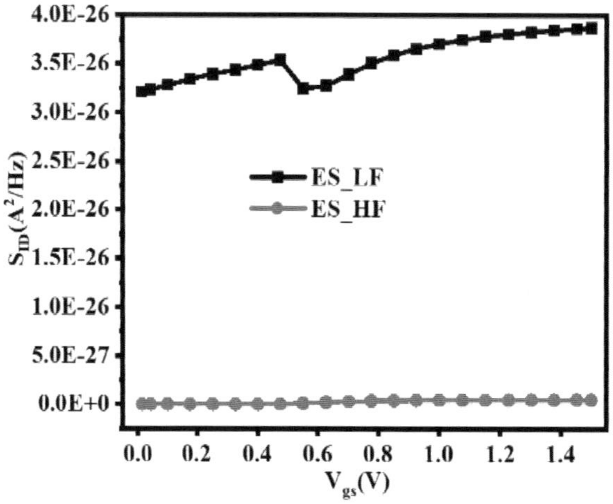

Fig.6(c)

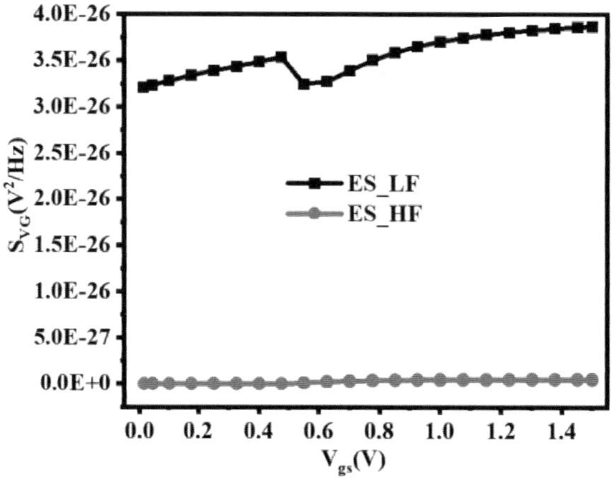

Fig.6(d)

Fig.6 Noise evaluation and difference between final and conventional device and also with low and high frequency (a) $S_{ID}V_g$ (b) $S_{VG}V_g$ (c) $S_{ID}V_g$_LF-HF (d) $S_{VG}V_g$_LF-HF

Fig.6(a) and 6(b) shows the variations in the S_{ID} and S_{VG} of both the conventional and final device along their noise variations and frequency enhancement throughout the device. It is observed that S_{ID} and S_{VG} of the device incremented crucially with the increment in extended source's length which is again due to the depletion of source and that of drain regions inherent for nano-scale FET devices.

Fig.6(a) and 6(b) also shows the effect on S_{ID} and S_{VG} because of the optimisation of the oxide width or height as the thickness of oxide was alter from 3 nm to 1 nm. It is apparent that the decrement in the thickness of the oxide , the S_{ID} and S_{VG} the device increases. As a result a steeper tunneling profile is observed on the source-channel junction.

Fig.6(c) and 6(d) also shows the effect on S_{ID} and S_{VG} due to the frequency variations; both at higher frequency and lower frequency. It is observed that on lower frequency the generation recombination noise dominates over all the other parameters of noise. As the frequency incremented the other parameters of noise get dominant over the generation and recombination noise. It can also be analysed that the presence and absence of trap-charges shows the dominance of number fluctuations & fluctuations of mobility.

5. CONCLUSION

A Hetero-Junction Single Gate Extended Source TFET is analysed and proposed in this paper. The paper deals with the noise, DC and AC analysis of the final device. The final device is optimised with different parameters of source doping, drain doping and channel doping. The extended source shaped structure hold on to a tunneling junction in the direction 90^0 to the channel which makes easier higher tunneling junction area without any sanction in device's area. The device design with two different materials (source-Ge and drain-Si) upgrade the ON-current squash the SS degradation behaviour with proper effect on it.The dominance of diffusion component of noise in the S_{VG} behaviour can be examined at high freq. as well as on the low frequency. The optimisation of the extended in the final device doesn't play a crucial role in alternation of capacitance (F), trans-conductance (g_m) and output conductance (g_d). The graph shows a significant and desired steep slope in I_dV_g graph plotting. BTBT tunneling also become efficient due to the optimisation across the extended source concept in the final device. The characteristics finalised for the selected Extended Source -TFET stressed its suitability for a broad area of upcoming generation low-power applications.

6. ACKNOWLEDGEMENT

This work is supported by DST-SERB, CRG grant, Govt. of India, CRG/2020/006229, dated: 05/04/2021.

7. REFERENCE

[1] W. Y. Choi, B. G. Park, J. D. Lee and T. J. K. Liu, "Tunneling field-effect transistors (TFETs) with subthreshold swing (SS) less than 60 mV/dec", *IEEE Electron Device Letters*, vol. *28, no.* 8, pp. 743-745, 2007.

[2] U. E. Avci, R. Rios, K. J. Kuhn and I. A. Young, (2011, August). Comparison of power and performance for the TFET and MOSFET and considerations for P-TFET. In *2011 11th IEEE International Conference on Nanotechnology* pp. 869-872, 2011.

[3] S. O. Koswatta, M. S. Lundstrom and D. E. Nikonov. "Performance comparison between pin tunneling transistors and conventional MOSFETs", *IEEE Transactions on Electron Devices*, vol.*5 no.* 3, pp. 456-465, 2009.

[4] J. Wan, C. Le Royer, A. Zaslavsky and S. J. S. S. E. Cristoloveanu, "Tunneling FETs on SOI: Suppression of ambipolar leakage, low-frequency noise behavior, and modelling", *Solid-State Electronics*, vol. *65*, pp. 226-233, 2011.

[5] W. Y. Choi and W. Lee, "Hetero-gate-dielectric tunneling field-effect transistors", *IEEE transactions on electron devices*, vol. *57* no. 9, pp. 2317-2319, 2010.

[6] J. S. Jang and W. Y. Choi, "Ambipolarity factor of tunneling field-effect transistors (TFETs)", *JSTS: Journal of Semiconductor Technology and Science*, vol. *11,* no. 4, pp. 272-277, 2011.

[7] N. Bagga and S. Dasgupta, "Surface potential and drain current analytical model of gate all around triple metal TFET", *IEEE Transactions on Electron Devices*, vol. *64,* no. 2, pp. 606-613, 2017.

[8] N. Bagga and S. K. Sarkar, "An analytical model for tunnel barrier modulation in triple metal double gate TFET", *IEEE transactions on electron devices*, vol. *62,* no. 7, pp. 2136-2142, 2015.
[9] H. G. Virani, R. B. R. Adari and A. Kottantharayil, "Dual-$ k $ spacer device architecture for the improvement of performance of silicon n-channel tunnel FETs", *IEEE transactions on electron devices*, vol. *57,* no. 10, pp. 2410-2417, 2010.
[10] S. Dash, B. Jena and G. P. Mishra, "A new analytical drain current model of cylindrical gate silicon tunnel FET with source δ-doping", *Superlattices and Microstructures*, vol. *97*, pp. 231-241, 2016.
[11] D. B. Abdi and M. J. Kumar, "In-built N+ pocket pnpn tunnel field-effect transistor", *IEEE Electron Device Letters*, vol. *35,* no. 12, pp. 1170-1172, 2014.
[12] W. Y. Choi and W. Lee, "Hetero-gate-dielectric tunneling field-effect transistors", *IEEE transactions on electron devices*, vol. *57,* no. 9, pp. 2317-2319, 2010.
[13] Y. Yang, X. Tong, L. T. Yang, P. F. Guo, L. Fan and Y. C. Yeo, "Tunneling field-effect transistor: capacitance components and modelling", *IEEE Electron Device Letters*, vol. *31,* no. 7, pp.752-754, 2010.
[14] H. Lou, L. Zhang, Y. Zhu, X. Lin, S. Yang, J. He, and M. Chan, "A junctionless nanowire transistor with a dual-material gate", *IEEE Transactions on Electron Devices*, vol. *59,* no. 7, pp. 1829-1836, 2012.
[15] S. Chander, S. K. Sinha, and R. Chaudhary, "Comprehensive review on electrical noise analysis of TFET structures", *Superlattices and Microstructures*, pp. 107101, 2021.
[16] S. Chander, S. K. Sinha, R. Chaudhary, and A. Singh, "Ge-Source Based L-shaped Tunnel Field Effect Transistor for Low Power Switching Application", *Silicon*, 2021.
[17] S. K. Sinha, S. Chander, and R. Chaudhary, "Investigation of Noise Characteristics in Gate-Source Overlap Tunnel Field-Effect Transistor", *Silicon*, pp. 1-8, 2022.
[18] A. Singh, S. K. Sinha, and S. Chander, "Impact of Fe Material Thickness on Performance of Raised Source Overlapped Negative Capacitance Tunnel Field Effect Transistor (NCTFET)", *Silicon*, pp. 1-8, 2022.
[19] S. K. Sinha, and S. Chander, "Investigation of DC performance of Ge-source pocket silicon-on-insulator tunnel field effect transistor in Nano regime", *International Journal of Nanoparticles*, vol. *13, no.* 1, pp. 13-20, 2021.
[20] I. A. Pindoo, S. K. Sinha, and S. Chander, "Performance analysis of Heterojunction tunnel FET device with variable temperature", *Applied Physics A*, vol. *127, no.* 10, pp. 1-10, 2021.

Non-Invasive Glucose Monitoring Using NIR Spectroscopy

Pulagam Sai Kumar Reddy, Darapu Mahesh, C Uday Teja, Muthuluri Janaki, Kiran Mannem

Department of ECE, Gokaraju Rangaraju Institute of Engineering and Technology, Hyderabad, India.

pulagamsaikumarreddy@gmail.com, darapumahesh79@gmail.com, chinnu.udaytej@gmail.com, janakimuthuluri@gmail.com, kiranmannem14@gmail.com

Abstract: Diabetes Mellitus (DM), a metabolic condition, can lead to major health problems if it is not adequately treated. Traditional blood glucose monitors are intrusive and give people pain and irritation. As a result, the study's goal was to find a non-invasive way to measure blood glucose. To determine blood glucose parameters, this system uses an NIR Spectroscopy-based system with a light source of wavelength 950nm. The light source illuminates the skin of the wrist, and a photodiode receiver captures the reflected beam. An Arduino UNO microcontroller receives the conditioned signal and digitalizes it. The spectrum is calculated by the Arduino board based on the subject's blood glucose levels. Pre-processing and segmenting the raw waveforms are required to obtain the peaks of the NIR spectrum and to compute the regression model.

Keywords: Photoplethysmography, CGM (Continuous Glucose Monitoring), Spectroscopy, Non-Invasive.

1. INTRODUCTION:

When a person's insulin level is out of control and the glucose concentration in the blood isn't processed properly, a slew of issues develops. Diabetes, a disease characterized by a high blood glucose level, develops because of this incapacity. The treatment of diabetes must be precise and managed carefully in-order to prevent sugar build-up in the blood and glucose intolerance, which increase the risk of serious vascular complication (such as heart attacks), peripheral vascular problems, kidney failure, and stroke. Diabetic neuropathy, also known as peripheral neuropathy and autonomic neuropathy, is also associated with increased risk of fatal complications. According to recent studies, there are 424.9 million diabetic people worldwide, with the number anticipated to rise to 628.6 million by 2045. Glucose monitoring on a regular basis using invasive method is expensive as well as time taking and may causes infections. Many diabetic patients are often attacked with cardiovascular strokes or diseases. There are many who doesn't monitor their glucose levels on a regular basis and remains undiagnosed and this leads to several health issues. Glucose monitoring on a regular basis is very important to avoid major health issues that are caused due to it. In this regard, many people are desiring for the non-invasive glucose measurement ahead of the invasive methods where we use painful methods. In this project, we are designing a non-invasive method for measurement of glucose value in the blood.

2. Existing Models, Proposed Model, Literature Survey

2.1 EXISTING MODELS:

Glucose is currently measured using standard methods and Continuous Glucose Measurement (CGM) and other devices which involves invasive method of measurement. To calculate the glucose level in invasive methods blood sample is required and if it is done regularly, it may cause infections and even time-consuming process. It motivates to develop the next generation devices, which are minimally invasive, and it involves small needle in the skin and Continuous glucose monitoring. In the next generation of devices where the glucose is measured totally noninvasively and real time continuous monitoring and remote monitoring, but this generation of devices are still in trails and development stages

2.2 PROPOSED MODEL:

In this proposed method, we measure glucose non-invasively using spectroscopic method. In this method, the glucose measurement is done using NIR spectroscopy. We use PPG technique and regression techniques for the measurement of concentration of glucose from the output obtained from the photodiode that is used to detect the light after absorption and transmission from glucose particles in the blood when LED light is passed through the fingertip or ear lobule.

2.3 LITERATURE SURVEY:

In the study, many researchers tried to analyse the different type of devices that are used for glucose measurement and try to summarize those into four generations. The classification is mainly done on invasiveness of the method. In the initial stages of developments in early 1970s, they have used reflectance technology which are made of heavy devices which requires very high amount of blood, and later due to availability of resources and latest technologies that made smaller devices with budget friendly prices where people are allowed to use personally.

One of the most well-known invasive methods is finger pricking where it only collects just 2-3 drops of blood, where people who frequently get tested may feel uncomfortable and become infectious. The next generation devices started being minimally invasive that includes small needles on skin which enables Continuous Glucose Monitoring (CGM).

Since these types of medical devices are still at an early stage of development, the emergence of the next generation of these devices is still more of an option than it is an actual application for them. In any case, we will represent them as fourth-generation medical gadgets since they provide real-time remote monitoring from a remote location and do not require traditional invasive approaches. The non-invasive techniques avoid invasive bodily contact and rely on techniques such as spectrometry and other measurements related to glucose levels.

Various types of non-invasive glucose testing methods are available. In addition to non-invasive glucose measurements that involve many approaches, other non-invasive glucose measurement methods include transdermal glucose measurement and optical glucose measurement. Mid-infrared spectroscopy, infrared spectroscopy, and near-infrared spectroscopy are used in optical non-invasive glucose testing. Raman spectroscopy and ocular spectroscopy are two types of spectroscopies.

We can determine the tissue's impedance. When an alternating current is passed through a tissue and the impedance frequency spectrum is measured in range of 100Hz to 100MHz, which is determined by the interaction of glucose with red blood cells. The results are a measure of the dielectric properties of that tissue.

A non-invasive optical glucose measurement method measures glucose in biological tissue by reflecting, scattering, and transmitting light in proportion to the sample's structure and chemical composition, which is the basis for many optical non-invasive glucose measurement methods. Based on the studied band of electromagnetic radiations and interpretation of the received spectrum and glucose levels, a combination of selectivity, interference to other compounds, multivariate calibration vectors, and several detection and multistage separation principles can be employed to distinguish these.

Using photoplethysmography or regression techniques, the glucose value is estimated in spectroscopy. Many factors influence non-invasive glucose measurement. The precision of the glucose reading is influenced by these factors. The measurement is imprecise due to the factors of water content, which can efficiently absorb a certain range of wavelengths, and tissues and their fluid, which impact the light passing through them. We learn through this study that there is a wavelength range (850nm-970nm) that is unaffected by the factors we considered.

Similarly, other researchers used ML and NN to create prediction models based on various criteria. However, some experiments have shown that the solution can be directly brushed on the sensor component to measure glucose levels, while others failed in error analysis, resulting in clinically significant inaccuracies. Despite the fact that some fourth-generation medical devices, such as Diasensor and MedOptix, were developed utilising optical spectroscopy, they have yet to be commercially viable.

3.Components Used, Block Diagram

3.1 Components Used

S. No	Component Name	Specification	Number
1	Arduino Uno R3	Atmega328P microcontroller	1
2	Amplifiers	LM358	2
3	Resistors	6.8k	3
		68k	2
		680k	2
		330	1
		47k	1
4	Capacitors	100nF	3
		1uF	2
5	IR LED and Photodiode Pair		1
6	O-LED Display		1

3.2 BLOCK DIAGRAM:

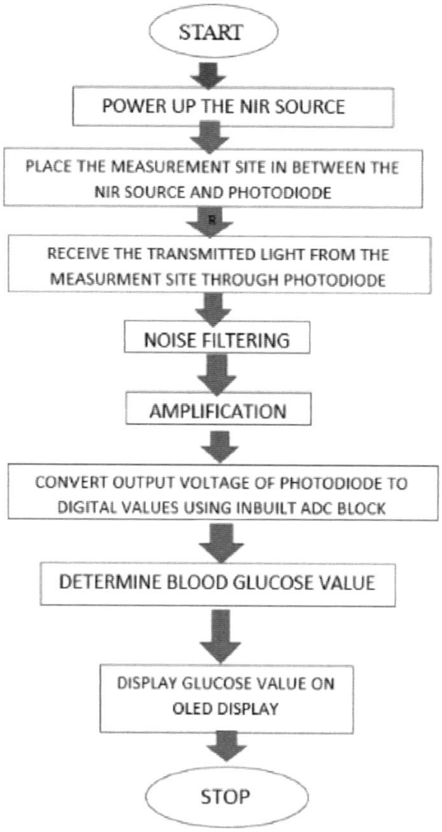

Figure 1: Block Diagram

4 IMPLEMENTATION TECHNIQUES:

4.1 Spectroscopy:

Spectroscopy is classified according to the wavelengths of the electromagnetic spectrum, such as IR spectroscopy, UV spectroscopy, and so on.

In the proposed model, we used NIR spectroscopy.

4.1.1 NIR Spectroscopy:

The concept of near infrared (NIR) spectroscopy is based on the absorption of electromagnetic radiation (EM) at wavelengths ranging from 780 to 2,500 nm. During light interaction with the sample, a detector measures its transmittance and absorbance. NIR spectroscopy is interested in both electronic and vibrational transitions. A NIR spectrophotometer measures light scattered off and through sample materials, enabling quick and accurate evaluation of their properties. The graph below shows an example of the absorption spectra of haemoglobin with and without oxygen as measured by NIR spectroscopy.

4.1.1.1 Versatility of NIR Spectroscopy

In many ways, NIR spectroscopy is extremely versatile. First and foremost, it is extremely versatile in its applications. NIR spectroscopy can be used in a laboratory, a factory, a hospital, a field, and a field laboratory.

4.2 REGRESSION:

Regression is a statistical method used to investigate and examine the nature and strength of the relationship between one variable (usually denoted by Y) and other variables (dependent variables) in finance, investments, and other fields. General form of relationship between variables using regression model:

$Y = b + a1X1 + ... + akXk + c$

Y is dependent variable, X1, ……Xk are independent variables, c is a residual constant

4.2.1 Polynomial Regression

The percentile of degree in a polynomial, which represents the coefficients of the relationship between an independent variable, x, and the dependent variable, y, is called Polynomial regression. Polynomial regression, denoted E(y |x), This type of regression fits a nonlinear relationship between values of x and mean y.

4.2.1.1 Polynomial Regression Model:

The least squares method is commonly used to fit polynomial regression models. Using the least-squares method, one can minimize the variance of the unbiased estimators of coefficients provided under the conditions of the Gauss-Markov theorem. It's a subset of multiple linear regression.

A polynomial regression's general equation is:

$Y = ao + a_1X + a_2X^2 + \ldots + a_mX^m + c$

4.2.2 Quadratic Regression:

Quadratic regression is a type of linear regression that is more complex than simple linear regression. It is a method which allows simulation of a relationship between two sets of independent variables. A quadratic regression model is used to determine which parabola equation is most conducive to fitting a set of data. In this data set, we have defined a predetermined set of graph points that take the form of a parabola.

The general form of a quadratic equation is y=a*x2+b*x+c, where a=0.

The R2 model is another name for it.

5 ILLUSTRATIONS OF PROCESS

5.1 CIRCUIT DIAGRAM:

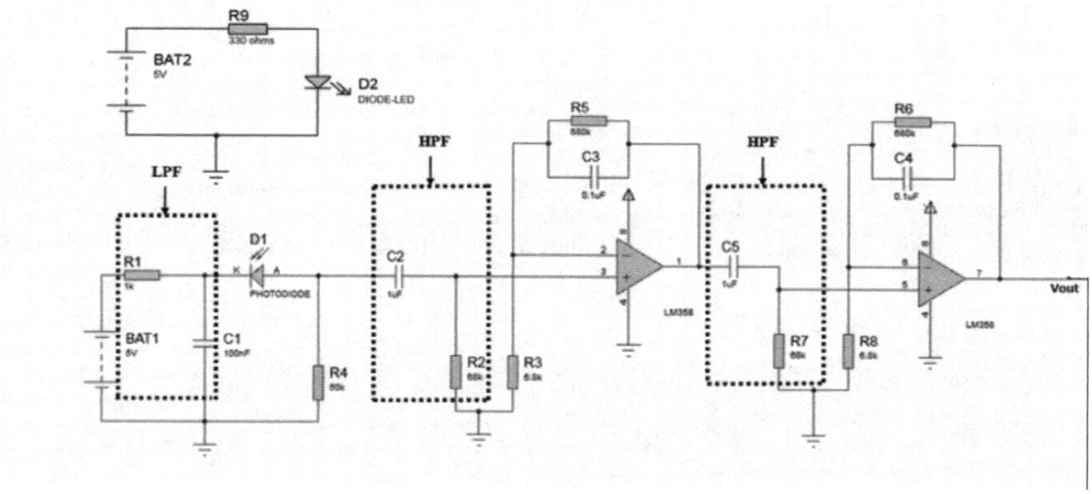

Figure 2: Circuit diagram

5.2 ILLUSTRATION:

To derive the polynomial regression equation, 80 diabetics of both genders were studied. These people' glucose levels are tested in the laboratory using a traditional invasive method, and simultaneously, the Analog voltage relative to their glucose level which is measured using the proposed hardware configuration as illustrated in Figure 3, with the results summarised in Table 1. The transmitted light from the IR LED is passed through the finger and the transmitted light is observed by the IR photodiode which is placed on the other side in the measurement site. The voltage is observed at the measurement site is then passed through the two amplifier circuits where voltage amplification is done. The output is given to the Arduino where the microcontroller digitizes the voltage value and it is used to calculate the glucose value.

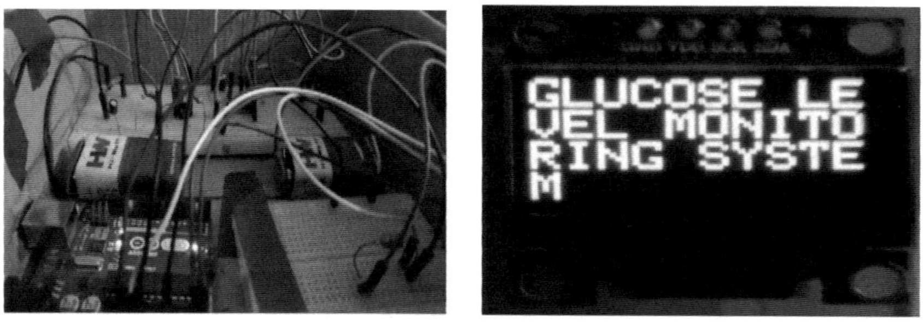

Figure 3: Designed model

The below table represents the voltage values that are calculated for corresponding invasively measured glucose values.

Journal of Physics: Conference Series **2325** (2022) 012021 doi:10.1088/1742-6596/2325/1/012021

Table 1: Analog Voltage vs Glucose Level of Samples.

S.no	Analog Voltage(mV)	Glucose Values(mg/dl)	S.no	Analog Voltage(mV)	Glucose Values(mg/dl)
1	499	142	41	609	163
2	509	146	42	602	174
3	519	156	43	423	111
4	519	157	44	584	177
5	548	177	45	502	145
6	524	159	46	524	163
7	543	209	47	541	172
8	568	133	48	517	153
9	573	179	49	511	147
10	583	224	50	522	158
11	592	175	51	489	137
12	597	187	52	473	125
13	607	196	53	507	151
14	627	191	54	597	178
15	695	167	55	583	189
16	735	220	56	503	132
17	612	244	57	595	178
18	847	247	58	598	184
19	833	248	59	567	165
20	867	276	60	495	143
21	935	302	61	749	225
22	999	321	62	630	234
23	636	178	63	798	267
24	538	162	64	481	139
25	950	350	65	834	248
26	821	241	66	967	315
27	611	199	67	633	193
28	929	301	68	746	221
29	864	272	69	629	183
30	512	154	70	637	167
31	593	179	71	511	156
32	603	186	72	866	276
33	577	174	73	957	309
34	741	221	74	831	237
35	589	183	75	597	179
36	599	191	76	964	310
37	532	157	77	735	217
38	497	141	78	721	209
39	854	233	79	599	195
40	477	132	80	594	187

y = (3*10^-5) *x^2 + 0.2903*x - 4.798 where x and y are Analog voltage (mV) and glucose level (mg/dl) respectively.

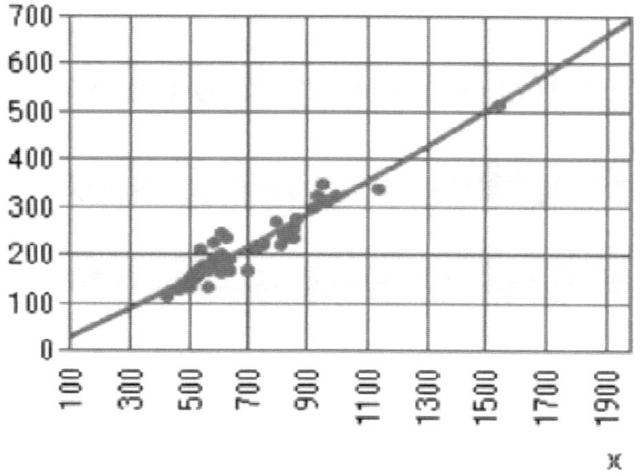

Figure 4: Quadratic Regression plot

The Arduino programme calculates the glucose level for a specific analogue voltage. The continuous Analog voltage readings obtained from the photo detector while positioning the finger between the Near Infrared LED and the photo detector are averaged and stored in an array. The glucose value corresponding to this average Analog voltage is computed by the microcontroller and shown on the O-LED display.

6 Result and VALIDATION:

The proposed model is assessed by collecting and comparing the glucose levels of 75 people using both the traditional invasive method and newly proposed model, and the results are tabulated in table 2. The proposed glucose measurement device is accurately evaluated with Clarke Error Grid Analysis.

Table 2: Comparison of Invasive Glucose and Non-Invasive Glucose Values

S.No	Invasive Glucose Values	Non-Invasive Glucose Values	Difference
1	117	120	-3
2	143	141	2
3	166	162	4
4	134	151	-17
5	145	139	6
6	120	129	-9
7	123	118	5
8	157	142	15
9	193	192	-1
10	106	103	3
11	245	214	31
12	134	148	-14
13	299	252	47
14	205	219	-14
15	164	170	-6
16	186	186	0

17	110	117	-7
18	157	142	15
19	120	129	-9
20	156	145	11
21	100	108	-8
22	250	235	15
23	92	85	7
24	195	181	14
25	145	154	-9
26	170	162	8
27	148	157	-9
28	192	220	-28
29	131	128	3
30	316	268	48
31	129	136	-7
32	220	198	22
33	130	104	26
34	149	160	-11
35	227	258	-31
36	165	152	13
37	110	152	-42
38	113	108	5
39	170	161	9
40	149	138	11
41	193	190	3
42	247	253	-6
3	301	289	12
44	231	245	-14
45	155	165	-10
46	278	254	24
47	130	141	-11
48	164	160	4
49	181	193	-12
50	196	190	6
51	252	267	-15
52	221	248	-27
53	190	187	3
54	175	182	-7
55	146	140	6
56	133	153	-20
57	157	160	-3
58	199	208	-9
59	137	122	15
60	189	176	13
61	172	190	-18
62	163	159	4
63	154	160	-6
64	183	211	-28
65	169	173	-4

66	123	123	0
67	175	165	`10
68	276	277	-1
69	321	319	2
70	247	260	-13
71	196	191	5
72	224	229	-5
73	187	189	-2
74	177	177	0
75	142	156	-14

The Clarke error grid analysis is used to assess and evaluate the clinical significance of precise value in the measurement of glucose level. It is performed in MATLAB giving the values of invasive and non-invasive values in table 2 as inputs.

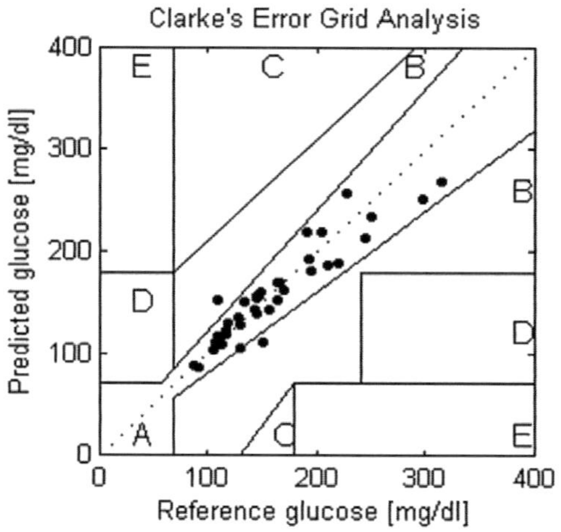

Figure 5: Result of Clarke Error grid analysis

Region A:

The difference between meter values and reference values is less than 20%.

Region B:

The difference between meter values and reference values is larger than 20%.

Region C:

Large correction and difference in values leads to unnecessary treatment.

Region D:

Values indicating a potentiality dangerous failure to detect hypoglycaemia or hyperglycaemia.

Region E:

Values that would confuse treatment of hypoglycaemia for hyperglycaemia and vice versa.

The results reveal that 82 percent of the anticipated glucose values fall within area A, which is a clinically acceptable range for glucose monitoring equipment. The results show that the proposed model or proposed approach when compared with the traditional method it gives 82% of accuracy in the values.

7. CONCLUSION:

The invasive glucose computation approach is uncomfortable, costly, and inconvenient. Furthermore, it does not provide long-term monitoring and there is a high risk of infection. This research presents a non-invasive approach for blood glucose testing utilising near-infrared LEDs to solve the drawbacks. The glucose level in the blood as detected by the photodetector is displayed on the O-LED display. By analysing error grids, it validates the proposed method. An innovative non-invasive portable blood glucose computing devices are an effective way to manage diabetic patients' health, this blood glucose monitor is portable and non-invasive. Patients can utilise the suggested model anywhere, and health care facilities can use it as well.

REFERENCES AND LINKS:

[1] Bruen D., Delaney C., Florea L., Diamond D. Glucose Sensing for Diabetes Monitoring: Recent Developments. Sensors. 2017:17:1866. Doi: 10.3390/s17081866. [PMC free article] [CrossRef]

[2] Gamble J.M., Clarke A., Myers K.J., Agnew M.D., Hatch K., Snow M.M., Davis E.M. Incretin-based medications for type2 diabetes: An overview of reviews. Diabetes Obes. Metab. 2015:17:649-658. Doi: 10.1111/dom.12465. [PubMed][CrossRef]

[3] International Diabetes Federation Guideline Development Group Global guideline for type 2 diabetes. Diabetes Res. Clinc. Pract. 2014:104:1-52. Doi: 10.1016/j.diabres.2012.10.001.

[4] Omar S.Khalil ,Spectroscopic and Clinical Aspects of Non-Invasive Glucose Meaasurments .10.1.1.492.3612 IEEE paper.

[5] Gayathri B., Sruthi k and K A Unnikrishna Menon, Non-Invasive Glucose Monitoring Using Near Infrared Spectroscopy IEEE paper uploaded on 09 June 2021.

[6] Mohan, D,C.Santosh Kumar "An efficient IKSVM based multi-parameter patient monitoring system", International Journal of Applied Engineering Research, Research India Publications, Volume 10,Number 9, p.22703-22710(2015).

[7] Non-Invasive Blood Glucose Monitoring Technology Article in https://www.ncbi.nlm.nih.gov/pmc/articles/PMC7731259/#__ffn_sectitle , by Liu Tang, Shwu Jen Chang, Ching-jung Chen, Jen-Tsai liu.

[8] Aleksandr Gorst, Kseniya Zavyalova, Aleksandr Mironchev., Journal in http://www.mdpi.com//journal/biosensors , titled Non-Invasive Determination of Glucose Concentration Using Near-Field Sensor published in 2021

[9] Balasubramaniam, Vivekanadam. "IoT based Biotelemetry for Smart Health Care Monitoring System." Journal of Information Technology and Digital World 2, no. 3(2020): 183-190.

[10] Shakya, Subarna, and P. P. Joby. "Heart Disease Prediction using Fog Computing based Wireless Body Sensor Networks (WSNs)." IRO Journal on Sustainable Wireless Systems 3, no. 1 (2021): 49-58.

[11] Kumbargeri, Vasavi, Nitisha Sinha, and Satyadhyan Chickerur. "A Quantum Machine Learning Classifier Model for Diabetes." In Innovative Data Communication Technologies and Application, pp. 603-613. Springer, Singapore, 2021.

[12] Harnsoongnoen.S, Wanthong A, Coplanar aveguide transmission line loaded with electric-LC resonator for determination of glucose concentration sensing. IEEE Sens.J.2017,12,1635-1640.

[13] R. W. Waynant "Overview of Non-Invasive Fluid Glucose Measurement Using Optical Techniques to Maintain Glucose Control in Diabetes Mellitus", A Publication Of The IEEE Lasers And Electro-Optics Society, 1998, vol.12.

[14] Mohan, D, C.Santhosh Kumar "An efficient IKSVM based multi-parameter patient monitoring system", International Journal of Applied Engineering Research, Research India Publications, Volume 10, Number 9, p.22703-22710 (2015).

[15] Swathi Ramasahayam, Sri Haindavi K, Bharat Kavala, Shubhajith Roy Chowdhury "Non Invasive Estimation of Blood Glucose using Near Infrared Spectroscopy and Double Regression Analysis", IEEE International Conference on Sensing Technology, 2013, pp. 627-631.

Design of Montgomery Multiplier – RSA Algorithm

K Venkatalakshmi, Pamuru Gayathri, Tadakaluru Sai Likhitha, Sushmitha Shinde, MOV. Pavan Kumar

Department of ECE, GRIET, Hyderabad, India.

kandulavenkatalakshmi04@gmail.com, gayathrireddy153@gmail.com, sailikhitha221@gmail.com, shindesushmitha1234@gmail.com, pavanomkar81@gmail.com

Abstract. In cryptography, during the process of encryption, we use multiplication modulo architectures (2^n- 1) and (2^n+ 1) are proposed that enable the implementation of very efficient combinational and pipelined circuits for modular arithmetic. This project is an improvisation of existing modulo multiplier architectures for higher speed and regularity. We use the Montgomery multiplier in this project for the RSA algorithm. Hence, the IDEA block cipher is represented as a high-performance modulo multiplier adder. The circuits that arise are compared qualitatively and quantitatively for the RSA algorithm.

Keywords— Cryptography, Encryption, RSA- Rivest, Shamir, Adleman algorithm, block cipher

1. INTRODUCTION

These days, data communication and internet services have increased rapidly. Thus, fundamental security standards for safeguarding sensitive information during the data transfer process are essential. Most of the systems use cryptography to provide such security services [3]. A public key is generated in the process of Cryptography, using Rivest, Shamir, and Adleman (RSA) algorithms, which is one among the most popularly used algorithms, public key algorithms have been adopted RSA is an unsymmetric type of algorithm that uses two kind of keys – public key, private key. The public key is used by everyone whereas, the private key is enclosed by a few clients. [4]. However, the RSA algorithm requires many modulo multiplications to be performed to achieve the required modulo exponentiation. The size of each modulus is usually 1024 bits for long-term security. Due to this reason, it's challenging to attain high data throughput rate without hardware acceleration. On the other hand, sensitive and Security requirements are crucial for private data transfer through mobile devices with internet connectivity, such as smartphones and tablets. Due to the limited battery power, and energy-efficient cryptosystem is required [5].

Therefore, an efficient architecture is needed to be implemented to perform quick modular multiplications with low usage of energy, high performance, high speed, and low complexity and delay parameters. A prominent approach to implementing this modulo arithmetic is Montgomery modular multiplication algorithm [2]. This Montgomery multiplier is a three-operand addition operation with an iteration loop that substitutes typical trial division with a sequence of addition and shift operations per modulus. But, in some cases, the addition of long operands and the carry propagations might be time-consuming, and this might influence the performance of the RSA algorithm in cryptosystems. There are few approaches for achieving high-speed rates of the Montgomery modular multiplications. One of which is representing the Montgomery modular multiplication's inputs and outcomes in binary format and the intermediate results in a carry-save representation by addition of the carry and sum terms, so that the carry propagation can be avoided. Each modular multiplication concludes by changing this carry-save representation format to the final binary representation. The second method involves combining all of the Montgomery multiplication's inputs and outputs in carry-save format by avoiding the final step to obtain the resultant, thus, eliminating the intermediate output to input format conversions. However, the number of operands must be increased in the modular multiplication resulting in the requirement of additional registers to store these operands.[9] This paper explains the efficient way of using the Montgomery multiplier of modular arithmetic. In this paper, the chapters that are involved are: Modular Arithmetic which gives a brief idea about different modular arithmetic algorithms and methods, Montgomery Multiplication, which includes the information regarding the usage and the process of functioning of the Montgomery multiplications, Implementation includes the process of implementation of Montgomery modular multiplications, Evaluation, and Results, showing the observed results and obtained values and finally Conclusion followed by their corresponding References.

2. MODULAR ARITHMETIC

Cryptography is a crucial aspect of secure communication. Because of the increased use of the internet for data exchange and electronic transactions, security has become the most important concern on the network [12]. Cryptography provides privacy, confidentiality, user authentication, integrity, and non-repudiation in addition to protecting data from theft or adaptation. To achieve these goals, two types of cryptographic algorithms are commonly utilized. It comprises symmetric cryptography, public-key cryptography, and asymmetric cryptography (RSA).[7] The asymmetric key algorithm uses two separate keys, one for encryption and the other for decryption. Data is encrypted using modular exponentiation of huge numbers, which is a time-consuming procedure because of the constant modular multiplications. Encryption, also known as ciphering, is the process of transforming plain text into an unreadable format. There are various mathematical operations involved in converting plain text to ciphertext. As a result, RSA encryption/decryption is essentially a modular exponentiation operation when both keys are generated. This mathematical operation is denoted by the symbol C = Ze encrypts plaintext block M to ciphertext block C. (mod z)[8]. Z = Cd (mod m) recovers the plaintext block, where C is ciphertext, Zis plain text, e is the public key exponent, and z is the modulus.

It involves a few modular operations such as modular multiplication, modular addition, and modular subtraction. To produce a private key, modular multiplication is required. As a result, the speed of modular multiplications determines the effectiveness of an RSA encryption system. In hardware, Montgomery multiplication is the best way for doing modular multiplications. This approach is quick and energy-efficient in hardware implementations. Allowing B.G mod N for modular multiplication, Montgomery multiplication replaces division with N right shifts to save division operations[13]. This approach also enables the multi-precision arithmetic while performing high-radix calculations. As a result, High-radix operations make it easier to construct modular multiplication systems.

Algorithm for Modular Exponentiation:
Input: R, j, p
Output: A = R*j mod p
Let p contain x bits If px-1=1 then A = R else A = 1
For d = x-2 down to 0
A = A×A
If j*d=1 then A=A×R.

3. MONTGOMERY MULTIPLICATION

Calculating X.Y (mod M) is a critical process in many cryptosystems, such as RSA [14]. Let M>0 be an even number. X.Y (mod M) depletion takes longer than X.Y multiplication without reduction. Montgomery invented Montgomery multiplication, a new formula for computing products (mod M) without time-consuming reduction (mod M). For a given integer n, Montgomery multiplication is defined as X.Y. 2 n (mod M). Precomputation allows for a hardware simplification of this algorithm. The values to be added to the intermediate result can be precalculated in the loop. It is critical to avoid any delays induced by propagation. As a result, Montgomery's multiplication is efficient because it has a conversion procedure among normal and Montgomery domains [11].

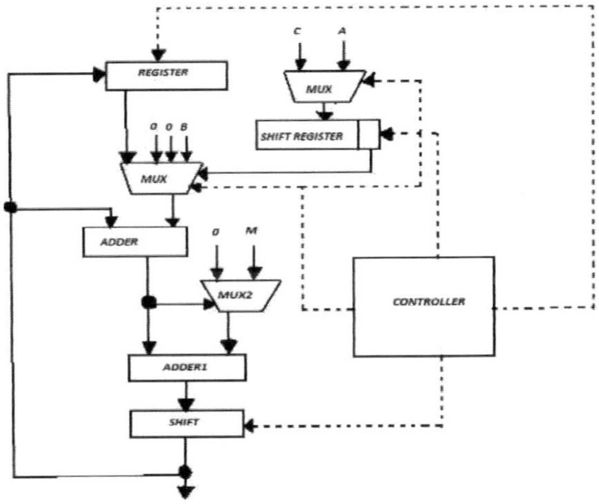

Fig.1: Block Diagram of Montgomery Multiplier

Table 1. Normal Domain to Montgomery Domain

Normal Domain	⇔	**Montgomery Domain**
B	⇔	**B`=B*2^q (mod N)**
G	⇔	**G`=G*2^q (mod N)**
B*G	⇔	**(B*G)`=B*G*2^q(mod N)**

4. IMPLEMENTATION:

Montgomery (modular) multiplication is a method for speeding up the computation of such multiplications [15]. Rather than dividing the product and subtracting n many times, it adds multiples of n to cancel out the lower bits before discarding them.

The core calculation of RSA encryption is nothing but calculating [6] $\boldsymbol{y^\wedge x\ (mod\ m)}$ for large x, y, m Modular multiplication and exponentiation can be performed in hardware correctly using Montgomery multiplication.

Inputs: x, y

$z = MP(x, y) = xy\left(2^k\right)^{-1}(mod\ m)$, done signal

$z = y^x(mod\ m)$, done signal

In this implementation, the code will be performed in two different blocks named mont_mul where the Montgomery multiplications are performed, and mont_expo where the Montgomery exponentiation is done. These implementations involve various stages of adders and shift registers. D-flipflops are used here as the shift registers [10]. The mont_mul block of code is substituted as a functional block in the mont_expo code block as shown in the figures below. This further involves the corresponding states where the process is defined stage-wise.

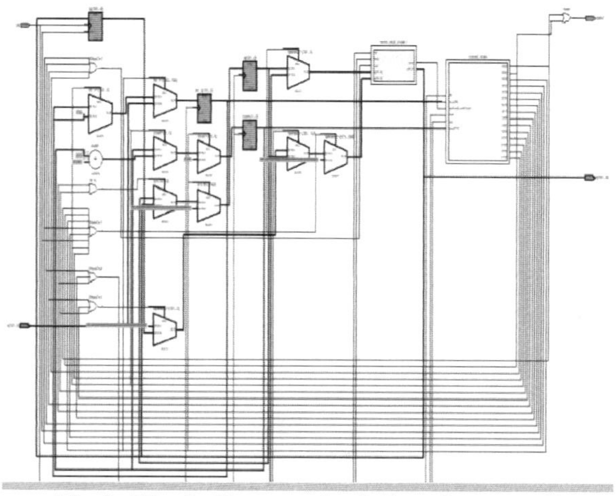

Fig.2: Block diagram depicting the overall operation of the Montgomery exponential modular arithmetic.

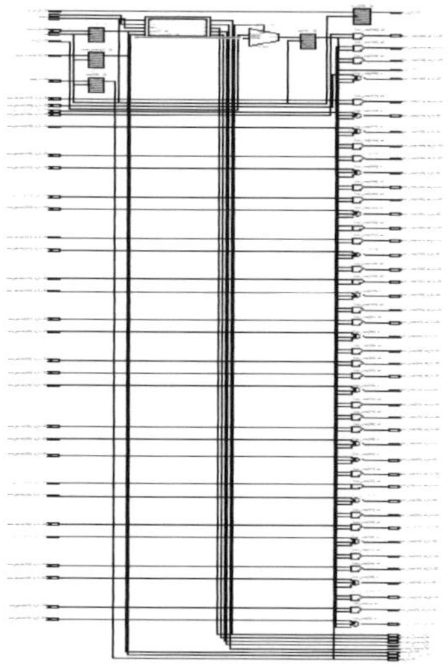

Fig.3: Internal operation done in the Montgomery multiplier block Substituted in the main operation block.

5. RESULTS:

After simulating the code which is being generated with the mathematical expression mentioned above and obtaining the output waveforms shown in the below figure as the result. This computation uses 5-bit inputs.

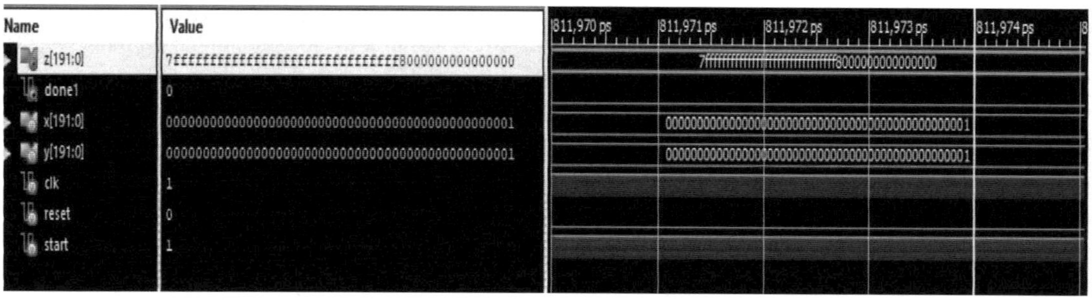

Fig 4: Figure showing the obtained simulated result as a waveform in Xilinx.

Table 2: Comparison table of results from various papers that are made on the Montgomery multiplications.

Montgomery algorithm	Scalable Diagram	N(bits)	Area	Time	Speed
RSA Algorithm	**Yes**	**1024**	**2106**	**4.3**	**269.8**
MWR2mm	**Yes**	**1024**	**1081**	**9.34**	**109.6**
Proposed IDDMM	**Yes**	**256**	**1102**	**17.03**	**60.23**
R 2k mm	**No**	**1024**	**1920**	**19.8**	**612.1**
Arch 2-ver 4 (k=32)	**Yes**	**1024**	**1612**	**18.9**	**86.8**
D-serial(D=4)	**No**	**1024**	**570**	**11.18**	**242.66**
Arch 2-ver 3 (K=64)	**Yes**	**1024**	**263**	**13.33**	**113.8**
Arch-2 ver-4 K=64	**Yes**	**1024**	**219**	**4.21**	**241.6**
Bit-serial	**Yes**	**512**	**957**	**9.64**	**53.11**
Ar-2-ver3 K=16	**Yes**	**512**	**208**	**8.29**	**61.70**
Arch2-ver 4 K=16	**Yes**	**512**	**10**	**8.67**	**59.02**

6. FUTURE SCOPE:

In this project, we have made the necessary changes to the existing Montgomery multiplier's architecture by changing some of the components, to enhance its features such as speed, time, and area. This project can be extended by changing the respective Libraries and Components in the architecture and observing the corresponding changes in the simulation. The features such as time and area can be enhanced by modifying the values of 'm' in $y^x(mod\ m)$ accordingly.

7. CONCLUSION:

The Montgomery Multiplier rested on CSAs, the throughput can be developed by maintaining a low complexity in the hardware. Instead of using more CSAs only one CSA can be used do the integration in the iteration loop based on the algorithm, the pre-computation of the values to be fed into the loop such as the B+N operation, and the long format conversion process by which the whole architecture will be having a small critical path and less hardware. The pre-computation and the format conversion process may lead to additional clock cycles which can increase the critical path, so if the CSA can do a three-input addition the additional clock cycles required for the mentioned processes can be made half just by using the modulo functions and computing multiplications and divisions just by power of 2. Thus, the Montgomery multiplication method speeds up the computation by implementing the modular arithmetic and the Montgomery multiplier will be having higher efficiency and the hardware takes a small area.

References

[1] Tenca, Alexandre F., and Cetin K. Koc. "A scalable architecture for modular multiplication based on Montgomery's algorithm." Computers, IEEE Transactions on 52.9 (2003): 1215-1221.

[2] Kaihara, Marcelo E, and Naofumi Takagi. "A hardware algorithm for modular multiplication/division." Computers, IEEE Transactions on 54.1 (2005): 12-21.

[3] Kuang, Shiann-Rong, et al. "Energy-efficient high throughput Montgomery modular multipliers for RSA cryptosystems." Very Large Scale Integration (VLSI) Systems, IEEE Transactions on 21.11 (2013): 1999-2009.
[4] Knezevic, Miroslav, Frederik Vercauteren, and Ingrid Verbauwhede. "Faster interleaved modular multiplication based on Barrett and Montgomery reduction methods. "Computers, IEEE Transactions on 59.12 (2010): 1715-1721.
[5] R. L. Rivest, A. Shamir, and L. Adleman, "A method for obtaining digital signature and public-key cryptosystems," Commun. ACM, vol. 21, no. 2, pp. 120–126, Feb. 1978.
[6] P. L. Montgomery, "Modular multiplication without trial division," Math. Comput., vol. 44, no. 170, pp. 519–521, Apr. 1985.
[7] C. K. Koc, T. Acar, and B. S. Kaliski, "Analyzing and comparing Montgomery multiplication algorithms," IEEE Micro, vol. 16, no. 3, pp. 26–33, Jun. 1996.
[8] Y. S. Kim, W. S. Kang, and J. R. Choi, "Implementation of 1024-bit modular processor for RSA cryptosystem," in Proc. IEEE Asia-Pacific Conf., Aug. 2000, pp. 187–190.
[9] V. Bunimov, M. Schimmler, and B. Tolg, "A complexity effective version of Montgomery's algorithm," in Proc. Workshop Complexity Effect. Designs, May 2002, pp. 1–7.
[10] R. Bhutada and Y. Manoli, "Complex clock gating with integrated clock gating cell," in Proc. Int. Conf. Design Technol. Integer. Syst. Nanoscale Era, Sep. 2007, pp. 164– 169.
[11] D. R. Sulaiman, "Using clock gating technology for energy reduction in portable computers," in Proc. Int. Conf. Comput. Commun. Eng., May 2008, pp. 839–842.
[12] J. Chao, Y. Zhao, Z. Wang, S. Mai, and C. Zhang, "Lowpower implementations of DSP through operand isolation and clock gating," in Proc. Int. Conf. ASIC, Oct. 2007, pp. 229–232.
[13] Vivekanandam, B. "Ideal Time-Based Voltage Control using Evolutionary Algorithm in Distributed Generator Centered Networks." Journal of Electronics and Informatics 2, no. 4 (2021): 233-238.
[14] A. Vilardo, A. Mazzeo, N. Mazzocca, and L. Romano, "A novel unified architecture for public-key cryptography," in Proc. Design, Autom. Test Eur. Conf. Exhibit., Mar. 2005, pp. 52–57.
[15] Cho, Koon-Shik, Je-Hyuk Ryu, and Jun-Dong Cho. "High-speed modular multiplication algorithm for RSA cryptosystem." Industrial Electronics Society, 2001. IECON'01. The 27th Annual Conference of the IEEE. Vol. 1. IEEE, 2001.

Damping Analysis in Si Torsional Micromirrors

Resmi R [1], V. Suresh Babu [2], M. R. Baiju [3]

[1] University of Kerala, LBS Institute of Technology for Women, Poojappura, Trivandrum, Kerala, India.

[2] APJ Abdul Kalam Technological University, College of Engineering Trivandrum, Kerala, India.

[3] University of Kerala; Kerala Public Service Commission, Kerala-695004, India.

resmilbs@gmail.com

Abstract. Micromirrors are extensively used in MEMS/NEMS based actuators and accordingly the design of superior performance mirror structures are prime requisite in MEMS/NEMS industry. Thermal and viscous damping of a dynamic vibrating micromirror are important factors degrading its performance by inducing energy dissipation. The primary sources of energy losses are viscosity and thermal conduction near the walls of the structure in the acoustic boundary layer. Hence it is necessary to accurately assess the bounds of energy dissipation owing to thermal and viscous damping. In this paper, the depths of the thermal and viscous boundary layers corresponding to thermal conduction and viscous drag at different eigen frequencies are investigated. Both the temperature and pressure distributions surrounding the vibrating micromirror are considered for the analysis. The penetration depths communicate the extent of thermal and viscous boundary layers and clearly indicate the extent of energy loss. In micromirrors, the thickness of boundary layer is a critical parameter since the dissipated energy is mainly distributed in it. The air domain surrounding the micromirror is modelled using thermoacoustics of COMSOL Multiphysics software along with the eigen frequency analysis. In the present study, the thickness of the boundary layers and quality factor are analysed for a micromirror vibrating in torsional mode. The thermal and viscous penetration depths decrease with eigen frequency and hence to develop high quality devices with low damping, the micromirror are verified to be operated at higher frequencies.

1. Introduction

Nowadays due to the fast growth in Micro-Electro-Mechanical Systems (MEMS), development of high quality and efficient microstructural elements are indispensable. Micromirrors are essential components in MEMS for integrated scanning in telecommunications, biomedical and consumer applications. MEMS technology processes can fabricate submicron features with high precision, low power consumption and very low unit cost[1].

The micromirrors are primarily used for the spatial manipulation of light in various applications[2]. Based on the operation modes, these devices are generally classified as torsional and piston mode micromirrors. In both types, the basic operation is to control the incident light in the requisite direction. In torsion mode, the deflection of light is accomplished by tilting and in piston mode, the phase of the light is altered through vertical displacement[3].

In order to render high performance and good quality micromirrors, utilization of low energy is a necessity condition, especially in applications with limited power sources[4]. However, in all types of vibrations, damping is associated and energy dissipation occurs in micromirrors. The main reason behind damping is the structural resonance of the vibrating microstructures. To improve the performance and to yield high profile devices with supreme efficiency and minute sizes, the design, modelling and fabrication of micromirrors have been a significant research area for the past decades.

Energy dissipation mechanisms existing in microstructures are classified as extrinsic and intrinsic. The main intrinsic dissipation mechanisms existing in micromirrors are Akhieser dissipation and acoustic thermal damping whereas extrinsic sources of damping include anchor loss, air-damping, squeeze film damping and surface loss[5] . To attain elevated Q factors, careful consideration of the extrinsic and intrinsic dissipation mechanisms are needed. Energy loss due to different damping mechanisms are controlled by changing the geometric design or working conditions. Air-fluid damping is reduced by operating the structure under vacuum. Anchor loss (support loss) is eliminated by designing nonintrusive supports. Energy loss like thermal damping thrusts a strict upper limit on the performance of the micromirrors[6].

Micromirrors regularly work in an atmosphere with surrounding air and hence its dynamics involve air–structure interactions to differing degrees[7]. The stiffness and damping of the micromirror continuously alter because of relative motion of thin layer of air between the planar structures. The abovementioned phenomenon leads to erratic dynamic characteristics and performance degradation. Hence, it is very essential to analyse the dynamic response of micromirrors which can be done proficiently by finite element modelling[8]-[10].

This paper deals with the analysis of the reciprocal actions between thin layers of surrounding air and vibrating MEMS micromirrors. Dissipation of energy in the neighbouring air of vibrating micromirror is presented which incorporates the impacts of viscous effects, thermal properties and compressibility. Finite element modelling is used which enables fully coupled acoustoelastic calculations to evaluate the dissipation of energy in the layer. Section 2 deals with the theory of thermal and viscous damping and Section 3 describes the finite element modelling of micromirrors. Section 4 elaborates the results and conclusions are provided in Section 5.

2. Theory of Thermal and Viscous Damping

Q factor (QF) of the micromirror is directly related to the energy dissipations and can be defined as the number of cycles needed for the amplitude of the motion to diminish to $e^{-\pi}$ of its original value. The existence of damping as a main thermal energy loss mechanism limiting the quality factor is first predicted by Zener[11]. In a torsional mode micromirror when vibration occurs, in the surrounding regions of micromirror a varying thermal resistance occurs which accounts for a temperature gradient across various parts of the vibrational micromirror. In torsional mode, vibration persists indefinitely when damping is absent. The local temperature gradients bring about irreversible flow of heat leading to substantial energy losses and attenuates the vibration amplitudes. This irreversible flow of heat and entropy generation leads to damping of the system. Thermal resistance exists surrounding the vibrating structure accounts for thermal power dissipation and this limits quality factor which is a measure of the energy loss mechanism [12]-[16] Similar to the thermal resistance, a viscous drag exists around the vibrating structure which is responsible for another type of energy loss surrounding the vibrating mirror.

This viscous drag is responsible for the viscous damping which also contributes to the total energy loss of the system. The two mechanisms are summarized as follows.

Thermal damping arises due to the varying thermal conduction between the gas surrounding the torsional micromirror and the vibrating structure. Similarly viscous damping exists in the vibrating micromirrors owing to the viscous forces at the interface. The total damping is explained by the losses originating from these two processes and the total damping constant is the sum of the damping constants due to the abovementioned energy loss mechanisms. The damping constant is inversely proportional to the Qfactor of the system.

$$\frac{1}{Q} = \delta = \delta_{TH} + \delta_{VISC} \quad (1)$$

where δ_{TH} is the thermal damping constant and $\delta_{VISC}$is the viscous damping constant.

2.1. Thermal and Viscous Boundary Layers and Prandtl Number

An important parameter in vibrating structures is the thickness of the viscous and thermal boundary layers known as the penetration depths. Along the thermal boundary layer, T varies in a significant way and its thickness can be defined as the distance from the vibrating structure at which the temperature is reduced to 99% of the original value.

The instantaneous local velocity is noted surrounding the vibrating micromirror which shows acoustic variations in the velocity. A high velocity region is seen near the edge of the micromirror. The extent of this region into the surrounding air is given by the scale of the viscous boundary layer also known as the viscous penetration depth.

In micromirrors, an important parameter associated with thermoacoustic interactions is denoted by the Prandtle number which depends on the viscous and thermal diffusion rates . It can be expressed as the ratio of kinematic viscosity (momentum diffusivity) to the thermal diffusivity which seems to be a dimensionless number and no length scale parameter is associated with it[17]-[18]. According to boundary layer theory, Prandtl number determines whether the thermal boundary layer is larger ($Pr \ll 1$) or smaller ($Pr \gg 1$) than the viscous boundary layer. When air at standard conditions are taken Pr is always less than 1 i.e. the thermal boundary layer is thicker than the velocity boundary layer. If the two layers have the same width, Pr =1. Thermal damping investigation of high performance resonating micro-mirrors have been done in [19]-[20]

For analysing both the velocity and thermal boundary layers, the boundary layer equation can be derived from the general energy equation. Viscous layers are formed in the vicinity of the wall, when fluid come across solid boundaries and the fluid in contact with the wall is not moving, viscous effects retard a layer in the vicinity of the wall. Thermal boundary layers are established when the wall is at a disparate temperature to the fluid, similar to viscous layers as discussed before. The boundary layer equations for an incompressible laminar flow with steady state characteristics in two dimensions can be expressed as:

$$\frac{\partial u}{\partial x} + \frac{\partial u}{\partial y} = 0 \quad \text{continuity} \quad (2)$$

$$u\frac{\partial u}{\partial x} + v\frac{\partial u}{\partial y} = -\frac{1}{\rho}\frac{dp}{dx} + \mu\frac{\partial^2 u}{\partial y^2} \quad \text{momentum} \quad (3)$$

$$u\frac{\partial T}{\partial x} + v\frac{\partial T}{\partial y} = \frac{\lambda}{\rho c_p}\frac{\partial^2 T}{\partial y^2} + \mu\left[\frac{\partial u}{\partial y}\right]^2 \quad \text{energy} \quad (4)$$

T, p, u and v are the temperature, flow pressure, and velocities along and perpendicular to the surface, respectively. μ, λ, cp and ρ are similarly the viscosity, thermal conductivity and specific heat and density. In the present analysis, Cartesian coordinate system is chosen and x and y represent the coordinates along and perpendicular to the surface respectively.

3. Design of Vibrating Micromirrors Using finite Element Method

3.1. Modeling of Micromirrors

A sketch of the idealized micromirror system modelled is shown in figure 1. The mirror is made up of Silicon and is surrounded by air. The modelled micromirror is 500μm by 500μm and only half is shown because of symmetry and thickness is 1μm.

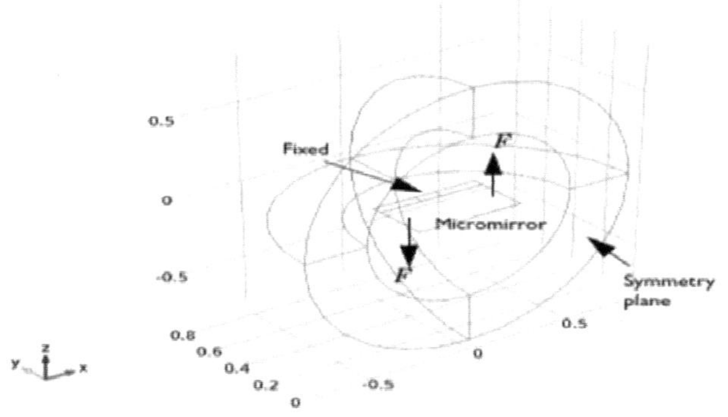

Figure.1. Geometry of the modelled vibrating micromirror

The frequency range selected for the analysis of vibrating micromirror is from 10KHz to 11 KHz in steps of 500Hz.The solid mirror is modeled using shell elements and the surrounding air is modeled using thermos acoustics. A pressure acoustics layer is used to limit the domain of computation for simplicity. The model is simulated to assess the damping the mirror experiences. Thermo acoustics is used to represent the air domain surrounding the micromirror.

3.2. Finite Element Analysis

The vibrating micromirrors can be analyzed by finite elements. Finite element modeling and simulations are performed using COMSOL Multiphysics to study the behaviour of the vibrating micromirror. The finite element model and simulated ones are shown in figure 2 and figure 3 respectively. The three physics interfaces –shell interface, the thermos acoustics, and the frequency domain interface. are set up and combined using multi physics couplings: The acoustic-thermoacoustic boundary coupling and the thermoacoustic-structure boundary coupling features, respectively. The thermal boundary layer thickness values and temperature variations are found out by Finite Element Analysis.

Finite element modelling is also used to evaluate the dissipation of energy in both the layers. The size of the elements selected for meshing the structure is shown in figure 3. The coarse mesh is used to resolve the acoustic boundary layer and the size of the mesh in the structure is a little bit larger to reduce the required memory.

A high velocity region is seen near the edge of the micromirror. The extent of this region into the surrounding air is given by the scale of viscous boundary layer which accounts for viscous penetration depth and corresponding energy loss. Heat transfer to and from the micromirror takes place within the thermal boundary layer and it is responsible for the thermal energy loss.

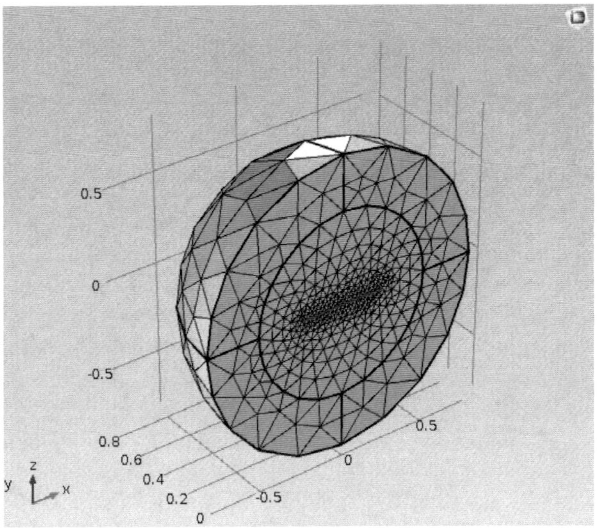

Figure.2. Finite element modelling of micromirror with meshing

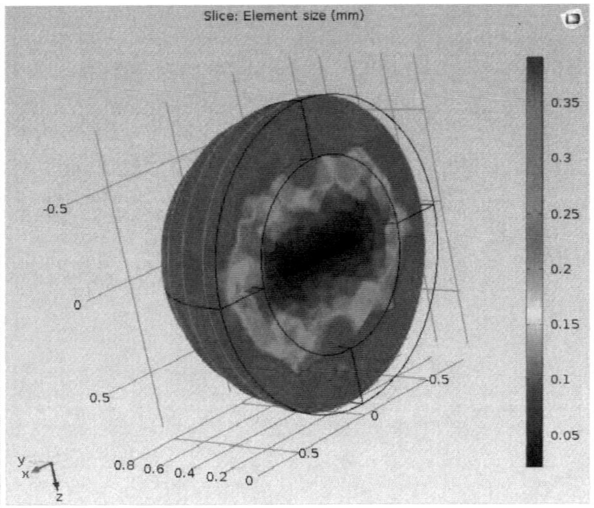

Figure 3. FEM simulated micromirror using COMSOL Multiphysics

4. Results and Discussion

4.1 Thermal and Viscous Penetration Depths and Damping Analysis

Inorder to find out the thickness of the boundary layers or penetration depths (δ) of thermal and velocity regions into the surrounding air eigenfrequency analyses are done and structural modes are investigated. The thickness of the thermal boundary layer is very crucial since heat is mainly dissipated in this boundary layer and energy is mainly lost. The extent of viscous region into the surrounding air is given by the scale of the viscous boundary layer. A high velocity region is seen near the edge of the micromirror. The energy loss in the velocity boundary layer is due to the viscous drag existing around the vibrating structure.

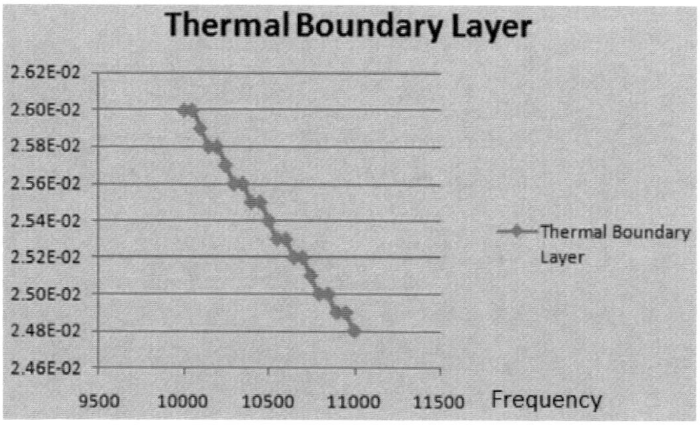

Figure 4. Thermal boundary layer of the micromirror

From the analysis, it is verified that as the eigen frequency increases the width of both the thermal and viscous boundary layers decrease and so energy loss decreases with high frequencies.

Operating the device under high frequencies substantially alleviates energy loss due to thermal and viscous damping since the penetration depths are reduced. Figure 4 shows the variation of thickness of thermal boundary layer and figure 5. illustrates the depth of viscous layer with hike in eigenfrequency of the structure

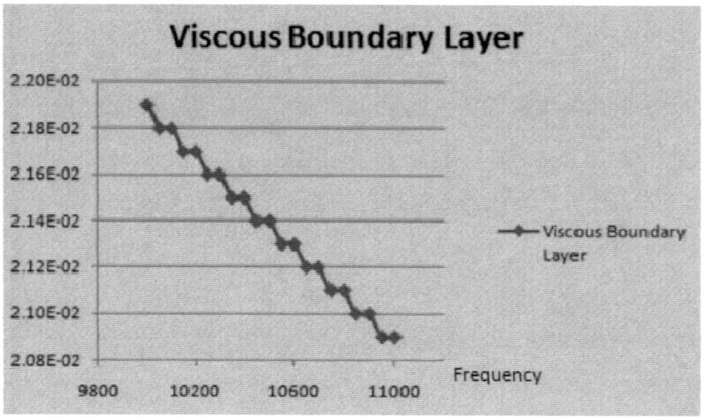

Figure 5. Viscous boundary layer of the micromirror

The viscous boundary layer thickness, δ is analytically given by the expression $0.22\sqrt{^{100}/_{f0}}$.For Prandtle number, Pr greater than 0.6, the thermal boundary layer width δT can be found out from the equation $\frac{\delta}{\delta T} = Pr\,^{1}/_{3}$.The simulated results shows good agreement with the analytical result.

The instantaneous local velocity surrounding the vibrating micromirror in the viscous boundary layer shows the acoustic variations in the velocity as a function of eigen frequency which can be calculated from the eigen frequency analysis.

4.2. Quality Factor and Damping Analysis

In order to attain high signal to noise ratio, high resolution and low power consumption, high Q factors are needed. Low values of Q factor shows degraded spectral purity, increased power dissipation and high noise.

Quality factor is related to the resonance frequency and can be expressed as Q= $f0/_{\Delta f}$ where $f0$ is the resonance frequency and Δf is the peak width at half power frequency.

The different energy loss mechanisms put an upper bound on the achievable Q factors and mechanisms to alleviate energy losses are very essential for high performance systems. The Qfactor is related to the attenuation rate α through $\alpha = \frac{\omega 0}{2Q}$ where $\omega_0 = 2\pi f_0$.

Figure 6.shows the Quality factor Vs. Eigen frequency of the modeled torsional micromirror which shows increase in value of Q for increased eigen frequencies. When the frequency of the system is 10000Hz, the Qvalue is 208.76 and as the frequency increases Quality factor is also enhanced. As the frequency of the system reaches 11000 Hz, Quality factor of the vibrating micromirror achieves a Q factor of 218.81. The values of Qfactor validate the analytical results obtained using Eq. (1).

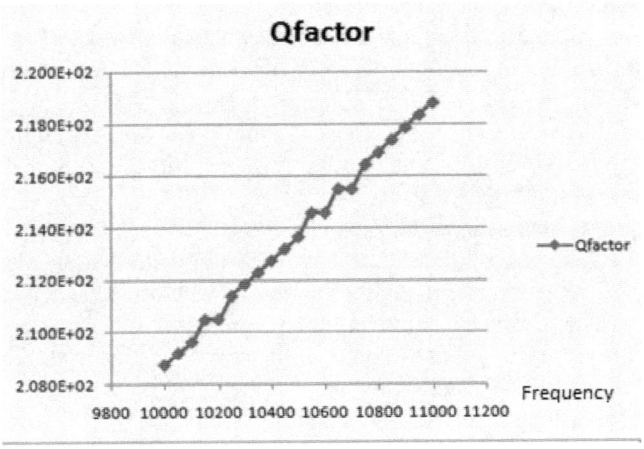

Figure 6. Quality factor vs. frequency of the micromirror

Conclusion

In this paper, the thermal and viscous penetration depths around the vibrating micromirror which accounts for the thermal and viscous dissipation power at different eigen frequencies are analyzed. It is found that as eigen frequency increases thermal and viscous boundary layer thickness or thermal penetration depth decreases. The thermal and viscous penetration depths are accountable for damping in a vibrating micromirror and estimation of its thickness is essential for designing high performance systems with minimum energy loss. The effect of thermal and viscous boundary layer thickness on quality factor is also analyzed and it is found that as eigen frequency increases the thickness of boundary layers decreases and dissipation also diminishes. As energy loss decreases with eigen frequency Qfactor of the vibrating mirror increases. In MEMS/NEMS industry, micromachined micromirrors are extensively used owing to its potential over conventional scanning mirrors with added advantages like high scan speed, minute size, and small cost with diffraction-limited optical performance.

References

[1] Olav Solgard, Asif A. Godil, Roger T. Howe, Luke P. Lee, Yves-Alain Peter, Hanes Zappe, "Optical MEMS:From micromirrors to complex systems," Journal of MicroElectroMechanical Systems, vol. 23, no.3, June 2014.

[2] R. Sulima and S. Wiak, "Modeling of vertical electrostatic comb-drive for scanning micromirrors," Int. J. Comput. Math. Electr. Electron.Eng., vol. 27, no. 4, pp. 780–787, 2008.

[3] XingguoXiong and HanyuXie, "MEMS dual-mode electrostatically actuated micromirror," Proceedingsof Zone 1 Conference of the American Society for Engineering Education (ASEE Zone 1), 2014.

[4] S. Finny and R. Resmi, "Material and geometry optimization for squeeze film damping in a micromirror," 2016 International Conference on Emerging Technological Trends (ICETT), 2016, pp. 1-5, doi: 10.1109/ICETT.2016.7873698.

[5] S. Finny and R. Resmi, "Quality factor and eigen frequency analysis of micromirrors," 2016 International Conference on Control, Instrumentation, Communication and Computational Technologies (ICCICCT), 2016, pp. 655-658, doi: 10.1109/ICCICCT.2016.7988032.

[6] Minikes A, Bucher I and Avivi G, "Damping of a microresonator torsion mirror in rarefied gas ambient," J Micromech Microengg 15:1762–1769K.,2005.

[7] Hamid Moeenfard, Mohammad TaghiAhmadian, and Anooshiravan Farshidianfar, "Modeling squeezed film air damping in torsional micromirrors using extended Kantorovich method lateral shift," Meccanica, 48:791–805, 2013.

[8] Shakya, Subarna. "Acoustic Features Based Emotional Speech Signal Categorization by Advanced Linear Discriminator Analysis." Journal: Journal of Soft Computing Paradigm December 2022, no. 4 (2021): 295-307.

[9] Shakya, Subarna. "Unmanned Aerial Vehicle with Thermal Imaging for Automating Water Status in Vineyard." Journal of Electrical Engineering and Automation 3, no. 2 (2021): 79-91.

[10] Tilmans H A C, "Equivalent circuit representation of electromechanical transducers: Lumped-parameter systems, J. Micromech. Microeng. 6157–76,1996.

[11] C.Zener, "Internal friction in solids II: General theory of thermoelastic internal friction," Phys. Rev., vol. 53, no. 1, pp. 90–99, Jan.

[12] T.V. Roszhart, "The effect of thermoelastic internal friction on the Q of micromachined silicon resonators", Tech. Dig. Solid-State Sens. Actuator Workshop, Hilton Head, SC, 13-16, 1990.

[13] Resmi R, M. R. Baiju, V. Suresh Babu "Thermoelastic damping dependent quality factor analysis of rectangular plates applying modified coupled stress theory" . AIP Conference Proceedings 2166,020029 (2019); https://doi.org/10.1063/1.5131616

[14] T. Veijola, A. Pursula and P. Raback, "Surface extension model for MEMS squeezed-film dampers" in Proc. DTIP'05, pp. 235,2005.

[15] R. Lifshitz and M. L. Roukes, "Thermoelastic damping in micro- and nanomechanical systems", Physical review B, Vol. 6, No 8, Feb. 2000, 5600-5609,2000.

[16] R. Resmi, M. R. Baiju and V. S. Babu, "Material dependent thermoelastic damping limited quality factor analysis of disc resonators," 2017 International conference of Electronics, Communication and Aerospace Technology (ICECA), 2017, pp. 675-680, doi: 10.1109/ICECA.2017.8212750.

[17] Yusuke Kawai, Jin-Hyeok Kim, "Parametrically Actuated Resonant Micromirror Using Stiffness Tunable Torsional Springs" Sensors and Materials,Vol.28 No.2131-139,2016.
[18] M. Gologanu, C.G. Bostan, V. Avramescu, O. Buiu, "Damping Effects in MEMS Resonators", Sensors and Wireless Lab Bucharest (SWLB), IEEE Conference,pp:67-74,2012.
[19] H.Tijdeman "Energy Dissipation in Thin Air Layers" ISMA 21 Conference on Noise and vibration Engineering, WB 26/TM 1582,1996.
[20] Farrugia, R., Portelli, B., Grech, I. et al. Air damping of high performance resonating micro-mirrors with angular vertical comb-drive actuators. Microsystem Technol (2019). https://doi.org/10.1007/s00542-019-04416-0

Semiconductor Technology based Air Sterilizer (STAS) for Indoor Air in Hospital/Clinic Rooms

Pallav Kumar Gopal[1], Dr. Vaishali Chandrashekhar Shelgaonkar[2] and Gaurav Upadhyay[3]

[1]B. Tech, Mechanical Engineering, Ghaziabad. [2]Prof. & HOD, Dept of Anesthesiology IGGMC Nagpur.[3]Prof. Mechanical Department, HRIT Ghaziabad, Uttar Pradesh, India.

pallavkumargopal@gmail.com, vas717@yahoo.com, gauravrakesh13@gmail.com

Abstract. Since the outbreak covid-19, the exposure of bacteria, fungi, and virus is a hazardous situation in outpatient clinics and hospital wards. Moreover, air pollution is a serious health issue that can impact the individuals differently. The major concern of these health issues is not only the polluted air of outdoor atmosphere but also contaminated air is spread inside the houses in urban as well as rural areas. In this regard, this paper proposes a Semiconductor Technology based Air Sterilizer (STAS) that integrated several layers of filtration, purification, sterilization and air freshening.Proposed STAS device is able to purify indoor pollutants and also lower down the amount of other harmful materials like, formaldehyde, moulds etc. simultaneously it can also kill organic substances and bacilli in air.The results have been tested in a hospital of Ghaziabad and the obtained data aftersterilization showed that the application of semiconductor technology used in STAS significantly reduced the concentration harmful gas particles, bacteria, and fungi in the installed environment and shifted the peak of the size distribution of particulate matter particles into coarser particles.

Keywords: STAS, Sterilizer, TiO2, HFL, CPL.

1. Introduction

Air pollution is the major cause of the deadliest and significant deaths in India in 2019, as stated by World Health Organization (WHO). According to WHO, 17.8% deaths in India were recorded on account of air pollution accounting for 1.67 million deaths on 95% certainty level. Moreover, WHO states that approximately 1/3rd of the deaths is due to air pollution which causes hazardous disease such as chronic respiratory disease, lung cancer, and stroke. Further more, as per WHO analysis, 92% population of the world lies in the exposure of air pollution that is beyond the legitimate air quality range. Besides, following smoking, dietary risks, and pressure, air pollution is put on forth greatest risk to life, according to the project "Global Burden of Disease". Having said that ventilation is a good way to maintain a clean and healthy environment, at the same time it can also be the way for the pollutants to enterin doors due to cross contamination, distribution, and inadequate system design etc. Since most people spend their most of the time in indoors, accumulation of fungi and bacteria is being widely studied since the last two decades [1-7]. Further more, owing to the outbreak of deadly infection such as COVID-19, exposure of bioaerosols in hospitals and clinic is a majorconcern. Some bacteria like Mycobacterium tuberculosis, Corynebacteriumdiphtheria,Neisseriameningitides, and Streptococcusprogenies are predominantly transmitted from infected patients via air borne droplets that can be account to nosocomial infection [8,9]. Thus, air filtration technology plays a key role in protecting human health by removing indoor air pollutions. Over the decades, several air purification and cleaning technologies have been propounded in order to clean indoor air. These technologies vary from Internet of Things (IoT) based technologies to filtration technologies. However, filtration-based technologies are becoming more popular [5]. Carbon nano tube filters are specifically seeming to gather aeroso lisednon-biological and biological particles [10]. The major concern is that these methods should also inactivate the biological particles after collecting them. In order to attain this purpose various techniques have been proposed that use UV-enhanced TiO2 cells [11], unipolarionemitters, and electrostaticfield [12].

Further more, air pollution is one of the major causes of indoor contaminated air [13-15], hence, in order to create a clean, dust free, virus free, and healthy environment and to contribute against the harm caused by covid19, we have designed an air purification device, STAS, having several filtration layers and utilizing the advancement of semiconductor filter. It is an air purification device that use semiconductor technology to represent a simple design and eliminates the constraints of technically skilled labor and advanced equipment. This work has also aimed to decrease the amount of ozone emitted from long hours of working of installed device.

The proposed STAS uses the semiconductor technology in order to trap the bacteria and fungi(Section 2.2) that were eliminated using the ionization process (Section 2.3), moreover HEPA filtering technique is used in order to eliminate the infections from the air (Section 2.4), and carbon filtering is also utilized in order to remove harmful gases from the air (Section 2.5), Furthermore many other layers are also used in STAS in order to remove and kill filamentous and unicellular fungi, the gram-negative, mammalian viruses bacteriophage, gram-positive bacteria, protozoa, CancerCells, algae, and othertypes of bacteria present in the air causing air borne diseases.

2. ResearchQuestions

This research has been proposed in order to give answer to the following research questions:

Q1: How the air contamination due to, pollutants & aerogenous spread of dieseses can be reduced, so as to improve patoients safety and prevent health hazardous to health care professionals.

Q2: To what extent the proposed model/device reduce the air contamination.

3. Device and Technology

STAS device is made up of 8 layers of filters that are implemented in order to improve the efficacy and for providing the fresh and pure air to living beings [fig 1].

- The very first filter that has been settled in STAS is called a pre-filter which has the basic function to obstruct the fallen hair and dust particles.
- The air obtained from pre-filtration process is passed through semiconductor filtration process followed by ionization process.

- Ionization environment is provided 8000-volt Direct Current (DC) power supply in order to ionize the neutral molecules and atoms and converting them into charged (positive and negative) molecules and atoms that work as a trap to kill viruses and bacteria from the contaminated air.
- Moreover, as aforementioned the air can also be contaminated with particulate matter particles having the size greater than or equal to 0.3 μm, in order to kill and trap such type of particles, medical grade HEPA filtration technology is used. Thus, the air coming from ionization process is inputted to HEPA filter where these particles are trapped by impaction, interception, and diffusion process.
- Furthermore, the air may contain Volatile Organic Compounds (VOCs) that result in pungent smell and odor in the atmosphere. In order to remove such unpleasant smell from air, Activation Carbon filtration technology has been used which is followed by photo catalyst filtration process that ends the gram positive, gram negative, and unicellular cells infection from the air.
- Thereafter, cold plasma is used to kill the bacterial infection by damaging the surface and microbial DNA structures of bacteria without harming the human tissues.
- Finally, the air is passed through Ultra-Violet (UV) filtration process that kills the moulds, microorganism and other pathogens from the air.

The resultant air obtained from all above filtration processes is sanitized and pure to take in by living beings. Each filtration layer of STAS is discussed in methodology section.

4. Methodology

Figure 1 presents the step by step working of the proposed STAS device in terms of various filtration layers embed in the device. Proposed STAS device will contribute to perform the intended work by means of the following 8 filtration layers.

4.1. Pre-Filtration Layer (PFL)

Pre-filtration layer is basically made up of a Woven Mesh Wire (WMW) filter that filters the dust particles, dander and fallen hair from the contaminated air. The filter used has a mesh like structure hence called WMW filter. This filter works in such a way that it restricts the air particles that are bigger the size than the size of pores of the filter, thus trapping the contaminants and letting pass the pure air particles.

4.2. Semi-Conductor Filtration Layer (SCFL)

The air extracted from the PFL is passed through a p-type semiconductor filter that is made up of a grade 304 stainless steel material & GI sheet. This works on Biot – Savart principle that says, a magnetic field is generated from moving charges carrying constant current. The generated magnetic field is in direct proportion to the current and is attributed to the small particles in the system. In our experiment, 8000V direct current supply is given owing to which the ions are generated thereby a magnetic field is induced. The produced magnetic field generates a force that is attractive force and binds the neutral atoms into molecules and these molecules further trap the viruses and bacteria.

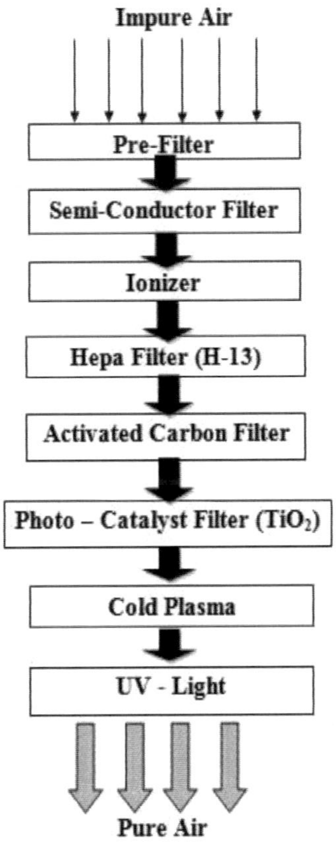

Figure 1. Layers of Filtration

4.3. Ionizer Layer (IL)

This layer is used to trap the bacteria, smoke, pollen, dust, and allergy carrying positive ions by adding the negative ions into the space. The negative ions are disclosed in the air with the help of electrical circuit which discharges the ions in the air. Further,this traps the positive ions thus resulting in improved quality of air. This purifier is ion based and so not harmful for the health and is completely safe, as the emitted negative ions do not harm the health. The particles present in air when attract the ions from the ionizer, also become charged. An electrostatic force of attraction works between the particles and the ions. The charged particles in turn attract either towards the nearest ceilings and walls, or towards the plates of the air purifier. This layer helps to reduce the infection rate to zero as the viruses and infections are eliminated by means of ionizer. Thus, the air of the ward becomes healthier to be breathed by the folks suffering from asthma, allergies, or other type of respiratory diseases. The use of ionizer in the proposed air purifier also helped in improving sleep patterns of patients by relieving the patients from tension suffering from Seasonal Affective Disorder (SAD). Moreover, it also helped the patients in escalating the infection resistant capacity, decreasing the skin temperature, and decreasing the high blood pressure.

4.4. HEPA Filtration Layer (HFL)

HFL layer deals with the particles of 0.21 μ and higher. This layer used H13 HEPA filtration that uses medical grade composed of a mat of randomly arranged paper fibers filter to remove these particles with ~99% efficacy. HEPA air filter captures submicron liquidaerosol(0.5 μm), virus, bacteria, moisture, dust, dirt, and pollen from the contaminated air. Photocatalytic Oxidation (PCO) technique of HEPA filter helps to remove the micro-organisms such as Bacillus subtilis, Staphylococcusepidermidis,Penicilliumcitrinum,and Aspergillus Niger. Further More, HEPA filtre can also capture floor dust which comprises of Bacilli, Clostridium, and Bacteroidia. HEPA Filter ismadefromfibersthatarearranged in a randommanner, ThesefibersarefurthermadefromFiber Glass and polypropylene of the size of diameter between 0.5 – 2.0 μm. These fibers are responsible for passing the air through a narrow-convoluted pathway. This pathway traps the small particles as the air in these pathways turns and twists and smaller air particles crush with the fibers during these turns and twists. Due to very less inertia possessed by small particles, these particles always revolve around the molecules of air.

4.5. Activated Carbon Filtration Layer (ACFL)

This layer uses carbon air filters that are intended to filter out the harmful gaseous particles. In this layer, a bed of activated charcoal (activated carbon) is used to eliminate the VOCs and unpleasant smells of pet and tobacco smoke released in air. Firstly, activated carbon is released with steam, carbon dioxide, or hot air, it makes the surface of carbon bigger as a lattice of tiny pores is formed. These pores trap more molecules thereby making the carbon filter more effective than usual.

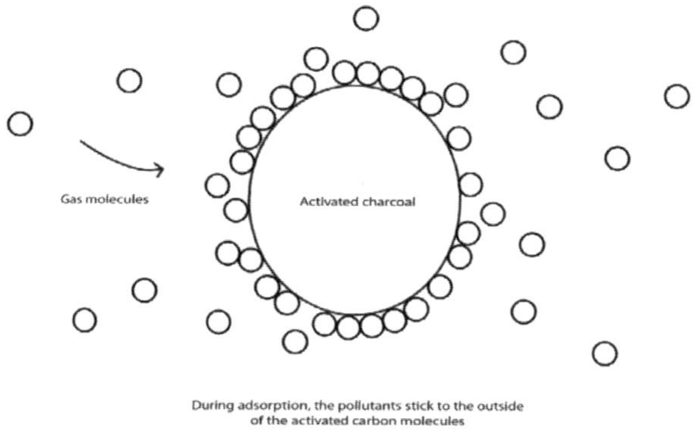

Figure 2. Adsorption Process

Figure 2 shows the adsorption process that shows how the pollutants are removed from the air using activated carbon filter. The process used by these filters is adsorption and not the absorption. In absorption process the pollutants are sticked inside the carbon whereas in adsorption they are at the outer surface. In this adsorption process the greater the area of the activated carbon will be, the greater the number of molecules with stick to the surface, hence the activation process is important here. Organic gases or chemicals are filtered from the air through the carbon air filters using adsorption process. VOCs can be effectively filtered only by carbon filters as other mechanical filters such as

HEPA filters cannot touch these substances. Moreover, carbon filter can also remove the harmful gases given off by cleaning and drying pain products or cigarette smoke.

4.6. Photo-Catalyst Filtration Layer (PCFL)

This layer uses Titanium Dioxide (TiO_2) as the filtration material. This catalyst is excited by ultraviolet (UV) light and is sometimes referred to as titania. This process of excitation of filtration material is known as photocatalytic oxidation or heterogeneous photocatalysis. This filter is made by substrate also called backing material covered with a thin film. Backing material is made of piece of metal or ceramic, when TiO_2 is hit by UV light, electrons are emitted from its surface. The water molecules present in atmosphere are broke up using these electrons into hydroxyl radicals(OH·). These radicals are uncharged hydroxide ions which are short-lived and highly reactive. OH· radicals attack the pollutant molecules that are bigger in size thus break their chemical bonds and turn them in to water and carbon dioxide molecules. Thus, these filters destroy molecules of harmful chemicals by transforming them. Thus, these photocatalyst filters kill the cancer cells, mammalian viruses bacteriophage, protozoa, Algae, filamentous and unicellular fungi, gram positive, and gram negative bacteria.

4.7. Cold- Plasma Layer (CPL)

CPL uses cold plasma is a state that is not in thermodynamic equilibrium. Cold plasma is the gas that is partially ionized and comprises of ground and excited state molecules, radicals, ultraviolet photons, electrons, and ionized gas's ions. The most effective plasma that can be effectively treated at room temperature and atmospheric pressure is, Cold atmospheric plasma (CAP), that can be used at both types of surfaces, highly featured and smooth. Plasma is sometimes considered to be the fourth state of the matter as it generally exists at thousands of °C. Surgical equipment are also sterilize using this much hot plasma. In addition to this, cold plasma is also used as the sterilizer that damages the microbial DNA and kills the bacteria without harming human tissue.

4.8. UV-Light Filtration Layer (UVFL)

UVFL layer is the last filtration layer that utilizes Ultraviolet germicidal irradiation (UVGI) in order to inactivate or kill micro-organisms. UVGI is the infection eliminating method that utilizes small-wavelength UVlight to destroy and disrupt the DNA of nucleic acids and stops their cellular functions. Several applications such as water, air, and food purification can be done using the UVGI method. UVGI device works in such a way that it creates in hospitable environments for pathogens, mould, viruses, bacteria, and other microorganisms by emitting strong UV light. In the proposed method, we have utilized the UV light of 222 nano meter wavelength in order to sterilize work facilities and for medical santitation in ward.

Structural design of proposed system is shown in figure 3.

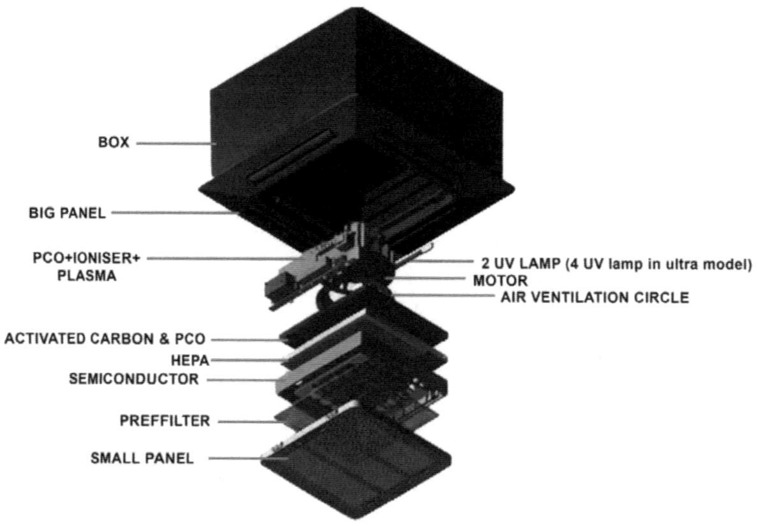

Figure 3. Structural Design of the Proposed System

5. Experimental settings

This section presents the experimental environmental settings for the installation of proposed STAS in that environment.

The experiment of the proposed STAS system has been done in a ward in a hospital in Ghaziabad where Covid patients were given treatment. Table 1 presents the characteristics of the experimented site/wardroom. The reason to select this ward is on purpose to give the answer to our proposed research question 1 (Q1).

Table 1: sampling site characteristics in hospital

Sampling-site	City	Volume (m^3)	Indoor Temp.	Rel. humidity (%)	Ventilation
Wardroom	Ghaziabad	120	32	65-50	Natural

6. Observation and results

The presented results are based on the machine microbiological as well as chemical parameters. The measured size of particulate matters is 2.5 and 10 μm, these are the size of air particles present in the air or you can also say that the presence of particulate matter in the atmospheric air is measured in these parameters. Table 2 presents the varied concentration levels of harmful gases, bacteria and fungi, owing to the setup of the STAS in the general ward of a hospital in Ghaziabad area that have the characteristics presented in section 3.1. It can be clearly seen from the results that the background concentration of bacterial aerosols and moulds is reduced to less than 1 when it was measured to be in the range of 9 to 12 CFM/90mm. Moreover, the concentration levels of fungi are dropped to 1 CFM/m^3 from 8 CFM/m^3.

Table 2. Before and after results in Ghaziabad area

Parameters	Before	Total	After 4 hours	Total	% leftafter STAS filtering
Particulate matter (2.5 μm) in μg/m^3 Particulate matter (10 μm) in μg/m^3	63.0 110.0	173	21.0 25.0	46	26.58%
Sulpher Dioxide(so_2) in μg/m^3 Nitrogen Dioxide (no_2) in μg/m^3 Carbon Dioxide (co_2) in % Ozone (o_2) in μg/m^3 VOCs in μg/m^3 Formaldehyde in μg/m^3	27.0 19.89 <0.9 <19.23 <0.1 <300	<367.1	<5.0 <10.0 <0.2 <18.23 <0.1 <120	<153.53	41.8%
Total Bacterial count in CFM/90mm	<12	<29	<1	<3	10.34%
Mould Count in CFM/90mm	<9		<1		
Fungi in CFM/m^3	<8		<1		

Figure 4 presents the graphical analysis of change percentage concentration of particulate matter, harmful gases, and bacteria and fungi after 4 hours of sterilization of wardroom by STAS machine. The change in percentage has been taken at the y axis, by considering the initial concentrations of all the three measured categories (taken on x-axis) to be 100%, y- axis shows the percentage drop in their respective concentrations. Also, the unintended system or human errors have been marked using the bars for each categories' value.

7. Discussion

STAS is able to kill and trap the bacteria with higher efficacy and performance. In addition, it also aids to sleep and breathe better, keeps the body fresher, and increases the oxygen level in brain for better functioning. The combination of these layers can make indoor atmosphere clean and free from all pollutants. The pollutants are trapped by the combination of impaction, interception, and diffusion mechanism. The larger particles follow the curved contours of the air and reluctantly attach to with another, this effect is called impaction, the mid-range particles are eliminated by interception process, and finally the particles of size < 0.3 μm are eliminated by diffusion process.

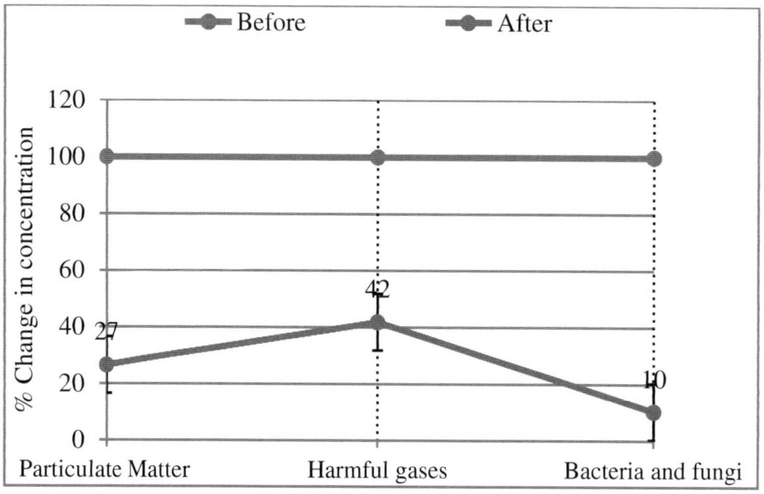

Figure 4. Before and after % concentration analysis of particulate matter, harmful gases, and bacteria and fungi.

Answer to research question 1

In order to give answer to our research question one, we have provided a systematic methodology of our proposed system in section 4, that is leading towards the complete answer of Q1 at each step of filtration. Our proposed air sterilization device STAS is the solution to reduce the air contamination due to, widespread covid pandemic and other polluting factors, and in order to provide fresh air to the patients of hospitals as well as care persons and doctors for effective diagnosing and treating the infected patients.

Answer to research Question 2

The results for the same are presented in section 6 and it is seen that proposed STAS device reduced the amount of particulate matter, harmful gases, and bacteria and fungi by 72%, 58%, and 90% respectively.

8. Conclusions and future research directions

The proposed STAS device can be utilized to control the air quality and to create a healthy environment. Air sanitization capability of STAS with smooth operation and high performance makes it rugged medical grade filtration device. Moreover, its long-life components make it the perfect to be fitted in the zones where there is high pollution. This device is user friendly as it has no difficulty in installation as well, along with this, it does not require much maintenance. Hence, this can also be installed in public medical shops as well as other shops.

This is effective against secondary, or hospital acquired infection in closed or indoor area, due to continuous circulation of air thereby providing safe atmosphere for patients and health care workers. We recommend the use of UVGI with semiconductors to prevent the secondary infection in the indoor room on the continuous circulation of the indoor air. It would also be helpful in eliminating the cross-infection, thereby providing more safety to patients as well as doctors. Simultaneously, it would be better if a disinfection system is also established in implementation process as per the requirements and norms by keeping in mind the principle of personnel and equipment protection, at the same time for conducting the effectiveness of disinfection and quality control management of compliance.

REFERENCES

[1] Pastuszka, J.S.; KyawThaPaw, U.; Lis, D.O.; Wlazło, A.; Ulfig, K. Bacterial and fungalaerosol in indoorenvironment in UpperSilesia, Poland. *Atmos. Environ.* **2000**, *34*, 3833–3842.

[2] Di Giulio, M.; Grande, R.; Di Campli, E.; Di Bartolomeo, S.; Cellini, L. Indoor air quality in universityenvironments. *Environ. Monit. Assess.* **2010**, *170*, 509–517.

[3] Blais-Lecours, P.; Perrott, P.; Duchaine, C. Non-culturablebioaerosols in indoorsettings: Impact on health and molecularapproachesfordetection. *Atmos. Environ.* **2015**, *110*, 45–53.

[4] Bra̧goszewska, E.; Mainka, A.; Pastuszka, J.S. Bacterialaerosols in an urban nurseryschool in Gliwice, Poland: A casestudy. *Aerobiologia (Bologna)* **2016**, *32*, 469–480.

[5] Liu, G.; Xiao, M.; Zhang, X.; Gal, C.; Chen, X.; Liu, L.; Pan, S.; Wu, J.; Tang, L.; Clements-Croome, D. A review of airfiltrationtechnologiesforsustainable and healthybuildingventilation. *Sustain. CitiesSoc.* **2017**, *32*, 375–396

[6] Kobza, J.; Pastuszka, J.S.; Bragoszewska, E. Do exposurestoaerosolspose a risk todentalprofessionals? *Occup. Med. (Chic. Ill)* **2018**, *68*, 454–458.

[7] Madsen, A.M.; Moslehi-Jenabian, S.; Islam, M.Z.; Frankel, M.; Spilak, M.;Frederiksen, M.W. Concentrations of Staphylococcusspecies in indoorair as associatedwithother bacteria, season, relativehumidity, airchange rate, and S. aureus-positiveoccupants. *Environ. Res.* **2018**, *160*, 282–291.

[8] Sarica, S.; Asan, A.; Otkun, M.T.; Ture, M. Monitoringindoorairbornefungi and bacteria in thedifferentareas of Trakya UniversityHospital, Edirne, Turkey. *Indoor Built Environ.* **2002**, *11*, 285–292.

[9] SubramanianSundarrajana, KwongLuck Tan, SoonHuatLim, SeeramRamakrishna, "ElectrospunNanofibersforAir Filtration Applications", ICMAT 2013,7th International Conference on Materialsfor Advanced Technologies, 30 June-5 July,2013, SuntecSingapore.

[10] Guan, T.; Yao, M. Use of carbonnanotubefilter in removingbioaerosols. *J. Aerosol Sci.* **2010**, *41*, 611–620.

[11] Grinshpun, S.A.; Mainelis, G.; Trunov, M.; Adhikari, A.; Reponen, T.; Willeke, K. Evaluation of ionicairpurifiersforreducingaerosolexposure in confinedindoorspaces. *Indoor Air* **2005**, *15*, 235–245.

[12] Yao, M.; Mainelis, G.; An, H.R. Inactivation of microorganismsusingelectrostaticfields. *Environ. Sci. Technol.* **2005**, *39*, 3338–3344.

[13] Ponmalar, Dr S. "Tungsten DiSulphide FBG Sensor for Temperature Monitoring in Float Glass Manufacturing." Journal of Information Technology and Digital World 2, no. 4 (2020): 191-200.

[14] Sungheetha, Akey, and Rajesh Sharma. "Fuzzy Chaos Whale Optimization and BAT Integrated Algorithm for Parameter Estimation in Sewage Treatment." Journal of Soft Computing Paradigm (JSCP) 3, no. 01 (2021): 10-18.

[15] Abhishek, M. B., S. Tejashree, R. Manasa, and T. G. Vibha. "Smart Agriculture Management System Using Internet of Things (IoT)." In Proceedings of International Conference on Sustainable Expert Systems, pp. 363-375. Springer, Singapore, 2021.

Journal of Physics: Conference Series **2325** (2022) 012025 doi:10.1088/1742-6596/2325/1/012025

Hybrid Precoder Design Using Alternating Minimization Algorithm in MultiCell Multi-User Millimeter-Wave Communications for Massive MIMO

Md. Ibrahim Al Imran*, Dr. Md. Farhad Hossain, Rashid Shabab

Department of Electrical and Electronic Engineering
Bangladesh University of Engineering and Technology, Dhaka, Bangladesh

Email: ibrahim.buet.eee@gmail.com*, mfarhadhossain@eee.buet.ac.bd, rashidshabab58@gmail.com

Abstract. In the field of upcoming generation in wireless communication technology there is an extreme demand for higher bandwidth which is expected to be fulfilled by the frequency spectrum ranging from 30GHz to 300GHz which in other words can be defined as millimeter wave (mmWave). But it is a fact that, the mmWave technology lacks the simplicity in the design of precoder. There is higher power loss and higher cost in the baseband precoding in the case of fully digital baseband precoder. As the supported number of analog-to-digital (ADC) converters is very few, the mmWave technology requires 'Hybrid Precoder' which may reduce the loss of extensive power and lessen cost. The cost-effective hybrid precoder transceiver formulation has not been standardized yet. In this paper, we proposed the design of an optimal hybrid precoder in fully connected structure which has a higher complexity but lower cost and lower power consumption in the case of multi-cell multi-user massive MIMO. We have used Alternating Minimization Algorithm for getting the optimal Reimannian manifold solution. Here the comparison between the Digital Baseband precoder performance and the Hybrid Precoder performance has been done using simulation tools.

1. Introduction

Every ten years or so, wireless communication technologies undergo a revolution. The nature of the system, pace, technology, frequency, data capacity, latency, and other factors vary significantly with each generation. From generation to generation, these technologies' standards, methodologies, and features change.

With the ever-increasing number of consumers, there is a demand for faster data rates, more bandwidth, and greater spectral efficiency. The availability of large unlicensed, semi-licensed, and licensed bandwidth in the mmWave frequency spectrum ranging from 30GHz to 300GHz piqued the interest of academics, business, and standards organizations in the hunt for significant capacity improvements [2]. The vast unlicensed spectrum between 30 GHz and 300 GHz is seen as a viable solution to the bandwidth shortage. Within this frequency band, exploratory research is evolving into a new mobile wireless technical standard known as fifth generation (5G). Commercialization is expected around 2020, with broad acceptance by 2025, according to industry opinion.

In comparison to earlier generations, 5G technology will need to serve a large range of ubiquitous applications. The major requirements of 5G technology to serve these diverse applications are

- *Higher data rate:* The maximum data rate is expected to be 20 Gbps, with user-experienced data rates ranging from 0.1 to 1 Gbps. [1].
- *Lower device power:* The device power of 5G devices is to be much lower than that of LTE devices.
- *Higher spectrum efficiency:* Spectrum Efficiency is expected to increase about 3-5 times in 5G networks compared with IMT-Advanced.
- *Lower latency:* From one end to another end, the latency in 5G networks is anticipated to be in millisecond level.
- *Higher energy efficiency*: Energy efficiency in 5G is expected to be increased to 100+ times of the energy efficiency of IMT-Advanced.
- *Traffic capacity per unit area:* The traffic capacity per unit area is anticipated to be 10^6 Mbit/s/m^2.
- *Greater mobility*: The mobility is anticipated to be 500 km/hr.

5G technology will have to support all these ubiquitous applications. So there is a requirement of potential enabling technologies that will support the standardization of fifth generation networks.

2. Complementary Enabling Technologies

The potential enabling technologies that will support the standardization of fifth generation networks are enumerated below:

2.1 Millimeter Wave:

The frequency band from 30GHz to 300GHz is termed to be the mmWave spectrum. The wavelengths range of this spectrum is spanning from 1mm to 10mm. Hence, the technology is termed as "Millimeter Wave Technology" [2].

2.1.1 Free-Space Loss: in case of mmWave technology we have several free space loss models.

$$FSL = 92.4 + 20\, log f + 20\, log(d) \ldots\ldots\ldots (1)$$

Where f is the frequency measured in GHz and d is the distance in km. Here the FSL is in dB unit in this equation.

$$\frac{P_r}{P_t} = \left(\frac{\lambda}{4\,\pi\,d}\right)^2 \ldots\ldots\ldots (2)$$

known as Friis' law of free space propagation loss. Where P_r is the power received, P_t is the power transmitted and the distance between the transmitter and the receiver is denoted by d.

2.1.2 Path loss model due to atmospheric attenuation: Modified Intercept

$$PL_{[dB]} = \alpha_p + 10\beta_p \log(d) + S_{\sigma_s} \ldots\ldots\ldots (3)$$

The modified version of the reference modified, which provide two parameters (αp and βp). In this paper we have used modified intercept path loss model.

2.2 Massive MIMO

The following are the most significant advantages of MIMO systems over traditional wireless systems:

- *Higher SNR:* Due to the coherent combining at the receiver, the SNR is improved. The improvement is up to the number of antennas at the receiver in uplink. On the other hands the downlink SNR is improved by increasing the beamforming gain [3].

- *Lower transmit power:* As data is sent via a beam, the signal is concentrated in a unique direction which results in a lower transmit power when compared to sending a signal in all the directions.
- *Increased data-rate and higher capacity*: Using the multiplexing gain, many orthogonal data streams are sent out at the same time. So the data rate may increase.
- *Interference reduction:* The signal-to-noise-ratio (SNR) is significantly improved as the number of BS antennas increases.
- *Reduction of fading effect:* Diversity is created by broadcasting the same data on different branches. As the branches are independently fading in nature connection dependability is improved a lot.

2.3 Hybrid Beamforming

Figure 01 shows the block diagram of a hybrid beamforming structures at the base station. Here a downlink transmission is shown which is applicable for both cmWave and mmWave bands [6]. F_{BB} is the baseband digital precoder which takes N_S data streams as the inputs and produces N_{RF} outputs. These are up-converted to RF. Then an analogue precoder maps them (F_{RF} to N_t antenna elements) for transmission.

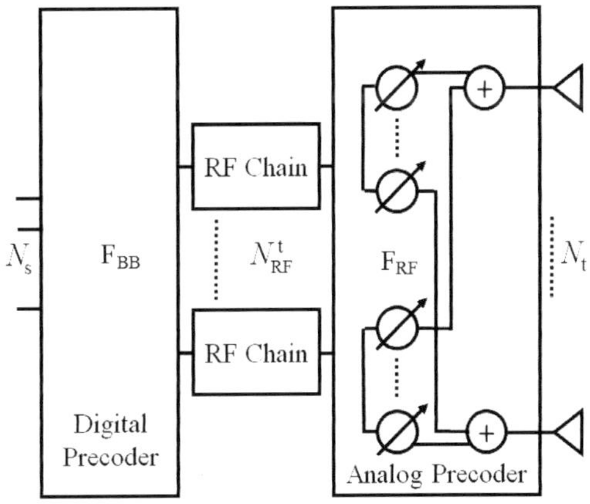

Figure 01: Hybrid beamforming structures for a downlink transmission at the base station

The structure at the receiver is similar as the transmitter. An analogue beamformer $\mathbf{W}_{RF}$ combines radio frequency signals from N_r antennas and generates N_{RF}^t outputs. These are then down-converted to baseband and again combined using a matrix $\mathbf{W}_{BB}$. Finally, the output signal y is produced for detection where the average power received is ρ and **H** is the channel matrix, **n** is the noise vector. So, the beamformer and the precoder are same device actually.

$$y = \sqrt{\rho}\boldsymbol{W}_{BB}^H\boldsymbol{W}_{RF}^H\boldsymbol{H}\boldsymbol{F}_{BB}^H\boldsymbol{F}_{RF}^H\boldsymbol{s} + \boldsymbol{W}_{BB}^H\boldsymbol{W}_{RF}^H\boldsymbol{n} \ldots\ldots\ldots (4)$$

3. System Model

A system having 7 cell cluster with 15 users per cell and cell radius 500m is shown in Figure 02.

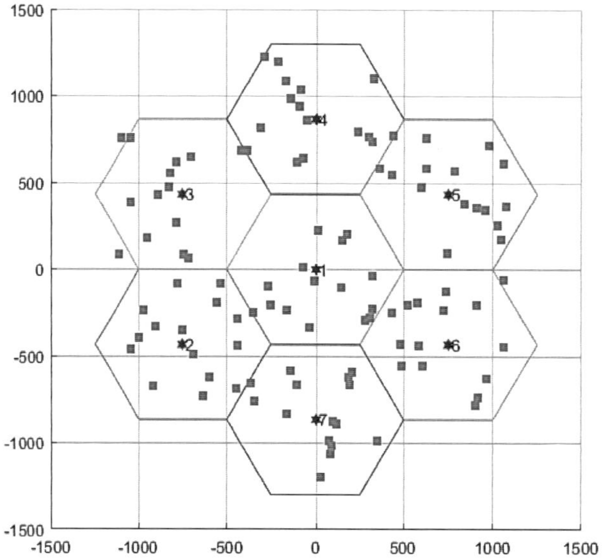

Figure 02: 7 cell cluster with 15 users per cell, cell radius 500m

3.1 Channel Modelling:

The propagation of mmWave is designed using a clustered channel model shown in Figure 03. The channel matrix $\boldsymbol{H}$ is as follows [2],

$$\boldsymbol{H} = \sqrt{\frac{N_t N_r}{N_{cl} N_{ray}}} \sum_{i=1}^{N_{cl}} \sum_{l=1}^{N_{ray}} \alpha_{il} \boldsymbol{a}_r(\phi_{il}^r, \theta_{il}^r) \boldsymbol{a}_t(\phi_{il}^r, \theta_{il}^r)^H \quad \ldots\ldots\ldots (5)$$

where N_{cl} is number of clusters, N_{ray} is the number of rays in each cluster, and α_{il} is the gain of the ray of a cluster. Here $\boldsymbol{a}_t(\phi_{il}^r, \theta_{il}^r)$ and $\boldsymbol{a}_r(\phi_{il}^r, \theta_{il}^r)$ and are the transmit and receive array response vectors respectively. Here the azimuth and elevation angles of arrival and departure are denoted by ϕ_{il} and θ_{il} respectively. Uniform square planar array (USPA) having number of antenna elements $\sqrt{N} \times \sqrt{N}$ is used for modelling. Therefore, the array response vector is [1]

$$\boldsymbol{a}(\phi_{il}, \theta_{il}) = \frac{1}{\sqrt{N}} \left[1, \ldots, e^{j\frac{2\pi}{\lambda}d(p\, sin\phi_{il} sin\theta_{il} + q cos\theta_{il})}, \ldots, e^{j\frac{2\pi}{\lambda}d((\sqrt{N}-1)\, sin\phi_{il} sin\theta_{il} + (\sqrt{N}-1) cos\theta_{il})}\right] \quad \ldots\ldots\ldots (6)$$

here, λ is the signal wavelength, d denotes spacing between two antennas, p and q are antenna indices.

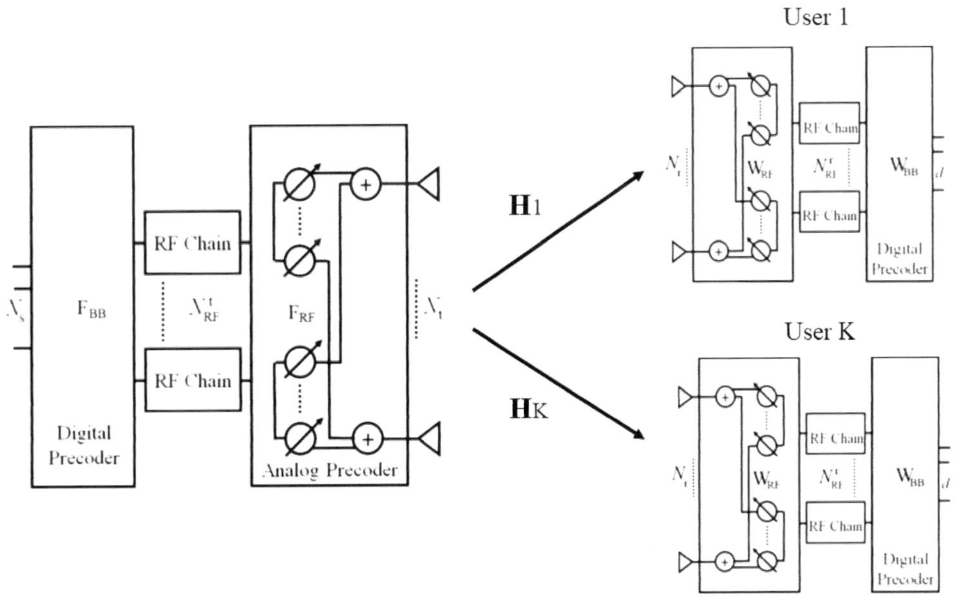

Figure 03: Hybrid beamforming structure of a multi-user MIMO system at the transmitter and at the receiver

- *Millimeter Wave Path Loss Model:* The path-loss models found in literature are summarized below [2]:

$$PL_{[dB]} = \alpha_p + 10\beta_p \log(d) + S_{\sigma_s} \ldots\ldots\ldots (7)$$

- *Random Distribution of Users*
- *Optimal Precoder Design*
- *Hybrid Precoder Design*
- *SINR Calculation*

$$\mathrm{SINR}_{k,l} = \frac{\frac{P_T}{\eta_l \mathrm{PL}_{k,l,l}} \left|\breve{\mathbf{h}}_{k,l,l}\mathbf{f}_{\mathrm{BB}_{k,l}}\right|^2}{\sum_{(m,i)\neq(k,l)} \frac{P_T}{\eta_i \mathrm{PL}_{k,l,i}} \left|\breve{\mathbf{h}}_{k,l,i}\mathbf{f}_{\mathrm{BB}_{m,i}}\right|^2 + N_o} \ldots\ldots\ldots (8)$$

where,

$$\hat{F}_{BB} = \frac{\sqrt{\mathrm{N_s}}}{\|\mathrm{F_{RF}F_{BB}}\|_\mathrm{F}} \mathrm{F_{BB}} \ldots\ldots\ldots (9)$$

3.2 Problem Formulation

In case of a single user in MIMO system, for a fully digital beamformer, the optimal beamformer is calculated from the singular value decomposition of the channel matrix **H**. The dominant left or right singular matrix is taken as the solution. Eigen decomposition is used as one approach. On the other hand, minimizing the Euclidean distance to this fully digital precoder provides the solution of an optimum hybrid beamformer. The objective functions are as follows.

Digital Base-Band Precoder Design

$$\underset{F_{BB}}{\text{minimize}} \left\| F_{opt} - F_{BB}F_{RF} \right\|_F$$

Analog RF precoder design

$$\underset{F_{RF}}{\text{minimize}} \left\| F_{opt} - F_{RF}F_{BB} \right\|_F^2$$
$$\text{subject to } \left|(F_{RF})_{i,j}\right| = 1, \forall_{i,j}$$

Hybrid Precoder Design

$$\underset{F_{RF}, F_{BB}}{\text{minimize}} \left\| F_{opt} - F_{BB}F_{RF} \right\|_F$$

$$\text{subject to } \begin{cases} F_{RF} \in \mathcal{A} \\ \|F_{BB}F_{RF}\|_F = N_s \end{cases}$$

Here $\mathcal{A}$ feasible set of unit modulus constraint $\left|(F_{RF})_{i,j}\right| = 1$

3.3 Alternating Minimization Algorithm

Jointly optimizing F_{RF} *and* F_{BB} *highly complicated* due to modulus constraint of F_{RF} [2]

$$\text{Channel Matrix, } \mathbf{H} = \mathbf{U\Sigma V^H}$$
$$\textbf{Precoder, } F_{opt} = [V]_{Ns\,x\,1}$$

Keeping F_{RF} fixed

$$\textbf{Solution, } F_{BB} = F_{RF}^{\dagger}F_{opt}$$

Where $F_{RF}^{\dagger}$ is the pseudo inverse matrices,
Keeping Fixed F_{BB}

$$\underset{F_{RF}}{\text{minimize}} \left\| F_{opt} - F_{RF}F_{BB} \right\|_F^2$$
$$\text{subject to } \left|(F_{RF})_{i,j}\right| = 1$$

Alternately minimizing F_{RF} *or* F_{BB} keeping the other fixed until a criterion is triggered.

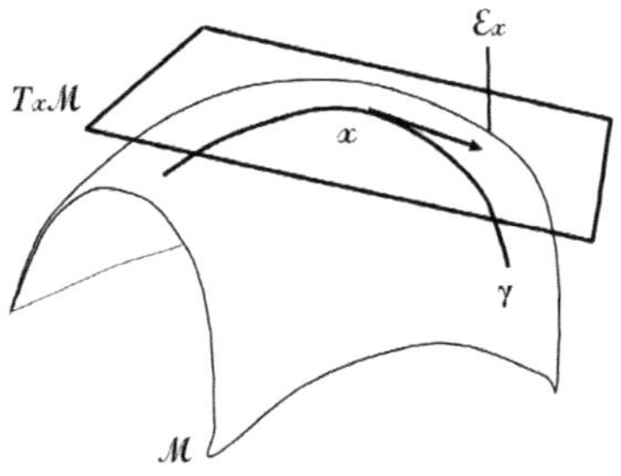

Figure 04: Tangent space of a Riemannian manifold and its tangent vector [2].

Conjugate Gradient Algorithm for Analog Precoding based on Manifold Optimization has been used to minimize F_{RF} *or* F_{BB} keeping the other fixed until a criterion is triggered [3].

3.4 Spectral efficiency:

It is assumed that both the transmitter and the receiver know perfect channel state information (CSI). The expected spectral efficiency can be expressed as

$$\mathbf{E}\,[\mathrm{R}_l] \approx \sum_{k=1}^{K} log_2(1 + \mathbf{E}\left[\mathrm{SINR}_{k,l}\right]) \ldots\ldots\ldots (10)$$

$$R = \log\det\left[\boldsymbol{I}_{N_s} + \frac{\rho}{\sigma_n^2}(\boldsymbol{W}_{RF}\boldsymbol{W}_{BB})^{\dagger}\boldsymbol{H}\boldsymbol{F}_{RF}\boldsymbol{F}_{BB} \times \boldsymbol{F}_{BB}^{H}\boldsymbol{F}_{RF}^{H}\boldsymbol{H}^{H}(\boldsymbol{W}_{RF}\boldsymbol{W}_{BB})\right] \ldots\ldots\ldots (11)$$

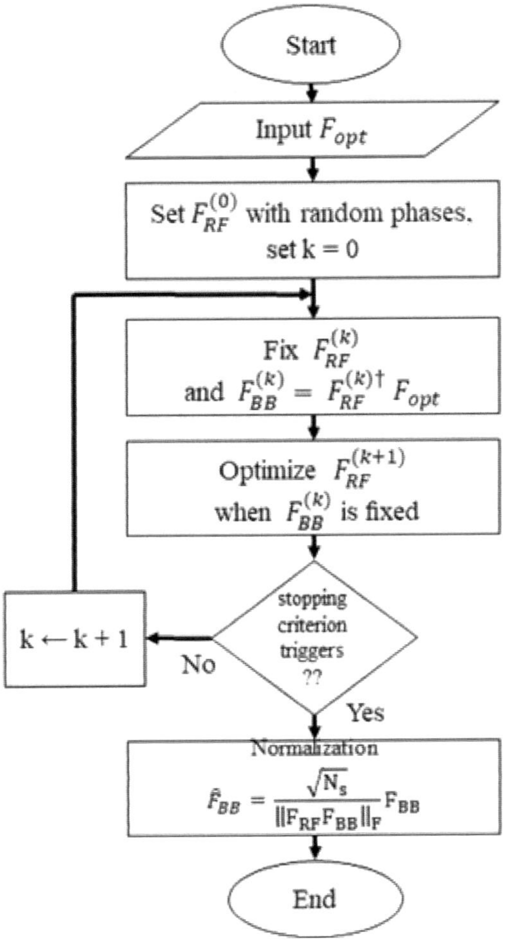

Figure 05: Flow chart of hybrid Precoder modelling for the Fully-connected Structure using Manifold Optimization

4. Simulation Results

The Alternating Minimization algorithm performance is shown here by plotting the performance graphs. We have taken number of data streams Ns=1 for simplicity of calculation. 7 cell cluster is taken each cell having radius of 0.5Km. We have taken uniform square planner array (USPA) both in the transmitter side and the receiver sides. $N_{cl} = 4$ and $N_{ray}=3$ is taken. The distance between the USPA antenna elements is $\lambda/_2$. The standard deviation of the elevation angle and azimuth angle is taken $\pi/_{128}$.

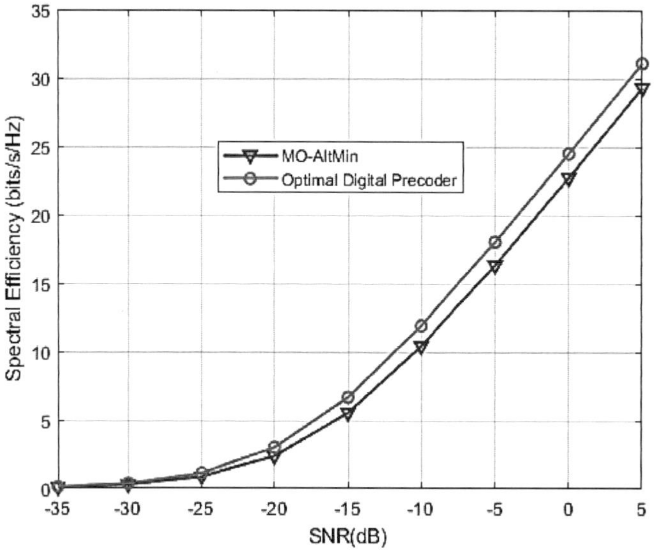

Figure 06: Spectral Efficiency(bits/s/Hz) vs SNR(dB) for both Optimal Digitial Precoder and MO-AltMin algorithm

Figure 06 shows Spectral Efficiency(bits/s/Hz) vs SNR(dB) for both Optimal Digitial Precoder and MO-AltMin algorithm. The red one is Optimal Digitial Precoder precoder and the blue one is designed using Alternating Minimization Algorithm.

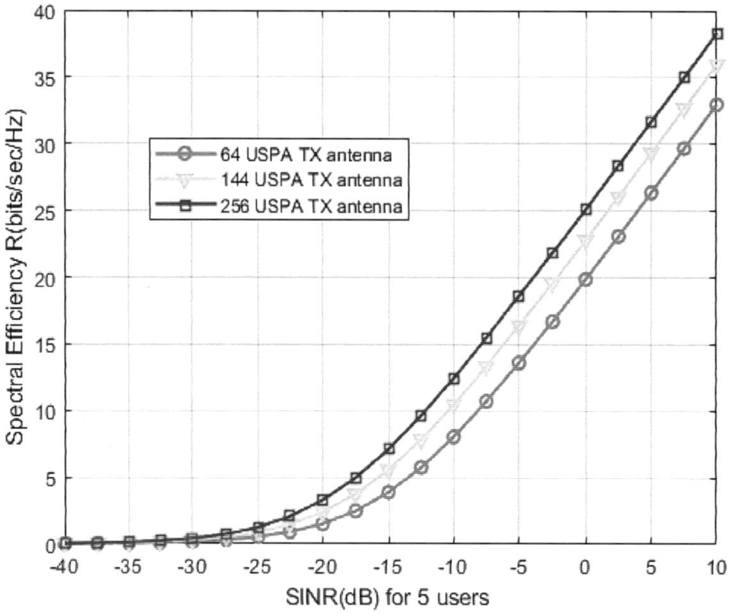

Figure 07: Spectral Efficiency(bits/s/Hz) vs SNR(dB) for MO-AltMin algorithm using 5 users varying number of USPA transmitting antennas (64, 144, 256 USPA antennas respectively)

Figure 07 shows Spectral Efficiency(bits/s/Hz) vs SNR(dB) for MO-AltMin algorithm using 5 users varying number of USPA transmitting antennas. Here 64, 144 and 256 USPA antennas have been used.

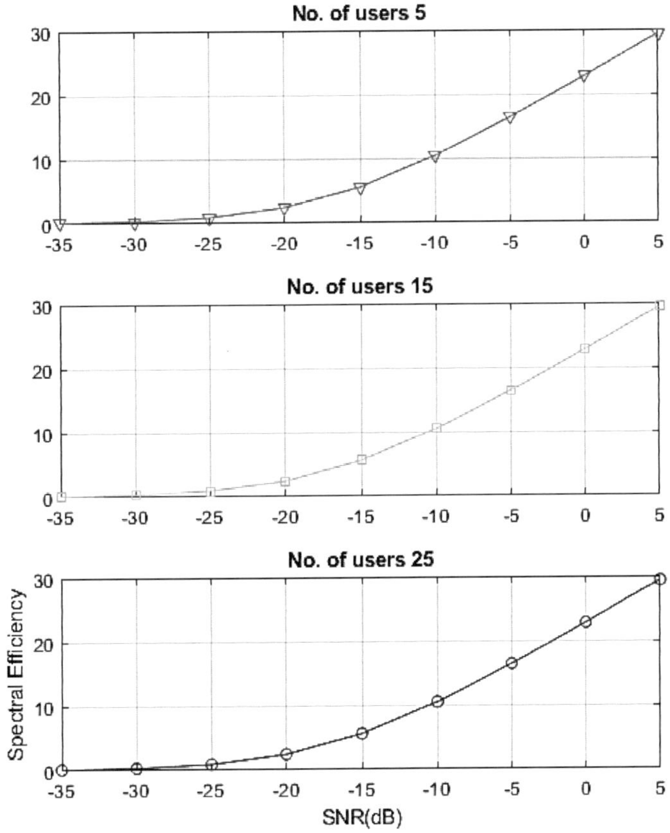

Figure 08: Spectral Efficiency(bits/s/Hz) vs SNR(dB) for MO-AltMin algorithm using 144 USPA transmitting antennas varying the number of users

Figure 08 shows Spectral Efficiency(bits/s/Hz) vs SNR(dB) for MO-AltMin algorithm using 144 USPA transmitting antennas varying the number of users. Here The number of users were taken 5,15 and 25 respectively.

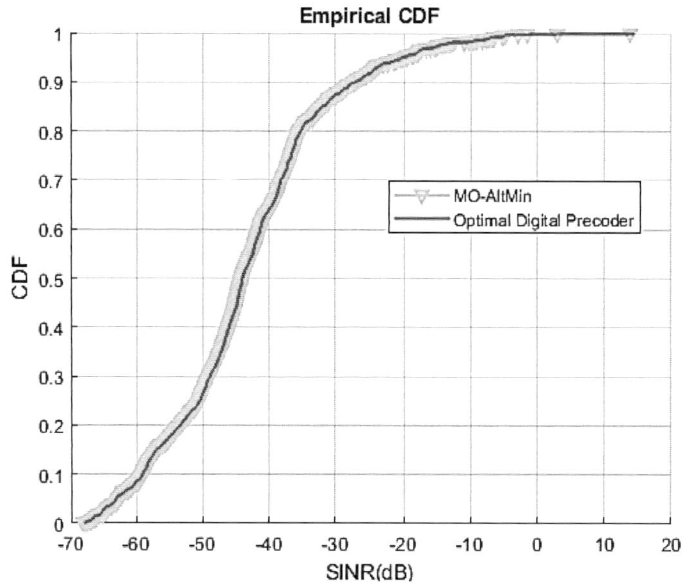

Figure 09: CDF plot of SNR(dB) for Optimal Digital precoder and Hybrid precoder for multi-Cell multi-user mmWave massive MIMO

Figure 09 shows a CDF plot of the SINR (dB) where the hybrid precoder is designed using two algorithms. The red one is fully digital precoder and the cyan one is designed by Alternating Minimization Algorithm. This demonstrates the vast superiority of the alternating minimization algorithm and our motivation for working with this algorithm.

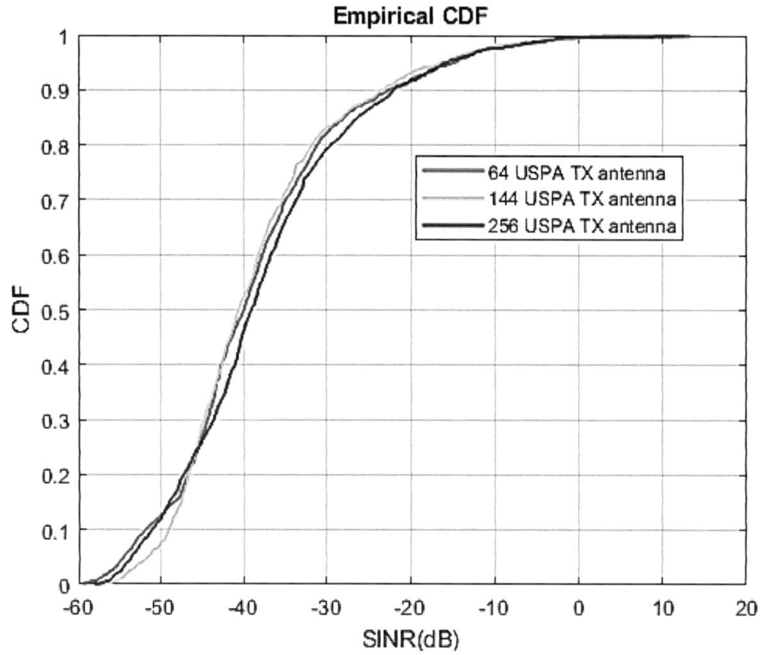

Figure 10: CDF plot of SNR(dB) for Hybrid precoder varying number of USPA transmitting antennas

Figure 10 shows a CDF plot of SNR(dB) for Hybrid precoder varying number of USPA transmitting antennas.

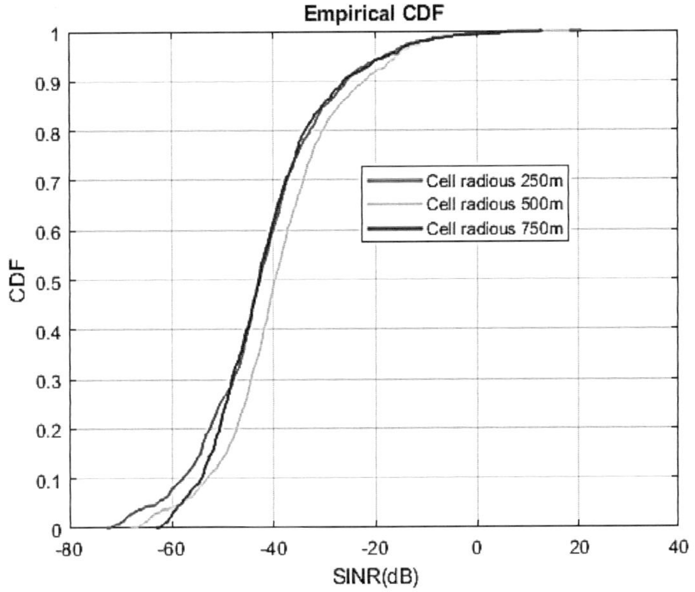

Figure 11: CDF plot of SNR(dB) for Hybrid precoder varying cell radious

Figure 11 shows a CDF plot of SNR(dB) for Hybrid precoder varying cell radious for multi-Cell multi-user mmWave massive MIMO.

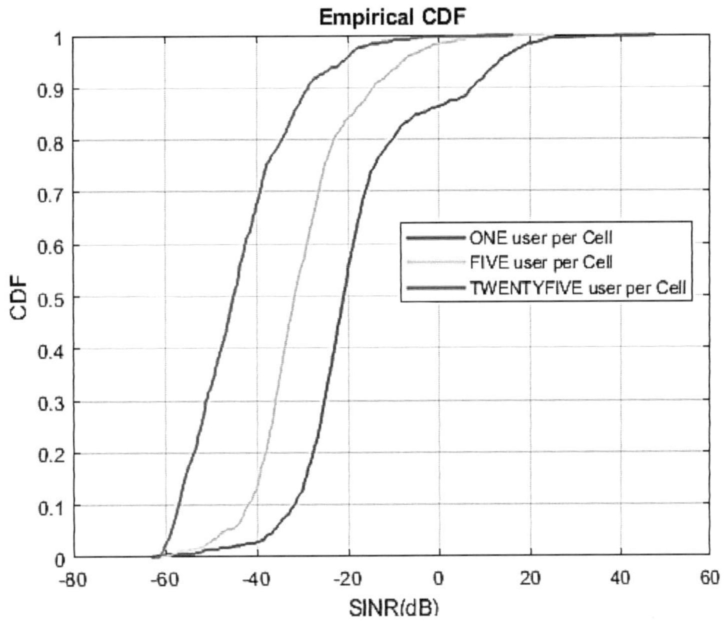

Figure 12: CDF plot of SNR(dB) for Hybrid precoder varying number of users

In Figure 12 a CDF plot of the SINR (dB) is obtained varying the number of users per cell. In the simulation we used 1, 5 and 25 users per cell. It is found that as the number of users increases the SINR decreases. For example, if a horizontal line is drawn along the 60% mark, then for K=1, SINR is approximately -20 dB, K=5, SINR is around -30 dB and K=25, SINR is around -43 dB.

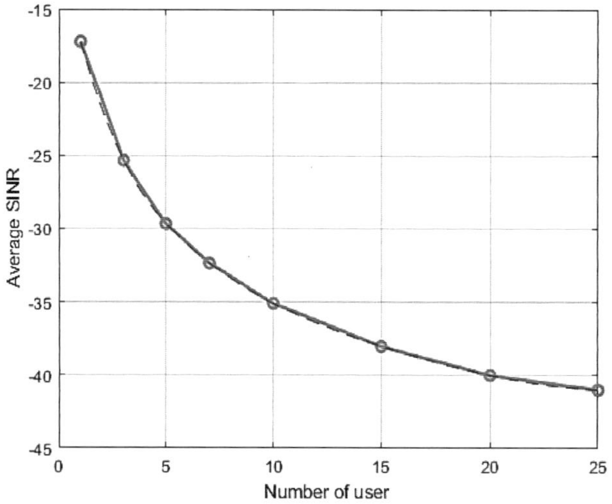

Figure 13: Average SINR(dB) for Hybrid precoder varying number of users

Figure 13 is the plot for average SINR(dB) Vs the number of users. It is found that as the number of users increases, the average SINR also decreases. As the number of user increases the interference also increases and hence the SINR decreases.

The expected value of spectral efficiency can be calculated using the equation (12) shown below [1].

$$\mathbf{E}\,[\mathrm{R}_l] \approx \sum_{k=1}^{K} log_2\left(1 + \mathbf{E}\left[\mathrm{SINR}_{k,l}\right]\right) \ldots\ldots\ldots (12)$$

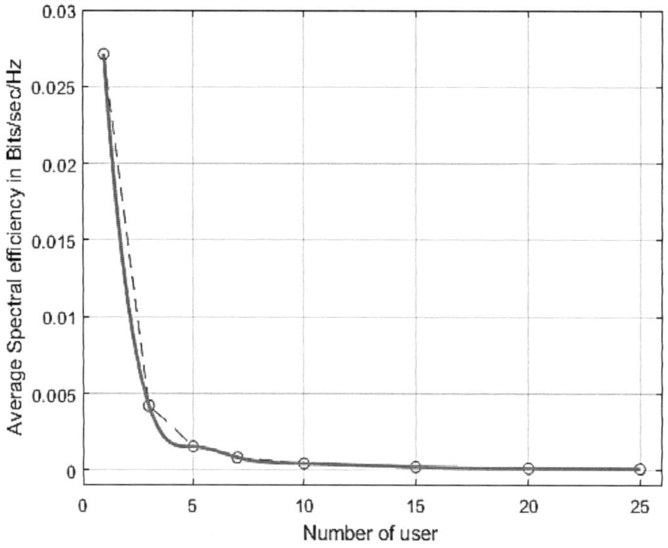

Figure 14: Average Spectral Efficiency (bits/sec-Hz) for Hybrid precoder varying number of users

Figure 14 is the plot of average Spectral Efficiency vs the number of users. It can be seen that as the number of users increases, the expected Spectral Efficiency also decreases significantly.

5. Conclusion

The next feasible step in fifth generation wireless communication is to move towards mmWave technology. The huge spectrum available in mmWave bands is subject to extensive attenuation. Hence, MIMO and hybrid beamforming are essential parts of mmWave communications. Hybrid beamforming optimization is a non-convex optimization problem, hence alternating minimization algorithm was employed to design hybrid precoders in mmWave scenario. Finally, simulations were carried out and the performance of the precoders varying different parameters were observed as part of the thesis work.
Several ideas for future work:
Accurate Channel Modelling: The accurate attenuations need to be incorporated in various channel models for different scenarios. Thus, measurements need to be carried out for proper modelling of mmWave channel.
Channel Estimation Techniques: In this paper we assumed perfect CSI. However, in practice the perfect CSI is not known. Hence, various channel estimation techniques may be explored to overcome these limitations.
In the modelling of the hybrid architectures and in the modelling of the channels we used ideal RF circuit components. But in practical circuit components non-ideal characteristics exist which needs to be considered in further studies.

References

[1] Sun, S., Rappaport, T. S., Shafi, M., & Tataria, H. (2018). Analytical framework of hybrid beamforming in multi-cell millimeter-wave systems. *IEEE Transactions on Wireless Communications*, *17*(11), 7528-7543.

[2] Hemadeh, I. A., Satyanarayana, K., El-Hajjar, M., & Hanzo, L. (2018). Millimeter-wave communications: physical channel models, design considerations, antenna constructions, and link-budget. *IEEE Communications Surveys & Tutorials*, *20*(2), 870-913.

[3] Yu, X., Shen, J. C., Zhang, J., & Letaief, K. B. (2016). Alternating minimization algorithms for hybrid precoding in millimeter wave MIMO systems. *IEEE Journal of Selected Topics in Signal Processing*, *10*(3), 485-500.

[4] Mumtaz, S., Rodriguez, J., & Dai, L. (2016). *MmWave Massive MIMO: A Paradigm for 5G*. Academic Press.

[5] Niu, Y., Li, Y., Jin, D., Su, L., & Vasilakos, A. V. (2015). A survey of millimeter wave communications (mmWave) for 5G: opportunities and challenges. *Wireless Networks*, *21*(8), 2657-2676.

[6] Heath, R. W., Gonzalez-Prelcic, N., Rangan, S., Roh, W., & Sayeed, A. M. (2016). An overview of signal processing techniques for millimeter wave MIMO systems. *IEEE journal of selected topics in signal processing*, *10*(3), 436-453.

[7] Rappaport, T. S., Xing, Y., MacCartney, G. R., Molisch, A. F., Mellios, E., & Zhang, J. (2017). Overview of millimeter wave communications for fifth-generation (5G) wireless networks—with a focus on propagation models. *IEEE Transactions on Antennas and Propagation*, *65*(12), 6213-6230.

[8] J. Walfisch and H. L. Bertoni, "A theoretical model of uhf propagation in urban environments," *IEEE Transactions on Antennas and Propagation*, vol. 36, no. 12, pp. 1788–1796, Dec 1988.

[9] ——, "A Millimeter-Wave Massive MIMO System For Next Generation Mobile Broadband," in *Conference Record of the Forty Sixth Asilomar Conference on Signals, Systems and Computers (ASILOMAR)*,Nov 2012, pp. 693–698

[10] M. Marcus and B. Pattan, "Millimeter Wave Propagation; Spectrum Management Implications," *Microwave Magazine, IEEE*, vol. 6, no. 2, pp. 54–62, June 2005.

[11] Rappaport, T. S. (1996). *Wireless communications: principles and practice* (Vol. 2). New Jersey: prentice hall PTR.

[12] ITU-R. (2013, Sep) Attenuation by atmospheric gases, P.676-10. International Telecommunication Union (ITU), Radiocommunication Sector. Geneva, Switzerland. [Online]. Available: http://www.itu.int/ rec/R-REC-P.676-10-201309-I/en

[13] ——. (2005, Mar) Specific attenuation model for rain for use in prediction methods, P.838-3. international Telecommunication Union (ITU), Radiocommunication Sector. Geneva, Switzerland. [Online]. Available: http://www.itu.int/rec/R-REC-P.838-3-200503-I/en

[14] ITU-R. (2013, Sep) Attenuation in vegetation, ITU-R P.833-8. International Telecommunication Union (ITU), Radiocommunication Sector. Geneva, Switzerland. [Online]. Available: http://www.itu.int/rec/R-REC-P.833/en

[15] M. A. Weissberger, "An Initial Critical Summary Of Models For Predicting The Attenuation Of Radio Waves By Foliage," ELECTROMAGNETIC COMPATIBILITY ANALYSIS CENTER (ECAC), ANNAPOLIS, MARYLAND 21402, Tech. Rep., July 1982. [Online]. Available: www.dtic.mil/cgi-bin/GetTRDoc?AD=ADA118343

[16] 3GPP, "Further Advancements for E-UTRA Physical Layer Aspects," 3rd Generation Partnership Project, Tech. Rep., Mar 2010. [Online]. Available: http://www.3gpp.org/DynaReport/36814.htm

[17] Jameel, F., Haider, M. A. A., & Butt, A. A. (2017, October). Massive MIMO: A survey of recent advances, research issues and future directions. In *2017 International Symposium on Recent Advances in Electrical Engineering (RAEE)* (pp. 1-6). IEEE.

[18] Molisch, A. F., Ratnam, V. V., Han, S., Li, Z., Nguyen, S. L. H., Li, L., & Haneda, K. (2017). Hybrid beamforming for massive MIMO: A survey. *IEEE Communications Magazine*, *55*(9), 134-141.

[19] Wang, W., Yin, H., Chen, X., & Wang, W. (2018, March). Performance Loss of Hybrid Beamforming with Imperfect Phase Shifters in Millimeter Wave Systems. In *2018 IEEE Vehicular Technology Conference (VTC) Fall*.

[20] Sohrabi, F., & Yu, W. (2016). Hybrid digital and analog beamforming design for large-scale antenna arrays. *IEEE Journal of Selected Topics in Signal Processing*, *10*(3), 501-513.

[21] Janisha, R. S., Vishnu, D., & Sheeba, O. (2021). Frequency Reconfigurable Circular Patch Antenna. In *Intelligent Data Communication Technologies and Internet of Things* (pp. 109-118). Springer, Singapore.

[22] Ramya, P., Valarmathi, R. S., Poongodi, C., & Hamsagayathri, P. (2019, August). Slotted Microstrip Patch 4× 4 MIMO Antenna for Ultra Wide Band Applications. In *International Conference on Inventive Computation Technologies* (pp. 11-17). Springer, Cham.

An Enhanced Control Procedure for that DVR to accomplish both LVRT and HVRT in the DFIG Wind Turbine

T. Hemanth [1], G. V. Nagesh Kumar [2], Vempalle Rafi [3]

[1]PG Student, Department of Electrical and Electronics Engineering, JNTUA College of Engineering, Pulivendula, Kadapa, India.
Email: hemanth64410@gmail.com

[2]Professor, Department of Electrical and Electronics Engineering, JNTUA College of Engineering, Pulivendula, Kadapa, India.
Email: drgvnk14@gmail.com

[3]Assistant Professor, Department of Electrical and Electronics Engineering, JNTUA College of Engineering, Pulivendula, Kadapa, India.
Email: vempallerafi@gmail.com

Abstract: In general, Double Fed Induction Generator DFIG-driven wind turbines are extremely sensitive to fluctuations in the grid voltage. For enhancing DFIG wind turbines, this paper proposes an improved procedure for controlling Dynamic Voltage Restorer (DVR) to simultaneously perform both Low Voltage Ride Through (LVRT) and High Voltage Ride Through (HVRT) operations. Consequently, both LVRT and HVRT capabilities are used to increase DFIG performance during grid disturbances. In this study, the Dynamic Voltage Restorer (DVR) has been presented to enhance the capabilities of LVRT and HVRT. MATLAB simulations show that a proposed strategy works as expected.

Index Terms — Double Fed Induction Generator (DFIG), Dynamic Voltage Restorer (DVR), Low Voltage Ride through (LVRT), High Voltage Ride Through (HVRT).

I. INTRODUCTION:

To accommodate the increasing amount of electricity supplied by decentralized renewable systems, such as wind power, the ambitious grid code standards are required for establishing a secure and safe energy network. For example, the grid code comprises rules for coping with faults as well as steady-state generation of watt and wattless electricity. To act like conventional power plants, real grid standards need wind pen to put-up to power network regulation, such as frequency and voltage sway. Grid regulations require wind turbines to stay connected and provide wattless power during grid voltage outages in order to sustain and restore a system's voltage as quickly as feasible. A DFIG (double-fed induction generator) is a significant innovation in wind turbines because of its variable-speed operation and individually adjustable watt and wattless power. As a result, the rotor side of the power electronic converter of DFIGs must be shielded from symmetrical or asymmetrical voltage dips [1-8].

In the case of high rotor currents or DC-link voltages, it is a usual practice to deactivate an RSC and connect the resistive network known as a "crowbar" to a motor circuit. When a crowbar is triggered, a machine consumes a copious of wattless power from the network, which is unsuitable while considering real mesh code demands, resulting in a significant short circuit current, as detailed in [9]. Different protection methods must be investigated in order to safely navigate the grid faults and comply with grid regulations for that fault ride-through of the DFIG, extra hardware such as the series dynamic resistance in a rotor [10] or a stator in the [11] or employing the series line side converter (LSC) architecture as in the [12] have been proposed. [12]Changing an RSC control to prevent a need for that extra hardware can reduce rotor currents during transient grid voltage fluctuations. Stator voltage feed forward [13], consideration of stator flux linkage [14], or use of observed stator currents as the backing for that a rotor current controller [15] can all be used to safeguard RSCs from damage. A number of other studies have focused on an enhanced performance when a grid voltage is unsymmetrical [16–19]. A converter in the [20] is protected by the demagnetizing current, but it becomes evident that the crowbar activation cannot be avoided during deep transient dips, and hence ongoing wattless power management cannot be assured. If a third-party power electronic gadget is hinged to rectify a grid voltage problem, the DFIG system's protective measures can be skipped. [21] describes the dynamic voltage restorer (DVR) system, which consists of a voltage origin follower coupled in series to a grid to correct faulty line voltage. As a result, the use of an external protection device simplifies a DFIG system. DVR has a number of disadvantages, including its high cost and complexity. The DVR may be used to safeguard any distributed load in the micro grid, such as the battery, in addition to protecting wind turbines that do not have adequate fault ride-along effort. [22] Compares the power and voltage ratings of several DVR configurations. [23] Describes the DVR operating at medium voltage. In [24] and [25], resonant controllers are used to compensate for those voltages that are not symmetric. In [26], the squirrel cage induction generator's fault ride-through capacity is provided by the DVR. [27] And [28] provide the DVR for that DFIG wind turbine protection, although only symmetrical voltage dips have been tested, and measurements do not cover transient grid failures and reactive power. In this study, the DVR linked to the DFIG powered by the wind turbine is used to investigate an unbroken fault ride-along of voltage dips that fulfils grid code standards. While a DVR adjusts for the erroneous line voltage, a DFIG wind turbine can retain its nominal functioning as required by actual mesh regulations. Measurement results under transient grid voltage dip on the 22 kW laboratory setup are reported here. This paper provides a thorough examination of DFIG effort and DVR sway, which was first described in the [19].

II. Modelling of Case Study DFIG

Fig. 1 depicts an examined wind turbine system, which comprises of main elements such as a turbine, gear, DFIG, and sequent voltage source converter with the dc connection (as indicated). DC chopper is used to reduce a voltage across a DC capacitor, as well as the crowbar. To decrease harmonics, the line filter connects a sequent follower to a grid, where an RSC and LSC are connected. A wind turbine is safeguarded by digital video recorder (DVR). A mechanical element of the turbine's dynamics will be omitted due to a brief duration of voltage disturbances, and a wind-driven mechanical torque will be considered to remain constant.

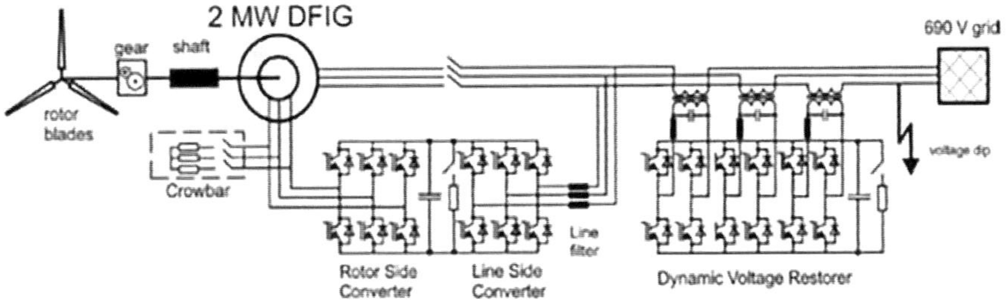

Fig. 1. DVR-enabled DFIG wind turbine system schematics

RSC design and operation necessitate an exact understanding of rotor voltage amplitude and frequency

$$v_s = R_s i_s + \frac{\partial \varphi_s}{\partial t} \quad (1)$$

$$v_r = R_r i_r + \frac{\partial \varphi_r}{\partial t} - j\Omega\varphi_r \quad (2)$$

$$\varphi_r = L_s i_s + L_h i_r \quad (3)$$

$$\varphi_r = L_r i_s + L_h i_r \quad (4)$$

There are three vectors: vamps (flush), verve (voltage), and vici (current). A stator and rotor quantities are denoted by subscripts s and r, respectively. The inductances of stator and rotor are represented by Ls = Ls + Ly and Lr = Lr + Ly, while a mutual inductance is represented by Ly. A stator and rotor resistances are represented by Rs and Rr, while an electrical rotor frequency is represented by.

$$\varphi_r = \frac{L_h}{L_s}\varphi_s + \sigma L_r i_r \quad (5)$$

A rotor voltage equation can be found by replacing (5) in (2).

$$v_r = \frac{L_h}{L_s}\left(\frac{\partial}{\partial t} - j\Omega\right)\varphi_s + \left(R_r + \sigma L_r\left(\frac{\partial}{\partial t} - j\Omega\right)\right) i_r \quad (6)$$

This is divided into two halves. A stator flux _s, which is normally provided by a constantly revolving vector, causes the first portion of the problem.

$$\varphi_s = \frac{v_s}{j\omega_s} e^{j\omega_s t} \quad (7)$$

The rotor current ire is to blame for that a second portion of (6). in many cases, a rotor resistance Rr and a leakage factor are insignificant, therefore a rotor voltage does not deviate significantly from a stator flux component. A normal rotor voltage amplitude Vr0 can therefore be calculated as

$$v_{r0} \approx v_s \frac{L_h}{L_s}\frac{\omega_r}{\omega_s} = v_s \frac{L_h}{L_s} s \quad (8)$$

s = 1 (/s) = r/s explains a slip, while r indicates a slip frequency. During full symmetrical stator voltage dips, a rotor voltage has induced a maximum.

$$v_s = \begin{cases} v_1 e^{j\omega_s t}, \text{for } t < t_0 \\ v_2 e^{j\omega_s t}, \text{for } t \geq t_0 \end{cases} \quad (9)$$

$$\varphi_s = \begin{cases} \varphi_1 = \frac{v_1}{j\omega_s} e^{j\omega_s t}, < t_0 \\ \varphi_2 = \frac{v_2}{j\omega_s} e^{j\omega_s t} \geq t_0 \end{cases} \quad (10)$$

As the continuous quantity, a stator flux cannot track a voltage step function. Assuming _ire = 0, due to its low impact on a rotor voltage, a differential equation (11) may be solved for that an evolution of stator flux.

$$\frac{\partial \varphi_s}{\partial t} = v_s - \frac{R_s}{L_s}\varphi_s \quad (11)$$

The answer is divided into two parts. Following the voltage dip, a Steady State stator flux is characterized by _s2 while a flux transition from _a too _s2 is described by _s1 (12)

$$\varphi_s = \varphi_{s0e^{-t\frac{R_s}{L_s}}} = \varphi_{s0e^{\frac{-t}{\tau_s}}} \quad (12)$$

Stator flow prior to and following a voltage dip is defined by (V1 – V2)/jess. A total flow of testator is a sum of two components.

$$\varphi_s(t) = \frac{v_2}{j\omega_s} e^{j\omega_s t} + \frac{v_1 - v_2}{j\omega_s} e^{\frac{-t}{\tau_s}} \quad (13)$$

Using a rotor voltage equation from (6) and a dynamic stator flux from (13) as input, a rotor voltage dynamic behavior under the uniform voltage dip is defined by (15)

$$v_s = \frac{L_h}{L_s}\left(\frac{\partial}{\partial t} - j\Omega\right)\left(\frac{v_2}{j\omega_s} e^{j\omega_s t} + \frac{v_1 - v_2}{j\omega_s} e^{\frac{-t}{\tau_s}}\right) \quad (14)$$

$$= \frac{L_h}{L_s}\left(s v_2 e^{j\omega_s t} - (1-s)(v_1 - v_2) e^{\frac{-t}{\tau_s}}\right) \quad (15)$$

It is possible to obtain the following rotor voltage in the frame revolving at rotor frequency:

$$v_r = \frac{L_h}{L_s}\left(s v_2 e^{j\omega_r t} - (1-s)(v_1 - v_2) e^{\frac{-t}{\tau_s}}\right) \quad (16)$$

The rotor voltage during the symmetrical voltage dip has two components, according to this research. Due to a fact that slip and residual stator voltage are inversely proportional to this initial component's value, it is tiny for that severe voltage dips and low slips. First, a second component of (16) is very large size and rotates at an electrical rotor frequency (= 60Hz) proportional to a (1/s) time constant. Stator time constant s causes a component to degrade exponentially. Rotor voltage will peak at a beginning of fault (t = 0) and during the full dip during symmetrical voltage dips. (V2 = 0)

$$v_{r,max} \approx \frac{L_h}{L_s}(1-s)v_1 \quad (17)$$

Notably, in this case, the highest rotor voltage is about 17 mill volts (mV). RSC of DFIGs is gauge for that part of stat0r ability as a rotor ability is roughly proportional to slip PR sap that is normally between 0.3.Ly/Ls = 1 indicates that a rotor voltage amplitude is likely to be determined by a Ly/Ls ratio.

$$v_r = \frac{sV_s}{N_{sr}} \qquad (18)$$

In this case, Nasr stands for a stator to rotor rotations ratio. In the practical wind-turbine-driven DFIGs, a turn's ratio is typically set at half or three-quarters to maximize a dc-loop voltage and narrow a converter's current rating. Determine a dc-loop voltage by using a formula

$$v_r = \frac{V_{dc}}{2} \qquad (19)$$

Where a pulse width modulation (PWM) modulation index m is used. When using carrier-based sinusoidal PWM, a modulation index is 1.0, but when using space vector modulation, it is 1.15. A extract is added in the traditional DFIG wind turbines to prevent an RSC from tripping owing to over-currents in a rotor circuit or overvoltage in a dc link when a grid voltage dips, which is the resistive network that is connected to a rotor windings of DFIG. Bypassing a rotor with a series of resistors, a crowbar regulates voltages and directs currents in the safe direction. There are no more RSC pulses when you use it, thus it acts more like the squirrel cage induction network linked directly to a mesh when you do. Reactive power absorption from a stator and a network can further reduce voltage levels, which is not permitted in the actual grid codes [9]. A crowbar circuit is also a sign of heightened stress.

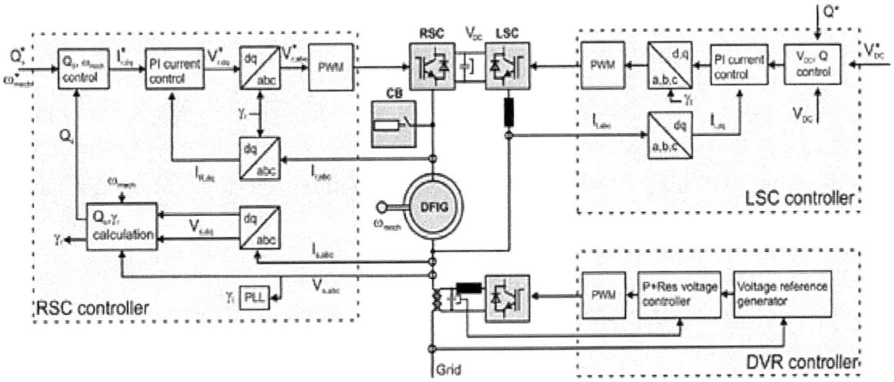

Fig2. DFIG DVR control system

Fig.2. to a system's mechanical components, such as a shaft and gear, as shown in the schematic picture. To learn more about voltage dips and crowbar protection, please refer to [6] and [9]. As the result, crowbar triggering should be avoided from the network and machine mechanical perspective. In any case, simulation results for that crowbar-protected DFIG wind turbines are used to compare a given technique here with a typical DFIG system. In the other words, this is where we design a crowbar resistance. Designing the extract resistance is also possible in the [7, 9]. An extract resistance has the maximum and minimum set by two limitations. A rotor short-circuits current Irma should be limited by the strong crowbar resistance as the first limitation. Crowbar resistance must be low enough to keep a rotor art from overheating, as a second limitation. Here will be the significant current flow through anti-parallel diodes when a rotor phase voltage over a crowbar crosses a larger follower voltage specified in the (18). Simulations employ the crowbar resistance of Crow = 150Rr. With the use of threshold control, it is feasible to inject the demagnetizing current and reduce the duration of the crowbar's operation so that a DFIG may quickly resume normal operation with active and reactive power regulation. The rotor over current triggers the passive crowbar circuit in a laboratory. As soon as safe conditions return, a crowbar can be manually deactivated by a user.

III. RSC Control

RSC plays a significant role in decoupled stator control. In this case, the vector sway map with inner current sway loops is hinged. Figure 2 depicts an overall control system. Control of stator voltage is decomposed into d and q components when stator-voltage oriented is hinged. A stator output active and reactive power are represented as and without taking into account a stator resistive voltage drop, respectively.

$$P_s = \frac{3L_h}{2L_s} V_{sd} I_{rd} \qquad (20)$$

$$Q_s = \frac{3V_{sd}}{2L_s}\left(\frac{V_{sd}}{\omega_s} + L_h I_{rq}\right) \qquad (21)$$

As the result, the d and q components of rotor current can be regulated independently of stator active and reactive power. Architecture of outer power control loops is based on (20) and (21).

Controls of LSC:

Reactive power is supported by an LSC, which regulates a dc voltage Vdc. A voltage drop across a line inductance LLC can be used to modulate a line current Il, with the following dynamics:

$$v_s = R_l I_l + L_l \frac{\partial I_l}{\partial t} \qquad (22)$$

To create a current controller, and to expressed c voltage dynamics, the following equation is used

$$C_{dc} \frac{\partial V_{dc}}{\partial t} = I_{dc} - I_{load} \qquad (23)$$

DC capacitance, Idc, and iLoad currents are utilized to create an outer DC voltage control loop, which can be referred to as CDC.

IV. DVR (Electrical System)

With the line filter, the DVR acts as the voltage source converter (usually LC type). For that sensitive loads or generators, the coupling transformer can be employed as the buffer between a grid and a sensitive load or generator. According to a DC link attach and power and voltage rating, many system topologies are examined in the [22]. A depth of voltage fault that needs to be adjusted heavily influences a DVR system rating. Active power needed by a DVR is simply given by when a DC-to-DC voltage fluctuates with the zero-phase angle jump.

$$P_{DVR} = \frac{v_1 - v_2}{v_1} P_{Load} \qquad (24)$$

V1 is a line voltage that is normal, while V2 is one that is malfunctioning. Keep in mind that voltage faults with phase slant leaps, which can result in a larger ability grade, require further attention [21].

Voltage controllers employed here are P+RES voltage controllers, which control a voltage across a filter capacitor directly without a need for that internal current controller.

$$v *_i (s) = v_c(s) + G_P + R_{es}(s).(v *_i (s) - v_c(s)) \qquad (25)$$

Where a voltage across a filter capacitor is fed forward. A PRs voltage controller's transfer function is defined as

$$G_P + R_{es}(s) = K_P + K_I \frac{s}{s^2+\omega_0^2} \qquad (26)$$

It has been discretized and results in a transfer function shown below:

$$G_P + R_{es}(s) = K_P + K_I \frac{zT_s - T_s}{z^2+(T_s^2\omega_0^2-2)+1} \qquad (27)$$

A new ant windup feature has also been included. A discrete controller's bode plots and root loci were analyzed to discover control parameters KP and KI. A discrete model of the inverter system includes the delay element in the series with an LC line filter's transfer function.

V. **Simulation Results**

Simulations of the 2-MW DFIG wind turbine system and the DVR have been done using MATLAB/Simulink and PLECS to demonstrate the effectiveness of the proposed approach, as shown in Fig. 1. Table I lists the parameters for that simulation. Simulink is used to create a control framework, which can be seen in Figure 2, while PLECS is used to represent all of the power electronic components. To demonstrate the effectiveness of DFIG, we used the two-phase, 37 percent voltage dip lasting 100 milliseconds, both of which were protected by a DVR (see Fig. 3(a) and 4(a)) and displayed in Figures 3 and 4. To a DFIG's response,

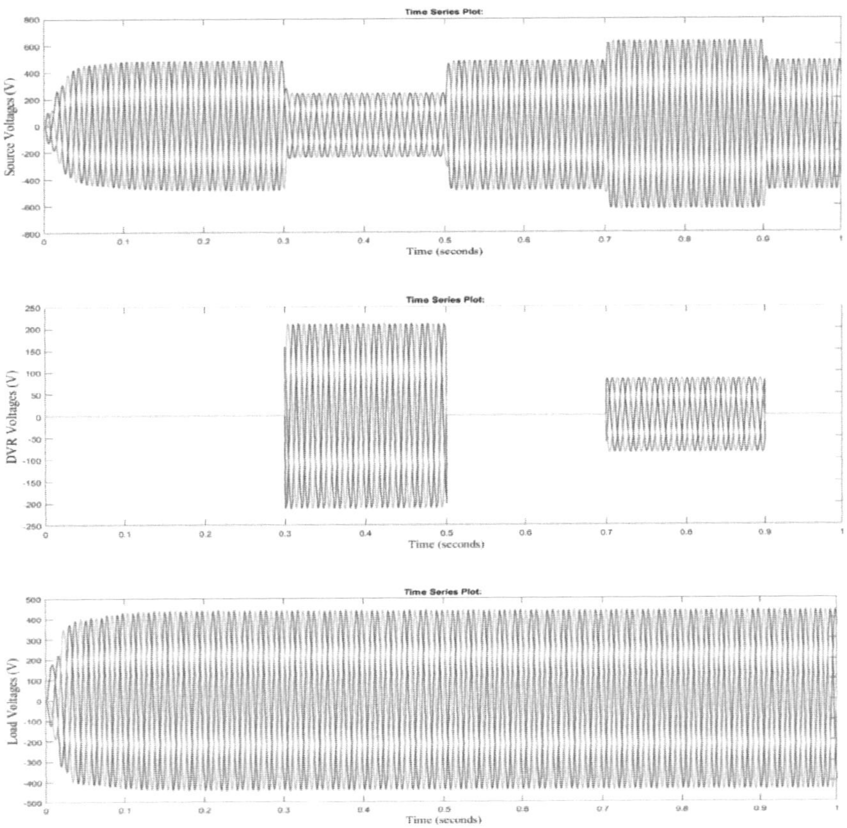

Fig3. Shows a DFIG Voltages, DVR Voltages, and Load Voltages

Journal of Physics: Conference Series **2325** (2022) 012026 doi:10.1088/1742-6596/2325/1/012026

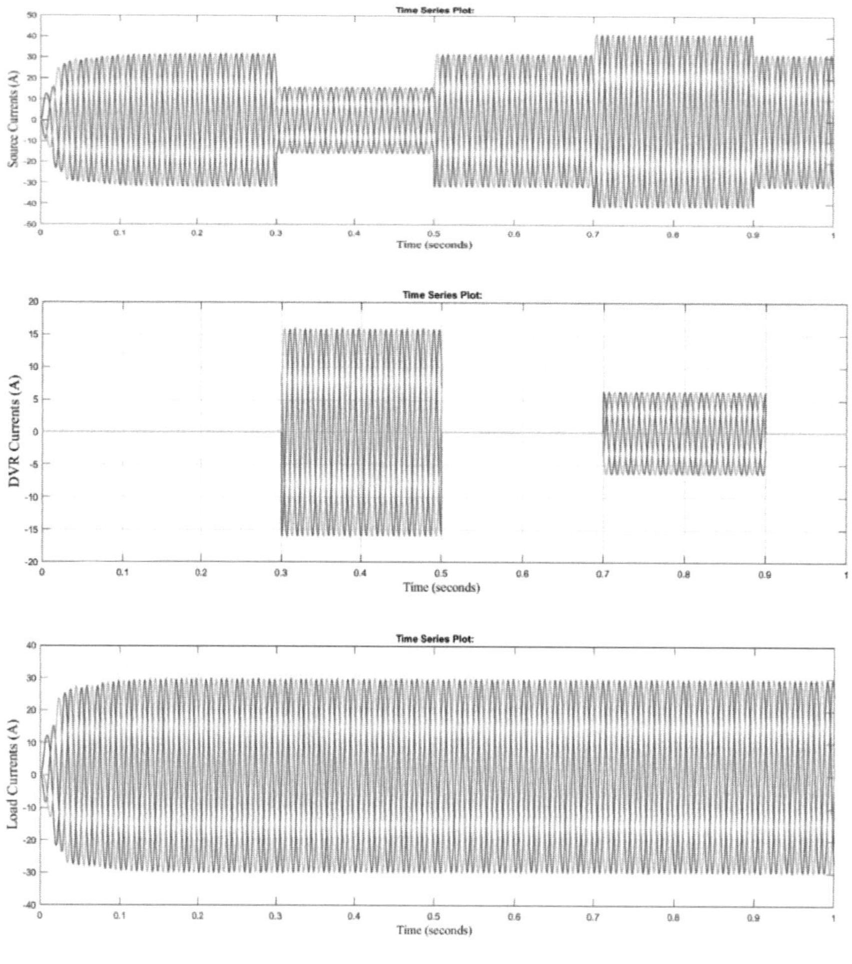

Fig4. DFIG Currents, DVR Currents, and Load Currents.

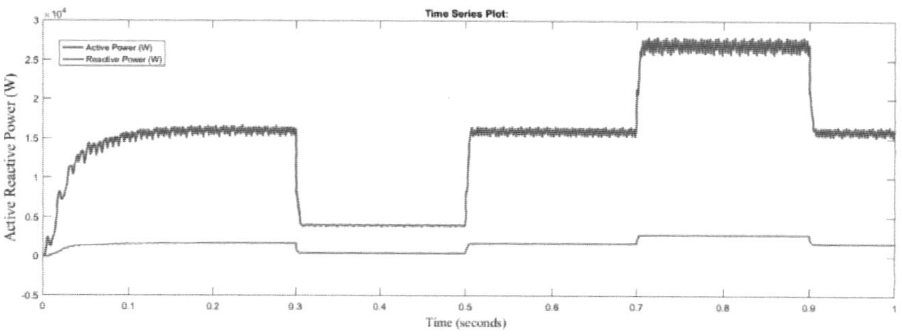

Fig5. Active Reactive power of DFIG.

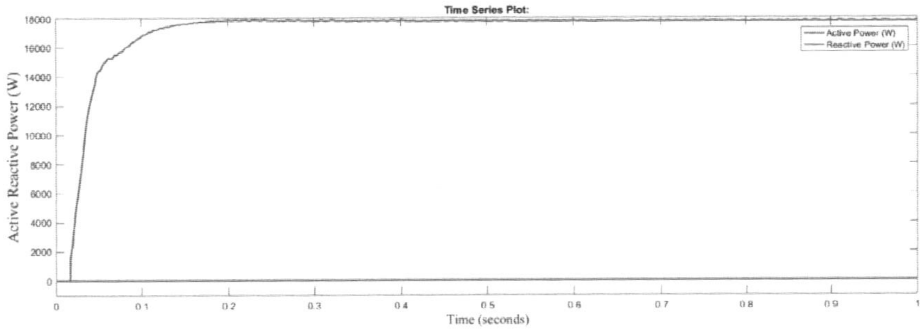

Fig6. Active Reactive power of Load

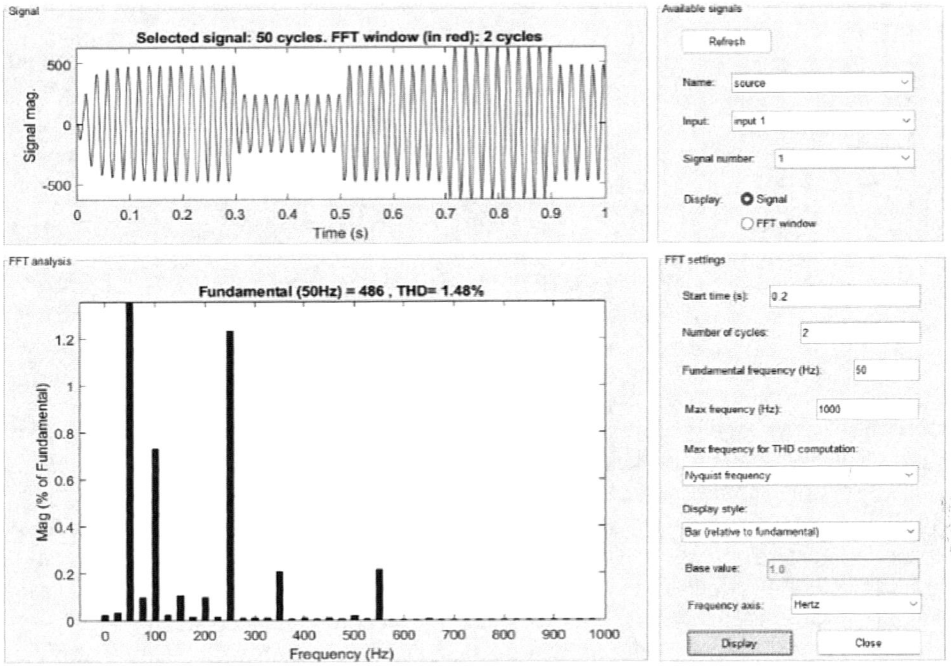

Fig7. THD of DFIG Voltages

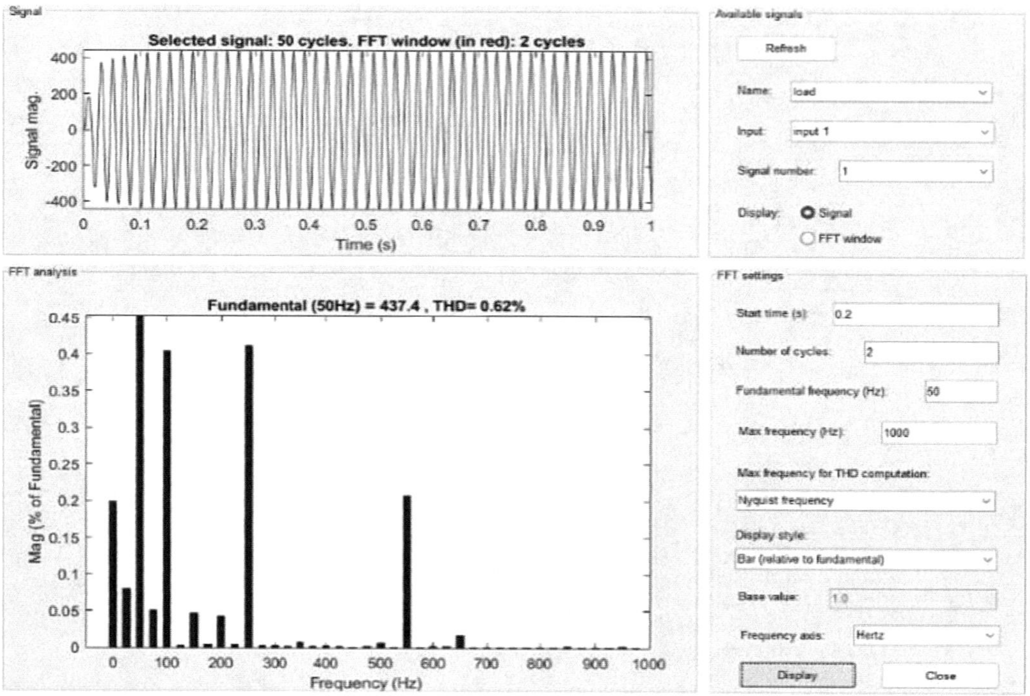

Fig8. THD of Load Voltages

High stator currents are a distinguishing feature. As a result, there are significant currents in the rotor map. A lever can be disengaged once a voltage level has been restored and transients have been reduced, although it is not evident here. When a crowbar is engaged, an RSC can be triggered and then disengaged, allowing a stator to provide machine excitation. It cannot be done because grid rules prevent wattless power management during power failure (see Fig. 3(h)). Owing to a lack of torque, a rotor is unable to accelerate, which may result in the turbine becoming separated due to over speed. Figure 3 shows that a DVR is not switched on in the simulations. A voltage drop may be practically entirely repaired when a wind turbine equipment is protected by a DVR, as demonstrated in Fig. 4 [see Fig. 4(c)]. Lower spindle and rotor overcurrent is generated as a result of DFIG's reaction, implying that a crowbar is not required. Despite the fact that a voltage drop has been totally addressed. An RSC, on the other hand, is still functioning and can independently manage a stator's watt and wattless power. During a grid fault, dynamic power generation (Qs=0.5Mvar) is therefore maintained at the required rate by grid codes. Keep in mind that DVR and DFIG must communicate. Figure 4 depicts the DVR power required to compensate for the voltage decrease (I). A DVR must use real and wattless power that cannot be fed into a damaged grid as a result of grid breakdown.

VI. Conclusion

The usage of a DVR in conjunction with a DFIG powered by wind turbines to provide uninterruptible fault ride-through of grid voltage problems is being investigated. A DVR can correct for faulty line voltage, allowing a DFIG wind turbine to operate normally and meet all grid code requirements without requiring any extra safeguards. Existing wind turbines with poor fault ride-through capability, as well as any distributed load in the micro grid, can be protected with a DVR. Simulated results for a 2 MW wind turbine under an asymmetrical two-phase grid breakdown demonstrate the usefulness of the proposed strategy as compared to a DFIG's low-voltage ride through deploying the crowbar when continuous reactive power output is hard. To confirm the results, measurements were taken during transient grid voltage, which decreases on a 22kW laboratory system.

References

[1] M. Stile and S. Papathanassiou, "A review of grid code technical requirements for that wind farms," Renewable Power Generat., IET, vol. 3, no. 3, pp. 308–332, Sep. 2009.

[2] R. Pena, J. Clare, and G. Asher, "Doubly fed induction generator using back-to-back pwm converters and its application to variable-speed windenergy generation," Electr. Power Appl., IEE Proc., vol. 143, no. 3, pp. 231–241, May 1996.

[3] S.Muller,M.Deicke, andR.DeDoncker, "Doubly fed induction generator systems for that wind turbines," IEEE Ind. Appl.Mag., vol. 8, no. 3, pp. 26–33, May/Jun. 2002.

[4] J. Lopez, E. Gubia, P. Sanchis, X. Roboam, and L. Marroyo, "Wind turbines based on doubly fed induction generator under asymmetrical voltage dips," IEEE Trans. Energy Convers., vol. 23, no. 1, pp. 321–330, Mar. 2008.

[5] M. Mohseni, S. Islam, and M. Masoum, "Impacts of symmetrical and asymmetrical voltage sags on dfig-based wind turbines considering phaseangle jump, voltage recovery, and sag parameters," IEEE Trans. Power Electron., to be published.

[6] S. Seman, J. Niiranen, and A. Arkkio, "Ride-through analysis of doubly fed induction wind-power generator under unsymmetrical network disturbance," IEEE Trans. Power Syst., vol. 21, no. 4, pp. 1782–1789, Nov. 2006.

[7] W. Zhang, P. Zhou, and Y. He, "Analysis of by-pass resistance of an active crowbar for that doubly-fed induction generator based wind turbines under grid faults," in the Proc. Int. Conf. Electr. Mach. Syst. (ICEMS), Oct., 2008, pp. 2316–2321.

[8] G. Pannell, D. Atkinson, and B. Zahawi, "Minimum-threshold crowbar for that the fault-ride-through grid-code-compliant dfig wind turbine," IEEE Trans. Energy Convers., vol. 25, no. 3, pp. 750–759, Sep. 2010.

[9] J. Morren and S. de Haan, "Short-circuit current of wind turbines with doubly fed induction generator," IEEE Trans. Energy Convers., vol. 22, no. 1, pp. 174–180, Mar. 2007.

[10] Vempalle, R., Dhal, P. Loss Minimization by Reconfiguration along with Distributed Generator Placement at Radial Distribution System with Hybrid Optimization Techniques. Technol Econ Smart Grids Sustain Energy 5, 18 (2020).

[11] A. Causebrook, D. Atkinson, and A. Jack, "Fault ride-through of large wind farms using series dynamic braking resistors (march 2007)," IEEE Trans. Power Syst., vol. 22, no. 3, pp. 966–975, Aug. 2007.

[12] P. Flannery and G. Venkataramanan, "Unbalanced voltage sag ridethrough of the doubly fed induction generator wind turbine with series grid-side converter," IEEE Trans. Ind. Appl., vol. 45, no. 5, pp. 1879– 1887, Sep./Oct. 2009.

[13] Suresh Kumar, L.V., Nagesh Kumar, G.V., "Power conversion in renewable energy systems: A review advances in wind and PV system", International Journal of Energy Research, 41(2), pp. 182–197, 2017.

[14] Ananth, D.V.N., Nagesh Kumar, G.V., "Tip speed ratio based MPPT algorithm and improved field oriented control for extracting optimal real power and independent reactive power control for grid connected doubly fed induction generator", International Journal of Electrical and Computer Engineering, 6(3), pp. 1319–1331,2016.

[15] Venkateswara Rao, B., Nagesh Kumar, G.V., "Sensitivity analysis based optimal location and tuning of static VAR compensator using firefly algorithm", Indian Journal of Science and Technology, 7(8), pp. 1201–1210, 2014.

[16] Deepak Chowdary, D., Nagesh Kumar, G.V., "Restoration of single phase distribution system voltage under fault conditions with DVR using sliding mode control", Indian Journal of Science and Technology, 1(5), 29353,2008.

[17] Venkateswara Rao, B., Nagesh Kumar, G.V., Ramya Priya, M., Sobhan, P.V.S., "Implementation of static VAR compensator for improvement of power system stability", ACT 2009 - International Conference on Advances in Computing, Control and Telecommunication Technologies, pp. 453–457,2009.

[18] Nageswara Rao, P., Siva Krishna Rao, G.V., Nagesh Kumar, G.V., "A Novel Technique for Controlling Speed and Position of Bearingless Switched Reluctance Motor Employing Sensorless Sliding Mode Observer", Arabian Journal for Science and Engineering, 43(8), pp. 4327–4346,2018.

[19] Nagesh Kumar, G.V., Amarnath, J., Singh, B.P., "Motion of free conducting particles in a single phase compressed gas insulated bus duct with electromagnetic field effect", Proceedings of 2008 International Conference on Condition Monitoring and Diagnosis, CMD, pp. 424–427, 4580317, 2008.

[20] J. Lopez, E. Gubia, E. Olea, J. Ruiz, and L. Marroyo, "Ride through of wind turbines with doubly fed induction generator under symmetrical voltage dips," IEEE Trans. Ind. Electron., vol. 56, no. 10, pp. 4246–4254, Oct. 2009.

[21] Murugan, Bala, and S. Manoharan. "Optimal power flow management control for grid connected photovoltaic/wind turbine/diesel generator (GCPWD) hybrid system with batteries." International Journal of Renewable Energy Research (IJRER) 3, no. 4 (2013): 819-826.

[22] J. Nielsen and F. Blaabjerg, "A detailed comparison of system topologies for that dynamic voltage restorers," IEEE Trans. Ind. Appl., vol. 41, no. 5, pp. 1272–1280, Sep./Oct. 2005.

[23] Rafi Vempalle, P.K. Dhal, Optimal analysis of time varying load radial distribution system with photovoltaic and wind generating system using novel hybrid optimization technique, Renewable Energy Focus, Volume 41,Pages 246-257,2022.

[24] Sekar, K. R., S. Yogapriya, N. Senthil Anand, and V. Venkataraman. "Ranking Diabetic Mellitus Using Improved PROMETHEE Hesitant Fuzzy for Healthcare Systems." In Intelligent Data Communication Technologies and Internet of Things, pp. 709-724. Springer, Singapore, 2021.

[25] Karuppusamy, P. "Design of Adaptive Estimator for Nonlinear control system in Noisy Domain." Journal of Electronics and Informatics 3, no. 3 (2021): 150-166.

Journal of Physics: Conference Series **2325** (2022) 012027 doi:10.1088/1742-6596/2325/1/012027

Design and Implementation of an Embedded System for Islanding Detection and Control Action in PV Inverter system

A. G. Hake[1], S. B. Chavan*[2], R. K. Kamat[3]
[1] Sr. Engineer, Agiliad Technologies Pvt Ltd, Pune
*[2] Corresponding author, Department of Technology, Shivaji University, Kolhapur, India, Email- sbc_tech@unishivaji.ac.in
[3] Department of Technology, Shivaji University, Kolhapur, India

Abstract. Many countries utilize hybrid energy systems for power generation. Transmission lines are used to deliver electricity to remote places from a variety of sources. During islanding, power lines are broken or damaged by natural or man-made causes. If not detected, animal life may be endangered. This paper presents anti-islanding technique that can be used to detect and prevent islanding. In this work, a prototype is developed using microcontroller for an anti-islanding PV inverter control system. The system detects islanding and disconnects the load from the source inverter and switches off the inverter. This technique continuously monitors the output power. A sudden drop in output power is indicative of islanding. The system immediately isolates the PV inverter from the output load and switches off all processing devices. A relay is used to isolate the PV inverter from the load. In the system developed here source isolation time is equal to relay switching time.

***Keywords**: Embedded System, Anti-islanding, PV inverter, Islanding detection, Hybrid Energy, Microcontroller*

1. Introduction

Natural or human-made causes of islanding can cause transmission/distribution lines to open (breaks/damages) and inverters to continue supplying power, resulting in a hazardous situation. It is necessary to detect islanding and implement anti-islanding measures. Several islanding detection and anti-islanding techniques have been developed by researchers. An overview of all existing islanding detection techniques is presented in [1] which provides guidelines regarding selection of suitable islanding detection method. In reference [1] the comparison of various methods is done based on parameters like detection time, error detection rate, power quality etc. Reference [2] provides another detail review on islanding detection techniques in which authors have shown more faith on active techniques. In a passive islanding detection method, Valentini et al [3] measured power mismatch in PV source and load to detect islands. Pigazo et al [4] suggested a hybrid technique for islanding detection which monitors and processes higher frequency components with wavelet transforms. Using inverter output current phase, Guo Kiang Hung et al [5] developed a programmed phase shift scheme for islanding detection; this method was tested for parallel RLC load. Riyasat Azim et al [6] described a hybrid islanding detection technique that uses active methods when passive methods fail to detect islanding. One active method is based on slip mode frequency shift [7]. Partha et al [8] presented a technique that is voltage independent and relies on current saturation of the controller. Manel et al [9] proposed a technique for fast detection of islanding using voltage sensors interfaced to the PCC for detection. Lei et al [10] presented a reactive power disturbance-based method with several advantages. Majid et al [11] presented a technique that uses impedance variation and inverter current phase change for islanding detection. In a recent paper, Skocil et al [12] reviewed islanding detection techniques for PV systems. Several active anti-islanding methods were discussed by Francesco De Mango et al. [13]. A design expression for boost converters is discussed in reference [14]. Fuzzy logic controller based approach is expressed in [16] to obtain maximum power from the PV system. Authors have reported higher speed compared with other few MPPT techniques.

The importance of islanding control schemes is examined in this work and a microcontroller based system is implemented to detect islanding mechanism. The prototype of PV inverter system is developed in the laboratory. The strategy presented here monitors the inverter output current. When inverter output current becomes zero, it is considered as the indication of islanding phenomenon. The controller system switches off the inverter and energizes the relay to isolate it from the transmission/distribution line.

2. System description

Figure 1 shows the system block diagram where the controller monitors PV voltage and current, inverter output voltage and current, and displays that data on a 16X2 LCD display. Furthermore, it compares the output voltage and current with the threshold to detect islanding and disconnects the inverter from the load using a relay. A cascaded boost DC/DC converter boosts the converter output to the required level to generate SPWM pulses after the inverter. These pulses are fed through a LC filter to obtain sine waves of 230V and 50Hz.

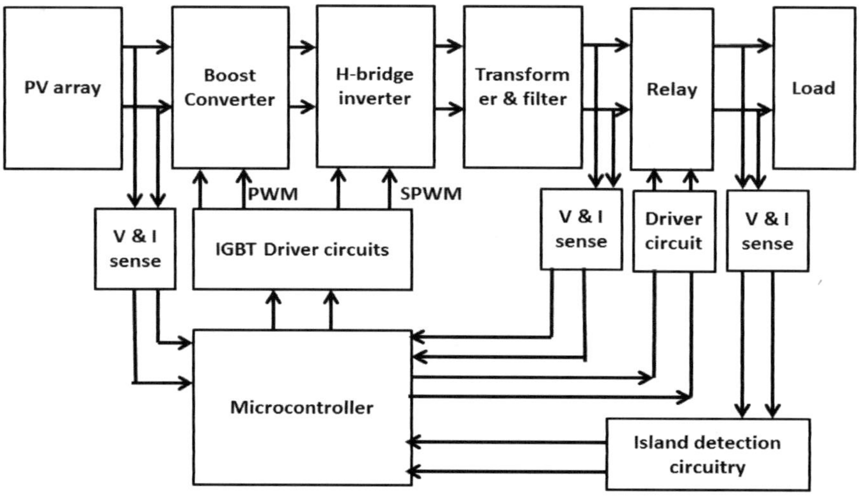

Figure 1. System block diagram

3. Prototype development

This section discusses design and development issues of hardware and software. System model is developed in MATLAB and first simulation results are tested and verified.

3.1 Hardware implementation

The system is developed using ARM 7, LPC 2148 microcontroller. Cascaded boost converter circuitry is used, it's PWM duty cycle is regulated using expression 1. Boost converter circuitry is designed for efficiency of 80 % for which the duty cycle of the driver circuitry is 70 %. The inductor and capacitor in the boost stage are selected using equations 2 and 3. The inductor of 1.50 mH and capacitor of 960 nF is selected for frequency of 10KHz. In cascaded boost converters and inverters, Fairchild semiconductors FGA 15N120 IGBT is used. A driver circuit based on IR 2113 is used. The

LC filter at output is designed using expression 4. Voltage divider networks and ACS 712 current sensors are used to sense input and output currents.

$$Duty\ cycle = 1 - \frac{[V_{inputMinimum} X\ Efficiency]}{V_{Output}} \quad (1)$$

$$L_B = \frac{V_{Input}[V_{Output} - V_{Input}]}{\Delta I_L\ X\ F_s\ X\ V_{Output}} \quad (2)$$

$$C_B = \frac{I_{OutputMaximum}\ X\ DutyCycle}{F_s\ X\ \Delta V_{Output}} \quad (3)$$

$$F = \frac{1}{2\pi\sqrt{L_f C_f}} \quad (4)$$

SPWM signals are generated using a look up table in program. For 50 Hz frequency, T=20 mS . The time period of 20 mS is divided into 200 parts, thus each period of 100 uS is obtained with step angle of 3.6 degrees for sine wave. For each step the time period is increased to create pulse of varying width. Thus 100 pulses are used for positive and remaining 100 are used for negative waveform. Sampling frequency of 10KHz is used with T= 100µS for each division. In program look up table and PWM channels are used to generate SPWM pulses. These pulses drives the IGBTs placed in H-bridge configuration. Laboratory prototype developed for experimentation is shown in figure 2.

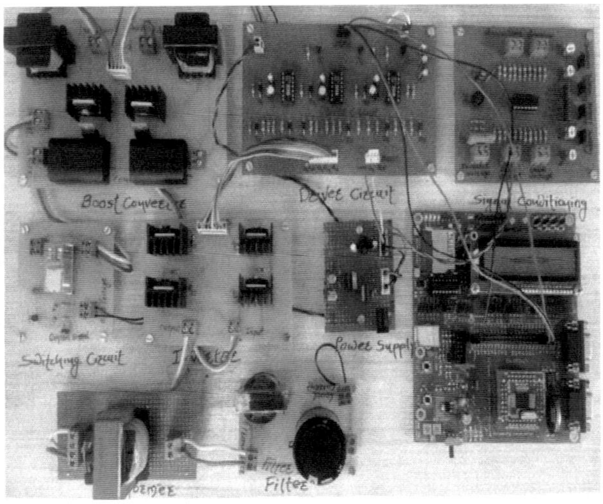

Figure 2. Prototype developed in laboratory

3.2 Software implementation

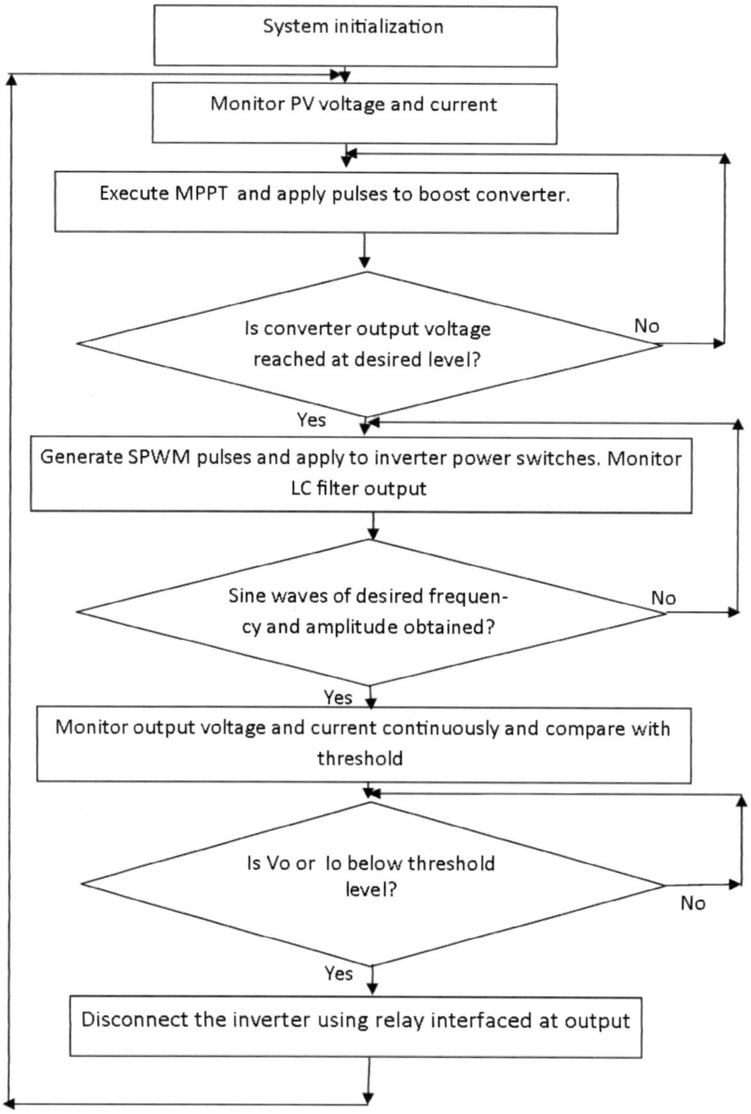

Figure 3. - Flow chart

Power switches are failure prone components [15], however, in this work all devices in power processing stages are assumed fault-free, and the system operation is described in flowchart in figure 3. The system monitors PV output power, execute MPPT routine and applies pulses to boost converter till the converter output voltage is reached at desired level. Then the SPWM pulses are applied to inverter power switches and LC filter output is monitored for desired amplitude and frequency. When islanding phenomenon occurs, the output current suddenly reduces to zero. In ideal working conditions zero output power indicates islanding phenomenon. When the controller system detects islanding phenomenon by monitoring the output current, it quickly switches off the gate pulses of IGBT's in converter and inverter stages and switches the relay OFF to isolate the transmission / distribution lines from the inverter system. Due to this mechanism the supply of power to distribution lines is stopped instantly. This avoids the possible situation of carrying power through the transmission / distribution line connected to the inverter system. Therefore any unintentional touch by living being to the broken transmission lines and to the current carrying objects will not cause the harm. The occurrence of islanding fault will be displayed on 16x2 LCD for further fault detection and corrective steps.

4. Results and discussion

In this work, inverter output voltage and current are monitored to detect islanding. Threshold of 1% current of peak current is considered, when output current drops below 1% of expected peak current, it is considered as faulty condition. A zero inverter output current indicates an OC transmission/distribution line, while a zero inverter output voltage indicates an SC line. When such conditions are detected, the inverter is turned off and the output relay is switched OFF , so the load is disconnected from the inverter.
The simulation model is developed in MATLAB. The SPWM pulses are applied to the gates of power switches connected in H bridge configuration. In figure 4, the H-bridge inverter output pulses are shown. In figure 5, the inverter output voltage in normal working condition is shown, and in figure 6, the inverter output current waveform in normal condition are shown.

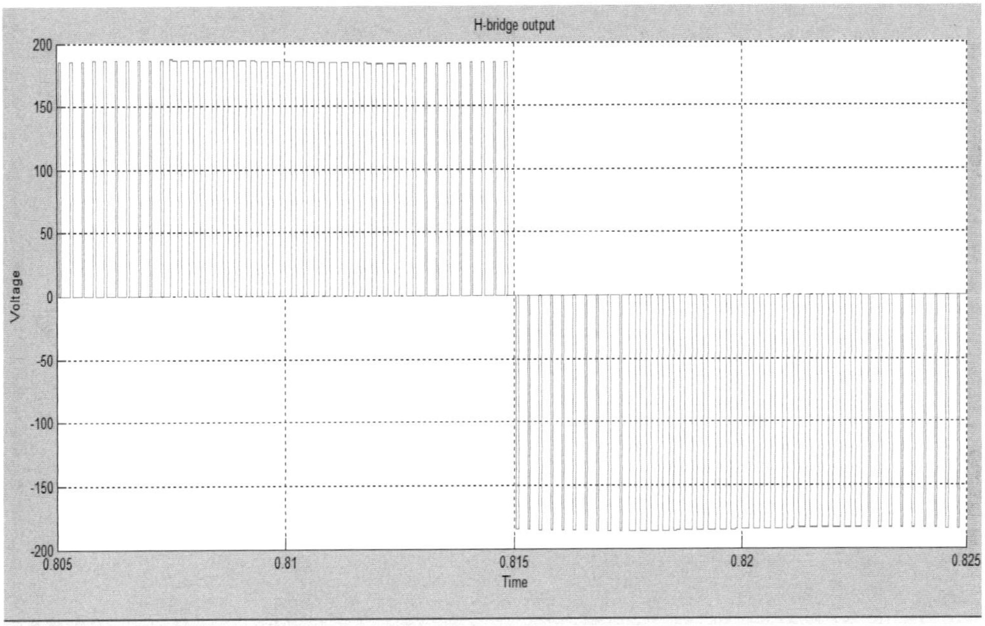

Figure 4. H-bridge inverter output pulses generated in simulation model

The output pulses of H-bridge inverter are fed to transformer and LC filter combination to get desired sine wave at the output. This output power can be applied to the transmission lines in the grid system. But for islanding detection and to initiate control action, relay is connected after filter. After detecting the islanding mechanism the relay is activated to break the supply and isolate inverter from the transmission lines. During this phase the IGBTs of boost converter and the inverter are also switched off.

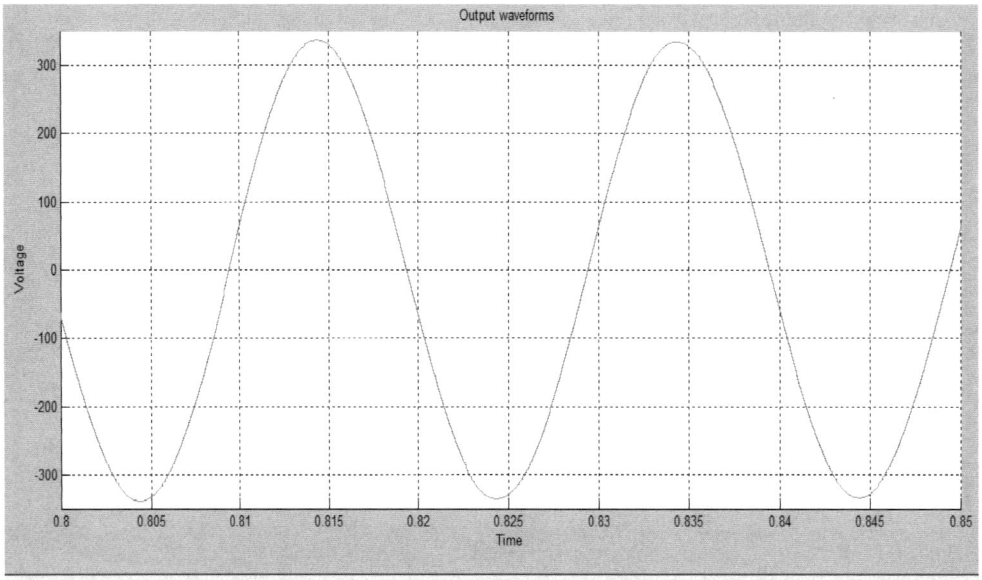

Figure 5. Inverter output voltage

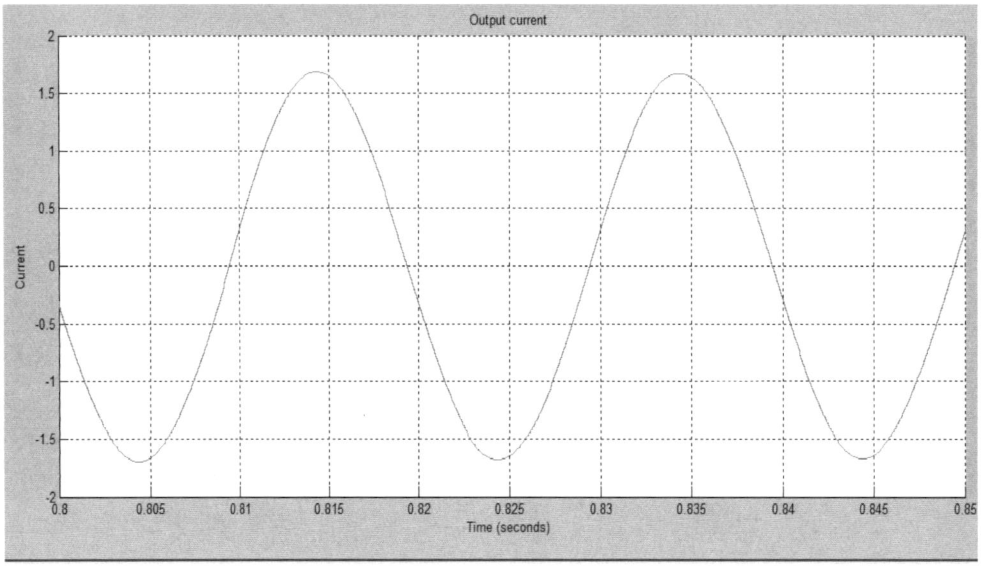

Figure 6. Inverter output current

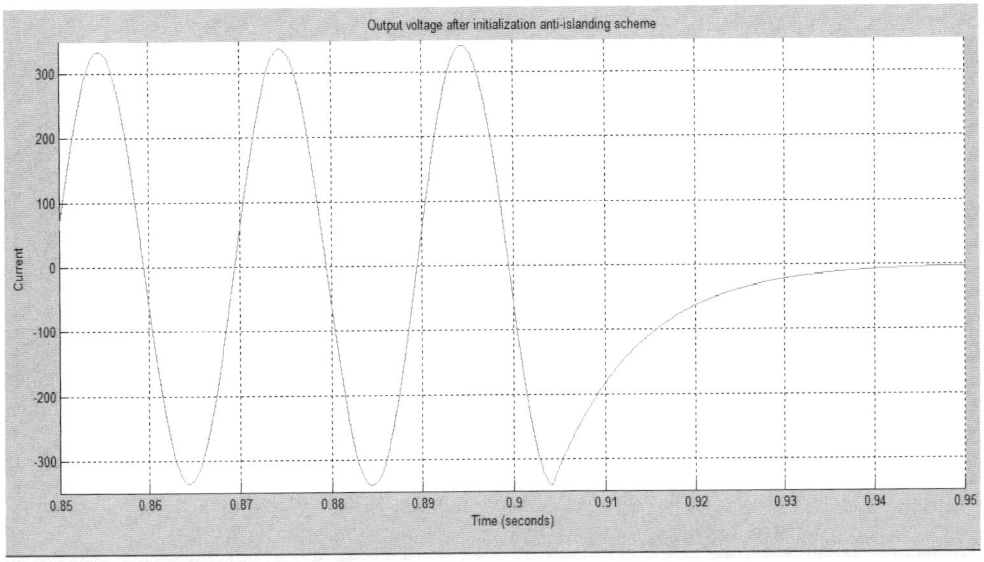

Figure 7. Output current after activation of anti-islanding scheme

A waveform of the inverter output voltage is shown in Figure 7 when the islanding phenomenon is detected and anti-islanding scheme is activated. After detecting the islanding, the inverter is turned off and the load is disconnected.

System prototype is developed in laboratory for experimentation. The PV inverter output waveforms in normal working condition are shown in figure 8. On the prototype provision is made to manually create the islanding phenomenon. When islanding is created for testing purpose, the developed system detects it and switches off the inverter and isolates it from transmission line. Therefore the line does not carry any power and thus the unintentional hazards to the living being are avoided. The output of the inverter after activating the anti-islanding strategy is shown in figure 9.

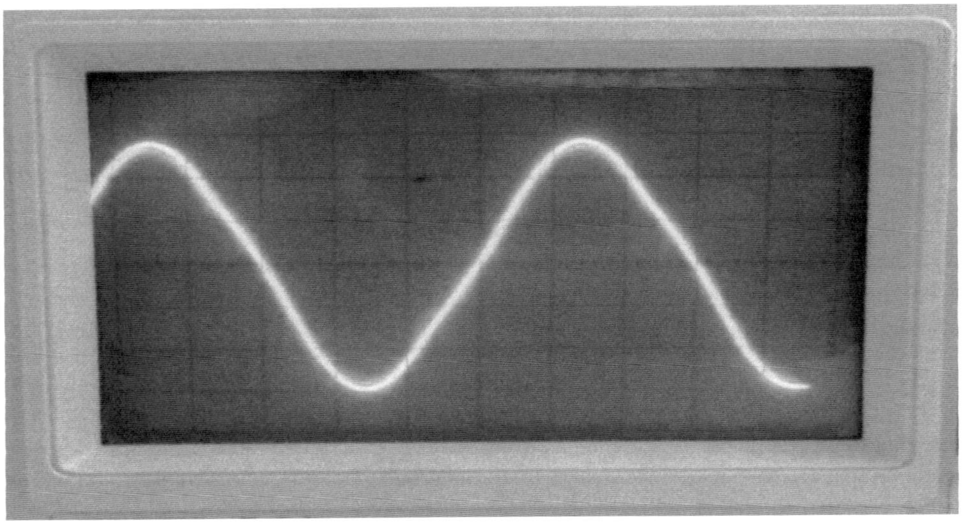

Figure 8. PV inverter output voltage waveforms

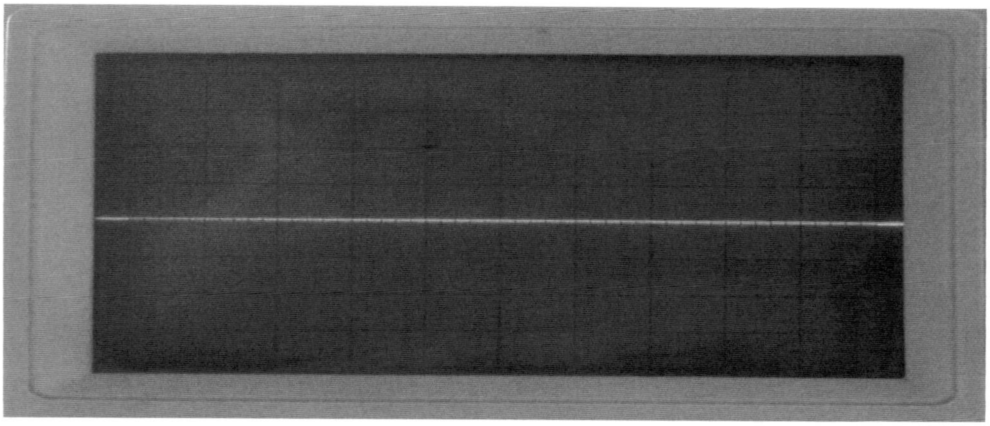

Figure 9. Output after activation of anti-islanding technique

5. Conclusion

This paper has presented a simple technique to sense islanding and to initiate anti-islanding measures. The system presented here monitors the output current and voltage for detection of islanding operation. Open circuit mechanism of transmission lines is detected and the inverter system is turned off and isolated from the load using a relay or circuit breaker. An inverter interfaced with such a mechanism would detect islanding and activate anti-islanding scheme. Quick detection of islanding mechanism and switching off the inverter avoids hazards to living being.

6. References

[1] Muhammed Worku, Mohamed A. Hassan, Luqman Maraaba, Mohammad Abibo, Islanding detection methods for microgrids: A comprehensive review, *Mathematics* , 9-3174, .pp.1-23, 2021

[2] J. A. Cebollero, David Canete, S. M. Arroyo, M. G. Gracia, H. Leite, S, A survey of islanding detection methods for microgrids and assessment of non-detection zones in comparison with grid codes, *Energies* 15 (460), pp. 1-30, 2022.

[3] Valentini M., Munk-Nielsen S., Valderrey Sanchez F., Martinez De Estibariz U. A new passive islanding detection method for grid connected PV inverters. In: *International symposium on Power Electronics, electrical drives, automation and motion*; p.223-228, 2008.

[4] Pigazo A., Moreno V.M.,Liserre M., Aquila A.D. Wavelet based islanding detection algorithm for single phase photovoltaic (PV) distributed generation systems. In: *IEEE international symposium on Industrial electronics*; p.2409-2413, 2007.

[5] Hung G.K., Chang C.C., Chen C.L. Automatic phase shift method for islanding detection of grid connected photovoltaic inverters, *IEEE transaction on Energy conversion*, Vol. 18, No. 1; p.169-173, 2013.

[6] Azim R., Li F., Zhao X. A hybrid islanding detection technique for inverter based distributed generators, In: *IEEE Electrical Power and Energy Conference (EPEC)*; p. 239-243, 2015.

[7] Mohammodpou B., Pahlevani M., Kaviri S.M., Jain P. Advanced slip mode frequency shift islanding detection method for single phase grid connected PV inverters, In: *IEEE Applied Power Electronics Conference and Exposition (APEC)* ; p. 378-385, 2016.

[8] Das P.P., Chattopadhyay S., Chakraborty S. A voltage independent islanding detection method and low voltage ride through of a PV inverter, *IEEE Transactions on Industry Applications*, vol. 54, no. 3, ; p. 2773-2783, 2018.
[9] Ghorbal M.J., Ghzaiel W.,Belkhodja I.S. Fast islanding detection of PV grid connected inverters, In: *International conference on electrical sciences and technologies in Magherb,* (CISTEM); p. 1-8, 2014.
[10] Lei L., Hu C., Hu J., Dong Y. Islanding detection method for PV grid connected inverter based on dynamic reactive power disturbance, In: *10th Conference on Industrial Electronics and Applications* (ICIEA); pp. 1471-1475 , 2015.
[11] Moghaddam M., Moaiyeri M.M., Afjei S.E. Islanding detection of solar power plants by considering the dynamics of PV and MPPT, In: *5th Conference on thermal power plants* (CTPP); p. 127-131 , 2014.
[12] Skocil T., Gomis- Bellmunt O, Miracle D., Arellano S., Duran J. Passive and active methods of islanding for PV systems, *13th European Conference on Power Electronics and Applications*; p. 1-10, , 2009 .
[13] De Mango F., Liserre M., Dell'Aquila A. Overview of anti-islanding algorithms for PV systems. Part-II: Active methods, *12th International Power Electronics and Motion Control Conference*; p. 1884-1889 , 2006.
[14] Brigitte H. Basic calculation of a boost converter power stage, Texas instruments, pp.1-19
[15] Chavan S.B., Chavan M.S. Power switch faults, diagnosis and tolerant schemes in converters of photovoltaic systems- A review, *International journal of advanced research in electrical, electronics and instrumentation engineering*; p. 11729-11737 , 2014 .
[16] A. Chandramouli, V. Sivachidambarnathan, Extract maximum power from PV system employing MPPT with FLC controller, *Journal of Electrical engineering and Automation*, 1 (2), pp. 58-71, 2019

Acknowledgement
Authors are grateful to Department of Technology Shivaji University Kolhapur India for providing required facilities to accomplish this work.

PARTIAL POWER PROCESSING BASED BIDIRECTIONAL CONVERTER FOR ELECTRIC VEHICLE FAST CHARGING STATIONS

[1]Geetha V M*, [2]S Saravanan, [3] M Swathisriranjani, [4]C S Satheesh, [5]S Saranraj

[1]P.G. Scholar, [2] Professor, [3]Assistant Professor, [4] Assistant Professor, [5]Assistant Professor, Department of Electrical and Electronics Engineering, Muthayammal Engineering College (Autonomous), Rasipuram 637408, Tamil Nadu, India.

mail2swathieee@gmail.com, geethaeee03@gmail.com

Abstract: This paper focuses on the design of a charging unit for an electric vehicle fast-charging station. Electric Vehicle is turning into a real alternative to conventional combustion-based vehicles due to their reduced greenhouse emissions. The conventional system experiences harmonic distortions, losses and takes more time to charge. To overcome these problems, new methods and techniques are established here.

With the help of a bidirectional-based converter, electric vehicle charging stations offer quick charging. When using a three-phase supply, phase voltage is given to the suggested converter. In conventional type converters, harmonic distortions are inherent; however, we can lessen voltage distortions by employing bidirectional converters. When tested with a prototype model featuring the suggested converter, the voltage ripple factor is quite low. Phase-shifting pulse width modulation is the topology employed in this procedure. The output voltages are controlled by a precise and simple algorithm. We can also use artificial intelligence gadgets to improve the circuit in the future. By comparing the input and output voltages in the feedback system, we can eliminate both steady-state and transient errors. The charging stations implemented in various areas are used effectively for efficiently charging Electric vehicles. The number of components is reduced compared to traditional converters and thus increasing the efficiency of bi-directional converters with less cost. This charger is also verified for fluctuations in power supply and supply voltages with higher magnitudes. The proposed charger gives us improved power factor and efficiency.

Keywords: E-Vehicle, converter, charger, green energy, pulse width modulation, charging station, efficiency.

1. Introduction

India is one of the world's most important vehicle markets. Traditional automobiles emit toxic pollutants into the atmosphere. These toxic gases contribute to increased pollution in the atmosphere, which has an impact on the ecosystem. The deterioration of air quality is one of the negative consequences of rising pollution in India. The amount of carbon emissions rises in lockstep with the number of automobiles on the road. The Air purifier is now available as a basic need in densely populated areas. It is the responsibility of every citizen to protect our environment. The Indian government is introducing many schemes to get rid of pollution. It is possible to replace conventional combustion-based vehicles with Electric Vehicles. We can reduce pollution levels effectively by employing Electric vehicles.

Another reason for buying Electric Vehicles is that we don't have the natural resource of conventional fuels in our country, hence we have to import petroleum fuels from other countries. During war and pandemic periods, the price of fuel will become higher. All levels of people might be affected by this price hike. But if we use electric Vehicles, We can solve this issue.

The national government is taking the initiative to renovate national highways and rural roads to encourage green energy by introducing electric vehicles by the Indian road service sector. The post-pandemic situation is handled effectively in a green environment. Attractive offers are given to customers who buy electric Vehicles as an initiative towards green and clean energy. Energy security has been improved by replacing combustion-type vehicles with Electric vehicles environment. Awareness about climatic change is created through online communications, Television channels, non-governmental organizations, and service sectors. Energy security has been improved by replacing combustion-type vehicles with electric vehicles. Increasing the number of electric Vehicles is the central component of zero-carbon emissions adopted worldwide. National governments are investing in the Research and development of Electric vehicle systems to provide a pollution-less environment. Eventually, citizens of India are also slowly investing in Electrical Vehicles. Regulations and Guidelines for a clean environment are introduced by the ministry of power. These regulations serve the following purposes: [1] Technical standards for the connection of generating resources are improved, [2] The safety of Electric supply in Electric Vehicles is ensured so that people can buy green vehicles without hesitation, [3] Providing charging stations at regular intervals. Most of the above guidelines are given by the Ministry of power and the Central Electricity Authority of India. Charging stations are installed by several organizations like National Thermal Power Corporation of India, Hindustan Petroleum, Bharat Petroleum, Power Corporation of India limited, and Energy efficiency Service limited.

We can witness the change in electric vehicle technological advancement over the last several years. The progress of Electric Vehicles is developing daily due to the depletion of non-renewable energy sources. In India, state governments are developing strategies and initiatives to encourage people to utilize electric vehicles. Tamil Nadu's goal is to attract more than Rs.55,450 crore in investment in the production of electric automobiles. Thus, we can create a comprehensive Electric vehicle ecosystem in the state. This investment also provides jobs for young people with attractive salaries. The objectives of these policies are as follows: [1] creating an adequate power supply by power generating capabilities, [2] Establishing more charging stations with favorable power tariffs. [3] Encouraging young talents to invent reliable technologies that support Electric Vehicles. Shared mobility is used by providing a suitable ecosystem and enhancing the infrastructures to make Tamil Nadu the electric hub of India. We can make Tamil Nadu the preferred destination for electric vehicles and manufacturing units for components required for pollution-free automobiles. From an economic standpoint, recycling and reusing batteries will save a significant amount of money. Several steps are taken to dispose of rejected batteries in an environment-friendly manner. Electrifying combustion-powered vehicles offer the most cost and emission savings. Expansion of road networks with basic infrastructure has led to the rapid increase in the number of Electric Vehicles in Tamil Nadu. The number of electric vehicle usage has seen a significant growth compared to previous years. According to the statistical analysis, approximately two lakh investments in electric vehicles were made in 1981, and this number has climbed to three crores by 2020. There is an annual growth rate of thirteen percent. There were three million transport vehicles and twenty-seven million non-transport vehicles sold in India. If we replace the combustion type vehicle with Electric Vehicles rapidly, we can make a clean environment in a short duration. Electric vehicles are now being produced with a high load hauling capacity. The electric vehicle's torque has also been increased to improve its performance. Electric cars have recently advanced to the point that programs are being conducted to electrify heavy-duty vehicles as well. Aside from the advantages listed above, the mobility of electric vehicles results in a higher level of energy gain. Electric vehicles can potentially be used as a means of storing renewable energy. People in India used to be hesitant to purchase electric vehicles, but that is changing, thanks to technological developments in electric energy systems. Although the future is bright with clean energy, there are some improvements we need to make to an existing system. Charging stations can be installed in many places like petrol bunk stations, but People don't have time to wait and charge their Electric Vehicles for a long time. This should be resolved with less battery charging time. If we do so, it will attract more people to advance towards green energy. The cost of electricity may increase if more people tend to use Electric Vehicles. This demand for an electric vehicle can also be rectified y using Renewable power generation. We can also install Photo Voltaic solar cells in the charging station for increasing power demand. The proposed converter can utilize the energy sources from renewable energy sources, utility grids, and also from other power generating methods.

2. Survey of Literature

Various literature is taken for reference. This literature is taken from various journals which are related to partial power processing based bidirectional converters for electric vehicle fast-charging stations as well as electrical and power electronics.

Wang, Haoxiang (2021) "Fault Diagnosis in Hybrid Renewable Energy Sources with Machine Learning Approach." Journal of Trends in Computer Science and Smart technology (TCSST) proposed Artificial neural network system based on machine learning to enhance the ability of hybrid vehicles. To meet the demand of electricity this paper explains the idea of integrating many renewable energy sources to form a hybrid-renewable energy source system.

Ranganathan, Dr G. (2021) "Energy Storage Capacity Expansion of Microgrids for a Long-Term" Journal of Electrical Engineering and Automation says that the electric vehicles that are used in charging stations are based on vehicle-to-grid wherein it is possible to regulate the charging rate and time and to transmit energy to the microgrid. This paper provides in depth utilization of charging stations by Performing some modifications in electric charges used for electric vehicles.

Jon Anzola (2018) "Partial Power Processing Based Converter for EV Fast Charging Stations" has proposed a paper on the optimization of electric vehicles and the algorithm to control the vehicle by using photovoltaic generation and stored energy in the charging station. The integration of the charging station in the building is explained in a stepwise manner.

3. Proposed System

The architecture of the project work is shown in Figure1 The components used in the block diagram are explained below.

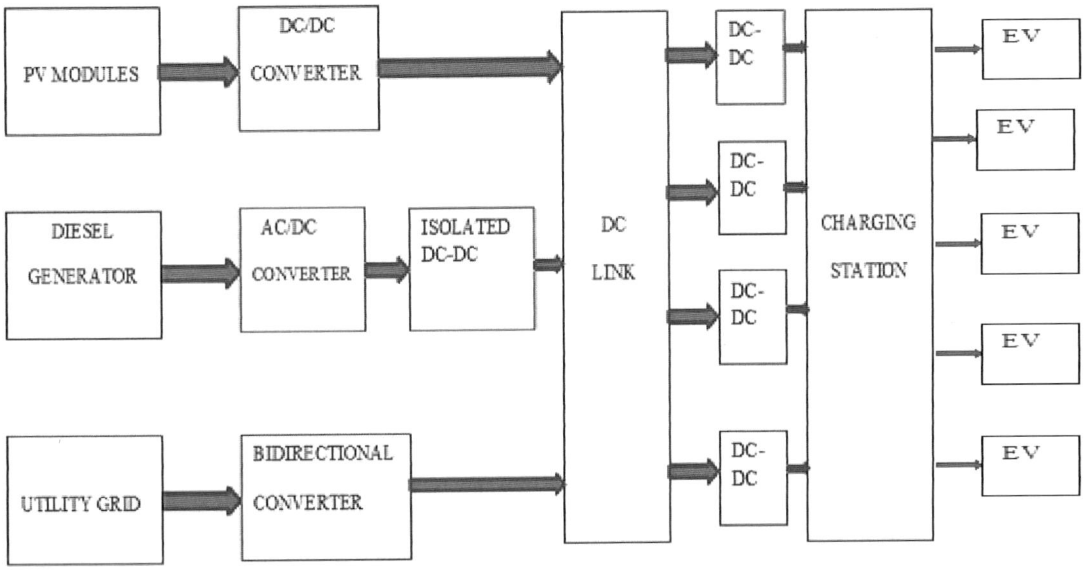

Figure1. Architecture of Project Work

3.1 Power Resource

The Renewable energy sources play a major role in power production. To meet the current energy demand, renewable energy is preferred. We can produce power by non-conventional energy sources without any interruptions if we have necessary natural resources. India can produce power using solar power panel without interruptions. Solar panels can be installed above the roof -top of charging stations. Among various states in India Thar desert is the best suitable location for solar radiation. People are already installing solar photovoltaic cells above their home for sufficient electricity. We can reduce the cost of electricity if we mount the solar panels above our house and use it for household purposes. Development of clean energy sources are increasing day by day to protect our environment. This solar panel cell produces direct current voltages and can be directly connected with the suitable converter if we want the power to be utilized by the batteries. On the other hand, we want to convert it into AC voltage by means of converter. Here the first method is implemented as charging stations need direct voltages. The diesel generator is used to produce alternating current. Although renewable energy is the preferred clean energy source, diesel generators are also utilized to charge electric vehicles via converters. Typically, this type of generator is used to supply peak loads. A baseload power producing station does not employ a diesel generator. Other fuels may be used instead of diesel in some machine models designed for diesel generators. The planned converter is linked to electric vehicle charging stations through a diesel generator.

To supply power to primary and secondary distributors, a utility grid is set up on the distribution side. It would be preferable if renewable energy sources were used as power generators and then transmitted to the utility grid. By comparing it to the water irrigation system, this power transmission mechanism can be clearly explained. We compare water with current because it exhibits similar properties. Water is allowed to flow through pipes and canals for irrigation purposes, indicating a resemblance in current flow from the generator to the utility grid via a transmission line.

3.2 Converters

Partial power converters or bidirectional converters are connected with the power source to charge the electric vehicles. Isolated DC-DC converters have separate return paths. The return path is not connected with high power supplies or with generating power stations. This connection enables the worker to clear fault in the DC-DC converter without any harmful effects of electric power. We can use any type of converter when we transfer power from DC-link to charging stations as our proposed converter is mainly used to convert voltage from the electric energy source to DC-link.

3.3 Smart Chargers

Electric vehicles in India will increase in the upcoming years. To digitalize electric vehicle charging stations, smart chargers are used. Smart chargers are installed in charging stations located near workplaces, apartments, standalone buildings, etc., to provide information about electric vehicles. Smart chargers are used to indicate the safety requirement of Electric Vehicles. Sharing real-time data is possible by using smart chargers. Charging stations protect the vehicle from overload conditions by using the information given by the mentioned charger. It gives the exact information about the availability of fuel. Accordingly, the charging station receives the required electric power in time.

Smart charging stations not only manage the charging decisions but also help the Electric Vehicle to meet the electric needs and reduce the pressure on power demand during peak hours. Discharging provides extra flexibility to the grid or other power resources by extending the range of power and saving the chargers from peak loading conditions. The charging systems with smart chargers give signals to the electric vehicles like where the charging station is located, the number of vehicles already charging in the particular charging station, etc., This smart charger can be implemented in Electric vehicles also and helps the electric vehicle to auto-drive, to indicate the temperature in the environment, to give automatic signals to nearby police stations or hospitals in case of any emergencies and to open up the airbag for safety purposes.

4. DC-DC CONVERTER

Charging stations utilizes DC-DC converters of different types.

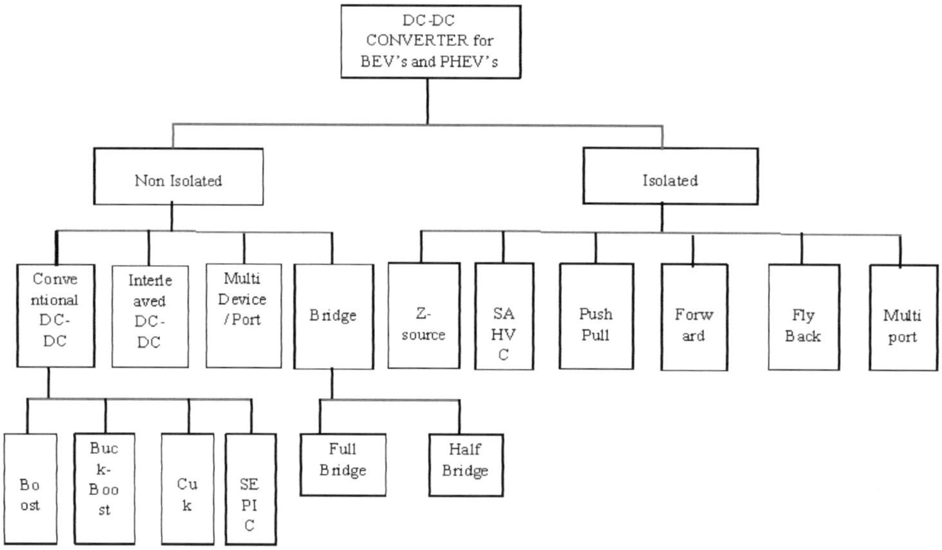

Figure 2. Classification of DC-DC converter topologies

4.1 Boost DC-DC Converter

The boost DC-DC converter increases the source voltage before passing it on to the charging station. While the magnitude of the voltage increases, the current is stepped down. This converter employs a thyristor as a switch and a diode to prevent negative voltage conduction. It consists of at least one storage element, such as an inductor which may be a reactor because an ideal inductor is not possible or both an inductor and a capacitor. The given voltage source is linked in series to the inductor, while the output voltage and capacitor voltage remain the same. Voltage ripple factors on the input and output sides are reduced by an inductor and a capacitor, respectively.

4.2 Interleaved Boost DC-DC Converter

The boost DC-DC converter increases the input voltage before passing it on to the charging station. While the magnitude of the voltage increases, the current is stepped down. This converter employs a thyristor as a switch and a diode to prevent negative voltage conduction. It functions as a switched-mode power supply with at least one storage element, such as an inductor or both an inductor and a capacitor. The source voltage is connected in series to the inductor, while the output voltage is connected in parallel to the capacitor. Voltage ripples on the input and output sides are reduced by an inductor and a capacitor, respectively.

4.3 Boost DC-DC Converter with Resonant Circuit in BEV and PHEV powertrains

Due to the delay in switching time, direct current to direct current boost converters suffers losses. The resonant circuit is connected to the regular converter, which solves the problem. When the switch is turned on and off with no discernible delay. We can reduce our losses by doing so. The use of a resonant circuit aids in the rapid openingand closing of thyristor switches.

4.4 Isolated ZVS DC-DC Converter

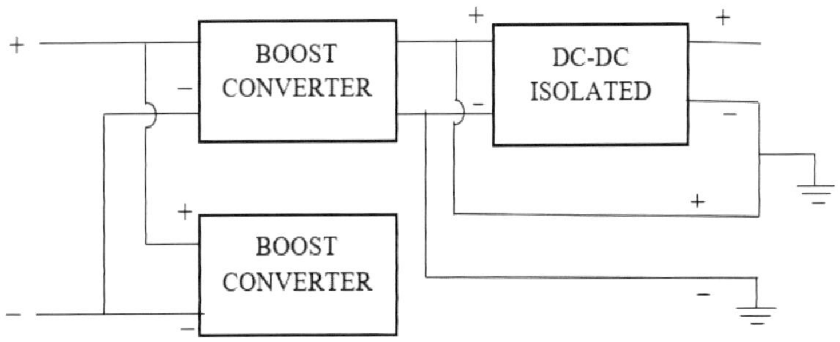

Figure 3. Isolated DC-DC Converter

An Isolated Zero-Voltage Switching direct current to direct current converter handles overcurrent issues.In conventional DC-Dc converter, the input and output shares same return path which leads to severe faults in other system also. In isolated DC-DC converter the input and output have different return path thus providing safety to the device and human. The load current varies from time to time when connected with different loads like resistive load, inductive load, capacitive load. To prevent significant losses, the device is switched on with a zero-voltage reference point. This conductor does not subject to heating frequently. The system is isolated from Dc mains for safety purposes. We can disconnect the affected path and rectify it easily when the converter is subjected to fault conditions. This is because the DC ground is connected to the local stations not directly with the power supplying sources. These are the reasons to select the above-mentioned converter.

4.5 Multiport isolated DC-DC Converter (MPC)

When we want to convert multiple input sources from direct current to direct current with increased magnitude multiport isolated converter is used. This converter is manufactured with galvanic isolation such that the faults in the converter is cleared easily. All the input sources are separated from one another so that it does not make over voltage issues. This technique reduces the voltage and current ripple factors in both input and output sides.Ports are also entirely isolated from each other. The proposed bidirectional converter consists of four diodes to restrict the negative voltage waveform when the converter acts as a rectifier. Active switches like Sa1 which consists of silicon-controlled rectifier also known as thyristors is connected as shown in figure 3, two capacitors C1 and C2 are connected in parallel to act as a direct current link, three-phase voltage source generally star connected is given to the converter. In addition to these components, the DC voltage source is also connected to provide bidirectional power flow. We can also use other switches instead of SCR but, the latter gives a better result. The topology used in this kind of arrangement is phase-shifting sinusoidal pulse width modulation to enhance the ability of the proposed converter. A large value of inductors provides ripple-free output current.

5. THREE – LEVEL STACKED NEUTRAL POINT CLAMPED CONVERTER

5.1 Circuit Configuration

A three-level stacked neutral point clamped converter in Figure 4. is widely used in medium to high voltage and high-power applications. We use a bi-directional three-level stacked neutral point converter as it minimizes the voltage ripples and eliminates the harmonics and power losses. An inductor is added to the proposed converter to increase the current value and to reduce the voltage ripple. If we convert the voltage from AC to DC then the waveform of both current and voltage becomes smooth. The power factor is increased by using a neutral point converter. All switches implemented in the above converters are capable of zero voltage switching due to the inductors. If we convert the voltage from DC to AC then the output waveform is sinusoidal without distortions. The power flow is bi-directional and it acts as both rectifier and inverter. The proposed converter gives better efficiency than the conventional type converters.

Reactors are connected with phase voltage which behaves like the input filter. The input supply to the proposed converter is given by either the utility grid or any other alternators. We use star connection in bi-directional neutral point clamped converter excess current flows through a neutral point that prevents the device from overcurrent faults. Phase voltage is 0.577 times the line voltage in the proposed converter. When the power flow direction is from AC to Dc the switching device is less prone to over-voltage fault due to less phase voltage. The phase-shifting PWM technique is also implemented for better performance of the converter.

DC-link is established by connecting two capacitors in series across the load. The capacitor is connected in parallel between the positive and negative terminals of the proposed converter. The transient is prevented from load side to distribution side when we connect the capacitor as dc link between the converters. Voltage ripple is prevented and continuous waveform is obtained. Dc-link capacitors are used to improve system energy density and the problem due to rapid switching scheme is resolved. We must be careful in selecting the type of capacitor to be implemented in the converter. We need high voltages to charge the batteries in the charging stations due to the physical limit of current. Magnitude of voltage can be raised easily compared to current. The dc-link to be connected should have very low equivalent series resistance, tolerance to high ripple current, low value of self-inductance because in some cases inductor is connected with capacitor to provide effective dc link system. Operating temperatures must be decided before the implementation. By using dc-link through capacitor the performance of the battery in charging station is improved.

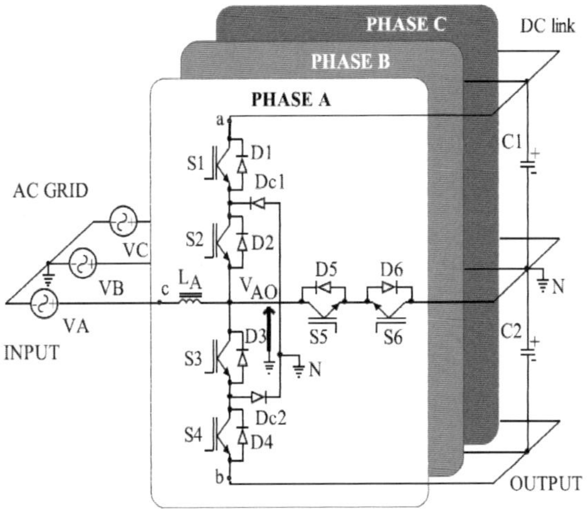

Figure 4. Bidirectional three level stacked converter

5.2 Methodology

The phase-shifting sinusoidal pulse width modulation is implemented in the proposed converter. The phase-shifting PWM technique is slightly modified to get bidirectional power flow and eliminates harmonics. This modification includes the addition of two IGBT switches compared to the traditional neutral point stamped converter. Zero Voltage Soft Switching of semiconductor switches introduced in this methodology reduces the switching losses and enables constant frequency operation. ZVS soft-switching range is prolonged by several modifications. One of the methods to extend the soft-switching range is by using additional components like passive auxiliary circuits in the conventional phase-shift power converter. Circulation current which may result in the dead short circuit is reduced by using high-frequency transformers or by additional active switch circuits. The controller used in the converter is Arduino Mega2500. As the control system is based on a microcontroller all the parameters can be adjusted through control switches. The proportional-integral controller is used to control the current loops. The Control board in the Arduino2500 microcontroller generates the driving pulse for the switches and the closed-loop of output voltages in a bidirectional converter.

The technique of phase-shifting with fuzzy logic and pulse width modulation controls the current using the semiconductor device INB154-Q2 by connecting it to the secondary side. This semiconductor device delivers the output current by detecting the magnitude of the load current. Voltage during Common mode excitation in the semiconductor device is more likely 40V but it becomes higher when it is connected to the virtual ground terminal. Voltage is measured by voltage divider or by using a potentiometer that has low voltage deviations and 2.2MHz of unity-gain bandwidth. Since the proposed bidirectional converter can flow in both directions the power is converted from analog to digital when flowing from source to load and from digital to analog when power flow is from load to source.

The phase difference is calculated by the information provided by the control system. The value of storage elements like the inductor is also required to identify the phase displacement in addition to output and input voltages. Power losses occur when we don't implement the zero-voltage switching loss scheme. If the switch is closed another switch is opened without any delay to prevent switching losses in the circuit. To achieve this condition high-frequency semiconductor switch is used. When we use the inductance of larger value voltage ripple factor is reduced that increases the efficiency of the converter. The proposed charger contains fewer components compared to normal converters. This calculates voltage phase differences easily. Due to over voltages, stress is created in the circuit that can be rectified using this methodology. Due to the zero voltage reference switching scheme, the information of voltage given to the battery charges has arrived very fast.

5.3 Control of Proposed EV Charger

The electric voltage charger is connected with Dc-link. The battery chargers are charged only up to a certain level due to the overloading effect. This required capacity is predicted by the Aurdino2500 controller. The voltage feedback path as shown in Figure 5. consists of multiple loops to gather information about the surroundings.

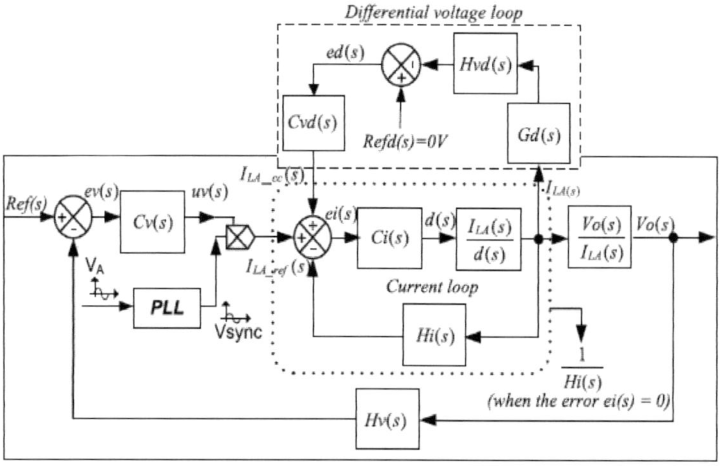

Figure 5. Phase shifting control

The equations for error and control signal are given as,

$$V_{Cuke}(n) = V_{Cukref}(n) - V_{Cuk}(n)$$

$$C_{vCuk}(n) = C_{vCuk}(n-1) + K_{pCuk}\{V_{Cuke}(n) - V_{Cuke}(n-1)\} + K_{iCuk}V_{Cuke}(n)$$

The gate pulses to DC-DC converter switches are obtained, after comparing the control signal to a sawtooth carrier waveform *Sc* with high frequency.

Where *Sc* represents the switching commutation pulses of the flyback converter.

The required inherent power factor correction, as well as, a well-regulated voltage waveform, is achieved by controlling the duty cycle ratio. The controller operates satisfactorily over a wide mains voltage range.

The converter is controlled using a negative feedback loop in current zero references. This provides the exact magnitude of voltage which is taken as the output. Normally, the negative feedback system is preferred over the positive feedback system as it is more accurate. Voltage can be measured by a proportional-integral controller. Error signals are eliminated by the phase-shifting method. The equations for current and voltage control loops are given by,

$$(n) = V^*(n) - V(n)$$

These control signals are passed through the phase-shifting PWM comparator to generate the required duty cycle to draw a unity power factor current from AC mains during battery charging in common current-common voltage mode for the steady-state as well as sudden variations in source voltage.

The battery charging capacity of each electric vehicle varies to its manufacturing procedure. So, Electric charging stations are implemented to provide the load of different capacities. Topologies included in the proposed three-level bi-directional neutral point clamped converter is phase-shifting pulse width modulation. This includes diodes, insulated bipolar junction transistor switches, and Inductor to facilitate bidirectional power flow. On the AC side, reactive power compensators are provided for smooth power flow characteristics. The output is thus controlled using Phase shifting pulse width modulation technique.

6. MATLAB SIMULINK

The simulink diagram is shown in Figure 6. The Simulink diagram is obtained by phase-shifting pulse width modulation technique which varies in both harmonic injection and reactive power consumption. The pulse width modulation technique injects high-level harmonics like fifth, seventh, and ninth harmonic currents. High-level harmonics do not affect the system. High-level harmonics are used to eliminate lower-level harmonics. Voltage ripples and distortions are caused by lower-level harmonics which are reduced by the mentioned technique. This methodology is easy to maintain because each valve is turned on naturally. The proposed converter is a two-quadrant converter that allows the power to flow in either direction. This gives the interesting feature of battery charging.

The proportional-integral controller is implemented to obtain the voltage reference required to drive the pulse width modulation with a signal generator. The mentioned controller is turned ON through the process of trial and error. The current reference is made equal to zero so that the main reference signal increases linearly. The simulation output obtained from the Simulink diagram gives the most accurate results with respect to prototype. From simulation diagram we can see the improvement in voltage profile and also the efficiency is improved. We can also sense the reduction in harmonics due to the proposed charger. Running a simulation generates a great deal of statistical data that must be analyzed with appropriate tools.

The simulation output obtained from the Simulink diagram gives the most accurate results with respect to prototype. From simulation diagram we can see the improvement in voltage profile and also the efficiency is improved. We can also sense the reduction in harmonics due to the proposed charger.

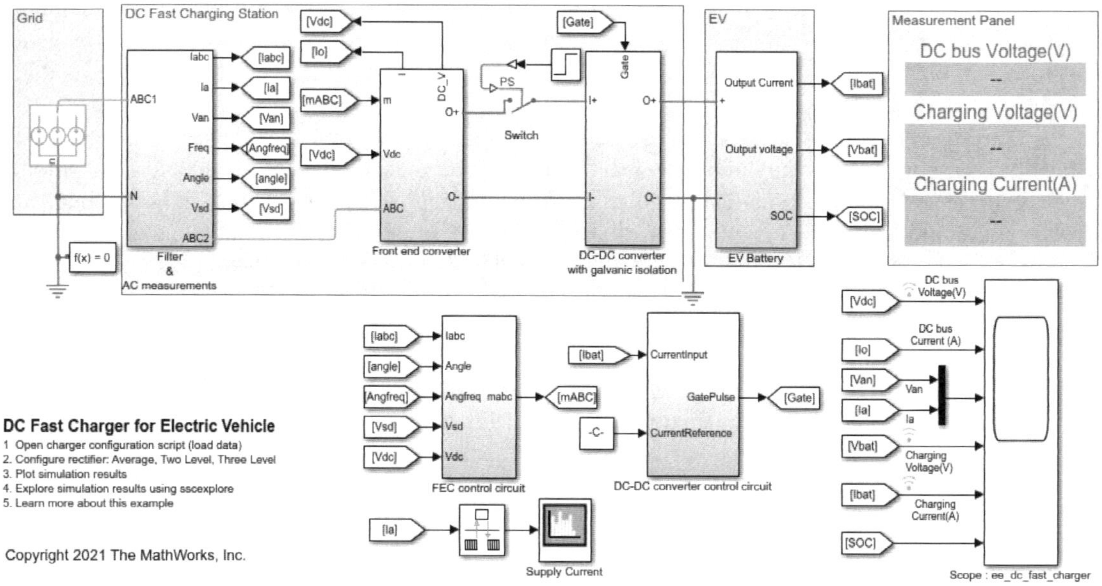

Figure 6. Simulink Diagram

6.1 Solar Power Output

The simulation output of solar power is shown in Figure 7. Voltage, current, and power are taken on the y axis whereas the x-axis represents the time. From the simulation output of solar power, we can ensure the capacity of the proposed bidirectional converter to be connected to the power grid. We can also use AC power as the bidirectional converter converts AC-DC and also DC-DC.

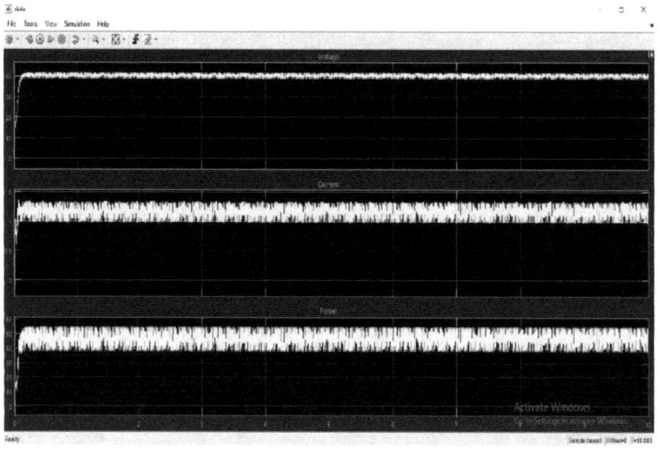

Figure 7. Solar Power Output

6.2 Harmonic Elimination

Pulse width modulation technique is used to eliminate harmonics. Figure 8. The output from pulse width modulation is shown. It eliminates the dominant lower-level harmonics by introducing higher level harmonics as higher-level harmonics does not influence the system much. From simulation output of pulse width modulation Technique, we can clearly see the reduction of harmonics in the prototype model.

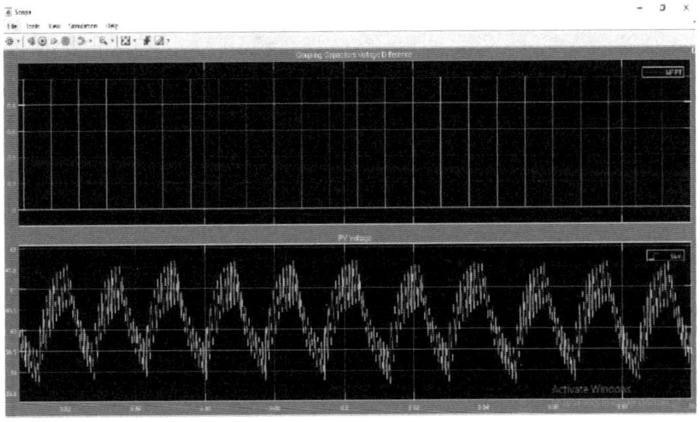

Figure 8. Pulse Width Modulation Technique

6.3 Performance of Battery

The bidirectional converter converts the voltage and current waveform such that the DC battery can be charged without many disturbances. The battery is charged in less time when we use the bidirectional converter to charge the battery compared to conventional converters. Figure 9. shows the battery performance of the prototype model. The state of charge and Voltage level is indicated in the y-axis whereas the time taken is represented in the x-axis.

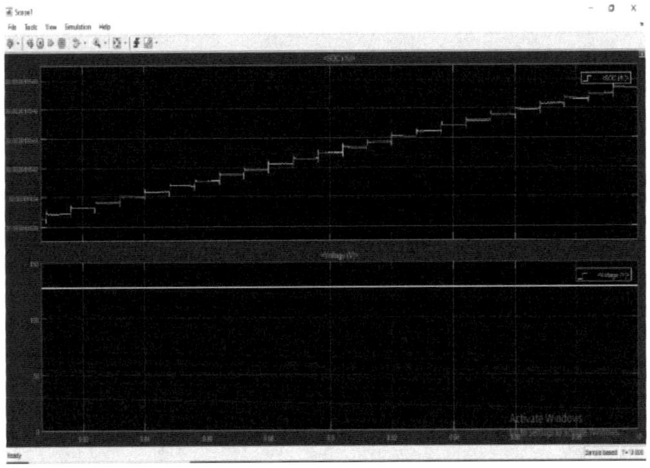

Figure 9. Battery Charging

From the simulation results, we can say that the performance of the proposed system gives the better charging features of Electric Vehicle charging stations. The time taken to charge the battery is reduced with increased efficiency. This simulation output is almost the same as the hardware prototype model we have designed and we can implement the real-time system by using the ideas of the prototype model.

7. CONCLUSION

The charging time in Electric Vehicle charging stations is reduced by the use of Partial power processing and bidirectional converter. This bidirectional converter gives the desired characteristics in direct current mode using single voltage feedback control. Electric charging station provides satisfactory steady-state characteristics and also withstand sudden fluctuations in voltage level. The charger shapes the waveform of output current and voltage. The total harmonic current is reduced to three percent in contrast to the conventional charger which accounts for about 55.3%. The efficiency of the controller is increased to ninety-three percent. Thus, the proposed converter gives maximum efficiency with fewer harmonic distortions. The control system control of Electric Vehicle power devices with a smaller number of elements conducting over a single shift cycle is proposed. The proposed converter offers excellent characteristics in DCM mode using single voltage feedback control. Therefore, the scale of the charger is reduced. The proposed charger shows satisfactory charging characteristics for steady-state and sudden fluctuations in supply voltage. The charger shapes the mains current to follow the mains voltage. The total harmonic distortion in output current is reduced as low as 2.8%, in contrast to the mains current total harmonic distortion of 55.3% in a standard charger. Therefore, the proposed charger gives the possible efficiency with less total harmonic distortion.

8. FUTURE WORK

In the future, we can design the proposed converter with slight modifications to be implemented in real-time applications. The cost of the electric chargers can be further reduced by using reusable batteries. The proposed charger can be further taken into research and development and benefit able projects can be proposed.

REFERENCES:

[1] Wang, Haoxing "*Fault Diagnosis in Hybrid Renewable Energy Sources with Machine Learning Approach*" Journalof Trends inComputer Science and Smart technology (TCSST) 3, no. 03 (2021): **222**-237.

[2] Ranganathan, Dr G "*Energy Storage Capacity Expansionof Microgrids for a Long-Term*" Journal of Electrical Engineering and Automation 3. No. 1: **55**-64.

[3] D.Ajithkumar, J.S.Akilan, K.Dileep, R.Lokesh, E.ViswanathanS.TamilselvanS.Saravanan, "*Design and Development of Electric Two Wheeler With Fast Charging*",. International Journal of Engineering Technology Research & Management, Vol.05, Issue.09, Pp.**94**-101, 2021.

[4] M. Ranjitha, N. Mohananthini, M. Swathisriranjani, "*Voltage balancing in diode clamped multilevel inverter using multi level boost converter*",International Journal of Advanced Research in Electrical, Electronics and Instrumentation Engineering, ISSN 2278 – 8875, volume 6, Issue 1,January 2017. pp. **182**-188, 2017.

[5] Swathisriranjani M, Mohananthini N, Ranjitha M, Baskar S and Kavitha D, "*Optimal Placement and Sizing of Active Power Line Conditioners for Minimizing Power Quality Problems*", Indonesian Journal of Electrical Engineering and Computer Science Vol. 5, No. 2, pp. 267 – 276, 2017 DOI: 10.11591/ijeecs.v5.i2.pp**267**-276.

[6] S.Arvinthraj, M.Arun, S.Inbhakumar, R.Sagayaraj, S.Saravanan, "*Multipurpose Hybrid Electric Vehicle for Agricultural Applications*", International Journal of Advanced Research in Electrical, Electronics and Instrumentation Engineering, Vol.10, Issue.10, Pp.**7366**-7371, 2021.

[7] J Mohammed siddi, A. Senthil kumar, S.Saravanan and M. Swathisriranjani, "*Hybrid Renewable Energy Sources For Power Quality Improvement With Intelligent Controller*", International Research Journal of Engineering and Technology (IRJET),Vol 7 Issue 2, pp.**1782**-1789,2020

[8] T.R. Vignesh, M.Swathisriranjani, R.Sundar, S.Saravanan and T.Thenmozhi,"*Controller for Charging Electric Vehicles Using Solar Energy*", International Journal of Engineering Research and Applications(IJERA), Vol 7 Issue 2, pp.**49**-53,2020.

[9] H. S. Youn, J. S. Park, K. B. Park, J. I. Baek, and G. W. Moon, "*A digital predictive peak current control for power factor correction with low-input current distortion,*" IEEE Trans. Power Electron., vol. 31, no. 1, pp. **900**-912, Jan. 2016.

[10] T.R.Vignesh, M.Swathisriranjani , Dr.S.Saravana, G.Nagarajan and R.Vinoth "*Implementation of Prototype Model For Charging Electric Vehicles Using Solar Energy*", Journal of Seybold Report, Vol-15 Issue-8, pp.**2838** -2846,2020.

Multi-Tasking Robot using Microcontroller for Agriculture Application

Dr. P. Sebastian Vindro Jude [1], Dr. C. Kathirvel [2], S. VenkateshKumar [3], R. MohanKumar[4]

[1]Assistant Professor (Sl. G), Department of Electrical and Electronics Engineering, Sri Ramakrishna Engineering College, Coimbatore, Tamil Nadu, India.

[2]Associate Professor, Department of Electrical and Electronics Engineering, Sri Ramakrishna Engineering College, Coimbatore, Tamil Nadu, India.

[3, 4]Assistant Professor (Sr. G), Department of Electrical and Electronics Engineering, Sri Ramakrishna Engineering College, Coimbatore, Tamil Nadu, India.

jude.panimayam@srec.ac.in, kathirveleee@srec.ac.in, venkateshkumar.s@srec.ac.in

Abstract. With innovations in every field of our lives, agricultural field also gets impacted by technological innovations on a huge scale. Agriculture is the backbone of India. In order to support this, a multi-functioning robot is required. Smart farming aids in boosting the quality and quantity of the yields. Currently, smart agriculture is implemented to save the manpower and time of work with the help of multitasking robots. This is the advanced booming technology in the agriculture sector. However, certain considerations and approaches should be taken when designing these types of robots, taking into account the agricultural setting in which they will be used. The main aim of this research is to provide smart farming at low cost without compromising on the efficiency of the processes involved. This paper discusses the design procedure and development of an independent multitasking robot that could be applied for agricultural purpose like soil digging, sowing seeds, harvesting as well as irrigation. This robot works on solar energy and will also help in conserving non-renewable energy resources on a large scale thereby contributing towards the sustainability of our planet. With help of technologies, this idea kindles the knowledge in developing more intelligent and smart robots for the agriculture activities.

Keywords – smart farming, multitasking agricultural robot

1. Introduction

Robotics is influencing our lives in many fields, some of which are Manufacturing, Education, Healthcare, Personal Technology, etc. Robotics is also impacting the agriculture field, and this can have a considerable impact on the farm produce especially since agriculture plays an important role for the Indian economy. Farmers today can benefit from the technological advancements. Automation can help in boosting the efficiency of the various processes involved right from sowing to harvesting the crops. Technology can bring in disruptive changes over the traditional laborious methods. Embracing new technologies will change the face of the agricultural sector. Smart farming can help bring in cost-effective solutions for the farmers. Agricultural robots help the farmers in automating the laborious and repetitive tasks, and therefore provides the farmer more time that can be devoted to other aspects of farming which in turn can help in improvising the overall production.

With the rising population across the globe, there is a huge gap between the demand and supply of agricultural produce. Automating can definitely help in reducing this gap. This paper discusses the design and development of a prototype of a smart robot that is capable of performing various operations like sensing the soil moisture content, ploughing, seed dispensing and irrigation. The system which is based on the ESP32 microcontroller can run in the auto/manual mode. In the manual mode, the wi-fi pairing application is used by the robots as a controlling device which supports the robot in the navigation around the field.

2. Literature Survey

The authors in [1] developed a robot that has capacity to accomplish the operations such as humidity sensing, seed dispensing and automatic ploughing. The AVR AtMega Microcontroller was used for monitoring and control. The complete field was tilled by the robot to perform the ploughing operation while concurrently performing the operation of dispensing the seeds on the tilled soil. The robot was operated in the "automatic" mode while it was on the field and in the "manual" mode while it was off the field. Paper [2] discussed a robot system used in the field of agriculture to cultivate the crops without using any manpower. The main objective was to reduce the work of the farm labourers and increase the productivity in the given span of time. The authors in [3] elaborated the use of new technologies to increase the farm yield. The paper discussed the use of latest electronic technology based on the use of Microcontrollers and GSM.

In [4], the authors have discussed the scenarios of decreased water tables, drying rivers and lakes, and unpredictable issues due to climate changes. The authors elucidated the dire need of effective utilization of water. The temperature sensors as well as moisture sensors were used at appropriate locations for monitoring the crops. A microcontroller-based system was evolved with the temperature threshold values and the soil moisture control, that was maintained on the amount of water dispensed. The solar photovoltaic cells were used to power the system. A duplex communication channel based cellular internet interface was included, which allowed the data for inspection and irrigation scheduling. Wireless sensor networks helped in the surveillance and controlling of the greenhouse parameters in smart agriculture.

In [5], the authors have used Arduino Uno ATmega 328p Microcontroller for controlling the robot. Esp3266 wifi module was used for the communication. A moisture sensor was used for checking the moisture level in the agricultural land. The node MCU ESP8266 microcontroller-based robot in [6] used various sensors such as PIR, soil moisture and flame sensor. The authors in [7], discussed a solar based agriculture robot that uses a solar panel with 12V DC battery and performs functions like grass cutting and fertilizer spraying. In [8] the authors have developed a multitasking robot using PIC 18f 4520 microcontroller which performs operations like sprinkling water, ploughing, and grass cutting.

3. Proposed System

The objective of this paper is to design and implement a system that is capable of performing most of the agriculture activities using a robot. The robot is controlled using an Android App specially designed for this system. It works over the Wi-Fi protocol. The system is solar powered thereby ensuring conservation of electricity. The block diagram of the system is shown in Fig. 1.

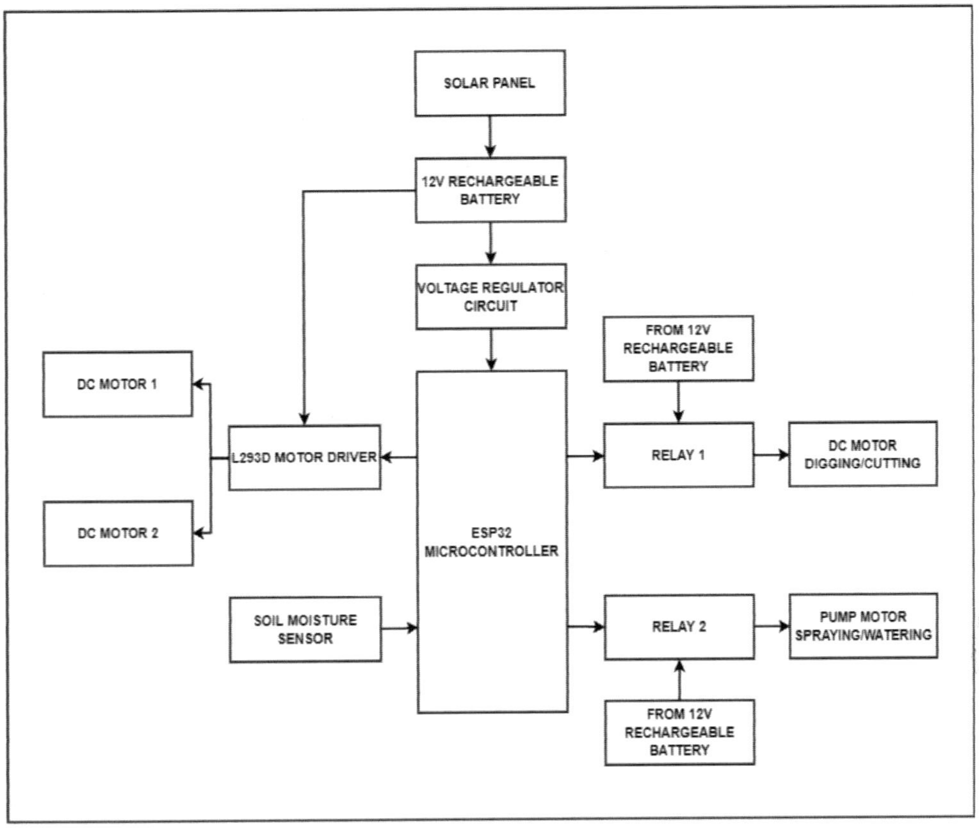

Fig. 1. Functional Block Diagram

The system is developed using the ESP32 microcontroller that also has an inbuilt wi-fi facility. A L293D motor driver is used for driving the two motors of the robot. ASG90 servo motor is connected for the function of seed dispensing. A soil moisture sensor aids in sensing the moisture content of the soil. For cutting and digging, a DC motor is connected via a relay.

A motor pump is used with a two-way sprayer nozzle for spraying and watering purpose. A mini solar panel is connected to a rechargeable lead acid batter for powering the robot. IC7805 fixed voltage regulator provides 5V DC regulated supply as required by the circuits. An Android app is designed using MIT App Inventor 2 for controlling the robot over wi-fi. Fig. 2 shows the simulation circuit diagram.

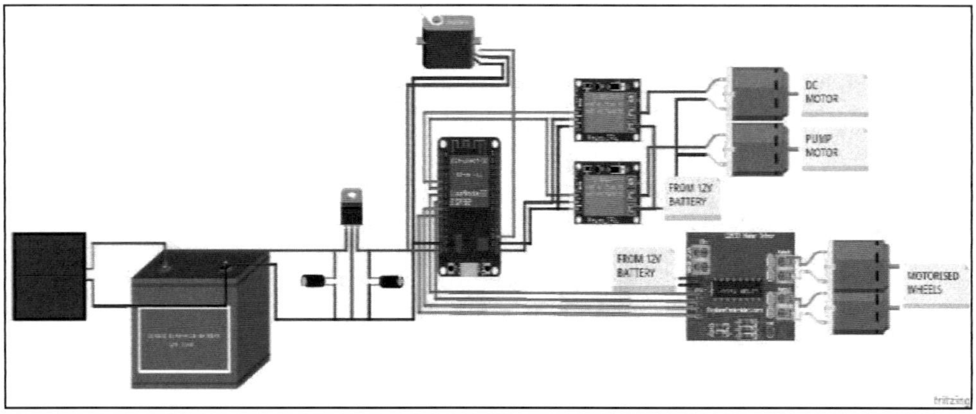

Fig. 2. Simulation Circuit Diagram

The flow chart of the proposed model is depicted in the following figure.

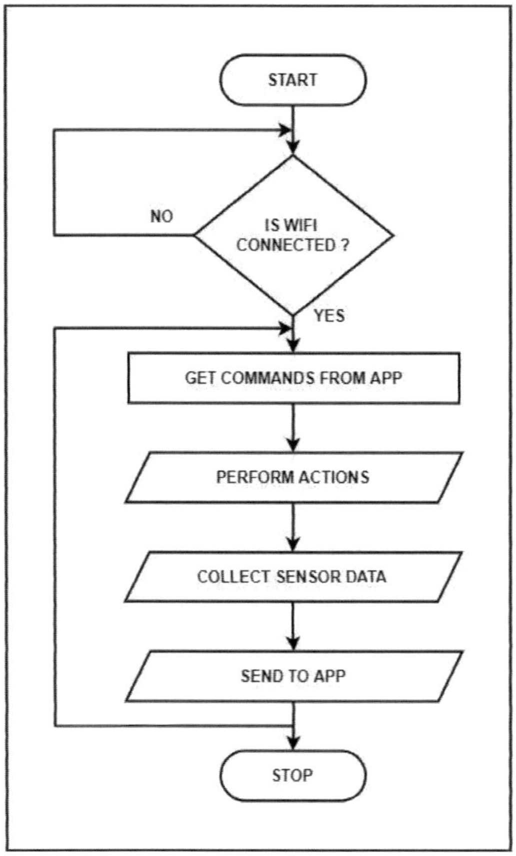

Fig. 3. Flow Chart

Fig. 4 and 5 show the solar panel installed on the robot and the base frame of the robot respectively.

Journal of Physics: Conference Series **2325** (2022) 012029 doi:10.1088/1742-6596/2325/1/012029

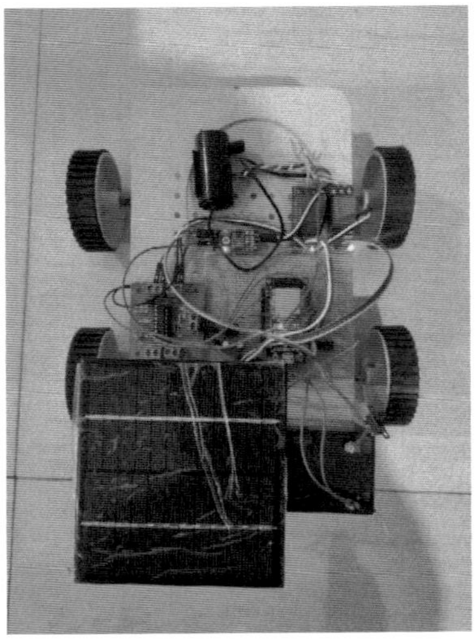

Fig. 4. Solar Panel installed on the Robot

Fig. 5. Base Frame of the Robot

4. Results

Software interface: The app is made using MIT app inventor 2 which is used to control all the activities of the robot using a mobile device with the help of wi-fi.

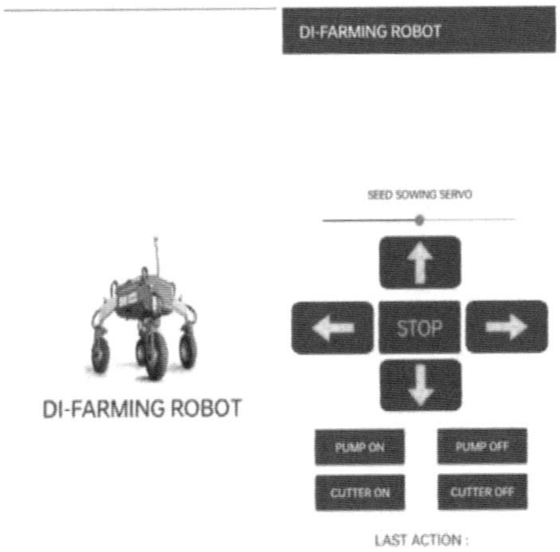

Fig.6. Mobile Application for the Robot Control

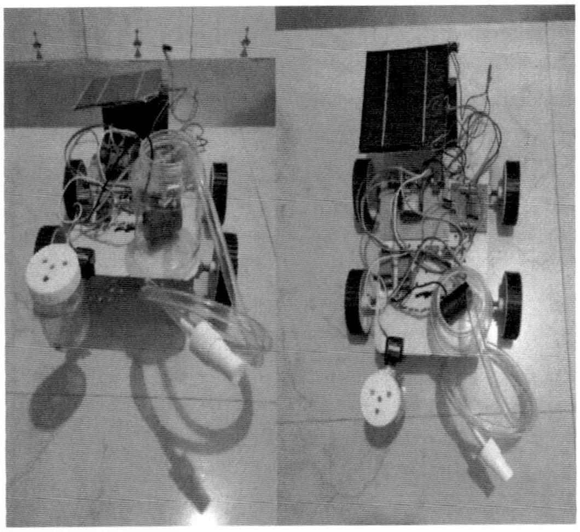

Fig.7. Hardware Interface and Implementation

Hardware interface:

Multi-tasking Robot- The moisture sensor senses the moisture level present in soil, and by using this information, appropriate decisions regarding watering the crop can be made. The various multitask such as ploughing, seed sowing, fertilizer spraying, water sprinkling, moisture sensing and grass cutting can be done on request of the user (Farmer).

Table 1. Details of the Estimated and Actual duration for various agricultural activities

Activities	Estimated Time	Actual Time
Seed Sowing	8 min	9.5 min
Watering	2 min	2.3 min
Spraying Pesticides	2 min	2.3 min
Grass Cutting	7 min	7.5 min

5. Conclusion

A prototype of a multi-purpose autonomous robot for agricultural purpose has been successfully implemented and has been tested for various functions like ploughing, dispensing seeds, watering and spraying. The robot can help the farmers in labour-intensive and monotonous tasks thereby providing more time for paying attention to other aspects of farming and the logistics involved. The future work can include incorporating Computer Vision for improving the efficiency of the robot.

6. References

[1] "Agricultural Robot for Automatic PloughingandSeeding",2015IEEE(TIAR2015),AmritaSneha.A, Ankita.A,Abirami.E.

[2] "DesignandImplementationofSeedingAgriculturalRobot",(JIRAS)(P.Usha,Dr.V.Nandagopal).

[3] "Automated Farming Using Microcontroller andSensors",InternationalJournalofScientificResearch and Management Studies (IJSRMS), 2015AbdullahTanveer, AbhishChoudhary,DivyaPal.

[4] "IOT Based Smart Agriculture", InternationalJournalofAdvancedResearchinComputerandCommunicationEngineering,2016NikeshGondchawar, Prof.Dr.R.S.Kawitkar

[5] "Digging and Seed Sowing AGRIBOT",2021,RenukaDhavale,ShwetaKalshetty,BhushanPatil,PrachiChoudhari,AjayPal Singh

[6] "SolarPoweredAutomatedMulti-TaskingAgriculturalRobot",2020,Jerosheja,Dr.MythiliC

[7] "Multi-TaskingAgriculturalRobot",2020,PratikshaK.Deshmukh,PrafullS.Mehetre,SuyashC.Kokane,PavanE.Joshi

[8] "SolarPoweredMulti-FunctionAgri-Robot",2020,ProfChandrakantD.Bhos,ShrutikaM.Deshmukh,PrajaktaA.Bhise,ShwetaB.Avhad

[9] "Multi-purposeAgriculturalRobot",2016NithinPV,ShivaprakashS:Special6.

[10] "SeedSowingRobot",2017Abdulrahman,MangeshKoli,UmeshKori,Ahmadakbar.

Implementation of Automatic Temperature Control for Confined Area

Manitha Samath[1], Surekha Kamath[1*], Varsha Nayak[1], Deeksha Tendulkar[1]

Department of Instrumentation and Control Engineering, Manipal Institute of Technology, Manipal, MAHE Manipal

surekha.kamath@manipall.edu

Abstract. The main aim of this paper is to develop a low cost system which keeps the room temperature at a desired temperature throughout without any human intervention irrespective of any changes in the external environment. The heating system is switched on if the surrounding temperature is less than the desired temperature or the cooling system is switched as the temperature sensed by the sensor goes above the surrounding temperature. Nevertheless, these kind of systems still needs some kind of improvement for being a better alternative.

Keywords: Temperature, Environment, Reference, Sensor, Switching

1. Introduction

May it be a man or a machine 100% performance is given only in idle conditions. One such factor which affects the performance of man and machine is temperature. Human beings have reached greater heights in the field of science and technology. The main aim of every human invention is to increase the level of comfort. Temperature control is one such field that is required. Automation empowers the commercial temperature controllers in order to modify the system specifications and characteristics as required over time without much human intervention. The wide range of the weather changes in the environment demands a simple heating and cooling system. This automatic temperature controller is designed such that it can be even operated by a person without any specific background according to the requirement the system need not be interfaced with the computer yet it can heat or cool the confined area provided.

2. Literature Review

A. Background Theory: With the ever growing technology every field is being automated, controlling temperature being one of the most important process in industries, medical, biotechnology field etc... it has seen many innovations

In [1] Rayan Mohamed Hamid et.al have discussed a cooling system which can control the temperature by using PIC microcontroller. Here the author has used GUI that is Graphical User

Journal of Physics: Conference Series **2325** (2022) 012030 doi:10.1088/1742-6596/2325/1/012030

Interface. The user can either control speed of fan manually or can set the temperature through the computer. The main disadvantage of this approach is that it has no heating system so the temperature of the surrounding can be reduced but can't be increased. Every time temperature has to be changed, the system has to be connected to a computer.

In [2] Mustafa Saad et.al have discussed an automatic fan speed controller. Here author uses pulse width modulation (PWM) technique to control the speed of the fan. Here as the temperature sensed by the sensor increases. the speed of the fan also increases. Here we cannot set any reference temperature according to our requirement and this system can be used only for cooling purpose.

In [3] A.L. Amoo et.al have used an AT89C51 microcontroller which receives the ADC signal which is sensed by the temperature sensor LM35. The sensed temperature is compared with the set value which is pre-defined by the user depending on this the speed of the fan differs. Control system can be classified as either open loop or closed loop. In open loop system output has no influence of the input action. Also open loop does not have feedback path. While closed loop system has a feedback path from the output of the controlled process to the input of the control system. So, closed loop system monitor the temperature and control the fan by switching to appropriate speed.

The need for automated temperature controlling system is a need of the hour. In industries, hospitals, laboratories etc… with rapidly changing temperature conditions it is severely needed. This could help the user to save a lot of time and energy for the measuring of temperature manually and striving hard to keep it at a stable level. Our work adds value to this research as we will be achieving the auto ranging of this temperature sensor which in turn can be applied to any similar sensor in the industries for very cost effective and easier operation of the apparatus consisting the sensor. This could help the user to save a lot of time and energy for the measuring of temperature manually and striving hard to keep it at a stable level.

3. Methodology

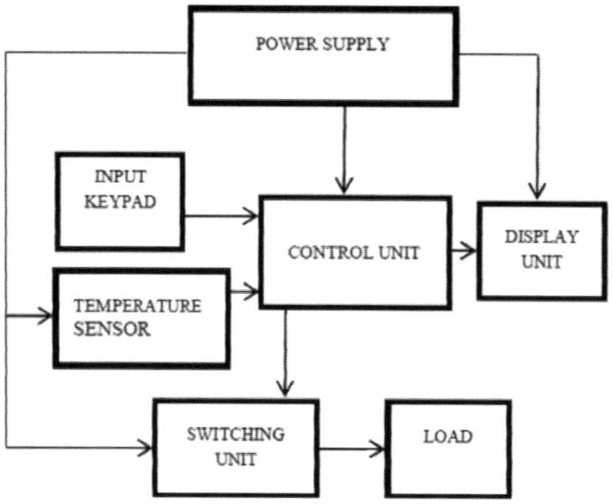

Figure 1. Block diagram of automatic control system

The block diagram of the proposed work is shown in Figure 1 which explains different components of the system. In this project we have used a PIC microcontroller to control the temperature of a confined area. The reference or the desired temperature is set using a keypad. LM35 temperature sensor sends the analog voltage values depending on the surrounding temperature which is fed into the

microcontroller, here temperature values are retained through calculations and then these values are compared with the reference values and necessary signals are sent. These signals either switches the fan or heater till the sensor temperature reaches the reference or desired temperature. Microcontroller scans the signals for every 10seconds and LM35 works with an accuracy of ± 1 /4 °C. Thus making the circuit very effective.
This Automatic control system is divided into 4 main parts

A. Power Supply Unit: Power Supply being one of the most important part of the circuit aims to meet the design specification of the main system. 12V heater and a 12V fan is used in our design it becomes necessary to step down the supply voltage of 240V (in India) to 12V this is achieved by using a step down transformer referring the relation between the power supply and turns ratio of a transformer. We can infer that the ratio of number of turns at primary winding to that in the secondary winding is equal to the ratio of voltage in primary winding to the voltage in secondary winding i.e.

$$\frac{Np}{Ns} = \frac{Vp}{Vs} \qquad \ldots\ldots(1)$$

$$\text{i.e. } \frac{Np}{Ns} = \frac{240V}{12V} = \frac{20}{1} \qquad \ldots\ldots(2)$$

Therefore a step down transformer with turn's ratio 20:1 is used. As this stepped down current is an alternating source to acquire a direct current we use rectifiers. Full wave rectifier's converts positive and negative halves of the ac waveform into single polarity waves as its output. The output from the rectifier will have ripples. To remove this ripples we use capacitor. Therefore, a step down transformer with turn's ratio 20:1 is used. As this stepped down current is an alternating source to acquire a direct current we use rectifiers. Full wave rectifier's converts positive and negative halves of the ac waveform into single polarity waves as its output. The output from the rectifier will have ripples. To remove this ripples we use capacitor

$$V_{pp} = \frac{I}{2fC} \qquad \ldots\ldots(3)$$

Where,
Vpp = peak to peak ripple voltage (in volts).
f= frequency of the AC power in Hertz (i.e. 50Hz)
C= Capacitance (in microfarad)
I= Current in Ampere (i.e. 1Amp)

$$V_{rms} = 12v$$

$$V_{pp} = V_{rms} \times \sqrt[2]{2} \qquad \ldots\ldots(4)$$

$$= 12 \times \sqrt[2]{2}$$

$$= 16.9705 \approx 17V$$

Assuming Ripple voltage of 20%

$$V_{pp} = \frac{20}{100*17} = 3.4V$$

$$C = \frac{I}{2fVpp} \qquad \ldots\ldots(5)$$

$$C = \frac{1}{2*50*3.4}$$

$$= \ 2.94 \times 1$$

2.94*10-3F is the calculated value which is approximately equal to 3300μF which is available easily. Thus the capacitor of 3300μF is used. Further to achieve precise value of 12V and 5V we use voltage regulators LM7812 and LM7805 respectively

***B. Input Unit*:** To control a parameter continuously monitoring it and getting feedback becomes necessary. Thus temperature sensor is used which continuously records the data from the surrounding. LM35 is used as a sensing element in this project. LM35 converts the change in temperature into change in voltage level. As the temperature is converted into voltage by the sensor. Now, the temperature is in the form of digital signal which is to be calculated and this is achieved from code. Keypad is used as an input device to set reference temperature this unlike analog knob, being a digital input eliminates human errors while setting up a reference temperature.

***C. Controlling and Switching Unit*:** In this project we have used PIC microcontroller (PIC 18F45K22) as controlling device, the main advantage using this microcontroller is its low power consumption property and ease of programming. These are more reliable compare to other microcontrollers having a low malfunctioning rate. PIC microcontroller uses RIC's architecture and thus it has a very fast performance. Relay circuit acts as a switching circuit which performs switching action between cooling and heating system. It is connected via transistor to the microcontroller because it provides fast DC switching (ON or OFF), hence that control relay coil.

***D. Output Unit*:** In this part of the circuit we have used a lamp as a heating system and a fan as cooling system. As it is just a prototype, the heat released by a filament lamp is sufficient to increase the temperature of a small closed area but for future applications effective heater and cooler can be incorporated.

Table 3.1 List of components

COMPONENT	SPECIFICATION	NUMBER
Transformer	Step down 20:1 1A	1
Diode	1N4001	4
Resistor	10k pot (variable resistor)	1
	10k ohm	6
	1k ohm	2
	220 ohm	1
Voltage Regulator	7805	1
	7812	1
LED	Red	2
	Green	1
Capacitors	3300 μF	1
	4700 μF	1
	22 pF	1
Pushbutton	Push Button	1

	Reset Button	1
Microcontroller	PIC series18	1
Crystal Oscillator	8MHz	1
Sensor	LM35 Temperature Sensor	1
Relay	12 V	2
Transistor	BC 107	2
Bulb	12V DC	1
Fan	12V DC	1
Display	LCD LM016L	1
Keypad	4*3	1

***F. Software Used*:** Simulation of Automatic Temperature Controller is performed using Electronic Design Automation software developed by Labcenter Electronics Ltd called Proteus Microcontroller is programmed using a hex file. MikroC Pro is the software used to convert the code from C language to a hex file.

4. Results and Discussions

A. Output of Power Supply Unit:

Design of the power supply unit is according to our requirement of 5V and 12V for the loads. We simulated power supply using Proteus software and tested output in it. The circuit connection was rigged on the breadboard for further verification, here we achieved a nearly approximate value of 12.23V and 4.95V as per our requirement.

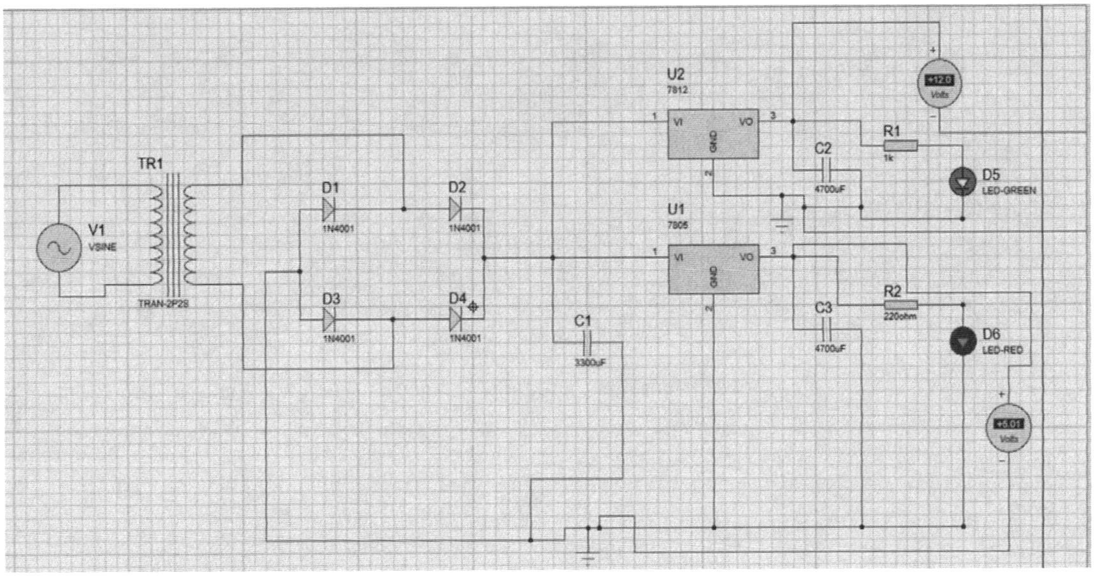

Figure 2. Simulation of automatic temperature control system

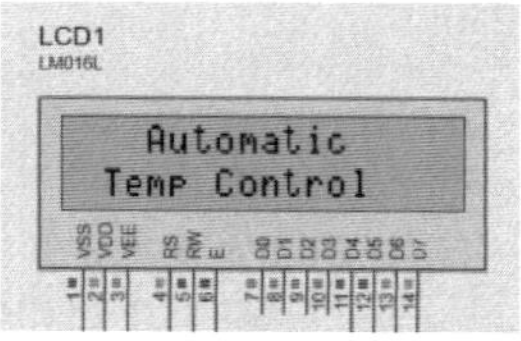

Figure 2.1

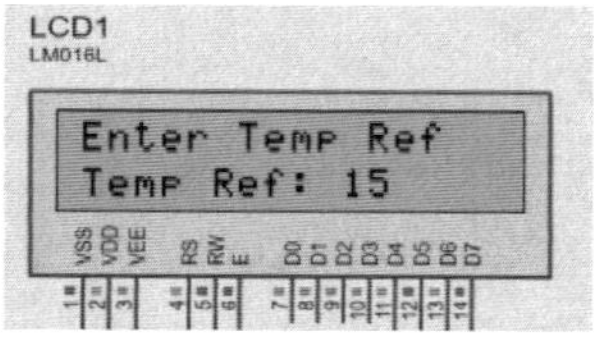

Figure 2.3

Figure 2.2

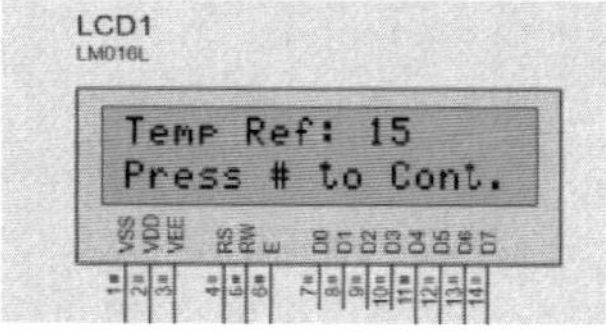

Figure 2.4

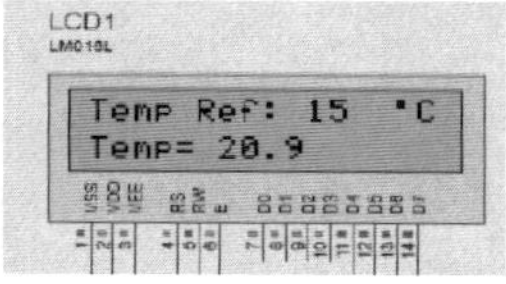

Figure 2.5

Figure 2.1: Automatic Temperature Control display
Figure 2.2: Prompt to enter reference temperature
Figure 2.3: Entering reference temperature
Figure 2.4: Moving to next process
Figure 2.5: Output screen

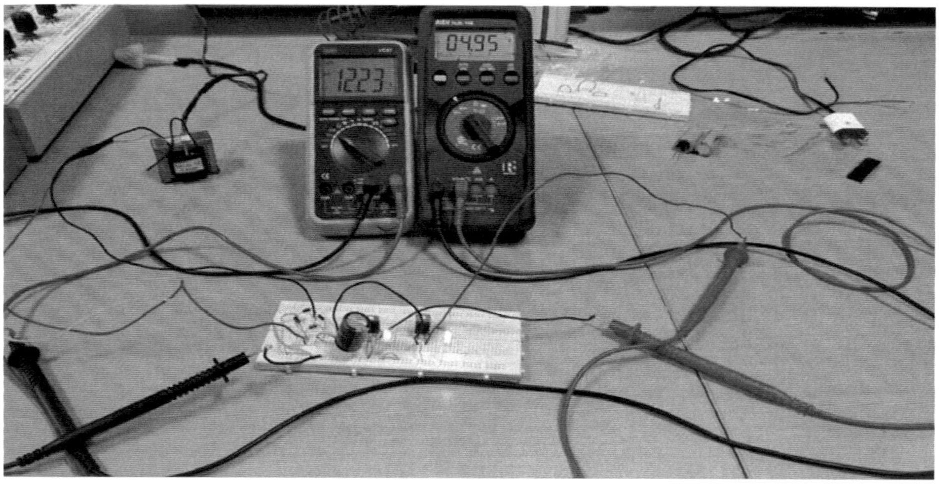

Figure 3. Breadboard connected power circuit

Power supply was designed according to our requirement of 5V and 12V for the loads. We simulated power supply using Proteus software and tested output in it. The circuit connection was rigged on the breadboard for further verification, here we achieved a nearly approximate value of 12.23V and 4.95V as per our requirement. Circuit was tested in oscilloscope for deviations and noise if any, we proceeded further as we attained a desired output. Figure 3 shows the breadboard connected power circuit.

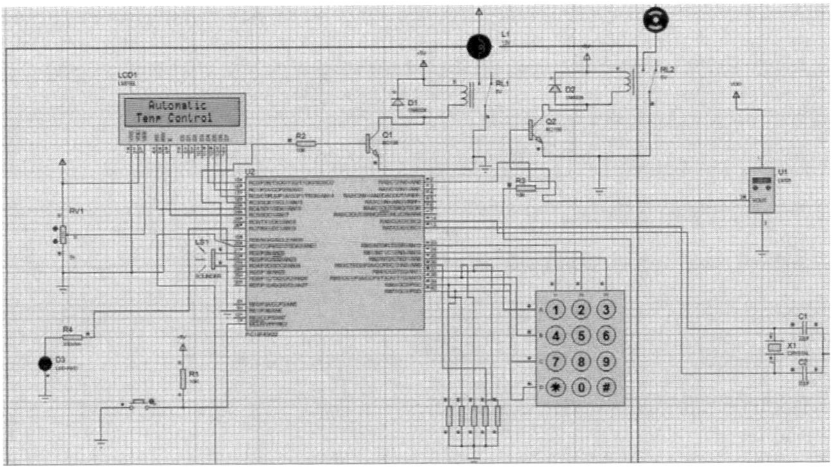

Figure 4: Whole circuit simulated in proteus

Figure 5 shows the circuit that is rigged up for automatic control of temperature of a confined area. Here a fan and bulb is used to indicate the highest and lowest temperature of an area. A keyboard is used to enter the required threshold values for highest and lowest temperatures.

Figure 5: Breadboard connection of automatic temperature controller

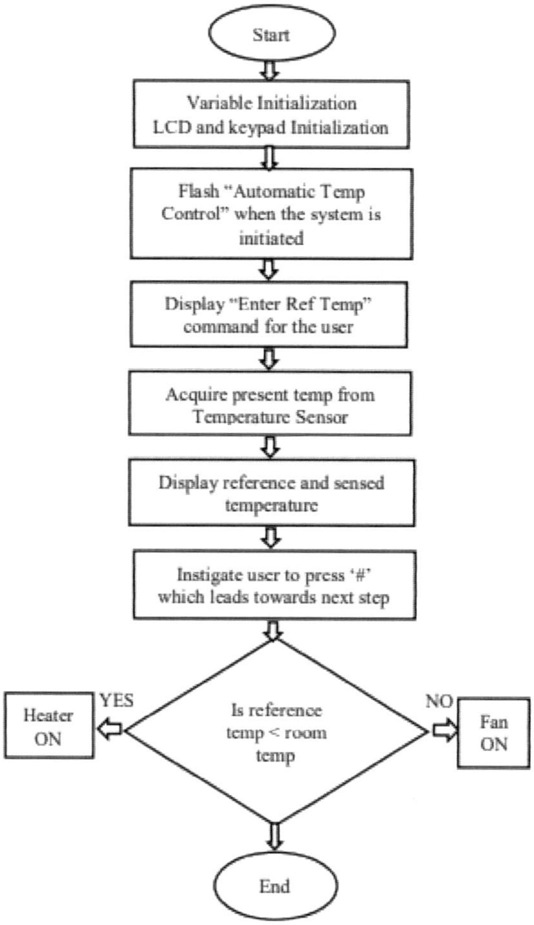

Figure 6: Flowchart of the program

B. Steps in Automatic Temperature Controller System

Step.1: The system first displays "Automatic Temperature Controller" in the LCD display as shown in Figure 2.1

Step.2: The system enables a user to enter reference temperature as in Figure 2.2. At this stage both the relays are disconnected from the loads (i.e. Lamp and Fan will be in OFF state).

Step.3: In this step the reference temperature is given with the help of keypad. The set temperature is displayed in the LCD as in Figure 2.3.

Step.4: Further in order to proceed '#' is pressed, which is initialized as 'ENTER' key as shown in Figure 2.4

Step.5: Reference temperature and the sensed present temperature of the confined area from LM35 are displayed which is shown in Figure 2.5

If suppose user has entered reference temperature as 200C and the temperature sensed is 21.90C, as the set temperature is lower compared to sensed temperature, turns the fan ON. Similarly, if it is the other way round then the microcontroller turns ON the heater. If suppose the sensed temperature is greater than 350C LED will glow and the buzzer will ring.

These steps are achieved by programming the microcontroller through the code which is illustrated in the flowchart shown in Figure 6.

5. Conclusion

We have designed implemented and verified an automatic temperature control system using a microcontroller. Our main objective is to make a system which fulfills the requirements in a cost-effective way. Components used in the implementation were easily available and economically affordable. This system maintains the temperature through its loads. This system can mainly use in medicals, homes, tissue culture, hospitals, industries etc. The devices which is being used in the project has a very simple form of structure. Hence these devices have a very low cost compared to other devices which is currently present in the market. The systems are capable of providing huge contribution in the various fields as mentioned above. Nevertheless, these kinds of systems still need some kind of improvement for being a better alternative.

References

[1] Rayan Mohamed Hamid and Eltahir Mohamed Hussein, "Cooling system temperature control using PIC microcontroller", International Journal of Science and Research, Volume 5, May 2016

[2] Mustafa Saad, Hossam Abdoalgader, and Muammer Mohamed, "Automatic fan speed control system using microcontroller",6th Int'l Conference on Electrical, Electronics & Civil Engineering, South Africa, Nov. 27-28,2014

[3] A. L. Amoo, H. A. Guda, H.A. Sambo and T.L.G. Soh, "Design and implementation of a room temperature control system: Microcontroller-based", 2014 IEEE Student Conference on Research and Development, 2014

[4] H.P. Thakre, Ahkil k. Gonde, Swapnil d. Balge, Shubham D. Dhawale, Ashwin P. Waghade, Akshay N. , Sabal, Prajkta K. "Automatic Temperature controller for various applications"Vol-3 , Issue-2 2017

[5] Aditya Srivastava; Anuj Upadhyay; Tanya Sharma & Vishal Srivastava "Automatic Temperature Control System using PIC microcontroller", International Journal of Research, April 2016,

[6] Muhammed Ali Mazidi, Rolind D. Mckinlay, Danny Causey, "PIC microcontroller and Embedded system" Pearson Publications, ISBN: 8131716759

IoT BASED INTELLIGENT LOW-COST ANTI-COLLISION CUM MONITORING SYSTEM FOR FISHING VESSEL

[1]Pranav S Wazarkar, [2]Neel S Bhurkunde, [3]Nitin S Yadav, [4]Sajal S Patil, [5]Poorvi K Joshi*

[1, 2,3,4,5] Shri Ramdeobaba College of Engineering and Management, Nagpur (MS), India.

[1]wazarkarps@rknec.edu, [2]bhurkundens@rknec.edu, [3]yadavns@rknec.edu, [4]patilss_2@rknec.edu, [5]joshipk@rknec.edu

Abstract. The collision of fishing vessels is the problem many fishermen face. Due to the lack of a proper system, this problem persists and inevitable danger needs to be solved. Even after detecting the objects that may collide with the ship, the captain may not be able to manoeuvre it to safety. The system proposed here would be able to overtake the control of the motor and the propeller, to slow down the ship. In this paper, an ultrasonic sensor is used to detect objects; and the system alerts the crew of the danger ahead. A DC motor is also used to replicate the operation of the propeller of the ship and emergency braking system. For control and sensing parameters, Arduino Uno is used as the microcontroller. To display the Real-time data on the web, ESP-S01 (8266) module is used. The relative distance of the object is logged on thingspeak through the ESP module. A machine learning model is implemented using Support Vector Machine (SVM) to predict the appropriate direction to avoid collisions, The testing accuracy of the model is 98%. Similarly, the logs and live updates help the port rescue team to help the crew under distress.

Keywords. Ultrasonic sensor, Espressif (ESP8266), Arduino, motor, Fishing vessel, Thingspeak, Machine learning.

1. Introduction

There are many anti-collision or collision avoidance systems in the automobile industry such as ABS, blind spot monitor, Emergency Braking Assist (EBA), etc. But there aren't many studies for monitoring ships. Nowadays the need for this system is rising rapidly due to the expansion of the fishing industry and maritime business. When fishermen go fishing, they are about 65nm (nautical) miles away from shore, therefore, they don't have many means of communication other than satellite phones or emergency transponders. But these are costly equipment and many of them can't afford it hence they have left with no option than to depend on self-safety means. Most of the time these small vessels are not detected in the RADAR system of big transport vessels (mostly at night) and as the fishing vessels do not have advanced systems to signal the transport vessels, they meet with an accident. Understanding the demanding need for this kind of system, many of the shipping companies try to get solutions for this problem. So to contribute to this, a system has been proposed that will detect neighbouring ships or other foreign objects which seem to have abnormal movement. In the proposed system, ultrasonic sensors are installed on boats as per boat architecture so that ultrasonic sensors cover maximum area. If any object is detected, then its relative distance is conveyed to the boat. For communication purposes, ESP module is used which will show the relative distance to the ship and also to the Thingspeak server, then it will raise an alarm on the boat so that the captain can control the boat manually. To make a system more interactive, the system provides the manoeuvring direction on the basis of relative distance extracted from Thingspeak and put into the Matlab SVM model, the testing accuracy of the model is 98%. In any circumstance, if the boat is not operated manually the system itself takes control of the motor and tries to avoid the collision. To control the speed of the ship, the motor will rotate in the opposite direction so that the opposite thrust will produce and the braking action would take place. In between this action, some reset time is given to reset the controller and shift the controls to manual if the ship is out of danger. All the above control action mentioned is controlled by Arduino to depict the prototype model of the system.

2. Related work

According to Yamin Huang et al., the system should propose the best possible solution to avoid the collision and if the interface does not work then the control system takes charge with the set of actuators that will control the ship automatically [1]. T D Subash et al. suggested that the system should detect and be on alert for incoming objects and the accident detection system should detect any accident-prone spot and alert a central team [2]. Arista Y. Wardani et al. studied the parameters concerning the architecture for making an anti-collision system for aquatic vessels [3]. P.Veerakumar et al. proposed prototype for collision avoidance using ultrasonic sensors, and a GSM module was used to provide distance, for foreign unidentified bodies [4]. Xiao-ling Liu, et al. proposed an anti-collision system-based design with a microcontroller (AT89C52 chip). The information from a constantly emitting laser pulse, is directly fed into the microcontroller to sound the alarm [5]. K. Ohtani et al. suggested using simple sensors to identify objects without getting in contact with them [6]. Ahmadhon Kamolov et al. proposed an IoT-based system automating linkage procedures by searching for locations using port-mounted sensors [7]. Megalingam et al. suggested a way to intelligently identify an accident at a place and specific time and report the same to the nearest Emergency Service Provider [8]. Y. Zhuo et al. suggested a way to provide an intelligent anti-collision decision-capable support system as it is very crucial to facilitate a precise anti-collision information capability [9]. Sivaganesan proposed that with the increasing AI, automation enabled with IoT application should use Edge computing rather than conventional cloud computing to reduce latency at end user and improve real time decision making capability of system [10]. Ubiquitous Computing Systems (UCS) for Wearable IoT devices especially for medical application was discussed by V.Suma and also how the parameters such as utility of bandwidth, error rate, storage management get improved using UCS was proved with results [11]. S.Unni et al. proposed the point-to-multi-point backhaul communication using long range Wi-Fi, and used the mesh topology where each boat acts as cluster for communication. They discussed various parameters with results at transmission power level, mechanical antenna tilt at base station, maximum channel bandwidth range [12]. K.Ohtani et al. proposed object detection system using simple ultrasonic sensors and neural networks for recycling waste segregation [13]. Kanarachos et al. discussed a new method for computation of optimal obstacle avoidance steering manoeuvres, with consideration of vehicles dynamics, the method is based on a reformulation of Pontryagin's Maximum Principle (PMP), in which the manoeuvre time was prescribed and the bang-bang control amplitude was minimised [14]. F.Kruse et al., discussed about collision detection system to detect all objects in front of train using multi-sensors and video technology for future automated train's safety [15].

3. Proposed System

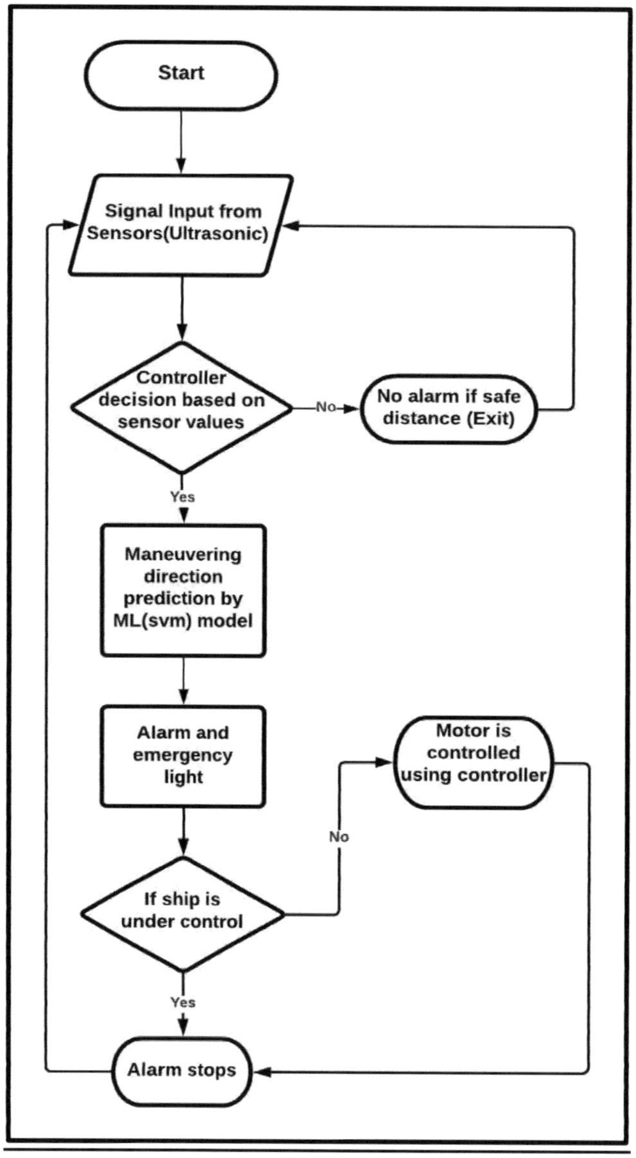

Figure1. Flowchart of the working model

The proposed system uses an ultrasonic sensor, alarm, emergency light, and Arduino as controller and motor (propeller) as an actuator. For communication purpose, the ESP8266-01 is used to send relative distance on the common interface (thingspeak) so that the base station of the ship as well as the other neighbouring ships are also aware of their relative positions, as Thingspeak provides a read API key through which other ships are also able to read the sensor values. The aim is to reduce the probability of the accidents happening as well as to control the ship, if it's not possible to control it manually, it attempts to take the ship to a safer position where the probability of collision is less or the damage about happen is reduced. Due to this communication, the rescue team at the nearest port can be alerted before an accident happens and more lives can be saved.

The flow of signals and the action to be taken are shown in figure 1, where first the input in terms of distance is taken from the ultrasonic sensors attached to the sides of the boat which sends continuous ultrasonic waves and measures the distance between the relative object and fishing he system makes decisions based on distance detected by the controller. If the boat is at a safe distance (greater than the threshold value configured in controller already) then there is no alarm. But if there is some kind of obstacle or object near the boat then the alarm and emergency light will be turned ON. The next step is the prediction of the direction which is done with the help of the SVM model; for this the values detected by sensors are uploaded on the Thingspeak via ESP Wi-Fi module, which are then used by SVM model for decision making. Then if the boat is taken under control by the captain, the alarm and emergency lights will be turned OFF. Consequently, if the captain is unable to take control of the ship, then automatic braking is done by controlling the propeller with the help of the controller. The architectural view of system is shown in figure 2 which significantly defines the type of sensors, actuators, controllers used in system. For each condition (type of input to sensor), the controlled action taken by the system is tabulated in the table 1.
Advantages of the proposed system:

- Object detection using ultrasonic sensors. The object can either be moving or stationary.
- Propeller control for automatic braking.
- A warning system to brake manually.
- A central information hub for data analysis and/or display.
- The Thingspeak server is used to send real-time values.
- Values received from Thingspeak are used to predict the manoeuvring directions using ML.

Table1. Working scenarios

INPUT	**OUTPUT**		**Communication**	**Actuator**
Ultrasonic sensors	**Alarm**	**Indicator**	**ESP**	**Motor**
Obstacle detected (manually controlled)	Sounds	Light	Relative distance send	No change
Obstacle detected (automatically controlled)	Sounds	Light	Relative distance send	Motor controlled to avoid collision
Obstacle not detected	No sound	No light	Relative distance send	No change

Journal of Physics: Conference Series **2325** (2022) 012031 doi:10.1088/1742-6596/2325/1/012031

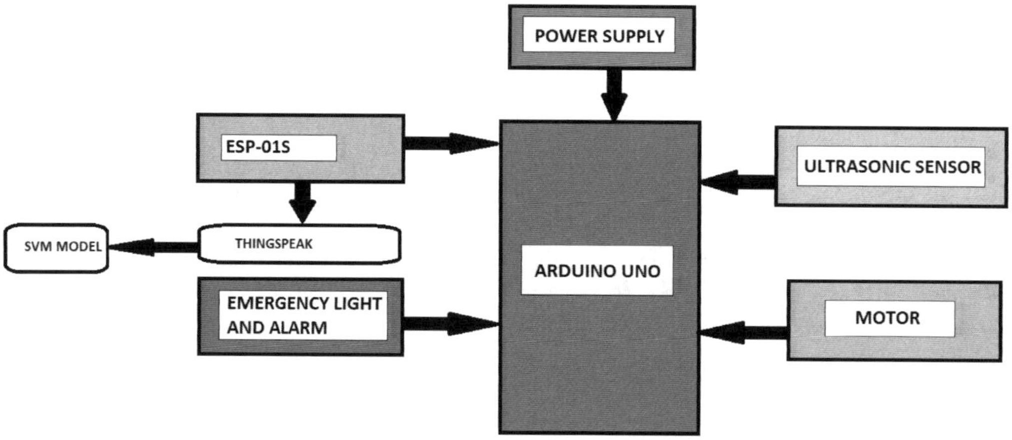

Figure 2. The architecture of the system

Circuit diagram and PCB layout

Figure 3, figure 4, and figure 5 are the circuit diagrams of the system, routed PCB layout, and actual 2D view of the PCB after manufacturing respectively. The PCB is designed to make the system robust to the harsh environmental conditions in the sea. PCBs can be easily mounted on the ship as per the ship's architecture as well as they can be embedded with other systems. Good packing also improves the life of the system.

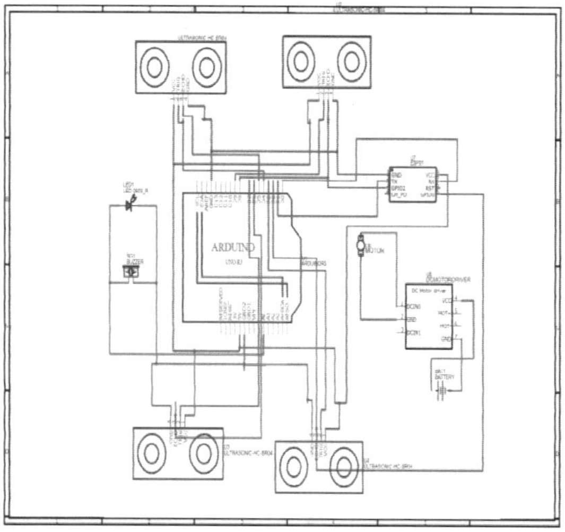

Figure 3. Circuit diagram

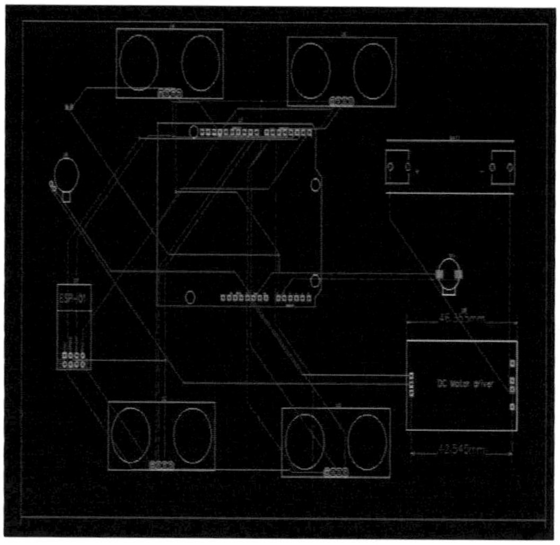

Figure 4. The routed layout of PCB

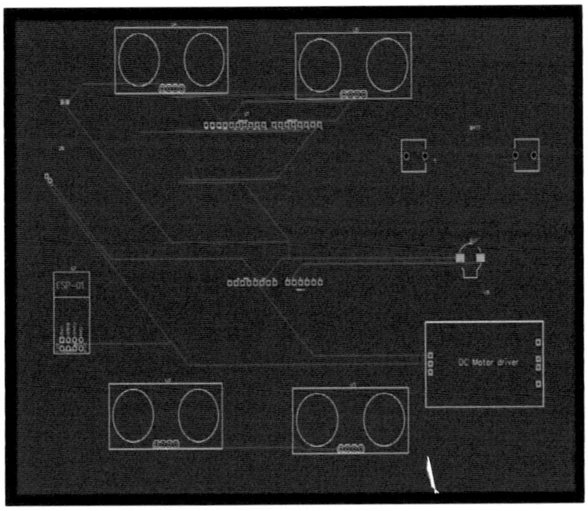

Figure 5. Physical view of PCB

Figure 6. Prototype model

4. Results

This work has successfully designed the prototype model as shown in figure 6 and it has been tested for various test cases and it is found that the outputs match with the ideal one. The SVM model is used for early decision making which is trained and tested on the dataset, and this will help the captain in taking appropriate decisions while manoeuvring in difficult situations. The testing and training accuracies of the model are 98% and 99% respectively. This model predicts direction based on values it gets from Thingspeak as shown in figure 7 which ensures the decision made is on the earliest set of data from the sensor. It provides the direction as 'Critical' which defines that the object is near to both the starboard (right side of ship) and port (left side of ship) side of the ship; 'left' indicates the object lies on the starboard side, 'Right' indicates the object is on the port side, the 'safe' indicates no object is in danger range, as shown in figure 8. The plus point of the system is that, in 'critical' conditions it will provide the automatic braking option using propeller control in the same way as a traditional braking system of the ship (motor will provide opposite thrust to increase the viscous drag), this process of providing the break via opposite thrust is continuous till the controller gets the next set of values in the safe region or the other case is that the controls are switched to manual. Figure 9 shows the testing accuracy of model using the confusion matrix, where the accuracy of the test data is about 99% with 0.2% false rate. Figure 10 shows the importance of each feature in percentage which is calculated by MRMR algorithm (Maximum Relevance – Minimum Redundancy).The main purpose of using this algorithm is to find how much is the contribution of each sensor to get the accurate decision regarding the direction of manoeuvring. On the basic of MRMR score some redundant values can be neglected while making decision.

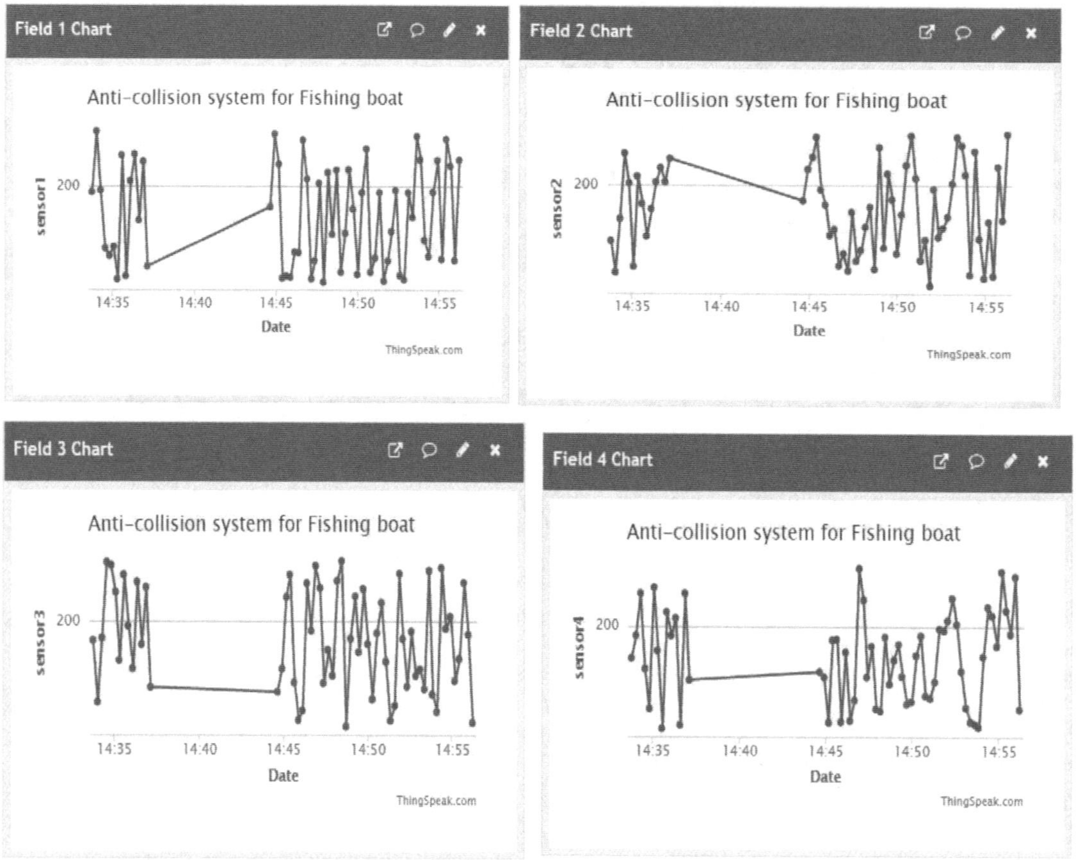

Figure 7. Real-time data transmission

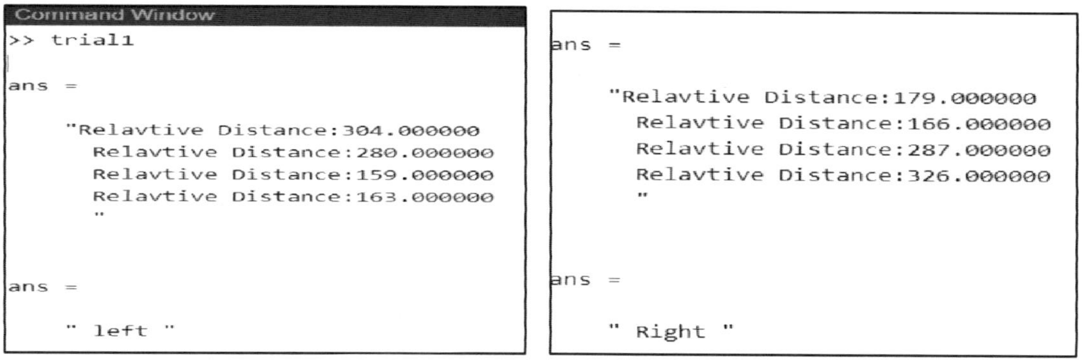

```
ans =

    "Relavtive Distance:179.000000
      Relavtive Distance:166.000000
      Relavtive Distance:187.000000
      Relavtive Distance:158.000000
      "

ans =

    " Critical "
```

```
ans =

    "Relavtive Distance:309.000000
      Relavtive Distance:296.000000
      Relavtive Distance:288.000000
      Relavtive Distance:316.000000
      "

ans =

    " Safe "
```

Figure 8. SVM prediction on Manoeuvring Direction

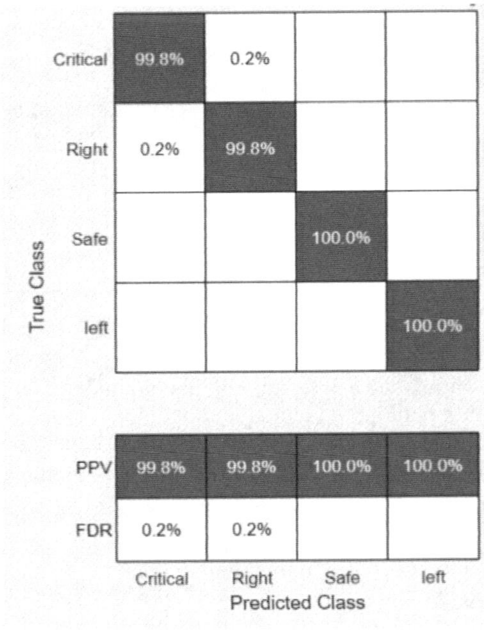

Figure 9. Confusion matrix for Test data

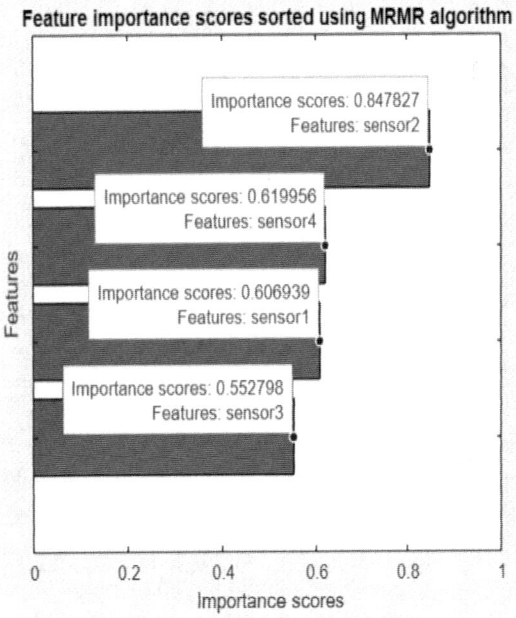

Figure 10. Feature importance comparison

5. Conclusion

A collision avoidance system is a great boon for the fishing industry and it will bring much relief for fishermen as they are the ones who risk their lives to fulfil the supply required related to seafood. This work successfully develops a system that can detect the ships nearby which in turn can avoid a collision. Real-time communication is very important for any system to make appropriate decisions; in the proposed system by transmitting the relative distance of the ship on Thingspeak and putting those values in the SVM model system, manoeuvring direction is provided so that the captain can take appropriate action to avoid the collision. In the condition where manual manoeuvring is not possible, then the system can take the propeller control and apply brakes by producing opposite thrust so that the damage can at least be minimized if not completely avoided. The experimental setup is tested against the different possible sets of conditions to ensure proper working and therefore it is concluded that the system can predict the collision and can even help avoid or minimize the damage.

6. References

[1] Y. Huang, L. Chen, R.R. Negenborn, P.H.A.J.M. van Gelder, *A ship collision avoidance system for human- machine cooperation during collision avoidance*, 2020 *Ocean Engineering* ***217***, *107913*. DOI:10.1016/j.oceaneng.2020.107913

[2] T. D. Subash, A.S. Pradeep, A.r. Joseph, A. Jacob, Jayaraj P.S, *Intelligent Collision Avoidance system for fishing boats*, 2020 *Materials Today: Proceedings* ***24*** *2457-2463*.
DOI: 10.1016/j.matpr.2020.03.776

[3] A.Y. Wardani, A.A. Masroeri, A.S. Aisjah 2015, *Vessel Collision Avoidance System (VCAS) Based on AIS Data, The 1st International Seminar on Science and Technology*.
DOI: 10.12962/j23546026.y2015i1.1140

[4] P.Veerakumar, R. Sundar, Dr. S. V. Saravanan, and T. Dhanya, *Automatic Collision Avoidance System Using Ultrasonic Sensor for Ship*, 2018 *International Journal of Management, Technology, and Engineering,* ***8,*** *Issue 9*. DOI: 10.1088/1742-6596/1011/1/012046

[5] X. Liu, F. Qin, H. Niu, *Design of Ship Anti-collision System Based On Microcontroller*, 2017 *7th International Conference on Applied Science, Engineering and Technology* ***122***.
DOI: 10.2991/icaset-17.2017.44

[6] K. Ohtani, M. Baba, *A Simple Identification Method for Object Shapes and Materials Using an Ultrasonic Sensor Array,* 2006 *IMTC -Instrumentation and Measurement Technology Conference*. DOI: 10.1109/IMTC.2006.328525

[7] A. Kamolov, S. Park, *An IoT-Based Ship Berthing Method Using a Set of Ultrasonic Sensors*, 2019 *Sensors* ***19****, Issue 5181*. DOI: 10.3390/s19235181

[8] Megalingam, R.N Nair, And R.K Pakya (2010*), Wireless Vehicular Accident Detection and Reporting system*, *International Conference on Mechanical and Electrical Technology*.
DOI: 10.1109/ICMET.2010.5598437

[9] Y. Zhuo, G.E. Hearn, *A Ship Based Intelligent Anti-Collision Decision-Making Support System Utilizing Trial Manoeuvres*, 2008 *Remote Sens*. DOI: 10.1109/CCDC.2008.4598079

[10] Sivaganesan*, D. Design and development ai-enabled edge computing for intelligent-iot applications, Journal of trends in Computer Science and Smart technology (TCSST)* **1**, no. 02 (2019): 84-94.
DOI: https://doi.org/10.36548/jtcsst.2019.2.002

[11] V. Suma *Wearable IoT based Distributed Framework for Ubiquitous Computing, Journal of Ubiquitous Computing and Communication Technologies (UCCT)* **3**, no.01 (2021): 23-32.
DOI: https://doi.org/10.36548/jucct.2021.1.003

[12] S. Unni, D. Raj, K. Sasidhar, and S. Rao 2015, *Performance measurement and analysis of long range Wi-Fi network for over-the-sea communication,13th International Symposium on Modelling and Optimization in Mobile, Ad Hoc, and Wireless Networks (WiOpt)*,**3**, Issue 11

[13] K. Ohtani, M. Baba (2006), *A Simple Identification Method for Object Shapes and Materials Using an Ultrasonic Sensor Array, IMTC 2006 - Instrumentation and Measurement Technology Conference.*

[14] Kanarachos, Stratis (2009). *A new method for computing optimal obstacle avoidance steering manoeuvres of vehicles ,International Journal of Autonomous Systems* **7** (1): 73–95

[15] F.Kruse, S.Milch, H.Rohling, *Multi Sensor System for Obstacle Detection in Train Applications*, Proc. of IEEE Tr., June, pp.42-46. (2003)

Computational Analysis of stochastic arithmetic computing and stochastic activation function

Ashok P, B. Bala Tripura Sundari

Department of Electronics and Communication Engineering, Amrita School of Engineering, Amrita Vishwa Vidyapeetham, Coimbatore, India.

E-mail: p_ashok@cb.students.amrita.edu

Abstract. Stochastic Number Generators (SNG) plays a significant role in designing a stochastic computing system. SNGs make the stochastic system comfortable for computing in the stochastic domain. The challenges in developing the stochastic computing system are correlation and hardware area occupancy. By considering these phenomena, we have considered Linear Feedback Shift Register (LFSR) based SNG and S-box based SNG in this work. Our contributions to this paper are stochastic computation for activation functions using the SNGs mentioned above and stochastic computation for arithmetic components in the stochastic domain. By considering the two SNG methods, the difference in the computation for accuracy has been analyzed for stochastic activation function and stochastic arithmetic computation. The better SNG method will be used as SNG for stochastic convolutional neural network design using this analysis.

Keywords: Activation function, Arithmetic computation, Stochastic number generation, Stochastic system, S-box - based SNG

1. Introduction

The vital part to be considered while designing a stochastic system is stochastic number generation. To have an accurate stochastic computer, input to the system should be free from correlation [1]. Conversely, the stochastic system cannot take binary digits as input to process its computation. Hence binary to stochastic conversion unit has to be used for transforming binary digits; this unit is otherwise referred to as a stochastic number generator. There are a few considerations in modeling a stochastic number generator (SNG) so that stochastic computing is accurate. The inputs should be uncorrelated, and the inputs shouldn't take up much area. Depending upon the accuracy requirements, the system can trade off power and area in future architectural implementations. As mentioned earlier, SNG is the input for the stochastic system. This input will affect essential factors in the computation process. The importance of SNG directly relies on the accuracy of the stochastic system. If the input (SNG) of the system is correlated, the system's precision is affected.

2. SNG and its related works

The survey has been focused on two aspects, namely stochastic number generation and arithmetic computation, and activation function. Different methods of random generation were studied. The basic computational components were analyzed in this literature.

The work on the generation of stochastic sequences had started with the use of Bernoulli sequences on a probability basis. With this phenomenon, a linear feedback network was formed for shift registers which were supported by EX-OR for generation [2]. A modulator-based pipelined steam generator had been proposed by Max et al. for fast operation with flexible design [3]. The generation of statistically independent sequences had been worked out by Jeavons et al., emphasizing the construction of stochastic neural network design [4]. Gupta et al. worked on generating random stochastic numbers that are uncorrelated, called weighted binary SNG. Jeavons et al. had focused on accuracy, which had led to the development of an SNG by cascading various modules that had replaced the use of comparators [5]. With the help of CMOS technology, a sampling-based number generator has been proposed by Buruk et al., which could be programmable. This work focused mainly on preserving the area of SNG.

There are two random number generator categories: true random number generator (TRNG) and pseudo-random number generator (PRNG). TRNG has the native entropy source usually used in device physics, whereas PRNG is based on deterministic algorithms. Yin Liu et al. had reported an approach to generate correlated bit streams [6]. This method aims to generate two stochastic streams. An effective sharing-based random number generator had been proposed to reduce the area by Bo yuan et al. This signifies binary and SC domain computing [7]. A new stochastic number generator CAPE had been proposed by Te-Hsuan et al. focusing on availing inputs for combinational circuits, and it supports progressive precision, thereby maintaining the accuracy of the system [8]. The down-sampling method was proposed by Hassan et al. as the deterministic approaches in SC have to be considered. It, therefore, has improved the progressive precision of the system [9].

To enhance the accuracy while simultaneously reducing the area and energy, a new method to generate a stochastic number generator was developed by Aidyn et al. [10]. In contrast to the above techniques mentioned, Sakamto et al. had proposed methods for constant generation of a stochastic number; this also addresses correlation and hardware area allocation. This method incorporates sharing of SNs focusing on hardware overhead. Yawen et al. had proposed a method of SN generation where one clock conversion had been used, which thereby focused on reducing hardware cost. LFSR as a random number source was used, which is considered for stochastic number generation. S-box, a cryptographic method, had been considered and developed, naming as SBoNG [11].

This work has two novel contribution methodologies: computation for the stochastic activation functions, namely the stochastic tanh and COS functions, and stochastic computation for arithmetic components. The activation functions and the arithmetic components have been computed using two different stochastic number generators as random number sources, LFSR- based SNG and S-box based SNG. A study has been done on the computations by comparing the errors obtained for two methods of stochastic number generators. The comparison is plotted in the graph for two different computation components for understanding. By considering the two SNG methods, the difference in the computation has been analyzed. From the analysis, the SNG that shows negligible error difference in the computation will be considered for the stochastic computing-based convolutional neural network design. This SNG will be further modified and will be extended in future for the development of VLSI-SC based neural networks using High Level Synthesis (HLS) [13], [14], [15], [16] .

3. Methodology

The proposed methodology, Linear Feedback Shift Register (LFSR) based SNG, and S-Box based SNG computation is based on the following Fig. 1

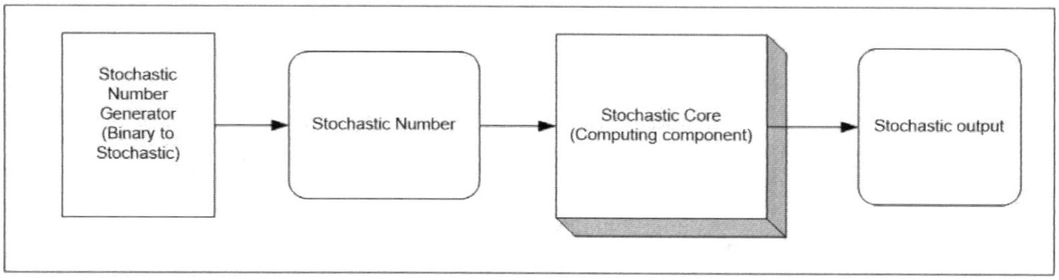

Figure 1. Stochastic Computing.

3.1. *Stochastic Number Generators*

The first method used here is LFSR-based computation; the following steps have been used for developing the appropriate stochastic computation.

(i) Random binary numbers are considered and placed in the register at the first iteration. Later after each cycle of Exor operation, the shifting values are prefilled in the register.
(ii) Perform EX-OR operation for values in the least significant bit
(iii) The resultant bit value is the required output at stage 1
(iv) Transfer its output to register by shifting 1 bit in the register
(v) Repeat the steps till the operation satisfies for $2^n - 1$
(vi) The collection of output obtained at step 3 is the required random number sequence

The second method used here is S-box based computation; the steps involved in the generation of stochastic number is as follows,

(i) Initialize the internal state 'S' and LFSR state, condition length of LFSR, internal state should be same and 'S' has to be multiple of 4.
(ii) The binary number must be considered as the value in the 'L' register. This value has to be shifted by one cycle.
(iii) The value in the LFSR performs EXoR with 'S'.
(iv) Split 'S' into 4 blocks.
(v) Each block performs S-Box operation.
(vi) Rotate the internal state by '1'.
(vii) At each instant of rotation, a binary number will be generated. This binary number is considered random.

3.2. *Stochastic Computation*

The following computations are performed using the above mentioned stochastic number generation methods.

Stochastic multiplication has been performed by logical AND gate using the expression $P(Z) = P(X)\, P(Y)$ Stochastic addition has been computed using scaled multiplexer in-order to scale the values between (0, 1). This method follows the expression $Z = (S\, A) + (1\, S)\, B$ Stochastic subtraction has been developed using scaled multiplexer with one input being negated considering the rule the values should be scaled between (0, 1). Stochastic subtraction is constructed using the equation $Z = (S \cdot A) + (1 - S) \cdot (1 - B)$. Stochastic division has

Table 1. Stochastic arithmetic computation - 4 bit

SNG 1 (INPUT 1)	SNG 2 (INPUT 2)	Stochastic output	SNG 1 (INPUT 1)	SNG 2 (INPUT 2)	Stochastic output
Stochastic Multiplication LFSR			**Stochastic Multiplication SBoNG**		
0001	0000	0000	0110	0100	0100
0011	0001	0001	1010	1011	1010
1010	1000	1000	0111	1011	0011
0000	0100	0000	1110	1111	1110
1000	1000	1000	0001	1000	0000
Stochastic Addition LFSR			**Stochastic Addition SBoNG**		
0001	0000	0000	0110	0100	0100
1111	0001	0001	1010	1011	1010
1010	1000	1010	0111	1011	1111
0000	0100	0100	1110	1111	1111
1000	1000	1000	0001	1000	1000
Stochastic Subtraction LFSR			**Stochastic Subtraction SBoNG**		
0001	0000	1001	0110	0100	0010
0011	0001	0010	1010	1011	0010
1010	1000	0011	0111	1011	0110
0000	0100	1000	1110	1111	1100
1000	1000	1110	0001	1000	0011
Stochastic Division LFSR			**Stochastic Division SBoNG**		
0001	0100	1001	0110	1111	0000
0011	1000	0011	1010	1011	0010
1010	0001	1010	0111	1011	0110
0000	0000	1100	1110	0100	1111
1000	0000	1110	0001	1110	0001

been developed by deducing the JK flip flop at toggled stage using the expression $P(Z) = \frac{P(X1)}{P(X1)+P(X2)}$.

Apart from arithmetic computing, complex computations like stochastic TANH and COS functions has been carried. The complex arithmetic system is approximated by piecewise linear approximation [12]. This will result in the optimized values where the coefficients function using Lagrange interpolation approximation. Therefore, for stochastic TANH function, $f(x) = 1 - C_i + a_i x$ has been used for computing. Similarly COS function has been developed by using $f(x) = a_i x + b_i$.

4. Experimental Analysis

The proposed methods LFSR-based SNG and S-box based SNG, have been used as stochastic number generators for stochastic arithmetic computation and stochastic activation function. The computed values are tabulated and comparison study has been represented in the graph below.

Table 2. Stochastic arithmetic computation - 8 bit

SNG 1 (INPUT 1)	SNG 2 (INPUT 2)	Stochastic output	SNG 1 (INPUT 1)	SNG 2 (INPUT 2)	Stochastic output
Stochastic Multiplication LFSR			**Stochastic Multiplication SBoNG**		
10100001	00100001	00100001	11111001	11100010	11100000
01000011	00000010	00000010	11001001	00110011	00000001
11000010	11000010	11000010	01010011	01000111	01000011
10010100	00010100	00010100	10110101	11001110	10000100
00101000	00001000	00001000	10110110	00000001	00000000
Stochastic Addition LFSR			**Stochastic Addition SBoNG**		
10100001	00100001	00100001	11111001	11100010	11101000
01000011	00000010	00000010	11001001	00110011	00000001
11000010	11000010	11000010	01010011	01000111	01000111
10010100	00010100	10010100	10110101	11001110	10000111
00101000	00001000	00101000	10110110	00000001	00110111
Stochastic Subtraction LFSR			**Stochastic Subtraction SBoNG**		
10100001	00100001	10100110	11111001	11100010	01011001
01000011	00000010	11000001	11001001	00110011	11001101
11000010	11000010	00000001	01010011	01000111	01110001
10010100	00010100	10010100	10110101	11001110	00110101
00101000	00001000	10110001	10110110	00000001	10111111
Stochastic Division LFSR			**Stochastic Division SBoNG**		
10100001	00010100	10100010	10110101	11001110	10110001
01000011	11000010	00000001	01010011	01000111	10011010
11000010	00000010	11000001	11001001	00110011	11001100
10010100	00100001	10011100	11111001	11100010	11011101
00101000	01000000	10111001	10101010	00100101	11101010

Table 3. Stochastic activation function - 4 bit

Ci	Ai	X	Output tanh	Ci	Ai	X	Output tanh
Stochastic TANH LFSR				**Stochastic TANH SBoNG**			
1111	1111	1100	1111	1001	1011	1010	1111
1111	1111	1110	1111	1001	1111	1100	0111
1101	1001	0001	1101	1111	1011	1111	1001
0111	0111	1110	1111	0111	1111	0001	0101
Stochastic COS LFSR				**Stochastic COS SBoNG**			
1111	1001	0111	1110	1001	1111	0100	1111
1001	1111	0011	1100	1011	1111	0010	0111
1111	1011	0001	1011	1101	1111	1110	0111
1111	1111	1000	0101	1111	1011	1010	0111

Table 4. Stochastic activation function - 8 bit

Ci	Ai	X	Output tanh	Ci	Ai	X	Output tanh
Stochastic TANH LFSR				**Stochastic TANH SBoNG**			
11111111	11111111	11010011	10101110	11111001	11111011	10100101	11101111
11111111	11111111	10111101	11100111	11111001	11111111	11101000	11011110
11111101	11111001	10001000	01011111	01111111	11111011	00110001	11011111
11110111	11110111	00011000	10111111	11110111	11111111	00110111	10011011
Stochastic COS LFSR				**Stochastic COS SBoNG**			
11111111	11111001	11001111	11111111	11111001	11111111	00001100	11111111
11111001	11111111	00001100	11110011	11111011	01111111	11111111	11011111
01111111	11111011	11111110	11111111	11111101	01111111	00000100	11111011
11111111	11111111	10001000	00111101	11111111	11111011	11000101	00111111

Table 5. Stochastic activation function - 16 bit

Ci	Ai	X	Output tanh	Ci	Ai	X	Output tanh
Stochastic TANH LFSR				**Stochastic TANH SBoNG**			
11111110 11111111	11111110 11111111	11110111 00110001	10011111 11101011	10000011 11111001	11010111 11111011	11101101 10100101	11111110 11111111
11111011 11111111	11011011 11111111	00011111 00110111	10111101 11011110	10001111 11111001	01001111 11111111	11110111 00110001	10111011 11011110
11011001 11111101	10000011 11111001	01110001 11110011	11111110 11101111	10001101 01111111	11001011 11111011	11110111 11101101	10111011 11111111
00111011 11110111	01000111 11110111	11101000 11001111	11111111 01111111	00101111 11110111	11101011 11111111	11101101 11110111	11000011 11111011
Stochastic COS LFSR				**Stochastic COS SBoNG**			
00111110 11111111	00000011 11111001	11111111 11111111	11111111 11111111	10000011 11111001	01001111 11111111	00010001 10001000	11111111 01110110
11110111 11111001	00100001 11111111	00110000 10000001	11111111 00111110	00010001 11111011	01101010 01111111	10111101 00011000	11111110 11100111
10011101 01111111	10111111 11111011	01001000 10100111	11111111 01011110	11001101 11111101	10001101 01111111	11001011 00010001	11111101 11111110
01100001 11111111	01011011 11111111	11111111 11111111	11111111 11111111	00111000 11111111	11000101 11111011	10010111 11101000	11111111 01101111

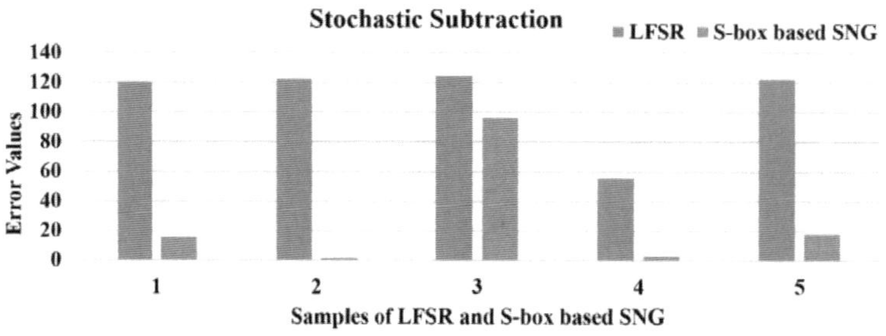

Figure 2. Error analysis in Stochastic subtraction.

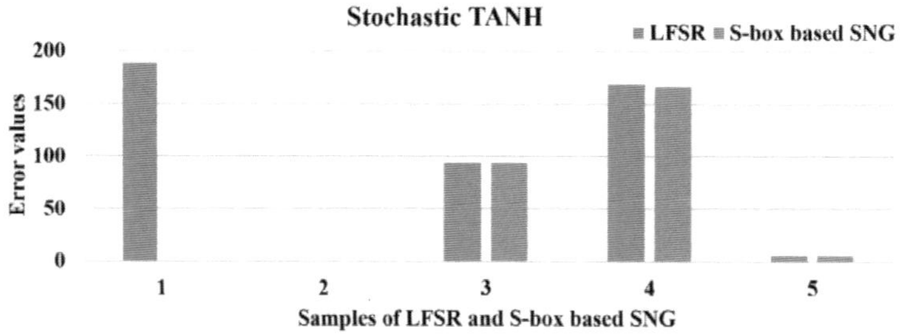

Figure 3. Error analysis in Stochastic TANH.

5. Discussion

The arithmetic computations have been performed in python and tabulated.The table 1 and table 2 represents the 4 bit and 8 bit stochastic computation of various arithmetic computation respectively. Similarly, the table 3, table 4 and table 5 represents the stochastic computation of TANH and COS activation functions respectively.The tabulated values have been analyzed, and random samples have been considered for error calculation. Figure 2 represents the error analysis for stochastic subtraction computation using both the methods of SNGs. Similarly, Figure 3 represents the error analysis for stochastic TANH computation using both the methods of SNGs. We considered the S-box-based number generator as a random number source in the contributed design. This stochastic number generator is considered as input for the arithmetic computation. Following this, the tanh and COS activation function has been computed using the S-box-based number generator as a random number source.We also have constructed LFSR-based stochastic activation and stochastic arithmetic computation. This has been done to perform an analysis on accuracy with respect to a stochastic number generator. Stochastic subtraction and stochastic tanh activation has been computed using both the methods of SNG separately. The accuracy difference in both methods is analyzed by error calculation. From the error calculation, graphs have been plotted considering both methods for comparison. By analyzing both the graphs, it is understood that LFSR-based method is prone to error and S-box based SNG method shows consistence in the performance across all the components with varying weights.

6. Conclusion

We have performed computational analysis for stochastic arithmetic computation and stochastic activation functions using two different stochastic number generators respectively. From the computation, the S-box based method seems to be advanced in generating stochastic numbers. The LFSR-based process appears to depend on the shifting from the previous cycle with which the sequences are generated. While analyzing the computation performed using S-box based method, due to the advance of each bit in the cycling process, there is no dependency in the generation of stochastic sequences. The design exploration based on accuracy and power trade- off has to be conducted for the VLSI architectural implementations. The computing will be incorporated into the VLSI-based convolutional neural network architecture and analyzed using high-level synthesis.

References

[1] Gaines, Brian R. "Stochastic computing." In Proceedings of the April 18-20, 1967, spring joint computer conference, pp. 149-156. 1967.

[2] Gaines, Brian R. "Stochastic computing systems." In Advances in information systems science, pp. 37-172. Springer, Boston, MA, 1969.

[3] Van Daalen, Max, Pete Jeavons, John Shawe-Taylor, and Dave Cohen. "Device for generating binary sequences for stochastic computing." ELECTRONICS LETTERS-IEE 29, no. 80-80 (1993): 22.

[4] Jeavons, Peter, David A. Cohen, and John Shawe-Taylor. "Generating binary sequences for stochastic computing." IEEE Transactions on Information Theory 40, no. 3 (1994): 716-720.

[5] Alaghi, Armin, and John P. Hayes. "Survey of stochastic computing." ACM Transactions on Embedded computing systems (TECS) 12, no. 2s (2013): 1-19.

[6] Liu, Yin, Megha Parhi, Marc D. Riedel, and Keshab K. Parhi. "Synthesis of correlated bit streams for stochastic computing." In 2016 50th Asilomar Conference on Signals, Systems and Computers, pp. 167-174. IEEE, 2016.

[7] Yuan, Bo, Yanzhi Wang, and Zhongfeng Wang. "Area-efficient scaling-free DFT/FFT design using stochastic computing." IEEE Transactions on Circuits and Systems II: Express Briefs 63, no. 12 (2016): 1131-1135.

[8] Chen, Te-Hsuan, Paishun Ting, and John P. Hayes. "Achieving progressive precision in stochastic computing." In 2017 IEEE Global Conference on Signal and Information Processing (GlobalSIP), pp. 1320-1324. IEEE, 2017.

[9] Najafi, M. Hassan, and David Lilja. "High quality down-sampling for deterministic approaches to stochastic computing." IEEE Transactions on Emerging Topics in Computing (2018).

[10] Zhakatayev, Aidyn, Kyounghoon Kim, Kiyoung Choi, and Jongeun Lee. "An efficient and accurate stochastic number generator using even-distribution coding." IEEE Transactions on Computer-Aided Design of Integrated Circuits and Systems 37, no. 12 (2018): 3056-3066.

[11] Neugebauer, Florian, Ilia Polian, and John P. Hayes. "S-box-based random number generation for stochastic computing." Microprocessors and Microsystems 61 (2018): 316-326.

[12] Luong, Tieu-Khanh, Van-Tinh Nguyen, Anh-Thai Nguyen, and Emanuel Popovici. "Efficient architectures and implementation of arithmetic functions approximation based stochastic computing." In 2019 IEEE 30th International Conference on Application-specific Systems, Architectures and Processors (ASAP), vol. 2160, pp. 281-287. IEEE, 2019.

[13] Krishnapriya, P. N., and B. Bala Tripura Sundari. (2019). High level synthesis for retiming stochastic VLSI signal processing architectures. Procedia computer science 143. 10-19.

[14] Nair, Neethu B., and J. P. Anita (2022). Design of Multistage Counters Using Linear Feedback Shift Register. In Inventive Communication and Computational Technologies, Springer, Singapore 161-173.

[15] Karthick, S., C. Kamalanathan, P. Sunita, S. Ananthakumaran, and E. Prabhu (2021). High speed energy efficient multiplier for signal processing. International Journal of Engineering Systems Modelling and Simulation 12, no. 4: 221-229.

[16] Ramesh, S. R., and R. Jayaparvathy (2019). Artificial neural network model for arrival time computation in gate level circuits. Automatika: ˇcasopiszaautomatiku, mjerenje, elektroniku, raˇcunarstvoikomunikacije 60, no. 3: 360-367.

Smart Framework for Automatic Control of Home Appliances Using IoT

Mallikarjuna Rao Gundavarapu [1], Srineeth Kumar Ineni [2], K. Sathvika [3], Gannavaram Shiva Keshava [4], Uppala Rishik Charan [5]

[1,2,3,4,5] Department of Computer Science and Engineering, Gokaraju Rangaraju Institute of Engineering and Technology, India.

gmallikarjuna628@grietcollege.com[1], srinithkumar35@gmail.com[2], sathvikarao09@gmail.com[3], chandrahaas.54321@gmail.com[4], rishikuppala@gmail.com[5]

Abstract: With the advent of the 21st century, comfort and efficiency started becoming a part of a common man's life. Virtual assistants like Amazon Alexa and mobile applications like Google Assistant have tremendously helped in making Home Automation Systems more effortless and orderly. Efforts are still being made for better intra-connection of our homes. Home automation has created a great impact in the field of the Internet Of Things. In this paper, we have used the pre-existing technology of Alexa Echo to make Home Automation more cost-effective by enabling non-smart devices to react to the virtual assistant. In this framework, we proposed a smart node (relay) that can remotely control the connected electrical gadget using text/voice commands. The status can be monitored and controlled by the Alexa app remotely without the need for physical voice commands to Alexa. Further, room ambience, as well as emotion-based lighting, can be provided.

Keywords: Home Automation, Virtual Assistants, Relay, Alexa Echo

1. Introduction

Home Automation makes it easier and more efficient to use home lighting, heating appliances, and other electronic appliances. It also provides users with increased levels of comfort and security. The resulting outcome is that of lesser energy consumption and a greener environment. [1] It might be as basic as controlling a few lights remotely or automatically, or it can be a comprehensive system that manages all key aspects of your home. Home automation is anything that turns your home into a smart environment by allowing you to control objects around the house automatically.

IoT stands for Internet of Things, which is defined as a network of physical objects that are embedded with circuits, software, network connectivity, electronics, and also sensors in order to connect and transfer data. Over an existing network infrastructure, the Internet of Things helps to sense the objects and operate them remotely, enabling more connections of the physical environment into virtual systems while also improving performance. [2]

To reduce "standby" power draws, simply enable the system to manage your energy consumption in the most effective way - switch lights off when rooms are vacant and lower the heating temperature when no one is home. Perhaps your key concerns are convenience and comfort, so your Smart Home can monitor air quality or operate your HVAC system to guarantee your home has a consistent flow of fresh, filtered air.

A wireless network for home automation is one example of a potential application. Consider a house fully equipped with sensors for controlling the lights with a remote control device as complicated as creating a network of items in your home like security system lighting which can be programmed with the help of a controller.[3]

2. Existing Approaches

2.1. Using E-mail

Interactive Home Automation System through E-mail using Raspberry pi, it helped to construct a home automation system on the Raspberry Pi by scanning the subject of an email and using Python, the Raspberry Pi's default programming language. LEDs represented the switching operations. Hence, this work limitation is known as tariff utilizing Dual Tone Multi-Frequency, and therefore it does not provide security for the system if the email is hacked, which means any unauthorized individual can access it.[4]

2.2. Using Raspberry pi 3

The component for providing the smart features to non-smart houses is a Raspberry Pi 3. It consists of Wi-Fi features, and the relay will activate and deactivate the appliances. [5]

2.3. Using Bluetooth Technology

In Home Automation using Bluetooth, instructions are delivered to the Arduino board's Bluetooth antenna with the help of an established Bluetooth connection from a mobile device; however, it is only practical for short distances. [6]

2.4. Using GSM

In Home Automation by using a Global System for Mobile Communication (GSM), GSM is used to send text messages to manage domestic equipment, such as lighting, climate control, and security systems.[7]

3. Proposed Architecture

3.1. Amazon Alexa

Alexa is an AI (Artificial intelligence) based virtual assistant invented by Amazon. Users can activate the device's skills through the wake word "Alexa" and these skills can be further deployed using specific voice commands. One can gather all the required information about weather, traffic, songs, and news. It can also help integrate and control several smart home devices by using itself as a hub.[8] Alexa can control such devices and these devices can consecutively access different content and services. When triggered, the recording of the voice command is sent to Amazon's cloud where the request is processed, and a response is given accordingly

3.2. Node MCU (ESP 8266)

Node MCU which is called ESP 8266, is used to connect objects and helps to transfer data through Wi-Fi protocol. In this project, Node MCU and Alexa need to be connected to the same Wi-Fi to establish a connection between the two. Node MCU is a micro-controller + Wi-Fi platform designed specifically for the development and execution of Wi-Fi and IoT projects. It consists of a USB port to upload the code, a 3.3V regulator, and a logic level converter circuit so you can upload the program and connect the circuits. It includes the ESP-12E Wi-Fi Module. The Arduino IDE is a software used to create and upload codes to Arduino compatible boards and other boards with third-party core compatibility. The ESP8266 community upgraded the Arduino IDE by adding a few features, that allow users to program the ESP8266 with the Arduino IDE and its programming language. There are numerous additional platforms that can be used to program the ESP8266 NodeMCU. [9]

An IoT application inherently requires the sensors to be interfaced with WiFi and remote communication also needs to be established through WiFi. Node MCU contains a built-in ESP-12E Wi-Fi Module and hence it does not require any external WiFi component as needed for arduino. Since WiFi is built in, the power consumption speed of operation and programming becomes much simpler with a node MCU.

3.3. Relay Board

A Relay board is an electromagnetically gadget switch that controls voltage supply and allows you to turn on/turn off circuits using a control switch/voltage barrier. The relay module has connectors with three sockets each - Common, Normally Closed(NC) and Normally Open(NO). The relay coil is directly energised and triggers NC and NO. The COM connects to the main terminal that is required to be controlled. The NC pin is connected to the COM pin. The secondary circuit is normally complete so by default, the current is flowing. The circuit is opened when a signal is sent from the ESP 8266 and disconnects the load. The NO pin is not connected to the COM pin so the circuit is complete only when a signal is sent to close it.

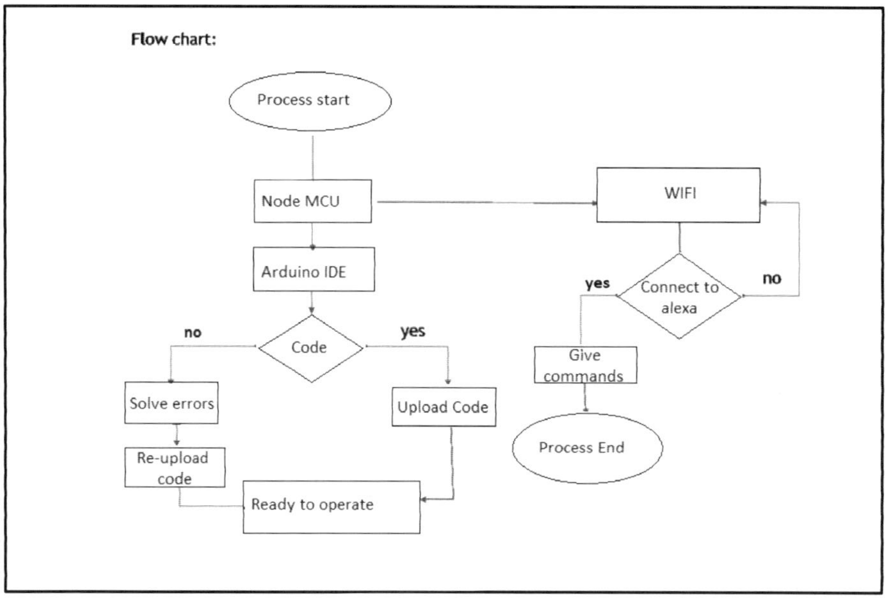

Fig 1. The architectural flow of the Framework

Electromagnetic relays are extremely responsive, minimizing needless delays. Relays are more efficient and depending on the number of devices to be controlled, multi-channel relays can be employed. [10]

4. Implementation

Before coding the program, it is necessary to install certain libraries like espalexa and select the required board from the board manager of IDE. The selected board for this project is NODE MCU 1.0(ESP-12E module). In the framework, all the devices used need to be registered with appropriate names and connected to NODE MCU digital pins (through which communication has to be established to the relay board). Alexa is connected to Node MCU with appropriate Wi-Fi credentials which need to be mentioned in the code. To trigger each individual device, a call-back function needs to be executed for each device. The call-back function code decides whether to switch on or switch off the device based on the input. To safeguard the devices appropriate relays, have to be used.

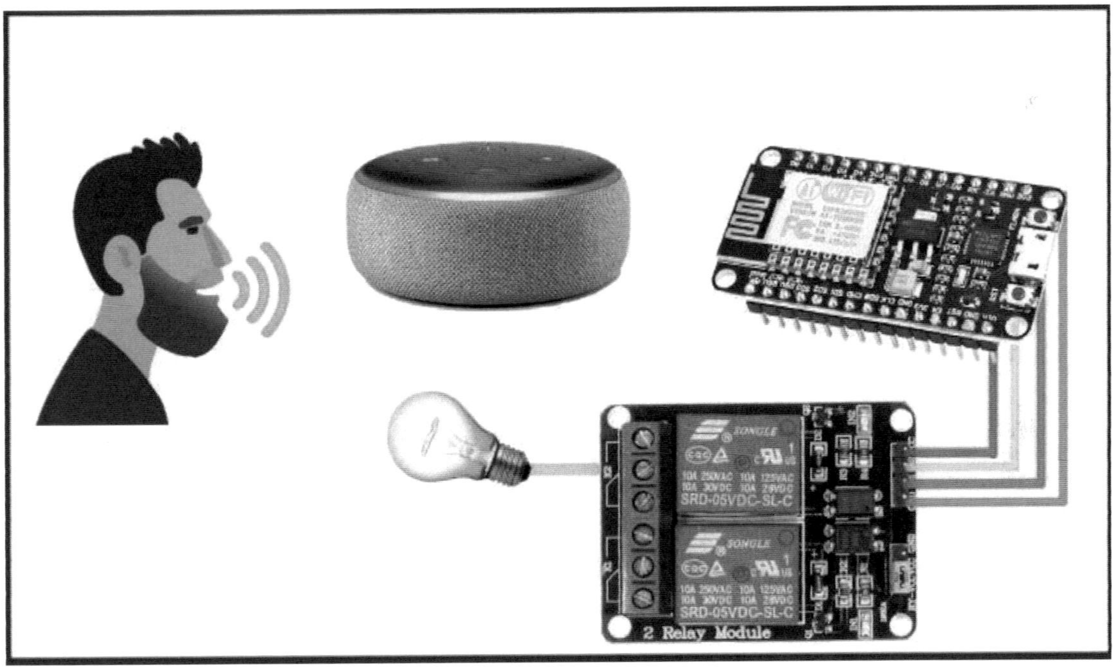

Fig. 2. Implementation

After deploying the code to Node MCU, the devices are required to be registered in the Alexa app to establish a connection between Alexa and Node MCU. In order to set up this connection, both the components need to be connected to the same network. After the connection is established, based on the commands, the Alexa will trigger the particular device through Node MCU with the help of a digital pin allotted to the particular device. Through digital pins, the input is sent to the relay board. According to the input, the relay board will switch the voltage barrier and turns on or off the device.

4.1. Technological stack

4.1.1. Physical design of IoT:

We have used the level-2 design of IoT where the single local node is responsible for sensing and controlling devices and the cloud is used to hold the database. We can monitor the power usage of connected appliances using the cloud with the remote app as shown in (Fig. 3).

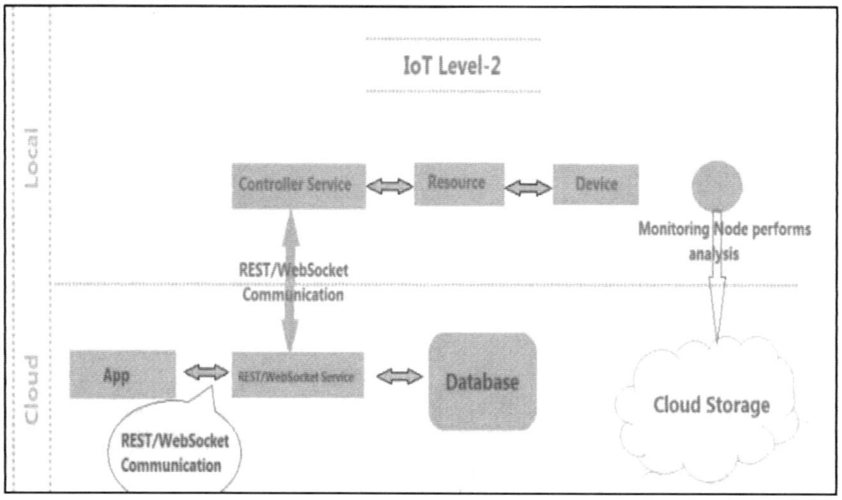

Fig 3. Level 2 Physical Design of IoT

4.1.2. Application Level

ASK stands for Alexa Skill kit, which is a simple Software Development Kit. ASK is used in this project to develop Amazon Echo skills. [11]

4.1.3. Programming Level

All the applications' source codes are programmed with C++, and the Arduino IDE was used to write and compile the code. It is a platform that is free to use. The first Arduino board was released in 2005, and there is now a range of options [12]. It is an Arduino-developed official program that is mostly used for editing, developing, and uploading code to IoT devices. The basic program in IDE is known as a sketch. It provides a HEX file as an output, and the output is uploaded into the board controller with the help of a USB.

5. Benefits

5.1. Energy Consumption

Home automation solutions reduce energy usage and thereby, reduce our carbon footprint too. It enables us to remotely control devices in order to minimise energy consumption by using the Alexa app on our phones. The system proposed by R S Hariharan et al [13] displays the energy consumed and generates an expected monthly bill through a mobile app. This is beneficial in controlling energy consumption.

5.2. Security

Surveillance cameras can be installed in various spots which can record footage, detect any movements and accordingly alert the owner. In the event of an individual being inaccessible to the location or forgetting to lock the entryways, this process can also be automated to increase security levels. Eric Fry and Shawn Johnson [14] used buzzers and motion sensors to implement the same which can be the future scope for this project.

5.3. Cost-effective

As energy consumption is decreased, the overall cost to pay for electricity also decreases. The prototype can also be built by individuals and the price of the components decreases when bought in bulk.

6. Pseudocode

1. *Define WiFi modules and output pins.*
2. *Declare the required functions.*
3. *Initialize the WiFi credentials SSID and password, and device names.*
4. *Set the flag as false to check for the initial WiFi connection.*
5. *Invoke the serial monitor.*
6. *Set the pin modes as output.*
7. *Invoke the connect WiFi function.*
8. *If WiFi is connected*

- *Add the devices to the esp Alexa module, with the parameters as the device name and function call to change the default state of the device.*

9. *If WiFi is not connected*

- *Print the message as "Cannot connect to WiFi. Please check data and reset the ESP."*

Function firstLightChanged (argument 1)
{
If the value of the argument 1 is a non-zero number
- *If the value of argument 1 is equal to 255*
  - *Set the first output pin to HIGH and print the message as "Device1 ON".*

Else if the value of argument 1 is zero
- *Set the first output pin to LOW and print the message as "Device1 OFF".*

}
End function

Function connectWiFi ( )
{
1. Initialize a flag, state to true and a variable i to 0.

Journal of Physics: Conference Series **2325** (2022) 012033 doi:10.1088/1742-6596/2325/1/012033

2. Set the WiFi mode to WiFi_STA.
3. Initialize the WiFi by passing the SSID and password.
4. Print "Connecting the WiFi" on the serial monitor
5. While status is not equal to WL_CONNECTED

- With a delay of 500 milliseconds, print "." to the serial monitor and increment the value of i.
- If the value of i is greater than 20
    - Set the state flag to false and break out of the loop.

6. If the state is true

- Print the SSID and IP address to the serial monitor

7. If the state is false

- Print "Connection failed" to the serial monitor.

8. Return state.
}
End function

The header file 'WiFi.h' is used to define the WiFi modules and create the necessary environment for WiFi interaction. WiFi.h also contains the function which enables the developer to authenticate ssid and password. On the occasion that user details match the aforementioned attributes, it will allow the user to access the cloud environment using WiFi.After connecting to the WiFi, we give voice commands to alexa, to switch ON or OFF any given device. From which the brightness is set to value based on the command (ON:255, OFF:0) and then call back function of the corresponding device is invoked by "Espalexa" module.

7. Screenshots of Experimentation Results

*7.1. Command: **Alexa, turn on light one.***

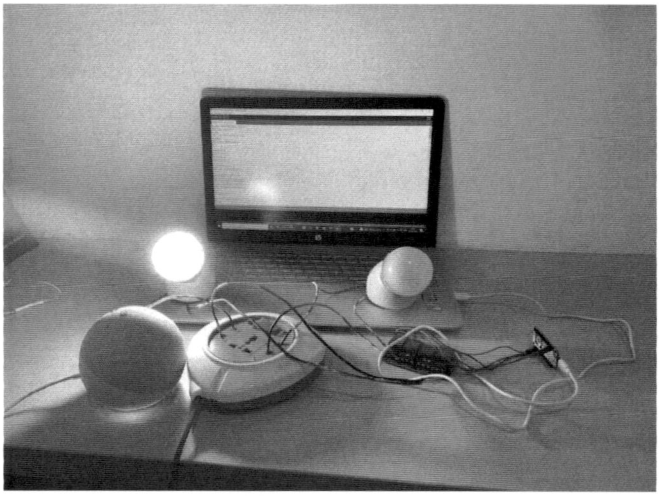

Fig4. Hardware Integration and controlling Left Light using Alexa

When Alexa Echo receives the command "Switch on the bulb 1", that particular digital pin in node MCU will send a high signal to the relay board. This turns on the bulb.
Similarly, a low signal is sent to turn off the bulb when the command "Switch off bulb 1" is given.

7.2. *Command:* ***Alexa, turn on light two.***

Fig5. Hard ware Integration and controlling Right Light using Alexa

When Alexa Echo receives the command "Switch on the bulb 2", that particular digital pin in node MCU will send a high signal to the relay board. This turns on the bulb.
Similarly, a low signal is sent to turn off the bulb when the command "Switch off bulb 2" is given.

7.3. *Command:* ***Alexa, turn on all lights.***

Fig6. Hardware Integration and controlling both lights using Alexa

When Alexa Echo receives the command "Switch on all bulbs", that particular digital pin in node MCU will send a high signal to the relay board. This turns on the bulb.
Similarly, a low signal is sent to turn off the bulb when the command "Switch off all bulbs" is given.

8. Conclusion and Future Scope Development

The 21st-century world seeks comfort and security at a more effective and personal level, hence the scope and market for home automation systems are greater than ever. But it is still a privilege enjoyed by few due to the presence of various products with exorbitant valuation. With our product, we aimed at bringing service into every home at a reasonable and affordable price that is compatible with any other appliance.

Presently, we have given a cost-effective solution by converting non-smart bulbs into functioning smart bulbs while incorporating a simple command given to a virtual assistant. We believe that this can be further developed. On a small scale, we can delve into regulating the speed of fans, dim the lights based on the external environment, and regulate the temperature of Air Conditioners.

The explosive boom of the production and deployment of smarter systems has raised concerns for privacy of the users. James L [15] presented a study to evaluate users' behavior when privacy is embedded into the home automation system. We wish to implement a middleware layer to provide protection and privacy for our users.

References

[1] Hamdan, Yasir Babiker. "Smart Home Environment Future Challenges and Issues-A Survey." *Journal of Electronics* 3, no. 01 (2021): 239-246 **3** 1

[2] Bhat O, Bhat S and Gokhale P 2018 Introduction to IoT *International Advanced Research Journal in Science, Engineering and Technology* **5** 41

[3] Shimi S. L, and Kumar M. 2015 Voice Recognition Based Home Automation System for Paralyzed People System *IOSR Journal of Computer Engineering* **4** 1

[4] Goyal L, Vaibhav A and Jain S 2014 Raspberry Pi based Interactive Home Automation System through E-mail *International Conference on Reliability, Optimization, and Information Technology* **1** 278-280

[5] T.S K and K M 2018 Voice Based Home Automation using Amazon Dot *Technical Research Organisation India* **4** 50

[6] Tazil M and Piyare R 2011 Bluetooth based home automation system using cell phone *IEEE 15th International Symposium on Consumer Electronics* **5** 28

[7] Ahmed S, Chan K, Teymourzadeh R and Hoong M 2013 Smart GSM Based Home Automation System *IEEE Conference on Systems, Process & Control* **1** 306-308

[8] Kaundinya A, Atreyas N, Srinivas S, Kehav V and Kumar N 2017 Voice Enabled Home Automation Using Amazon Echo *International Research Journal of Engineering and Technology (IRJET)* **4** 682-683

[9] Parihar Y 2019 Internet of Things and NodeMCU *Journal of Emerging Technologies and Innovative Research* **6** *1087*

[10] Kumar A, Senthilnathan N, Rishikeshanan R, Muthusamy S, Somesh S, and Bala, V. 2020 Real-time Implementation of Home appliance control using ALEXA *IOP Conference Series: Materials Science and Engineering* **937** 5

[11] Matriaya S and Purwar R 2018 Voice Controlled Home Automation Using Amazon Echo Dot *IOSR Journal of Computer Engineering* **20** 17

[12] Melin P, Amestica O, Lagos G and Duran-Faundez C 2019 An Experimental Comparison of Arduino IDE Compatible Platforms for Digital Control and Data Acquisition Applications *IEEE CHILEAN Conference on Electrical, Electronics Engineering, Information and Communication Technologies* **10** 7575

[13] R S Hariharan, Reema Agarwal, Madhurya Kandamuru and H Abdul Gaffar 2021 *IOP Conf. Ser.: Mater. Sci. Eng.* 1085 012026 9

[14] Fry E and Johnson S Embedded Home Security System *Computer Science, United States: College of Natural Sciences Colorado State University-Fort Collins* 2-3

[15] James L 2020 Smart Privacy for IoT: Privacy Embedded Design for Home Automation Systems *College of Computing and Engineering Nova Southeastern University* 1

Design and implementation of power efficient clock gated dual-port SRAM

Mohammad Waqar Bhat[1], Kaartik R [2], Dr. Sowmya K B[3]

[1] Department of ECE, RV College of Engineering, Bengaluru, India.

[1,2,3] {mohammadwb.lvs21,kaartikr.lvs21,sowmyakb }@rvce.edu.in

Abstract. Synchronous clock gated Dual-port RAM has been designed in this paper. To increase the design's power, a negative latch-based clock gating approach was used. On Xilinx Vivado ML Edition, the design was implemented on an XC7Z010ICLG225-1L device with a -1-speed grade and the Zynq-7000 FPGA family. The Verilog HDL was used to create the design. A 2-bit AND gate is used to realize clock gating which takes the clock and enable signal as input thus clock become only high when enable is high which reduces the switching frequency and thus power consumption by the clock is reduced as power is directly proportional to $(\text{frequency})^2$. Apart from clock gating, the SRAM is Dual-Port based, allowing multiple reads or writes to occur simultaneously, or at approximately the same time, unlike single-port RAM, which permits only one access at a given time. The power consumption value of the clock signal with clock gating at 100 GHz is 627.61 mW as compared to power consumption without clock gating which is 1405 mW, at a clock frequency of 100 GHz, a 55 percent reduction in total clock power was achieved. Xilinx Vivado ML Edition's XPower Analyzer tool was used to calculate the device's power.

Keywords: Clock Gating, Dual Port Ram, XPower Analyzer.

1. Introduction

Random Access Memory (RAM) is a type of semiconductor peripheral device (memory element) used in VLSI circuits to store data or communicate with other peripheral devices. There are two forms of RAM: static RAM (SRAM) and dynamic RAM (DRAM) (DRAM). Each bit is saved in static random-access memory (SRAM) utilising a bistable latching circuit with both logic high and logic low stable states. SRAM is used for cache and internal registers in CPUs. SRAM requires a constant power supply for data storage and transfer as they are volatile, i.e., data is permanently lost when power is disconnected.

In contrast to single-port RAM, which enables only one access at a time, dual-ported SRAM allows several reads or writes to occur at the same time, or almost at the same time. Dual-port SRAMs allow two CPUs or peripheral devices to access the same memory in the same device at the same time. As a result, two peripheral devices can communicate with each other via memory. Finally, it improves the memory's efficiency because two peripheral devices are acquiring data from it at the same time.

Clock signals are one of the vital power contributors in VLSI design and are almost responsible for 30%- 70% of the entire dynamic power of the peripheral device [1]. Several methods have been invented to regulate the dynamic power of a peripheral device; clock gating is notable among those. In the primary clock gating approach, the input clock signals of the device are AND 'ed or ORed with designer input enable (EN) signals. Thus, gating of the clock can actually save power in sequential circuits. It simply disables the clock signal where it consumes unnecessary power [2], [3]. In the field of clock gating, the Negative latch-based strategy has also been proven to be the best [8]. NOR-based, positive latch-based, and AND-based clock gating approaches were investigated. The AND-based clock gating approach was developed on RAM in 2013 [10]. The paper's main contribution is that we used a negative latch-based clock gating technique to improve the power efficiency of our 8-bit dual-port SRAM. The layout is well-organized to handle write-read disagreements. The concept uses a negative latch-based clock to reduce power consumption. We're also working on putting this design into action on the reconfigurable system, as seen in [4-6] [9] [11,12].

It is equipped with two central processors (CPU). Each CPU has its own set of signals, including read, write, enable, semaphore, address, and data lines. The Peripheral Device has two ports, allowing several peripheral devices to connect with one another without the use of any other special communication routes between two peripherals.

On clock edges, both peripheral devices are gaining memory. The clock is generated by the Gated Clock, and the usage of clock gating at the clock has enhanced the circuit's power. The enable signal is used to activate the dual-port RAM clock. Figure 4 shows a negative clock-gated SRAM circuit. The RTL perspective of a negative edge-triggered dual-port SRAM cell is shown in Figure 3. The RTL view of the architecture makes the circuit relationships more obvious. When left enable is set to high, it will send the address bits it wants to access, as well as a read and write signal that indicates whether the peripheral device wants to write or read on that address. It will try to read the memory if it wants to it will give the address bits only. It will transmit the address bits alongside the data bits if it wishes to write. The read write actions will cause the control logic to set the semaphore of this peripheral device. Peripheral device B, on the other hand, can access the memory space at the same time by providing the control block with the necessary inputs. No inputs will reach the control block if enable is flipped logic low, or the peripheral device will be reviewed in an OFF state.

2. Literature Review

Authors describes the implementation of different clock gating strategies for different sequential circuits in Paper [1] Benini, L.; Siegel, P.; De Micheli, G., "Saving power by synthesising gated clocks for sequential circuits, "Design & Test of Computers, IEEE, vol.11, no.4, pp.32,41, Winter 1994. Clock gating (CG) and Power gating (PG) are the most widely used low-power design concepts for reducing dynamic and static power, respectively. Aside from the significant energy savings they give, their popularity stems from their high compatibility with common CMOS technologies and scaling, flexibility for automation applications, and ease of interface with commercial design frameworks provided by major EDA suppliers. Unfortunately, their installation is not free, and scheduling and area issues must be carefully examined. Furthermore, both require dedicated power-managed devices that are responsible for detecting idle conditions and generating the appropriate control signals. Such control units, which are frequently built as independent yet different entities, could be a major source of space and timing overheads, complicating the design verification step. As a result, it's easy to see how, despite being highly desirable in terms of energy savings, combining clock-gating and power-gating under a single control domain could be a good solution for lowering implementation costs.

In Paper [2] {V. G. Oklobdzija, Digital System Clocking—High-Performance and Low-Power Aspects. New York, NY, USA: Wiley, 2003.} It is observed the differences in the modelling approach can affect the design process and explore some of the precautions that designers can take to minimise increased power dissipation and potential difficulties during the layout phase and final timing closure.

3. Techniques and Methodology

Clock signals are one of the most significant power contributors in any VLSI design, accounting for almost 70% to 30% of the device's total dynamic power [1]. Several ways for controlling the dynamic power of a peripheral device have been developed, with gating at the clock signal being one of them. The device's input clock signals are AND ‘ed / OR ‘ed with the designer input enable signal in a basic clock gating mechanism. In sequential circuits, clock gating can therefore conserve power. It simply suppresses the clock signal in areas where it is wasting energy [2], [3]. In the field of clock gating approaches, the negative latch-based strategy has been demonstrated to be one of the best [8]. The technique was compared to clock gating strategies based on NOR, AND, and positive latch. The AND-based clock gating approach was developed on RAM in 2013 [10].

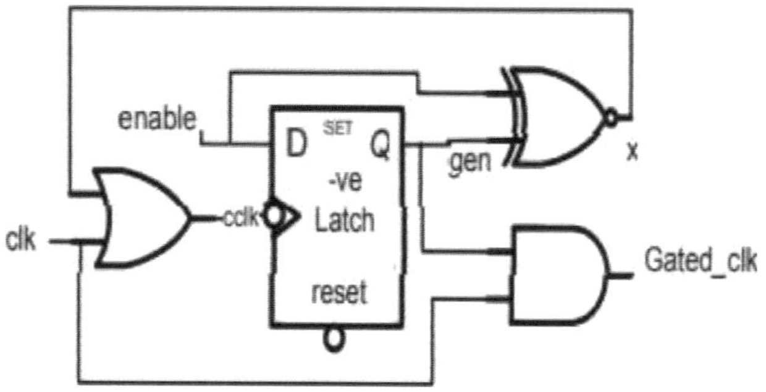

Figure 1. Negative latch-based Clock Gated Circuit.

3.1 CLOCK GATING TECHNIQUE USING NEGATIVE LATCH

The clock gating circuitry in Figure 1 is based on a negative -latch. An Enable input is utilised here to command the clock of the controlling peripheral device and assign it to the Gated CLK output, which is then used as the clock of the target peripheral device. Gen is initially set to '0,' Enable to '1,' and x to '0.' As a result, when this x is OR ‘ed with a clock, a clock signal is generated for the controlling peripheral device, which is Latch. When the clock is LOW, the latch is driven, and the output Q is received in the next clock cycle. The Gated CLK is derived by ANDing the global clock and Gen, which is logic LOW in this case. As a result, the latch is held in place and no switching actions are performed. Gen now changes to logic HIGH on the following clock cycle, which is estimated by a clock AND. As a result, it generates a clock pulse that activates the targeted peripheral device. OR gate will always be high because Gen is logic high and Enable is also logic high. As a result, the latch will remain in its current state and can only be toggled when Enable changes. As a result, when Enable is set to logic low, Gated CLK is set to logic low as well. That is how the target gadget will carry out its tasks.

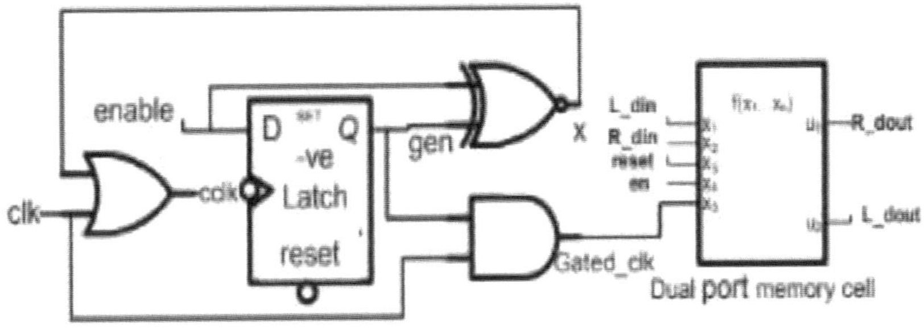

Figure 2. Circuit diagram negative latch-based Dual-port SRAM.

The Clock Gating circuit gives this clock as output, the power of our circuit is optimised by using clock gating circuitry at the clock input. The clock signal for the dual-port RAM circuit is enabled using the enable signal. The circuit of negative clock-based dual-port SRAM is exhibited in figure 2. The RTL diagram of the negative latch-based dual-port SRAM cell is displayed in figure 3. The RTL schematic of the design the circuit connections are more understandable.

When the left enable is turned high, it will send address bits together with the read-write signal, indicating whether the device wishes to write or read on that specific address. If it wants to read, it will merely give the address. If it wishes to write, it will provide the address as well as the data. The semaphore of this device is set accordingly using the control block. Device B can access the memory locations, at the same time, by giving required inputs to the control logic. No inputs will reach the control logic, when the enable is low or the device will be evaluated to be in OFF state.

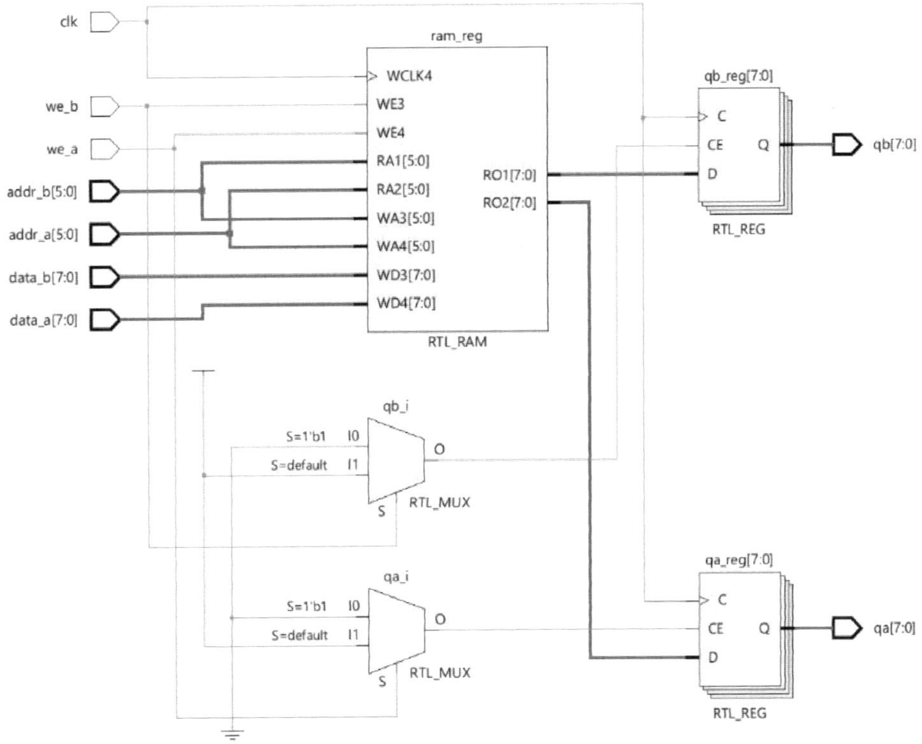

FIGURE 3. Clock gated 8-bit dual-port SRAM RTL schematic

4. SIMULATION RESULTS AND DISCUSSION

This section defines the behavior of the proposed clock gated 10-bit dual-port SRAM circuit using the Xilinx XPower Tool. Results of the Proposed clock gated 10-bit dual-port SRAM are compared with results of 10-bit dual-port SRAM without clock gating. Both the dual-port SRAMs are simulated in the Xilinx Vivado simulator tool, using the XC7Z010ICLG225-1L device family.

Verilog HDL is used to synthesize the design of the circuit. XPower Analyzer tool of XILINX VIVADO calculates the power consumption of the peripheral devices. The two designs' power has been tested at 1 THz, 1GHz, 100 GHz, and 1000 GHz frequencies, with the results compared. Without using the clock gating technique, Table 1 shows the clock, logic, BRAM, signal, dynamic, static, and total power levels of the 10-bit dual port.

The clock, logic, BRAM, signal, dynamic, static, and total power values of the 10-bit dual port with the clock gating technique are shown in Table 2. At 1 THz, 1GHz, 100 GHz, and 1000 GHz frequencies, the various power values were examined.

TABLE 1. 8-BIT dual port sram without clock gating on-chip power consumption values

POWER ON-CHIP	8-bit Dual Port SRAM Without Clock Gated			
FREQUENCY	1 THz	1 GHz	10 GHz	100 GHz
CLOCK	14050	14.05	140.5	1405
LOGIC	0.25	0.01	0.01	0.07
IOs	2875.78	3.34	29.22	288
BLOCK RAM	17978.8	18	179.81	1797.9
STATIC	433.88	45.33	45.96	53.31
DYNAMIC	34905.43	35.44	349.6	3491.12
SIGNAL	0.59	0.05	0.06	0.16
HIERARCHY	0.97	0.97	0.97	0.97
TOTAL	35339.31	80.77	395.56	3544.43

TABLE 2. 8-Bit dual port sram with clock gating technique on-chip power consumption values

POWER ON-CHIP	8-bit Dual Port SRAM with Clock Gated			
FREQUENCY	1 THz	1 GHz	10 GHz	100 GHz
CLOCK	6276.01	6.28	62.77	627.61
LOGIC	58.03	0.41	1.37	6.73
IOs	11.42	11.42	11.42	11.42
BLOCK RAM	3316.61	3.32	33.17	331.66
STATIC	75.29	45.3	45.48	47.28
DYNAMIC	9663.06	21.76	109.28	978.41
SIGNAL	0.99	0.34	0.55	0.99
HIERARCHY	0.65	0.65	0.65	0.65
TOTAL	9738.35	67.07	1025.68	1025.68

Table 2 shows the clock powers of the two architectures at frequencies of 1 THz, 1GHz, 100 GHz, and 1000 GHz, respectively. At 1 THz, 1GHz, 100 GHz, and 1000 GHz frequencies independently, a 55 percent reduction in total clock power was found. The power of the clock is found to be considerably lowered when the clock gating approach is applied to the device.

The addition of clock gating circuitry increased the logic and signal power in the design, but this increase was found to be insignificant.

Table 2 shows the power consumption of the two designs by input and output power. Due to the installation of more circuitry, the power increases by 70.75 percent at 1 GHz, but drops by 96.03 percent, 99.60 percent, and 60.92 percent at 100 GHz, 1 THz, and 10 GHz, respectively.

At 1 THz, 1GHz, 100 GHz, and 1000 GHz frequencies, overall BRAM power was reduced by around 81.55 percent. The power of the hierarchy has decreased by 32.99 percent.

The static power of the two peripheral devices nearly remains identical up to 100 GHz frequency but has been found to decrease considerably at 1 THz frequency. At 1 THz, 1GHz, 100 GHz and 1000 GHz frequencies, 82.65%, 0.07%, 1.04% and 11.31% decrease in static power has been observed respectively. An considerable reduction in dynamic power has been observed. The dynamic power decreases by 72.32%, 38.60%, 68.74%, and 71.97% at 1 THz, 1GHz, 100 GHz, and 1000 GHz frequencies respectively. This reduces the supply power by 72.44%, 16.96%, 60.88%, and 71.06% respectively at 1 THz, 1GHz, 100 GHz, and 1000 GHz frequencies.

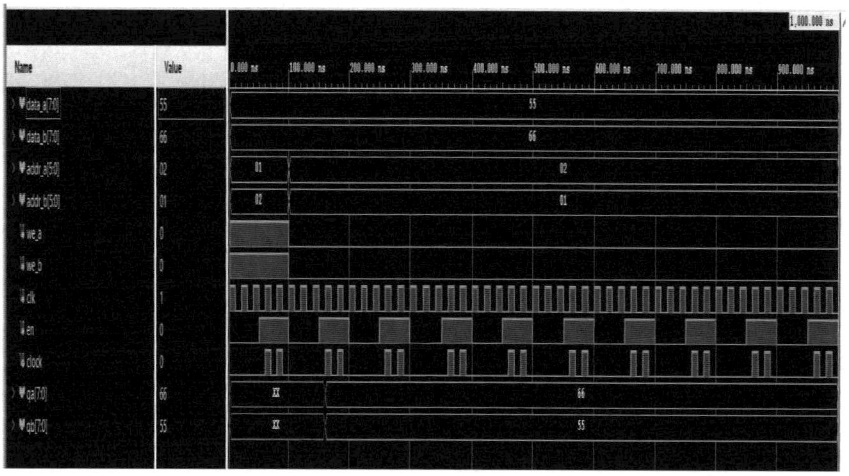

FIGURE 4. Simulation window of 8-bit dual-port SRAM

The clock gated 8-bit dual-port SRAM simulation waveform created by Vivado is shown in Figure 4.

5. Conclusion

Clock Gated Dual Port SRAM has been successfully designed, simulated, and optimised. The design was created using Verilog HDL. The Total Power Consumption of Clock Gated Dual Port RAM is reduced by 16.96 percent (on 1GHz), 60.87 percent (on 10GHz), 71.062 percent (on 100GHz), 72.44 percent (on 100GHz) as compared to the clock power consumption of dual-port SRAM without employing negative latch-based clock gating Techniques (on 1THz). At 1GHz, 1THz 100 GHz, and 1000 GHz frequencies, total supply power was reduced by 16.96%, 72.44 percent, 60.88 percent, and 71.06 percent, respectively. Cadence design tools at the transistor level can lower the proposed design's power even more.

6. References

[1] Benini, L.; Siegel, P.; De Micheli, G., "Saving power by synthesizing gated clocks for sequential circuits," Design & Test of Computers, IEEE , vol.11, no.4, pp.32,41, Winter 1994.

[2] V. G. Oklobdzija, Digital System Clocking—High-Performance and Low-Power Aspects. New York, NY, USA: Wiley, 2003.

[3] M. S. Hosny and W. Yuejian, "Low power clocking strategies in deep submicron technologies," in Proc. IEEE Intll. Conf. Integr. Circuit Design Technol., Jun. 2008, pp. 143–146.

[4] Ashutosh Gupta and Kota Solomon Raju, "Design and Implementation of 32-bit Controller for Interactive Interfacing with Reconfigurable Computing Systems" International Journal of Computer Science and Information Technology (IJCSIT), Vol.1, No.2, pp 80-87, Nov 2009. ISSN: 0975-3826(online); 0975-4660.

[5] Gupta, A., Duhan, M., & Raju Kota, S. (2009). HDL Implementation of Sine-Cosine Function Using CORDIC Algorithm in 32-Bit Floating Point Format. The Icfai University Journal of Science & Technology, 5(2), 40-48.

[6] Sharma P and Gupta A. (2009), "Design, Implementation and Optimization of Highly Efficient UART" , The IUP Journal of Science and technology, Vol: 5, No. 4, pp. 21-30.

[7] Sterpone, L.; Carro, L.; Matos, D.; Wong, S.; Fakhar, F., "A new reconfigurable clock-gating technique for low power SRAM-based FPGAs," Design, Automation & Test in Europe Conference & Exhibition (DATE), 2011 , vol., no., pp.1,6, 14-18 March 2011.

[8] Jagrit Kathuria, M. Ayoubkhan, Arti Noor, MIT International Journal of Electronics and Communication Engineering, "A Review Of clock Gating Technique", MIT Publications, ISSN 2230-7672, Vol 1, No.2, Aug 2011.

[9] Agarwal, C.; Gupta, A.: 'Modeling, Simulation based DC Motor Speed Control by Implementing PID Controller on FPGA', IET Conference Proceedings, 2013, p. 9.07-9.07, DOI: 10.1049/cp.2013.2358 IET Digital Library.

[10] Pandey, B.; Singh, D.; Baghel, D.; Yadav, J.; Pattanaik, M., "Clock Gated Low Power Memory Implementation on Virtex-6 FPGA," Computational Intelligence and Communication Networks (CICN), 2013 5th International Conference on , vol., no., pp.409,412, 27-29 Sept. 2013

Moving Object Detection and Speed Estimation by Digital Image Processing

K N V Khasim[1], Anilkumar G[2], Vamshi G[2], Sai kumar Ch[2], Nikhil Yadav M[2]

[1] Professor, ECE, Gokaraju Rangaraju Institute of Engineering and Technology, Bachupally, Hyderabad, India
[2,] UG Students, ECE, Gokaraju Rangaraju Institute of Engineering and Technology, Bachupally, Hyderabad, India

E-mail: khasim.knv@griet.ac.in, gugulothuanilkumar09@gmail.com, vamshi1628@gmail.com, saitejajake@gmail.com, nikhilyadavm092001@gmail.com

Abstract. Intelligent network systems and security surveillance are used for smart city development. Manual and heritage technology is used in surveillance systems to detect speeds higher than usual in developing countries. This process can be automated with ease by leveraging the potential of image processing. Moving object detection is one of the essential tasks in image processing because of its prominent role in many real-world applications. Vehicle speed detection can be achieved by employing image and video processing methods. In this work, a unique approach is proposed for moving object detection and speed estimation based on an integrated approach in video sequence frames as opposed to the most commonly used RADAR and LIDAR devices for traffic law enforcement. Video data is collected and analyzed for speed in real-time without any sensor calibrations,thereby it removes any external hardware dependency requirements. Moving vehicles are segmented out by using frame-subtraction and masking techniques. The proposed algorithm tracks the time taken by the car to cover a predetermined fixed distance in order to calculate its speed.

Keywords: Image processing, object detection, OpenCV, Python, Image Subtraction, Thresholding, Masking

1. Introduction

Owing to the more traffic accident rate in many countries, speed regulations have become the need of the hour. Traffic management and enhancing road safety, can be one course of action to avoid road accidents. On the other hand, another approach to prevent road accidents are to prevent civilians from speeding by penalizing such people in the form of fines. In order to identify such cases, the traffic department makes use of hardware-based speed detection systems which require the physical presence of the traffic officials at the time of over speeding. While effective, such methods are limited by the fact that there are numerous cases which often go undetected.

Some researchers are developing sensor based technology in networks that provide consumers with traffic information, thus reducing the number of traffic violations. Many researchers often use global positioning systems in combination with sensors that operate on RADAR or even digital cameras to determine vehicle speed and volume of traffic flow on the roads to monitor over speeding. Significant resources have been spent on the calibration of Large-Scale Traffic Control Systems operation, which is both costly and time consuming in terms of monetary value, time and energy.

Intelligent speed adaptation providing efficient monitoring for its operator, registration and recording of the vehicle speeds that exceed the speed limit, is therefore an essential priority. In this work we suggest an automated, lightweight speed detection system for video surveillance using image processing that can be used by law enforcement.

2. Related work

One of the most fundamental and critical tasks in computer vision is to detect moving objects along with their video sequence movement. This provides the basis for a variety of automated applications in a range of domains, including monitoring, increased reality and movement capture. Object monitoring is a key component of the IVS or Intelligent Video Surveillance program and can further be replicated for other suspicious activity detection systems. There are ample methods and algorithms for object tracking, which are the basis for implementing an IVS with a mean shift algorithm.

Nearly every vehicle detection system comprises two basic stages: [1] a generation of hypotheses, hypothesizing the position in the picture of potential vehicles and hypothesis verification (HV), verification of the assumptions [2]. The aforementioned paper presents a motion-based approach for generating hypotheses. Motion-based methods measure the presence of the vehicle using the relative movement between the sensor and the scene obtained through optical flow measurements [3]. It also works on designing and implementing highly adaptive algorithms and systems based on field-specific road, vehicle, and control knowledge in real time [4]. Automated tracking and motion detection is a challenging task in traffic monitoring. Cheng et al. proposed a digital image correlation-based, high precision vehicle tracking algorithm [5]. Experiments carried out by Cheng et al. have shown that the approach works well in long-range monitoring. With the scaling value of the tracking device, the high precision relative speed can be determined.

For some computer vision implementations, object recognition is a crucial step. These included three dimensional reconstruction, compression, medical imaging, augmented reality, image reconstruction and monitoring [6].In any traffic management system, moving object detection is the first and foremost step. Pawar et al . explained a variety of target detection techniques [7].A large variety of models and theories of traffic flow have been developed over the past 50 years [8].

Rahmaniar et al. introduced a new and efficient technique for the detection of moving objects in the environment in real time by unmanned aerial vehicles. In order to estimate the camera movement and hence to stabilize the image sequence, the first two successive frames are found. Then, the moving object candidate is detected in regions of interest of the objects. This is the image in the front. In addition, moving objects, based on the motion vectors, in the foreground and the background are classified as the dynamic and static objects. These techniques have surpassed many state of the art techniques, as this technique achieves the speed of measurement as 30 frames per second and also maintains a high precision of 95% [10]. The standard or current system for monitoring or calculating vehicle speed is through the use of RADAR instruments and other variants such as the continuous G-factor, series and acceleration median. The constraints of this device or models are that it is not as precise and is expensive. Furthermore, it does not capture images over long distances. For speed estimation, various techniques have been used. A combination of Gaussian techniques for speed estimations is proposed by Dhulavvagol et al. [11].

In order to detect high speed vehicles in the Vehicle Ad Hoc Network, Nayak et al. propose a Position Based Algorithm for High Speed Car Detection (PHVA), using a vehicle cloud server, used as computing as a service [12]

A crucial step for traffic safety applications is the automated regulation of vehicle speed by sensing speed sign labels from speed boards on the roadside. In addition, the driver will be notified of the changes in the speed limit with an alarm message [13]. Thus, the driver is informed. In a situation where, even after the warning sign the driver does not lower the limit, the authorities are informed. Automatic control of the speed of the vehicle by detecting the speed signs labels from speed sign boards, which are laid on the road side is a critical step for traffic surveillance applications. Furthermore, necessary steps must be taken to alert the driver by sending a caution notification [13]. When the driver does not lower the speed even after the sign of caution, details of the vehicle are notified to the authorities.

3. Proposed Method

The proposed work is structured as a three-tier application. This work is tailored to work for traffic video data, consequently, the input for the scope of this work is restricted to traffic surveillance videos.During the process of streaming the input video, each stream is captured using cv2.VideoCapture() method of OpenCV and each frame is treated as an individual image on which the tracking algorithm is applied.

The Image Detection and Segmentation follows the Image Acquisition module. This module consists of extracting the different ROI's, elimination of any background if present and pre-processing steps.

The processing steps are as follows:

1. Convert the color video frame to grayscale. This is done because images in RGB format increase complexity of the algorithm as the additional information provided by the three-color channels are not of any use to the algorithm. It also reduces the amount of noise in the image frame.

2. Gaussian blurring is performed to remove high-frequency noise which aids in contour detection which is performed in a later stage.

3. Apply fastNIMeansDenoising() method provided by OpenCV to remove remnant noise which uses an underlying Non-Local Means denoising algorithm.

After the preprocessing steps have been applied, Segmentation is carried out. Segmentation is the process of partitioning an image into chunks of pixels called image objects, thereby simplifying the process of analysis. The classifier is then used to carry out the object detection in each frame.

3.1. Image Detection and Segmentation

The first step in the algorithm is to capture each frame from the input surveillance video. After the video frame has been captured, we first apply the above preprocessing steps to reduce the background noise and any noise which might be incurred during image processing stages. We treat each detected object as a different ROI in each frame, on which we operate independently.

There are 3 basic steps which are used :

1. Edge Features are used to detect edges
2. Line Features are used to detect lines
3. Four Rectangle Features detect slanted lines

One or more objects of interest are detected, we draw a bounding box around it to signify a moving object in the frame. The frame is then written to the output video, allowing us to visualize the detected object. Since this procedure is done on a frame by frame basis, the detection of vehicles appears in the video output.Video is a collection of frames. Each frame is subtracted from the previous frame in the video to remove the static background and only identifies the moving objects in the frame.

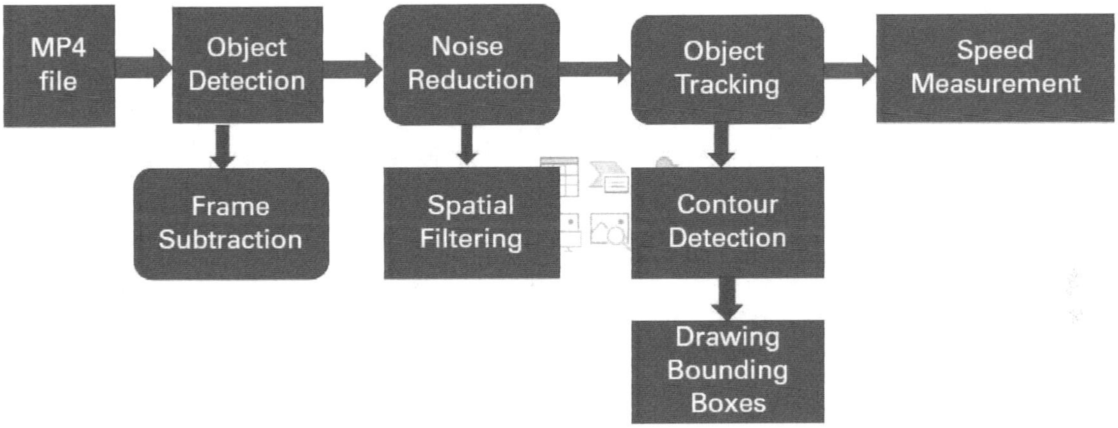

Fig.1 Block Diagram for Implemented Algorithm

3.2. Speed Estimation

The final Stage is to evaluate the speed of each moving object in the video stream. The accurate measurement of displacement between exactly 2 reference points on the path being captured in the video stream. When a car crosses the first point of reference, we start a timer which is specific to that vehicle. Similarly, when the vehicle crosses the second reference point, we stop the timer and calculate the time difference, i.e. the time taken by the vehicle to move from the first reference point to the second.

We then use the simple formula for speed calculation to estimate the speed for each moving vehicle:
Speed = Distance/Time ....(1)

This must be done on a per vehicle basis to effectively track the speed of each car. Finally, we compare the estimated speed of each car to the threshold speed. In the event of over speeding, we highlight the required ROI.

4. RESULTS AND CONCLUSION

Original Video

Fig.2 Original input Video frame

Video after Image subtraction

Fig.3 Background Elimination

Processed Video

Fig.4 Processed Video

Precision

It can be defined as the ratio between true positives and the sum of the true positives and false positives is defined as precision. Eq. (3) depicts the formula for precision.

Precision = TP /(TP + FP) ....(2)

= 32/(3 + 3)

= 0.94

Accuracy

Accuracy is generally defined as the percentage of correctly classified objects, i.e., the percentage of correctly detected vehicles.

Accuracy = (TP + TN)/(TP + TN + FP + FN) ...(3)

= (31 + 36)/(31 + 38 + 3 + 2)

= 0.90

5. CONCLUSION

The use of high-resolution cameras has surpassed conventional sensor-based speed detection techniques. Thus, this research work is the result of an innovative technology integrated with a video monitoring system for speed detection and the enforcement. The system identifies moving objects on the road, extracts its other parameters, labels all its related elements, which essentially helps to obtain the speed of the vehicle using different python functions, based on the image's geometry and camera position. The system works well to assess the level of both local and highway traffic. The system uses various image processing methods such as background subtraction, binary image formation, gray scale image formation, etc. It can further be enhanced by using OCR technique for license plate identification. Another future enhancement would be to detect two-wheelers in addition to the 4 wheelers which it currently detects. Based on the results obtained, the research work is proven that it can be used as a real time solution for commercial traffic control systems.

REFERENCES

[1] G. Mathur, D. Somwanshi and M. M. Bundele, "Intelligent Video Surveillance based on Object Tracking," 2018 3rd International Conference and Workshops on Recent Advances and Innovations in Engineering (ICRAIE), Jaipur, India, 2018, pp. 1-6.

[2] Vehicle Detection by Independent Parts for Urban Driver Assistance "IEEE transactions on intelligent transportation systems, vol. 14, no. 4, (2013) December

[3] IEEE transactions on intelligent transportation systems, vol. 12, no. 2, (2011), June

[4] FLEXYS: Motion-based Traffic Analysis and Incident Detection (IBBT/VUB-ETRO)

[5]. T. Cheng, J. Wang, B. Wu, Y. Jiang, X. Zhu and Q. Du, "A High Precision Vehicle Tracking Algorithm Based on Digital Image Correlation," 2019 3rd Conference on Vehicle Control and Intelligence (CVCI), Hefei, China, 2019, pp. 1-6.

[6]. Al Najjar M., Ghantous M., Bayoumi M. (2014) Object Detection. In: Video Surveillance for Sensor Platforms. Lecture Notes in Electrical Engineering, vol 114. Springer, New York, NY

[7]. B. Pawar, V. T. Humbe and L. Kundnani, "Morphology based moving vehicle detection", 2017 International Conference on Big Data Analytics and Computational Intelligence (ICBDAC), Chirala, 2017, pp. 217-223.

[8] A. P. Shukla and M Saini, "Moving Object Tracking of Vehicle Detection": A Concise Review, International Journal of Signal Processing, Image Processing and Pattern Recognition Vol.8, No.3 (2015), pp.169-176

[9] Signal & Image Processing: An International Journal (SIPIJ), vol. 5, no. 1, (2014) February.

[10] W. Rahmaniar ,W. Wang and H. Chen, "Real-Time Detection and Recognition of Multiple Moving Objects for Aerial Surveillance", Digital Signal, Image and Video Processing for Emerging Multimedia Technology, Electronics, 2019.

[11]. P. M. Dhulavvagol, A. Desai and R. Ganiger, "Vehicle Tracking and Speed Estimation of Moving Vehicles for Traffic Surveillance Applications," 2017 International Conference on Current Trends in Computer, Electrical, Electronics and Communication (CTCEEC), Mysore, 2017, pp. 373-377.

[12]. R. P. Nayak, S. Sethi and S. K. Bhoi, "PHVA: A Position Based High Speed Vehicle Detection Algorithm for Detecting High Speed Vehicles using Vehicular Cloud," 2018 International Conference on Information Technology (ICIT), Bhubaneswar, India, 2018, pp. 227-232.

[13]. A. Reddy K., S. Patel, K. P. Bharath and R. Kumar M., "Embedded Vehicle Speed Control and Over-Speed Violation Alert Using IoT," 2019 Innovations in Power and Advanced Computing Technologies (i-PACT), Vellore, India, 2019, pp. 1-5.

[14]. Dhungel, Prasanga, Prashant Tandan, Sandesh Bhusal, Sobit Neupane, and Subarna Shakya. "Video Compression for Surveillance Application using Deep Neural Network." Journal of Artificial Intelligence and Capsule Networks 2, no. 2 (2020): 131-145

Solving Wave Equation and Heat Equation by Rohit Transform (RT)

***Rohit Gupta[1], Dr. Rakesh Kumar Verma[2], Dr. Sanjay Kumar Verma[3]**

[1*]Lecturer, Dept. of Applied Sciences (Physics), Yogananda College of Engg. and Tech., Jammu, India.
[1*]Email id: guptarohit565@gmail.com

[2]Asstistant Professor, Dept. of Applied Sciences, Yogananda College of Engg. and Tech., Jammu, India.
Email id: rakesh.verma189@gmail.com

[3]Associate Professor, Dept. of Physics, Dean Faculty of Sciences, Cluster University of Jammu, India.
Email id: sanjay66_verma@yahoo.co.in

*Corresponding Author: Rohit Gupta (Email id: guptarohit565@gmail.com)

Abstract: The wave equation and the heat equation are widely known differential equations coming to light in engineering, basic and material sciences. The differential equations which represent the wave equation and the heat equation are usually solved by the exact technique or by the approximate technique or by the purely numerical technique. Since the implementation of these techniques is very complex, computationally vigorous, and requires elaborate computations, therefore, for finding the solutions of differential equations depicting the wave equation and the heat equation, there is a need to ask for integral transform techniques. Integral transform techniques render productive means for finding the solutions of problems coming to light in engineering, basic and material sciences. The Rohit transform (RT) is a new integral transformation put forward by the author Rohit Gupta recently in the year 2020 and has been utilized for finding the solutions of problems coming to light in engineering, basic and material sciences like other transform techniques. In this study, the RT is brought in for finding the solutions of the heat equation and the wave equation expressed in terms of differential equations which are generally partial in nature.

Keywords: RT, Heat equation, Wave equation, Differential Equations.

1. Introduction

The wave equation and the heat equation coming to light in engineering, basic and material sciences are expressed in terms of differential equations which are generally partial in nature. There are different techniques like exact technique or approximate technique or purely numerical technique for finding the solutions of differential equations depicting the wave equation and the heat equation. Since the implementation of these techniques is very complex, computationally vigorous, and requires elaborate computations, therefore, for finding the solutions of differential equations depicting the wave equation and the heat equation, there is a need to ask for integral transformation techniques. Integral transformation techniques render productive means for finding the solutions of problems coming to light in engineering, basic and material sciences [1-13]. The RT has been put in for finding the solutions of problems coming to light in engineering, basic and material sciences [14-20]. In this paper, the differential equations depicting the wave equation and the heat equation have been solved by the RT to obtain their exact solutions. The organization of the paper is given below:

1. Section-1 explains the RT and its basic properties.
2. Section-2 explains the applicability of the RT to the differential equations depicting the heat equation and the wave equation
3. Finally, the discussion and the conclusion are furnished.

2. Rohit Transform (RT)

The RT of f (z), z ≥ 0, denoted by $G(r)$, is defined as $G(r) = r^3 \int_0^\infty e^{-rz} f(z)dz$, provided the integral is convergent, where r is a parameter which may be real or complex [15-19].
The RT of few basic functions are

- $R\{z^n\} = \frac{n!}{r^{n-2}}$
  - $where\ n = 0,1,2,3 \ldots \ldots$
- $R\{e^{az}\} = \frac{r^3}{r-a}$,
- $R\{sinaz\} = \frac{a\,r^3}{r^2+a^2}$,
- $R\{sinhaz\} = \frac{a\,r^3}{r^2-a^2}$,
- $R\{cosaz\} = \frac{r^4}{r^2+a^2}$,
- $R\{coshaz\} = \frac{r^4}{r^2-a^2}$

2.1 Properties of RT

The RT $\frac{\partial f(x,y)}{\partial y}$ w.r.t. y, is as follows

$$R\left\{\frac{\partial f(x,y)}{\partial y}\right\} = r^3 \int_0^\infty e^{-ry} \frac{\partial f(x,y)}{\partial y} dy$$

$$R\left\{\frac{\partial f(x,y)}{\partial y}\right\} = r^3\left\{-f(x,0) - \int_0^\infty -re^{-ry} f(x,y)dy\right\}$$

$$R\left\{\frac{\partial f(x,y)}{\partial y}\right\} = r^3\left\{-f(x,0) + r\int_0^\infty e^{-ry} f(x,y)dy\right\}$$

$$R\left\{\frac{\partial f(x,y)}{\partial y}\right\} = r^3\{-f(x,0) + rR\{f(x,y)\}\}$$

$$R\left\{\frac{\partial f(x,y)}{\partial y}\right\} = rG(x,r) - r^3 f(x,0)$$

Hence $R\left\{\frac{\partial f(x,y)}{\partial y}\right\} = rG(x,r) - r^3 f(x,0)$

Since $R\left\{\frac{\partial f(x,y)}{\partial y}\right\} = rR\{f(x,y)\} - r^3 f(x,0)$, therefore, on changing f(x, y) by $\frac{\partial f(x,y)}{\partial y}$ and $\frac{\partial f(x,y)}{\partial y}$ by $\frac{\partial^2 f(x,y)}{\partial y^2}$, we find

$$R\left\{\frac{\partial^2 f(x,y)}{\partial y^2}\right\} = rR\left\{\frac{\partial f(x,y)}{\partial y}\right\} - r^3\frac{\partial f(x,0)}{\partial y}$$

$$R\left\{\frac{\partial^2 f(x,y)}{\partial y^2}\right\} = r\{rR\{f(x,y)\} - r^3 f(\mathrm{x},0)\} - r^3\frac{\partial f(x,0)}{\partial y}$$

$$R\left\{\frac{\partial^2 f(x,y)}{\partial y^2}\right\} = r^2R\{f(x,y)\} - r^4 f(\mathrm{x},0) - r^3\frac{\partial f(x,0)}{\partial y}$$

$$R\left\{\frac{\partial^2 f(x,y)}{\partial y^2}\right\} = r^2G(x,r) - r^4 f(\mathrm{x},0) - r^3\frac{\partial f(x,0)}{\partial y}$$

Hence $R\left\{\frac{\partial^2 f(x,y)}{\partial y^2}\right\} = r^2G(x,r) - r^4 f(\mathrm{x},0) - r^3\frac{\partial f(x,0)}{\partial y}$

From above study, we found that

$$R\left\{\frac{\partial f(x,y)}{\partial y}\right\} = rG(x,r) - r^3 f(x,0),$$

$$R\left\{\frac{\partial^2 f(x,y)}{\partial y^2}\right\} = r^2G(x,r) - r^4 f(\mathrm{x},0) - r^3 f'(\mathrm{x},0),$$

$$\mathrm{R}\left\{\frac{\partial^3 f(x,y)}{\partial y^3}\right\} = \mathrm{r}^3\mathrm{G}(\mathrm{x},\mathrm{r}) - \mathrm{r}^5\mathrm{f}(\mathrm{x},0) - \mathrm{r}^4\mathrm{f}'(\mathrm{x},0) - \mathrm{r}^3\mathrm{f}''(\mathrm{x},0)$$

3. IMPLEMENTATION OF RT TO THE HEAT EQUATION AND THE WAVE EQUATION

Integral transform techniques are widely used for finding the solutions of initial value problems expressed in terms of differential equations which are generally partial in nature [8-12, 21, 22, 23]. In this study, the RT has been put forward for solving the heat equation and the wave equation expressed in terms of differential equations, coming to light in engineering, basic and material sciences.

Problem 1: Considering the heat equation [23]

$$\frac{\partial g(x,t)}{\partial t} = \frac{\partial^2 g(x,t)}{\partial x^2} \qquad (1),$$

$$here\;\; g(x,0) = 3\sin 2\pi x,\, g(0,t) = 0,\, g(2,t) = 0,\, 0 < x < 2 \; and \; t > 0.$$

Solution: The RT of (1) w.r.t. t, gives

$$rG(x,r) - r^3 g(x,0) = \frac{\partial^2 G(x,r)}{\partial x^2}$$
$$rG(x,r) - r^3\, 3\sin 2\pi x = \frac{\partial^2 G(x,r)}{\partial x^2}$$
$$\frac{\partial^2 G(x,r)}{\partial x^2} - rG(x,r) = -r^3\, 3\sin 2\pi x \quad (2)$$

The solution of homogeneous equation: $\frac{\partial^2 G(x,r)}{\partial x^2} - rG(x,r) = 0$ is written as

$$G(x,r) = Pe^{\sqrt{r}x} + Qe^{-\sqrt{r}x} \quad (3)$$

Here P and Q are constants and are set on by initial conditions.

Since $g(0,t) = 0,\ and\ g(2,t) = 0$ [23], therefore, $G(0,r) = 0\ and\ G(2,r) = 0.$

Using $G(0,r) = 0$, we can write

$$0 = P(1) + Q(1)$$

Or

$$P = -Q$$

Using $G(2,r) = 0$, we can write

$$0 = Pe^{2\sqrt{r}} + Qe^{-2\sqrt{r}}$$
$$0 = -Qe^{2\sqrt{r}} + Qe^{-2\sqrt{r}}$$
$$0 = Q(-e^{2\sqrt{r}} + e^{-2\sqrt{r}})$$

Since $\left(-e^{2\sqrt{r}} + e^{-2\sqrt{r}}\right) \neq 0, therefore, we\ can\ write$

$$Q = 0.$$

The particular solution of (2) is given by

$$G(x,r) = L\cos 2\pi x + M\sin 2\pi x \quad (4)$$

To determine the constants L and M, the technique of undetermined coefficients has been applied.

$$\frac{\partial G(x,r)}{\partial x} = -2\pi L\sin 2\pi x + 2\pi M\cos 2\pi x \quad (5)$$
$$\frac{\partial^2 G(x,r)}{\partial x^2} = -4\pi^2 L\cos 2\pi x - 4\pi^2 M\sin 2\pi x \quad (6)$$

Substitute equations (4) and (6) in (2), we can write

$$-4\pi^2 L\cos 2\pi x - 4\pi^2 M\sin 2\pi x - r\,L\cos 2\pi x - rM\sin 2\pi x = -r^3\, 3\sin 2\pi x$$

Or

$$-\left(4\pi^2 + r\right)L\cos 2\pi x - (4\pi^2 + r)M\sin 2\pi x = -r^3\, 3\sin 2\pi x$$

Comparing the coefficients of $\cos 2\pi x\ \ and\ \sin 2\pi x\ \ on\ both\ sides, we\ have$

$$-\left(4\pi^2 + r\right)L = 0$$

Or

$$L = 0$$

And

$$\left(4\pi^2 + r\right)M = r^3\ 3$$

Or

$$L = \frac{3r^3}{(4\pi^2 + r)}$$

Hence the solution of (2) is written as

$$G(x,r) = (0)\cos 2\pi x + \frac{3r^3}{(4\pi^2 + r)}\sin 2\pi x$$

Or

$$G(x,r) = \frac{3r^3}{(4\pi^2 + r)}\sin 2\pi x \quad (7)$$

The inverse RT of (7) is given by

$$g(x,t) = 3e^{-4\pi^2 t}\sin 2\pi x$$

Hence, the solution of (1) is written as

$$g(x,t) = 3e^{-4\pi^2 t}\sin 2\pi x$$

It is checked that this is certainly the solution of the heat equation, as for t = 0, $g(x,0) = 3\sin 2\pi x$

Problem 2: Consider the wave equation [23]

$$\frac{\partial^2 g(x,t)}{\partial t^2} = \frac{\partial^2 g(x,t)}{\partial x^2} \quad (8)$$

$$here\ \ g(x,0) = \sin x, \frac{\partial g(x,0)}{\partial t} = 0, g(0,t) = 0, g(\pi,t) = 0, 0 < x < \pi\ and\ t > 0$$

Solution: The RT of (8) w.r.t. t, gives

$$r^2 G(x,r) - r^4 g(x,0) - r^3\frac{\partial g(x,0)}{\partial t} = \frac{\partial^2 G(x,r)}{\partial x^2}$$

$$r^2 G(x,r) - r^4 \sin x = \frac{\partial^2 G(x,r)}{\partial x^2}$$

$$\frac{\partial^2 G(x,r)}{\partial x^2} - r^2 G(x,r) = -r^4 \sin x \quad (9)$$

The solution of homogeneous equation: $\frac{\partial^2 G(x,r)}{\partial x^2} - r^2 G(x,r) = 0$ is given by

$$G(x,r) = Le^{rx} + Me^{-rx} \quad (10)$$

Here L and M are constants and are set on by initial conditions.

Since $g(0,t) = 0\ and\ f(\pi,t) = 0$, therefore, $G(0,r) = 0, G(\pi,r) = 0.$

Applying the boundary condition $G(0,r) = 0,$ we have

$$L(1) + M(1) = 0$$

Or

$$L = -M$$

Applying the boundary condition $G(\pi,r) = 0,$ we have

$$0 = Le^{\pi r} + Le^{-\pi r}$$
$$0 = -Me^{2r} + Me^{-2r}$$
$$0 = M(-e^{2r} + e^{-2r})$$

Since $\left(-e^{2r} + e^{-2r}\right) \neq 0, therefore, we\ can\ write$

$$M = 0.$$

The solution of (9) is written as

$$G(x,r) = P\cos x + Q\sin x \quad (11)$$

To determine the constants P and Q, the technique of undetermined coefficients has been applied.

$$\frac{\partial G(x,r)}{\partial x} = -P\sin x + Q\cos x$$
$$\frac{\partial^2 G(x,r)}{\partial x^2} = -P\cos x - Q\sin x \quad (12)$$

Substitute equations (11) and (12) in (9), we have

$$-P\cos x - Q\sin x - r^2 P\cos x - r^2 Q\sin x = -r^4\sin x$$

Or

$$-(1+r^2)P\cos x - (1+r^2)Q\sin x = -r^4\sin x$$

Comparing the coefficients of $\cos x\ and\ \sin x\ on\ both\ sides, we\ have$

$$-(1+r^2)P = 0$$

Or

P = 0

And

$$(1+r^2)Q = r^4$$

Or

$$Q = \frac{r^4}{(1+r^2)}$$

Hence the solution of (9) is written as

$$G(x,r) = (0)\cos 2\pi x + \frac{r^4}{(1+r^2)}\sin x$$

Or

$$G(x,r) = \frac{r^4}{(1+r^2)}\sin x \quad (13)$$

The inverse RT of (13) gives

$$g(x,t) = cost\sin x$$

Hence, the solution of (8) is written as

$$g(x,t) = cost\sin x$$

It can be checked that this is certainly the solution of the wave equation, as for t = 0, $g(x,0) = \sin x$

4. Discussion

Since the implementation of techniques like exact technique or approximate technique or purely numerical technique is very complex, computationally vigorous, and requires elaborate or vigorous computations for finding the solutions of differential equations depicting the heat equation and the wave equation, coming to light in engineering, basic and material sciences. The RT, explained in the paper, presented a compelling system for finding the solutions of widely known differential equations depicting the heat equation and wave equation. The exact solutions heat equation and the wave equation have been derived by the exertion of RT and the result accorded a confirmation for the RT which took part a great contribution for deriving the exact solutions of differential equations depicting the heat equation and the wave equation by streamlining their vigorous computation.

5. Conclusion

The authors victoriously explained the applicability of the RT to find the solutions of widely known differential equations depicting the heat equation and the wave equation, coming to light in engineering, basic and material sciences.

Conflict of Interest: There is no conflict of interest.

References

[1] Sadikali Latif Shaikh, Some Applications of the New Integral Transform For PartialDifferential Equations,Math. J. Interdiscip. Sci., 7(1), 45–49 (2018)

[2] Ali E. Abaoub, Abejela S. Shkheam, The New Integral Transform 'Abaoub-Shkheamtransform', IAETSD Journal for Advanced Research in Applied Sciences,7(6), 8-14 (2020)

[3] Zill, D.G., Advanced Engineering Mathematics, Jones & Bartlett (2016)

[4] Rachid Belgacem, Ahmed Bokhari, Mohamed Kadiand DjelloulZiane,Solution of non-linear partial differential equations by Shehu transform and its applications, Malaya Journal of Matematik, 8(4), 1974-1979 (2020)

[5] Mohmed Zafar Saber and Sadikali L.Shaikh, Solution of Some Non Linear Partial Differential Equation by New Integral Transform Combined with ADM, Global Journal of Pure and Applied Mathematics, 16(5), 741-748 (2020)

[6] Eman A. Mansour , Emad A. Kuffi , Sadiq A. Mehdi, The new integral transform 'SEE transform and its applications', PEN, 9(2),1016- 1029 (2021)

[7] Eman A. Mansou, Emad A. Kuffi , Sadiq A. Mehdi, The New Integral Transform SEE Transform of Bessel's Functions, Turkish Journal of Computer and Mathematics Education, 12(13), 3898-390 (2021)

[8] Rahul Gupta, Rohit Gupta, Dinesh Verma, Propounding a New Integral Transform: Gupta Transform with Applications in Science and Engineering, International Journal of Scientific Research in Multidisciplinary Studies, 6(3), 14-19 (2020)

[9] Zain UL Abadin Zafar, ZZ Transform method, IJAEGT, 4(1), 1605-1611 (2016).

[10] Rahul Gupta, Rohit Gupta, Dinesh Verma, Analysis of series RL and RC networks with sinusoidal potential sources by Gupta transform, ASIO Journal of Engineering & Technological Perspective Research, 5(1), 28-30(2020)

[11] Chaudhary, R., Sharma, S.D., Kumar, N., & Aggarwal, S. Connections betweenAboodh transform and some useful integral transforms. International Journal of InnovativeTechnology and Exploring Engineering, 9(1), 1465-1470 (2019)

[12] Rahul Gupta, Rohit Gupta, Dinesh Verma, Application of Novel Integral Transform: Gupta Transform to Mechanical and Electrical Oscillators, ASIO Journal of Chemistry, Physics, Mathematics & Applied Sciences, 4(1), 4-7 (2020)

[13] Rahul Gupta, Rohit Gupta and Loveneesh Talwar, Response of Network Circuits Connected To Impulsive Potential Source via New Integral Transform: Gupta Transform, ASIO Journal of Engineering & Technological Perspective Research, 5(1), 18-20 (2020)

[14] Rahul Gupta, Rohit Gupta,Loveneesh Talwar, Gupta Transform Approach to the Series RL and RC Networks with Steady Excitation Sources, Engineering and Scientific International Journal, 8(2), 45-47 (2021)

[15] Rohit Gupta, On Novel Integral Transform: Rohit Transform and Its Application to Boundary Value Problems, ASIO Journal of Chemistry, Physics, Mathematics & Applied Sciences, 4(1), 08-13 (2020)

[16] Neeraj Pandita, Rohit Gupta, Analysis Of Uniform Infinite Fin via Means Of Rohit Transform, International Journal of Advance Research and Innovative Ideas in Education, 6(6), 1033-1036 (2020)

[17] Rohit Gupta, Rahul Gupta, Dinesh Verma, Solving Schrodinger equation for a quantum mechanical particle by a new integral transform: Rohit Transform, ASIO Journal of Chemistry, Physics, Mathematics & Applied Sciences, 4(1), 32-36 (2020)

[18] Rohit Gupta, Rahul Gupta, Analysis of RLC circuits with exponential excitation sources by a new integral transform: Rohit Transform, ASIO Journal of Engineering & Technological Perspective Research, 5(1), 22-24 (2020)

[19] Anamika, Rohit Gupta, Analysis Of Basic Series Inverter Via The Application Of Rohit Transform, International Journal of Advance Research and Innovative Ideas in Education, 6(6), 868-873 (2020)

[20] Loveneesh Talwar, Rohit Gupta, Analysis of Electric Network Circuits with Sinusoidal Potential Sources via Rohit Transform, International Journal of Advanced Research in Electrical, Electronics and Instrumentation Engineering, 9(11), 3020-3023 (2020)

[21] E. Butkov. Mathematical physics, Addison-Wesley Pub Co, 1968

[22] N.P. Bali and Dr. Manish Goyal. A Text Book of Engineering Mathematics, S Chand and Company Limited, 2014

[23] *R. Gupta, N. Pandita and R. Gupta, "Solving One-Dimensional Heat and Wave Equations Via Gupta Integral Transform,"* 2022, International Conference on Sustainable Computing and Data Communication Systems (ICSCDS), *2022, pp. 921-925, doi: 10.1109/ICSCDS53736.2022.9760823.*

Design and implementation of zeta converter for LED driver applications

J. Gnanavadivel [1, a)], M. Shunmathi [2, b)], I. Raghul [3], R. Mani Bharathi [3], E. Maneesh [3]

[1]Associate professor, Department of Electrical and Electronics Engineering, Mepco Schlenk Engineering College, Sivakasi, Tamil Nadu, India.

[2]Research Scholar, Department of Electrical and Electronics Engineering, Mepco Schlenk Engineering College, Sivakasi, Tamil Nadu, India.

[3] UG Scholar, Department of Electrical and Electronics Engineering, Mepco Schlenk Engineering College, Sivakasi, Tamil Nadu, India.

[b)] Corresponding author: shunmathi@mepcoeng.ac.in

[a)] gvadivel@mepcoeng.ac.in

Abstract: The objective of this study is to analyze the design and operation of low power, stand-alone active led driver applications. The system is powered by solar PV panels with a maximum input power of 125 W. The chosen converter supports continuous conduction mode. The device seems to have a switching frequency of 20 kHz. The duty cycle of a converter is controlled by the MPPT controller, which keeps the dc bus voltage constant. MPPT employs an incremental conductance method. The intention of this converter used to control the voltage at 48 V. The preferred converter is verified in both Matlab/Simulink and its performance is evaluated on hardware prototype of 25 W. While PV module efficiency is low and power output is dependent on solar irradiation and ambient temperature, maximizing power output while increasing efficiency is of key significance. The resulting system is highly efficient, low cost and able to track power faster.

Keywords: Continuous conduction mode, DC-DC converter, Maximum power point tracking, Incremental conductance algorithm.

1. Introduction

Solar power sources are gradually being integrated into power system networks to allow for continuous increases in power generation given the shortage of oil and coal as well as to mitigate negative environmental impacts [1]. Solar energy may be considered a very essential and sustainable energy among fuel reserves owing to this wide usage, vast number, and sustainability [2]. PV systems, on the other hand, are particularly sensitive and require a large boost in performance to reach a much higher power efficiency impactful energy system. MPPT is a system level technique for ensuring that the PV system consistently delivers maximum power to the load. Because of the constant variations in temperature variation, the Photovoltaic system can exploit and respond to emerging irradiance and temperature if the MPPT algorithm is used [3]. The main goal of the MPPT algorithm is to determine the output power differential by measuring the ripple voltage or maximum power from PV module. The output voltage of a PV module is affected by temperature and the amount of energy emitted from the sun. In order to maintain maximum power, a charge controller in the Photovoltaic system must use a DC-DC converter especially in combination for the optimization technique [4]. To improve the efficiency of solar PV systems, a considerable literature and a range of research strategies are being conducted throughout the years. The study presents a maximum power point control model called on a genetic technique to solve the inadequacies and limits of current controllers. The developed controller uses a DC-DC converter and can charge a variety of batteries and the system is expensive due to the numerous circuit breakers [5]. They designed a fuzzy controller with an adjustable output step function for a maximum power point tracker for solar systems. As a result of this research, an MPPT charge controller for a rechargeable battery able to integrate with a solar PV system is designed [6]. The system, which includes a PV module, a boost converter, a MPPT controller, and a resistive load, can increase PV performance by tracking the MPPT with an incremental conductance method. The proposed system, however, is expensive to build because to the numerous switches and passive components [7]. The zeta converter, and that is a kind of buck–boost converter, can be used to continuously increase the output voltage. The maximum power point tracking of an PV array, this property gives an unlimited region [8]. If MPP occurs within prescribed limitations, MPPT can be accomplished with buck [9] and boost [10] converters. The MPPT technique, which uses an Artificial Neural Network to forecast the global MPP in a solar pv subjected to continuous and fast evolving shade patterns, accurately and fast, however the system's shortcoming was its slow tracking speed [11]. In addition, the relevance of a Photovoltaic array fed BLDC motor driven water pump employing zeta converter is discussed in [12]. The BLDC motor and the zeta converter both have benefits that can aid in the designing of a PV system fed water pumping system that can work effectively under dynamically changing environmental conditions. A zeta converter, on the other hand, provides certain advantages over standard buck–boost, cuk converters employed in PV based systems. To achieve the voltage gain, a split inductor boost converter is used. Low energy usage, lower voltage strains on active switches and diodes, and reduced control complexity are all advantages of the selected converter [13]. The high gain voltage conversion characteristic is more appealing, and it is used to capture energy from highly efficient energy resources like fuel cells and solar panels. The recommended converter's performance has been proven by experimental data, with a high efficiency [14]. Because traditional boost converters never provide adequate gain, this research presents a transformer-free DC-DC converter with increased gain [15]. The converter provides a considerable voltage gain by reducing the duty cycle and lowering the conduction loss. References [16-17] describe sepic and cuk converter topologies that combine any of the voltage enhancing approaches with the classic converter model to improve gain. With a higher component count, the boosting voltage options stated above, as indicated in reference [18-19], improve the converter's gain. In Reference [20], high power LED lightning system discussed. It has essentially twice the voltage gain of a standard zeta converter. It will be able to handle problems such as power dissipation and reverse recovery. The variance MPPT algorithm is employed in this system, which reduces the Maximum power point tracking period and PV array energy losses, especially when irradiance varies. An MPPT technique will be utilized to design and build the zeta dc-dc converter, which will operate as the system's charge controller.

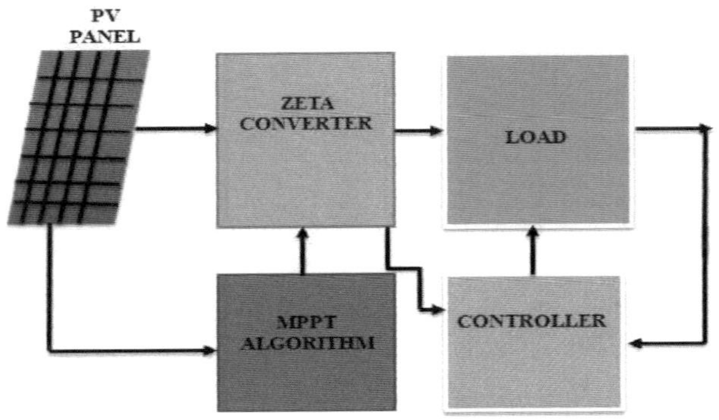

Figure 1. Solar Photovoltaic System Block Diagram

2. Energy Management System

Researchers are focusing their efforts on solar and wind power generation to fulfill expanding global energy demand. Foremost prevalent method for harvesting solar energy is to employ photovoltaic panels, which take photon energy from the sun and convert it to electrical energy. The goal of this study is to create a high-efficiency DC-DC converter using simply a solar cell as a power source. To address this issue, various energy methods are available, with solar energy seeming to be a more impactful vital source of renewable energy. It appears to have numerous advantages over alternative energy sources, including being a clean source of energy, allowing for the use of free raw materials and being a more commercial energy source. A simplified block diagram of a photovoltaic power management is shown in figure 1. The entire system is built and modeled to assess its effectiveness. The suggested converter further reduces switching losses by eliminating voltage rise over the switching devices. Simulated data is used to illustrate the MPPT controller's and intended system's performance under various atmospheric conditions and load disturbances.

3. Zeta Converter

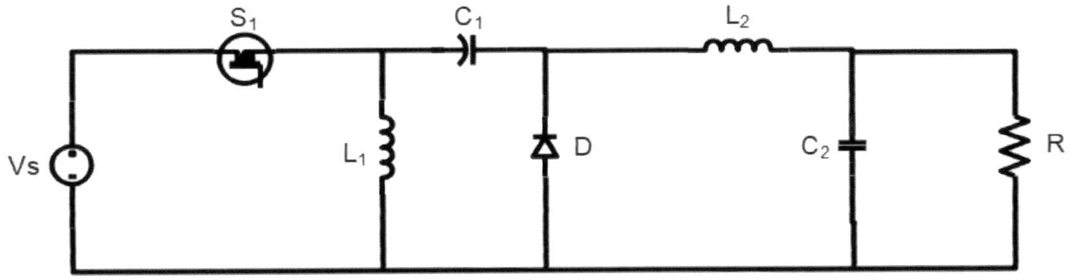

Figure 2. Zeta Converter Circuit Diagram

The zeta converter block diagram is shown in Figure 2. It includes a DC power supply, a load and a zeta converter. The power source provides the input voltage to the converter. The R-load is fed the output of the zeta converter. The zeta converter's output voltage can be adjusted to suit the load. Other components of a zeta converter include switch (S_1) two inductors (L_1, L_2) and two capacitors (C_1, C_2). The switching device is provided the gate pulse. The duty cycle of the gate pulse is used to regulate the converter's output. Buck operation happens whenever the duty cycle is less than 0.5, whereas boost operation occurs when the duty cycle is more than 0.5. From an input voltage that varies above and below the output value, the ZETA converter topology produces a positive output voltage.

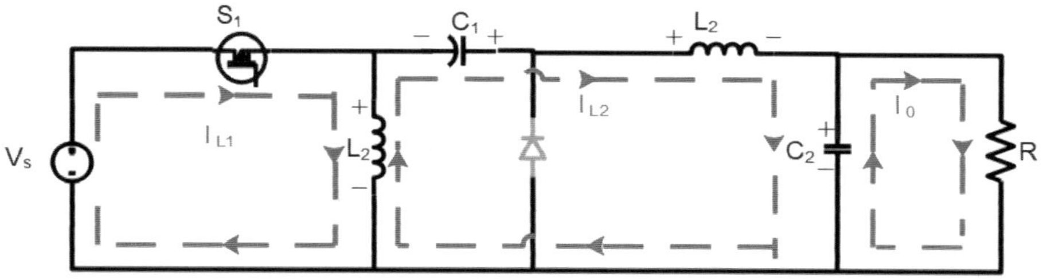

Figure 3. Mode-1 Diagram

Mode 1

The switch S_1 is on condition during this mode of operation depicted in figure 3. The input source's energy was stored across the inductors L_1 and L_2, resulting in a linear increase in current across the inductors. The voltages across capacitors C_1 and C_0 are exactly equal to V_0. The reverse voltage across the diode D_1 at this time is calculated as follows:

Where $T = T_{ON} + T_{OFF}$

The inductor L_1 is parallel to the voltage source,

By KVL,

$V_s = V_{L1}$

For inductor we know that, $V = L\, di/dt$.

$$V_s = V_{L1} = L_1(\Delta I_{L1}/T_{on}) \quad \ldots\ldots\ldots\ldots.(1)$$

By KVL on loop 2,

$$V_{l2} = V_s + V_{c1} - V_o$$

$$L_2\,(\Delta I_{L2}/T_{OFF}) = V_s + V_{c1} - V_0 \quad \ldots\ldots\ldots\ldots\ (2)$$

By applying KCL,

$$I_{L2} = -I_{C1}$$

$$C_1(dV_{c1}/T_{ON}) = -I_{L2} \ldots\ldots\ldots\ldots (3)$$

$$I_{L2} = I_{C2} + I_0$$

$$I_{L2} - I_0 = C_2(dV_0/T_{ON}) \quad \ldots\ldots\ldots\ldots\ldots (4)$$

MODE – II

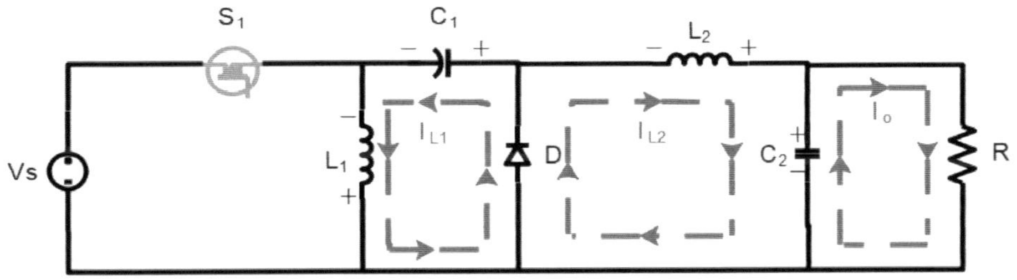

Figure 4. Mode-2 Diagram

MODE 2: The diode D_1 begins to conduct the switch S_1 is switched off. As a result, no energy passes via the existing path in this mode of operation depicted in figure 4. The load receives energy from the stored energy in the inductors i_{L1} and i_{L2}. As a result, the current flowing through the inductor current drops. The voltage applied to the switch is calculated as follows:

By KVL in loop 1,

$$V_{L1} = V_{c1}$$

$$L_1(\Delta I_{L1}/T_{OFF}) = V_{c1} \ldots\ldots\ldots\ldots\ldots\ldots (5)$$

By KVL in loop 2,

$$V_{L2} = V_0$$

$$L_2(\Delta I_{l2}/T_{OFF}) = V_0 \quad \ldots\ldots\ldots\ldots\ldots\ldots (6)$$

By applying KCL,

$$I_{L1} = I_{C1}$$

$$I_{L1} = C_1(dV_{c1}/T_{OFF}) \quad \ldots\ldots\ldots\ldots\ldots\ldots (7)$$

Equating $L_1(\Delta I_{L1})$ from equation (1) and (5)

$$V_s T_{ON} = V_{c1} T_{OFF} \quad \ldots\ldots\ldots\ldots\ldots\ldots(8)$$

Equating $L_2(\Delta I_{L2})$ from equation (2) and (6)

$$V_0 T_{OFF} = V_s T_{ON} + V_{C1} T_{ON} - V_0 T_{ON} \quad \ldots\ldots\ldots\ldots\ldots\ldots(9)$$

Duty cycle (D) the ratio of on time period and off time period of gating signal.

$$D = T_{ON}/T \quad \ldots\ldots\ldots\ldots\ldots\ldots (10)$$

$$T_{ON} = D\,T \quad \ldots\ldots\ldots\ldots\ldots\ldots (11)$$

$$T_{OFF} = (1 - D)T \quad \ldots\ldots\ldots\ldots\ldots\ldots(12)$$

Substituting equation (8) in equation (9)

$$V_0(T) = V_s T_{ON} + V_s (D^2T/(1-D)$$

$$V_0 = V_s (D/(1 - D)) \quad \ldots\ldots\ldots\ldots\ldots\ldots(13)$$

Equation (13) is the output equation of zeta converter.

By power balance equation,

$$P_{in} = P_o$$

$$V_s I_s = V_0 I_0$$

$$I_o = V_s I_s / V_s$$

Substituting equation (13)

$I_o = (1 - D)\, I_s / I_0$

The Design equation for inductors,

From equation (1)

$L_1 = V_s\, D\, T / \Delta I_{L1}$

$T = 1/T_{OFF}$

$L_1 = V_s\, D / (\Delta I_{L1}\, f)$

$L_2 = V_s\, D / (f\, \Delta I_{L2})$

The Design equation for capacitors:

The capacitors C_1 and C_2 for a maximum allowable voltage ripple, ΔV .where Q is the electric charge. By rewriting to an average rate of change, we get $\Delta V = \Delta QC$. Now, for C1 if we assume negligible current ripple, it is discharged by I_{L2}=I0.

$$\Delta V_{C1} = \frac{\int_0^{DT_s} I_0\, dt}{C_1}$$

$$\Delta V_{C1} = \frac{I_o D T_s}{C_1}$$

$$C_1 = \frac{I_o D}{\Delta V_{C1} f_s}$$

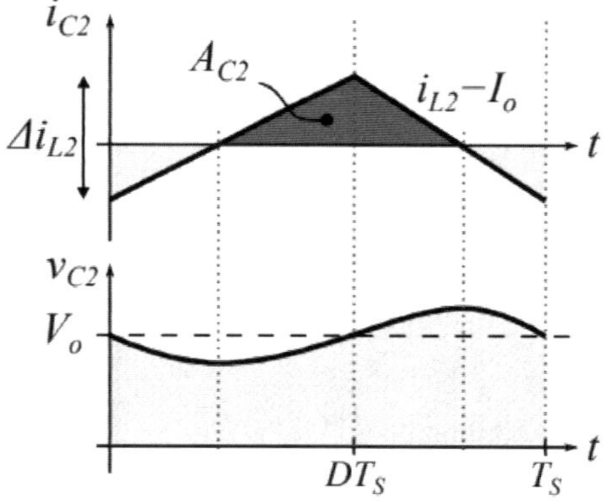

Figure 5. Switching Waveform

Now, the charge which determines the voltage change of C_2, is shown as the area A_{C2}. Geometrically we obtain that

$$A_{C2} = \frac{1}{2}\frac{T_s}{2}\frac{\Delta i_{L2}}{2}$$

Collecting the terms and using (12) for ΔiL2, we get that

$$A_{C2} = \frac{T_s}{8} \cdot \frac{V_o(1-D)T_s}{L_2}$$

However, this only represents the charge. To size the capacitor, we utilize C = ΔQ ΔV, where ΔQ = A_{C2}. For C_2 this gives

$$C_2 \geq \frac{A_{C2}}{\Delta V_{C2}} = \frac{V_o(1-D)}{8 \cdot L_2 f_s^2 \Delta V_{C2}}$$

Table 1. Converter design specifications

S.NO	PARAMETERS	VALUE
1	Inductor L_1	0.29 mH
2	Capacitor C_1	100 µF
3	Capacitor C_2	100 µf
4	Resistance	9.216 Ω

TABLE 1 shows the designed converter specifications.

4. Result Analysis

At a variety of load levels and temperatures, the converter produces 48 V. In this study, the solar PV cell is paired with a DC converter to achieve the best voltage regulation. The voltage level at the output end is determined by an incremental conductance algorithm method. FIGURE 6 depicts the modeling of the closed loop zeta converter.

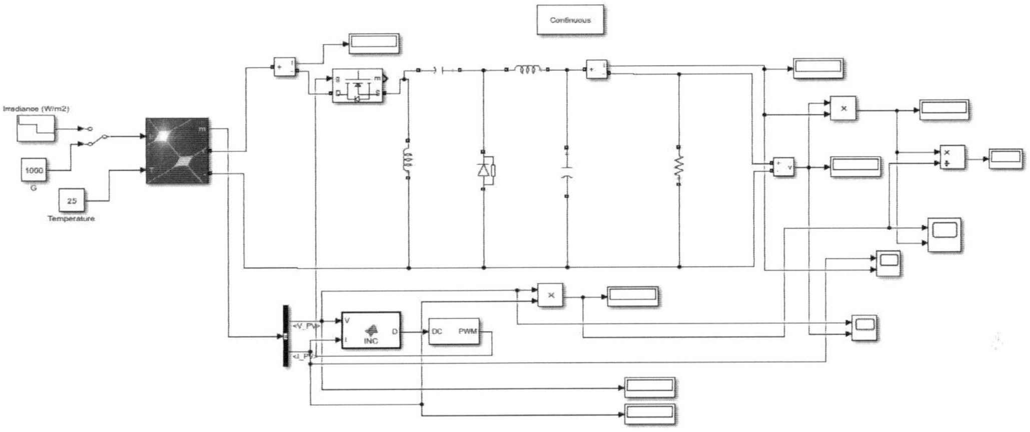

Figure 6. Simulation Diagram

Type 1: With a constant irradiation of 1000W/m^2 and a constant temperature of 25°C. The results are achieved by changing the load, as shown in TABLE 2. FIGURES 7 and 8 depict the output current and voltage. The results of adjusting the load while keeping the irradiation level and temperature at 1000W/m^2 and 25°C, respectively, are presented in this section.

Table 2. Load Regulation at 25°C and 1000W/m^2 Irradiation

Power (W)	Output Voltage (V_0)	Output Current (I_0)	Efficiency (%)
100	48	2.535	93.56
75	48	2.38	94.62
50	48	2.09	95.17

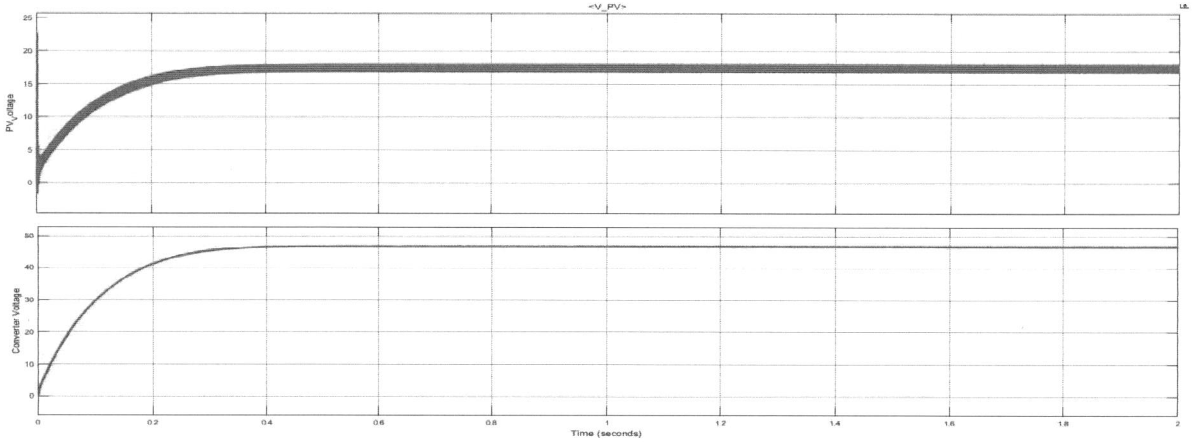

Figure 7. Waveform of Output Voltage at 1000W/m^2 and 25°C

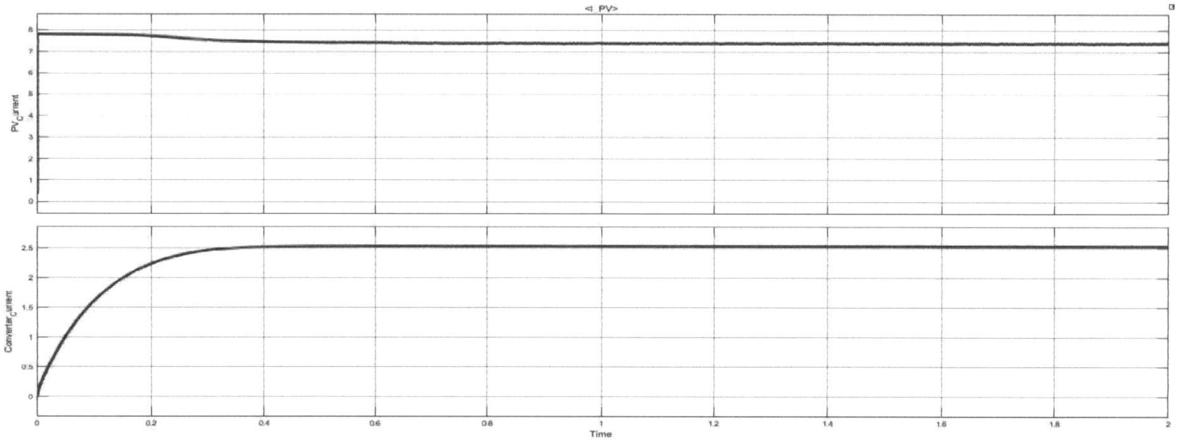

Figure 8. Waveform of Output Current at 1000W/m^2 and 25°C

Table 3. Load Regulation at 20°C and 850W/m^2 Irradiation

Power (W)	Output Voltage (V_0)	Output Current (I_0)	Efficiency (%)
100	48	2.52	93.43
70	48	2.43	94.86
50	48	2.03	95.74

Type 2: For a constant 850W/m^2 of irradiation and a constant temperature of 20°c

TABLE 3 emphasizes on changing the load while maintaining irradiance 850W/m^2 and temperature at 20°C, accordingly. FIGURES 9 and 10 depict a solar panel's voltage waveform. This converter can also maintain 93% efficiency, which is suitable for converters that use renewable energy. Figure 11 depicts the prototype of zeta converter fed LED driver setup.

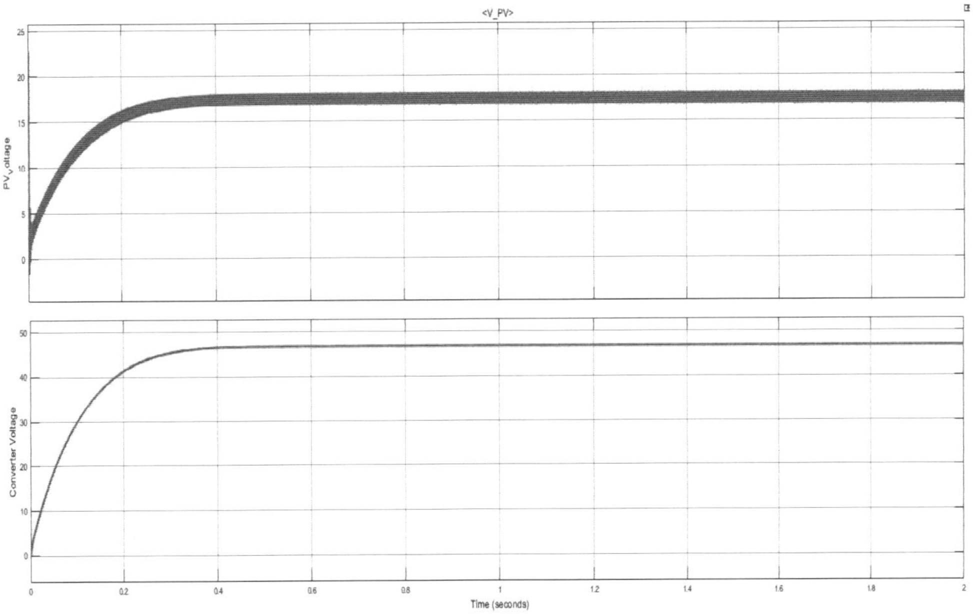

Figure 9. Waveform of Output Voltage at 850w/M^2 and 25°C

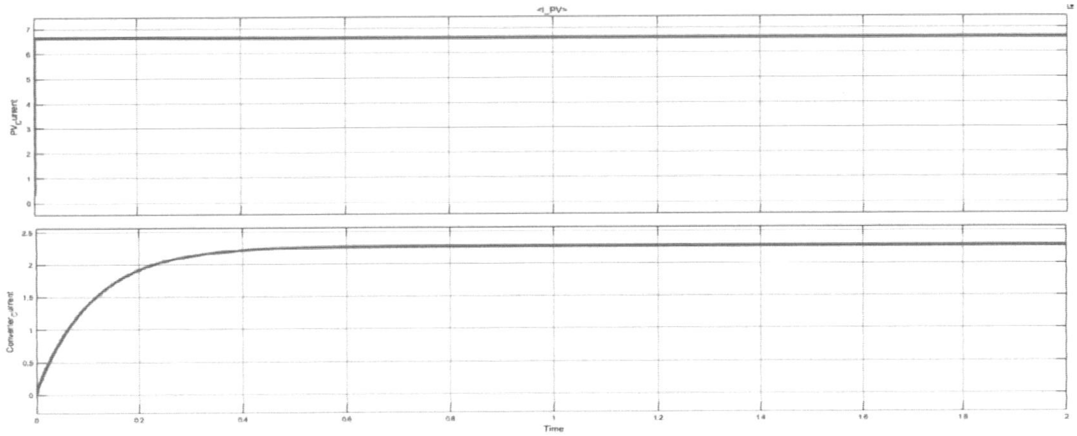

Figure 10. Waveform of Output Current at 850W/m^2 and 25°C

5. Hardware Implementation

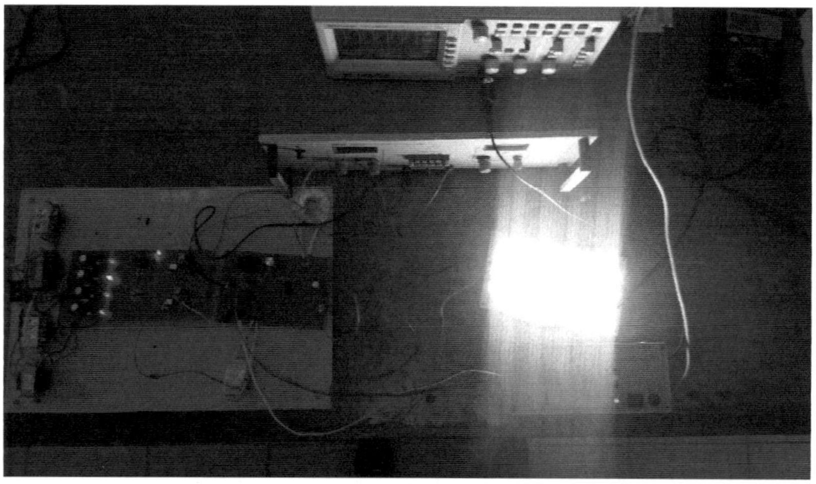

Figure 11. Hardware Result

6. Conclusion

The Zeta converter is simple to use and provides a constant output current with no ripples. The output voltage of the zeta converter can be increased or decreased. For integrating the entire structure, the subsystems of the overall system, such as the PV panel and the zeta converter, were separately simulated. The proposed system then uses closed loop control. The output voltage's steady state inaccuracy is decreased with the adoption of a MPPT control algorithm. The zeta converter's output has been used for led driver applications. The experimental setup finally built, and the results are compared to those obtained from simulation.

7. References:

[1] Hossain MZ, Rahim NA. Recent progress and development on power DC-DC converter topology, control, design and applications: A review. Renewable and Sustainable Energy Reviews. 2018 Jan 1;81:205-30.
[2] S. Alagammal, Dr. N. Rathina Prabha, I. Aarthy "Centralized Solar PV Systems for Static Loads using Constant Voltage Control Method", Circuits and Syst. 2016, pp 4213-4226.
[3] N Rathina Prabha, J Gnanavadivel, KS Krishna Veni, "Performance Investigation of Single Switch Dual Output DC-DC SEPIC Converter for PV Applications", Int. J. Adv. Sci, 2020, pp 4676-4683.
[4] J.Gnanavadivel, M.Shunmathi, P.Muthu Thiruvengadam, S.Narthana, "Analysis of DC-DC Converter with High Step-up Gain for Alternative Energy Sources", Int. J. Eng. Technol. 2021, pp 162-168.
[5] Hussein AA, Fardoun AA. Design considerations and performance evaluation of outdoor PV battery chargers. Renewable Energy. 2015 Oct 1;82:85-91.

[6] Ananda-Rao K, Matar Y, Baharudin NH, Ismail MA, Abdullah AM. Design of MPPT charge controller using zeta converter for battery integrated with solar Photovoltaic (PV) system. In Journal of Physics: Conference Series 2020 (Vol. 1432, No. 1, p. 012058). IOP Publishing.
[7] Kumar R, Singh B. BLDC motor-driven solar PV array-fed water pumping system employing zeta converter. IEEE Transactions on Industry Applications. 2016 Jan 28;52(3):2315-22.
[8] Lu, D.D. and Nguyen, Q.N., 2012. A photovoltaic panel emulator using a buck-boost DC/DC converter and a low cost micro-controller. *Solar Energy*, *86*(5), pp.1477-1484.
[9] Elgendy MA, Zahawi B, Atkinson DJ. Assessment of the incremental conductance maximum power point tracking algorithm. IEEE Transactions on sustainable energy. 2012 Jul 25;4(1):108-17.
[10] Kumar R, Singh B. Buck-boost converter fed BLDC motor drive for solar PV array based water pumping. In 2014 IEEE International Conference on Power Electronics, Drives and Energy Systems (PEDES) 2014 Dec 16 (pp. 1-6). IEEE.
[11] Bendib B, Belmili H, Krim F. A survey of the most used MPPT methods: Conventional and advanced algorithms applied for photovoltaic systems. Renewable and Sustainable Energy Reviews. 2015 May 1;45:637-48.
[12] Norouzzadeh E, Ale Ahmad A, Saeedian M, Eini G, Pouresmaeil E. Design and implementation of a new algorithm for enhancing MPPT performance in solar cells. Energies. 2019 Jan;12(3):519.
[13] Singh B, Bist V. Power quality improvements in a zeta converter for brushless DC motor drives. IET Science, Measurement & Technology. 2015 Apr 23;9(3):351-61.
[14] Y. Jiao, F.L. Luo and B.K. Bose, "Voltage-lift spilt-inductor-type boost converters," IET Power Electronics, vol. 4, no. 4, pp. 353-362, Apr 2011.
[15] Yu Tang, Ting Wang and Yaohua He, "A Switched-Capacitor- Based Active-Network Converter With High Voltage Gain," IEEE Transactions on Power Electronics, vol. 29, no. 6, pp. 2959 – 2968, Jun 2014.
[16] Ramaprabha, N., J. Gnanavadivel, N. Senthil Kumar, and ST Jaya Christa. "Power quality improvement in single phase AC to DC zeta converter." (2013): 533-537.
[17] Naveena M, Gnanavadivel J, Christa SJ, Kumar NS. Bridgeless Zeta Converter for Power Quality Improvement. In International Conference on Innovations in information, Embedded and Communication Systems (ICIIECS) 2016.
[18] Perianayagi M, Prabha NR, Gnanavadivel J. Design and analysis of single switch dual output DC-DC SEPIC converter for PV applications. IEEE International Conference on Innovations in Communication, Computing and Instrumentation 2019 Mar 23 (pp. 112-116).
[19] Gnanavadivel J, Yogalakshmi P, Kumar NS, Muralidharan S. High efficient single stage Cuk LED driver for universal input voltage applications with improved power quality. Optoelectron. Adv. Mater. Rapid Commun. 2018, 1;12(11-12):694-9.
[20] Gnanavadivel J, Kumar NS, Christa ST, Yogalakshmi P. Design and Implementation of High Power LED Lighting System for Health Care Applications. Current Signal Transduction Therapy. 2019;14(1):31-37.

Comparison of Deep Learning & Adaptive Algorithm Performance for De-Noising EEG

Md. Ibrahim Al Imran*, Mamun Rabbani
Department of Biomedical Physics & Technology, University of Dhaka, Dhaka, Bangladesh

Email: ibrahim.buet.eee@gmail.com*, mamunrabbani@du.ac.bd

Abstract. Various forms of artifacts can readily contaminate an electroencephalogram recorded using surface electrodes. A comparison of several electroencephalogram (EEG) de-noising methods is shown here. Five distinct forms of noise are reduced using three different strategies, and the results are compared. These three procedures are Recursive Least Squares (RLS) adaptive algorithm, Least Mean Squares (LMS) method, and Fully Connected Neural Network (FCNN). The results are shown using time-domain plots of the real EEG signal, noisy EEG signal, and forecasted EEG signal. For comparing the performance of the three de-noising techniques here relative-root-mean-square-error (RRMSE) and signal-to-noise-ratio were used. Here, exploring the values of the parameters, we find that FCNN predicts a better result than other two algorithms.

1. Introduction

EEG is recorded by placing electrodes on the scalp in a non-invasive EEG recording method. The procedure of non-invasive recording isn't extremely exact. Using surface electrodes, the amplitude of a clean EEG signal varies from 10μV to 100μV. Even with precise electrode placement and high gain amplifiers, there is still a background noise of $2\mu V_{P\text{-}P}$, which is considerable [2]. The EEG signal is critical for neurological disorder diagnosis and therapy. So, the EEG signal must be devoid of noise. However, because numerous forms of noise may readily contaminate the signal, it must be de-noised.

2. Related Work

EEG signals are commonly utilized to investigate human brain function. There hasn't been any research comparing the deep learning neural network with the adaptive method for de-noising the EEG signal yet. A method is proposed for the precise acquisition of EEG signal in [20]. But the results show that the acquisition system needs to be further improved and de-noising should be performed. Artifacts in EEG can be classified into two types, physiological artifacts and non-physiological artifacts [21]. Ocular noise and muscular noise are physiological artifacts introduced during the acquisition of the EEG signal. On the other hand, White noise, power line noise and baseline wandering noise are due to non-physiological artifacts. Among all the artifact signals, usually the most troublesome artifacts observed in the EEG signals of the frontal lobe are due to the ocular-activity, which are called Electro-ocular artifacts [22]. EEG signal corrupted by the ECG signal, EOG signal and mixed with the 50Hz sinusoidal power line interference is filtered by using cascaded adaptive filters in [4]. Although this work does not consider other noise signals and uses only LMS algorithm. Adaptive filter has "self-regulation" and "tracking" capacities. The most important property of adaptive filter is that it can work effectively in unknown environment [23]. The RLS algorithm has a great advantage that it converges very fast and it is free from the eigenvalue spreading problem. So, to remove the artifacts from the noise corrupted EOG signal the RLS adaptive filter is very suitable [24].

For evaluating the performance of the proposed filter structures the signal-to-noise ratio improvement (ΔSNR) is measured in [25]. The average percentage improvement of 2.9078% in SNR value is obtained for the five subjects in [13] where EOG is the only considered noise. A novel convolutional neural network model based on deep learning has been proposed in [6]. Here EOG and EMG artifacts are taken into consideration in two separate ways.

3. Method

In this paper, both simulated noise and recorded noise signals were used. The sampling frequency of the EEG signal is 256Hz. So, every signal used in this thesis must have a sampling frequency of 256Hz. MATLAB 2021a software has been used for the performance analysis of adaptive filter and FCNN model has been developed, trained and evaluated using Python 3.

3.1 Generation of the Noise-corrupted EEG

White noise is always present in every recorded biomedical signal [3]. Power line noise of 50Hz and its harmonics also affect the recorded EEG signal [4]. Baseline wandering noise affects the signal whose origin is usually unknown [5]. While recording EEG of the frontal lobe of a subject, EOG acts as a noise to the EEG [6]. Movement of various muscles of the subject generates Electromyogram (EMG). This EMG also affects the recording of the EEG signal [7]. All these types of noise get added to the actual EEG signal and are responsible for the change in the amplitude of the EEG.

The recorded EEG signal is taken from the website [27] having a sampling frequency of 256Hz (Figure 01). So, every signal used must have a sampling frequency of 256Hz.
Simulated Noise corrupted signal is generated by combining the EEG epochs and the epochs of the noise signals linearly according to formula (1),

$$\mathrm{y} = \mathrm{x} + \sum_{i=1}^{n} \lambda_i n_i \quad \ldots\ldots\ldots (1)$$

where x is the pure EEG signal, n_i is the noise signal and λ_i is the factor of contribution of the unwanted signals and y is the combined signal of EEG and noise which is actually the noise-corrupted EEG signal.

$$\lambda\,\mathrm{n} = \sum_{i=1}^{n} \lambda_i n_i \quad \ldots\ldots\ldots (2)$$

So, from the formula (1) and (2), SNR of the noise corrupted signal can be varied by varying the factor λ using the formula (3) [6],

$$\mathrm{SNR\ (dB)} = 10 \log \frac{\mathrm{RMS(x)}}{\mathrm{RMS}(\lambda\,\mathrm{n})} \quad \ldots\ldots\ldots (3)$$

The linear mixture of the EEG signal and the combination of different noise signals is done using the value of the percentage ratio value λ as in [30].

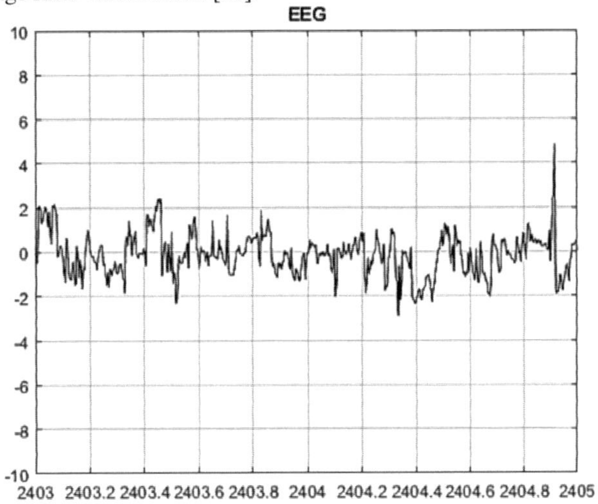

Figure 01: Clean EEG Signal for a Short Duration

The plot of a noise-corrupted EEG signal is shown in Figure 02.

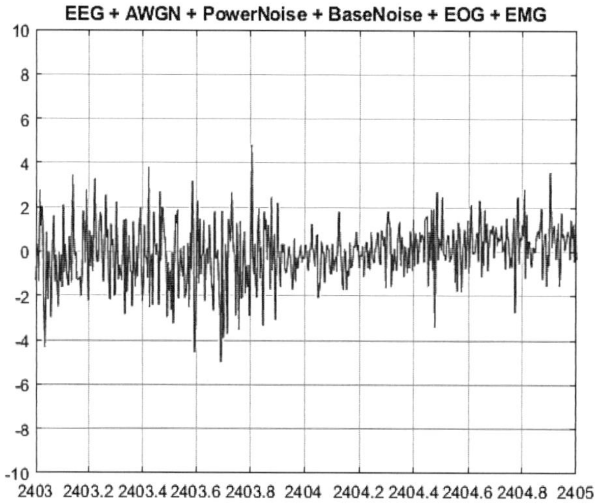

Figure 02: Noise-corrupted EEG Signal for a Short Duration

3.2 De-Noising Techniques

The de-noising techniques used here are the fully connected neural network, the cascaded recursive least square adaptive algorithm and the cascaded least mean square adaptive algorithm.

3.2.1 Fully Connected Neural Network

Any type of filter can be considered as a function which takes contaminated signal in the input and produces clean signal in the output. In Fully Connected Neural Network (FCNN), the function is designed in such a way that, it approximates that function automatically. It contains the input layer and the output layer and a combination of hidden layers between them. The network contains some neurons in each layer and all the neurons of one layer is connected with the neuron of other layers. That is why, the network is named Fully Connected Neural Network[6].

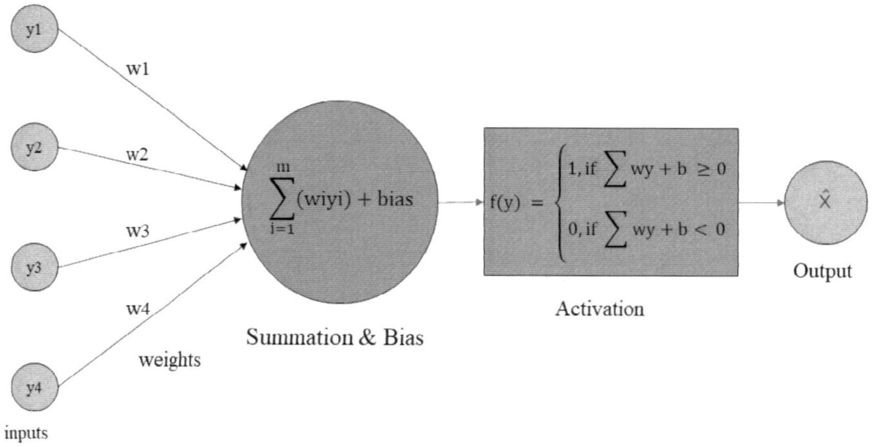

Figure 03: A Single-Layer Perceptron

It takes an input vector **y** and multiplies it some weights (w1, w2, w3, w4) and adds a bias. Finally, the products are summed and the result goes through an activation function. The activation function imposes a nonlinearity in the signal. Finally, the output $\hat{x}$ is generated after taking the output of the activation function.

In this paper, Rectified Linear Unit (ReLU) is used as the activation function which can be expressed as,

$$f(y) = y^{+} = \max(0, y) \dots\dots\dots (4)$$

Where y is the input to the neuron. ReLU imposes a nonlinearity in the signal by allowing only positive inputs. Dropout regularization is also used in between the layers to overcome the problem of overfitting.

3.2.1.1 Preparing the Required Simulated Epochs

To create the noise-corrupted data, 3800 pairs of EEG and noise epochs were utilized. The training set is made up of 3000 pairs, the validation set is made up of 400 pairs, and the test set is made up of the left 400 pairs. At first, randomly combined 3000 pairs of EEG and noise data are generated ten times according to formula (1) and (3) and used to train the model. The SNR's value was distributed uniformly between -7dB and 2dB in this case. The training network grew to 30000 epochs as a result. Similarly, 400 pairs of epochs with 10 different SNR values (-7dB, -6dB, -5dB, -4dB, -3dB, -2dB, -1dB, 0dB, 1dB, 2dB) were randomly paired for training purposes. The testing dataset grew to 4000 epochs as a result.

3.2.1.2 The Architecture of the Fully-Connected-Neural-Network (FCNN)

At first the network takes the contaminated signal as input and in each layer multiplies it with some random weights and adds biases. After traveling through all the layers, a signal appears in the output. Then the signal is compared with the actual ground truth signal and a large error is observed. The network's weights and biases are then modified as a result of this error propagating across all the preceding levels. The output error is decreased by modifying the weights and biases. The network is therefore trained in such a way that the error is reduced to the smallest value achievable.

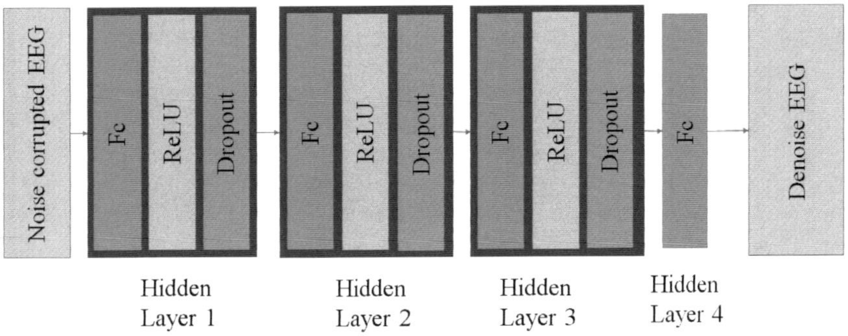

Figure 04: The FCNN Architecture

Once, this network is trained, it can be tested with some unknown signals using the trained weights and biases.

We constructed a FCNN with 4 hidden layers (Figure 04). The number of neurons in each FCNN layer is equal to the input signal's number of data points. Here, the number of neurons is 512. To avoid the problem of overfitting, we randomly drop 30% of the neurons of each layer.

3.2.1.3 Learning Process

A nonlinear function f should be designed in such a way that it would map the noise corrupted EEG signal y to the clean EEG signal x. This can be expressed simply as,

$$f: y \rightarrow x \dots\dots\dots (5)$$

We use y to represent a sample from the noise corrupted EEG signal distribution P(y), and x to represent a sample from the clean EEG signal distribution P(x). Function f is used to represent the denoising network. As a result, f maps samples from the P(y) distribution to the P(z) distribution. The aim now is to minimize the difference between P(z) and P(x).

For the simplification purpose, signals are divided by the standard deviation of the noise corrupted EEG signal, σ_y.

$$\hat{x} = \frac{x}{\sigma_y} \ldots\ldots\ldots (6)$$

$$\hat{y} = \frac{y}{\sigma_y} \ldots\ldots\ldots (7)$$

After the denoising process the signals are multiplied by the standard deviation. The loss function we utilize is the mean squared error (MSE). $L_{MSE}(f)$ is calculated using the formula below [6],

$$L_{MSE}(f) = \frac{1}{N}\sum_{i=1}^{N} \|f_i(y) - x_i\|_2^2 \ldots\ldots\ldots (8)$$

Here, $f_i(y)$ denotes the ith sample of the output epoch of the network, and x_i is the ith sample of the ground truth epoch x and N represents the number of samples in each epoch. For the training purpose of our FCNN we use 400 epochs.

3.2.2 Adaptive Filter

Adaptive filter is based on the optimization theory [16]. Adaptive filter has the capacity to modify its properties as the filter coefficients change according to selected features of the signal which is being passed through the filter.

In Figure 05, y(n) is the input primary signal and s(n) is the secondary signal which is actually the noise signal. The output $\hat{s}(n)$ is generated by the linear filter H(z), which is the estimated noise signal. Then $\hat{s}(n)$ is subtracted from y(n) to compute the error signal e(n). The error signal e(n) is used to update the coefficient of the filter according to the optimization algorithm [8].

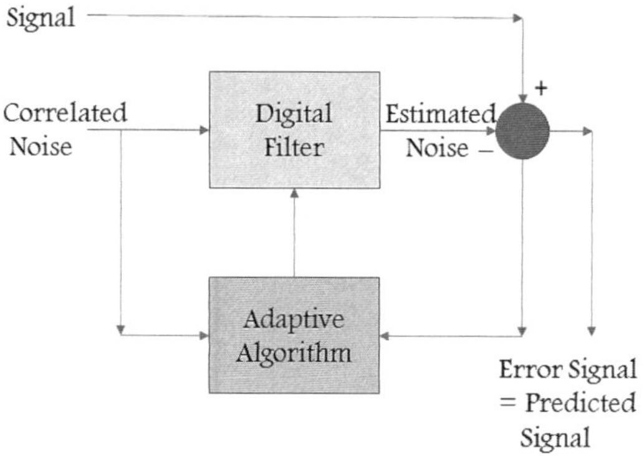

Figure 05: Single-Stage Adaptive Filter

For a single-stage Adaptive filter, it can be written simply for the FIR digital filter that,

$$\hat{s}(n) = \sum_{k=0}^{L} w_k\, s(n-k) \ldots\ldots\ldots (9)$$

Here, w_k is the filter coefficient, L is the filter order, and $\hat{s}(n)$ is the filter output [8]. So, error signal can be calculated by the formula (10),

$$e(n) = y(n) - \hat{s}(n) \ldots\ldots\ldots (10)$$

So,

$$e(n) = y(n) - \sum_{k=0}^{L} w_k\, s(n-k) \ldots\ldots\ldots (11)$$

3.2.2.1 Cascaded RLS Adaptive Filters

The weighted least square error,

$$\varepsilon(n) = \sum_{i=0}^{n} \lambda^{n-i} |e(i)|^2 \ldots\ldots\ldots (12)$$

Here λ is the exponential weighting factor. Taking the derivative of equation (3-12) with respect to $w_k^*(n)$ the formula can be derived [8],

$$\mathbf{R}_s(n)\mathbf{w_n} = \mathbf{r}_{ys}(n) \ldots\ldots\ldots (13)$$

Here $\mathbf{R}_s(n)$ is the autocorrelation matrix at time n and $\mathbf{r}_{ys}(n)$ is the cross-correlation matrix between y(n) and s(n). The inverse of autocorrelation matrix $\mathbf{R}_s(n)$ is taken as,

$$\mathbf{P}(n) = \mathbf{R}_s^{-1}(n) \ldots\ldots\ldots (14)$$

Filter order is L. The weighting vector $\mathbf{w}_0 = \mathbf{0}$ is taken for initialization. δ is the value used to initialize $\mathbf{P}(0)$ using the formula (15),

$$\mathbf{P}(0) = \delta^{-1}\, \mathbf{I} \ldots\ldots\ldots (15)$$

The gain vector is $\mathbf{g}(n)$ is calculated according to the following the formulae,

$$\mathbf{z}(n) = \mathbf{P}(n\text{-}1)\, \mathbf{s}^*(n)$$

$$\mathbf{g}(n) = \frac{1}{\lambda + \mathbf{s}^T(n)\mathbf{z}(n)} \mathbf{z}(n)$$

$$\alpha(n) = y(n) - \mathbf{w}_{n-1}^T \mathbf{s}(n)$$

So, the updated weighting vector is calculated according to formula (16),

$$\mathbf{w_n} = \mathbf{w}_{n-1} + \alpha(n)\mathbf{g}(n) \ldots\ldots\ldots (16)$$

Figure 06 shows the flow chart of cascaded adaptive filter following RLS adaptive algorithm. For each of the filters, filter order is taken 20.

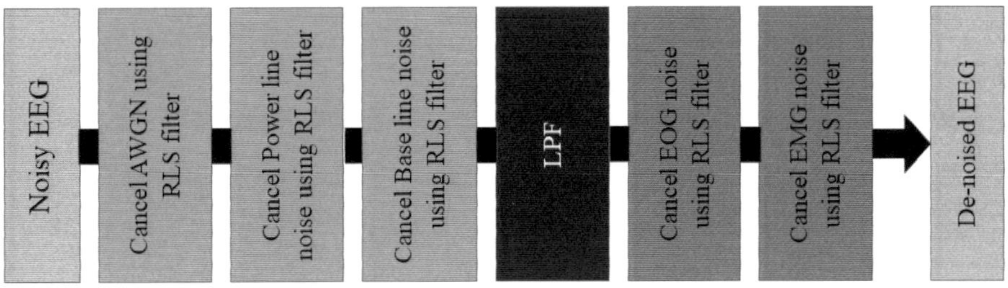

Figure 06: Flow Chart of Cascaded RLS Adaptive Filters

As white noise of high frequency still prevails, an equiripple Low pass filter is used. As a result the high frequency noise can be filtered out. The cut off frequency for the LPF is taken 50Hz.

3.2.2.2 Cascaded LMS Adaptive Filters
In the adaptive LMS filter, initially the weights are taken to be very small. The filter weights are updated at each of the steps based on the gradient of the mean square error. From the equation (11) we derive the LMS algorithm updated coefficient [4],

$$e^2(n) = y^2(n) - 2y(n)\sum_{k=0}^{L} w_k s(n-k) + \left[\sum_{k=0}^{L} w_k s(n-k)\right]^2 \quad \dots\dots\dots (17)$$

The filter coefficient is updated by the formula (18),

$$w_k(n+1) = w_k(n) + \mu(-\nabla_k) \quad \dots\dots\dots (18)$$

where μ is the coefficient used to control adaptation rate.

$$\nabla_k = \frac{\partial\{e^2(n)\}}{\partial w_k(n)} \quad \dots\dots\dots (19)$$

Combining equations (18) and (19) we get,

$$w_k(n+1) = w_k(n) - \mu\frac{\partial\{e^2(n)\}}{\partial w_k(n)} \quad \dots\dots\dots (20)$$

Finally, we can write,

$$w_k(n+1) = w_k(n) - 2\mu\, e(n)s(n-k) \quad \dots\dots\dots (21)$$

Figure 07 shows the flow chart of cascaded adaptive filter following LMS adaptive algorithm. For each of the filters, filter order is taken 20. The value of step size is taken μ = 0.005.

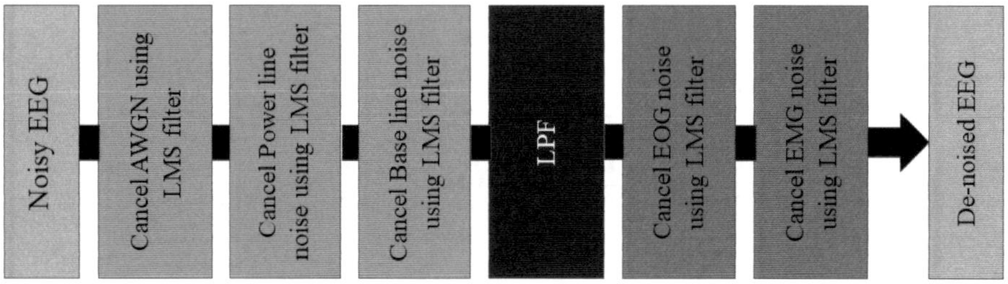

Figure 07: Flow Chart of Cascaded LMS Adaptive Filters

As white noise of high frequency still prevails, an equiripple Low pass filter having a cut off frequency of 50Hz is used to filter the high frequency noise.

4. Result
The performance of different de-noising schemes has been compared using ΔSNR and RRSME. The value of these parameters for clean EEG has been taken as the benchmark.

4.1 Noisy EEG Signal as The Input
To calculate the performance, we use the noise corrupted EEG signal as the input signal in our algorithms and generate the predicted output. And then compare the predicted output with the actual clean EEG signal.

4.1.1 Time-Domain Plot Using the Noisy EEG
Time-domain plot of the ground-truth EEG signal and Predicted EEG signal are illustrated.

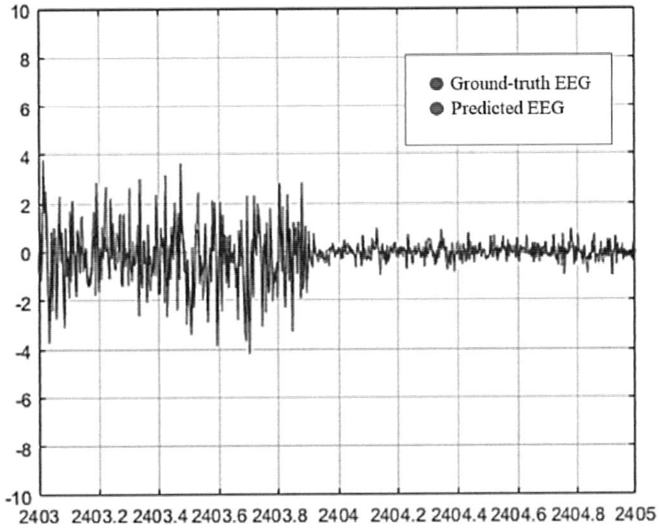

Figure 08: Ground-truth EEG Signal & FCNN Predicted Signal

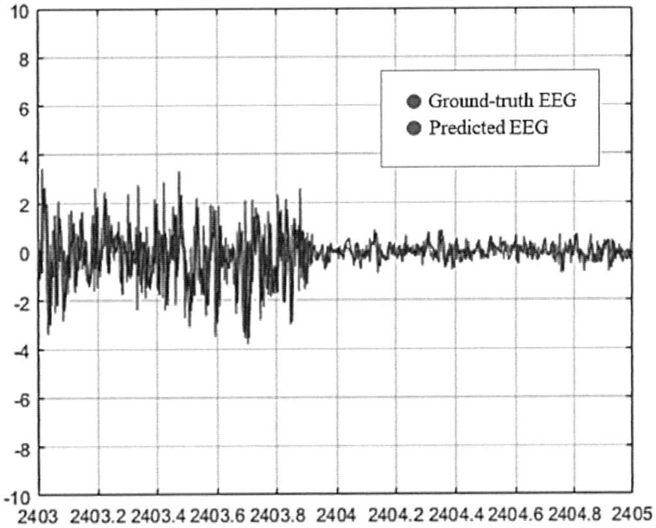

Figure 09: Ground-truth EEG Signal & RLS Predicted Signal

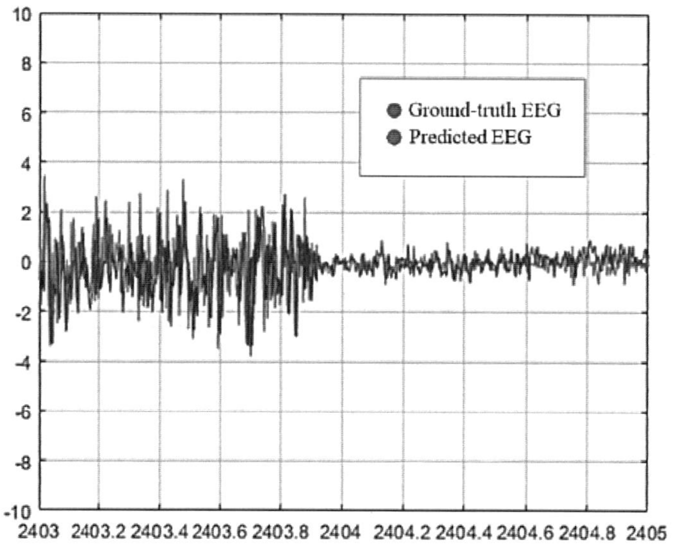

Figure 10: Ground-truth EEG Signal & LMS Predicted Signal

The FCNN and RLS predicted signals follow the ground-truth EEG signal better than LMS.

4.2 Comparison of performance

To calculate the performance we use the ground-truth EEG signal as the input signal instead of the noisy signal. The parametric calculation is done so that the performance can be evaluated. The signals are illustrated in the same plot (Figure 11).

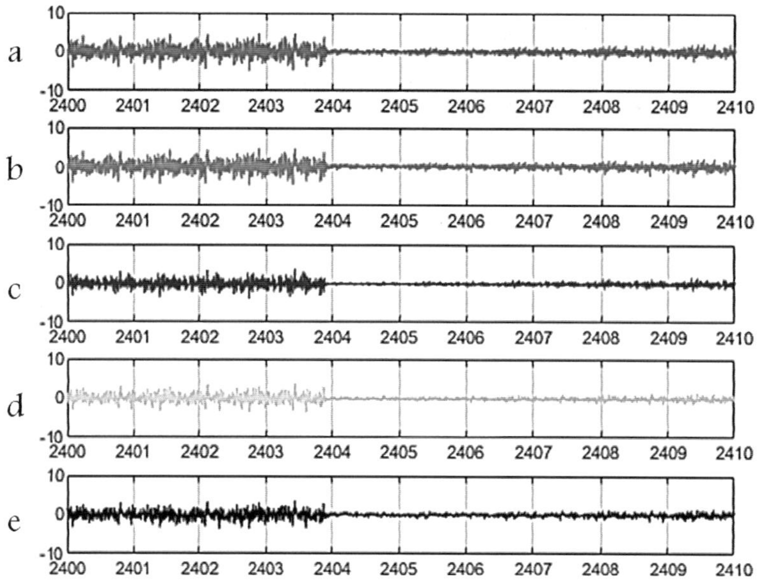

Figure 11: Ground-truth EEG signal in the input, (a) Actual EEG Signal, (b) clean EEG signal as input instead of noisy EEG Signal, (c) FCNN Predicted EEG Signal, (d) RLS Algorithm Predicted EEG Signal & (e) LMS Predicted EEG Signal

Now the noisy EEG signal is taken as the input and the parametric calculation is done. All these signals are illustrated in the same plot (Figure 12) to make a visual illustration.

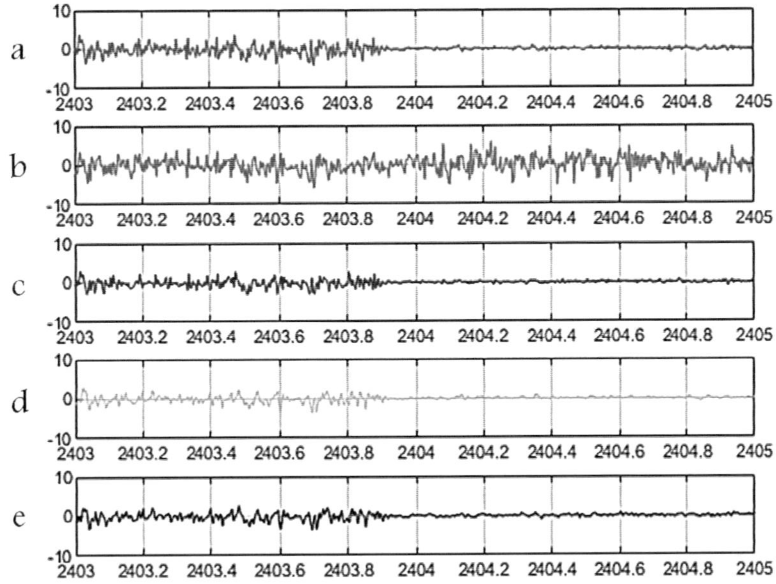

Figure 12: Noisy EEG Signal in The Input (a) Actual EEG Signal, (b) Noise Corrupted EEG Signal, (c) FCNN Predicted EEG Signal, (d) RLS Predicted EEG Signal & (e) LMS Predicted EEG Signal to Make a Visual Comparison

4.2.1 Performance Calculating Parameters

Three parameters are calculated here to compare the performance of the algorithms. The formulae to calculate these parameters are described below.

ΔSNR is the improvement in signal-to-noise ratio. Here to calculate the ΔSNR, variance of the signal is used in equation (22) [17].

$$\Delta\text{SNR} = 10\log_{10}\left[\frac{\text{variance(actual signal)}}{\text{variance(predicted signal)}}\right] - 10\log_{10}\left[\frac{\text{variance(actual signal)}}{\text{variance(noisy signal)}}\right] \quad \dots\dots\dots (22)$$

Relative RMS error is calculated with respect to the RMS value of the predicted signal using equation (23) [6].

$$\text{RRMSE} = \frac{\sqrt{\text{mean(predicted signal} - \text{actual signal)}^2}}{\sqrt{\text{mean(predicted signal)}^2}} \quad \dots\dots\dots (23)$$

Table 01 shows the values of ΔSNR and RRMSE for both clean EEG signal and the noisy EEG signal.

	ΔSNR (dB)		Relative RMS error	
	Clean EEG in the input	Noisy EEG in the input	Clean EEG in the input	Noisy EEG in the input
FCNN	2.9171	8.2863	0.7037	0.7083
RLS	2.8028	7.9535	1.7767	1.8055
LMS	2.4232	7.3257	1.8074	1.8526

Table 01: Clean EEG & Noisy EEG as Input Signal, Values of ΔSNR and RRMSE Clarifies that FCNN gives a better result

As, in the case of clean EEG signal in the input, the predicted signal is almost similar as the clean EEG signal. So, there is little scope to improve the SNR. But In case of noisy EEG signal in the input, there is a high scope of improving the value of the SNR. The ΔSNR is the greatest for FCNN and least from LMS for both clean EEG and noisy EEG.

For clean EEG signal, RRMSE always has a lower value than for noisy EEG signal which is expected. But actually, the relative RMS error has almost the same value for both clean EEG and noisy EEG. In both cases, the Relative RMS error gives much less value for FCNN than others. So, it can be said that, FCNN algorithm has better performance.

5. Discussion

A comparison among different algorithms of EEG signal De-noising is conducted to bring out the effect of the noise minimizing techniques over the simulated noisy EEG signal. If the origin of the noise is unknown, the adaptive algorithms fail to de-noise the signal. The origin of the baseline wandering noise signal is unknown in various cases. So, the output of the adaptive filters is erroneous. On the other hand, to minimize the noise using the neural network algorithm, we need a dataset to train the network and also there exists a chance of overfitting. We found that, the output is fine enough to ensure that the algorithms are valid. Exploring the parameters we say, FCNN gives the best output.

References

[1] Malmivuo, Jaakko, and Robert Plonsey. Bioelectromagnetism: principles and applications of bioelectric and biomagnetic fields. Oxford University Press, USA, 1995.

[2] Brown, B.H., Smallwood, R.H., Barber, D.C., Lawford, P.V. and Hose, D.R., 2017. Medical Physics and Biomedical Engineering: Medical Science Series. CRC Press..

[3] Karagiannis, A. and Constantinou, P., 2009, November. Noise components identification in biomedical signals based on empirical mode decomposition. In 2009 9th International Conference on Information Technology and Applications in Biomedicine (pp. 1-4). IE.

[4] Correa, A.G., Laciar, E., Patiño, H.D. and Valentinuzzi, M.E., 2007, November. Artifact removal from EEG signals using adaptive filters in cascade. In Journal of Physics: Conference Series (Vol. 90, No. 1, p. 012081). IOP Publishing..

[5] Reddy, A.G. and Narava, S., 2013. Artifact removal from EEG signals. International Journal of Computer Applications, 77(13)..

[6] Zhang, H., Wei, C., Zhao, M., Liu, Q. and Wu, H., 2021, June. A novel convolutional neural network model to remove muscle artifacts from EEG. In ICASSP 2021-2021 IEEE International Conference on Acoustics, Speech and Signal Processing (ICASSP) (pp. 1265-1.

[7] Abdul-Latif, A.A., Cosic, I., Kumar, D.K., Polus, B. and Da Costa, C., 2004, December. Power changes of EEG signals associated with muscle fatigue: the root mean square analysis of EEG bands. In Proceedings of the 2004 Intelligent Sensors, Sensor Networks.

[8] Nefian, A.V. and Hayes, M.H., 1998, May. Hidden Markov models for face recognition. In Proceedings of the 1998 IEEE International Conference on Acoustics, Speech and Signal Processing, ICASSP'98 (Cat. No. 98CH36181) (Vol. 5, pp. 2721-2724). IEEE..

[9] Böcker, K.B., van Avermaete, J.A. and van den Berg-Lenssen, M.M., 1994. The international 10–20 system revisited: Cartesian and spherical co-ordinates. Brain Topography, 6(3), pp.231-235..

[10] Lin, J., Sun, X., Wu, J., Chan, S.C. and Xu, W., 2016, November. Removal of power line interference in EEG signals with spike noise based on robust adaptive filter. In 2016 IEEE Region 10 Conference (TENCON) (pp. 2707-2710). IEEE..

[11] https://www.probabilitycourse.com/chapter10/10_2_4_white_noise.php.

[12] Marino, M., Liu, Q., Koudelka, V., Porcaro, C., Hlinka, J., Wenderoth, N. and Mantini, D., 2018. Adaptive optimal basis set for BCG artifact removal in simultaneous EEG-fMRI. Scientific reports, 8(1), pp.1-11..

[13] Kher, R. and Gandhi, R., 2016, April. Adaptive filtering based artifact removal from electroencephalogram (EEG) signals. In 2016 International Conference on Communication and Signal Processing (ICCSP) (pp. 0561-0564). IEEE..

[14] Liu, G.T., Volpe, N.J. and Galetta, S.L., 2019. The neuro-ophthalmic examination. In Liu, Volpe, and Galetta's Neuro-Ophthalmology (pp. 7-36). Elsevier..

[15] Potvin, J.R., 1997. Effects of muscle kinematics on surface EMG amplitude and frequency during fatiguing dynamic contractions. Journal of Applied Physiology, 82(1), pp.144-151..

[16] Islam, S.M.M. and Farid, M.S.U., 2015. Denoising EEG signal using different adaptive filter algorithms. Int. Journal of Enhanced Research in Science, Technology & Engineering, 4..

[17] Väisänen, O. and Malmivuo, J., 2009. Improving the SNR of EEG generated by deep sources with weighted multielectrode leads. Journal of Physiology-Paris, 103(6), pp.306-314..

[18] Abdullah, M.M.B., Uli, J. and Tari, J.J., 2008. The influence of soft factors on quality improvement and performance: Perceptions from managers. The TQM Journal..

[19] Roy, V., Shukla, S., Shukla, P.K. and Rawat, P., 2017. Gaussian elimination-based novel canonical correlation analysis method for EEG motion artifact removal. Journal of healthcare engineering, 2017..

[20] Chen, J., Li, X., Mi, X. and Pan, S., 2014, October. A high precision EEG acquisition system based on the CompactPCI platform. In 2014 7th International Conference on Biomedical Engineering and Informatics (pp. 511-516). IEEE..

[21] Mannan, M.M.N., Kamran, M.A. and Jeong, M.Y., 2018. Identification and removal of physiological artifacts from electroencephalogram signals: A review. Ieee Access, 6, pp.30630-30652..

[22] Yang, B., Duan, K., Fan, C., Hu, C. and Wang, J., 2018. Automatic ocular artifacts removal in EEG using deep learning. Biomedical Signal Processing and Control, 43, pp.148-158..

[23] He, Y., He, H., Li, L., Wu, Y. and Pan, H., 2008, December. The applications and simulation of adaptive filter in noise canceling. In 2008 International Conference on Computer Science and Software Engineering (Vol. 4, pp. 1-4). IEEE..

[24] Yang, B., Duan, K. and Zhang, T., 2016. Removal of EOG artifacts from EEG using a cascade of sparse autoencoder and recursive least squares adaptive filter. Neurocomputing, 214, pp.1053-1060..

[25] Sudha, N.S. and Dodda, R.K., 2017, May. Design of error normalized LMS adaptive filter for EEG signal with eye blink & PLI artefacts. In 2017 International Conference on Trends in Electronics and Informatics (ICEI) (pp. 54-58). IEEE..

[26] Falkner, T., Abedi, A., Hayes, M. and Paul, J., 2010, May. Improving the signal-to-noise ratio of event-related eeg signals in high risk newborns. In 2010 25th Biennial Symposium on Communications (pp. 28-31). IEEE..

[27] https://github.com/ncclabsustech/EEGdenoiseNet/blob/master/EEGdenoiseNet%20da taset.pdf.

[28] https://physionet.org/content/eegmmidb/1.0.0/.

[29] https://console.cloud.google.com/storage/browser/emgdb-1.0.0.physionet.org?pli=1.

[30] Iriarte, J., Urrestarazu, E., Valencia, M., Alegre, M., Malanda, A., Viteri, C. and Artieda, J., 2003. Independent component analysis as a tool to eliminate artifacts in EEG: a quantitative study. Journal of clinical neurophysiology, 20(4), pp.249-257..

[31] Jafarifarmand, A. and Badamchizadeh, M.A., 2013. Artifacts removal in EEG signal using a new neural network enhanced adaptive filter. Neurocomputing, 103, pp.222-231..

[32] Akin, M., 2002. Comparison of wavelet transform and FFT methods in the analysis of EEG signals. Journal of medical systems, 26(3), pp.241-247..

[33] Tang, Q. and Wang, X., 2019, August. A Periodogram-Based Method to Identify Forced and Natural Oscillations Using PMUs. In 2019 IEEE Power & Energy Society General Meeting (PESGM) (pp. 1-5). IEEE..

[34] Vijayakumar, T., R. Vinothkanna, and M. Duraipandian. "Fusion based feature extraction analysis of ECG signal interpretation–a systematic approach." Journal of Artificial Intelligence 3, no. 01 (2021): 1-16.

[35] Dhaya, R. "Deep net model for detection of covid-19 using radiographs based on roc analysis." Journal of Innovative Image Processing (JIIP) 2, no. 03 (2020): 135-140.

[36] Bhuvaneshwari, M., and E. Grace Mary Kanaga. "Classification of VEP-Based EEG Signals Using Time and Time-Frequency Domain Features." In Intelligent Data Communication Technologies and Internet of Things, pp. 737-747. Springer, Singapore, 2021.

Artificial Intelligence & Deep Learning: Pet and Spect Imaging and Classification on Brain MRI Images List

[1]Amit Kumar Goel, [2]Sagar Yadav, [3]Md. Zubair, [4]Naman Jain

[1]Professor, School of Computing Science and Engineering, Galgotias university, India.

[2]M.Tech Student, School of Computing Science and Engineering, Galgotias University, India.

[3,4]B.Tech Student, School of Computing Science and Engineering, Galgotias University, India.

amit.goel@galgotiasuniversity.edu.in

Abstract: Cognitive abilities commonly linked with human brains include learning and problem solving. Another option is just a preprogrammed rule that instructs the computer to perform a given action under specified circumstances. Voice recognition and image recognition, for instance, are both widely used in daily life and may be found everywhere. "Deep learning" (DL) approaches, in particular,due to their ability to come up with creative solutions to complex situations throughout the past decade. Both cancerous and non-cancerous tumors can be seen in the human brain (benign). Google developed the open-source deep learning library Tensor Flow. It can be combined with traditional machine learning. It was created on top of PIL that the Python Pillow module exists (Python Image Library). This collection of modules is critical to Python's ability to perform image processing. "A deep learning API developed in Python" is powered by "the TensorFlow-based machine learning platform TensorFlow." NumPy is a Python library for manipulating arrays, as the name suggests. Also included are linear algebra, the Fourier transform, and matrix functions. As far as the ResNet50 model goes, it has 48 convolutional layers, a maxpool and an averagepool layer. Up to a scale of 3.8 x 109, floating point operations can be performed. ResNet-50 is a type of neural network with 50 layers. You can import a pre-trained version of the network from the ImageNet database.

Keywords: SPECT, Image Processing, Image Resizing, Image Segmentation

1. Introduction

Deep learning (DL), a type of artificial intelligence (AI) approach, has gotten a lot of interest recently due to its potential to provide new answers to challenging problems. Additionally, clinical and scientific contexts are increasingly incorporating the use of artificial intelligence and deep learning (AI/DL). Medical imaging using many imaging modalities has demonstrated modern AI/DL algorithms to be exceptionally good at learning from large and complicated data sets.

Picture segmentation and categorization in medical imaging have been re-examined by deep learning approaches that have improved image segmentation. Data adjustments were also reviewed in light of the new information (such as removing artefacts or noise). Additionally, images were investigated and altered (prognosis, diagnosis, and monitoring of treatment response). In order to overcome some of the fundamental limitations of current molecular imaging techniques, an AI-based strategy has been devised. Deep learning's superior performance over previous methods makes a paradigm shift plausible.

Radiologists and clinical experts may rely on existing information to execute this difficult and time-consuming operation, but MRI scans pose substantial challenges in finding and extracting tumor locations. Human organ anatomical structures can be seen utilizing image processing techniques. Basic imaging techniques have a hard time detecting aberrant structures in the human brain.

A brain tumor develops when the brain or central spinal cord grow abnormally. Secondary (metastatic) tumors develop outside of the brain, as opposed to primary (localized) tumors. benign and malignant (cancer-causing) brain tumors are both possible (benign). More than 120 different types of brain tumors are classified by the WHO based on the type and location of tumor cells. They are split into groups based on the cells from which they originate. In contrast to grades I and II, grades III and IV are considered to be "higher-grade tumors." An inaccurate diagnosis of a brain tumor can have life-threatening effects, hence it is critical that the patient's diagnosis and treatment be as precise as possible.

MRI scans of patients have previously been used to identify and locate brain tumours. Preprocessing, edge detection (Sobel, Prewitt, and Canny edges) and segmentation are some of the methods suggested (thresholding). The tumor's location was finally pinpointed with the help of k-means clustering. As a result, the output image contains more regions than just the tumour, which could lead to confusion.(Jia & Chen, 2020)

"Applications of deep learning in SPECT and PET imaging"

a) "Instrumentation and image acquisition/formation"

In PET scanners, detector modules play a crucial function. If you're interested in PET imaging, the ideal detector should be able to precisely locate events while also exhibiting exceptional energy and time resolution capabilities. In order to distinguish between random and scattered photons and genuine coincidences, Incoming photons must be able to be precisely identified by the detector. "The sensitivity, spatial resolution and signal-to-noise ratio" of the scanner is affected by these parameters (true coincidence versus scatters or random). There has been a lot of progress in PET equipment, despite this, machine learning algorithms may be employed to handle complex and multi-parametric challenges.

PET scanners' total spatial resolution is enhanced by precise localisation of the interaction location inside the crystals. With optical photons being distributed in a random manner and numerous Compton scattering and reflection occurring at the crystal edges, precise placement of the interaction is difficult. Machine learning approaches surpassed Anger logic and correlated signal enhancement in finding the crystal's centre of gravity.

PET instrumentation's ability to precisely measure time is also a factor in TOF imaging and random and scatter rejection efficiency. Noise introduces substantial statistical uncertainty and bias into the data. Simple strategies like feeding detector signals into a CNN model have shown promising results for estimating TOF information. For a training dataset, 68Ga point sources were scanned at 5-mm intervals over the PET scanner's field of view to collect data for a deep learning algorithm (reference).

Journal of Physics: Conference Series **2325** (2022) 012039 doi:10.1088/1742-6596/2325/1/012039

A TOF resolution of 185 picoseconds was found to be significantly higher than that of conventional techniques, which typically have resolutions ranging from 210 to 527 picoseconds.

The speed at which photons go through a crystal and their interactions inside it may be precisely predicted using deep learning algorithms, according to current study. Significant experimental measurements are required to provide the labelled data (which is used as a reference in these models), which is a big problem. A single pencil beam and a series of measurements are required for location estimation training. Similar tasks can be achieved utilizing uniform or fan-beam sources, as well as the application of "clustering to the training dataset".Data from hundreds of point source sites must be collected for TOF training purposes. The temporal modification of PET detector waveforms has been suggested as a means of creating fake ground truth data.

ASICs should be used to build powerful machine learning-based algorithms for event placement, timing, and calibration in the next generation of SPECT and PET systems (FPGAs). It's a good idea for nuclear imaging systems to build a single model that can extract the time, location, and energy from photo detector outputs.(Arabi et al., 2021)

2. Review of Literature

(Zaidi & El Naqa, 2021)A wide range of medical imaging applications, including better scanning and disease diagnosis, have benefited from deep learning. Deep learning algorithms for various imaging modalities have been developed as a result of these developments. If you'd want to learn more about the most prevalent methods and how to implement them, you may read this article. Next, we'll take a closer look at neural networks and their potential applications in medical imaging. This article covers the radiology pipeline in the industry and compares more traditional approaches to more in-depth learning in the field of radiology.

(Arabi & Zaidi, 2020)Such research is greatly aided by AI, particularly deep learning techniques. Low-dose PET imaging, image denoising, radiation dosimetry, and computer-aided diagnostics are all included in this category. To provide a better understanding of the triumphs and challenges AI and deep learning in healthcare face, this study briefly examines these technologies.

(Nensa et al., 2019)For many medical practitioners, the term "artificial intelligence" (AI) and its inner workings remain a mystery, leading to both unrealistic expectations and unfounded fears. This study provides a conceptual taxonomy and a brief introduction to AI's technological principles. Artificial Intelligence's potential impact on nuclear medicine is also examined, as well as the challenges it may pose. The basic architectures and methods utilized in deep learning applications are also briefly described. There are many hurdles to overcome before full-scale adoption of AI-based solutions for PET and SPECT pictures in the clinic can be achieved, but there are also many potential".

(Arabi, Hossein; AkhavanAllaf, Azadeh; Sanaat, Amirhossein; Shiri, Isaac; Zaidi, 2021)An increasing amount of AI is being applied in SPECT and PET imaging, notably in the form of deep learning algorithms. The goal of this research is to outline the most significant applications of AI in these imaging modalities in detail. So, several imaging modalities are mentioned, followed by a description of AI-based solutions to address the underlying limitations and limits of these imaging techniques. AI-based solutions for PET and SPECT images in the clinic face a number of challenges before they can be widely adopted, but there is also a great deal of potential.

(Tahrir Ibraq Siddiqui, 2021)In the Siemens SPECT-CT devices, two hefty detector heads (each weighing approximately 1500 pounds) can be moved into various positions for radioactive imaging. Large torque motors in the gantry drive these detectors, allowing for both linear and rotational motion.

There is a high probability that the detectors will damage the items they collide with (stools, tables, patient limbs, etc.) and themselves if they come into contact with them. Deep learning object detection is proposed to prevent collisions between detector heads and objects in the course of detector motion by creating a complete end-to-end deep learning object detector.

(Chrysostomou et al., 2019)For SPECT imaging, we've come up with a novel way to rebuild tomographic images. The new "CNN Reconstruction CNNR" approach is based on deep convolutional neural networks (CNNs), which are utilised in the new reconstruction method. The CNNR Projection was trained using data from software phantoms. Both software and hardware phantoms were employed to test the CNNR method's effectiveness. This "Maximum Likelihood Expectation Maximization" (FBP) technique is used to obtain the tomographic images (MLEM) as well as the maximizing of ordered subset expectation (OSEM)

3. Objectives

- To describe the most important uses of AI in these imaging modalities.
- To identify the best object detection framework, collect and process an image dataset of target objects.
- TO examine the tomographic images based on CNN in AI and deep learning.

4. Methodology

4.1. Implementation Details

Libraries

Import all the necessary libraries like Pillow, os, Numpy, matplotlib, etc.

a) Pillow

To display an image, the Python pillow library's image class is utilized. There are a few built-in features in the pillow image modules, such as the ability to load photos or generate new ones.

b) Os

Operating system-related functions are provided by the Python OS module. Python's standard utility modules include the operating system. This module makes it possible to take advantage of operating system-specific features when on the go. File system functions are provided by the operating system and operating system path modules.

c) Numpy

For the Python programming language, NumPy (Numerical Python) provides an open-source library called NumPy. It is utilized in scientific computing and array manipulation. Beyond the array object, high-level working tools for arrays are also provided by this framework.

d) "Matplotlib"

"It is a cross-platform, data visualization and graphical charting package for Python and NumPy". A feasible open source alternative to MATLAB is thus provided by this package. For developers, matplotlib provides APIs that may be used to incorporate plotting into GUI programs.

e) Loading Dataset

The Load method can be used to load data from an IDataReader instance into a single DataTable. In this method, you can load several result sets from an IDataReader into multiple tables within a DataSet.

Data Description
"Total number of images in the training set": 5712
"The total number of images in the testing set: is": 1311
Total number of classes: 4
Class names: glioma, meningioma, notumor, pituitary

f) Sample Images of Dataset
In computer vision, a dataset is a collected collection of digital images used to test, train, and evaluate algorithms.

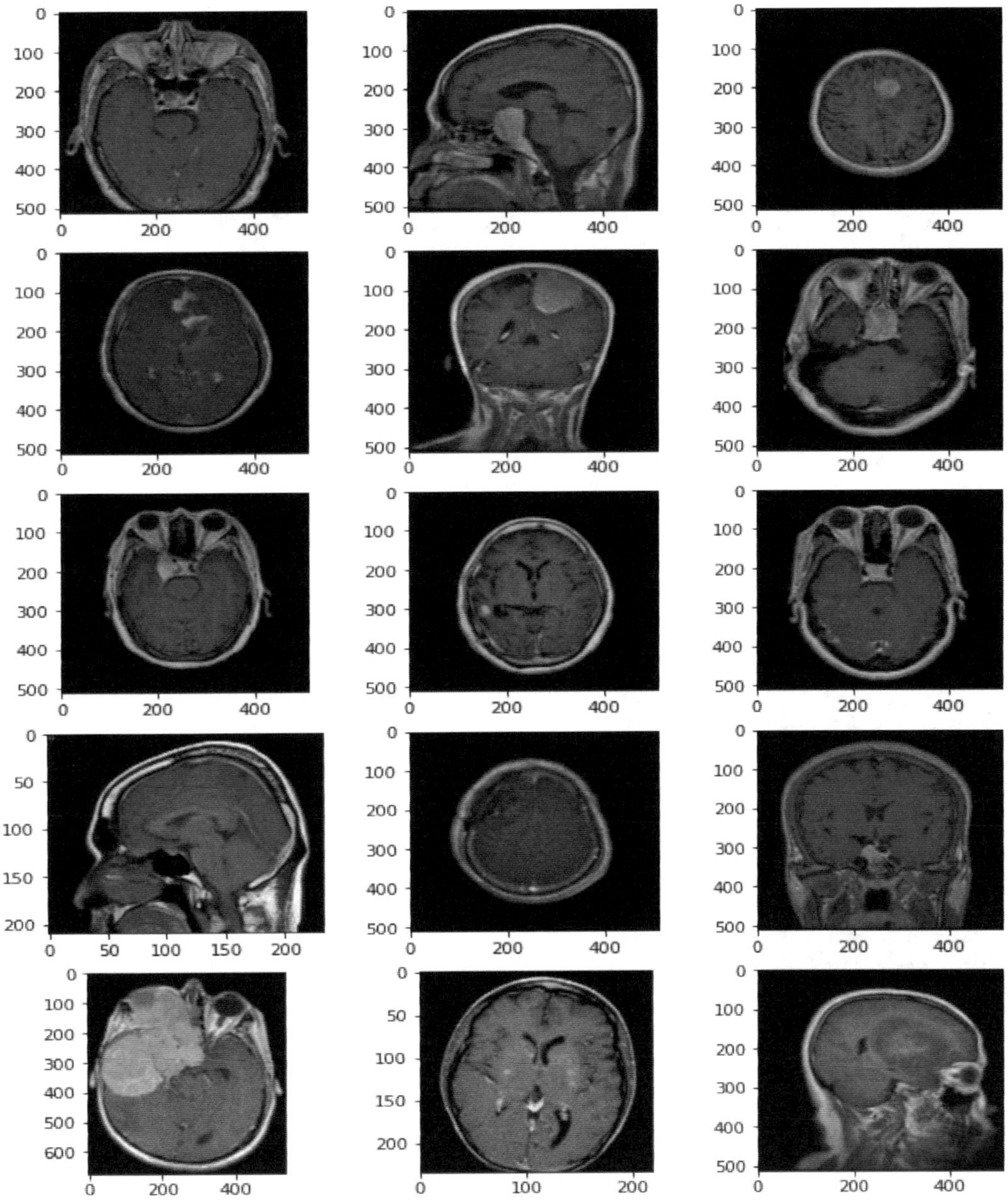

Fig.1: Sample Image of Dataset

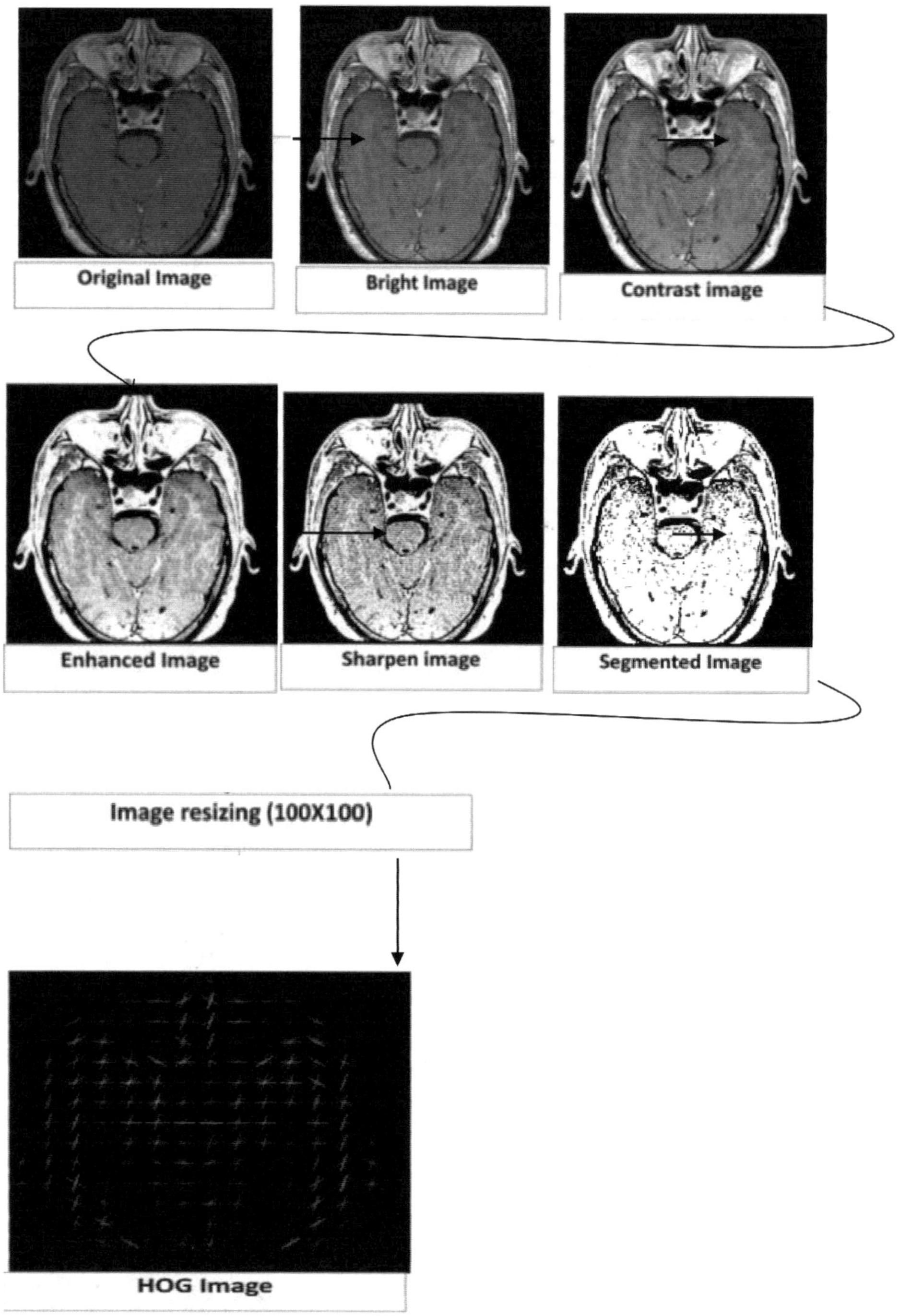

Fig 2: Applying Image Processing

Here, we use pillow library as we already explained.
Firstly, we increase the brightnesss of an Original image. We increase the Contarst of an image. Now, we enhanced the contrast image . Then, we increase the sharpness of an image so that we have a sharp

image.We increase segmentation of an image the image then we resizing the segmented image in 100 length× 100 width.In the end, we put image in the Histogram of Oriented Gradients (HOG).

4.2. Image Resizing:
When you resize an image, you can make it smaller or larger without removing any of the original content. File size and quality are often impacted by resizing an image because the dimensions are altered. A typical reason to resize images is to lower the file size so that they may be emailed or shared on the internet more easily.

4.3. Image Segmentation:
An image is segmented into smaller subsets called Image segments, which simplifies further processing or analysis of the image by lowering its overall complexity. In a nutshell, segmentation is the process of labeling individual pixels. We used OTSU in this. When it comes to image segmentation, the Otsu approach uses a global adaptive binarization threshold.

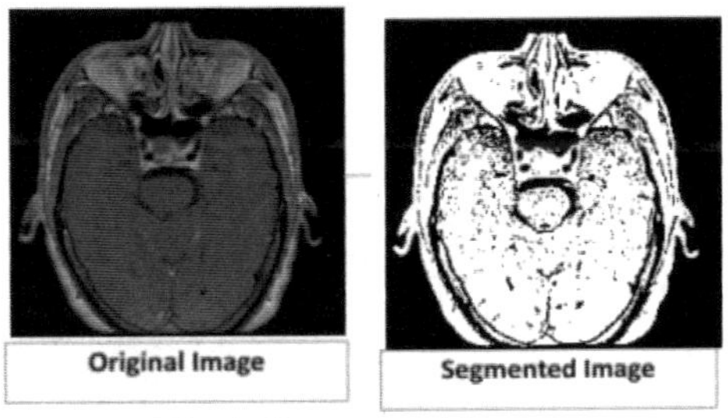

Fig 3: Segmented image

4.4. **HOG**:
As a feature descriptor, the Histogram of Oriented Gradients, or HOG, is similar to Canny Edge Detector (CED) and SIFT (Scale Invariant Feature Transform). Computer vision and image processing are two areas where it is used to identify and locate objects. Gradient orientations are counted in a specific area of an image. Edge Orientation Histograms and Scale Invariant a Feature Transformation (SIFT) are also examples of related techniques (SIFT). Focusing on the object's structure or form, the HOG descriptor is used. It is superior than all other edge descriptors because it calculates the features based on both the magnitude and the angle of the gradient.

4.5. Pre-Processing
In the pre-processing, we make the features and labels of the dataset. And appending all the data and labels into the list.
Assigning labels 0 as 'glioma', 1 as 'meningioma', 2 as 'notumor', and 3 as 'pituitary'

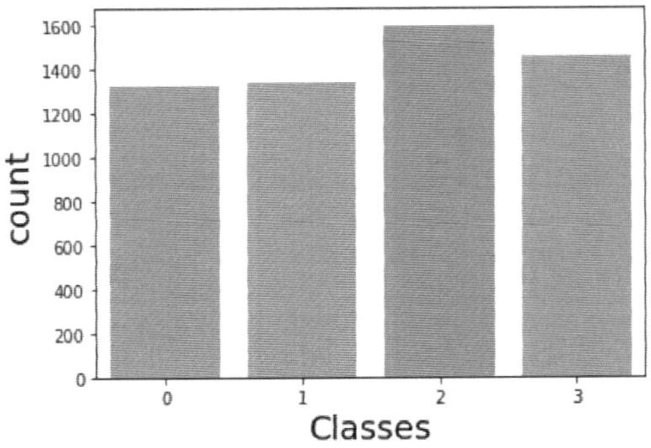

Fig 4: data distribution graph of training data

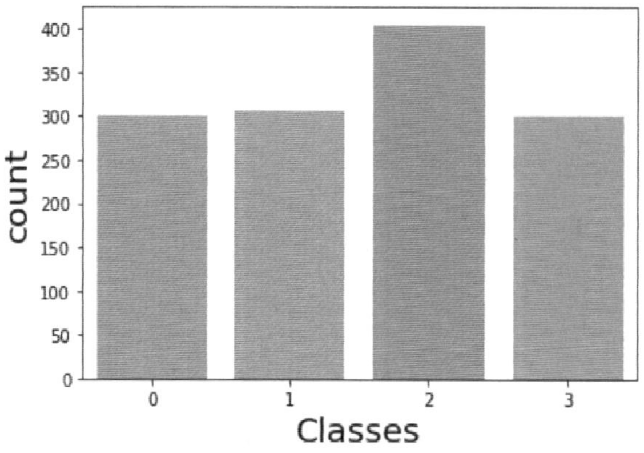

Fig 5: data distribution graph of testing data

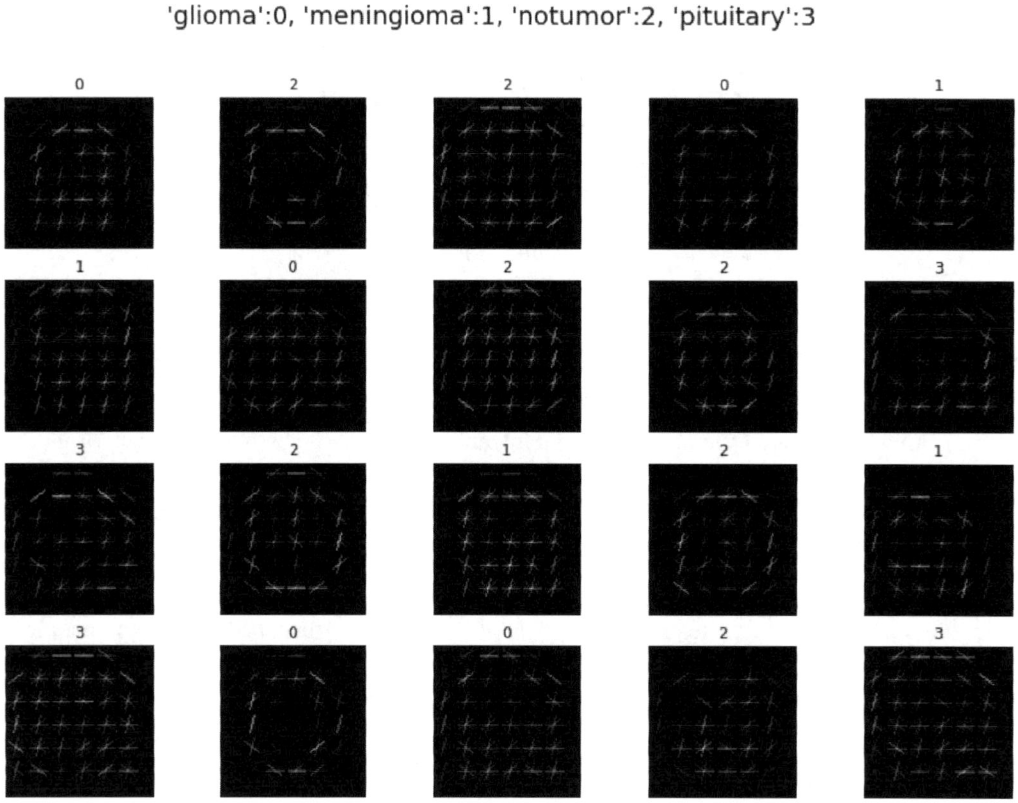

Fig 6: images after the preprocessing and image processing techniques and feature extraction.

4.6. Model Building & Training

We use transfer learning for the model building and training:
Resnet50 is the model we use in this approach.

Layer (type)	Output Shape	Param #
resnet50 (Functional)	(None, 4, 4, 2048)	23587712
flatten_1 (Flatten)	(None, 32768)	0
dropout_2 (Dropout)	(None, 32768)	0
dense_2 (Dense)	(None, 64)	2097216
dropout_3 (Dropout)	(None, 64)	0
dense_3 (Dense)	(None, 4)	260

Total params: 25,685,188
Trainable params: 3,152,196
Non-trainable params: 22,532,992

Fig 7: the model layer architecture:

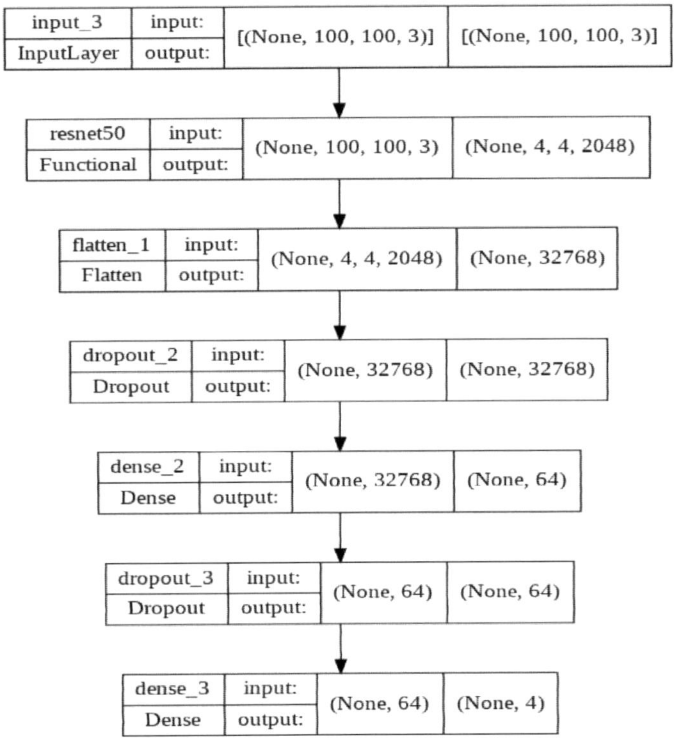

Fig 8: Model: "Resnet50 with layer"

4.7. Parameters

- Optimizer: Adam

A Neural Network algorithm optimization solver, Adam takes little memory and is well-suited to problems with a vast amount of data or parameters or both. Adam is computationally efficient. A prominent variation on stochastic gradient descent is the Adam extension.

Journal of Physics: Conference Series **2325** (2022) 012039 doi:10.1088/1742-6596/2325/1/012039

- Loss function: Categorical cross-entropy

In multi-class classification tasks, a loss function known as categorical crossentropy is employed. These are challenges in which the model must decide which of many possible categories an example belongs to. Formally, it's meant to measure the difference between two probability distributions and express it as a numerical value.

- Learning rate: 0.0001

The learning rate determines how quickly the model is able to adapt to a new situation. Smaller learning rates necessitate more training epochs because the weights are updated less frequently, whereas higher learning rates produce quick weight changes and require fewer training epochs.

- Batch size: 32

Before updating the weights of a model, a batch size of 32 will be utilized to estimate the error gradient from the training data set.

- Epochs: 50

Epochs represent how many times the training dataset has been traversed in its entirety. The number of samples in the training dataset must be fewer than or equal to the batch size. It's possible to set the number of epochs to anything from one to infinite here.we use 50 epochs.

- Activation: softmax

The softmax function is used as the activation function in the output layer of neural network models that predict a multinomial probability distribution. employing a softmax activation function for multi-class classification challenges in which more than two class labels are required.

5. Results

As a result, Uniformity and machine understanding of acquired data are critical considerations is supplied with the algorithm's right inputs and outputs. Here we Applying Image Processing and Feature Extraction process.

Table 1: The accuracy is given below:

Method	"Train Acc."	"Train Loss"	"Test Acc."	"Test Loss"	"No. of Epochs"
Proposed Method	0.9838	0.046	0.8794	0.5354	50

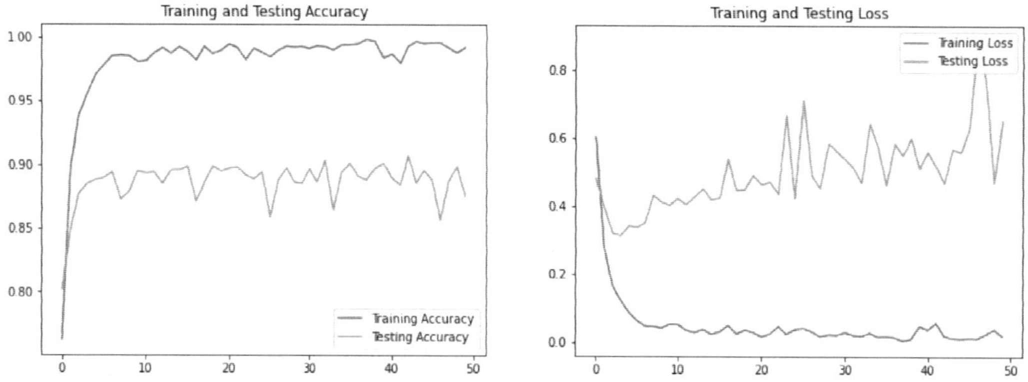

Fig 9: Graph of Training accuracy and Testing accuracy

Prediction:
Taking a picture of glioma and get the correct output

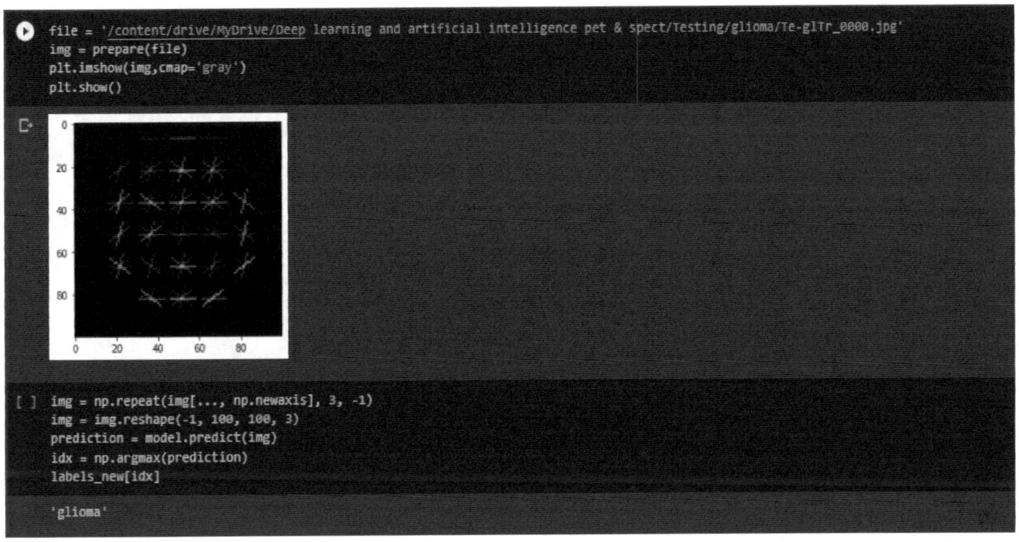

Fig 10: Taking another image which is from meningioma it is showing the correct output:

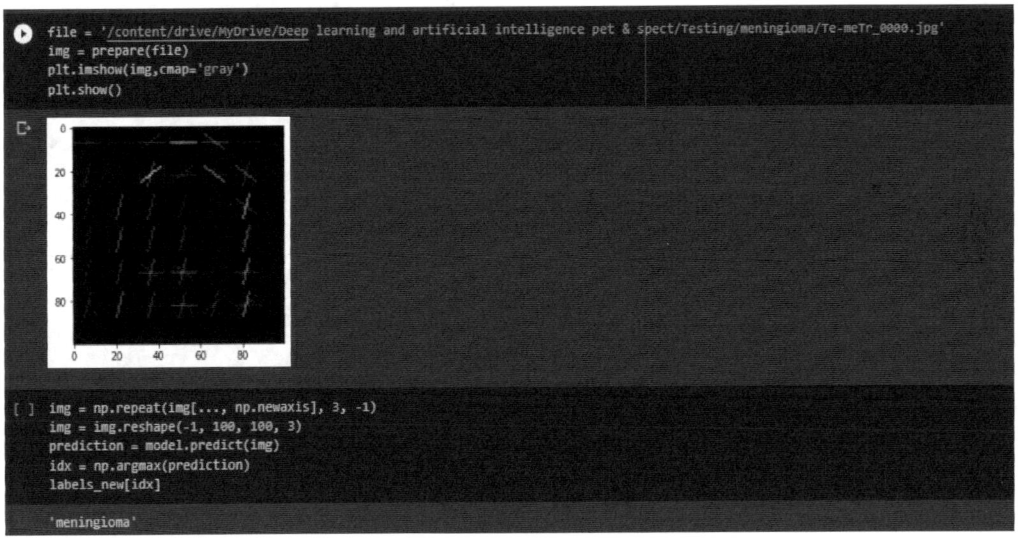

Fig:11 Third picture from the no tumor

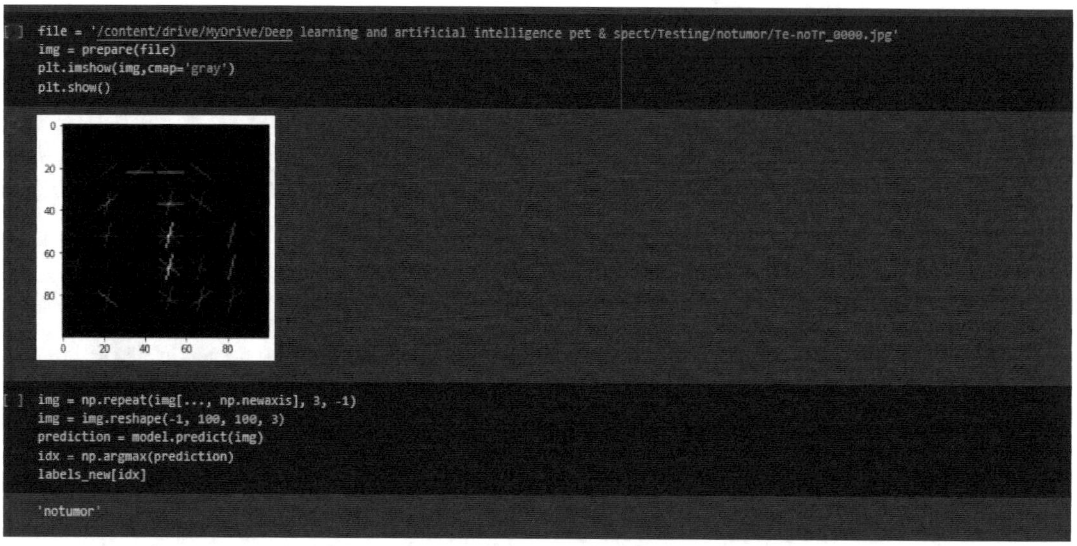

Fig 12: Fourth picture from the pituitary class also predicted correct prediction

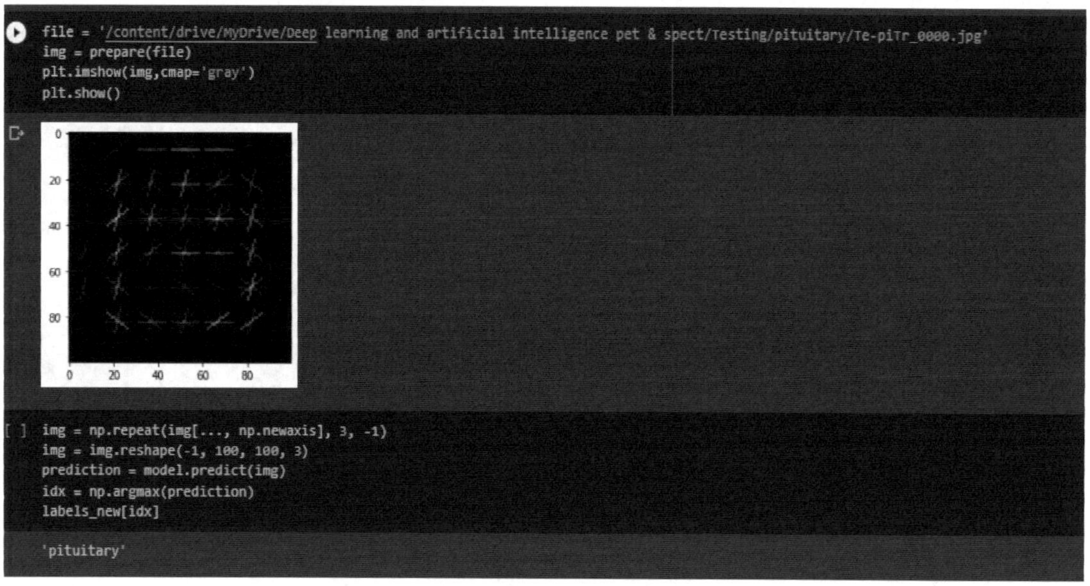

Fig 13: Output sample

6. Conclusion

Many areas of the imaging and radiology pipeline can benefit from deep learning, typically exceeding standard methods in terms of speed or accuracy. There has to be more study into explainable AI because these AI technologies can have a direct impact on diagnosis and treatment plans. Some progress has been made, but many deep learning algorithms are still viewed as black boxes and it is difficult to understand how and why the algorithm makes certain predictions and under what situations it can fail.

References

[1] Arabi, Hossein; AkhavanAllaf, Azadeh; Sanaat, Amirhossein; Shiri, Isaac; Zaidi, H. (2021). The promise of artificial intelligence and deep learning in PET and SPECT imaging. -European Journal of Medical Physics. https://pure.rug.nl/ws/portalfiles/portal/177460937/1_s2.0_S1120179721001241_main.pdf

[2] Arabi, H., AkhavanAllaf, A., Sanaat, A., Shiri, I., & Zaidi, H. (2021). The promise of artificial intelligence and deep learning in PET and SPECT imaging. Physica Medica, 83, 122–137. https://doi.org/10.1016/j.ejmp.2021.03.008

[3] Arabi, H., & Zaidi, H. (2020). Applications of artificial intelligence and deep learning in molecular imaging and radiotherapy. In European Journal of Hybrid Imaging. https://doi.org/10.1186/s41824-020-00086-8

[4] Chrysostomou∗, C., , Loizos Koutsantonis, Member, IEEE, Christos Lemesios, Member, IEEE, A., & Costas N. Papanicolas. (2019). SPECT Imaging Reconstruction Method Based on Deep Convolutional Neural Network. IEEE. https://arxiv.org/pdf

[5] Jia, Z., & Chen, D. (2020). Brain Tumor Identification and Classification of MRI images using deep learning techniques. IEEE Access, 1–1. https://doi.org/10.1109/access.2020.3016319

[6] Nensa, F., Demircioglu, A., & Rischpler, C. (2019). Artificial intelligence in nuclear medicine. Journal of Nuclear Medicine. https://doi.org/10.2967/jnumed.118.220590

[7] Tahrir Ibraq Siddiqui. (2021). INTELLIGENT COLLISION PREVENTION SYSTEM FOR SPECT DETECTORS BY IMPLEMENTING DEEP LEARNING BASED REALTIME OBJECT DETECTION. Master of Science in Engineering. https://www.pfw.edu/departments/etcs/depts/ece/graduate/theses/2021_Tahrir_Thesis.pdf

[8] Zaidi, H., & El Naqa, I. (2021). Quantitative Molecular Positron Emission Tomography Imaging Using Advanced Deep Learning Techniques. In Annual Review of Biomedical Engineering. https://doi.org/10.1146/annurev-bioeng-082420-020343

Effective and Efficient Honey Harvest Alert System for Bee Farms

[1]Dr. A. Albert Raj, [2]Kavin, [3]T. Jeeva, [4]Kiran L.K

[1] Professor, Sri Krishna College of Engineering and Technology,
Coimbatore, Tamil Nadu, India.
albert@skcet.ac.in

[2] UG Student, Sri Krishna College of Engineering and Technology,
Coimbatore, Tamil Nadu, India.
18euec047@skcet.ac.in

[3] UG Student, Sri Krishna College of Engineering and Technology,
Coimbatore, Tamil Nadu, India.
18euec074@skcet.ac.in

[4] UG Student, Sri Krishna College of Engineering and Technology,
Coimbatore, Tamil Nadu, India.
18euec081@skcet.ac.in

Abstract: The technology described in this paper could be used to support the rural people who practice apiculture. The present innovation helps to increase the quality attributes of the honey by automating the process of monitoring the honey broods in order to obtain the first quality honey. The proposed invention continuously monitors the honey brood using the webcam and image processing technique is used to identify the capping level with absolute accuracy. We also monitor the humidity level inside the brood box using the humidity sensor ensuring the quality of honey produced and quality of the environment for the sustaining of bees. Image processing and machine learning is used to identify the level of capping occurred in the beehive and first quality of honey could be extracted if at least 75% of cells of the hive are capped.

Keyword: Image processing, Machine learning, Capping, Humidity, Apiculture

1. Introduction:

Traditionally honey is harvested from mountain cliffs and trees using the bare hands, hanging from the ropes and this method is not safe. But the modern method incorporates the usage of honey broods and practice apiculture for large honey production. According to National Agricultural Statistics-2020, it is observed that amount of pure honey production has gradually reduced over the past few years, mainly due to improper handling of honey broods. When Apiculture is practiced in a rural environment, the entire environment will get pollinated by the bees which in turn will improve the ecosystem of the village / rural area. As the honey harvest alert system will improve the productivity of bee farms, more farmers will get attracted to develop bee farms which in turn will improve the livelihood of the rural population. Pollination plays an important role for the bee keeping process. Water, wind and Animals contribute for the proper pollination on right time.

Modern method of honey harvesting is done using honey broods that are specially designed wooden boxes normally take about a month for the comb to become ripe for harvesting. At Present ripens of honey comb are varied by physically removing the frames and inspecting. In case the harvest is delayed, the honey comb will yet be solidified due to the formation of sugar crystals. Hence the challenge lies in correctly identifying the correct time for removing the honey broods so that honey can be extracted to the maximum extent, the average honey extraction per box is about 4kg per month out of which about 1kg is in the form of sugar crystals. Hence the net extraction is only about 3kg with the prevailing price of Rs.1000 per kg, the loss per box is Rs.1000/-. This loss can be avoided by the honey harvest alert system.

2. Objective of the Invention:

To provide an "Effective and Efficient Honey Harvest Alert System for Bee farms"' capable of supporting the productivity of honey which is being practiced by an individual, a group and village people with the following features:

- Increased Productivity of honey farms.
- Reduction in wastage of honey due to formation of sugar crystals.
- Increased pollination in surrounding area will improve the ecological condition.
- Reduced man power for maintaining bee farms.
- Higher production of honey will act as substitute for sugar and will improve the overall well-being of consumers, due to the anti-oxidant content in the honey.
- Improved standard of living of small and rural farmer.

3. Literature Survey:

In the past few years, few articles have been published in the journals. Each survey tackles smart bee keeping boxes with a different angle of analysis. The authors of [Zacepins and Karasha, 2013] describe the progress of the research in temperature monitoring of bee colonies, from simple thermometers and data collected manually to infrared imaging and the start of embedded solutions. The survey paper presented in [Meikle and Holst, 2015] reflects on the advancement of PB data analysis by listing comprehensively the types of data that a beehive can offer. The research described in [Zacepins et al., 2015] is also close to our survey. The authors of [Marchal et al., 2020] examine the range of data collected in an apiary and the expansion to smart technologies directly on flowers, bees, and nests.
Jalled et.al[2] aimed their project to incorporate state-of-the-art techniques for object detection with the goal of achieving high accuracy with real-time performance. In this paper, they use a completely deep learning-based approach to solve the problem of object detection in an end-to-end fashion. D.Joy et al[1] proposes the idea of teaching a computer to detect and define a colour well enough to have useful applications. Ettore Baglio[3] in his book "Chemistry and technology of honey production. Springer, 2017 explain the composition of honey involves carbohydrates, water, nitrogenous substances, and minerals with the additional presence of vitamins, organic acids, flavour aromatic substances, and polyphenols. The production flow of commercial honey involves six different steps, from the initial extraction to the final packaging. He also describes the influence of honey composition on physical properties like color and density etc. Hoover et.al [4] presented "Pollen collection, honey production, and pollination services in a seminar:" managing honey bees in an agricultural setting" – 2018" .The process of managing honey bees in the farm field is explained where both the farmer and the person who practice apiculture gets benefited, as the bees increases the pollination rate of the crops increasing

the yield for the farmers and also produce honey as a by-product. During this process the brood box is moved too different environment with different level of vegetation thus allowing us to study the behavioural changes of bees and the change of the quality in honey. "Quality of Honey Harvested and Processed Using Traditional Methods in Rural Areas of Kenya"- 2019 [5,6] The difference in the honey harvested using the traditional method and the honey collected using the brood box setup is compared in this study. Samples of the honey collected using the traditional method seems to have no loss in its nutrition content and mineral value though some samples seem to have compromised in aroma due to the excessive smoking of the hives is explained [7-10]. The quality of the honey is analysed using the quality attributes such as sugar level, moisture level, HMF and Diastase level. Quality of honey harvested using the brood box mostly depends on the environment that is placed and the flora present around the box [11-14]. In view of the above problems, this paper designs a smart beehive system, which can better realize scientific beekeeping and intelligent management [15-20]. The system adopts distributed data acquisition and monitoring structure, which consists of a single beehive, a beehive group node and a monitoring center three-level structure.

4. Methodology

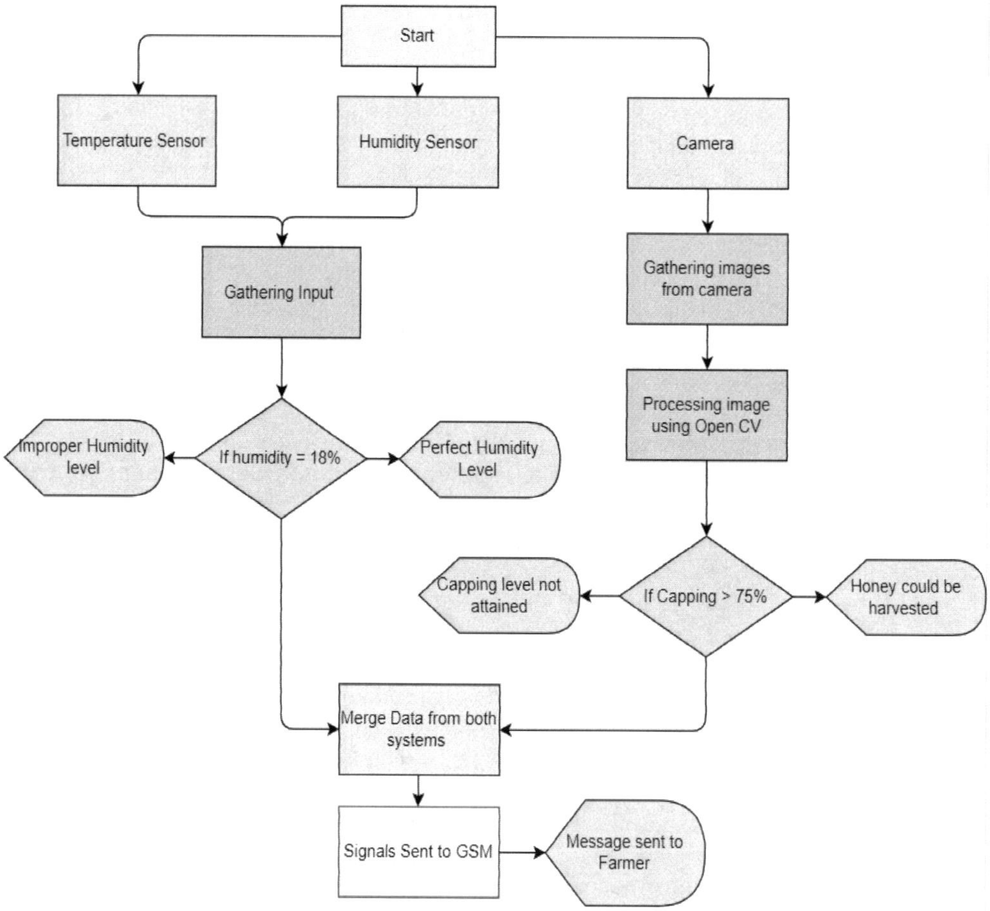

Fig 1: Over all Architecture

The proposed innovation consists of three major parts one of them is sensor mechanism to collect various data from the environment and the camera to take the picture of the brood to identify the level of capping occurred, microcontroller which acts as the brain of the prototype, here we have used raspberry-pi as the microcontroller and a separate mechanism is used to send the messages to the farmer in the real-time which consist of GSM-Module and 16*2 LCD display.

The functioning of our proposed innovation as given in (fig 1) begins with the collection of various data like humidity, temperature and moisture level from their respective sensors. Parallelly the web cam turns on when the process begins and captures the shoot of the brood at the regular interval period and sends it to the raspberry-pi for the image processing of the bee-hive in the real-time.

The microcontroller used in the innovation uses the data collected from the various sensor and the images from the camera to calculate the perfect time for harvesting honey in the first quality without compromising in its quality attributes. Harvesting the honey in perfect ripping time helps us to avoid the crystallization of the honey into sugars thus retaining its chemical composition perfectly. Capping of the honey comb is the process where the bees cover their hive using a special substance after the single hexagonal compartment is completely filled with the honey and the difference between the capped hive (single hexagonal compartment) and the uncapped hive could be visually identified. For the formation of the first quality of the honey the humidity level of the environment must be around the 18-20%, if this condition is not maintained that is more likely that honey will lose its viscous nature and texture of the honey will be different or sugar crystals will be formed in the honey affecting the chemical attributes of the honey.

In addition to these conditions honey should not be harvested unless at least 75% of the cells in the honey comb are capped i.e. for every four hexagonal cells at most one hexagonal cell could be uncapped (without completely filled with honey) if this condition is not met the honey will be in the unripe stage and the chemical compound formed due to the fermentation of the honey is more likely to be not formed and thus resulting in the lesser quality honey which in turn reduce the profit for the person who practice apiculture as the market price for this honey is very low comparing to the honey harvested in the perfect time.

If the honey comb attains the perfect capping percentage and the humidity level of the environment is also maintained at the constant level between 18% - 20%, with these conditions achieved and thus the honey formed in the honey comb which is monitored using our prototype is ready to be harvested in first quality. This message is sent to the person who practice apiculture using GSM modules attached in the prototype and the registered phone number of the farmer. The benefit of this method that the farmer could be in any part of this world but could get the update about the farm in the regular interval in real time. Also, we have attached an LCD 16 × 2 display in the brood box for the convenient monitoring of the level of capping attained in the comb and the humidity level of the environment in the real time whenever needed.

5. Hardware System Structure Design

The specific system hardware architecture is shown in Fig.2.The intelligent bee box is mainly consisting of signal acquisition circuit, humidity and Temperature sensor, attitude detection sensor, recording device, communication interface circuit and charging control circuit. When the measured load is loaded on the scale body, the load cell installed under the scale body generates a voltage signal proportional to the measured load. After the conditioning circuit is amplified, filtered, and digital-to-analog converted, it is transmitted to the MCU to complete the weighing signal acquisition. The single-chip computer uses the ambient temperature and humidity signal collected by the digital temperature sensor to obtain the current temperature and humidity value. On the other hand, according to the temperature compensation algorithm, the temperature compensation of the measured load weighing result is completed. The system uses MMA7361 to obtain the attitude of the box. When the box inclines or collides, the acceleration along the X, Y and Z axes will change, and the voltage value proportional to the acceleration will be output. After filtering, digital-analog conversion, inverse triangulation calculation and attitude calculation, the current attitude of the box can be obtained. The Air800_M4 packages the obtained data value and the GPS and sound information collected by itself to the client to complete the function of the whole system

Fig 2: The hardware system structure

6. Novelty

The innovation we bought in the existing method is that we have automated the process of monitoring the honey brood every day manually in order to extract honey in first quality by using the image processing technology and the also extracting the use of various sensors to monitor and provide the feedback of the environmental condition to the person who practice apiculture to obtain the honey in the first quality. The use of the humidity sensor in this prototype is to monitor the humidity level of the environment in the real time and send the measured data to the microcontroller for the further processing. If there are any unfavourable conditions like the humidity level going beyond or above the ideal condition then the microcontroller sends an alert message to the farmer. We have used the OpenCV software and the python to process the image obtained from the Camera in the real time to identify the level of capping occurred in the honey brood.

7. Implementation

Fig 3 shows the hardware circuit which is implemented in the boxes. For programming we have used the Python and OpenCV (A Python library) in the backend to perform the image processing, to obtain the percentage of the capping and also to check various conditions to obtain first quality honey based on the data collected from the various sensors based to the various predefined conditions obtained from the various researches.

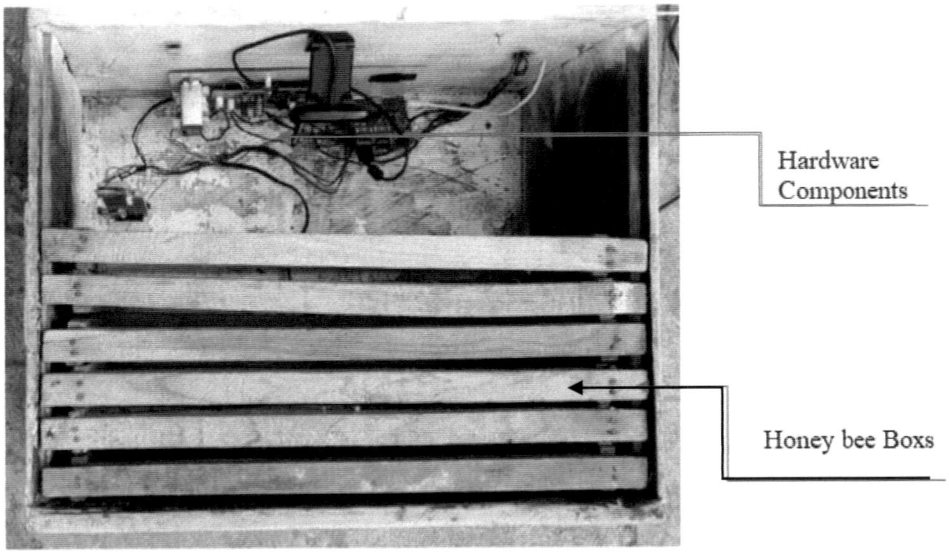

Fig 3: The implementation of hardware in Boxes

8. Conclusion:

The detection of the prefect time to harvest the honey in the first quality includes many challenging factors that should be done manually by the human. As we humans are more liable to perform parallax errors, the proposed method have automated the process of monitoring the honey broods using the image processing technology over traditional human method. By the use of this method result analysis have shown that the honey production has increased and the quality of the honey produced is also increased. Thus, resulting in the increase in the income for the farmer. By using this proposed method overall bee count of the colony is also increased which results in the increased pollination rate in the surroundings and reduction in the rate of global warming.

Reference:

[1] T Joy, D., Kaur, G., Chugh, A., & Bajaj, S. B. (2021). Computer Vision for Color Detection. International Journal of Innovative Research in Computer Science & Technology (IJIRCST) ISSN, 2347-5552.
[2] Jalled, F., & Voronkov, I. (2016). Object detection using image processing. arXiv preprint arXiv:1611.07791.
[3] Baglio, Ettore. Chemistry and technology of honey production. Springer, 2017.
[4] Muli, Elliud, A. Munguti, and S. K. Raina. "Quality of honey harvested and processed using traditional methods in rural areas of Kenya." Acta Veterinaria Brno 76.2 (2007): 315-320.
[5] Hoover, Shelley E., and Lynae P. Ovinge. "Pollen collection, honey production, and pollination services: managing honey bees in an agricultural setting." Journal of Economic Entomology 111.4 (2018): 1509-1516.
[6] Magazine on "The importance of bees and other pollinators for food and agriculture" by Food and Agricultural organization of United nations, May 2018
[7] Deborah Ruth Amulen, Marijke D'Haese, Eline D'Haene, James Okwee Acai, Jacob Godfrey Agea, Guy Smagghe , Paul Cross," Estimating the potential of beekeeping to alleviate household poverty in rural Uganda" DOI: https://doi.org/10.1371/journal.pone.0214113 March 2019.

[8] Tosi, Simone; Nieh, James C.; Sgolastra, Fabio; Cabbri, Riccardo; Medrzycki, Piotr, "Neonicotinoid pesticides and nutritional stress synergistically reduce survival in honey bees" by (2017) in proceedings of the Royal Society B: Biological Sciences.
[9] Zacepins, A., Stalidzans, E., and Meitalovs, J. (2012). Application of information technologies in precision apiculture. In Proceedings of the 13th International Conference on Precision Agriculture (ICPA 2012), Indianapolis, IN, USA.
[10] Huang, Shaokang, et al. "A novel method for the detection and diagnosis of virus infections in honey bees." Journal of Virological Methods 293 (2021): 114163.
[11] J.Vijila ,A.Albert Raj 2020 "Discovering Virtual Reality Framework for Rescuing Emergency Response Training during Natural Disaster using Wireless Networks" Vol 45 No1 pp 327-343.
[12] P.Maria Jessi, A.Albert Raj 2019 "A neoteric approach for preserving infrastructure and nodes mobility through weight based clustering and distributed scheduling" Journal of Intelligent & Fuzzy Systems 36 (4), 3345-3356.
[13] R.Bommi,A,Albert Raj 2018 An Efficient optical Pilot Scheme for Channel Estimation in MIMO OFDM System " International Journal of Printing and Packaging Allied Sciences Vol 5, no 1, pp 67-
[14] Gnana jebadhas , A.Albert raj 2017 An Image processing frame work for white matter segmentation in suspected Alzheimers disease Bio medical research 2017 ;28(6)2836-2842.
[15] Maria Jessi ,A,Albert Raj 2017 A newfangled method to maintain Infrastructure and Mobility of Nodes by weigh based Clustering and Distributed Scheduling , International Journal of Printing & Packaging Allied Science Vol 5, no 1 ,pp 24-33 ISSN 2320-4287.
[16] Kebila , A,Albert raj 2017 Pilot Design Channel estimation in MIMO OFDM system using Cyclic Delivery Scheme " International journal of Advances in Engineering & technology ISSN 22311963.
[17] Huang Wei, Yang Wei, Zhang Wendong. Design of beehive bee monitoring system based on single chip computer [J]. China Test, 2014, 40 (02): 82-85.
[18] Jiang Xuan, Lin Bingfang, Liu Nacui. Design of scientific apiculture management platform based on Internet of Things [J]. Fujian Computer, 2018, 34 (09): 33 + 62.
[19] Cheng Wei, Zeng Baiwei, Sun Xianzhong, et al. Intelligent beehive management system - Honeybee House [J]. Internet of Things Technology, 2014, 4 (04): 11-12.
[20] Meikle, W. G., Weiss, M., and Stilwell, A. R. (2016). Monitoring colony phenology using within-day variability in continuous weight and temperature of honey bee hives. Apidologie, 47(1):1–14.
[21] Marchal, P., Buatois, A., and Kraus, S. e. a. (2020). Automated monitoring of bee behaviour using connected hives: Towards a computational apidology. Apidologie

Intrusion Detection System using Machine Learning and Microwave Doppler Radar

A Ushasree, Atluri Shanmukha Datta, Varikallu Sai Krishna, Pingili Manish Reddy and Rampelly Sanjay Kumar

Department of Electronics & Communication, Gokaraju Rangaraju Institute of Engineering and Technology, Hyderabad, Telangana, India.

ushasreemishra@gmail.com, shanmukdatta12345@gmail.com

Abstract: A security system is an essential part of a safe home as thefts and intrusions are increasing in recent times; moreover, keeping the home safe is a major concern while being away. This paper proposes an innovative way of security alert system using a microwave doppler radar and integrating it with other peripherals such as camera, alarm, and raspberry pi, implemented using machine learning techniques. The machine learning technique used is the deep neural networks. This structure helps in monitoring the target space while receiving notifications on WhatsApp and Email to improve the security.

Keywords: Microwave RADAR, Doppler Effect, Raspberry Pi, Machine Learning, Computer Vision.

1. INTRODUCTION

Security is one of the important topics in every facet like surveillance, industrial applications, offices, and home. A security system that monitors the entire area and reacts effectively is needed. There are many security systems for both indoor and outdoor such as CCTV, infrared detectors, ultrasonic detectors, and PIR sensor [1]. However, one or the other system have limitations of being expensive, more electrical consumption, more memory space utilization of the recording system and complex circuitry. In common, the drawback in all these systems is that there is a good possibility of false intruder detection based on 'line of sight' cut by any entity and not necessarily an intruder. Many security systems such as RFID, OTP based system, and biometrics are used to prevent unauthorized access [2][3][4]. But merely preventing unauthorized access is not enough. This project involves in using microwave technology because they are more stable and less prone to false alarms. Here, microwave radar plays an important role. Microwave radar generates an electromagnetic (RF) field between the transmitter and the receiver [5]. The microwave radar transmits microwave signals and evaluates returned signals for a moving object [6]. By using this microwave technology, the detector is less sensitive to false alarms. Microwave radar utilize Doppler principle to detect motion of an object [7]. If a target is moving towards the radar system, the frequency of its radar echo is slightly higher than the frequency of the radar system itself; and if it is moving away, the frequency is slightly lower [7]. This change in frequency with target motion is the doppler effect. Machine learning is a method by which a machine can imitate a human in decision making process by training with required datasets. Depending on the type of dataset given to a machine learning algorithm, it can learn to make decisions close to human like. Machine learning can also be considered a type of artificial intelligence as it can make human like decisions artificially. Machine learning is a very important part of the current era, as it can decrease human dependence on iterative and valuable human resources, and talent can be put into much more important tasks requiring innovation and creativity.

2. LITERATURE SURVEY

In closed-circuit television (CCTV), signals are not broadcasted publicly, but are monitored for security and surveillance purposes. In order to use CCTV, you must strategically place the cameras and observe their output on a monitor. CCTV does not give notice in real-time when recording interference [9]. CCTV implementation faces some challenges. To start with, it does not produce notifications or warnings instantly when it captures any suspicious activity. Additionally, CCTV systems are designed to record continuously, so even if nothing suspicious occurs in a home, they still capture what is happening. Streaming and storing continuous video consumes a lot of bandwidth and storage [9][10][11].

Gayathri P. Sonawane conducted research on the different types of video surveillance systems using ARM9 and ARM11 for the implementation [12][13]. H.264, MPEG, and JPEG are just a few examples of video compression techniques. Among those techniques, H.264 is a better choice for usage without losing the original image properties. A cloud-based system stores images and videos and sends SMS and email notifications to their owners. For better monitoring purposes live streaming is also used. Rajat Kumar used an Ultrasonic sensor [14][15]. It worked better than CCTV. An ultrasonic sensor can detect the object in only one direction. It can be achieved with the help of a DC motor. The motor rotates the sensor in a step manner so as to cover 360 degrees. Due to this, the cost is high, and implementation is difficult. Ying-Wen Bai developed a system to detect intruders [16]. If any person passes through the field sensors, there will be a fluctuation in the output and the system identifies the change. Activating the video camera is done using the Major Voting Mechanism (MVM). The basic functionality of the system depends on PIR sensor. Infrared radiation, which comes from the body, is what makes the system work. According to studies, the body temperature of a normal individual can radiate infrared light at wavelengths between 10 and 12 micrometers [17].

Infrared sensors detect motion by detecting fluctuations in infrared light. A new method (UGS) for detecting strangers and the motion of pedestrians was proposed by Xin Jin [18]. Detection and classification were carried out by Pema Chodon using wavelet-based techniques and PIR (Passive Infrared) sensors [19][20]. Rakesh et al., proposed a real time system using beagle board, ZigBee, and FTP Server [21]. Ansari et al., proposed a security system using raspberry Pi. Raspberry Pi has higher clock speed, multiple connectivity options and more flexibility at affordable price compared to other development boards [22]. E. Baraneetharan has done a survey on various techniques that can be used to integrate with Wireless Sensor Networks for intrusion detection. WSNs were used for reason such as biodiversity, ecosystem protection, surveillance and other military applications. Machine learning enabled this system to perform a wide range of applications such as, securing the network, increased efficiency, identification, analysis and other such features [23].

Table: 1 Comparison between various sensors

Radar Type	CDM324	PIR motionsensor	Ultrasonic sensor
Type of communication	wireless	wireless	wireless
Distance of communication	15m	10m	4m
Obstacles	Can be placed behind walls orcasing	Needs to be open	Needs to be open
Weather conditions	Can work inany weatherconditions	Performanceimpacted byweather conditions	Performanceimpacted byweather conditions
Identification parameters	Speed and approximatesize	No parameterscan be obtained	Distance formsensor

3. PROPOSED SYSTEM

In the proposed system, a microwave radar module uses the Doppler frequency discrepancy between transmitted waves and the returned waves which hits on the target when the target is moving and output its returned signal as a voltage offset in DC. In radar, readable output is given in voltage which is measured by calculating the frequency shift Δf, which is then amplified, filtered, and processed by a signal conditioning circuit in a homodyne transceiver configuration. As a result, it is possible to detect, locate, track, and recognize objects at considerable distances using this technology.

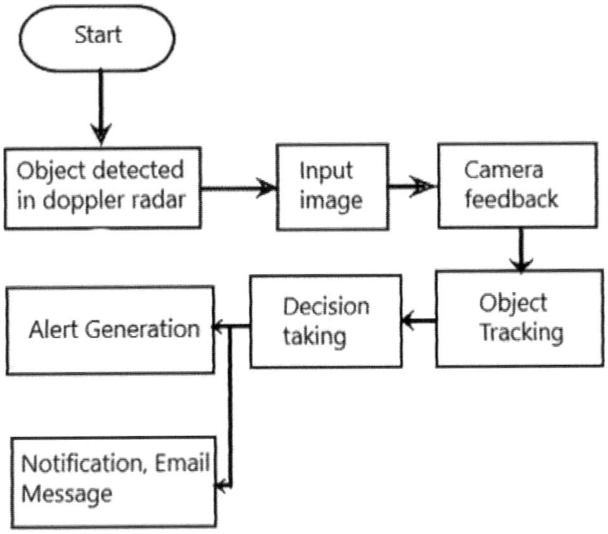

Fig:1 Block diagram of the proposed system's operation

The security system uses a Raspberry PI as its CPU. Once an object or person is detected, deep learning methods are used to identify the person and the numbers are identified using Haar cascade, and YOLO algorithms.

4. HARDWARE AND SOFTWARE USED

The project is done with **Raspberry Pi 3B+** board. Because raspberry is a tiny low-cost development board that can also be considered as a stand-alone computer, it supports many programming languages like python, and java. It has many IDE tools included in OS installation. Other development boards such as Arduino, ESP32, Intel Galilu, and Nvidia Jetson either lack modules such as direct camera support, WIFI and Bluetooth support or too expensive. As such Raspberry Pi 3B+ is used which has all required components at affordable cost.

A Raspberry Pi is a development board that has all the components required to run as a standalone computer and is researched and developed by RaspberryPi foundation. It is a tiny computer that can use a standard keyboard and mouse. Raspbian OS is used in itas the most popular operating system which has to be uploaded into a micro-SD card and inserted into the Raspberry Pi.

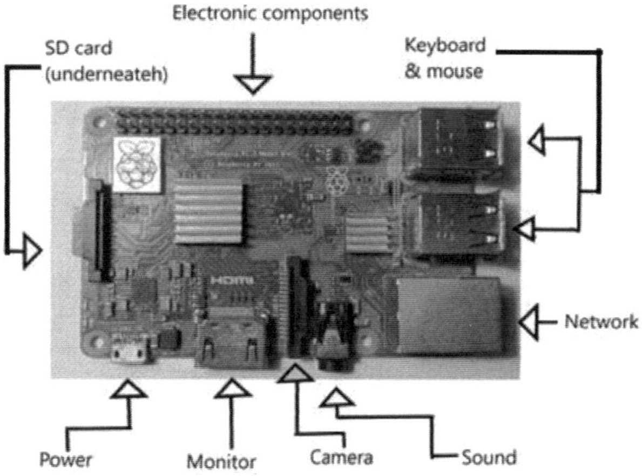

Fig: 2 Raspberry pi 3B+

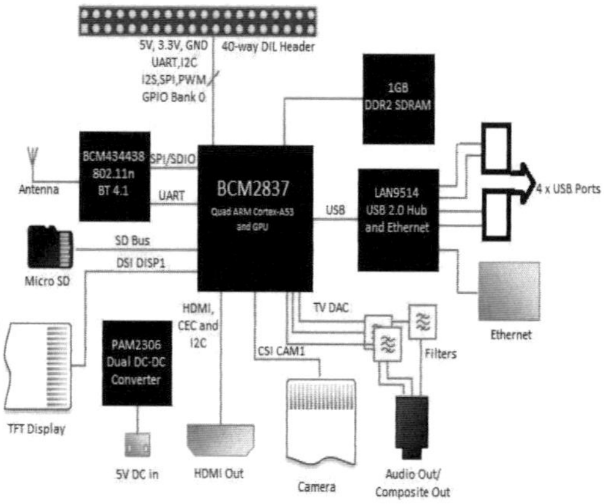

Fig:3 Raspberry Pi 3B+ Architecture.

The Raspberry Pi 3 Model B+ Variant has a 1.4 GHz clock speed and is built using 64-bit quad-core Broadcom Arm Cortex A53-architecture. It has 15% performance improvement compared to its predecessor model 3B. It has Bluetooth 4.2 and Bluetooth low energy are supported. It has dual-band wireless local area networks at 2.4 and 5 GHz. Additionally, it features four USB 2.0 ports, a 40-pin GPIO header, a full-size HDMI port, port for displays and cameras, and composite video and audio outputs.

Doppler radar is used in this project to detect the motion of objects. This project uses the dopplerRADAR CDM324.

Fig: 4 CDM324 Doppler Radar

The **CDM324** is a single-channel 24 GHz band radar transceiver. A planar microstrip antenna structureis used in it, it has an independent transmitting and receiving channels and a single IF output. The main purpose of CDM324 is for velocity information detection of short-range targets. Radial sensors candetect, locate, track, and recognize a wide range of objects from a distance. During operation, electromagnetic energy is sent toward targets, and then echoes are observed at the targets. It works on 30ma of current and has an operating voltage of 5v. Its output power is 16dbm and gives an output of300mv

The term **Amplifier** refers to any circuit that produces an amplitude increase in the signal it receives. Because of their ability to amplify relatively small input signals, small signal amplifiers are commonly used devices. In this project, an LM324 op-amp that boosts the 30mV output of the Doppler radar to 5V for the convenience of the Raspberry Pi microcontroller has been used.

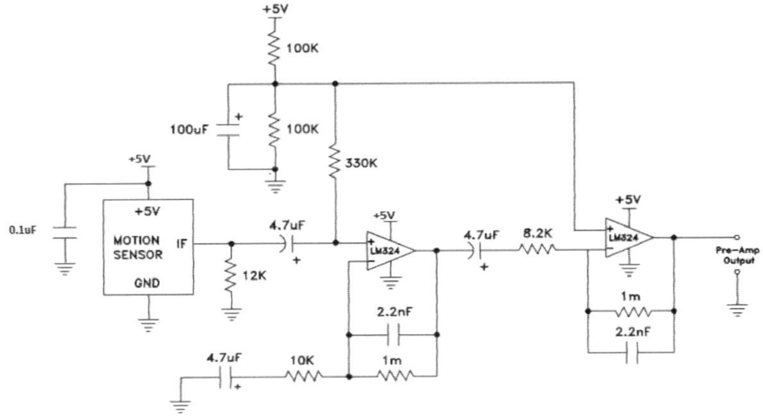

Fig: 5 Amplifier circuit diagram.

This project uses High Current Active Alarm **Buzzer** Driver Module, that will buzz at a frequency (2300 ±300Hz) on its own even when you just apply steady DC power.

Fig: 6 Alarm

Fig:7 Raspberry Pi Cam.

Raspberry PI cam is an original camera module designed for Raspberry PI development board. The original Raspberry PI cam is a 5-megapixel camera module and has both normal and infrared versions which is used for face detection purposes.

YOLO (You Only Look Once) is an algorithm to detect and recognize objects in real time. It is the most usedalgorithm due to its high speed and accuracy. YOLO is used to detect and identify multiple objects in a single image. This algorithm works by following three simple steps: -

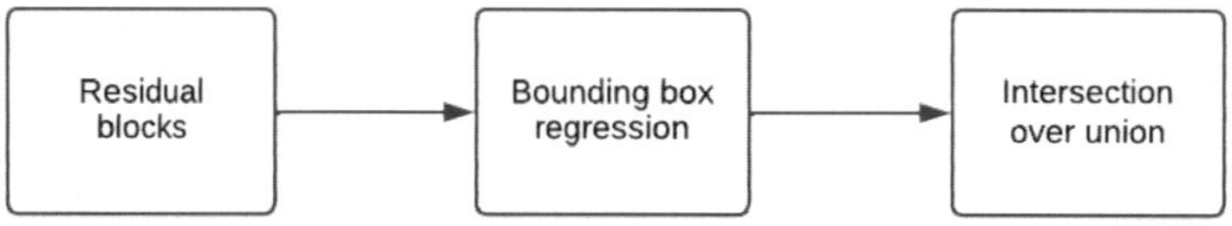

Fig:8 Yolo Workflow.

YOLO workflow starts by dividing image into various grids called **Residual blocks,** which has a dimension of I x I and every cell will detect objects that appear within them. The outline which highlights the object is called **Bounding box** and every box in the image consists of width (bw), height (bh), class (c), and with center (bx, by). **Intersection over Union** is an evaluation metric used to find an object from boundingbox regression. The formula of IoU is given as the area of overlap divided by the area of union.

Fig: 9 Detection by YOLO

Haar Cascade is also known as Viola-Jones Face identification technique. It identifies faces in real time or in an image by using Object Detection algorithms. The algorithm uses edge or line detection feature. It helps in recognizing facial highlights and disregards everything else like, trees, structures, background objects etc.

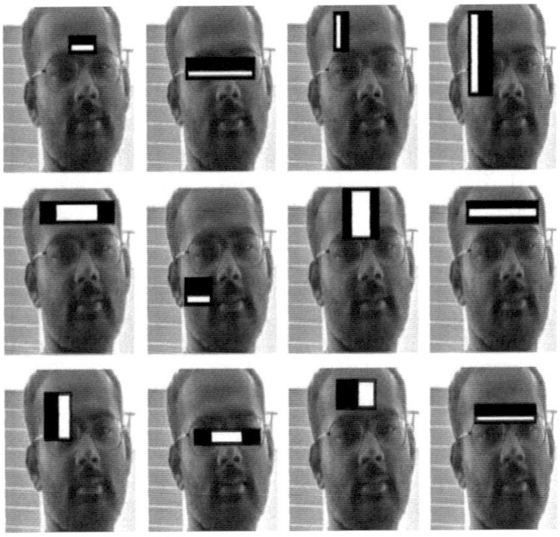

Fig: 10 HAAR Cascade representation.

Here, the live feed from camera must be verified with the dataset. It is done for the identification of known and unknown individuals. This can be achieved using computer vision. **OpenCV** which isan open-source computer vision library can be used to identify objects and faces. When features need to be added, NumPy can be included. **NumPy** is a library used for mathematical computations. It provides multidimensional arrays which arefaster than python lists.

Time helps to get local time or current time, which helps in finding the duration of a process. It returns floating values as output. **OS** is a module used to get files or content present in a folder. It also has functions which can help in creating or removing folders. **RPi.GPIO** is a python module to control the GPIO interface (pins) of the Raspberry Pi.

5. METHODOLOGY USED FOR OBJECT DETECTION AND FACE RECOGNITION

Deep Neural Networks (DNN): Neural networks perform numerous calculations, like human brain`s neural structure, and is inspired after the brain neural structure working principle, which might be meant to understand designs. DNN is used to design the workflow similar to human neural networks. Unlike ANN, a concept credit assignment path can be used, which makes it easier for autonomous works. As such, the nodes can be programmed to perform specific tasks. These nodes are used to integrate Harr cascade and YOLO for verification from dataset.

YOLO: Compared to fast RCNN which carries out detection on numerous area proposals and makes multiple predictions in more than one instance for numerous areas in a photo, Yolo structure is greater and passes the photo (M x M) through the FCNN and the output obtained is (N x N) prediction. This photo M x M is divided into separate grids for identification. For every grid, this structure divides the photo into m x m bounding boxes, and each bounding box is assigned a class probability. It is to be noted that bounding box is much more likely to be larger than the grid itself. This helps in accurate identificationof the objects.

Haar cascade: Face identification is an innovative method that can classify the human faces according to the data given in objective pictures with respect to human faces in real-time pictures. It compares facial highlights and disregards other things such as structures, trees, vehicles and bodies. In face recognition, the errand is to find out the regions and sizes of a known quantity of appearances; for example, the distance between eyebrows, forehead size, cheeks measurement etc., Face recognition may be regarded as a step into the future, as it can help in quickly recognizing people and reducing redundant tasks for many applications.

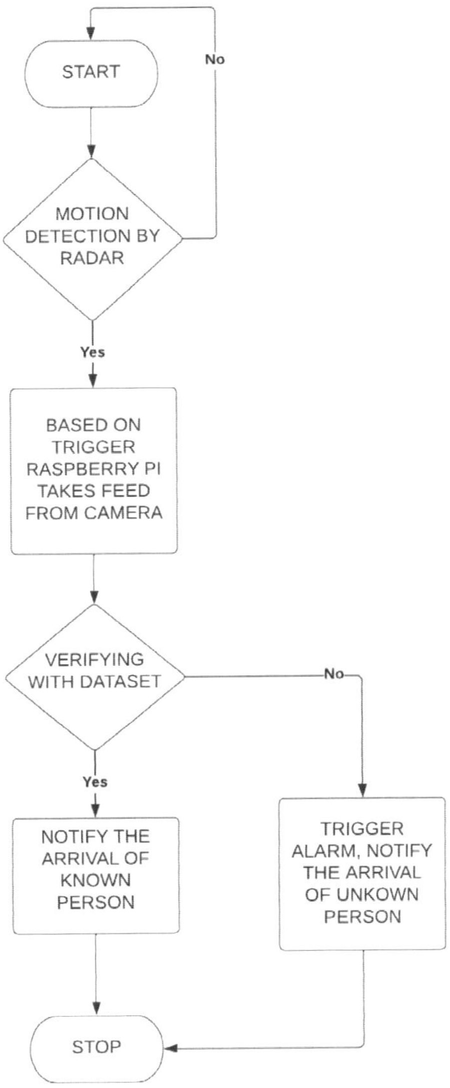

Fig: 11 Flow chart

6. IMPLEMENTATION

The below schematic diagram of home security using radar explains the interfacing section of each component. An amplifier is built using a series of resistors and capacitors to amplify the signal received from radar using LM324 operational amplifier which is connected to raspberry pi. An alarm and camera are also connected for video and surveillance.

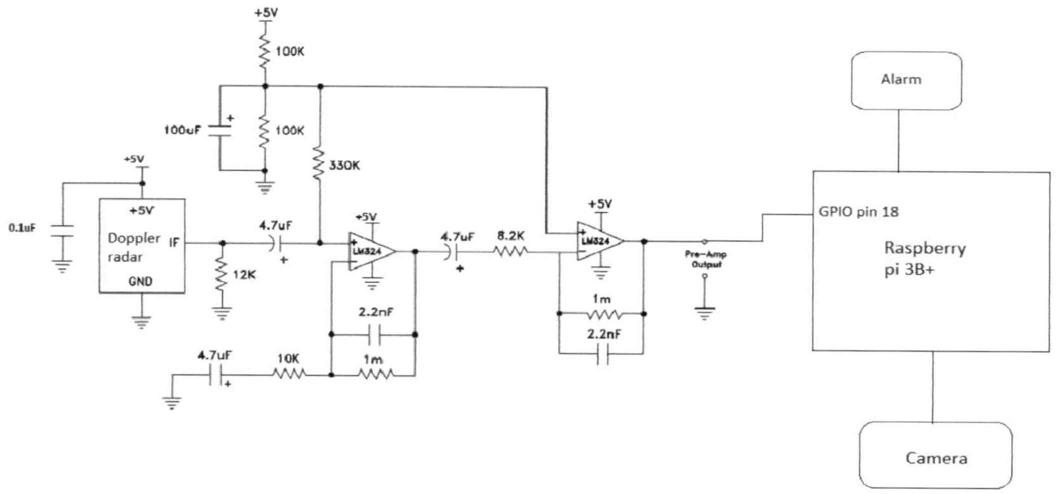

Fig: 12 Schematic diagram of the project

The circuit shown in the figure 12 is constructed using the IC LM324AD, motion sensor, and passive components like resistors and capacitors.

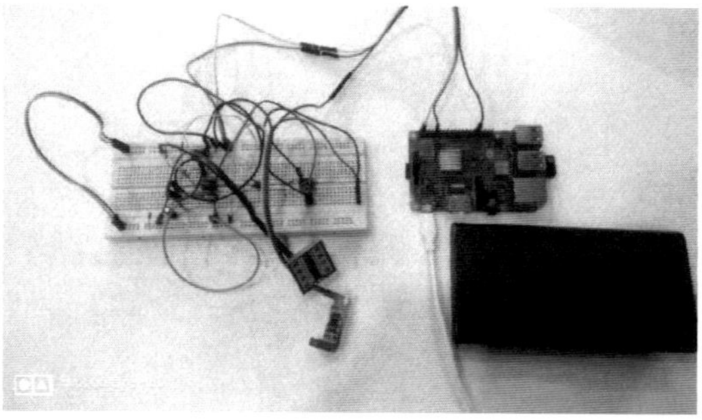

Fig: 13 Components setup.

This security system is implemented using various hardware and software components. The base of this system is CDM324 Doppler radar which detects the motion of any object within its specified range using Doppler effect. Once an object enters the range of the doppler radar, there will be a change in the received frequencies. Radar calculates the frequency difference and gives an output in mV. The security system uses a Raspberry PI as its CPU. But the output voltage from the radar is too small. So, an amplifier is used which amplifies the output voltage into a more readable signal. The output signal is sent to Raspberry Pi. The output signal will show 0 for no detection and above 1, if any object is detected and a noise is generated. So, the values are subtracted by a threshold which makes these values nearly 0 and 1. This process eases out the testing phase. Based on trigger, feed is taken from the camera. Once an object or person is detected, deep learning methods are used to identify the person and the numbers using Haar cascade and YOLO algorithms. The first in the flow is YOLO which identifies the numbers of intruders and check if it is human or not. YOLO classifies the image into grid cells and divides those grid cells into bounding boxes by using another program called Intersection Over Union. This process is used to eliminate the background and other useless objects, thus giving the bounding boxes with objects within them. After that, those objects are compared with the dataset. The dataset has images of known persons, animals, and vehicles in different angles. Vehicles, animals, and small objects are ignored. From this dataset, if the intruder is classified as unknown, then immediately a notification is sent through Email and WhatsApp along with a loud alarm.

7. RESULTS

The testing is done in a real-world scenario in an office. Once individuals enter the detecting range, the radar raspberry pi camera is activated and starts recording. The raspberry pi then analyses the recording in real- time and identifies the number of intruders and known individuals using YOLO and Haar cascade. Once intruders try to enter, the alarm starts ringing and the event is notified through WhatsApp and email.

Fig: 14 Unknown person identification

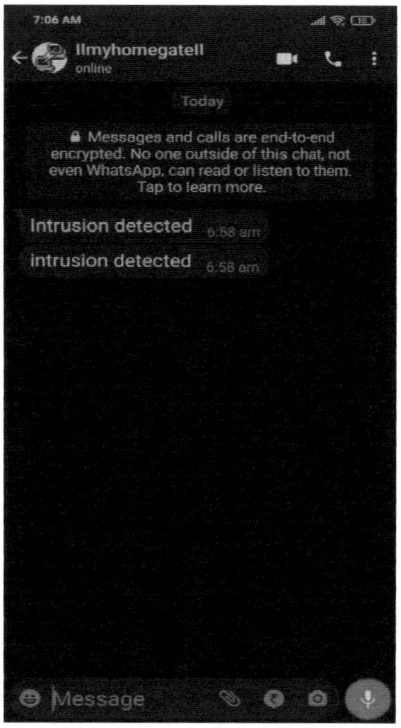

Fig: 15 WhatsApp notification

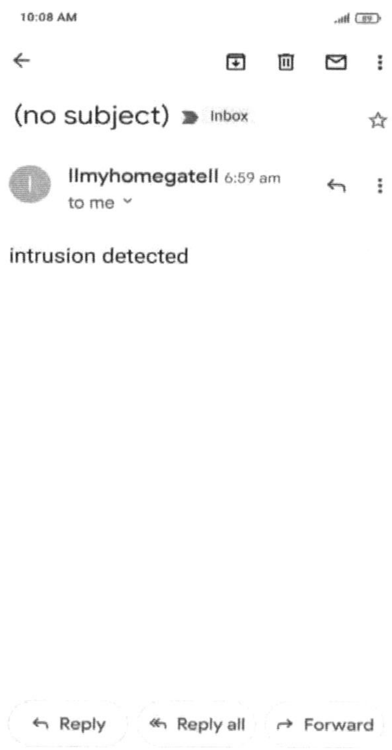

Fig: 16 Email notification

8. CONCLUSION

The project has been designed to enhance the home security system by alerting the household members. In this paper, an innovative security system is proposed using microwave radar and machine learning techniques. The sensor gets activated when there is an object in its range, and when an intruder is identified, the system alarms and notifies the owner. Despite the cost of the radar sensor, the proposed system gives the accurate results in comparison with all other existing solutions. It is believed that the chosen technical and technological approach is better than the former solutions in terms of accuracy.

REFERENCES

[1]. N. Surantha and W. R. Wicaksono, “Design of Smart Home Security System using Object Recognition and PIR Sensor,” in Procedia Computer Science, vol. 135, pp. 465-472, 2018.

[2] A. Jules, "RFID security and privacy: A research survey", IEEE Journal on Selected Areas in Communications, vol. 24, no. 2, pp. 381- 394, 2006.

[3] J. Jeong, M. Y. Chung and H. Choo, "Integrated OTP-based user authentication scheme using smart cards in home networks," In Proceedings of the 41st IEEE Annual Hawaii International Conference on System Sciences, pp. 294-294,2008.

[4] A. K. Jain, A. Ross, and S. Prabhakar, "An introduction to biometric recognition," IEEE Transactions on Circuits and Systems for Video Technology, vol. 14, no. 1, pp. 4-20, 2004.

[5]. Rajat Kumar Dwibedi,L Pattathu Rani, P.Vinnarasi Ponnury, "Security Alert System Using Radar", International Journal of Engineering Development and Research (IJEDR), ISSN:2321-9939, Vol.6, Issue 1, pp.723-727, March 2018, Available :http://www.ijedr.org/papers/IJEDR1801125.pdf

[6]. Skolnik, M. I. introductiontoradarsystems-merrillskolnikiiirded-140506103641-phpapp02. pdf.

[7]. Pozar, D. M. (2012). Microwave Engineering. Fourth Editions, University of Massachusetts at Amherst, John Wiley & Sons, Inc, 26-30.

[8]. Kumar, Sushant, and S. S. Solanki, "Remote homesurveillance system," IEEE Int.Con.Advances in Computing, Communication, and Automation, 2016, pp. 1-4.

[9]. R. Chandana, S. a K. Jilani, and S. J. Hussain, "Smart Surveillance System using Thing Speak and Raspberry Pi," Int. J. Adv. Res. Comput. Commun. Eng., vol. 4, no. 7, pp. 214-218, 2015.

[10]. J. Bangali and A. Shaligram, "Design and Implementation of Security Systems for Smart Home based on GSM technology," International Journal of Smart Home, vol. 7, no. 6, pp. 201-208, 2013.

[11]. U.RAMAKRISHNA and N. SWATHI, "Design and Implementation of an IoT Based Smart Security Surveillance System", International Journal of Scientific Engineering and Technology Research, Vol. 5, Issue 4, pp 697-702, Feb-2016.

[12]. Gayatri.P.Sonawane and Dr A.J. Patil, "A Survey on Different Digital Video Monitoring Systems", International Journal of Computer Science Trends and Technology, Volume 3 Issue 3, pp. 132-134, May-June 2015.

[13]. S. N. Jyothi and K. V. Vardhan, "Design and implementation of real-time security surveillance system using IoT," In International Conference on Communication and Electronics Systems (ICCES-2016), pp. 1-5,2016.

[14]. Rajat Kumar Dwibedi, L Pattathu Rani, P.Vinnarasi Ponnury, "Security Alert System Using Radar", International Journal of Engineering Development and Research (IJEDR), ISSN:2321-9939, Vol.6, Issue 1, pp.723-727, March 2018, Available :http://www.ijedr.org/papers/IJEDR1801125.pdf

[15]. Y. W. Bai, L. S. Shen and Z. H. Li, "Design and implementation of an embedded home.

[16]. Ying-Wen Bai, Li-Sih Shen and Zong-Han Li, "Design and Implementation of an Embedded Home

Surveillance System by Use of Multiple Ultrasonic Sensors", IEEE Transactions on Consumer Electronics, Vol. 56. Issue No. 1, February 2010.

[17]. S. Yuvaraj Prof. and Ramesh S., "Improved Response Time on Safety Mechansim Based on PIR sensor," International Journal of Emerging Technology and Advanced Engineering, vol. 2, no. 4, April 2012.

[18]. Xin Jin, Soumalya Sarkar, Asok Ray, Shalabh Gupta, and Thyagaraju Damarla,"Target Detection and Classification Using Seismic and PIR Sensors", IEEE Sensors Journal, VOL. 12, NO. 6, JUNE 2012

[19]. P. N. Saranu, G. Abirami, S. Sivakumar, K. M. Ramesh, U. Arul and J. Seetha, "Theft Detection System using PIR Sensor," 2018 4th International Conference on Electrical Energy Systems (ICEES), 2018, pp. 656-660, doi: 10.1109/ICEES.2018.8443215.

[20]. Chodon, P., Adhikari, D. M., Nepal, G. C., Biswa, R., & Gyeltshen, S. (2013). Passive Infrared (PIR) Sensor Based Security System. International Journal of Electrical, Electronics & Computer Systems, 14(2). surveillance system by use of multiple ultrasonicsensors," IEEE Transactions on Consumer Electronics, vol. 56, no. 1, pp. 119-124, 2010.

[21]. Rakesh, V. S., P. R. Sreesh, and Sudhish N. George, "An improved real-time surveillance system for home security system using BeagleBoard SBC, Zigbee and FTP webserver," IEEE Int.Con, 2012, pp. 1240-1244.

[22]. Ansari, Aamir Nizam, Mohamed Sedky, NeelamSharma, and Anurag Tyagi, "An Internet of things approach for motion detection using Raspberry Pi," IEEE Int.Con.Intelligent Computing and Internet of Things, 2014, pp. 131-134.

[23]. Baraneetharan, E. "Role of Machine Learning Algorithms Intrusion Detection in WSNs: A Survey." Journal of Information Technology 2, no. 03 (2020): 161-173.

Routing Protocols and Security Issues in Vehicular Ad hoc Networks: A Review

Pimal Khanpara, Stuti Bhojak

Department of Computer Science and Engineering, Nirma University, Ahmedabad, India.

pimal.khanpara@gmail.com

Abstract. Vehicular Ad hoc Networks, also known as VANETs, are temporary wireless networks consisting of different types of automobiles as network nodes and the connections among these vehicles as links. VANET is a booming technology that has become quite popular in research, academia, and industry domains. A VANET can be considered as a moving network in which the nodes are automobiles. As VANETs are basically ad hoc networks, they have all the fundamental characteristics of ad hoc networks including computational and storage constraints. A network of VANETs can also be formed to enhance the safety and other services for vehicle drivers. VANETs are proven to be efficient in many real-life scenarios including traffic control and driver assistance. Information communicated through VANETs can definitely help to improve the user driving experience and avoid collisions. This paper discusses various deployment scenarios for VANETs along with the communication requirements and security concerns.

Keywords – Ad hoc Networks, Vehicular Ad hoc Networks, Routing, Security, Attacks

1. Introduction

Vehicular Ad hoc NETworks, abbreviated as VANETs, are considered a special class of Mobile Ad hoc Networks (MANETs) to facilitate communication among moving vehicles. As VANETs are derived from MANETs, they possess most of the fundamental characteristics of ad hoc networks. However, VANETs have their distinguishing characteristics which make them different from regular ad hoc networks. For example, in self-configuring VANETs, nodes (vehicles) move continuously and at a faster rate, usually from 0 to 40 m/s compared to MANET nodes. One of the basic objectives of VANETs is to increase the safety and comfort of driving for passengers while on the road. Vehicles can join or leave the network at any time.

VANETs are widely regarded as one of the most important technologies for enhancing the efficiency and safety of today's transportation systems. Due to the high rate of mobility of automobiles, connectivity in VANETs is constantly changing, generating rapid changes in the network architecture. A variety of vehicles ranging from light-weight personal or passenger vehicles to heavy-weight transportation vehicles can be considered to form a vehicular network. Because vehicles operate at varying speeds depending on the given traffic situations such as sparse or dense traffic, road conditions, rural or urban regions, traffic rules, etc., handling the dynamic mobility of the vehicles in an efficient manner is one of the biggest challenges in vehicular networks [1].

VANET increases the safety of transportation by informing drivers about road conditions, traffic bottlenecks, and weather conditions. Vehicles are increasingly being equipped with On-Board Units (OBUs), GPS (for navigation), wireless communication devices, and sensors, according to current developments (RADAR, LADAR). The primary goal of VANET applications is to gather and modify data so that drivers may make decisions. If 802.11p technology is employed, VANETs can connect to automobiles within a range of 100 to 900 meters. Vehicular Communication is seen as a key enabler for future autonomous transportation systems. However, the privacy and confidentiality of drivers' as well as vehicles' information must be maintained. Attackers might target to exploit and violate the secrecy of personal information such as a driver's name, address, license plate number, etc. by implementing different types of disclosure attacks. Confidentiality, Authenticity, Integrity, Availability, and Non-Repudiation are the key security requirements in VANETs. Therefore, Vehicle-to-Vehicle (V2V), as well as V2I (Vehicle to infrastructure) communications, are also required to be protected to make the adversaries unable to tamper or drop the control or data packets transmitted over the network [11]. The general architecture of a VANET is shown in figure 1.

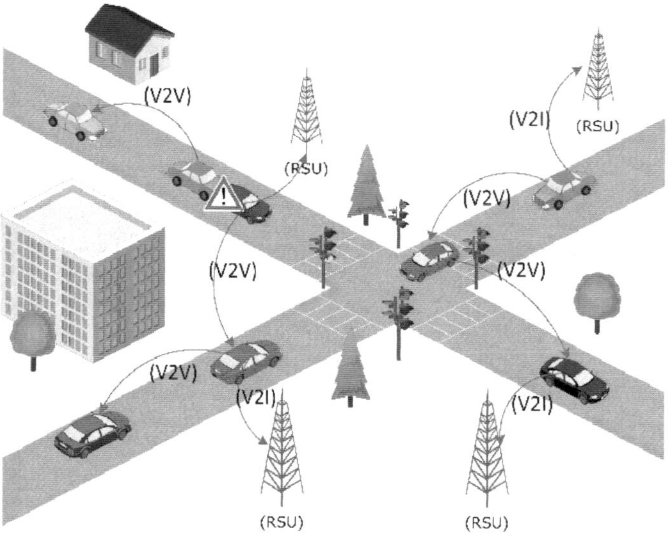

Figure 1. General Architecture of a VANET [2].

2. Characteristics of VANETs

A VANET can be targeted by adversaries in a variety of ways by launching various types of attacks. Hence, securing the nodes, links, and communication sessions is one of the major requirements in VANETs. As network connections in VANETs are not always consistent, the following issues need to be addressed [2]:

- *Absence of centralized infrastructure* – Just like MANETs, in VANETs also, there is no fixed, centralized infrastructure available [8]
- *Network Management* – The topology of the network can change rapidly due to the high mobility of the participating nodes
- *Security* – Security breaches or violations can result in severe safety issues and can also be transformed into life-critical emergency situations

- *Frequent Disconnections* – In VANETs, vehicles travel at random speeds, especially depending on dynamic traffic conditions which results in frequent link breakages and a very dynamic network topology
- *Congestion and Collision Control* – Unbounded network size creates a lot of challenges. The region of network deployment also plays an important role to determine collision avoidance and congestion control rules [25]
- *MAC Design* – VANETs use a wireless communication medium, i.e., air to transmit or receive messages. As the communication medium itself is not secured, adversaries might take an advantage of this fact to disrupt the network operations at various layers, especially the MAC layer
- *Environmental Impact* – VANETs use electromagnetic waves for communication. It is important to consider this fact while evaluating the environmental effects of VANET communications

The key characteristics of VANETs can be summarized as [2]:

- *Mobility:* In VANETs, vehicles often travel at a rapid rate. Predicting the position of the vehicle is difficult
- *Rapidly Changing network topology:* In VANETs, as vehicles travel at a random pace and in diverse directions, the location of vehicles will vary greatly due to their speed and motion. Within a short amount of time, nodes might leave/join the network and it directly affects the network topology
- *Unlimited battery power:* Vehicles in MANETs travel with their batteries, so the network components don't have to rely on a restricted power source to perform effectively
- *Unbounded Network size*: The size of a network is not restricted to a certain range
- *Frequent Exchange of Information*: Real-time information captured in the network is required to be exchanged among vehicles and Road Side Units (RSUs) on a frequent basis to take important runtime decisions with minimum errors
- *Wireless communication*: Vehicles are wirelessly connected with one another and with the RSUs to facilitate the exchange of information within the network
- *Time Critical:* As vehicle driving is a time-sensitive application, it is required to minimize the delay in data transfer and hence, the information must be provided to the vehicles within a certain time frame
- *Physical protection*: Vehicles are considered network nodes in VANETs and hence, their physical safety is also important for uninterrupted network operations [9]. Adversaries might attempt to make a physical compromise by launching physical attacks on the nodes

3. Security in Vehicular Ad hoc Networks

The rapid expansion of vehicular networks has resulted in a rise in security concerns also. Users of the VANETs, like users of other networks, need security in terms of data integrity, confidentiality, and availability, among many others. Every vehicular user requires privacy, which means that information about the vehicle and driver must be shared with other lawful users in a secure manner. The communication medium has a significant impact on the security aspects of VANETs. Messages should be communicated in a protected environment [12]. Delays in transmission can cause a slew of security risks. Due to delayed processing in VANETs, attackers find sufficient time to launch attacks or tamper with the standard network functionalities.

3.1. Security of VANET Entities

There are various types of entities involved in the overall functioning of vehicular networks. These entities participate in many network operations and hence, their safety is quite important for continuing the desired functioning of the network [17]. Key entities for VANETs are described below:

- *Vehicles:* Vehicles include all types of vehicles that can take part in the formation of VANETS i.e. cars, buses, etc.
- *RSU:* The roadside infrastructure is also crucial for any VANET to ensure correct and efficient V2V or V2I communications in a timely manner
- *Vehicle Drivers:* As drivers of vehicles are responsible for vehicle movements, their integrity and correct decision-making abilities are very important for any vehicular network
- *Third-Party Users:* It includes traffic control personnel like traffic police, transport regulatory authorities
- *Attackers:* It includes internal as well as external attackers who try to violate the network rules to disrupt network services with malicious intentions

3.2. Security Requirements in VANETs

The major security requirements for vehicular ad hoc networks are depicted in figure 2.

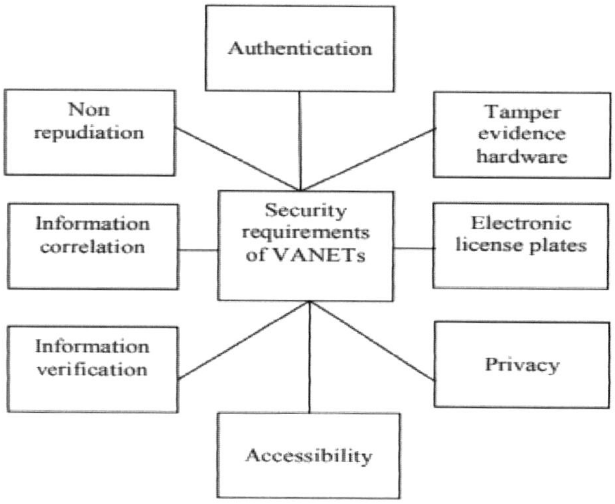

Figure 2. Security Requirements in VANETs [4].

- *Authentication:* It ensures that the message was sent by a legitimate sender to the intended recipient. Before processing any received information, the recipient verifies the authenticity of the sender via different techniques.
- *Accessibility:* It states that only legitimate users should be able to access confidential information or ongoing communications [15]. Attacks on access rights policies should cause the system to shut down in order to safeguard the data and prevent any kind of data sharing.
- *Message Integrity:* It is very important from the security perspective as the overall functioning of a VANET relies on the messages exchanged among different network

entities. Fake or false messages can result in degraded network performance [22]. Bogus or tempered information sharing can be the first step towards launching some attack in the network and hence, must be detected and prevented immediately.

- *Privacy:* There must be an assurance that the privacy of users' messages is protected against unauthorized users. It also addresses the overall confidentiality of past/ongoing communications in the network.
- *Event information recording:* The information regarding the vehicles must be recorded in case of any accidents or other fatal events.
- *Non-Repudiation:* It illustrates that a user cannot deny the fact that it has transmitted a message. Basically, it prevents any user from not taking responsibility for the transmissions it has imitated.
- *Tamper Detection:* The most important prerequisite to making information more secure is to identify the attempt of tempering through some hardware/software means. It is ultimately needed to safeguard data from unauthorized access.
- *Electronic license plates:* The government-issued vehicle number plates contain some useful information like unique identification of the vehicle or the vehicle owner, date of vehicle registration, etc. this kind of information is useful in tracking down the vehicles or their owners in case of any adverse or accidental events.
- *Information Correlation:* information correlation is an important process in vehicular networks to determine the importance, essentiality, and consistency of data.
- *Reliability:* It is one of the most important requirements in a vehicular network domain. The users of the network should be able to rely on the information received and the overall network operations.
- *Scalability:* Scalability refers to the ability to accommodate a large number of network entities as per the demand of the situation. In scalable networks, when the size of the network or the number of nodes grows, the performance of the network should not suffer significantly [16]. It should be simple to expand the network in terms of users, traffic, and other factors without affecting its functioning and services.

4. Security Issues in Vehicular Ad hoc Networks

Various types of attacks are possible in the vehicular ad hoc environment. Different types of attacks have a different impact on the network operations depending on their characteristics and severity. A few major attacks that can target VANETs are shown in figure 3.

4.1. Network Attacks

These are high-priority attacks that have the potential to disrupt the whole network (communication, vehicles, RSUs). In this type of network, attackers can have an impact on all users connected to it. Network attacks might take one of the following forms [3]:

- *Denial of service attack (DoS):* This attack attempts to disrupt the network operations by sending a huge quantity of useless/fake messages. The main goal of this type of attack is to prevent users from accessing the available network services.
- *Distributed Denial of Service attack (DDoS):* A distributed Denial of Service attack operates in a similar way as DoS attacks. However, in this attack, multiple entities or mediums are used to target the legitimate node or network operations [18].
- *Sybil attack:* Sybil attacks are used in peer-to-peer networks when a single hostile attacker has control over several nodes. A single attacker claims many identities in this attack. A large number of fake or bogus identities are generated in order to carry out this attack.

- *Node impersonation attack:* All VANET nodes contain unique identification numbers. An adversary uses some legitimate node's identity with malicious intentions, especially to gain authorization or access to network resources or functionalities. If successful, it becomes very difficult for the network to identify such attackers as they use legitimate nodes' identities [26]. After impersonating a valid user, an adversary can initiate a variety of damaging actions in the network.

4.2. Application Attacks

These attacks are carried out against the system's apps and services. The attacker either changes the behaviour of network applications or tampers with the information transmitted via system applications. The receiver has no notion that the communication has been tampered with by the attacker [7].

4.3. Timing Attacks

This type of attack is used to cause a lag in communication between the sender and the recipient. The attacker's primary goal is to cause delay and lengthen the time it takes for a message to be sent.

4.4. Social Attacks

The purpose of social attacks is to annoy or provoke people. In this form of attack, the attacker sends offensive communications to legitimate users.

4.5. Monitoring Attacks

These attacks are used to track the network nodes or users. It can be a passive form of attack with an attempt to observe the activity and behaviour of the users for the network operations. The information gathered by launching this attack can be used to initiate other types of attacks in the network. Popular forms of the monitoring attacks are Man in the Middle attack and Traffic analysis attack [3]:

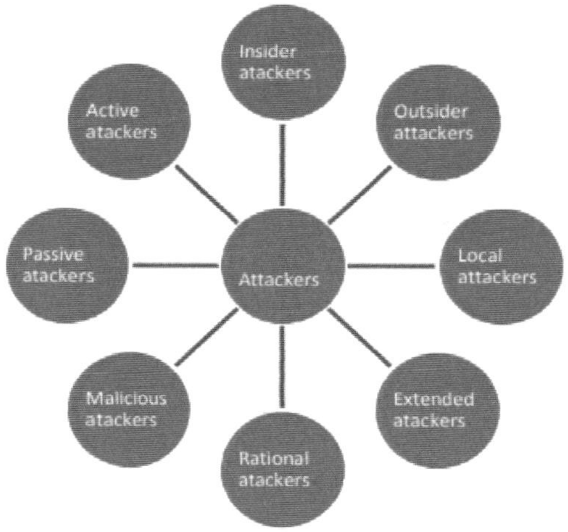

Figure 3. Types of Security Attacks in VANETs.

5. Routing Protocols in Vehicular Ad hoc Networks

The mechanism of routing is very important for any network and VANETs are not an exception [14]. In VANETs, it is possible to use different types of routing protocols to exchange node information with as little overhead as possible between nodes in a network. Different components of the VANET routing protocols can be defined, such as the routing algorithm utilized, routing information specifics, protocol similarities, network architecture, and so on. Topology-based, Position-based, Cluster-based, Geo-cast, and Broadcast are the five basic types of routing protocols used in VANETs [26].

5.1. Topology-based Routing Protocols

VANETs employ the topological routing process that delivers the essential information stored in the routing tables of nodes for communication. There are three types of topological routing protocols available for VANETs: proactive (table-driven), reactive (on-demand), and hybrid (combination of proactive and reactive) [10].

- *Proactive Routing:* The exact routes and other relevant information are stored in the routing tables in the proactive routing protocols. Proactive routing is independent of the need for message requests used for finding the available routes in the network. This kind of routing takes the advantage of previously recorded routes for various destinations. However, because of the low latency, it cannot be utilized for applications with temporal limitations. Moreover, the overhead of maintaining an up-to-date routing table in order to reflect frequent network changes is a tedious task. FSR, DSDV, and OLSR are some well-known proactive routing protocols for VANETs.
- *Reactive Routing:* The reactive routing mechanism lightens the load on the network while keeping the present routes in place. It operates on an on-demand or as-needed basis, lowering the entire network's overhead. The main advantages of reactive routing are less storage overhead and the absence of frequent routing table updates [20]. However, reactive routing protocols suffer from longer delays in establishing the required routes compared to proactive protocols. If the number of nodes in the network is small, reactive routing strategies can be implemented with maximum efficiency [13].

5.2. Cluster-based Routing Protocols

A cluster is a collection of nodes having comparable properties. Cluster-based routing is utilized in clusters, which are made up of a collection of nodes. In a cluster-based routing mechanism, one node serves as the cluster head, broadcasting information to the rest of the cluster nodes [19]. The benefit of cluster-based routing is scalability, but the disadvantage is high network latency caused by VANET's highly mobile nodes.

5.3. Geo-cast Routing Protocols

Geo-cast routing is a method of routing that is dependent on the physical locations of the participating nodes. Nodes available in specific zones are to be considered for routing. In geo-cast routing, the zone of relevance is also important which refers to the nodes that are at a specific place.

5.4. Broadcast-based Routing Protocols

These are the most widely used routing protocols in vehicular networks for traffic allocation, assessment of climatic conditions and risks, and communicating roadway conditions between vehicles and roadside units. BROADCOMM, UMB, V-TRADE, and DV-CAST are some popular examples of broadcast routing protocols.

5.5. Position-based Routing Protocols

In this category of routing, the selection of the node for delivering detailed information is dependent on the geographic location of the node. The packet information is sent via the nearest neighbour node to the destination node [6]. The benefit of using a position-based route is that there is no need to establish or maintain a global route for the communicating nodes. The greedy V2V protocol based on geographic locations is an example of position-based routing. The following description presents a brief of well-known routing mechanisms under this category.

5.5.1 GPSR

One form of position-based routing is the Greedy Perimeter Stateless Routing (GPSR) algorithm. Greedy forwarding is a mechanism in GPSR where nodes transmit packet information to an adjacent node that is at the shortest distance from the target node. The current position of the node and its immediate neighbour should be known for GPSR implementation. For transmitting information, GPSR uses the beacon idea. The notion of delivering hello messages to its neighbour is used by the node beacon to determine its destination. They are valid for a fixed amount of time; if a node does not reply, its item in the routing table is deleted. The planar traversal strategy and the greedy forwarding utilized by the GPSR routing protocol are described in Zone of Relevance (ZoR) Information packets are provided from a given source to all other nodes within the Zone of Relevance in Geo-cast based routing. For any encountered routing errors, warnings are broadcasted within ZoR.

Because GPSR is based on neighbour position data along with the destination location information, a VANET has a high mobility of nodes, hence no permanent structure of the nodes exists. Due to the dynamic architecture of a VANET, the greedy technique occasionally fails because the destination node's immediate neighbour is not found. Because of these issues with the Greedy technique, GPSR employs the idea of recovery mode [3].

The greedy methodology and perimeter packet forwarding mechanism are used by GPSR. When networks are stable, the greedy technique is employed, but when they become dynamic, the perimeter packet forwarding method is used to recover the situation, and when the network is stable again, the greedy method is used and the same procedure is followed. To eliminate routing loops, the perimeter approach is used on planar networks [23]. The same is shown in figure 4. When the GPCR reaches its local maximum, the recovery mode is activated. The packets are backtracked for other solutions to adopt a greedy mode during the recovery phase. The right-hand rule is used to forward packets.

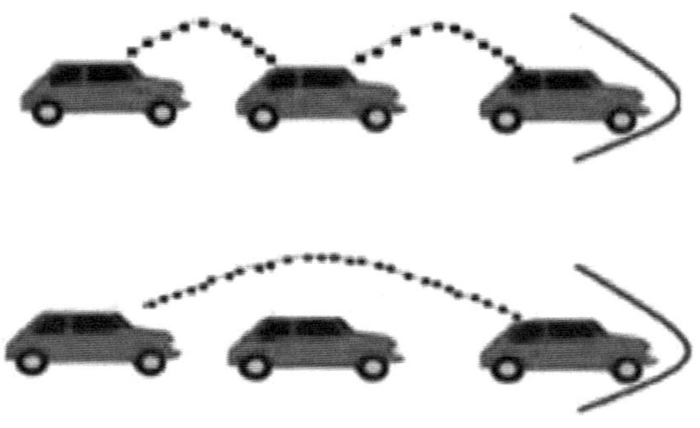

Figure 4. Greedy Approach: Planar Traversing Graph [3].

5.5.2 GPSRJ+

Another example of a position-based routing protocol is GpsrJ+. GpsrJ+ is categorized as a beacon. The nodes in GpsrJ+ are virtually linked. For packet forwarding, GpsrJ+ supports two modes: greedy mode and recovery mode. GpsrJ+ packets are sent to the nearest destination in Greedy forwarding [10]. If the forecast status indicates that the packet information will be sent onto a road in a different direction, the packet is transferred to the junction node. When the stage reaches the point where it points to the local maximum, the recovery mode is activated. Packets are resuming their journey down the highway in recovery mode. If the nodes are not required, GPCR can surpass them and route the packets to the junction [4]. The classification of routing protocols for VANETs is presented in figure 5. The analysis of these routing protocols along with their key characteristics is shown in table 1. The appropriateness of using a routing protocol of a particular type along with the advantages and drawbacks is presented in table 2.

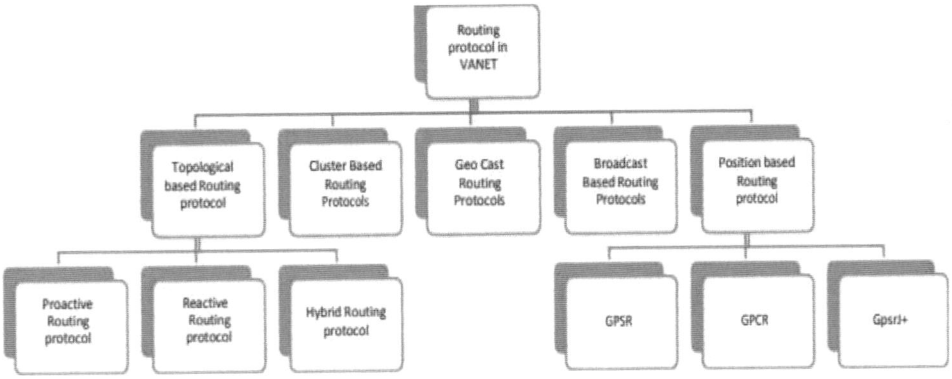

Figure 5. Classification of VANET Routing Protocols.

Table 1. Key Characteristics of VANET Routing Protocols

Type of Protocol	Forwarding Technique	Location Information Requirement	Traffic Flow	Recovery Mechanism	Deployment Scenario
Proactive	Multi-hop	No	Realistic	Multi-hop Forwarding	Urban/dense
Reactive	Multi-hop	No	Realistic	Store-Carry-Forward	Urban/dense
Position-based	Heuristic-based	No	Realistic	Store-Carry-Forward	Urban/dense
Delay-bounded	Store-Carry-Forward	No	Not realistic	Multi-hop Forwarding	Sparse
Cluster-based	Multi-hop	Yes	Not realistic	Store-Carry-Forward	Urban/dense
Broadcast	Multi-hop	No	Realistic	Store-Carry-Forward	Highway/sparse
Geo-cast	Multi-hop	No	Realistic	Flooding	Highway/sparse

Table 2. Advantages and Drawbacks of Routing Protocols

Protocol	Advantage	Disadvantage
Topological-based routing	Can have the advantages of background paths	Poor performance in high mobility scenarios
Cluster-based routing	High scalability	Delay is proportional to the mobility of nodes
Geo-cast routing	Minimum packet overhead	Network partitioning
Broadcast-based routing	Guaranteed packet delivery	Under-utilization of bandwidth
Position-based routing	Can be used in highly challenging environments	Relies on routing data received from GPS

6. Security Mechanisms for Vehicular Ad hoc Networks

Authentication of participating entities and non-repudiation are major requirements for securing vehicular communications. Many researchers have proposed mechanisms to guarantee security in the VANET environment. The most commonly used security mechanisms for VANETs are public key infrastructure, ID-based cryptosystems, and situation modeling-based mechanisms. These mechanisms are discussed in this section.

6.1. Public-Key Infrastructure (PKI)

In public-key cryptography, each node uses two keys, a public key and a private key. The role of the private key is to provide authentication and non-repudiation. Before transmitting a message, each sender signs it with their private key. Upon receiving this signed message, the recipient can verify the digital signature of the sender by using the sender's public key. This is how authentication and non-repudiation are achieved. Valid digital signature verification ensures that data is not tampered with during transmission. The public key of each node must be authenticated by the other nodes in the network for proper authentication, hence a PKI method is necessary for secure communication in VANETs. The certificate authority (CA) is in charge of creating authenticated certificates in the PKI [5].

The following are the responsibilities of the CA:

- Issuing public key certificates
- Disseminating and binding the certificates to the respective entities
- Renewing certificates for public keys
- Revoking public key certifications

CAs can revoke certificates using Certificate Revocation Lists (CRLs), which include the malicious or suspicious nodes' certificates. As a result, before authenticating a message received from any node, the CA checks to see if the sender's certificate is listed in the CRL. Although PKI is an effective method for protecting VANETs, it has significant flaws that can lead to insecure VANETs in specific situations. Despite the use of anonymous certificates for identity privacy, location privacy is not always given in VANETs, and attackers can attack the vehicle by tracking a specific vehicle between two locations specified by the attacker. If the vehicle is moving at the same pace and on the same path which is being monitored by the attacker, then the adversary can link the anonymous certificate for that vehicle to it and target it. Therefore, it is necessary to apply PKI location privacy techniques.

6.2. Identity-based Cryptography

The basic concept underlying ID-Based cryptography is to utilize any known information that represents the user's identity to authenticate the digital signature. This public information might include an email address, a network address, a user name, or a combination of these. Shamir [21] introduced id-based cryptography for the first time in 1984. The ID-Based cryptosystem's key advantage is that it generates pseudonyms that may be modified as needed for security reasons. Additionally, the ID-Based approach allows users to use many pseudonyms for increased anonymity. Because a trusted third party is not necessary to store, fetch, or validate the public key certificates of various road safety situations, an ID-based approach might be an efficient alternative for the PKI technique in VANET contexts. As a result, the ID-Based method beats the PKI system in terms of time, communication bandwidth, and storage, especially when CRL costs are included. The primary aim of ID-Based cryptography is to ensure the entity's privacy. For example, pseudonym formation should be done using an irreversible method that generates the pseudonym using the entity ID while assuring that the pseudonym is only available for the same entity [6]. As a result, several studies have been done to address the ID-Based mechanism's privacy issue.

6.3. Situation-based Modeling

Situation Modeling-based techniques are based on the generation of VANET models for vehicle driving patterns and behavior in order to improve VANET security through node management. A small number of models can be used to summarize various VANET settings and circumstances [5]. However, the issue is based on the obvious unpredictability of vehicle movements. The major concern with situation modeling-based techniques is producing numerous models based on various vehicle placements and movements in the terrain area. Furthermore, produced models should be able to check themselves with respect to unexpected events. Many attempts have been made by the researchers to resolve the issues pertaining to situation modeling-based techniques that entail generating a large number of models to match various VANET deployment scenarios and network conditions [6]. The above-mentioned VANET security mechanisms each have their own set of benefits and drawbacks, making them suitable for a variety of scenarios. Table 3 summarizes the issues faced by VANET security methods.

Table 3. Challenges for VANET Security Mechanisms

VANET Security Architecture	Challenges
Public Key Infrastructure	Location privacy, Maintenance of CRL, ensuring the availability of CA in the network, Security of CA
Identity-based Cryptography	Handling identity privacy-related issues
Situation-based Modeling	Complexity due to the dependence on various possible deployment scenarios

7. Conclusions and Future Directions

Mobile nodes, which are equipped with wireless communication and serve as network entities form a temporary vehicular network. By maintaining V2V and V2I communications, the network self-organizes and enables greater communication and coordination among vehicles and the available infrastructure. In VANETs, malicious users might attempt to mislead the legitimate vehicles by sending incomplete or tampered information. VANET communication differs significantly from MANET communication scenarios. The emphasis of this paper is on how to deal with different types of probable security violations while yet maintaining the desired efficiency and Quality of Service in the dynamic VANET environment. The paper also discusses the drawbacks or loopholes of the existing security mechanisms for vehicular networks. The categorization of routing is also presented in the vehicular ad hoc network context in order to provide a better understanding of possible routing schemes. The efficiency of routing algorithms is an important parameter to determine the overall performance of the network, even in adverse conditions. There are numerous characteristics that distinguish VANETs from other areas of research, as this work has addressed. There are many aspects of this field that make it worth researching in its way, whether it's its distinct features, the inherent services it can provide, or the difficulties that this technology faces. By looking into these, this paper highlights the benefits of VANET research and encourages further research in the field. The paper then continues further to detail the recent research being done to deploy VANET technologies, as well as how such advances may assist in improving the solutions in this domain.

References

[1] Mejri MN, Ben-Othman J, Hamdi M. Survey on VANET security challenges and possible cryptographic solutions. Vehicular Communications. 2014 Apr 1;1(2):53-66.
[2] Saini M, Singh H. VANET its characteristics attacks and routing techniques: a survey. International Journal of Science and Research. 2016 May;5(5):1595-9.
[3] Jain M, Saxena R. Overview of VANET: Requirements and its routing protocols. In2017 international conference on communication and signal processing (ICCSP) 2017 Apr 6 (pp. 1957-1961). IEEE.
[4] Kaur R, Singh TP, Khajuria V. Security issues in vehicular ad-hoc network (VANET). In2018 2nd International conference on trends in Electronics and Informatics (ICOEI) 2018 May 11 (pp. 884-889). IEEE.
[5] Tomar R, Prateek M, Sastry GH. Vehicular adhoc network (vanet)-an introduction. International Journal of Control Theory and Applications. 2016;9(18):8883-8.
[6] Bariah L, Shehada D, Salahat E, Yeun CY. Recent advances in VANET security: a survey. In2015 IEEE 82nd vehicular technology conference (VTC2015-fall) 2015 Sep 6 (pp. 1-7). IEEE.
[7] Sheikh MS, Liang J. A comprehensive survey on VANET security services in traffic management system. Wireless Communications and Mobile Computing. 2019 Sep 15;2019.
[8] Khanpara P. A review on fuzzy logic based routing in ad hoc networks. International journal of advanced research in engineering and technology. 2014;5(5):75-81.
[9] Talpur A, Gurusamy M. Machine learning for security in vehicular networks: A comprehensive survey. IEEE Communications Surveys & Tutorials. 2021 Nov 18.
[10] Hota L, Nayak BP, Kumar A, Sahoo B, Ali GG. A Performance Analysis of VANETs Propagation Models and Routing Protocols. Sustainability. 2022 Jan;14(3):1379.
[11] Nazib RA, Moh S. Reinforcement learning-based routing protocols for vehicular ad hoc networks: A comparative survey. IEEE Access. 2021 Feb 10;9:27552-87.
[12] Trivedi R, Khanpara P. Robust and secure routing protocols for MANET-based internet of things systems—A survey. In Emergence of Cyber Physical System and IoT in Smart Automation and Robotics 2021 (pp. 175-188). Springer, Cham.
[13] Channakeshava RN, Sundaram M. A Study on Energy-Efficient Communication in VANETs Using Cellular IoT. In Intelligence Enabled Research 2021 (pp. 75-85). Springer, Singapore.
[14] Khanpara P, Valiveti S, Kotecha K. Routing in ad hoc network using ant colony optimization. In International Conference on Future Generation Communication and Networking 2010 Dec 13 (pp. 393-404). Springer, Berlin, Heidelberg.
[15] Shah M, Khanpara P. Survey of techniques used for tolerance of flooding attacks in DTN. In Information and Communication Technology for Intelligent Systems 2019 (pp. 599-607). Springer, Singapore.

[16] Khanpara P, Trivedi B. Survivability in MANETs. International Journal of Advanced Research in Computer Engineering & Technology (IJARCET). 2018 Jan;7(1).
[17] Maddiboyina HV, Sankar Ponnapalli VA, Naresh Kumar A. A Study on the Implementation of Secure VANETs Using FPGA. In Recent Innovations in Computing 2022 (pp. 179-187). Springer, Singapore.
[18] Shetty SR, Manjaiah DH. A Comprehensive Study of Security Attack on VANET. In Data Management, Analytics and Innovation 2022 (pp. 407-428). Springer, Singapore.
[19] Teixeira D, Ferreira J, Macedo J. Systematic literature review of AI/ML techniques applied to VANET routing. In Future of Information and Communication Conference 2022 Mar 3 (pp. 339-361). Springer, Cham.
[20] Qasim OA, Noori MS, Dabag A, Ahmad ML. Performance evaluation of ad-hoc on-demand distance vector protocol in highway environment in VANET with MATLAB. Telkomnika. 2022 Feb 1;20(1).
[21] Chaudhry SA. Comments on" A Secure, Privacy-Preserving, and Lightweight Authentication Scheme for VANETs". IEEE Sensors Journal. 2022 Apr 18.
[22] Khanpara P, Trivedi B. Security in mobile ad hoc networks. In Proceedings of International Conference on Communication and Networks 2017 (pp. 501-511). Springer, Singapore.
[23] Singh G. Performance Evaluation for Ad hoc Routing Protocol in Vehicular Ad hoc Network (VANET). Global Journal of Computer Science and Technology. 2022 Mar 12.
[24] Talukder S, Ira SK, Khanom A, Sneha PB, Saleh W. Vehicle Collision Detection & Prevention Using VANET Based IoT With V2V.
[25] Hassan A, Attia R, Rizk R. Internet of Vehicles and Intelligent Routing: A Survey-Based Study. InInternational Conference on Advanced Machine Learning Technologies and Applications 2022 (pp. 517-531). Springer, Cham.
[26] Sireesha K, Malladi S. A Survey of VANET Security Models and Its Issues on Node Level Data Transmission. In2022 Second International Conference on Artificial Intelligence and Smart Energy (ICAIS) 2022 Feb 23 (pp. 1409-1417). IEEE.

Diabetic Retinopathy Detection Using Automated Segmentation Techniques

S. Prabha[1], S. Sasikumar[1] and Ch. Leela Manikanta[1*]

[1]Department of ECE, Hindustan Institute of Technology and Science, Chennai, India.

sprabha@hindustanuniv.ac.in, ssasik@hindustanuniv.ac.in, manimanikanta.ch@gmail.com

Abstract. This paper contains a brief discussion about Diabetic Retinopathy. As the name indicates, it's a medical complication present in diabetic patients which affects the retina. Diabetic Retinopathy acronymed as DR is a medical circumstance where the high glucose levels in the blood start affecting the blood vessels in the retina. The paper discusses the non-invasive technical method to detect diabetic retinopathy involving various algorithms in every phase of the process. The input fundus images are taken from STARE Database. The methodology conveyed in this paper involves contrast-limited adaptive histogram equalization for noise cancellation purposes and enhancing the base contrast of the image. The Segmentation consists of 2 steps and the first step consists of the Fuzzy C-Means clustering primarily to find the coarse vessels present in the retina. Additionally, the Region-based active contour is used to select the region of interest which is to highlight the blood vessels. As a result, Our proposed segmentation method extracts the blood vessels accurately, resulting in the similarity measure value of 85%. Furthermore, these segmented retinal blood vessels are given as the input to CNN classifiers in order to detect Diabetic Retinopathy. For our proposed method, an overall accuracy to detect DR was 92%. This methodology can be used for mass screening processes in the field of ophthalmology.

Keywords: Diabetic Retinopathy, Contrast-Limited Adaptive Histogram Equalization (CLAHE), Segmentation, Fuzzy C Means, Region based active contour, retinal images.

1. INTRODUCTION

Human eyes are commonly known as the organ of vision. It's the organ that initially detects light and transfers the signals through the optic nerves to the brain in order to perceive the visionary image. Eyes are an essential part of the human body as a common fact. It's also a highly sensitive organ, with 23 neurons that transform light into electrochemical impulses. Sclera, Iris, Cornea, Pupil, Lens, Ciliary Body and Muscle, Conjunctiva, Retina, Vitreous Body, Optic Nerve, Macula, and Retinal Blood Vessels are the components of the eyes. These are the essential and common parts of the eyes and can be further segmented if each organ is studied specifically. As it is known that the human body is an interconnected system that is connected and functions by relying on other organs, the same it's an exclusive feature that the nerves in the eyes are to other organs of the body as well. The retina is a fragile portion of the eye that is located in the base of the eye and is the innermost, light-sensitive layer. The retina's job is to give the sharp, centre vision needed for tasks like reading, driving, and anything else that requires minute concentration.

The world is evolving towards something new on a daily basis, leading to advancements in every sector be it technical, business or even the overall life style of human beings. Along with these advancements, an essential sector with regular advancements is the medical sector. There are researches, inventions and discoveries of new medical circumstances often. As a result, the medical world and advanced technology work hand-in-hand to resolve problems in an effective manner. The same way diabetes is a medical condition wherein the sugar levels present in the blood are higher than required. It is a disease that can be found in human beings of all ages certainly, including new born babies.

Diabetes is further 48 divided into Prediabetes, Type 1 Diabetes, and Type 2 Diabetes. Prediabetes is a phase wherein the glucose present in the blood is slightly higher than the amount needed. Classification of Type 1 diabetes is as production of minute or nill amount of insulin by pancreas. When the body's ability to metabolize, blood sugar is hampered by excessive amounts of glucose, Type 2 Diabetes develops. Patients with diabetes commonly face other problems like foot ulcers, cardiovascular complications, gum diseases, fatigue, stroke, nerve damage, chest pain, vision complications. Diabetic Retinopathy is a familiar stage of diabetes. It is a medical disorder in which excessive blood sugar levels impact the eyes in a number of ways. And one such condition is Diabetic Retinopathy also known as DR. It is a medical condition wherein high blood sugar levels begin to harm the blood vessels in the eyes, creating serious problems. The most prevalent concerns that individuals with Diabetic Retinopathy may have include vessel obstruction or leakage, as well as the initiation of new blood vessel growth in the Retina. When one end of a blood vessel becomes clogged due to excessive sugar levels, a blockage occurs. And this further tends to lead to the next complication of leakages of the blood vessels when there is continuous supply of blood to the vessel and when there is a blockage, the blood cannot be present within the closed space, hence leading to the leakage of blood as a result of the pressure. Sometimes there can even be new blood vessels formed. In the short or long run, these conditions may result in various issues like blurred vision, impaired color vision, eye floaters, etc., which restricts vision, poor night vision, dark or empty spots in vision.

There are 4 stages in Diabetic Retinopathy, and they are:

Stage 1- Mild Non-proliferative Retinopathy: In this stage, microaneurysms start to show up. Microaneurysm is the swelling of blood vessels located in the retina. The blood vessels are in the initial stage of bulging at this point of time.

Stage 2- Moderate Non-proliferative Retinopathy: The swelling of the blood vessels starts to increase. This further starts to severely affect the blood flow in the retina. And this also prevents the proper nourishment required in the eyes.

Stage 3- Severe Non-proliferative Retinopathy: In this stage the blockages in the blood vessels start taking place in large amounts. This significantly reduces the amount of blood flow in the thin areas.

Stage 4- Proliferative Retinopathy: This is the critical stage where the birth of new blood vessels occur, causing severe complications in the delicate areas.

These four stages of DR have their own complications and treatments for cure but this can be done only if the problem can be detected accurately. In order to prevent all these complications, it is advised for all the patients with diabetes Type 1 and Type 2 to take a DR test. Nevertheless, not every patient agrees to do so as there are various factors involved in taking the DR test medically. The medical DR tests are extremely time consuming they require the physical contact in the eyes which is a method prone to infections if done without proper sanitization, and they are also high in cost. That is why there are various technical methods proposed and carried out with computational algorithms to detect DR. The engineers' methods so far are extremely time effective so that treatment can be started out in an early stage. They are non-invasive, meaning there is no physical contact required so safety is always insured. And finally, they are also cost effective. Another added advantage of the engineer's methods is that the stages of DR can also be configured so that the diagnosis can be done accordingly.

2. RELATED WORKS

Using robust hybrid features Aslani [1] demonstrated a new supervised retinal vessel segmentation method They took the hybrid feature vector to merge the supporting complementary local data. There are thirteen Gabor filter responses that make up the 17D feature vector used. It is also noteworthy that as the numbers are reduced, the accuracy also decreases. The proposed method tends to be designed and executed to achieve the highest average accuracy in the DRIVE and STARE datasets. Furthermore, the progress and scope of the development were discussed. Azzopardi [2] discusses trainable COSFIRE filters for vascular description with attributes of retinal imaging. The dominant feature here is that the COSFIRE filter has been structured. B in this case means bar, which is a construction for a container. The implementation of the filter named COSFIRE depends on the approximate values of required parameters. Christodoulidis[3] Multiscale Tensor Voting Framework (MTVF) is the reference for the methodology proposed in this paper. It is added up with multi-scale line detection to get rid of line detection limitations when managing smaller vessels. Although the proposed methodology has produced excellent results, there were some issues that needed to be explored and addressed. If the MTVF is attached to the false positive (FP) vessel-like structure of the main vasculature, the main vessel may be lost.

Zhu[4] The document ultimately aims to improvise the standards of the segmented image. Using State-of-the-art technology they enhanced 39 discriminating feature vectors for fundus imaging. The binary retinal vessel segmentation is the acquired output by the classifier. As a result, the diagnosis of retinal diseases is difficult. Therefore, it is crucial to use the right image segmentation algorithms to accurately find all retinal blood vessels. Geetha[5] proposed his work based on the Principal component analysis (PCA) that was used to create the feature vector in this paper. Additionally, k-mean clustering is applied to this result to group the pixels into cup or non-cup groups. Vessel group classification improved accuracy but reduced sensitivity in some cases. Hassanien[6] In this article, the automated approach to segmentation is based on two-step optimization principles; retinal blood vessels were created. was demonstrated. The suggested method combines artificial bee colony optimization with a fitness function for fuzzy cluster compression with partial membership at first level and approximate vasculature at aggregation. Roy Chowdary[7] conducted the research based on blood vessel segmentation from fundus photographs by extracting major vessels and classifying sub-images. In this document, a set of eight features was introduced to distinguish between vascular and nonvascular pixels. The fine pixels of the vessels are found with both Gaussian and GMM classifiers. Sil Kar [8] discussed this research study to track the numerous thresholds, they used their understanding of the maximum response of the matched filter and the fuzzy conditional entropy. And differential evolution calculations are used to find the ideal values. The discussed procedure is also effective in removing the irregular and thin blood vessels in pathological retinal images.

Sreejini [9], this article discusses a PSO-based parameter that was introduced to determine the exact values of the multi-scale coordinated filter parameters. The results generated by the multiscale matching filter take precedence over the single scale matching filter for vessel segmentation. Singh[10] this work analyzes the methodology for this particular experiment consisting of PCA or principal component analysis during the pre-processing phase. In the next stage, CLAHE is employed in order to improve the retinal picture. To obtain the segmented image, after running the proposed combined filter, the post-processing phases include optimal entropy-based threshold and length filtering. The results suggest that this method performs relatively well and efficiently compared to other prominent Gaussian distribution functions. Saranya[11] The Convolution Neural Network (CNN) is the predominant methodology utilized to observe and classify the images according to the four phases of DR as it is capable of managing the pre-processing of images and normalization as well. Chaudhary[12] The proposed methodology contains the raw fundus images which are carried out for processing for sound reduction and further changed into gray image. Later the segmentation of the optic disc and the retinal nerves take place. The extracted features will be used for the classification. Post processing consists of erosion and dilation using a structuring element on the pre-processed images. The CNN and the Fuzzy classifier are used to identify the status of the image. This will confirm the presence of DR or a normal eye condition.

Imran Qureshi[14], demonstrated the computer assisted diagnosis (CAD) systems for treating Diagnosis Retinopathy. It's a combined study on the effectiveness of the different methodologies used to detect DR. The segmentation and the method of extraction using the image combination, pre-processing, improvement and segmentation methods have been briefly explained in the paper [18]. Akinboro[15] The project is carried out specifically to classify the stage of DR with the proposed approach. The methodology chosen is by first generating the digital fundus image through the convolution neural network CNN and 3662 datasets. The CNN model is built with the Keras infused. Author [13][16][17] did the research work on threshold based on an iterative algorithm that has been implemented in the research paper focusing on the accuracy of the segmented images. The proposed methodology is made up of feature extraction, followed by image segmentation, image pre-processing, and finally the classification based on CNN classifier which provides the results of whether DR is present. The proposed methodology is designed in such a way to detect diabetic retinopathy in an earlier stage in order to take appropriate diagnosis. Accuracy, precision, specificity, and the sensitivity rates of this proposed system stays above 85% in all the four aspects. The overall procedure focuses on maintaining the accuracy rates of the results and every algorithm chosen, carefully aids to it.

3. PROPOSED WORK

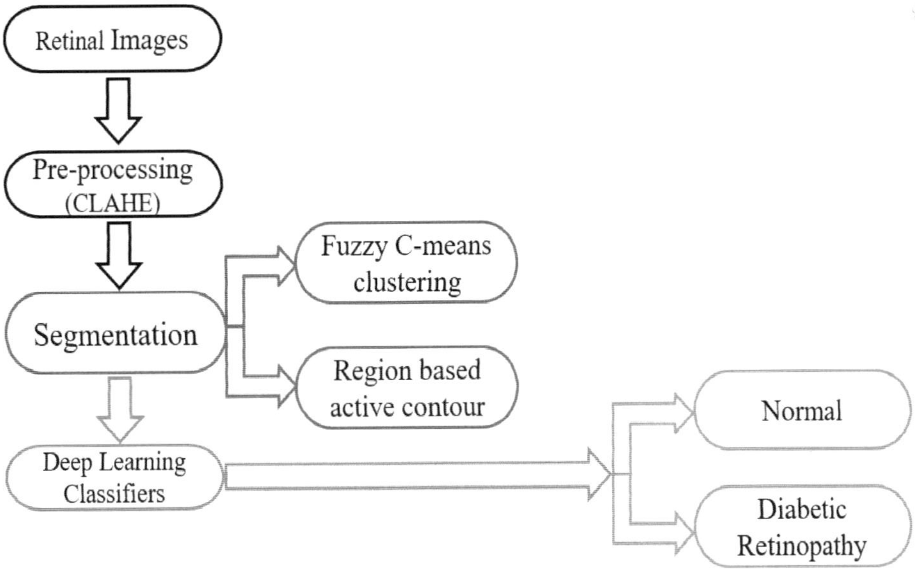

Figure 1: Proposed work architecture

The innovative technique flows through the following six phases as shown in figure 1.

1. The proposed method uses a fundus camera as an input image (STARE database).

2. For the first step of pre-processing, the input image is transformed to the green channel. The picture is then subjected to CLAHE to boost the image's local contrast.

3. Fuzzy C-Means (FCM) is utilized to detect the coarse vasculature during the segmentation phase.

4. The region based active contour method is used to select the region of interest & aids in highlighting the blood vessels.

5. The segmented retinal blood vessels are seen in the output image.

6. The segmented image is given as an input image to the deep learning classifier to find whether the respected image is affected with DR or NOT.

The process of detecting DR will first be started out by first acquiring the retinal fundus image from the STARE Database. STARE stands for Structural Analysis of the Retina. In an image, second channel is the green channel, that second channel alone is extracted to correct non uniform illumination. The pre-processing will be done through the CLAHE algorithm. This algorithm is utilized to improve the image's contrast level and for noise cancellation characteristics. As a result, the image is additionally enhanced so that the unseen features are also visible. The CLAHE has three parameters: block size, histogram bins and max slope through these three methods the image will undergo pre-processing. Then the optimization stage takes place in two steps. The first stage of optimization includes the involvement of Fuzzy-C Means clustering to specifically spot the coarse vessels present in the retina. Later the region of interest is selected utilizing the region based active contour algorithm for better quality and higher accuracy levels of the blood vessels that exist in the particular image. Now the output image will show the segmented retinal blood vessels which will be sent as an input image through the CNN classifier and this is where the presence of diabetic retinopathy is observed, reviewed and confirmed.

Resulted in the creation of the Designed analysis of the Retina database (STARE). The STARE database photos were taken using a 35degree narrow field view camera and have a resolution of 700*605 pixels. Retinal imaging is a frequent procedure performed during an eye examination. To the posterior inferior surface of the eyeball in order to see the pupil of the eye an optical camera is assisted. The retinal layer, the optic nerve, the fovea, and the surrounding arteries are all imaged. The ophthalmologist can then use this image as a reference point for evaluating any findings or to examine the eye condition.

AHE (adaptive histogram) is a computer image processing approach for enhancing image contrast. Traditional histogram equalization differs from the adaptive technique. It's designed to build a number of histograms, each corresponding to a different portion of the image, and then use them to disperse the image's brightness values. This makes it great for increasing local contrast and sharpening edge sharpness in different areas of a photo. AHE also tends to exaggerate noise in relatively similar areas of an image. By limiting gain, a variant of AHE known as CLAHE was developed to minimize or prevent noise. Microaneurysm pixels were enhanced with the CLAHE filter. The CLAHE filter produces adequate vein enhancement while removing excess noise. Other global enhancement methods such as traditional contrast stretching and universal histogram equalization work less efficiently compared to the CLAHE filter. The use of an independent elemental analysis is the second way to increase the contrast of the veins. Various image enhancement techniques have been added in recent research at the pre-processing stage, including grayscale conversion of image, spectrum normalization, brightness enhancement using CLAHE, and gamma modification to improve overall appearance. The CLAHE histogram has been clipped to a standard clipping value to avoid excessive contrast enhancement that could result in an image with an odd appearance and unwanted detail. Also, incorrect approaches to implementing contrast enhancement could result in poor appearance in missing areas, such as small veins lead. By summing up a global threshold to the fixed point of contrast the above consequence could be solved. Edge detection is a technique for detecting frequent changes in pixel values. The brightness level of neighbouring pixels is used to examine the edge information of a given target pixel. Without a clear difference in brightness, there can be no image edge.

The FCM (Fuzzy C-Means) clustering technique allows a single data item to be assigned to two or more clusters. Pattern Recognition also runs on the basis of the above Strategy. This algorithm assigns the link to each data point corresponding to each group centre based on the distance between the group and the data point. The closer the data is to the cluster centre, the more likely it is to belong to that cluster centre. The iterative unsupervised Fuzzy C Means (FCM) method is the most extensively used clustering technique for picture segmentation. Pattern recognition and clustering are used to analyze and segment most medical images. Compared to the k-Means algorithm, fuzzy C-mean clustering is considered a better option. The fuzzy C-means algorithm, unlike the k-means technique, which requires data points to belong to only one group, allows data points to belong to many groups with probability.

The fuzzy C means that the clustering process starts first by outputting the input image first. Then the size of the image is retrieved. The possible range is then calculated. In addition, the possible number of iterations is identified. Then the given dimension of the images is calculated. Then the iteration begins. When the maximum number of iterations is reached, the procedure is repeated, and the picture is submitted for iteration once more until the maximum number of iterations is achieved. This is the complete procedure of the fuzzy pooling process C.

Active Contouring is a Segmentation approach that separates relevant pixels from an image for further processing and analysis using forces and power constraints. An active contour is a model used in the segmentation process. Contours are the lines that define the area of interest in an image. There's a need of converting segmented images into multiple similar images to get represented in the region-based technique. The basic idea behind active contour models, sometimes referred to as snakes, is to generate a curve to group objects in a given image while constrained by these constraints. The curve approaches its interior normal and must end at the edge of the object. In classic active snake and contour models, To come to a halt, an edge detector employs the curve evolution at the boundary of the target element based on the gradient of the image. The following steps are used to evacuate the papilla: i. To capture the retinal image mask, first select the underlying raw, apply the region-based active contour to the segmented image mask. I. The mask from the input picture is removed, resulting in a new mask. ii. The segmented image without a mask is the result of subtracting the new mask from the previous one. The region based active contour methos uses statistical information to construct a region stopping function which stops the contour evaluation between different regions. This method is not sensitive to initialization of level set function and can recognition object boundaries efficiently.

The Structural Similarity Indexing Method (SSIM) is a tool that compares the structural similarity of two photographs. If the second image is of perfect quality, the SSIM index can be regarded as a quality indication for one of the photos under consideration. It is an improved version of the first proposed Global Image Quality Index. The SSIM tool is used in this case to compare the actual terrain image with the segmented image. The relevant output image for the fundus image retrieved from the STARE database is the actual image of the floor. And the segmented image is the starting picture that is received after the CNN after the final processing of the image. The SSIM tool will be used to compare the differences present in the ground truth image and the segmented image which will prove the accuracy and the efficiency of the methodology. Keras was used to build and compile the Convolutional Neural Network (CNN) models. Except for their last layer, both models have the same architecture; the first model, which was used to classify the severity of DR, has 5 outputs, while the other model, which was used to identify DR, has 2. The models are divided into four stages:

i. Input

ii. Feature Learning

iii. Classification

iv. Output

The retinal fundus image is the CNN model's input, which was read by the computer as a matrix of 64 * 64 * 3 pixels, that represents the height, width, and dimension. The RGB values are represented by the three dimensions of the images. The input phase's outcome is subsequently passed on to the feature learning phase. The input image is passed through a series of convolution layers with filters in the feature learning phase, and the model extracts and learns features about the images. The extracted features are then forwarded to the classification phase, in which the image is categorized with probabilistic values 26 between 0 and 1. The pre-processed image is fed into the input layers, which produce an array of pixels as output, which is then transferred to the first convolution layer. The link between/among of the pixels exists by learning visual attributes of minute squares and convolutions. The first convolution layer uses a 64 * 64 * 3 picture as input and produces an output of 62 * 62 * 128 feature map with relu activation.

The output was fed into the second convolution layer as input. The second convolution layer, which likewise uses a (3 * 3) filter matrix with relu activation, produced a (60 * 60 * 128) output that was transferred to the pooling layer. By picking the largest element from the feature map, the max pooling layer minimizes the feature map's dimensionality, producing an output of size (30 * 30 * 128) and passing it to the dropout layer. Over fitting was reduced by using the dropout layer. During the training, it switches off some neurons randomly. The output of the dropout is flattened using a flatten layer, which transforms the entire feature map into a single column of size (115200). The dense layer contains densely linked neurons and gets input from the prior layer's neurons. To reduce overfitting, a dropout layer was added between the two thick layers. The last dense layer employed the SoftMax activation function to assign probabilistic values between 0 and 1 to each class in the image.

4. RESULTS OBTAINED

4.1 Normal Images:

The normal images refer to the pictures of healthy eyes received from the STARE database. These images do not contain or have the symptoms of DR. The retina will be clear in this image with normal and regular functioning blood vessels without any complications or abnormalities. Images considered to be normal will provide the image of the healthy retina when the image is fully processed.

4.2 Abnormal Images:

Abnormalities refer to images that contain Diabetic Retinopathy. These images will commonly show up as abnormalities while retrieving from the STARE database. The images classified as abnormalities will contain the image of the affected retina where the blood vessels may have blockages or leakages.

In other cases, there might also be new blood vessels formed or growing depending on the severity of the stage as shown in figure 2.

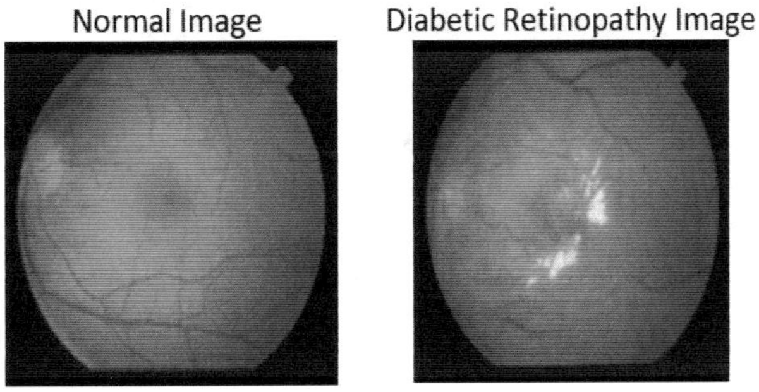

Figure 2. Describes the Normal and the Diabetic Retinopathy Image.

4.3 CLAHE Images:

As we know that having the perfect contrast to an image will give the best results. To obtain best results we are using Contrast Limited adaptive histogram equalization, The contrast to an image shouldn't be low meanwhile it shouldn't be high. So, it should be in the limit. To maintain that certain contrast limit to an image we are using CLAHE.

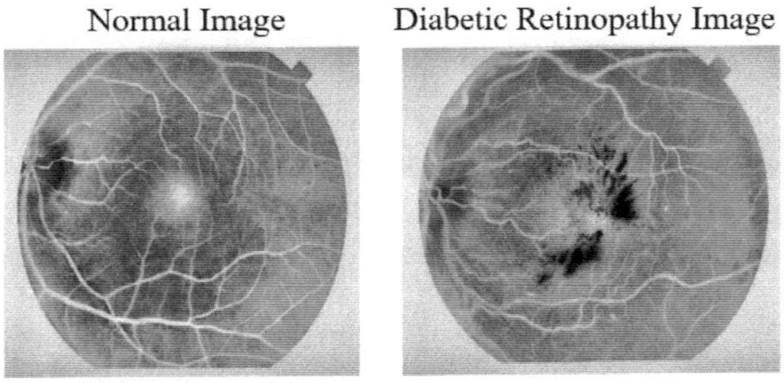

Figure 3. Describes the Normal and the Diabetic Retinopathy Image after pre-processing.

Here, by using the CLAHE we can clearly observe the Blood vessels in the image compared to the input image as shown in figure 3.

4.4. Segmented Images:

Segmented images are the images received after the raw fundus image goes through the whole segmentation process. This is the image that will be received from the Segmentation process. The image of the blood vessels present in the retina can be seen in this image. The segmented image can also be considered as the proof for the presence of Diabetic Retinopathy for the particular processed image as shown in figure 4.

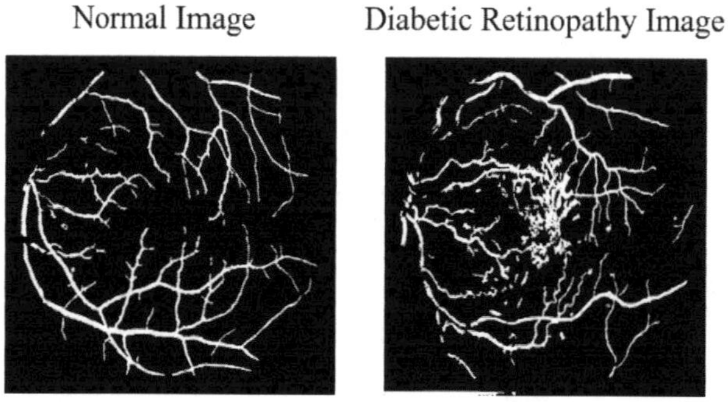

Figure 4. Describes the Normal and the Diabetic Retinopathy Image after segmentation.

4.5 Ground truth Images:

These images are the segmented Outputs of the STARE project. Ground truth is data that's famed to be real or true, provided by direct observation and measurement as shown in figure 5.

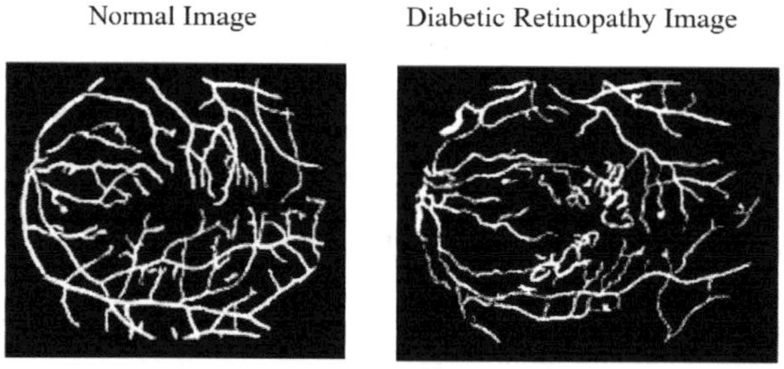

Figure 5. Describes the Normal and the Diabetic Retinopathy Image of ground truth images.

4.6 SSIM

SSIM stands for structural similarity index measure. The Structural Similarities Indexing Method (SSIM) is a technique for assessing structural similarity between two photographs. If one of the photos being compared is of perfect quality, the SSIM index can be considered as a quality indicator for that image. It's an improved version of the global picture quality index that was first suggested. In the present scenario the SSIM tool will be used to compare the ground truth image and the segmented image. The ground truth image is the respective output image for the fundus image that is retrieved from the STARE Database. And the segmented image is the output image received from the CNN after the final processing of the image. The SSIM tool will be used to compare the differences present in the ground truth image and the segmented image which will prove the accuracy and the efficiency of the methodology.

Our proposed segmentation method extracts the blood vessels of each image accurately, resulting in the similarity measure value of 85%. The primary reason for the high accuracy rates is the combination of the Fuzzy C-means clustering and the Region based active contour algorithms.

Journal of Physics: Conference Series **2325** (2022) 012043 doi:10.1088/1742-6596/2325/1/012043

4.7 Classification:

S. No	Input Image	CLAHE Image	Segmented Image	Ground truth Image	Accuracy	Classification
1	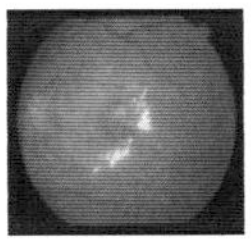	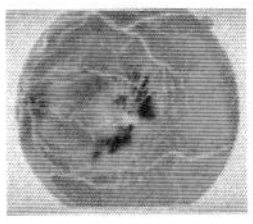	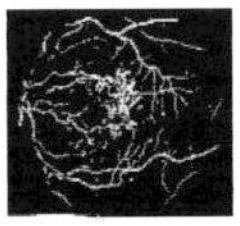	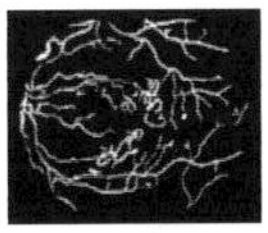	0.96	Affecting with DR
2	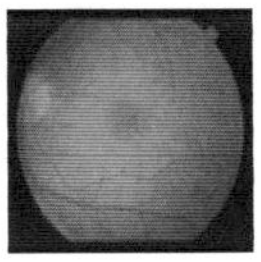	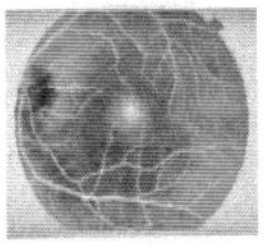	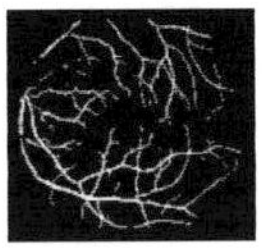	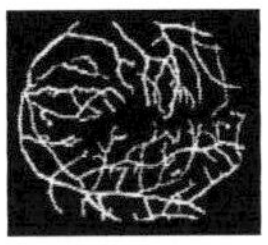	0.93	Not Affecting with DR
3	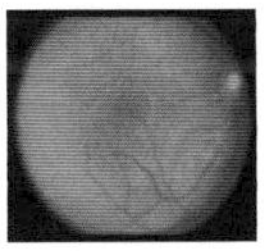	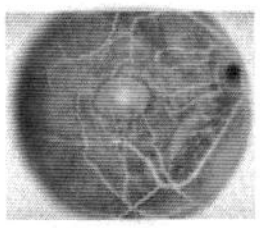	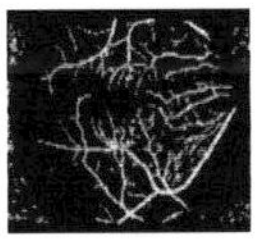	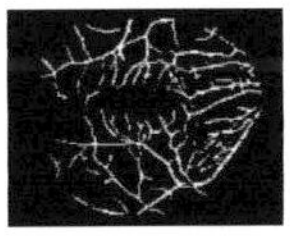	0.92	Not Affecting with DR
4	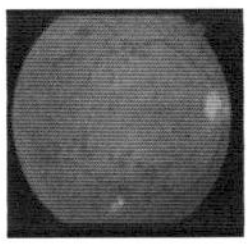	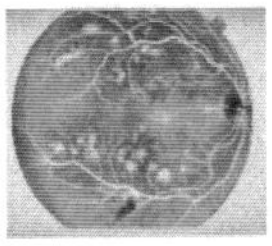	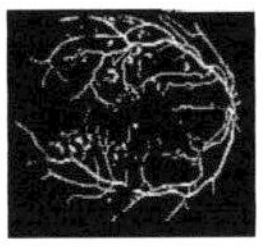	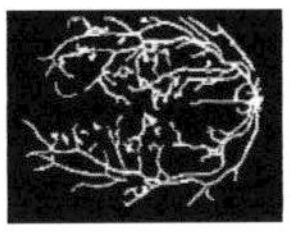	0.95	Affecting with DR

Table 1: Classification

The output images are the segmented images. The structural information in the segmented image and the ground truth image are compared pixel by pixel using SSIM metrics. If there is a very good similarity between segmented and ground truth image, it displays the value between 0.9 to 1. Those segmented images are given as the input images to the CNN classifier to classify whether the image is affected with diabetic retinopathy or not.

Total we have 88 segmented images. All these 88 images are divided into two stages, one is training dataset and another stage is testing dataset. In that 88 images, 67 images are given to train the classifier and the remaining 21 images are given to test the classifier. Each stage is divided into 2 classes, class0 and class1. Class0 represents the normal segmented images which are not affected with Diabetic retinopathy. Class1 represents the segmented images affected with diabetic retinopathy.

The classifier will undergo the classification process with the help of training dataset and testing dataset to provide the accurate result. For our proposed method, an overall accuracy to detect DR was 92%.

4.8 Performance measure:

The performance measure of our proposed method includes sensitivity, specificity and accuracy are shown in the figure.6. The primary reason for the high accuracy rates is the combination of the Fuzzy

C-means clustering and the Region based active contour algorithms along with CNN classifier. The visual representation of the performance measure has been provided in figure 6.

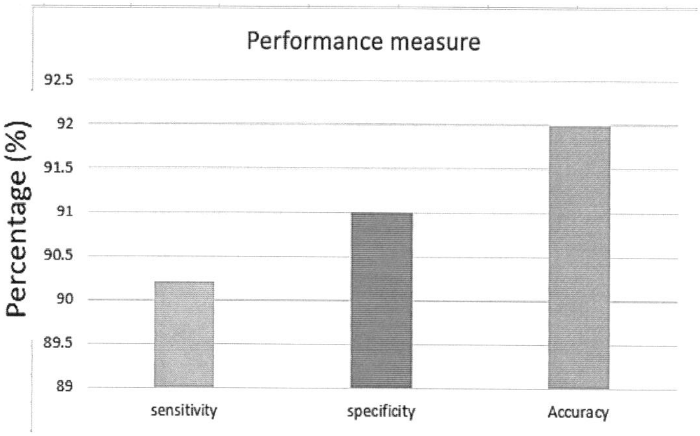

Figure 6. Shows the Performance Measure

5. CONCLUSION

In order to determine the severity level of DR, it is critical to accurately identify the retinal blood vessels during the ophthalmological examination. Many algorithms aren't capable of differentiating the depigmented abnormal retinal images from the retinal blood vessels. The research findings have emerged from each of the STARE database's 88 photos. The results indicate that the proposed methodology consisting of FCM clustering along with Region based active contour accurately identifies all the blood vessels. Both these algorithms aid in increasing the accuracy rate of the segmented blood vessels. There are no discontinuities between the minor vessels and they are also identified in this process.

CLAHE minimizes the level of noise in depigmented retinal images. By examining the segmented vessel structure through the proposed method, it is evident that it is capable of minimizing the ophthalmologists' effort in analyzing the diabetic retinopathy affected blood vessels. The proposed retinal blood vessels segmentation approaches can be used for datasets with similar attributes. Our proposed segmentation method extracts the blood vessels of each image accurately, resulting in the similarity measure value of 85%. For our proposed method, an overall accuracy to detect DR was 92%.

6. FUTURE SCOPE

In the long-run, the proposed methodology can be implemented in the field of ophthalmology for retinal screening if it is further developed. There is still room for advancements in the present procedure wherein there can be progressive algorithms designed to be involved in order to classify the stage of Diabetic Retinopathy. If this advancement is added in this methodology, it will be a procedure that will specifically convey the stage of DR the patient is currently in so that appropriate diagnosis or treatment can be started at the earliest. Once the methodology is developed with the classification, the procedure can be implemented for ophthalmological campaigns carried out especially in rural areas.

Journal of Physics: Conference Series **2325** (2022) 012043 doi:10.1088/1742-6596/2325/1/012043

As the method discussed is time efficient, it can be introduced in campaigns and other screening areas as the retinal examination can be done in a short span of time and the results can also be retrieved soon.

7. REFERENCES

[1] Aslani, Shahab, and Haldun Sarnel. "A new supervised retinal vessel segmentation method based on robust hybrid features." *Biomedical Signal Processing and Control* 30 (2016): 1-12.

[2] Azzopardi, George, et al. "Trainable COSFIRE filters for vessel delineation with application to retinal images." *Medical image analysis* 19.1 (2015): 46-57.

[3] Christodoulidis, Argyrios, et al. "A multi-scale tensor voting approach for small retinal vessel segmentation in high resolution fundus images." *Computerized Medical Imaging and Graphics* 52 (2016): 28-43.

[4] Zhu, Chengzhang, et al. "Retinal vessel segmentation in color fundus images using an extreme learning machine." *Computerized Medical Imaging and Graphics* 55 (2017): 68-77.

[5] GeethaRamani, R., and Lakshmi Balasubramanian. "Retinal blood vessel segmentation employing image processing and data mining techniques for computerized retinal image analysis." *Biocybernetics and Biomedical Engineering* 36.1 (2016): 102-118.

[6] Hassanien, Aboul Ella, Eid Emary, and Hossam M. Zawbaa. "Retinal blood vessel localization approach based on bee colony swarm optimization, fuzzy c-means and pattern search." *Journal of Visual Communication and Image Representation* 31 (2015): 186-196.

[7] Roychowdhury, Sohini, Dara D. Koozekanani, and Keshab K. Parhi. "Blood vessel segmentation of fundus images by major vessel extraction and subimage classification." *IEEE journal of biomedical and health informatics* 19.3 (2014): 1118-1128.

[8] Kar, Sudeshna Sil, and Santi P. Maity. "Retinal blood vessel extraction using tunable bandpass filter and fuzzy conditional entropy." *Computer methods and programs in biomedicine* 133 (2016): 111-132.

[9] Sreejini, K. S., and V. K. Govindan. "Improved multiscale matched filter for retina vessel segmentation using PSO algorithm." *Egyptian Informatics Journal* 16.3 (2015): 253-260.

[10] Singh, Nagendra Pratap, and Rajeev Srivastava. "Retinal blood vessels segmentation by using Gumbel probability distribution function based matched filter." *Computer methods and programs in biomedicine* 129 (2016): 40-50.

[11] P.Saranya,K.M.UmaMaheswari(2021), "Classification of Different Stages of Diabetic Retinopathy using Convolutional Neural Networks", 2nd International Conference on Computation, Automation and Knowledge Management (ICCA KM), pp 59-64.

[12] Chaudhary, Sonali, and H. R. Ramya. "Detection of Diabetic Retinopathy using Machine Learning Algorithm." *2020 IEEE International Conference for Innovation in Technology (INOCON)*. IEEE, 2020.

[13] Prabha, S., et al. "Machine Learning Technique for Analysing the Diabetic Retinopathy." *2022 International Conference on Sustainable Computing and Data Communication Systems (ICSCDS)*. IEEE, 2022.

[14] Qureshi, Imran, Jun Ma, and Qaisar Abbas. "Recent development on detection methods for the diagnosis of diabetic retinopathy." *Symmetry* 11.6 (2019): 749.

[15] Fayemiwo, M. A., S. A. Akinboro, and O. A. Adepegba. "Identification and Classification of Diabetic Retinopathy Using Machine Learning." *Adeleke University Journal of Engineering and Technology* 1.1 (2018): 245-259.

[16] Prabha, S., et al. "Retinal Image Analysis Using Machine Learning." *2020 International Conference on Communication and Signal Processing (ICCSP)*. IEEE, 2020.

[17] Bindhu, V., and Villankurichi Saravanampatti PO. ""Semi-Automated Segmentation Scheme for Computerized Axial Tomography Images of Esophageal Tumors."" Journal of Innovative Image Processing (JIIP) 2, no. 02 (2020): 110-120.

[18] Rao, Smaran S., Gajanan Maske, and Antara Roy Choudhury. ""Iris Image Segmentation and Localization using Dynamic Reconfigurable Processor."" Journal of Innovative Image Processing 2, no. 3 (2020): 147-155.

Friend Recommendation System using Image Encryption and Deep learning

M.Mohammed Yaseen[1], M.Akilan[2], A.G.Ranjith[3], S.Soundariya[4], S.V.Manisekaran[5], S.Ramakrishnan[6]

[1, 2, 3] Student B. Tech IT, Department of Information Technology,
Dr. Mahalingam College of Engineering & Technology, Pollachi, Tamil Nadu, India.

[4] Assistant Professor, Department of Information Technology,
Dr. Mahalingam College of Engineering & Technology, Pollachi, Tamil Nadu, India.

[5] Assistant Professor, Dept. of Information Technology,
Anna University Regional Campus, Coimbatore, Tamil Nadu, India.

[6] Professor, Dept. of Information Technology,
Dr. Mahalingam College of Engineering & Technology, Pollachi, Tamil Nadu, India.

Email: - mohammedyaseen2001@gmail.com, akilanmaruthachalam5678@gmail.com, agranjith8@gmail.com, soundaricse@gmail.com, manisekaran@aurcc.ac.in, ram_f77@yahoo.com.

Abstract. The recommendation system functions similarly to a content filtering system, providing information depending on the preferences and behavior of the user. People use image sharing platforms such as Facebook, LinkedIn, and Instagram to share photographs of various elements to convey their sentiments with linked peers. As image sharing has been increasingly popular among individuals in recent years, it plays an important part in our daily lives. However, because of photo leaks and misuse, people are becoming increasingly concerned about their privacy. Encrypting photographs before publishing to social media is a better approach but decrypting the images and then searching for comparable images is computationally costly. The prior model advocated encrypting pictures using AES encryption and classifying them with deep autoencoders which is only a 2-step compression technique and takes quite more time for classifying. We provide a feasible privacy-protected friend recommendation system in proposed model by encrypting images using the Gaussian blur technique and a deep learning strategy for feature extraction and image classification for user recommendation. For training the dataset, the deep learning system utilizes the Keras model with CNN method, and a new model is generated from the features extracted from the blurred images. After that, the test dataset is compared to the model developed for similar image prediction. The accuracy measurements used to plot the graph are val acc and val loss. The accuracy result for the original image is 91.8%, while the accuracy result for the blurred image is 86.2% percent, demonstrating that our model is highly efficient and results in lower data loss.

Keywords: Friend recommendation system, Deep learning, Encryption

1. Introduction

Recommender systems, a sort of information filtering system, are used to forecast the bias or ratings that a user is likely to give to an item. Consumers are using recommendation systems to help them deal with contradictory information. The recommendation system is mostly used in digital entertainment platforms, and several companies, like Amazon, and Spotify use recommendation algorithms to generate significant engagement and profit. Collaborative, content-based, and other forms of available information recommendation systems. Currently, virtually all social media companies provide a friend recommendation system, which creates the way for the creation of a smart community among various users. It's more possible that users who look alike have anything in common to have a connection between them. Keras is a Python-based high-level neural network API. In the backend, it uses TensorFlow or Theano. Keras was created to make neural network implementation easier. A Convolutional Neural Network is a form of neural network that consists of numerous convolution layers followed by fully connected layers. In terms of complexity and the weights of convolution layers used during feature extraction, CNN uses fully connected layers during classification. As a result, image recognition, speech recognition, and video analysis are all common uses for these networks. In this paper, we use the Keras model and CNN method for feature extraction and classification purposes for user recommendation. The dataset comprises the user's personal data, with 80% of the dataset being utilized for training and 20% for testing. However, the accuracy of the feature retrieved from the photos is determined by the quality of the privacy-protected friend recommendation system.

2. Related Work

The recommending system gathers data from a variety of sources, including the content seen by the user, feedback forms submitted, and website cookies. There have been several studies on recommendation systems in this area. By safeguarding the image against sensitive data being stolen, the suggested system is related to a privacy-protected friend recommendation system. A suitable recommender system should gather data and filter depending on users' preferences.

For masking private portions of images and splitting them into private and public parts, an image blurring technique [5] was proposed. Additionally, by selecting them, the user may generate personalized content.

A combined image encryption [8] was utilized, in which the photos are obfuscated by blurring them with AES Encryption is used for security. A deep autoencoder is used to extract characteristics from photos and then construct user profiles using binary data for classification.

The Facebook friend recommending system [7] was developed based on graph mining confusion matrix. The algorithm's performance matrix is created using the confusion matrix, and classification is performed using the random forest approach. A recommender system based on lifestyle [6] was developed which collects users' lifestyles using mobile phone sensors and then recommends them to one another. The user's lifestyle is retrieved using a probability distribution algorithm, and the activity recognition approach is employed for classification.

Recently, a recommender system was applied in medical research [18]. Using the k-clique and deep learning approaches, an effective RS for patient diet suggestion was designed. The k-clique is used in conjunction with DRS for deep learning classifiers and better precision on patient datasets based on their health conditions.

Location and orientation-based RS [19] is used to provide personalized suggestions based on data obtained from different IoT devices such as smart watches, iPads, and smartphones. Outdoor evaluations were only conducted on both content-based and collaborative-based algorithms.

Traditional data mining, neighborhood construction, and dimensionality reduction are used to classify an e-commerce-based [15] recommendation system based on customer purchasing preferences. For analysis purposes, a dataset containing Fingerhut Inc web-purchase transactions and movie lens dataset is utilized. The results obtained from reduction technique [20] for RS reveal that the extracted features may accurately reflect the picture quality after the development of a system using visual features and a series of convolution operations to finish the deep learning.

For feature extraction and image classification, an image-based RS [14] employs multiple CNN models such as vgg16, vgg19, and xception. For assessments, several versions of Movielens dataset are employed. E-commerce sites achieved higher accuracy.

A collaborative based filtering [13] to compare similar images uploaded between users for classification is performed based on location which also improves the scope of recommendation.

3. Methodology

3.1 Image Encryption

Image obfuscation is the technique of removing critical information from images through non-linear pixel conversion to create an image that is unrecognizable[13,16]. Images may be encrypted in a variety of ways to improve security. A detailed image is nothing more than sharp edges that make the image flawless, but image blurring reduces the sharpness of the image edges, making the image indistinct. Only a portion of the data is encrypted. The pixels are changed randomly in the scrambling approach, disrupting the association between nearby pixels by performing discrete cosine modification across picture pixels but the loss of data is more. As a result, further Gaussian Blurring research works have been carried out. Figure 1 depicts the flow diagram:-

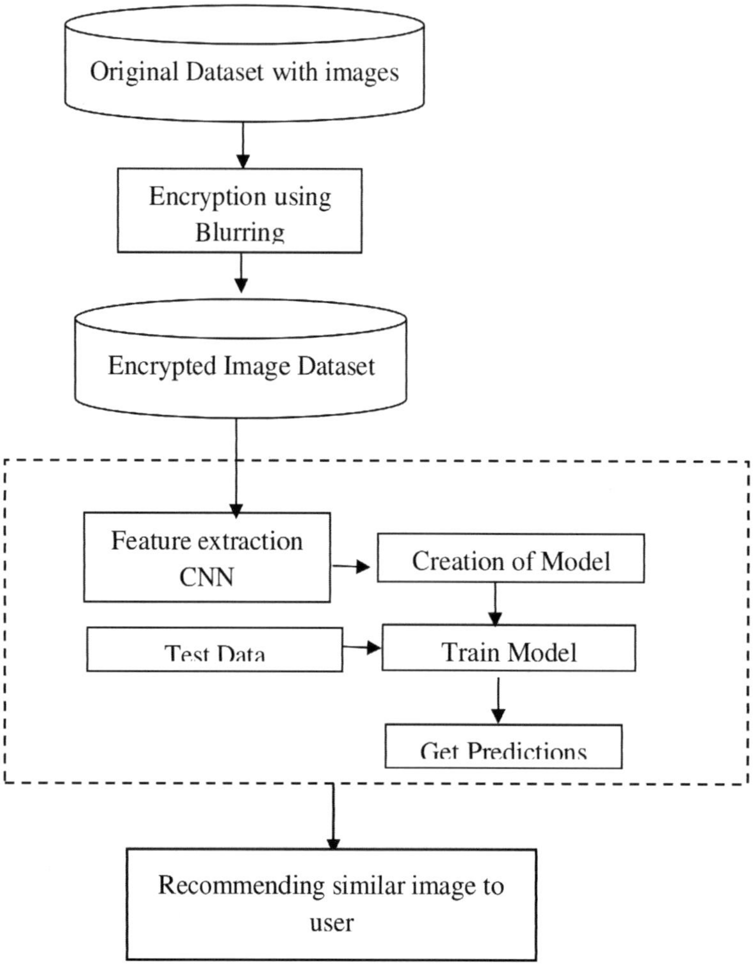

Figure 1: Process Flow Diagram

The formula for gaussian blur is as follows

$$G(x,y) = \frac{1}{2\pi\sigma^2} e^{-\frac{x^2+y^2}{2\sigma^2}} \quad (1)$$

The mask creates a character with the dimensions [x, y] and standard deviation of a Gaussian low pass filter. For example, setting [x, y] to [4 4] determines how the kernel weights and sets the extent to which the Gaussian filter is employed as a definition of the number of rows and columns. We employed a Gaussian blur mode dependent on image size for encryption.

Figure 2: Original image

Figure 3: Blurred image

3.2 Feature Extraction and Image Classification Using Keras

RS uses implicit data to improve the quality of recommendations. Furthermore, significant feature extraction allows the system to learn more precisely and effectively. Convolutional neural networks were used to extract features in this investigation (CNN). Convolutional models are within the domain of deep learning approaches. The inclusion of numerous hidden layers beyond standard basic neural networks is referred to be "deep." CNN models are used by the developers to detect and classify images, and if the system does not consider the network's last layer, the model appears as a feature extraction tool. As a result, the suggested technique extracts visual characteristics using CNN vgg16 model. Keras functions are data learning models that are used to detect and optimize feature values. The 16-step compression technique for vgg16 model is show in Figure (vgg16-implementation-in-keras). Outcome is really accurate. Model with high efficiency. To cope with large-scale photos,[10] the vgg16 function explores depth layer with modest filter size 3 x 3. Block formats are available for all vgg16 functions. A sequential layer is followed by a mass consolidation tier in each vgg16 block. The total sequential layer employs the same kernel size 3 x 3 and 1 stride to keep the result size consistent. After each consecutive layer blending 2 × 2 size with 2 steps, the resolution is reduced by half after each block. The exit layer is coupled to two hidden layers in every vgg16 model. Vgg16 is a consecutive neural network structure that focuses on having a 3 by 3 filtered layer with stride size 1 and a 2x2 max-pool tier with stride size 2 whenever it employs the exact padding value. It maintains the sequential way and 2x2 maxpool filter throughout the design, with two linked layers and softmax for output. The 16 in vgg16 indicates that it has 16 weight tiers, and this network is the biggest, with over 138 million frames.

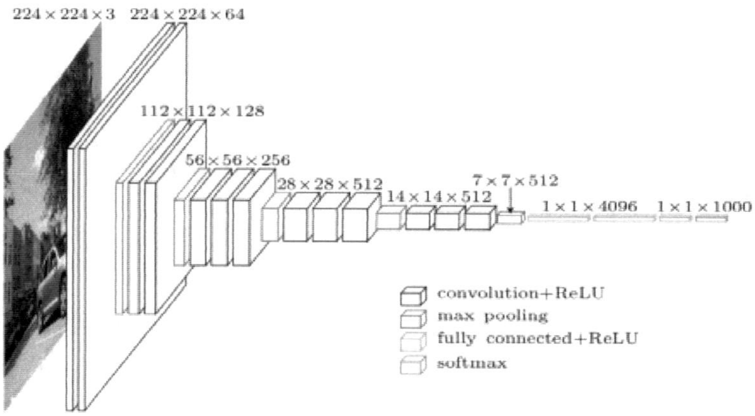

Figure 4: VGG16 step compression

4. Results and Discussion

The proposed framework is implemented in two parts: first, images are blurred; second, we utilize a deep learning search strategy to extract features and obtain similar image results. For all implementations, we use a laptop with an i5-8th generation CPU, 8 GB RAM, 1 TB HDD, and a Nvidia 1050ti 4GB GPU.

The vgg16 function in Keras handles accuracy analysis. The vgg16 function is a type of sequential neural network [9] with improved accuracy. We use two datasets for calculating accuracy which are fruits 360 dataset and vegetable Image dataset. The fruits 360 image dataset is used with 9345 images approx. each of size 100x100 px where 6231 images are used for training phase and 3114 are used for validation phase. The vegetable image dataset with 15000 images for training phase and 3000 images for validation phase each of size 224x224 px is used. The accuracy analysis is done for totally three methods where original image, AES Encrypted image and Blurred images are employed. The classification report provides precision, recall and f1score values from which the accuracy is obtained.

Precision is a number that represents the number of information digits and reflects the value of the number. The formula for Precision is as follows: -

$$Precision = \frac{True\ Positive}{True\ Positive + False\ Positive} \tag{2}$$

The ability of a model to track down all relevant examples within a data collection is referred to as recall. The formula for recall is as follows: -

$$Recall = \frac{True\ Positive}{True\ Positive + False\ Negative} \tag{3}$$

The F1-score combines a classifier's accuracy and recall into a single statistic by calculating their harmonic mean. Its primary application is to compare the performance of two classifiers. The formula for f1score is as follows: -

$$f1\ score = \frac{2x(Precision.Recall)}{Precisin + Recall} \tag{4}$$

The following graph plotted using matplotlib depicts the accuracy and loss for blurred image dataset for both fruits and vegetable datasets: -

Fruits 360 Dataset **Vegetable Image Dataset**

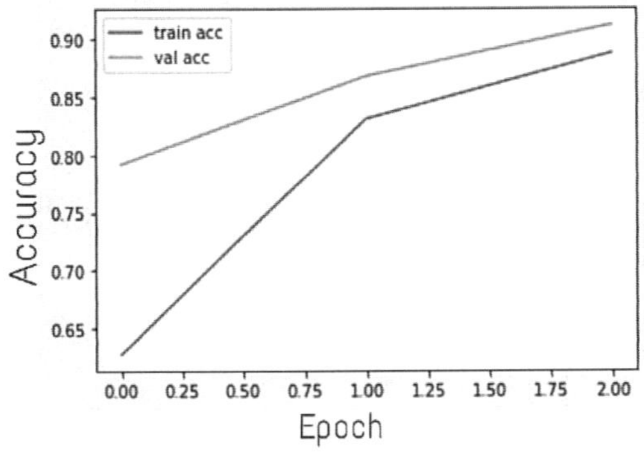

Figure 5: Accuracy Graph

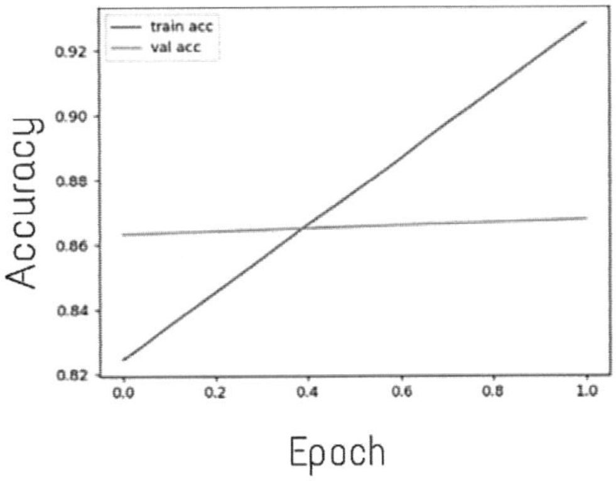

Figure 7: Acuuracy Graph

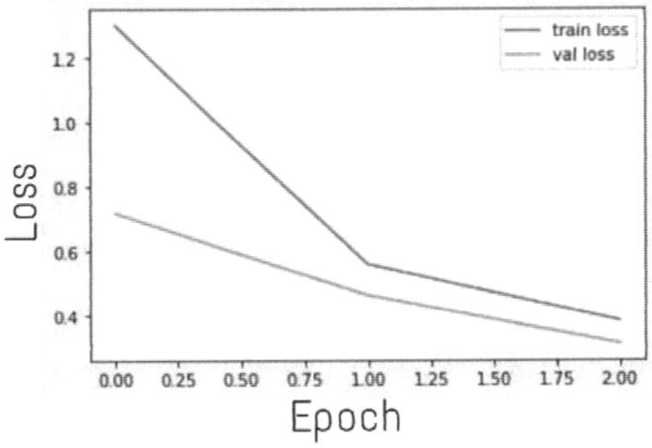

Figure 6: Loss Graph

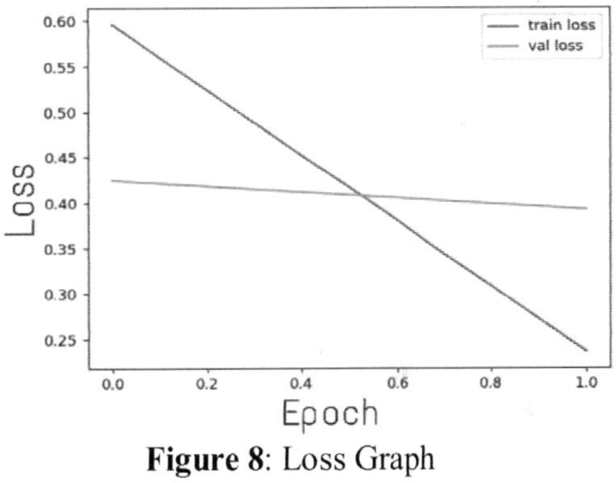

Figure 8: Loss Graph

We can observe from the experimental outcomes that the accuracy graph is climbing exponentially, indicating a good model, however the loss chart shows a considerable amount of loss, whereas the training phase clearly represents an efficient and accurate model. The following table shows the accuracy comparison for training data results between the original, AES, blurred image: -

Table 1. Fruits 360 Dataset

Method	Apple_6			Cucumber_1			Pear_1			Zucchini_1			Carrot_1		
	Precision	Recall	F1score	Precision	Recall	F1score	Precision	Recall	F1score	Precision	Recall	F1score	Precision	Recall	F1score
Original	0.79	0.99	0.88	1.00	0.96	0.98	0.98	0.94	0.96	0.99	1.00	0.99	0.97	1.00	1.00
AES	0.001	0.00	0.001	0.002	0.001	0.012	0.002	0.001	0.001	0.09	0.07	0.081	0.07	0.04	0.06
Blur	0.69	0.87	0.77	0.89	0.84	0.89	0.82	0.94	0.87	0.94	1.00	0.97	0.86	0.85	0.85

Table 2. Vegetable Image Dataset

Method	Bitter guard			Bottle guard			Brinjal			Broccoli			Bean		
	Precision	Recall	Flscore	Precision	Recall	Flscore	Precision	Recall	Flscore	Precision	Recall	Flscore	Precision	Recall	Flscore
Original	0.99	0.99	0.92	1.00	0.96	0.98	1.00	0.95	0.96	0.98	1.00	0.99	0.98	1.00	1.00
AES	0.00	0.00	0.00	0.08	0.17	0.11	0.00	0.00	0.00	0.00	0.00	0.080	0.07	0.05	0.06
Blur	0.88	0.79	0.90	0.95	0.90	0.92	1.00	0.37	0.54	0.82	0.93	0.87	0.97	0.79	0.87

Table 3. Result of Dataset with accuracy

Dataset		Original	AES	Blur
Fruits 360	Training Accuracy	0.887401223	0.0043473847	0.9318097865
	Validation Accuracy	0.911726832	0.0067685948	0.8790954045
Vegetable Image Dataset	Training Accuracy	0.9294494986	0.0057878798	0.9281800985
	Validation Accuracy	0.9453421145	0.0067876346	0.8676077553

5. Conclusion

According to the results of the experiment, the deep learning technique in this approach efficiently predicts images. The proposed design addresses two problems: blurring images for encryption and classifying encrypted images using deep learning. According to the data, the classification of the dataset improves considerably as the number of times it is training increases. Furthermore, the quality is dependent on the size of the dataset, and this strategy is a good way of securing privacy.

Reference

[1]. L.A. Adamic, E. Adar, Friends, and neighbours on the web, Social. Network. 25 (3) (2003) 211–230

[2]. Cheng, Y., Liu, J., & Yu, X. Online social trust reinforced personalized recommendations. Personal and Ubiquitous Computing, 2016, 20(3), 457–467.

[3]. Vincent, P., Larochelle, H., Bengio, Y., & Manzagol, P.-A. Extracting and composing robust features with denoising autoencoders. Proceedings of the 25th international conference on machine learning, ACM, 2008, 1096–1103.

[4]. Yuan, L., Korshunov, P., & Ebrahimi, T. Privacy-preserving photo sharing based on a secure JPEG. IEEE conference on computer communications workshops IEEE, 2015, 185–190.

[5] Zhang, L., Jung, T., Liu, C., Ding, X., Li, X.-Y., & Liu, Y. Pop: Privacy-preserving outsourced photo sharing and searching for mobile devices. 35th International conference on distributed computing systems (ICDCS), IEEE, 2015, 308–317

[6] Dagadu M Jadhavar and V R Chirchi. Friend Recommendation System for Online Social Networks. International Journal of Computer Applications November 2016, 153(12):1-3.

[7] Friend recommendation using graph mining on social media Kosaraju Naren Kumar, Kanakamedala Vineela, ijetms.in Issue:5, August-2020, Volume No.4.

[8] Obfuscated image classification for secure image-centric friend recommendation Kazi Wasif Ahmed, Omit Chanda, Noman Mohammed, Yang Wang, Sustainable Cities and Society 41, 2018, 940-948.

[9]. Simonyan, K., & Zisserman, A. Very deep convolutional networks for large-scale image recognition, 2014, arXiv:1409.1556.

[10]. Simonyan, K., Vedaldi, A., & Zisserman, A. Deep inside convolutional networks: Visualizing image classification models and saliency maps, 2013, arXiv:1312.6034.

[11] Souza Araujo, C., Correa, D., Paulo, L., Couto da Silva, A. P., Oliveira Prates, R., & Meira, W. It is not just a picture: Revealing some user practices in Instagram. 2014 9th Latin American Web Congress (LA-WEB), IEEE, 2014, 19–23.

[12] Strizhov, M. Towards a practical and efficient search over encrypted data in the cloud, 2015, IEEE international conference on cloud engineering (IC2E), IEEE, 496–498. Tella, A. Social media strategies for dynamic library service development. IGI Global.

[13]. Soni, M. T. S. S., & Singhai, P. (2020). Friend recommendation system using machine learning method. Journal of Scientific Research & Engineering Trends, 6(5).

[14] Malekpour Alamdari, P., Navimipour, N. J., Hosseinzadeh, M., Asghar Safaei, A., & Darwesh, A. (2022). Image-based Product Recommendation Method for E-commerce Applications Using Convolutional Neural Networks.Sarwar, B., Karypis, G., Konstan, J., & Riedl, J. (2000, October). Analysis of recommendation algorithms for e-commerce. In Proceedings of the 2nd ACM Conference on Electronic Commerce (pp. 158-167).

[15] Yuan, X., Wang, X., Wang, C., Squicciarini, A., & Ren, K. (2014). Enabling privacy-Preserving imagecentric social discovery. 34th international conference on distributed computing systems (ICDCS), IEEE, 198–207.

[16] Ning, H., Dhelim, S., & Aung, N. (2019). PersoNet: Friend recommendation system based on big-five personality traits and hybrid filtering. IEEE Transactions on Computational Social Systems, 6(3), 394-402.

[17]. Wang, W., Vong, C.-M., Yang, Y., & Wong, P.-K. Encrypted image classification based on a multilayer extreme learning machine. Multidimensional Systems and Signal Processing, 2016, 1-15

[18]. Manoharan, S. (2020). Patient diet recommendation system using K clique and deep learning classifiers. Journal of Artificial Intelligence, 2(02), 121-130.

[19]. Joe, M. C. V., & Raj, J. S. (2021). Location-based Orientation Context Dependent Recommender System for Users. Journal of trends in Computer Science and Smart technology (TCSST), 3(01), 14-23.

[20]. Sulthana, A. R., Gupta, M., Subramanian, S., & Mirza, S. (2020). Improvising the performance of image-based recommendation system using convolution neural networks and deep learning. Soft Computing, 24(19), 14531-14544.

Journal of Physics: Conference Series **2325** (2022) 012045 doi:10.1088/1742-6596/2325/1/012045

SOFT COMPUTING TECHNIQUES FOR DRIVER ALERTNESS

Penuballi Phani Srikar Reddy, Gunturu Sasi Pavan Khadyoth, Vennalakanti Sri Ganesh Raman, Gadikota Sai Dheeraj Reddy, Hima Bindu Valiveti

Department of Electronics & Communication Engineering, Gokaraju Rangaraju Institute of Engineering and Technology, Hyderabad, Telangana, India.

ABSTRACT: There has been an alarming increase in the number of accidents that occur due to drowsiness while driving. In order to reduce roadside accidents, the detection of driver fatigue or drowsiness is crucial. Detecting fatigue during driving is crucial for reducing accidents, as well as improving the safety of both the driver and the passengers. Various methods can be used to detect drowsiness among drivers, but fuzzy logic-based detection stands out for its ability to avoid false alarms. As part of the proposed system, we are using eye-tracking in combination with methods such as Haar cascade to identify the level of drowsiness of the driver. This system has been tested in real-time.

1. INTRODUCTION:

Driver drowsiness can occur due to reasons such as fatigue, sleeping disorders, sleep depravity, over work, alcohol consumption, improper lighting conditions, food intake before driving, and medications. When a driver suffers from drowsiness or falls asleep it often results in accidents. 21% of road accidents are a result of drowsiness of drivers. About 60% of drivers working in logistics and transport sector suffer from the said issue. About 30,000 deaths and 1.5 lakh people are disabled due to road accidents each year in India. Several methods to detect the driver drowsiness exists such as Pressure sensitive steering wheel, Heart monitoring techniques like ECG, and brain activity monitoring techniques like EEG, Head movement tracking etc. It is the aim of this study to determine whether the Haar cascade method is an effective method of detecting and reducing driver drowsiness.

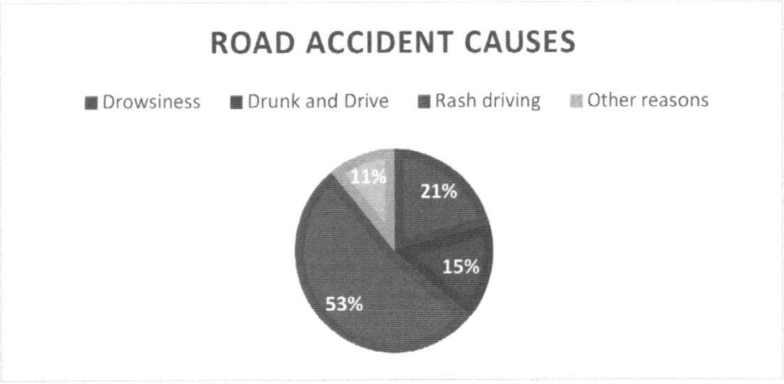

2. LITERATURE SURVEY:

Since drowsiness is one of the most common causes of vehicle accidents, researchers and automotive manufacturers have developed several solutions ranging from finding patterns in driving habits to analysing brain waves and vitals while driving. The majority of these solutions use machine learning and statistics to support their predictive algorithms. In general, there are three main categories to be considered as described below.

Rafi Ahmed et al proposed drowsiness detection using lane departure and head tilting. Although this is an innovative approach it has severe drawbacks such as being too late in waking up the driver. Often times there is not much time to wake up and react [1]. Mohammad Amin Assari et al developed a hardware-based system for the detection of drowsiness in drivers through external infrared lights and camera to make the system independent of lighting. However, this system has its own flaws complex circuitry and interfering with drivers' field of vision [2]. Md. Tanvir Ahammed Dipu et al proposed an improved driver drowsiness system based on yawning however this system has limitations such as false alarm being wrongly timed [3]. Eddie E. Galarza et al proposed a system using a mobile application to detect driver drowsiness. This system has limitations due too much dependence on mobile phone. If the mobile phone is not in a condition to capture all the symptoms of the drowsiness, the system fails to warn the driver [4]. Charlotte Jacobé de Naurois et al developed a technique for predicting driver drowsiness by utilizing Artificial Neural Networks (ANN). This system is effective in detecting eye lid movement, gaze, and head movement however due to the model being prediction type it is proven to false alarm and Artificial Neural Network (ANN) require large dataset which is not practical for mobile applications [5]. Rateb Jabbar et al studied a system involving Convolutional Neural Network (CNN) to be implemented in embedded systems. This system has few limitations which include necessity of a large dataset which makes the system significantly slower and also requirement of good lighting conditions [6]. Elena Magán et al proposed a system where the drowsiness model is trained using a series of 60 seconds videos rather than images. Fuzzy logic has also been implemented into the system. This system is flawed as it takes a full 60 seconds to detect driver drowsiness as it cannot effectively alert the driver to prevent accidents [7]. T. Vesselenyi et al proposed a system using Artificial Neural Networks (ANN) this system is highly efficient and is compact enough to use in mobile applications. However due to using low number of neural networks this system has very low learning ability [8]. Amna Rahman et al proposed a similar model as mentioned earlier which monitors blinking of the driver. The demerits of this model are bad illumination and lack of corner detection [10]. Marco Javier Flores et al implemented a model that uses computer vision and AI. This model has been tested in real time under various scenarios and circumstances. This model poses a higher error rate under the influence of interruptions [11][12][13]. CNN, SVM and HMM algorithms are used for gaining a highly accurate model but it is very difficult to deal with a larger dataset. The accuracy of the model is defined by the size of the dataset. In the above-mentioned model, a small dataset has been used due to which the accuracy might appear higher than the original value [14][15]. Qaisar Abbas et al developed a hybrid fatigue detection system using multiple cameras and ECG sensors. This model has a higher accuracy rate due to the presence of ECG sensors. The downside is that a lot of hardware is required for continuous monitoring and this might obstruct the driver's field of vision [16]. Dr. Samuel Manoharan et al developed a brain disorder model using MRI scans. Using hard and soft clustering along with super pixel regions of the image are segmented and segregated so that required features can be extracted from the resultant images. H. The authors used a two-level convolution neural network to classify the driver behavior and the emotion in short durations [17]. The current work can be incorporated along with heterogenous networks based on the principles of software defined networking wherein the energy efficient algorithms can be successfully implemented [18][19].

3. PROPOSED SYSTEM:

The proposed system is done in four phases face capturing, feature mapping, drowsiness detection, and alert. Cameras are used for capturing face in real time. This capturing is done over a period. Haar Cascade uses recorded or real time faces and uses line or edge detection features to track open and closed eyes. Clustering is an unsupervised machine learning technique. This method involves training a machine learning model on unsupervised data to detect drowsiness. It has been used to classify the real-time images acquired by the camera of the driver and track the drowsiness pattern. This pattern is used in real-time to identify whether the driver is feeling drowsy.

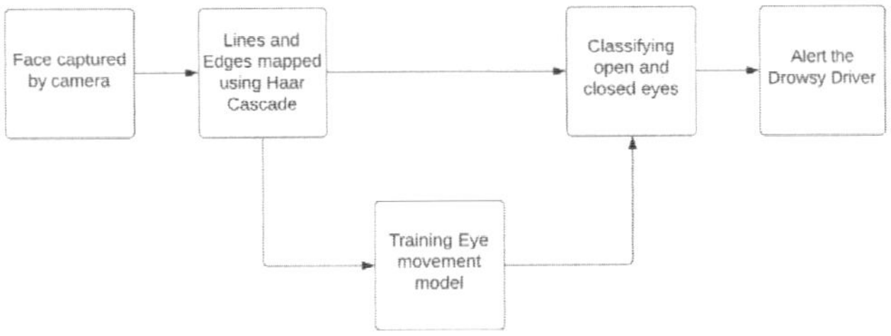

Fig 1 Block diagram of the working model

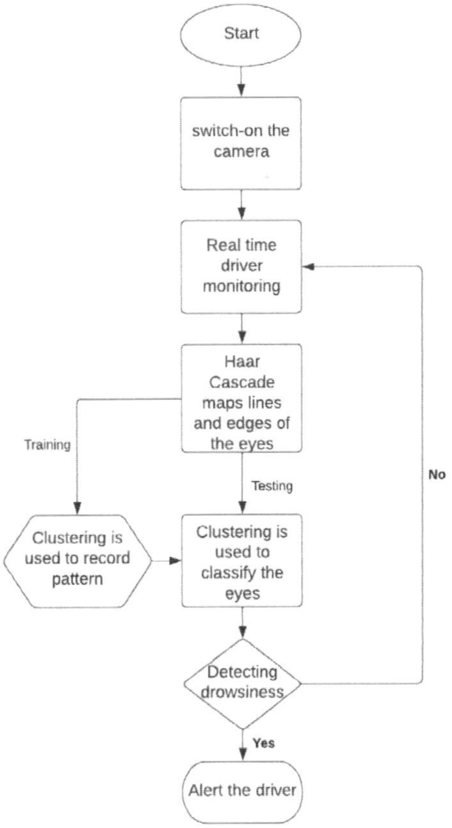

Fig 2 Flowchart of the working process

4. FRAMEWORK AND APPLICATIONS:

OpenCV library is a collection of programming functions primarily targeted at real-time computer vision. Its purpose was to facilitate the development of computer vision applications and accelerate the integration of machine perception into commercial products. As it is a cross-platform library, OpenCV runs on Windows, Linux, Android, iOS, and Mac operating systems. Some of the most popular programming languages are supported, including C++, C, Python, and Java. OpenCV uses GPU acceleration for real-time operations.

Haar Cascade was first introduced by Paul Viola and Michael Jones in 2001. In this approach, a cascade function is trained by analysing a large number of positive and negative images. Objects in other images are then detected based on this function. Haar cascade divides the picture into several grids of pixels which are used to calculate lots of simple features. Features such as eyes, nose, forehead, cheeks, chin, mouth, space between eyebrows, etc., are used to calculate lines and edges of the face.

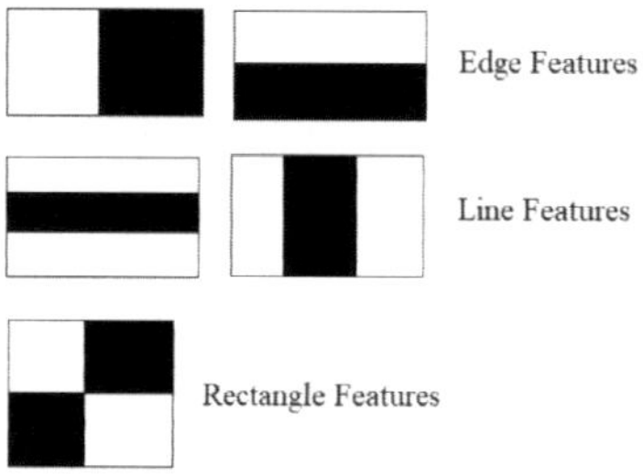

Fig 3 Types of Haar features

As shown in the above figure there are three types of masks to extract features. The two types we implemented in this model are edge and line. The masks when applied onto the image reveal the required portion of the image and cover the rest. The image portion under the black region of the mask undergoes feature extraction and the portion under the white region is covered for the time being. When required the black and white regions are swapped so that all the important features are extracted.

Clustering is a method in which large cluster of data is used to train an ML model. Clustering Large Applications also known as **CLARA** is used to speed up the training algorithm by picking small samples of data. The criterion for sampled data is the average of the dissimilarities in every object in the data cluster available. It is an extension to the PAM method for large data.

A pattern is generally defined as a vector of certain number of measurements knowns as features. Pattern recognition systems usually use training samples from known categories to form a judgement for unknown patterns. Unlike pattern recognition systems, clustering does not require supervision. It classifies images into clusters based on similarities. In this model clustering recognizes images which are similar to each other and forms multiple clusters based on them. These clusters are then used to recognize differences and provide a decision rule to classify other unknown images.

5. IMPLEMENTATION:

Face Capturing:

The image is taken without any infrared tracking and then an infrared image is taken with normal image as background. The infrared light helps in highlighting the lines and edges of faces which is crucial for face detection.

Face Mapping:

Haar cascade is used for face mapping. Haar cascade divides the picture into several grids of pixels which are used to calculate lots of simple features. Features such as eyes, nose, forehead, cheeks, chin, mouth, space between eyebrows, etc., are used to calculate lines and edges of the face. Now this cascade method is used for drawing the lines and edges of eyes. To map the aspect ratio of eyes we have used 25 fixed points.

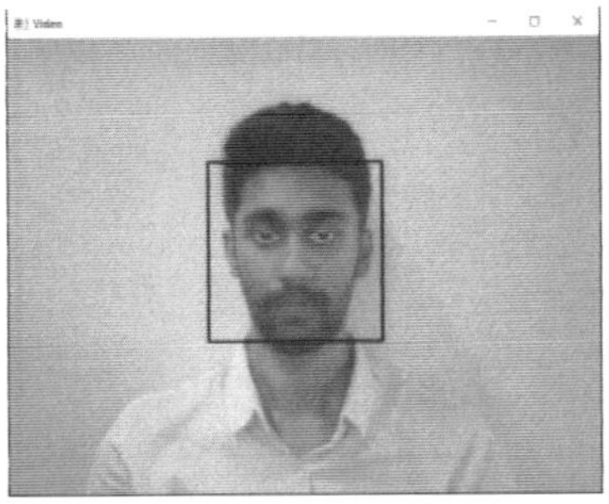

Fig 4 Face mapping

Drowsiness Detection:
To detect drowsiness, we have used CLARA clustering method which involves in selecting small samples of data from a large available dataset. By using the mapped images from Haar Cascade we use a simple formula to calculate eye aspect ratio which is the difference between points of upper eye lid and points of lower eye lid respectively.

$$Eye\ Aspect\ Ratio = \frac{|A1-B1|-|A2-B2|+\cdots+|An-Bn|}{2*|C|} \quad (1)$$

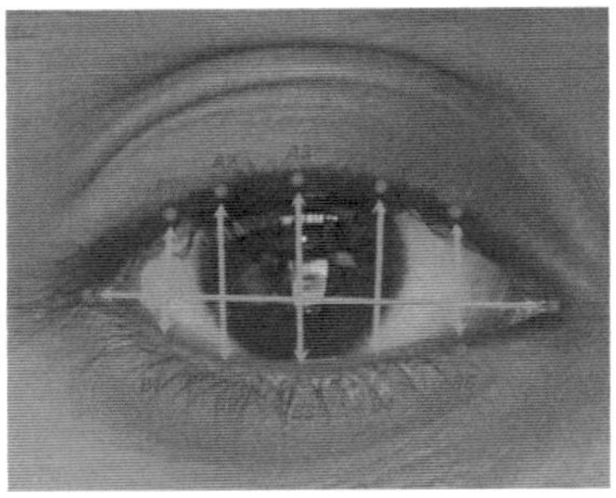

Fig 5 Eye landmarks for EAR

This aspect ratio assists us in determining if someone is drowsy or not, depending on the pattern of drowsiness.

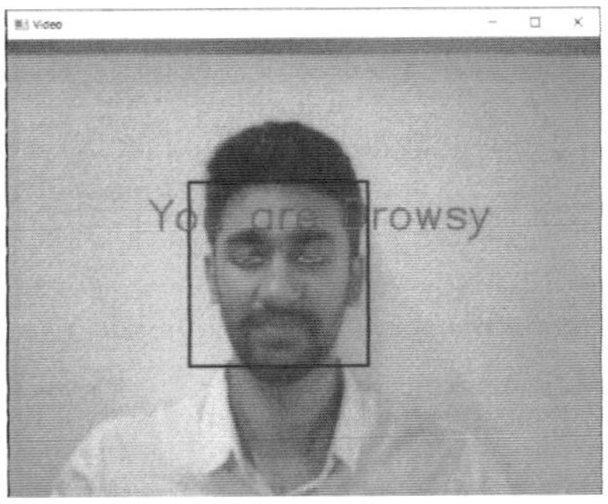

Fig 6 Drowsiness detection

Using OpenCV classification of eyes is performed i.e., if the eyes are open or closed. In this model a threshold value of EAR has been set to differentiate between open and closed eyes. When the EAR is below the given threshold value it is determined that the eyes are closed and vice versa. If the model detects that the eyes are closed for a certain time interval it is determined that the driver is drowsy and needs to be alerted immediately.

Alarm Alertness:
Whenever the person is detected feeling drowsy according to the model an alarm is activated to notify the person and wake up the driver.

6. RESULTS:

The testing is done in a real-world scenario. Whenever a person is driving the car if he is showing the patterns of drowsiness for more than three seconds the model triggers an alarm to get rid of the drowsiness. In about 80% of the circumstances this model was able to prevent the accidents.

Table 1 Comparative analysis of drowsiness tests

Test	Number of Observations	Number of Hits	Success rate
Eyes open without Glasses	60	53	88.3%
Eyes open with Glasses	60	48	80%
Eyes closed without Glasses	60	50	83.3%
Eyes closed with Glasses	60	47	78.3%

7. CONCLUSION:

This paper describes about the drowsiness detection model using clustering method. By using clustering, we have gathered data from a large number of videos and images. The cluster dataset we used to train our model is UTA Real-Life Drowsiness Dataset [9] which was created by University of Texas and has about 30 hours of videos as a common dataset for drowsiness detection systems. Through CLARA clustering we have taken a small number of frames from every video in the dataset thus effectively reducing the size of the dataset and making the robustly trained system more efficient. In this model, a major focus is on developing a highly accurate model and preventing drowsiness in drivers. This model can be integrated into dashcams to avoid additional hardware and to also avail for existing vehicles without major modifications. However, this model is not without limitations, the most prominent being, detection while wearing sunglasses, masks, or other face coverings.

References

[1] Ahmed, Rafi, Kazi Emrul Kayes Emon, and Md Foisal Hossain. "Robust driver fatigue recognition using image processing." 2014 International Conference on Informatics, Electronics & Vision (ICIEV). IEEE, 2014.

[2] Assari, Mohammad Amin, and Mohammad Rahmati. "Driver drowsiness detection using face expression recognition." 2011 IEEE international conference on signal and image processing applications (ICSIPA). IEEE, 2011.

[3] Dipu, Md Tanvir Ahammed, et al. "Real-time Driver Drowsiness Detection using Deep Learning." International Journal of Advanced Computer Science and Applications 12.7 (2021).

[4] Galarza, Eddie E., et al. "Real time driver drowsiness detection based on driver's face image behavior using a system of human computer interaction implemented in a smartphone." International Conference on Information Technology & Systems. Springer, Cham, 2018.

[5] de Naurois, Charlotte Jacobé, et al. "Detection and prediction of driver drowsiness using artificial neural network models." Accident Analysis & Prevention 126 (2019): 95-104.

[6] Jabbar, Rateb, et al. "Driver drowsiness detection model using convolutional neural networks techniques for android application." 2020 IEEE International Conference on Informatics, IoT, and Enabling Technologies (ICIoT). IEEE, 2020.

[7] Magán, Elena, et al. "Driver Drowsiness Detection by Applying Deep Learning Techniques to Sequences of Images." Applied Sciences 12.3 (2022): 1145.

[8] Vesselenyi, Tiberiu, et al. "Driver drowsiness detection using ANN image processing." IOP conference series: materials science and engineering. Vol. 252. No. 1. IOP Publishing, 2017.

[9] arXiv:1904.07312 [cs.CV]

[10] A. Rahman, M. Sirshar and A. Khan, "Real time drowsiness detection using eye blink monitoring," 2015 National Software Engineering Conference (NSEC), 2015, pp. 1-7, doi: 10.1109/NSEC.2015.7396336.

[11] Flores, M.J., Armingol, J.M. & de la Escalera, A. Real-Time Warning System for Driver Drowsiness Detection Using Visual Information. J Intell Robot Syst 59, 103–125 (2010). https://doi.org/10.1007/s10846-009-9391-1

[12] Ngxande, Mkhuseli, Jules-Raymond Tapamo, and Michael Burke. "Driver drowsiness detection using behavioral measures and machine learning techniques: A review of state-of-art techniques." 2017 Pattern Recognition Association of South Africa and Robotics and Mechatronics (PRASA-RobMech) (2017): 156-161.

[13]Valiveti, Hima Bindu, et al. "Soft computing based audio signal analysis for accident prediction." International Journal of Pervasive Computing and Communications (2021).

[14]Koduru, Anusha, Hima Bindu Valiveti, and Anil Kumar Budati. "Feature extraction algorithms to improve the speech emotion recognition rate." International Journal of Speech Technology 23.1 (2020): 45-55.

[15] Mankena, Santosh Vardhan Reddy, et al. "SICU Ambience and Patient Health Monitoring System with IOT principles." 2021 Second International Conference on Electronics and Sustainable Communication Systems (ICESC). IEEE, 2021.

[16] Abbas, Qaisar. "HybridFatigue: A real-time driver drowsiness detection using hybrid features and transfer learning." International Journal of Advanced Computer Science and Applications 11.1 (2020): 9.

[17] Chand, H. Varun, and J. Karthikeyan. "CNN Based Driver Drowsiness Detection System Using Emotion Analysis." INTELLIGENT AUTOMATION AND SOFT COMPUTING 31.2 (2022): 717-728.

[18] Valiveti, Hima Bindu, et al. "Energy efficient ambience awake routing with openflow approach." CMC-COMPUTERS MATERIALS & CONTINUA 67.2 (2021): 2049-2059.

[19] Valiveti, Hima Bindu, and Chaitanya Duggineni. "Software Defined Device to Device Communication Handover-Latest Advancements." 2021 6th International Conference on Inventive Computation Technologies (ICICT). IEEE, 2021.

Quality of Service Aware Performance Improvement in Wireless Body Area Sensor Networks

Ashley Anoop[1], Nazrin Jariya[1,*] and P Ashna Aziz[1]

[1]Computer Science Department, T. K. M College of Engineering, Kerala, India.

nazrinasaf@gmail.com

Abstract. Modern healthcare quality has been under increasing pressure. Wireless Body Area Networks have enabled better monitoring of health and reduced costs. A wireless body area network comprises miniature sensors that can be worn or implanted to record a person's health data. The uninterrupted and quick access to a patient's health status makes it easier for the doctor to diagnose in an emergency. The network uses C.S.M.A/C.A during data transmission. With packet collision and loss being two significant issues, we designed an enhanced algorithm to improve the overall performance and Quality of Service of the Wireless Body Area Networks by introducing aging factors. The algorithm avoids unnecessary long backoff periods and brief backoff periods that lead to performance degradation, improving various Quality of Service parameters like packet reception rate, packets received per node, consumed energy, estimated network lifetime, and packet loss rate.

Keywords. Healthcare, MAC protocols, Quality of Service, Wireless Body Area Networks

1. Introduction

The Wireless Body Area Networks (W.B.A.N.) consist of wireless battery-operated sensor nodes patients wear to monitor their health uninterruptedly. These nodes are responsible for bio-physiological data collection and transmission, initiating actions, and triggering alarms during an emergency. Since these networks are used in time-critical applications, receiving data from every node is essential, and no data reception will result in reduced performance. The W.B.A.N. architecture supports intra-B.A.N communications, inter-B.A.N communications, and beyond- B.A.N. communications. The transmissions between the wireless sensor nodes and between the wireless sensor nodes and the personal device are intra-B.A.N communications. The transmissions between the personal device and other access points are inter-B.A.N communications. The transmissions that involve the internet are beyond-B.A.N communications. The Medium Access Control (M.A.C) protocol in W.B.A.N. has to be energy-efficient, scalable, fair, and support a high data rate.

Quality of service or Q.o.S is the measurement of the complete performance of a service. The Q.o.S parameters measured as observations are objective, whereas those based on human perceptions are subjective. The Q.o.S requirements are different for different applications. Some of the Q.o.S parameters include packet loss rate, packet reception rate, energy, estimated network lifetime, and transmitted packets.

Following is a summary of the remainder of this article. The access mechanism of I.E.E.E 802.15.4 M.A.C also known as ZigBee M.A.C is detailed in the next subsection. The currently existing M.A.C protocols that support Q.o.S are reviewed in section 2. An enhanced version of C.S.M.A/C.A algorithm with aging factors is proposed in section 3. The simulation details and results are discussed in sections 5 and 6 respectively. The paper concludes with section 7.

1.1. 802.15.4 M.A.C Access Mechanism

Under I.E.E.E 802.15.4, a group drafted the Wireless Personal Area Network (W.P.A.N) standard. The standard defines the M.A.C and physical layers. The M.A.C layer responsibilities include managing

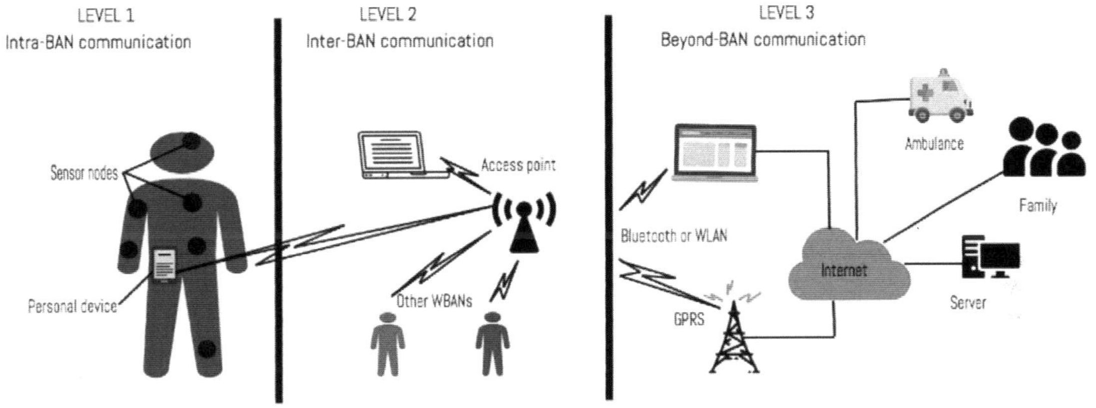

Figure 1. Wireless Body Area Network Architecture.

beacon usage, channel access, Guaranteed Time Slot (G.T.S), frame validation, etc. The physical layer uses the following frequency bands in the range 2.4GHz to 2.4835 GHz, 902MHz to 928 MHz, and 868MHz to 868.6 MHz for operation. Band1 supports sixteen channels, band2 supports ten channels, and band3 supports one channel [3]. The topology and bandwidth divide ZigBee M.A.C layer operation into the non-beacon mode and beacon mode. In a beacon-less way, a coordinator always waits for data by default. Any device with data to send can perform data transmission when the channel is free. Otherwise, it remains for a random time as per protocol definition. The coordinator waits for a data request from a device whenever it wants to send data to that device. The coordinator acknowledges the request. The coordinator performs data transmission using C.S.M.A/C.A algorithm when there is data waiting and transmits an empty data frame in case of no pending data. This mode of operation lacks device synchronization by the coordinator and G.T.S reservation. The absence of superframe and slot synchronization allows only random channel access. The coordinator uses periodic beacon transmission to ensure device synchronization in the beacon-enabled mode. This notifies the device about the superframe duration and data transmission time.

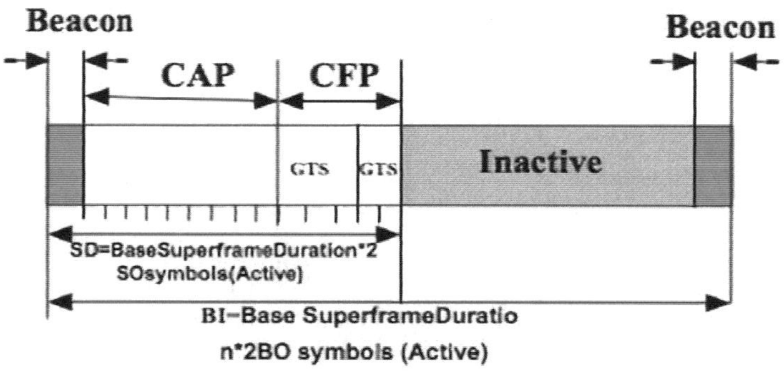

Figure 2. Superframe Structure.

The superframe structure [6] used in this mode starts with a beacon. It also has active and inactive periods, as presented in figure. 2. The transmission between nodes occurs in the active phase. During the inactive stage, nodes power down for energy conservation. The beacon interval or *B.I* denote the length of the superframe. The superframe duration or *S.D* represents the length of its active period. *B.I* and *S.D* calculated as follows:

$$B.I = aBaseSuperframeDuration * 2^{BO}$$
$$S.D = aBaseSuperframeDuration * 2^{S0}$$

The value for aBaseSuperframeDuration is 0.187ms. The variable *B.O* refers to the beacon order, and *S.O* refers to the superframe order. The coordinator chooses its values from $0 < S.O. < B.O < 14$. The active period includes the Contention Access Period (C.A.P) and Contention Free Period (C.F.P). The C.A.P resembles the beacon-less mode operation where devices perform random transmissions within the time slot. It does not allow transmissions in the middle of a time slot. The C.F.P. guarantees medium access to nodes for a certain number of slots known as G.T.S. The nodes use C.S.M.A/C.A for transmission during C.A.P.

The backoff exponent (B.E) computes the backoff delay (B.D). The backoff period is the time needed to send twenty symbols, where one symbol is equal to four bits. A device before assessing the channel waits for random backoff periods ranging from 0 to 2B.E. − 1. Initially, the value of variable B.E is equal to macMinB.E. The variable N.B (number of backoffs) indicates how often the C.S.M.A/ C.A algorithm back-offed while trying to access the medium. Its value is zero for every new transmission try. The channel sensing mechanism makes sure that the medium is free for the duration of contention window (C.W) backoff slots before any device begins to send data. As per the standard algorithm, nodes remain idle throughout the backoff period to conserve energy. The I.E.E.E. 802.15.4 non-beacon operational mode uses unslotted C.S.M.A/C.A.

The backoff exponent (B.E) computes the backoff delay (B.D). The backoff period is the time needed to send twenty symbols, where one symbol is equal to four bits. The variable N.B (number of backoffs) indicates how often the C.S.M.A/ C.A algorithm back-offed while trying to access the medium. The channel sensing mechanism makes sure that the medium is free for the duration of contention window (C.W) backoff slots before any device begins to send data. During step one, N.B and C.W is initialised to 0 and 2 respectively. If the battery extension is enabled B.E takes the minimum of 2 and macminBE. Otherwise, the value of variable B.E is equal to macMinBE. Its value is zero for every new transmission try. The Clear Channel

Assessment (C.C.A) is performed to check whether the channel is free for transmission. C.W is decremented by one if the channel free to be bus and this process repeats until C.W becomes zero. When C.W becomes zero, the device has successfully obtained access to an idle medium for data transmission. On the other hand, if the channel is found to be busy, the number of backoff is incremented by one. The backoff exponent takes the minimum of B.E + 1 and macmaxBE. This device before assessing the channel again waits for random backoff periods ranging from 0 to 2B.E. − 1. This process repeats until the N.B remains less than macmaxCSMAbackoffs. As per the standard algorithm, nodes remain idle throughout the backoff period to conserve energy. When the number of backoff attempts becomes greater than the default value assigned by the M.A.C protocol it results in a failed transmission. The I.E.E.E. 802.15.4 non-beacon operational mode uses unslotted C.S.M.A/C.A algorithm. The I.E.E.E. 802.15.4 beacon operational mode uses the slotted C.S.M.A/C.A algorithm given below

Algorithm 1 Slotted CSMA/CA Algorithm

```
Begin
NB ← 0
CW ← 2
if BatteryLifeextension = enabled then
    BE ← min(2, macMinBE)
else
    BE ← macMinBE
end if
Step A: Locatebackoffboundary
Step B: Delayforrandom(2^{B.E} − 1)
Step C: PerformCCAonbackoffboundary
if Channel = Idle then
    CW ← CW − 1
    if CW = 0 then
        Output : Successfultransmission
    else
        GotoStepC
    end if
else
    CW ← 2
    NB ← NB + 1
    BE ← min(BE + 1, macMaxBE)
    if NB > macMaxCSMABackoffs then
        Output : Failedtransmission
    else
        GotoStepB
    end if
end if
End
```

Figure 3. Slotted C.S.M.A/C.A Algorithm.

2. Literature Review

[5] ensures data transmission without collision using a packet scheduler and two queues, Collision Resolution Queue and Data Transmission Queue. Despite its ability to provide quality-of-service features such as energy consumption, and reliability, it has considerable overhead in maintaining the queues and scheduler.

In [8], throughput and delay are met through a superframe structure that combines advertisement, C.A.P, Beacon, Data Transmit Slot (D.T.S), and Emergency Data Transmit Slot (E.T.S). The advertisement passes primary network data to new nodes. The nodes make data transmission requests during C.A.P and perform C.C.A. The beacon broadcasts information to synchronize all the network nodes. During D.T.S, high-priority data transmission occurs. In case of insufficient D.T.S slots, data transmission follows in E.T.S. Even though it prioritizes time-critical data transmission, it lacks energy efficiency.

[9] guarantees reliable and timely data transmission by upgrading ZigBee's acknowledgment packet and using eight priority parameters that prioritize packets in the application layer. The usage of access categories also ensures Q.o.S. The experimental results proved [9] to perform better than ZigBee.

[12] uses a dynamic superframe where slots are allocated according to traffic classification based on delay and reliability factors, ensuring energy efficiency. This traffic classification assigns priority class backoff values that follow suitable slot allocation for data transmission.

[13] uses a T.D.M.A-like scheduler and Markov-like channel to ensure Q.o.S features like energy efficiency. The ordering of transmissions to ensure low packet loss and high-reliability transmission before slot allocation makes it suitable for heterogeneous W.B.A.N networks.

The two-level network architecture in [6] includes inter-B.A.Ns and intra-B.A.Ns. It supports communication across tiers in hospital networks to ensure reliability and timeliness. The M.A.C, routing, and scheduling modules are drafted and tuned to meet the domain-related requirements and deliver the best results. The reliability and timeliness services are provided by using a multiple-level packet scheduling system. The cluster head in a B.A.N gathers and combines data in a cluster to send it to the base station following a consistency check.

3. Proposed Approach

The network nodes use C.S.M.A/C.A for sensing the channel and determining whether it is idle before transmitting the data. The packet collision and loss are the two main issues in these wireless networks. A backoff algorithm decides how long to wait before sending data when the channel is occupied or following a collision. A node that backs off unnecessarily long periods can severely hamper the throughput, cause delay, and greater energy consumption. Whereas a node that backs off too soon will perform as many more Clear Channel Assessments (C.C.A), find the channel to be still busy, and eventually declare a failure to transmit. Hence it is crucial to choose the proper method or algorithm for backoff to improve the performance.

3.1. Aging Factors

In this work, we introduced two new parameters that account for aging. The aging factors avoid both indefinite waiting of nodes with failed transmissions and short backoff periods with successful communications. Successcount accounts for successful transmission and failcount indicates a failed data transfer. The introduction of aging factors imposes certain conditions, which are as follows:

- If more than one successful transmission takes place, then both macMaxBE and macMinBE are incremented by one until they reach their respective upper limits, MaxBElimit + 1 and MinBElimit + 1.
- If more than one failed transmission occurs, both macMaxBE and macMinBE are decremented by one until they reach their respective lower limits, MaxBElimit - 1 and MinBElimit - 1.
- On every successful transmission, failcount will be decremented by one if the value is positive and successcount will be incremented by one.
- On every failed transmission, failcount will be incremented by one and successcount will be decremented by one if the value is positive.

Journal of Physics: Conference Series **2325** (2022) 012046 doi:10.1088/1742-6596/2325/1/012046

A lower limit of MinBElimit - 1 is chosen so that the node waits for a small-time interval before packet transmission in case of repeated failed transmissions. An upper limit of MinBElimit + 1 prevents nodes from waiting for overly long periods to transmit packets if they have been successful in their previous transmissions. The value limit of macMaxBE allows the node to have a different B.E. value each time it finds the channel busy. The macMaxBE value is limited for nodes to have different B.E. values every time the channel is active. In this work, the main focus is on enhancing the various Q.o.S parameters for the W.B.A.N networks, ensuring the effective transmission of time-critical data across the network. The focus is on the Contention Access Period. The design mainly consists of an improved algorithm, a variant of the existing standard, slotted C.S.M.A/C.A. The Q.o.S parameters considered while designing this algorithm include packet loss rate, packet reception rate, energy consumption, estimated network lifetime, and transmitted packets.

4. Enhanced Slotted C.S.M.A/C.A Algorithm

4.1. Attempt Transmission

In this phase, a transmission (or retransmission) attempt occurs. A transmission attempt occurs if the internal M.A.C state is not set up or inactive. We check for available retries and delays if a packet queues for transmission. If the M.A.C state is G.T.S, the packet transmits in G.T.S. The successcount, failcount, and number of backoffs (N.B) are zero for each new transmission attempt. The value of the contention window in the slotted C.S.M.A scheme is two. The value of macMinBE and macMaxBE depends on the value of successcount and failcount. The maximum allowed C.S.M.A backoff's value will also change accordingly. The value of the binary exponent depends on the battery life extension feature. If battery life extension is present, B.E takes the lesser one among macMinBE and two; else, it takes macMinBE. At this stage, C.S.M.A/C.A executes.

Algorithm 2 Attempt Transmission

```
Begin
NB ← 0
CW ← 2
if successcount > 1 then
    if macMaxBE < MaxBElimit + 1 then
        macMinBE ← macMaxBE + 1
        macMaxCSMABackoffs ← macMaxCSMABackoffs + 1
    end if
    if macMinBE < MinBElimit + 1 then
        macMinBE ← macMinBE + 1
        macMaxCSMABackoffs ← macMaxCSMABackoffs − 1
    end if
else if failcount > 1 then
    if macMaxBE > MaxBElimit − 1 then
        macMaxBE ← macMaxBE − 1
        macMaxCSMABackoffs ← macMaxCSMABackoffs − 1
    end if
    if macMinBE > MinBElimit − 1 then
        macMinBE ← macMinBE − 1
        macMaxCSMABackoffs ← macMaxCSMABackoffs + 1
    end if
end if
End
```

Figure 4. Enhanced Slotted C.S.M.A/C.A Algorithm

4.2. Perform C.S.M.A/C.A

The channel access in C.A.P is in the form of slotted C.S.M.A. All node synchronizations and transmissions can begin only at the boundaries of time units called backoff slots in the slotted algorithm. A node that has a packet ready for dispatch first backs off for a random number of backoff slots, chosen uniformly between 0 and 2 B.E − 1, before sensing the channel where the parameter B.E is the backoff exponent. This random backoff reduces the probability of collisions within contending nodes. The channel-sensing algorithm ensures that the channel is free for a contention window duration, expressed in the number of backoff slots before the node tries to send data. As per 802.15.4 standard, two backoff slots constitute C.W duration. The backoff exponent is increased by one when the channel is busy, and new backoff slots are allotted for the node to wait for the medium to become available again. The process continues, and once B.E equals the aMaxBE, the process is frozen at aMaxBE or until a certain number of backoff stages is reached. At this point, the upper layer is declared inaccessible. The parameter macMaxCSMABackoffs determines the maximum number of allowed random backoff stages. For each successful or failed transmission, the successcount and failcount adjust accordingly. At each of these stages of implementation, the algorithm tries to overcome the drawbacks of the existing standard and improve the Q.o.S parameters, thus ensuring efficient transmission. With the introduction of aging factors, the algorithm can avoid short and long backoff periods.

Algorithm 3 Perform CSMA/CA

```
Begin
if BatteryLifeextension = enabled then
    BE ← min(2, macMinBE)
else
    BE ← macMinBE
end if
Step A: Locatebackoffboundary
Step B: Delayforrandom(2^(B.E) − 1)
Step C: PerformCCAonbackoffboundary
if Channel = Idle then
    CW ← CW − 1
    if CW = 0 then
        successcount ← successcount + 1
        if failcount > 0 then
            failcount ← failcount + 1
        end if
        Output : Successfultransmission
    else
        GotoStepC
    end if
else
    CW ← 2
    NB ← NB + 1
    BE ← min(BE + 1, macMaxBE)
    if NB > macMaxCSMABackoffs then
        failcount ← failcount + 1
        if successcount > 0 then
            successcount ← successcount + 1
        end if
        Output : Failedtransmission
    else
        GotoStepB
    end if
end if
End
```

Figure 5. Perform C.S.M.A/C.A – Algorithm

5. Body Area Network Simulation

The algorithm is tested using the Castalia simulator. It supports body area networks, wireless sensor networks, and low-energy embedded device networks. The research scholars use this OMNeT++ platform to test their distributed algorithms or protocols in wireless and radio channel models with natural node characteristics, mainly those related to accessing radio resources. Since this simulator is highly parametric, it can be applied to analyze the characteristics of various platforms for specific applications. Castalia's basic units are composite modules. The modules exchange messages through nodes. The nodes communicate with each other via wireless channels and the physical processes that they monitor. Figure. 6 illustrates the contents of a composite node module. The message passes are shown using solid arrows, and the function calls are shown using dashed arrows.

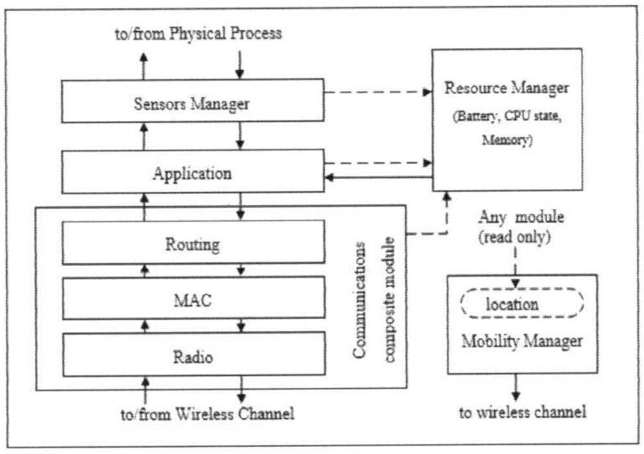

Figure. 6. Castalia composite node module.

The Castalia configuration files dynamically select four different module types: M.A.C, Routing, Application, and MobilityManager. This simulator allows the creation of new user modules in the application module and other modules to implement and test new protocols and algorithms. It supports four types of application modules, namely BridgeTest, ConnectivityMap, ThroughputTest, ValuePropagation, and ValueReporting. The communication module of each sensor network node (S.N) contains a radio module. The variety of low-power radio features that are captured by the radio module is expected to be used in W.S.N. It also supports four M.A.C modules, namely TunableM.A.C, T.M.A.C, I.E.E.E 802.15.4/ZigBee M.A.C, and I.E.E.E 802.15.6 M.A.C. Some parameter tuning on T.M.A.C results in S.M.A.C. The parameters like transmission power, retransmission attempts, scheduled or random access, and polling can be modified in one or more nodes in this module for performance evaluation. The Mobility Manager maintains the location state of a node accessed by other modules with function calls. It provides periodic updates about a node's position to the wireless channel. The resource manager module tracks the node's energy consumption, clock drift, and baseline power consumption. The configuration setup supports both temporal variation and G.T.S during simulation. A wireless channel with temporal variability is represented by the general configuration setup. The G.T.S, also known as Guaranteed Time Slots, is a T.D.M.A-based scheme that 802.15.4 supports. All the 16 slots in 802.15.4 use a contention-based approach for channel access while the G.T.S is in the off state.

The Castalia configuration file contains the default parameter initializations simulation requires. The configuration file includes a general section where the simulation time is set to 100 seconds. The number of sensor network nodes is set to 15. The files needed for the wireless channel module parameters pathLossMapFile and temporalModelParametersFile are also included. The S.N module consists of submodule nodes which are referenced using arrays. For example, to collect the trace information of a node with index six SN.node[6].Application.collectTraceInfo is assigned the value 'true'. The operational characteristics of the B.A.N radio module are defined inside RadioParametersFile.txt. The power value for radio packet transmission is set using TxOutputPower. Since we are simulating a B.A.N that uses ZigBee M.A.C the communication module parameter MACProtocolName is set to ZigBee M.A.C. The communication module parameter RoutingProtocolName assumes the default value "Bypass routing". The application module is set to ThroughputTest using the ApplicationName parameter. In the ThroughputTest module, all the nodes send traffic to the default sink node 0. The packet rates are assigned values ranging from 10 to 15. The sensor nodes are deployed based on the default custom configuration and the default static node NoMobilityManager configuration. The simulations are carried out under general configuration with G.T.S off. The simulator will run all combinations of parameters when there exist multiple values for each parameter. All other parameters whose values weren't changed assumed the default values given in Castalia.ini files.

6. Results and Discussion

The general simulation parameters used are:

- macMinBE and macMaxBE represent the minimum and maximum values that the backoff exponent can take, respectively.
- macMaxCSMABackoffs indicates the maximum number of retries the packet will receive before it aborts and attempts a retry.
- macMaxFrameRetries indicates the maximum number of attempts until the upper layer gets notified during packet loss.
- sim-time-limit denotes the simulation time.
- numNodes denotes the number of nodes in the network.

Table 1. Simulation parameters and their values.

Parameters	*Values*
macMAXBE	7
macMINBE	5
macMaxCSMABackoffs	3
sim-time-limits	100s
num-nodes	15

The general simulation parameters and the values considered while designing this enhanced version of C.S.M.A/C.A algorithm for the IEEE 802.15.4 standard is shown in the table. 1. The outputs considered for comparison include packet loss rate, packet reception rate, packets received per node, consumed energy, estimated network lifetime, and transmitted packets. The simulations were performed at packet rates ranging from 10 to 15 to evaluate ZigBee M.A.C performance.

The performance of various Q.o.S factors under the general configuration when the G.T.S is off is visualized in figures 7-12. The x-axis represents the rate at which the node sends packets measured in packets per second. The y-axis of figure 7 represents the lifetime of the sink node measured in days. The y-axis of figure 8 represents the energy consumed by node 0 measured in joules. The y-axis of figure 9 represents TXed packets, i.e., the total number of packets transmitted by node 2. The y-axis of figure 10 represents the packet loss rate at the application layer expressed in percentage. The y-axis of figure 11 represents the average of packets received by the sink node from all nodes. The y-axis of figure 12 represents the packet reception rate at the application layer expressed in percentage.

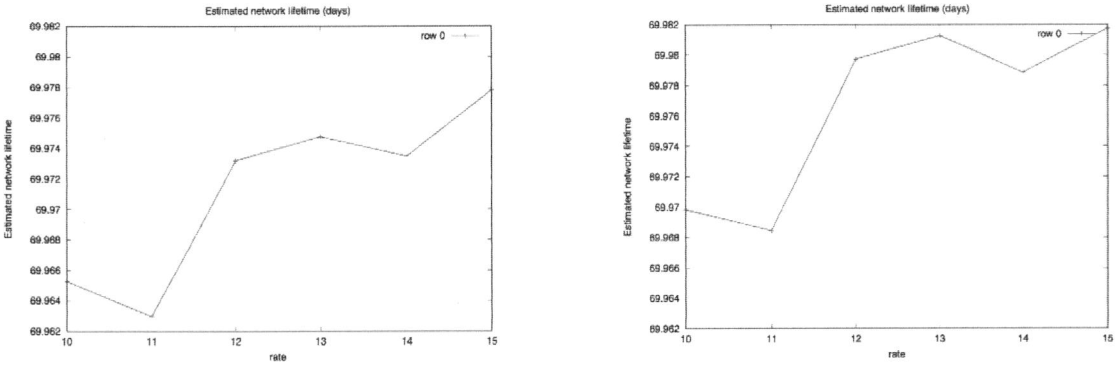

Figure. 7. Network Lifetime

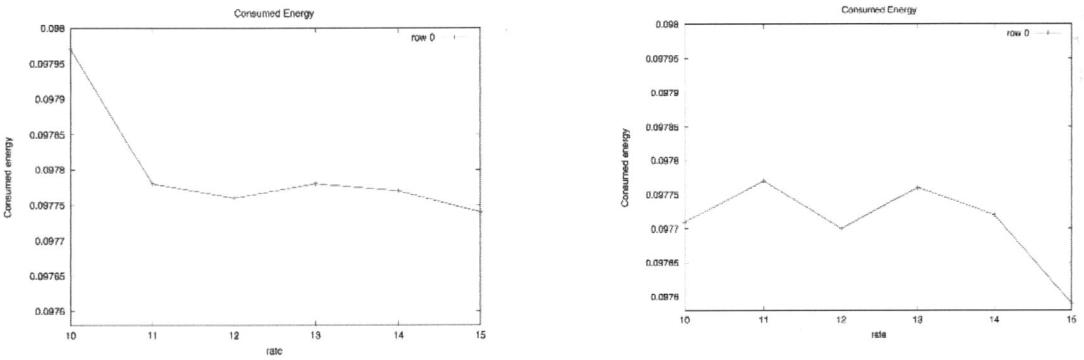

Figure 8. Energy Consumed

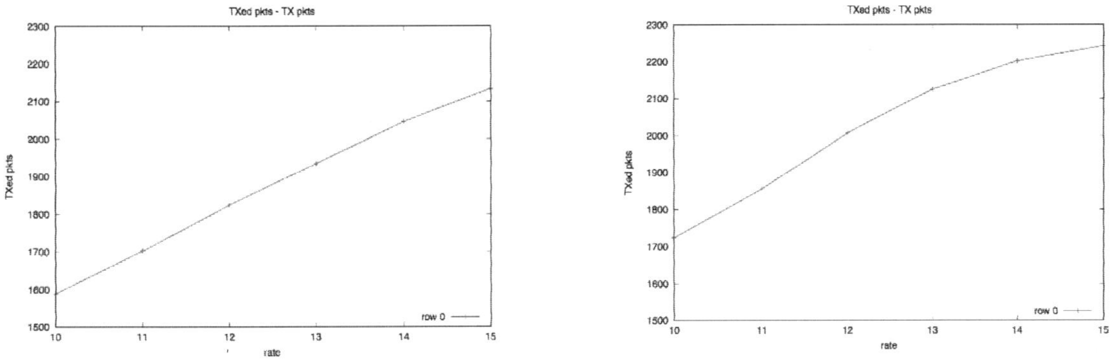

Figure 9. Packets Transmitted

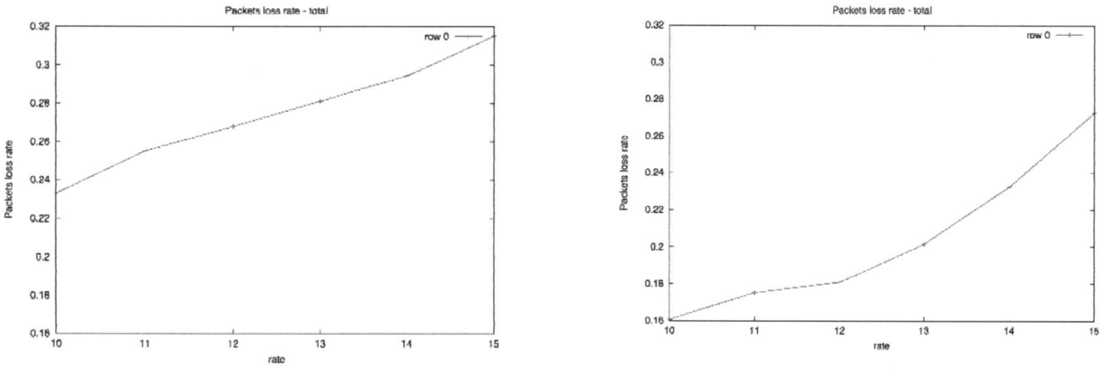

Figure 10. Packet Loss Rate

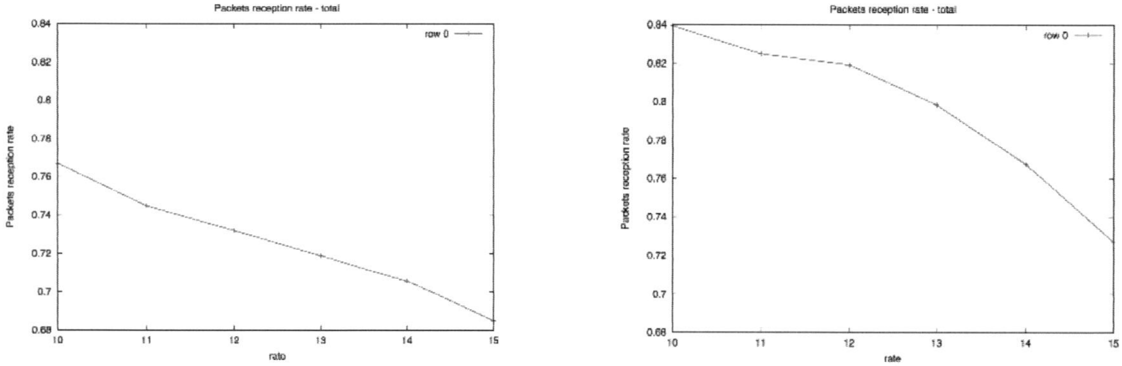

Figure 11. Packet Reception Rate

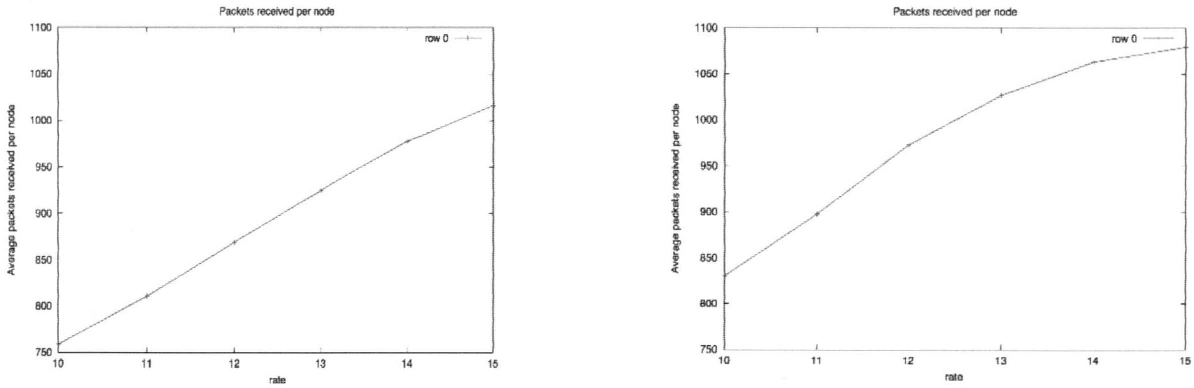

Figure 12. Packet Received Per Node

The Q.o.S needs to be maintained and improved to transmit the time-critical data effectively. After performing simulations with rounds of the slotted algorithm with different packet rates, we obtained improved results for all the Q.o.S parameters considered. ZigBee M.A.C with enhanced C.S.M.A/C.A algorithm works well with efficiency and effectiveness. It achieves the targeted task with a better performance rate. The results are satisfactory and comparable with the already existing I.E.E.E 802.15.6 ZigBee standard.

7. Conclusion

The Wireless Body Area Networks are a new generation of highly localized body area networks used for many medical applications, from monitoring implants to tracking the health of the elderly. But if this specialized area matures into a thriving industry, it will need a better definition of its scope and potential and a clearer vision for its future. In this work, we have considered the slotted scheme of the C.S.M.A/C.A algorithm, where we have laid focus on the Contention Access Period of the superframe structure. We added two aging factors to the existing default parameters to improve the current algorithm. There are a variety of possible directions related to this field that experts can explore for future research work. An extension for this work can focus on high packet rate networks as this work supports low rate networks. Also, one of the main focuses could be extending the network capacity, which would help analyze a large number of nodes, for transmission of time-critical data, without collision. Q.o.S would always be the most critical factor for transferring time-critical data. Researchers can incorporate security, privacy, ease of use, compatibility with other B.A.N's, improving the battery technology for energy conservation, etc. The application range of W.B.A.N's can be extended to services such as deployment and control of prosthetics or remote assistive devices, deep brain stimulation, in the military field, for sea and space explorers, etc.

References

[1] Frank Agyei-Ntim and Kimberly E Newman 2013 Lifetime estimation of wireless body area sensor networks using probabilistic analysis Wireless Pers. Comm. **68** 1745-59

[2] Negra R, Jemili I and Belghith A 2016 Wireless Body Area Networks: applications and technologies Procedia Computer Science **83** 1274-81

[3] Bradai N, Fourati L C and Kamoun L 2014 Investigation and performance analysis of MAC protocols for WBAN networks Journal of Network and Computer Applications **46** 362-73

[4] Jung, J I 1996 Quality of service in telecommunications in proposition of a QoS framework and its application to b-isdn IEEE Comm. Magazine **34** 108-11

[5] Otal , Alonso L and Verikoukis C 2009 Highly reliable energy-saving MAC for wireless body sensor networks in healthcare systems IEEE Journal on Selected Areas in Comm. **27** 553-65

[6] Ali K A, Sarker J H and Mouftah H T 2010 Urgency-Based MAC Protocol for Wireless Sensor Body Area Networks IEEE Int. Conf. on Comm. Workshops 1-6

[7] Cao H, Gonz´alez-Valenzuela S and Leung V C M 2010 Employing IEEE 802.15.4 for Quality of Service Provisioning in Wireless Body Area Sensor Networks IEEE Int. Conf. on Advanced Information Networking and Applications 902-09

[8] Yoon J S, Ahn G, Joo S and Lee M J 2010 PNP-MAC: Preemptive Slot Allocation and Non-Preemptive Transmission for Providing QoS in Body Area Networks IEEE Consumer Comm. and Networking Conf.1-5

[9] Garcia J J and Falck T 2009 Quality of Service for IEEE 802.15.4-based Wireless Body Sensor Networks Int. Conf. on Pervasive Computing Technologies for Healthcare 1-6

[10] Barua M, Alam M S, Liang X and Shen X 2011 Secure and quality of service assurance scheduling scheme for WBAN with application to eHealth IEEE Wireless Comm. and Networking Conf. 1102-06

[11] Monowar M M, Hassan M M, Bajaber F, Al-Hussein M and Alamri A 2012 McMAC: towards a MAC protocol with multi-constrained QoS provisioning for diverse traffic in wireless body area networks Sensors **12** 15599-627

[12] Anjum I, Alam N, Razzaque M A, Hassan M M and Alamri A 2013 Traffic priority and load adaptive MAC protocol for QoS provisioning in body sensor networks Int. Journal of Distributed Sensor Networks **9** 205192

[13] Yan Z, Liu B and Chen CW' 2012 QoS-driven scheduling approach using optimal slot allocation for wireless body area networks IEEE Int. Conf. On E-health Networking, Applications and Services 26-72

[14] Su H and Zhang X 2009 Battery-dynamics driven tdma mac protocols for wireless body-area monitoring networks in healthcare applications IEEE Journal on Selected Areas in Comm. **27** 424-34

[15] R Sruthi 2016 Medium access control protocols for wireless body area networks: A survey Elsevier, Procedia Technology **25** 621–28

[16] Gang Zhou, Jian Lu, Chieh-Yih Wan, Mark D Yarvis and John A 2008 Stankovic BodyQoS: adaptive and radio-agnostic QoS for body sensor networks IEEE Conf. on Computer Comm. 565-73

[17] Huq M A, Dutkiewicz E, Fang G, Liu R P and Vesilo R 2012 MEB MAC: improved channel access scheme for medical emergency traffic in WBAN Proc. of the Int. Symp. on Comm. and Information Technologies 371-76

[18] Chen B and Pompili D 2011 Transmission of Patient Vital Signs Using Wireless Body Area Networks Mobile Network and Applications **16** 663-82

[19] Abdullah Sevin, Cuneyt Bayilmis and Ismail Kirbas 2016 Design and implementation of a new quality of service-aware cross-layer medium access protocol for wireless body area networks Computers and Electrical Engineering **56** 145-56

[20] Garcia J and Falck T 2009 Quality of service for IEEE 802.15.4-based wireless body sensor networks Proc. of pervasive computing technologies for healthcare

[21] Athanassios Boulis 2011 Castalia A simulator for Wireless Sensor Networks and Body Area Networks

[22] Chen J I Z and Yeh L T 2020 Data forwarding in wireless body area networks Journal of Electronics **2** 80-87

Wireless Communication for Optical Link in Network

Dr. P. Sebastian Vindro Jude[1], Dr. C. Kathirvel[2], S. Venkatesh Kumar[3]

[1]Assistant Professor (Sl. G), Department of Electrical and Electronics Engineering, Sri Ramakrishna Engineering College, Coimbatore, Tamil Nadu, India.
[2]Associate Professor, Department of Electrical and Electronics Engineering, Sri Ramakrishna Engineering College, Coimbatore, Tamil Nadu, India.
[3] Assistant Professor (Sr. G), Department of Electrical and Electronics Engineering, Sri Ramakrishna Engineering College, Coimbatore, Tamil Nadu, India.

jude.panimayam@srec.ac.in, kathirveleee@srec.ac.in, venkateshkumar.s@srec.ac.in

Abstract- Laser, as a correspondence medium can give a decent substitute to the present-day correspondence frameworks as the issue of connection point looked on account of EM waves, isn't there and a high arrangement of mystery is accessible. Optical correspondence offers a practical option in contrast to RF interchanges for between satellite connections and different applications where superior execution joins are a need. This paper is planned to give an outline of advancement that has been achieved with the assistance of lasers to send sound and explains on the way things are being utilized alongside future turns of events. In the transmitter, the power of the laser point of support is managed by the consequence of the code oscillator. Utilizinga push to on switch the tone oscillator is initiated to alarm theindividual at the collector end prior to beginning a voice correspondence utilizing the mouthpiece. The beneficiary getsthe power changed. We will dissect the outcome utilizing sun powered chargers and standard opticalidentifiers
(Light-dependent resistors).

Keywords: Optical Communication, Tone Oscillator, Light-dependent resistors.

1. INTRODUCTION:

Laser Communication is one of the arising areas of remote correspondence frameworks. Its low commotion proportion makes it one of the appropriate correspondence media for the trading of data. Presently, laser substitution is taken on in satellite correspondence for space research exercises and because of its proficiency on low clamor proportion, reasonable, low power and adaptability, and protection from the radio obstructions make lasercorrespondence is one of the examination regions in remote correspondence. In this interaction, this paper includes one such use of laser correspondence for data trade between any two devices.[1]

This paper proposes a vital idea of lasercorrespondence which is one of the immature and impending fields of correspondence. Along these lines, this kind of correspondence will be additionally carried out in the future in numerous areas, like PC correspondence, satellite correspondence, and so forth. This will diminish the use of transports inside the hardware, which may likewise decrease the utilization of metal, subsequently lessening e-squander. Laser correspondence will play a significant part in giving exceptionally high information rates and limit. Optical laser transmission enjoys numerous innate benefits contrasted andRF transmission.

2. WORKING PRINCIPLE:

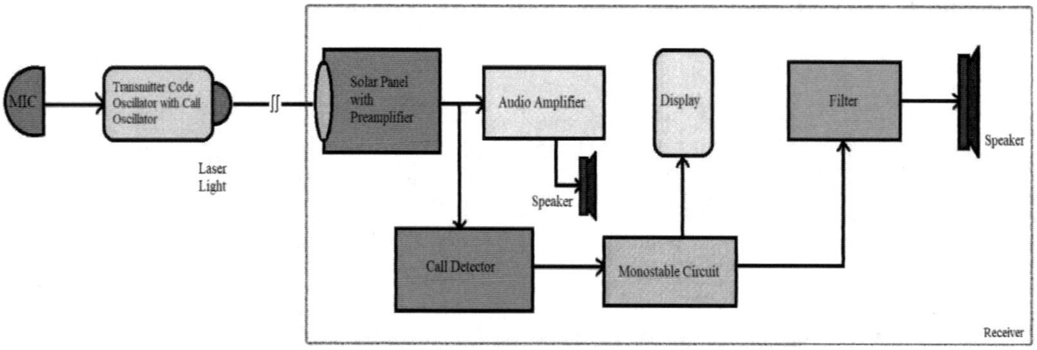

Fig.1 Block diagram

The mic is moreover called a recipient. [2].

The transmitter area contains a condenser mouthpiece. In transmitter, the recurrence is created which isto match the recurrence in the PLL utilized in the recipient segment.

The recipient circuit utilizes a NPN phototransistor or a sunlight powered charger the light sensor that is trailed by two-stage semiconductor preamplifier[4].

IMPLEMENTATION:

a. Transmitter Circuit

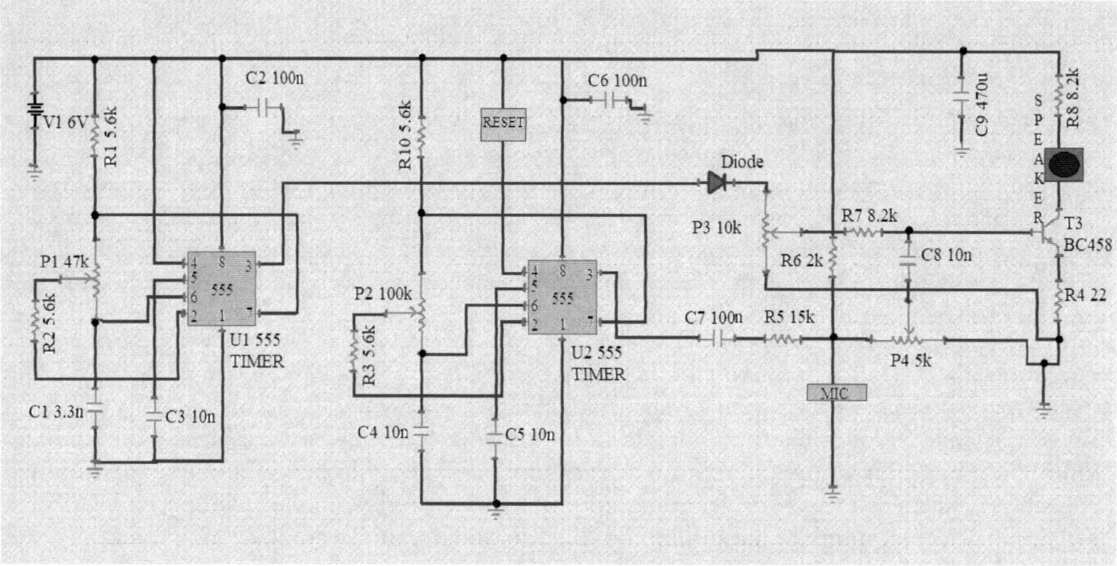

Fig.2 Transmitter circuit

b. Code Oscillator:

This circuit comprises of an IC NE555 clock.. here, the swaying recurrence is chosen by the resistors (R1, R2), a variable Resistor P1, Capacitor C1. VR1 is changed in such a manner, that its wavering coordinates with the middle recurrence of PhaseLock Loop circuit PLL1 in the recipient end. Then the result from this IC is taken care of to the blendercircuit (Transistor T3) by means of diode 1N4001 and pot meter VR3. We'll change variable resistor P2 to change the tone recurrence to coordinate it with the middle recurrence of Phase lock circlecircuit PLL2 in the recipient end.

c. Condenser Microphone:

The output from U2 is coupled with mixer circuit via capacitor C7 and C8, resistor R5 and Potresistor P4. The main use of Pot P4 is to vary the biasing signals (to adjust the audio signal when someone speaks into microphone).

d. Audio Frequency Mixer circuit:

The results of IC NE555 clock U1, U2 and the voice signal is blended by the semiconductor T3. **Laser Diode:**

A semiconductor laser diode creates a lot higher and exact result power and exceptionally directional shafts contrasted with the LEDs.

e. Receiver Circuit:

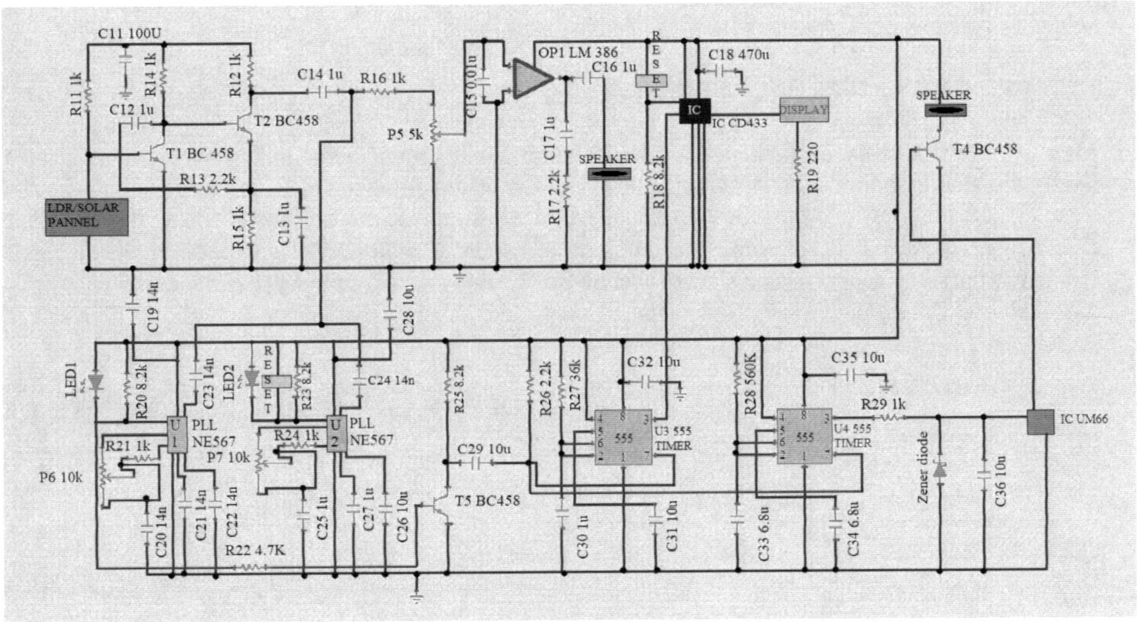

Fig.3 Receiver circuit

- PLL IC NE567 (U1, U2):

The outside parts around PLL chooses the middle recurrence of activity and the result delay. The preamplifier yield is associated with PLL IC NE567 U1 and U2 through capacitors C23 and C24 individually. ICNE567 U1 is utilized to distinguish the 10khz - 15khz sign. Pre-set variable resistor P6 is utilized for tuning PLL IC U1 to its ideal reach. The consequence of PLL IC U1 is coming up short in case any repeat between the scope is distinguished. LED1 illuminates to demonstrate congruity of the optical way of correspondence. At the point when the laser shaft is interfered with, the result of the IC U1 goes high to semiconductor T5 and its gatherer voltage falls, which sets off a monostable circuit worked aroundIC U3 and U4 (both are IC NE555) individually. Accordingly, the result of these ICs goes high for some dominating time (that is settled by resistor and capacitors blends present in the monostable circuit). Since the resultof IC NE555 U3 is associated with decade counter IC CD4033, it gives a clock heartbeat to counter IC to addition its count, showing the quantity of hinders causedin the correspondence link[5][. Here, the ongoing includeis displayed in a 7-section show. After continuous interferences, the count begins from the start in the wakeof counting 9 successive intrudes. Likewise, it tends to be reset physically utilizing the RESET switch. As we probably are aware, IC U3 and U4 are set off all the while, the result of IC U8 is beat high for a foreordainedtime frame to give voltage to Zener diode SD1 1N5817 to a tune IC UM66. This IC produces a pleasant tune at whatever point a light shaft is interfered, then, at that point, the result is taken care of to one more speaker through a semiconductor T4.

For starting a call, the individual at the transmitter will pressthe RESET change to caution the remote end individual of forthcoming voice correspondence. Along these lines, the regulated light result will have both 1khz - 2khz tone part and 10khz - 15khz code oscillator yield. the 1khz - 2khz tone is decoded by PLL U2, whose middle recurrence is acclimated to in the transmitter. Subsequently, PLL IC U2 go about as the call indicator. In this manner, When the tone part is distinguished by IC U2, its result goes low to illuminate LED2. Since the tone part at theresult of the preamplifier is additionally gone through enhancer LM386, the tone is heard from both speakers [6].

Consequently, the discourse is imitated at the result of LM386 by means of an amplifier [7]. The code partis recognized by PLL IC new one it is shown by one as madesense of before to imply continuous light way which.

3. RESULTS AND CONCLUSION:

After a complete analysis of the circuit and testing for its implementation, we have arrived at a conclusion that lasershave the potential to transmit the information signal from the transmitter to the receiver in a quite efficient manner since it's based on the principle of spontaneous emission. This circuit of transmitter and receiver is designed for a **range of 3-4km**. The reception of the signal is efficient atthe receiver and its conversion is done by obtaining good throughput of the circuit.

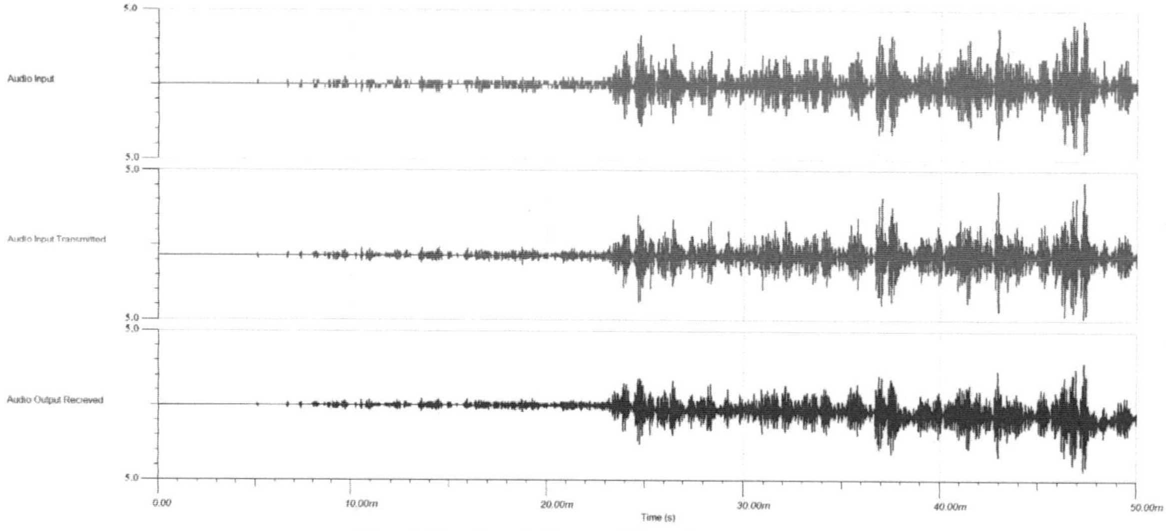

Fig. 4 Simulated Output Waveforms

The above picture is the simulated output of the signal transmitted and received. Using the TINA software, an input signal of 15V peak-to-peak is given and an output signal of same strength is obtained with a frequency of 500 Hz. The signal is pre-recorded audio message that is converted to a .wav file as TINA software doesn't support .mp3 files. The maximum voltage is set to 15V peak-to-peak. Similar output is obtained at the receiver end, thus establishing a successful communication channel between the transmitter and the receiver.

4. APPLICATIONS:

a. Defence and Sensitive Areas: Laser Defense frameworks laid outacross the world can identify foe boats and rockets while simultaneously incapacitating them.
b. Airport Runways: Remote optical correspondence frameworks on air terminal runways can convey messages back to base camp telling officials when andwhat planes have landed.
c. NASA: Nasa has made heaps of various advances which incorporate remote optical correspondence. One late NASA achievement was a fruitful trade of laser beats with the MESSENGER rocket and an Earth- based observatory which set a standard for laser transmission in space.
d. Satellite-to-Satellite Communication: Nasa has alsodeveloped Satellite-to-Satellite communications usinglaser communications through wireless media.

5. ADVANTAGES AND DISADVANTAGES:

Advantages:
1. Transmission of information through longer distances without any further boosters.
2. The circuit is modular, we can increase therange by using an appropriate laser.
3. The Same phase will be there for input and output signals, and it's maintained by PLL IC NE567.
4. Security and no loss of data [9].

Disadvantages:
1. Circuit must be in line-of-sight condition.
2. Distance of transmission can be variable by a few things like glass, fog causes a decrease in light intensity.
3. Interference from background light sources [10]

REFERENCES:

[1] Rohini Deshpande, Gagan Ganapathy A A, Govind Kumar R, Lokesh Guptha, Jayant Verma, "Free-Space Laser Communication", 2021 International Conference on Electronics, Communication, Information Technology, and Management (ICECIM) School of Electronics and Communication Engineering, REVA University, Rukmini Knowledge Park, Yelahanka, Bangalore, ISBN: 978-81-952307-0-9.

[2] Mohammad Dehghani Soltani, Elham Sarbazi, Nikolaos Bamiedakis, "Safety Analysis forLaser-based Optical Wireless Communications: A Tutorial", IEEE Journal,2021

[3] Chao Shen, Chicheng Ma, Dong Li, JunhuiHu, Guoqiang Li, Peng Zou, Junwen Zhang, Ziwei Li, Nan Chi, "High-speed visible laser light communication: devices, systems, andapplications", Event: SPIE OPTO, 2021

[4] CAI Xiang-Peng, "A Cost-Efficient RGB Laser-Based Visible Light Communication System", Navigation college, Quanzhou Normal University, Quanzhou, Fujian, China 26 August 2021

[5] Anshul Arjariya; Lijo P Jose, "Laser-Based Free Space Optical Communication System for Future Space Application", 2021

[6] Audre Arlene A, Jagadeesh Basavaiah, Chandrashekar Mohan Patil, "An Approach to End- to-End Audio Transmission Using Laser Communication", Vidya Vardhaka College Of Engineering, May 2021

[7] Giuseppe Schirripa Spagnolo, Lorenzo Cozzella, and Fabio Leccese,"Underwater Optical Wireless Communications", 2020

[8] Chao Shen, "Laser-based high bit-rate visible light communications and underwater optical wireless network",2020 Photonics North (PN)

[9] Vijayashri Belgaonkar, "Performance Evaluation of Laser-Based FSO Communication System in Turbulent Atmosphere", International Journal of Advanced Trends in Computer Science and Engineering, Sir M. Visvesvaraya Insititute of Technology

[10] Daria Stepanova, Valentin Pryanichnikov, Sergey Khandorin, Andrey Kulchitsky, and Alexey Kuznetsov, "Developing A Highly AccuratePointing System for FreeSpace Optical Communications". Moscow Institute of Physics and Technology, Moscow 141701, Journal of Communications Vol. 14, No. 12, December 2019

Brain Tumor Segmentation and survival prediction using 3D U-net

Kotha Srinidhi [1], Musthyala Ashwini [1], Banala Sowjanya Sree [1], Pinumalla Apoorva [1], Vadla Akshitha [1], K.N.V. Khasim [2]

[1] Department of ECE, GRIET, Hyderabad, Telangana, India.
[2] Assistant Professor, Department of ECE, GRIET, Hyderabad, Telangana, India.

kotha.srinidhi123@gmail.com

Abstract. Detecting Brain Tumor is a critical problem in the medical field and analyzing its location and size has been a challenging task for doctors. Designing automated brain tumor systems for MRI is one of the important factor in this decade, to avoid high expenses, low accuracy due to parallax errors and time consumption. Automated systems assist in diagnosis and early detection of tumors, which allows physicians to focus more on treating a patient. A method based on Convolutional Neural Network (encoder decoder architecture) captures 3D Flair (MRI) for glioblastomas as an input, which could be anywhere, any shape or size of the brain. These factors made us to investigate for an artificial intelligence system that utilizes the flexibility of a highly efficient neural network with an astonishing efficiency. The U-Net model is essential for achieving optimal performance in segmenting brain tumor and patient stage. The current approach uses 3d image instead of 2d image for proper diagnosis.

Keywords: Convolutional Neural Network, U-Net architecture, 3D volumes, Brain Tumor (Gliomas), Segmentation.

1. Introduction

Brain is the most important part the human as it manages all other organs to function properly. So, for us it's very important to care and protect the brain properly. A brain tumor is a medical condition or disease that occurs as a result of abnormal cell growth in the brain. Before any kind of treatment, Diagnosis and detection are must for brain tumor. MRI generates images with a detailing insight, high resolution on various body parts and used to diagnose abnormalities. MRI is a best tool for brain tumor Diagnosis, analysis, monitor and surgery plan. In image processing, segmentation is the process of dividing a digital image into distinct parts to retrieve meaningful information for easy analysis and investigation. Segmentation is used to identify the regions of the brain whose cells differ from normal in order to detect brain tumors in an MRI. Segmentation is the difficult part of detecting tumor, this is because the perfectness of tumor detection is determined by the accuracy with which they are segmented. This paper focuses more on proficiently segmenting the regions of the brain tumor from 3D MRI images by using a semantic segmentation algorithm. The brain scan from all the angles is represented by 3D images to provide a 3D view for detection. Gliomas is a type of tumor, that grows rapidly and requires surgery, radiotherapy and have poor survival condition. The survival period of a gliomas effected patients is 12-18 months and 25% of them survive for more than a year, 5% of them for about 5 years.

Now-a-days, the traditional computer vision segmentation techniques are surpassed by Deep learning-based, as the Convolutional neural networks (CNN) of Deep learning technique can learn from examples, performs segmentation for 2D and 3D MRI accurately. In this Paper, the volumetric 3D brain's tumor segmentation is described from the 3D MRIs using semantic segmentation. Encoder-decoder structure of CNN is followed, which extracts deep image features reconstructs segmentation masks, which is called as asymmetrically large encoder, as a work.

2. Related Work

In recent years, many approaches have been developed to the brain tumor detection, segmentation and analysis. [1] T. kalaiselvi and P. Nagaraju, in the year 2015, that combines the two techniques FCM and SRG for the efficient detection of the brain tumor region. This [13] semi-automatic method extracts the tumor region from T2-weighted MRI brain images. [6] R. B Dubey, M. Hanmandlu, S. K. Gupta (2009) proposed an automated system for brain tumor volume measurements based on MR imaging. [10] Lu Y, Jiang T, Zang Y (2003) demonstrated an algorithm of volume extraction for segmentation of cerebrovascular structure on brain MRA data sets. [4] A. Mustaqeem, A. Javed and T. Fatima (2012) performed segmentation that is done using a technique based on threshold segmentation, watershed segmentation and morphological operators which was experimented by MRI scanned images. [12] Heath LM, Hall LO, Goldgof DB and Murtagh FR (2001) presented the first automatic segmentation method that seperates Non-enhancing brain tumors from Healthy tissues in MRIs to aid in the task of tracking tumor size overtime. In addition to [1], [7] T. kalaiselvi and P. Nagaraja developed a rapid automatic method to detect brain tumor from T2-weighted MRI brain images using the principle of modified minimum error thresholding (MET) method.

With reference to base papers, this paper specifies the optimization of the advantages of using U-net architecture for identifying the region of tumor in the brain and its stage which helps doctors to predict the severity of the condition. It can be obtained by performing semantic segmentation and morphological operations.

3. Dataset

The System is Trained with a Dataset of 12 3D MRI Images acquired from the Kaggle Public dataset under Brain Tumor Segmentation (BraTS20). Imtool is used for visualizing these 3D image slices. Imtool should be downloaded and call this Imtool function in the code, this tool helps in viewing the image by scrolling through the slices, and images will covert to image formats such as .png, .jpg, .tif or nifti(.nii) file. In order to read the 3D format data niftiread command is used which reads the nifti image file which is named by the file name in the current folder and it will output's the volumetric data(V). Figure 3.1 shows the centered slice of the 3D image and its respective ground truth labels of the dataset. Center slice is considered because the tumor visibility will be more. This is the main reason for selection of center slice.

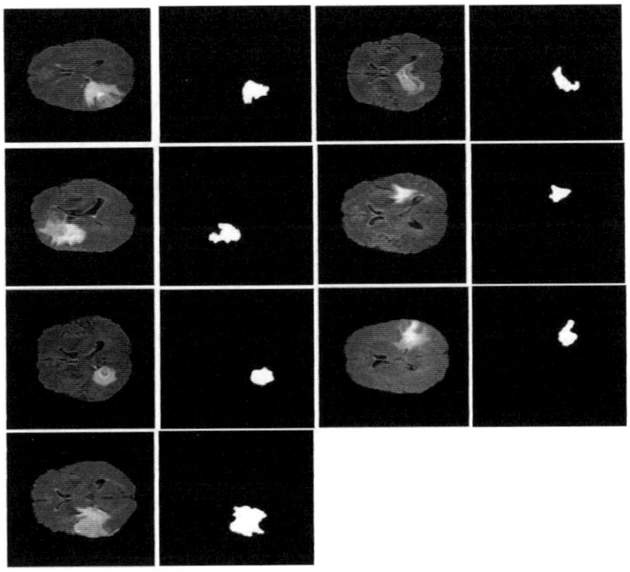

Fig 1 - Slices and its Ground Truth labels

4. Methodology

This paper incorporates U-net Architecture which is used to segment the tumor out of 3D MRI dataset. The system is designed using Matlab 2020a. The whole idea is implemented using Deep learning concepts. The Dataset contains 12 3D volume of MR Images acquired from the Kaggle Public dataset under Brain Tumor Segmentation (BraTS20).

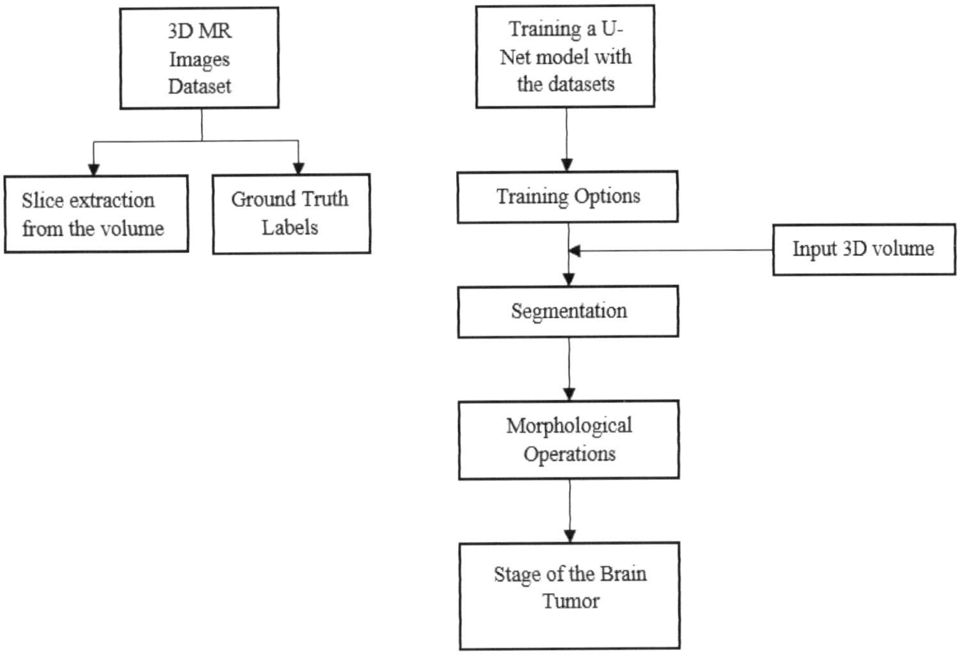

Fig 2 - Block Diagram

Journal of Physics: Conference Series **2325** (2022) 012048 doi:10.1088/1742-6596/2325/1/012048

4.1. Input Image:

A 3D image is read from the file folder at the very first step. The image is in .nii format.

4. 2. Conversion of 3D volumes into 2D slices:

In order to have further processing easier, the 3D images have to split. Hence, as soon as the workspace is loaded with that 3d volume, it is converted into (number of) 2D slices.

4.3 Extraction of Center Slice:

Out of those (number of) slices of the volume, the center slice is selected where the tumor can be clearly visible. Center slice can be obtained by assigning volumetric 3D data dimensions to rows, column, and slices. Now the obtained slice value is divided by 2 and rounded it off. This value is nothing but centre slice.

4.4 Creating Ground Truth Labels:

The middle slice from the tumor image is preprocessed using morphological operations like Erode, dilate, open, close masks. The output image is Ground truth image.

4.5 Ground-truth-image:

It is the proposed model's target image that reflects the actual nature of the problem which is or assumed to be true. The datasets of slices and ground truth labels are then created.

4.6 Training Options:

Using 3D U-net Architecture the machine is trained. It is a supervised learning method used mainly in medical analysis. It consists of encoder and decoder paths for analysis and merging.

- **Encoder:** It contains the left analysis method. In this process, each and every layer have two 3 × 3 × 3 convolutions, each and every layer followed by ReLU and max-pooling of 2 × 2 × 2 large combinations in two steps in each dimension. Shortcut connections from the equal adjustment layer in the analysis method provide a way to integrate with the required features of high resolution in the merging path.

- **Decoder:** It contains right merging method. In the merging process, each layer contains a 2 x 2 × 2 up-convolution with two steps on each side, followed by two 3 × 3 × 3 convolutions, each followed by ReLU. In the final layer of Unet architecture, the convolution 1 × 1 × 1 reduces the number of output channels into the number of 3 labels. And batch normalization before each ReLU.

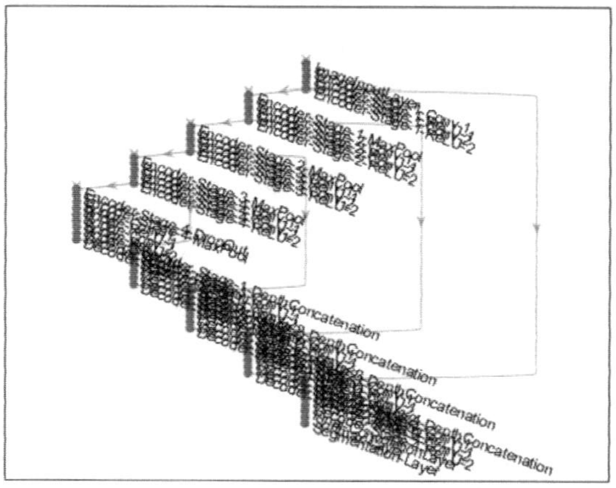

Fig 3 - Architecture of U-net layer graph.

4.7 Semantic segmentation:

The test image is then made to undergo a process called semantic segmentation. It is the method in which each and every pixel of an image is classified from a set of classes which are predefined i.e., a label is assigned to each pixel in the image.

4.8 Morphological operations:

These operations will process the digital images depending on their size. Each pixel corresponds to the value of other pixels in its neighborhood, in morphological operation. You can build a morphological operation by choosing the size and shape of neighboring pixels. These operations enables to create the same size of an output image by applying a structuring element to an input image.

- **Erode:** Shrink foreground mask and expand foreground holes. It is used to shrink the erosion of an image P by a structuring element Q i.e., removing the pixels from the boundaries of object. The below equation describes the erosion of an image.

$$P\ominus Q=\{z|(Q^{\wedge})Z\in P\}$$

- **Dilate:** Expand foreground mask and shrink foreground holes. Dilate function for more information. The tumor region in the sliced image is expanded by using the process of dilation i.e., adding pixels to the edges of the object in an image. In Dilate operation the highest pixel value in the neighboring input image which can be analysed by below equation.

$$P\oplus Q=\{z|[(Q^{\wedge})Z\cap P]\in P\}$$

- **Open Mask:** This removes forefront objects which are smaller than the structuring element.

- **Close Mask:** This process remove holes which are smaller than the structuring element.

By performing these morphological operations, we get the segmented output of tumor. In an image, boundaries and skeleton of objects are identified. A window, namely, Structure element is created by morphological operations to compare corresponding pixel with neighborhood by getting placed in all locations.

4.9 Finding stage of tumor and survival prediction:

For finding the area of tumor, first the output image is converted into gray-scale image and then to binary image. The outline of the tumor is depicted, by identifying the area covered by white pixels (binary) and black pixels (~binary) based on which the stage of tumor of human survival is predicted.

5. Results

This paper implemented a model and trained in Matlab on AMD Ryzen 3 3250U with Radeon Graphics using BraTs training dataset. Training is carried out for 7 three dimensional images and on methodology around twelve 3D images were tested, for which we considered the centered slice which consists of maximum tumor area which results in better accuracy.

The below set of images illustrate the centered slice and segmentation operation at various stages. These slices are taken on random basis.

Journal of Physics: Conference Series **2325** (2022) 012048 doi:10.1088/1742-6596/2325/1/012048

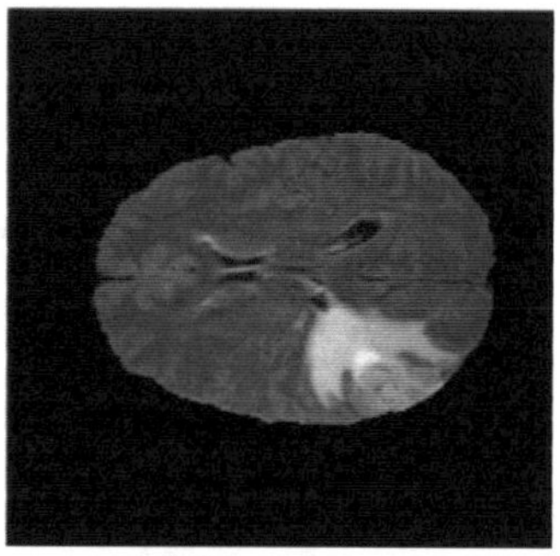
(a) Input image

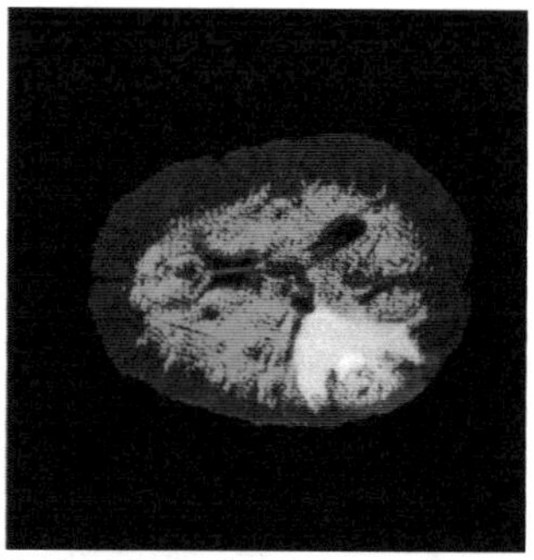
(b) Overlayed segmentation regions

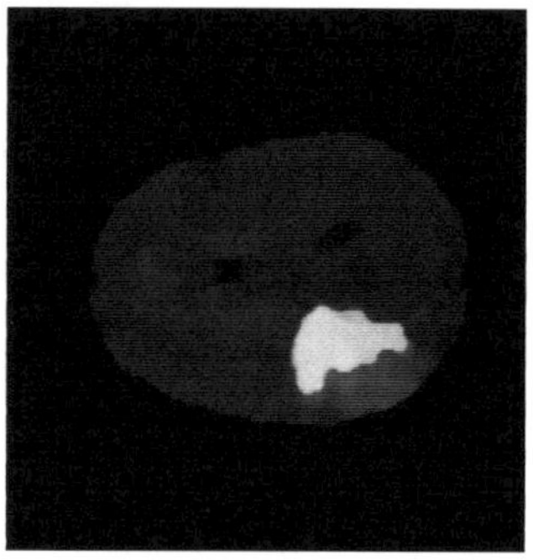
(c) Segmented Tumor

(d) Outline of Detected Tumor

Fig 4 - Slice 1 and its output at various stages.

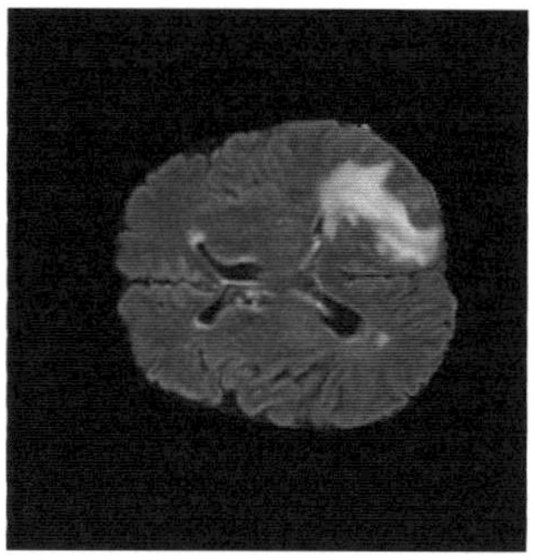
(a) Input Image

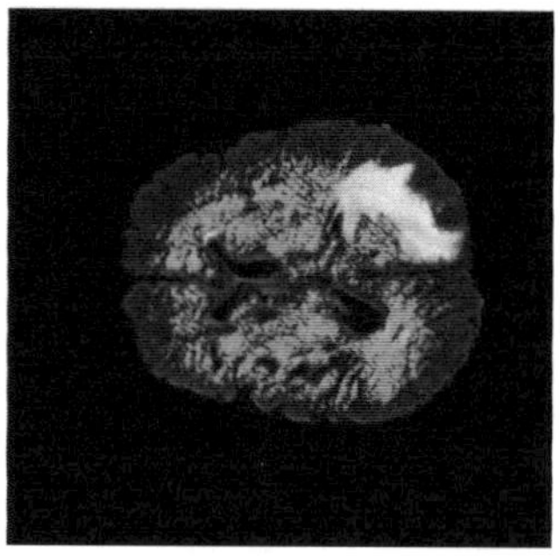
(b) Overlayed Segmentation Regions

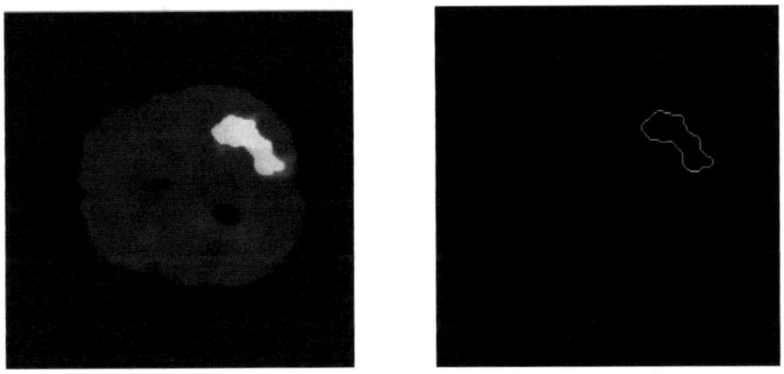

(c) Segmented Tumor (d) Outline of Detected Tumor

Fig 5 - Slice 2 and its output at various stages.

The graph in Fig 6 Shows the accuracy obtained for the tested image and also the time required for the model to train the network.

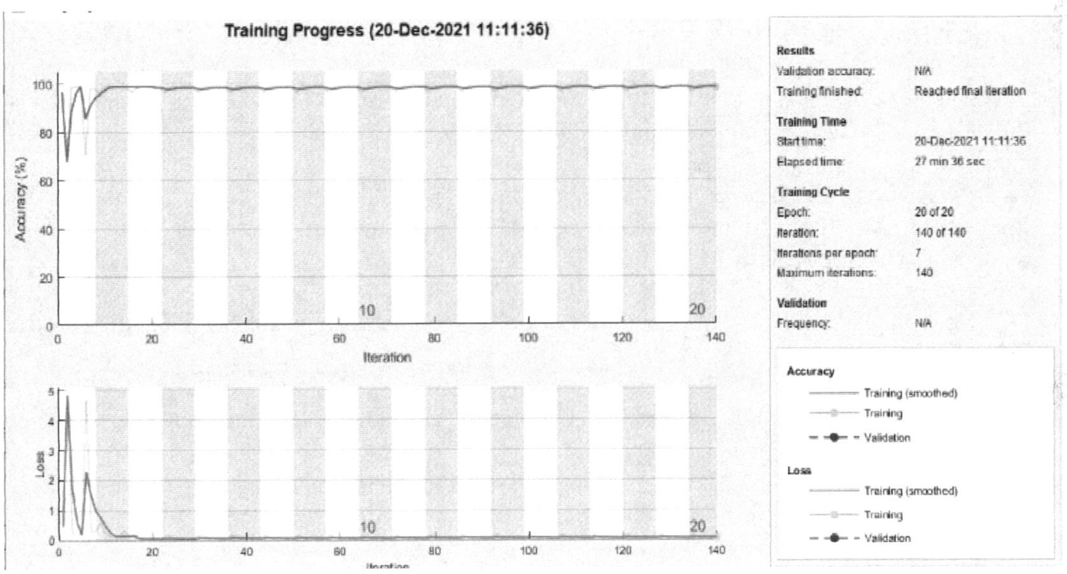

Fig 6 - Training Progress

6. Conclusion

This paper describes a network for semantic segmentation to detect brain tumor from three-dimensional Magnetic Resonance Images. The data is collected from BraTs website. The proposed method helps to detect the Gliomas, the most common Brain Tumor using U-Net modal network. The U-net modal provides better performance with very few training samples. However due to the processer limitations it was difficult to increase the size of the data set as the time required to train the network increase as the data set is increased, due to which 7 three dimensional images are used to train the network. This model works excellent for brain tumor medical images that helps doctor to detect the severity of the tumor.

7. References

[1] Kalaiselvi, T., and Nagaraja, P., “Brain Tumor Segmentation of MRI Brain Images through FCM clustering and Seeded Region Growing Technique”, International Journal of Applied Engineering Research, vol.10, no.76, 2015, pp. 427-432.

[2] Sungheetha, Akey, and S. R. Rajesh. "Comparative Study: Statistical Approach and Deep Learning Method for Automatic Segmentation Methods for Lung CT Image Segmentation." J. Innov. Image Process 2 (2020): 187-193.

[3] Fletcher-Heath LM, Hall LO, Goldgof DB, Murtagh FR, “Automatic segmentation of non-enhancing brain tumors in magnetic resonance images”, Artificial Intelligence in Medicine, Vol.21, pp.43-63, 2001

[4] A. Mustaqeem A. Javed and T. Fatima, “An Efficient Brain Tumor Detection Algorithm Using Watershed & Thresholding Based Segmentation”, International Journal of Image, Graphics and Signal Processing, Vol.10, pp.34-39, 2012.

[5] S. Sivanand, “Adaptive Local Threshold Algorithm and Kernel Fuzzy C-Means Clustering Method for Image Segmentation”, in proc of IJLTET International Journal of Latest Trends in Engineering and Technology, 2013.

[6] R. B. Dubey, M. Hanmandlu, S. K. Gupta, “Region growing for MRI brain tumor volume analysis” Indian Journal of Science and Technology, Vol.2 No. 9, pp.26-31, 2009.

[7] T. Kalaiselvi and P. Nagaraja, “A Rapid Automatic Brain Tumor Detection Method for MRI Images using Modified Minimum Error Thresholding Technique” International Journal of Imaging Systems and Technology, 2014.

[8] J.C. Bezdek, “Pattern Recognition with fuzzy objective function algorithms”, Plenum Press, New York, 1981.

[9] N. Moon, E. Bullitt, K. Van Leemput, and G. Gerig, “Automatic brain and tumor segmentation”, MICCAI, Vol.2489, pp. 372–379, 2002,

[10] Lu Y, Jiang T and Zang Y (2003), “Region growing method for the analysis of functional MRI data”, NeuroImage. 20, 455-465.

[11] Yang Y, Yan X, Zheng C and Lin P (2004), “A novel statistical method for segmentation of brain MRI”, IEEE, 946-949.

[12] Heath LM, Hall LO, Goldgof DB and Murtagh FR (2001), “Automatic segmentation of non-enhancing brain tumors in magnetic resonance images”, Artificial Intelligence in Med. 21, 43-63.

[13] Guha, Shetu Rani, and S. M. Rafizul Haque. "Performance comparison of machine learning based classification of skin diseases from skin lesion images." In International conference on communication, computing and electronics systems, pp. 15-25. Springer, Singapore, 2020

[14] Bu Lee, Soo-Min Song, Jae-Sung Lee and Myoung-Hee Kim (2005), “Tumor segmentation from small animal PET using region growing based on gradient magnitude”, Dept. of Computer Sci. Engg., Ewha Woman’s University.

[15] Wafa M, Zagrouba E. "Tumor extraction from multimodal MRI." Computer Recognition Systems 3. Springer Berlin Heidelberg, 2009. 415-422

Machine Learning approach for Software Effort Estimation using Combination of Principal Component Regression and Neural Network

Priya Varshini A G[1], Anitha Kumari K[2], Bhalavaishnavi V S[3], Kalpana Devi C[4], Pranesh S[5]

[1]Assistant Professor, Department of Information Technology, Dr Mahalingam College of Engineering and Technology, Pollachi, Coimbatore 642 003, India.

[2]Associate Professor, Department of Information Technology, PSG College of Technology, Coimbatore 641 004, India.

[3,4,5]Department of Information Technology, Dr Mahalingam College of Engineering and Technology, Pollachi, Coimbatore 642 003, India.

Corresponding Author: priyavarshini.a.g@gmail.com

Abstract. For software projects that deploy vital tasks, it is difficult to estimate the effort of a project. In order to anticipate a few hours of labour effort (either time or personal) to deploy or maintain the software programme, software measurement points must be used. It is difficult to predict the behavior of an application that is engaged in software development during the initial phases of the effort. The hybrid model technique is used in this paper. It is necessary to apply Supervised Learning techniques in the Machine Learning algorithm. It is further subdivided into different types, such as linear regression, logistic regression, SVM (Support Vector Machine) algorithm, Naive Bayes algorithm, PCR(Principal Component Regression) algorithm, (Neural Network)NNET algorithm, KNN (K-Nearest Neighbour) algorithm, K-means, Random Forest algorithm, Dimensionality reduction algorithms, Gradient boosting algorithm, and Ada Boosting algorithm, to name a few examples. PCR algorithm and the nnet algorithm have been utilised for hybrid method, as shown above. Predictions from JM1/Software have been used to create this data collection. With 10886 unique instances and 18 unique traits, this is a very large amount. Metrics for evaluating this system include Mean Absolute Error (MAE), Mean Relative Error (MRE), Mean Magnitude of Relative Error (MMRE), Percentage of Predictive Accuracy (PRED), and R-squared. Compared to single model approaches based on machine learning algorithms techniques, the proposed hybrid using principal component regression and neural networks produced the best results, as demonstrated by the results.

Keywords: Machine Learning, Software effort estimation, regression, Hybrid approach, PCR and NNET.

1. Introduction

The impact of high accuracy estimates on the success of software development projects as the cost of software development expands due to rising prices and rising wages. The increase in software expands the span of more domains, architectures, and so on. The demand for smart and more innovative software makes the drive cheaper and more faster software solution. To estimate approximate predictions, it is necessary to mold the initial multiple uncertain and unstable state data. [1] We conducted planning, investment analysis and budget analysis. The iterative process for planning for constant input from estimates attracts a lot of attention. The estimate of the identification and size of the efforts, such as personnel skills, manpower, resource saving system or building. Software from the early stages of data uncertainty and unstable state estimation is difficult to tackle task development. The time or effort of a person's software is estimated. Software efforts were made to estimate the following techniques [2]: expert judgement based on similarity law estimation function point analysis, machine learning techniques [15], including Fuzzy-based approach in deep learning model using regression method, classification method, clustering method, neural network, and ensemble method, and expert judgement based on similarity law estimation function point analysis

On the basis of similar historical projects, a hybrid model-based estimation was used to predict the amount of effort needed. Moreover, because it was based on historical data, it delivered accurate results. [3] When it comes to software voltage prediction, different machine learning techniques such as regression, classification and grouping approaches have a considerable impact. [4] Estimating linear regression (both single and multiple regressions) and logistic regression (including flexible network regression, ridge regression, lasso regression, and progressive regression) requires the application of a variety of regression techniques that include

The decision tree method, the random forest approach [5], the SVM classification algorithm, the NNET algorithm, the KNN algorithm, and the Naive Bayes method are all examples of classification methods that are used to estimate the results. The best hybrid model, according to a comparison of multiple methods, is PCR with NNET. If you want the most efficient algorithms for software effort estimation, the Hybrid Model, which incorporates PCR and NNET, is your best option.

With PCR, we can easily reduce the dimensionality of a large dataset and fit a linear regression model to a smaller number of variables while retaining the majority of the original predictors' variability. Because only a few of the principal components are used, the number of variables in the model is reduced, which is always an advantage.

2. Related Work

Consequently, the comparative evaluation of effort prediction methods described by Whigham, Peter A., et al. [6] has become a typical strategy when examining how to forecast effort over a variety of different project types. Using an automatically transformed linear model (ATLM) as a baseline model for comparison with (Software Effort Estimation) SEE approaches, this article proposes that SEE methods be used. With mixed numeric and categorical data, ATLM can be used without the need to adjust any parameters.

Rather than relying on the accuracy of individual models to determine their applicability in anticipated situations, Walkerden, Fiona, and Ross Jeffery [7] asserted that the accuracy may be defined based on knowing how to calibrate software data. Empirical estimate models are discovered to be potentially accurate when compared to alternative estimation strategies among the diverse set of models investigated. FP analysis techniques are common in practice in the empirical category, while the datasets COCOMO (Constructive Cost Model), SLIM, and SEER-SEM are popular in theory. For determining stable ranks of diverse predictors by Keung, Jacky, and colleagues This is done through the use of a new evaluation algorithm called CLUSTER [8], which is focused on one single error measure (MRE) and uses it in conjunction with a newly designed evaluation method known as CLUSTER [8]. In this effort estimation, the MRE data set was employed.

Several single models were combined into an ensemble technique by Priya Varshini, A.G., and others. Averaging, weighted averaging, bagging, boosting, and stacking were among the ensemble estimation strategies explored [11]. Random Forest-Based Stacked Ensembling Approach is the name of the method. Albrecht, China, Desharnais, Kemerer, Kitchenham, Maxwell, and Cocomo81 were the datasets that were examined for estimation. The mean absolute error, the root mean squared error, and the R-squared were the evaluation metrics that were utilized.

The Neural Network toolbox22, which is included in the MATLAB software, is used to develop the proposed upgraded Neural Network Model by Sehra et al. [9]. Simulation and modelling are made easier with Matlab tools. One hidden layer and a variable number of processing elements, or neurons, are used to construct neural network models. COCOMO II, MMRE, and MRE are the datasets that have been taken into account.

3.Proposed Work

Hybrid approach of Principal Component Regression and NNET was used for effort estimation.

3.1 PCR Algorithm

Principal component regression (PCR) is a principal component analysis (PCA) based on the regression method. The basic idea of PCR is to calculate its impact and to understand the main elements that are used as predictors for linear regression models [12]. It uses the typical least squares method. Some of the most significant advantages of PCR are as follows: Dimensionality reduction, Prevention of multicollinearity and Over-fitting mitigation.

- Dimensionality reduction:

Using the PCR method, it is easy to perform dimensional reduction, high-dimensional data as a fitty linear regression model with a small setting of variables at the same time to keep the variation of the original predictors [13]. All the use of the main components of each variable, this model allows you to expect to reduce the complexity of the model, which is a plus. If you need many key elements and explain the variation of the data, about as many key components as the variables in the number of data sets, PCR may be performed. In this scenario, it would also be better to use the usual vanilla linear regression is positive if the key components are sufficient to explain the changes in predictors.

One of the most common linear dimensionality reduction approaches is Principal Component Analysis. This method performs a direct mapping of the data to a lower-dimensional space that maximizes the data's variance.

It's a statistical process that orthogonally turns a dataset's 'n' coordinates into a new set of n coordinates known as the main components. The first principal component with the greatest variation is created as a result of this conversion. Each subsequent principal component has the greatest potential variance if it is orthogonal (not correlated) to the ones before it.

The relative scaling of the original variables has an impact on PCA conversion. As a result, before using the PCA approach, the data column ranges must be standardized first.

- Prevention of multicollinearity

If there is a certain degree of multicollinearity between variables in the data set, then this procedure can be done. The problem to avoid is that the raw data is straight and the combination of predictors is uncorrelated.

- Over-fitting mitigation

The least squares method to fit a model of the main components will produce better results than the least squares method to fit a model of the original data if all of the PCR's basic assumptions are met, because the majority of the changes and related information of the relevant variables is condensed in the main components, and by estimating the smaller coefficient, the risk of overfitting can be reduced significantly.

In evaluation and regression, the least squares method is frequently applied. This method is described as a standard strategy in regression analysis for approximating sets of equations with more equations than unknowns. When we talk about principal component regression, we're talking about a pattern that is created by the composition of an independent variable from a linear combination of independent variables. However, unlike principal component analysis, this method involves assembling components that are meant to best describe a dependent variable. In many cases, this allows you to employ fewer model variables while improving what is typically acceptable for your model. Because the principal component analysis (PCA) method relies on variables whose influence has not been well studied, it is an unsupervised method that is very dependent on them. Unless you are constructing a component, you will not be able to provide a meaningful description of the dependent variables.

3.2 NNET Algorithm

From a variety of perspectives, neural networks have long been one of the most intriguing machine learning models, not only because of the fancy backpropagation algorithm, but also because of their complexity (such as deep learning with many hidden layers) and a structure inspired by the brain. It is possible to forecast or presuppose in the production of neural networks using artificial intelligence (AI). Each node neural network has an input port that accepts a collection of nodes based on the output that determines the activation function for that node neural network's input port. Finally, a regression model to a modified neural network that can be controlled is used as the final activation function. The use of 'Keras" allows you to benefit from the addition of a neural network of relevant python library structures.

The ability of neural networks to compare to humans has always existed [17]. Similarly, to how information travels via a person's nervous system, trabecular information travels through an animal's nervous system. The first layer of the neural network gets the original input, processes it, and then applies the information from the hidden layer to complete the processing process. The information from the previous layer is passed to the hidden layer, which is responsible for producing the outputs of the previous layer. Natural neural networks have the advantage of being well-suited to their surroundings. It learns from the information provided, i.e., the train learns from its own data, which has a known result, and optimizes its weight is a better predictor of the case with unknown results than the train learning from its own data. One of the most basic neural networks is the recognizer, often known as a single-layer neural network. With the addition of the activation function, one of the IDs received three-dimensional inputs and actions using a weighted approach. Using labelled data and a learning method, it is taught how to improve the weights of the chemicals in the processor. Another important determinant of model identification is the inability to deal with non-linear models. It is possible to circumvent this constraint by using a multi-layer neural network, which can aid in the solution of nonlinear issues. A connection is made between the input and hidden layers, which in turn connects to the output layer. In addition, the link weighting is enhanced by the use of rule learning.

Neural networks can be trained using a number of different learning methods, including least mean square (LMS), gradient descent (ND), and Newton's rule (CN). Improved performance of backpropagation error procedures can be achieved by employing learning principles. It is utilized in the calculation of the error at the output device. It is the error propagating backwards from each unit in such a way that the error at each unit is proportionate to how much of the overall error at the output unit was contributed by that unit.

3.3 Algorithmic Approach
We use the combination of two algorithms PCR and nnet.

Step 1: dataset has been obtained from the JM1/Software prediction model. With 10886 unique instances and 18 unique traits, this is a very large amount.

Step 2: Pre-processing and summarization of the dataset values occur in Step 2. Once the data has been compiled, mean, median, quantile and maximum values can be determined.

Step 3: Depending on the algorithms being used, you will need to install the library functions and packages.

Step 4: Using the pre-processed data, normalization is performed in order to convert the values of numeric columns in the dataset to a similar scale without distorting the difference in range of values between columns.

Step 5: To complete Step 5, create a function that accepts as inputs a training and a test set as well as train labels and the number of neighbors you wish to locate using the algorithm. Each row of the test data is represented by a factor vector, and the function guesses which class each row represents.

Step 6: The dataset is fed into a collection of two base algorithms, which are then tested against each other. In terms of evaluation metrics, the algorithms are compared. It is the Mean absolute error (MAE), Mean Squared Error (MSE), Root Mean Squared Error (RMSE), prediction (pred), and R-squared that are important to consider (R-squared).

After evaluating the metrics in step 6, the best algorithm is predicted. R programming is used to execute the code.

4. Result Analysis
Promise repository has been used to obtain these datasets for this project. The data set was collected directly from the JM1/Software prediction model. With 10886 unique instances and 18 unique traits, this is a very large amount.

In regression analysis, the MSE, MAE, RMSE, and R-Squared metrics are primarily used to assess the accuracy of predictions and the performance of models [10]. When averaging the absolute difference over a data collection, MAE (Mean absolute error) is used to assess the average error between the original values and the anticipated values. The mean of the absolute mistakes is called the standard deviation. When the expected deviation differs from the actual deviation, the RMSE (Root Mean Squared Error) measure [14] is used.

If the values are compared to the original values, the R-squared coefficient indicates how well they fit. From 0 to 1, the value can be regarded as a proportional change in the number. When the model's value is higher, the model's quality is higher as well. Several machine learning algorithms are utilized in the hybrid model approaches [16], each of which is applied on top of the baseline classifiers. It was projected that all hybrid model approaches would have similar mean absolute error, mean relative standard error, and R-squared value. It was decided that the hybrid model was superior if it produced less errors (MAE and RMSE) and if the R-squared value was closer to one.

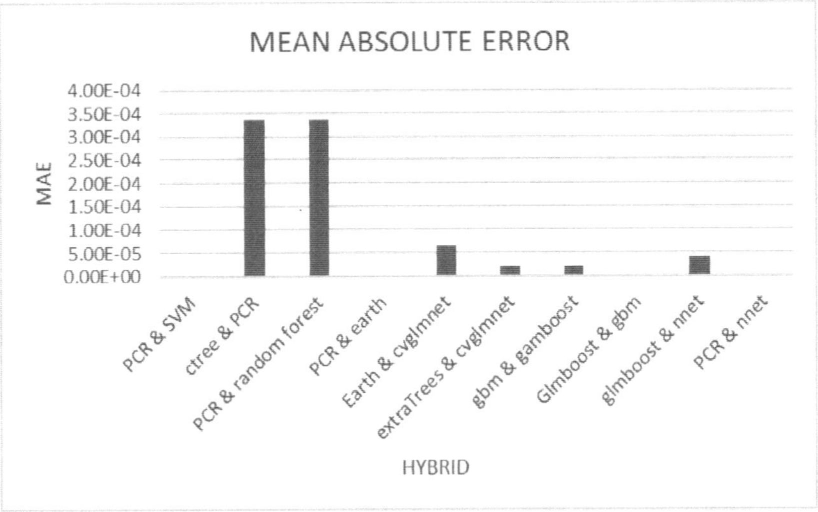

Figure 1. Error rate vs. hybrid models for MAE.

Figure 1 shows that hybrid using PCR and NNET produces less error, MAE (1.79E-09) compared with other algorithms.

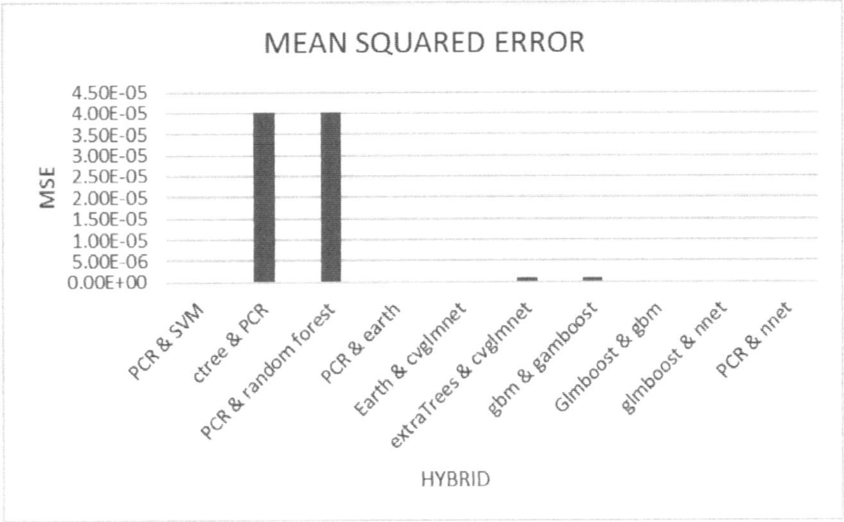

Figure 2. Error rate vs. hybrid models MSE

Figure 2 shows that the hybrid using PCR and NNET produced less error, MSE (5.49E-18) compared with other algorithms.

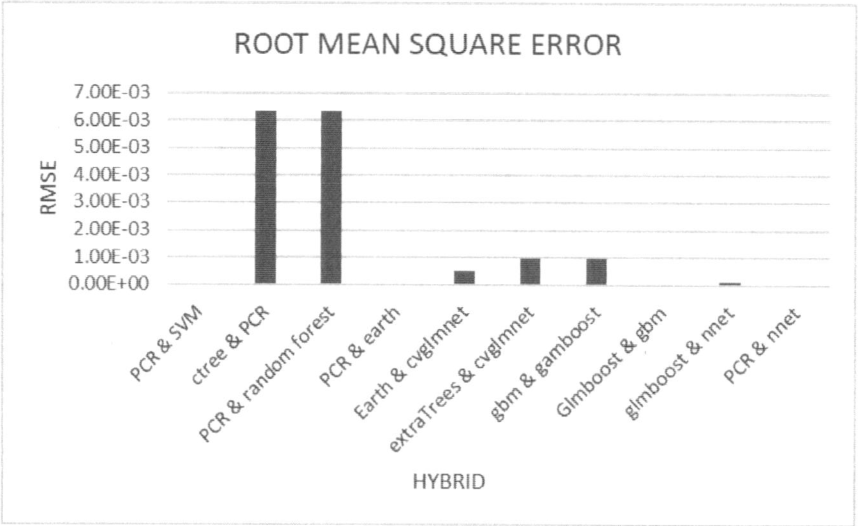

Figure 3. Error rate vs. hybrid models for RMSE.

Figure 3 shows that hybrid using PCR and NNET produces less error, RMSE (2.34E-09) compared with other algorithms.

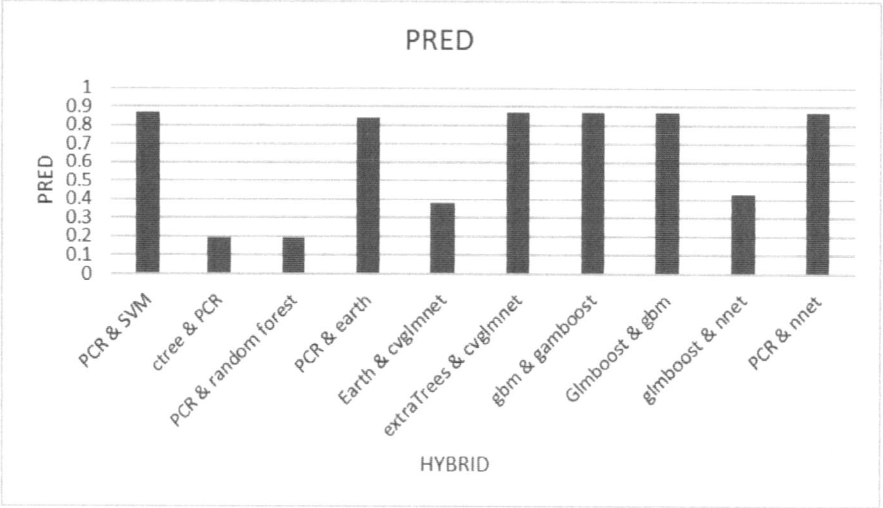

Figure 4. Prediction vs. hybrid models for PRED.

Figure 4 shows that hybrid using PCR and NNET produces better predictive model whose value should be closer to 1; PRED (0.8740469) compared to other hybrid algorithms.

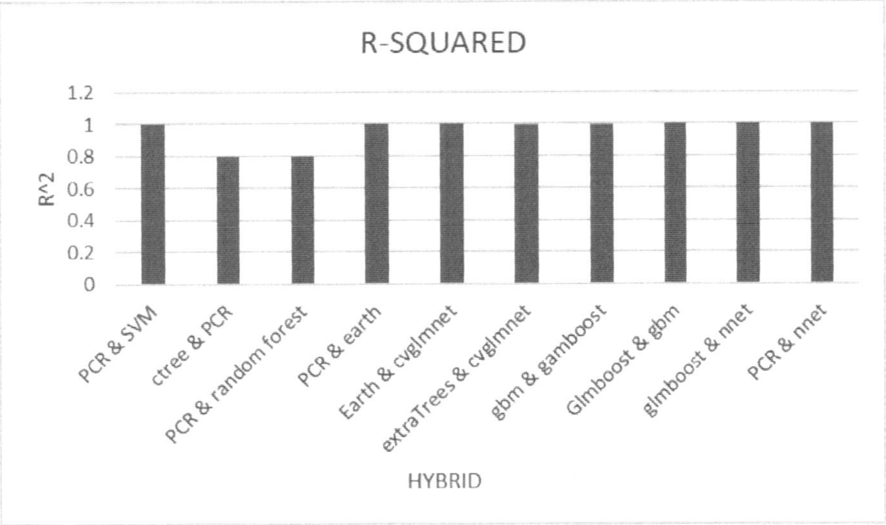

Figure 5. R-Squared vs. hybrid models for R-squared.

Figure 5 shows that hybrid using PCR and NNET produces better predictive model the R-squared value should be closer to 1; R-squared value (1) compared to other hybrid algorithms. It works by calculating how much variance in the dataset's predictions is explained. Simply expressed, it refers to the disparity between the dataset's samples and the model's predictions.

From Figure 1-5, it can be concluded that hybrid models based on principal component regression and nnet create less errors (MAE and RMSE) than other hybrid models; R-squared values were also closer to one in this case.

5. Conclusion

Calculating software effort estimations is a difficult undertaking to accomplish. The cost of the project is estimated by comparing it to the costs of previous initiatives. In this aspect, PCR and nnet are more efficient than all other algorithms when compared to one another. Five real datasets are employed in this work, with the performance measures MAE, MSE, RMSE, PRED, and R-SQUARED being the most notable. On the basis of the metrics used to compute the values for this model, it can be inferred that the suggested model is more accurate than the already available development estimation model. This estimation is utilized as an input for the pricing process, project planning, iteration planning, budgeting, and investment analysis, among other processes. In the future, a hybrid model for software work estimation will be created in order to make more accurate predictions.

References

[1] Phannachitta, Passakorn. "On an optimal analogy-based software effort estimation." *Information and Software Technology* 125 (2020): 106330.
[2] Jordan, Michael I., and Tom M. Mitchell. "Machine learning: Trends, perspectives, and prospects." *Science* 349.6245 (2015): 255-260.
[3] Carleo, Giuseppe, et al. "Machine learning and the physical sciences." *Reviews of Modern*

Physics 91.4 (2019): 045002.

[4] Molokken, Kjetil, and Magne Jorgensen. "A review of software surveys on software effort estimation." *2003 International Symposium on Empirical Software Engineering, 2003. ISESE 2003. Proceedings.* IEEE, 2003.

[5] Dejaeger, Karel, et al. "Data mining techniques for software effort estimation: a comparative study." *IEEE transactions on software engineering* 38.2 (2011): 375-397.

[6] Whigham, Peter A., Caitlin A. Owen, and Stephen G. Macdonell. "A baseline model for software effort estimation." *ACM Transactions on Software Engineering and Methodology (TOSEM)* 24.3 (2015): 1-11.

[7] Walkerden, Fiona, and Ross Jeffery. "An empirical study of analogy-based software effort estimation." *Empirical software engineering* 4.2 (1999): 135-158.

[8] Keung, Jacky, Ekrem Kocaguneli, and Tim Menzies. "Finding conclusion stability for selecting the best effort predictor in software effort estimation." *Automated Software Engineering* 20.4 (2013): 543-567.

[9] Sehra, Sumeet Kaur, et al. "Research patterns and trends in software effort estimation." *Information and Software Technology* 91 (2017): 1-21.

[10] Rijwani, Poonam, and Sonal Jain. "Enhanced software effort estimation using multi layered feed forward artificial neural network technique." *Procedia Computer Science* 89 (2016): 307-312.

[11] Priya Varshini, A.G.; Anitha Kumari, K. Predictive analytics approaches for software Effort Estimation: A review. *Indian J. Sci. Technol.* 2020, *13*, 2094–2103. [CrossRef]

[12] Kocaguneli, Ekrem, and Tim Menzies. "Software effort models should be assessed via leave-one-out validation." *Journal of Systems and Software* 86.7 (2013): 1879-1890.

[13] Tellinghuisen, Joel, and Andrej-Nikolai Spiess. "Comparing real-time quantitative polymerase chain reaction analysis methods for precision, linearity, and accuracy of estimating amplification efficiency." *Analytical Biochemistry* 449 (2014): 76-82.

[14] Shepperd, Martin, and Chris Schofield. "Estimating software project effort using analogies." *IEEE Transactions on software engineering* 23.11 (1997): 736-743.

[15] Shakya, Prerana, and Subarna Shakya. ""Critical Success Factor of Agile Methodology in Software Industry of Nepal."" Journal of Information Technology 2, no. 03 (2020): 135-143.

[16] Shakya, Subarna, and S. Smys. ""Reliable Automated Software Testing Through Hybrid Optimization Algorithm."" Journal of Ubiquitous Computing and Communication Technologies (UCCT) 2, no. 03 (2020): 126-135.

[17] Burgess, Colin J., and Martin Lefley. "Can genetic programming improve software effort estimation? A comparative evaluation." *Information and software technology* 43.14 (2001): 863-873.

Copy Move Forgery Detection: The Current Implications and Contemporary Practices

Pranshav Gajjar[1*], **Aayush Saxena**[2], **Het Shah**[1], **Nandish Kikani**[1], **Karan Lakhani**[1], **Pooja Shah**[1], **Ankit Sharma**[1], **Krishn Limbachiya**[1]

1. Institute of Technology, Nirma University
2. SRM Institute of Science and Technology

E-mail: 19bce060@nirmauni.ac.in (*)

Abstract. Copy move forgeries are a type of digital retouching that takes place in the field of image modification. This type of retouching occurs when one element of a picture is replicated over to a different segment of the photograph. This is typically done in an effort to mask or cover up undesirable qualities of the photograph thus copy move forgeries are a form of image modification. When it comes to digital picture modalities, copy move forging is by far the most prevalent form of picture manipulation that may take place. This is mainly observed when certain elements or segments of the recorded file are duplicated in the recorded file. The identification and localization of forgeries is a prominent problem that has piqued the attention of scholars working in the field of digital forensics and continues to do so. The identification of photo forgery has previously been the subject of several methodologies and uncountable publications, all of which were created and published. An abridged description of the present methodology and approaches is provided in this document, along with a survey and condensed summaries of current research and successes, among other things.

Keywords: Copy Move Forgery Detection, Image Processing, Cyberforensics, Image Manipulation, Forgery Detection.

1. Introduction

Digital picture forgery is the creation of fabricated images by modifying the original image information. Copy move counterfeiting is among a number of image manipulation methods that may be used to fabricate a photograph. Forgery committed using the copy move technique involves cutting off a section of a picture, copying and pasting it into a different location or locations inside the same shot (by whether or not altering the copied element of the image) [1]. The primary purpose of this form of forgery is to hide certain components or areas of the picture by replicating them [2]. As a by-product of the technological advancements and the reach of the internet, humans may now instantly get intriguing multimedia from the internet and recreate or tamper with it as they see appropriate. It is a type of forgery in which portions of a photograph are replicated and then inserted back into the original [3]. As a result, image forensics and copy-move forgery detection have become increasingly important in our networked world, and sufficiently developed algorithms and techniques can be used to identify image fabrication, forged imagery [4] and also for other related practices [5]. To address the issue of fabricated images or digital information, various research has been conducted on how to detect corrupted media or information, both manually and automatically. Splicing and copy-move fraud are the two

Figure 1. This figure (a) follows the original, unedited version of the picture by the one (b) which is the altered image [8].

most common forms of digital picture forgeries nowadays [6]. Splicing is a tampering method that involves the method of swapping a certain number of elements from the test-target digital picture with picture blocks or pieces from a different origin. To detect a spliced image a common procedure is followed to check for differences in the test-target image's traits, values, or attributes [7], the underlying methodologies have substantial implication in the generalized domain of forgery detection. Copy-move forgery (CMF), in contrast to splicing, copies pieces from the test-target picture, then inserts them into someplace in the same image to induce misleading or misunderstanding [7], having suitable algorithms for both these related paradigms is important, however, the scope of this paper pertains to copy move forgery detection. A lot of considerations should be addressed while developing certain copy move forgery detection technique. To begin, the detection approach can be deemed necessary to deliver detection findings that are both accurate and reliable. In practice, however, the suggested approach should be efficient when it comes to speed and computational complexity [7]. As a direct consequence of this, finding a solution to the conundrum posed by the trade-off between efficiency and agility is presently challenging. In addition to this, a successful CMFD strategy ought to be invulnerable to a variety of kinds of assault and strategies for modification, including the addition of rotation, resizing, compression, and noise. [3]. In the figure 1, a demonstration of copy move forgery that might help illustrate the concept is mentioned.

After image manipulation, the primary goal of image (CMFD) is to locate particular regions of the picture that would be similar to the other regions of the image in the same way that the copied fraction is reproduced from the same image. This is accomplished by comparing the original and duplicate portions of the image. Certain operations, such as scaling, filtering, noise embedding, and so on, are often performed on the whole forged picture or a duplicated segment of the image before pasting it on the image in order to make the forgery undetectable. This is done so that the forgery cannot be detected. Detecting forgeries in pictures is essential in light of the fact that visuals play such a significant role in the communication process in a variety of contexts. In situations involving law enforcement, images need to be credible, and their validity has to be verified. In situations like this, the use of an image forgery detection system is essential [9]. There are two types of image forensics technology: passive detection and active detection. The active detection method determines image authenticity by using prior information obtained from the image, such as watermarking. Passive detection methods, as opposed to techniques that include active detection, would not require previous familiarity with a picture. Passive detection systems can employ the detective method to find tampering zones [10]. Thus the resulting great majority of image forgery detection algorithms employ a passive-based method to detect the tampering discussed in this study. There are two forms of passive detection technology: block based methods and key point based strategies [8]. The novel study as presented in this paper is divided into multiple sensations which further elaborates on the conventional approaches, the current standard practices and cutting edge methodologies, and finally the conclusion and the possible future work and related implications.

2. Conventional Methodologies

This section pertains to the conventional methodologies and techniques as presented in the available literature. Lin et al. [11] provided an efficient solution for a CMFD system which relied on radix sort, however the model that has been presented significantly reduces the amount of time complexity, while the adopted features increase the capacity to resist different assaults. The method discovers duplicated areas when rotated through specifically defined angles, but not when rotated through random angles, and only a few tiny copied regions are discovered. Classifiers used included Log-polar Coordinates, SIFT, PCA, SDA, Descriptors, Connected Component Analysis, and Radix Sort. Shivakumar et al. [12] examine the detection of copy move forging as a field and talk about certain challenges that it is now facing, such as how difficult it is to identify forgery in tampered images with compression, noise, and rotation, and how difficult it is to build a model for robust forgery detection.

The execution and empirical performance of a plethora of widely-used and accepted features for CMFD is comprehensively examined by Christlein et al. [13], which analyze various characteristics inside a common processing pipeline. Their method consisted of two steps. They conducted a more extensive per-pixel study in the first stage, when just a tampered/original classification was done, and a more comprehensive inquiry on a per-image basis in the second stage. They use Threshold Determination, Detection at the image-centric level, and Detection at the pixel-centric level as classifiers. Gupta et al. [14] have stated that employing the Discrete Cosine Transform rather than Principal Component Analysis is a better method for identifying copy-move forgeries in textured images. They divided the picture into blocks and then examined the image for duplicated blocks. Cozzolino et al. [15] have proposed a novel model for copy-move forgeries based on patch match, a randomized iterative nearest neighbor search approach. This approach effectively exploits the regularity of natural pictures to converge extremely rapidly to a near-optimal and smooth field, and the algorithm has been improved to be more efficient vs spinning images while preserving computing efficiency.

Uliyan et al. [16] a Hessian-based method was proposed using CSLBP, also known as a centre symmetric binarization pattern, was the one that was suggested. The explained architecture is comprised of four exhaustive steps: first, object identification through the utilisation of normalised cut segmentation; second, the localization of the regional interest points of each item through the utilisation of the Hessian method; third, the extraction of the features corresponding to CSLBP; and fourth, the accurate detection of repeated or duplicated areas within the paradigm of picture forgeries. The results of the testing indicate that the method can withstand post processed copy move fraud even when it is scaled and JPEG compressed. D'Amiano et al.

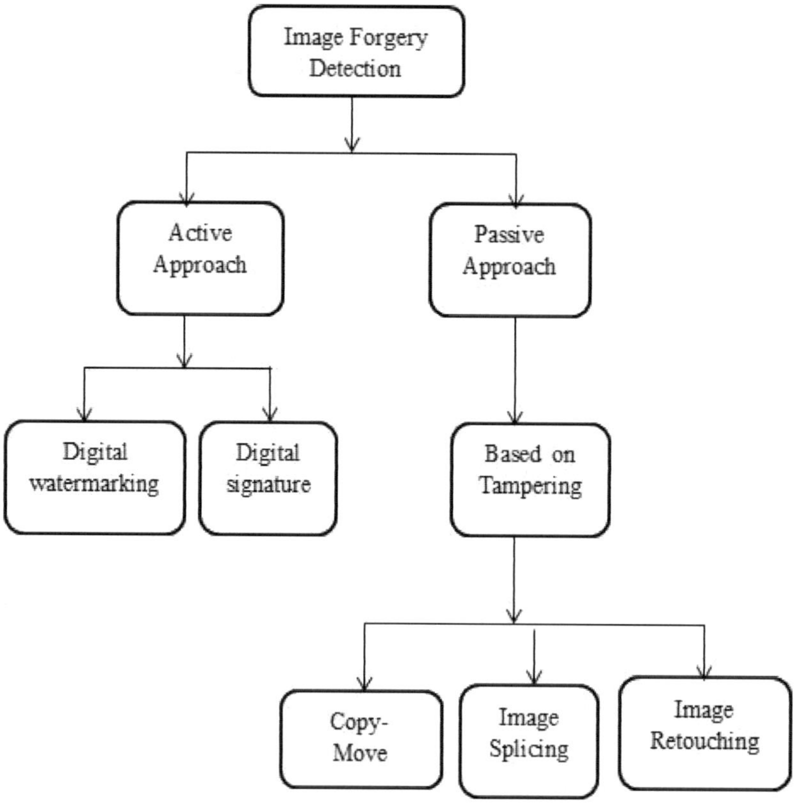

Figure 2. The generalized classification of various forgery detection techniques [8].

[17] have proposed an algorithm based on a modified version of patch match extended for video copy move forgery detection and accurate localization. The explained architecture is robust to noise and has invariant rotation characteristics; it also has a quicker approach for closest neighbor search owing to dense field matching throughout the video; and lastly, it has a fast post-processing for detection.

Li et al. [18] has suggested a new detection technique for copy move forgery designed to extract interest points for comparison. One of the significant distinctions between the previous technique and this model is that the model here first segments the image into independent semantic patches before extracting interest points, which improves its performance over its predecessors. The generalized classification and description for the totality of forgery detection approaches is mentioned below in figure 2.

Bayram et al. [19] have studied various block matching algorithms for copy-move forgery detection, as well as many different block-based features proposed in this field, on the basis of time complexity, robustness to distinct types of image modifications such as compression, rotation, scaling up or down, and so on. Sreelakshmy et al. [20] have built a hybrid approach that uses both keypoint-based and block-based forgery detection algorithms. The image is segmented into various superpixels using adaptive overlapped segmentation, after that, important points are retrieved from either of these sections using the SURF technique, and characteristics are correlated with one another. If the benchmark is surpassed, the picture is considered of being forged, and fused areas are produced. Finally, morphological operations are performed on the image to indicate detected forgery. Al-Hammadi et al. [21] have suggested an enhanced SURF-based copy-move forgery detection system for detecting small-scale forgeries. They applied the single image super-resolution (SISR) technique to improve keypoint recognition in processed images, allowing them to successfully identify forgery of small size.

Dixit et al. [22] have proposed a method that uses statistical feature extraction to locate duplicate areas in the image. They split the image into blocks of pixels based on mean and variance. Mean tells the contribution of each block while variance tells how much the block differs from the neighboring block. The following approach has been compared and evaluated against other methods. Tarman et al. [23] have used a Modified SIFT (MSIFT) algorithm which is an enhanced version of SIFT, for copy-move forgery detection. The proposed algorithm uses clusters and their mean values to locate forgery and hence it is significantly better than the standard SIFT procedure and is proven as a more efficient version than the previous methods. These addons succesfully present a way to improve upon the standard SIFT algorithm. Zhang et al. [24] explore the field of copy-move forgery focusing on the two traditional techniques of copy-move forgery i.e keypoint and block-based, along with the two frameworks of copy-move forgery detection. Lastly, the conclusions and research directions given could prove useful for the researchers in this field.

Kaur et al. [25] have proposed a keypoint-based approach to calculate the performance analysis of different features. The method that was utilised in this study includes pre - processing, the recognition of feature points, the extraction of features, and lastly the pattern recognition of a number of different feature vectors. Khan et al. [26] have proposed a hybrid technique that combines the block-based and non block-based methods of copy-move forgery detection along with this, the performance of various image features like SIFT, MSER, SURF, FAST, Harris, and MinEigen are evaluated on the basis of precision, accuracy, recall, execution time and F-1 score. Chen et al. [27] have evaluated the performance of Fractional Zernike Moments (FrZMs) by using colored robust image copy-move forgery detection. The model used the Fractional Quaternion Zernike moments (FrQZMs) as a feature and for matching it used a modified version of patch match algorithm. Kaur et al. [28] have suggested a passive method for detecting copy-move forgery. It uses Kernel Principal Component Analysis (KPCA) for feature extraction and then sorts it in alphabetical order, then similar blocks are identified to construct the replicated blocks. This methodology has good accuracy and precision as well as time taken to detect forgery is reduced.

3. Contemporary Advancements

This section functions on the abridged illuminations and descriptions of the extremely recent articles and studies. Aspects of the research and technological advancements discussed in this article relate to the simultaneous identification of multiple copy-move forgeries, the theory of single counterparts, and their detection using a combination of techniques based on block and keypoint data [29]. This is achieved through the combination of block-centric detection techniques, an example of which is AS or the adaptive over-segmentation mechanism, as well as keypoint-based detection methods, an example of which is AKAZE or the accelerated Kaze paradigm or SIFT, which can be elaborate as the scale-invariant feature transform technique, which leverages the suggested methodology as additionally robust against different geometrical attacks while also being computationally reasonable. Furthermore, the input picture is transformed into the distinctive color channel entities, and the 'Cr' channel is then utilized for further computing as it notices tampering based artifacts that are not perceptible to the naked eye in the original image [29].

Furthermore, because a suitable number of key point entities are extracted, Even in smooth areas, the copy move forgery may be identified with greater precision, resulting in more accurate detection of the forgery. On the COVERAGE, MICC-F220, GRIP, and the IMD datasets, the experimental results are carried out, and various performance based measurement criteria and metrics, such as the precision metric and the recall metric, F1 and F2 metrics, are measured and evaluated [29]. The observed derivatives demonstrate that the proposed technique in the aforementioned paper is resistant to a variety of geometrical attacks, including rotation, scaling, JPEG compression, and noise addition, and that it performs significantly better than the other existing techniques when compared to them. In addition, the article described a statistical estimation test that utilized ANOVA and interpretation related to the cross-dataset paradigm and demonstrated sufficient utility to be promoted as a deployable test measure in the future.

Numerous copy-move forgeries have been found lately that make use of the homogeneous regions of an image in conjunction with the extensive-scaling attacks to accentuate or hide the target elements. These forgeries are easy to construct but challenging to detect. On the other hand, conventional CMFD or Copy-Move Forgery Detection approaches are incapable of detecting these sorts of forgeries because they are still unable to extract a significant number of quality key points from the homogeneous region(s), resulting in inferior detection accuracy and efficiency [30]. As a result of our research, we suggest the following new CMFD scheme: It is intended to increase the capacity of extracting effective key points from the homogenous area by using a more advanced SIFT structure with inherent scaling invariance. 2). When huge keypoints extraction in the homogeneous area is improved, the computational strain on feature matching is significantly increased, a common issue in CMFD. A novel FLM or Feature Label Matching approach is presented as a means of breaking down the enormous key points into separate little labeling groups, each of which comprises just a small number of key points, resulting in greatly increased matching efficacy and efficiency [30].

In terms of performance, identifying genuine key points during matching is a vital topic to consider. It is suggested in this paper that the HSF or Hierarchical Segmentation Filtering method be used to filter out potentially dangerous anomalies based on statistics from coarse to fine hierarchical segmentation, as the HSF methodology can be perceived as non-trivially superlative in comparison with the other related methodologies. The fusing of coarse-to-fine hierarchical segmentation maps, as the last step, exactly fills the forgery areas. After a number of tests, we found that the suggested approach provides outstanding detection performance for a variety of assaults, particularly when detecting homogenous region(s) under a massive scale attack (s) [30]. On the IMD dataset (which can be considered a comprehensive collection of data), and the CMH datasets, extensive experimental findings reveal that the suggested approach obtains the highest F1 scores while incurring the least amount of computing cost as perceived in the modality of forgery sample entities under various geometric attacks. According to our results, the suggested strategy improves F1 scores by at least 20% and 25%, respectively, when validated against the current state-of-the-art methods and when scaling factors of 50% and 200 percent are used to large-scale sub-datasets of IMD [30]. The inherent working is graphically elaborated below in figure 3.

Compared to standard approaches the paper [31] implemented the related and corresponding techniques on a stream of images for video CMFD. A critical and very tough problem, video copy-move forgery detection (VCMFD) is made more complex by a number of challenges, including a large quantity of video information, a range of forgery kinds, and a rich set of forgery objects, and homogeneous forgery sources. These obstacles pose four major issues that have not yet been handled in VCMFD: Forgery videos are frequently detected with the ineffectiveness in several popular forgery cases, as well as with inefficient matching in computing a multitude of video pixels with multi-dimensional elements under dozens of corresponding iterations, as well as with high false positive (FP) rates [31].

Filling forgery regions is also inefficient, with some forgeries failing to be detected at the pixel level. It is suggested in this study that a new VCMFD approach be used to overcome the following issues: A unique quick keypoint-label matching or the FKLM technique is devised that creates many keypoint-label groupings for each multidimensional feature. Instead of leveraging

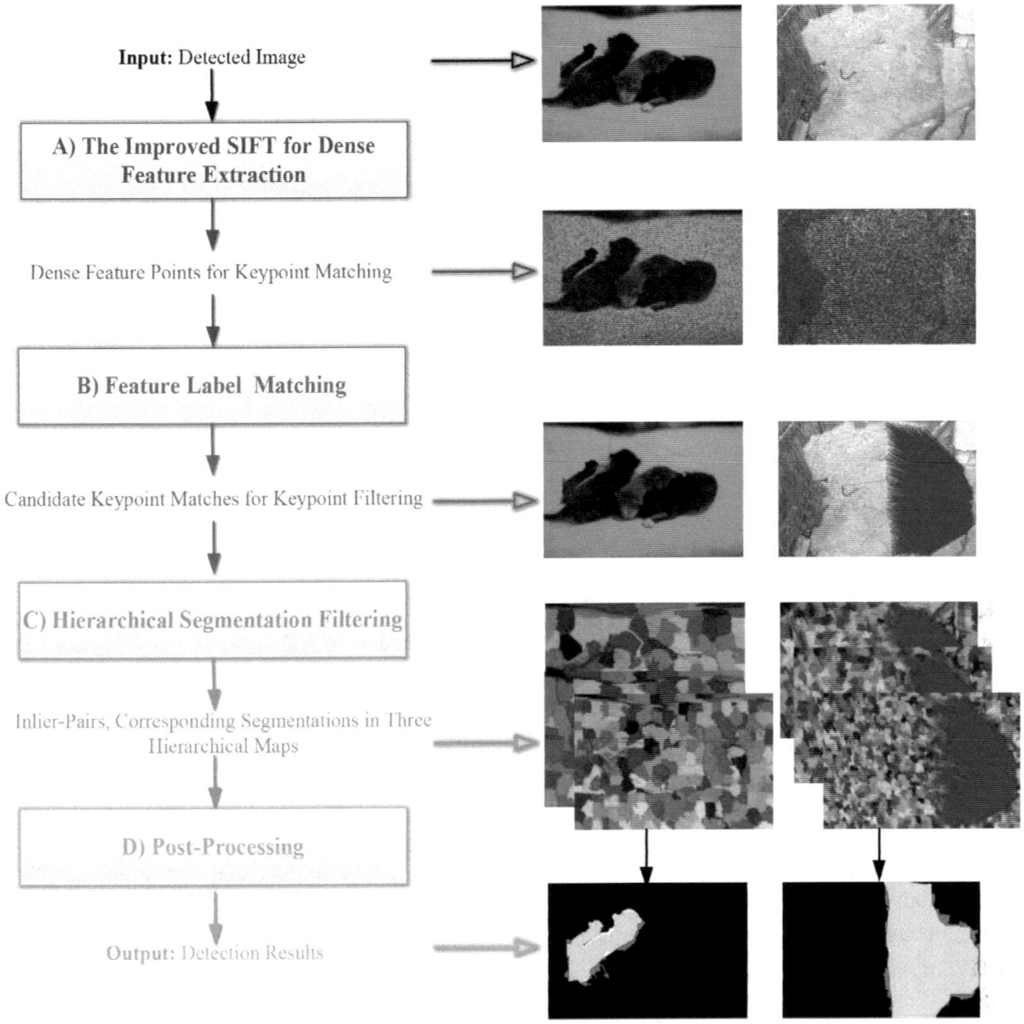

Figure 3. The underlying working of the proposed techniques and methodologies in the aforementioned description [30]. The illustration also showcases sample forged images and the corresponding final detection results.

the standard near-neighbouring test (2NN) for obtaining the correlations for matching keypoints, this paper utilizes the standard deviation metric; ii) an enhanced SIFT structure is provided that is capable of extracting all features in all instances of video copy-move fraud; ii) A one-of-a-kind coarse to fine filtration approach that is aimed to further cut down on the total amount of counterfeit keypoint matches is being developed. This technique depends on the inherent qualities of precise keypoint matches.; iii) responsive block stuffing based on absolute keypoint-matches contributes to the valid and a computationally efficient filling of suspicious regions, even at high frame rates; iv) a novel coarse-to-fine filtering approach based on the inherent properties of precise keypoint matches is proposed to more efficiently calculate the needed processes [31]. After that, the suspicious area locations that are associated with the forgery vision permanence concept suggest the presence of forgery videos [31].

The experiments show that, when validated against the current state-of-the-art methods or the aforementioned conventional approaches, the proposed method in the original paper [31] achieves the most elevated detection accuracy, the lowest false positive rate (FP), and improves at least '16 percent' and '8 percent' of F1 scores on the GRIP 2.0 dataset and a mixture of 'SULFA 2.0' and 'REWIND' datasets, respectively. The suggested technique, in addition, has a low computational time (4.45 s/Mpixels), which is approximately half to three-quarters the time required by the most recent 'DFMI-BM (8.02 s/Mpixels) and PM-2D (13.1 s/Mpixels) approaches' [31]).

The article's [32] primary goal was to offer a powerful digital picture copy-move forgery detection algorithm that can be implemented in real-time, as suggested. Initially, the textural form of the input picture is extracted using the suggested approach. Because the SIFT keypoints and descriptors were derived from textual pictures, more resilient keypoints and descriptors will be used in the analysis [32]. Keypoint matching is done in order to identify whether or not the picture is fabricated, and suspicious areas are identified and the localization of fabricated pixels is accomplished by the use of Ciratefi-based techniques [32].

Through the use of Connected Component Labeling and morphological operations, the post-processing stage is used to improve the accuracy of the labeling pixels in the image. The 'GRIP' and 'CMH' collections of data are exploited to illustrate the efficacy of the current state-of-the-art and suggested methods, respectively [32]. The approach is impervious to geometric distortion and picture degradation assaults, as well as other types of attacks. The findings show that the suggested technique outperforms all other methods in terms of performance, particularly when it comes to geometric distortion assaults such as image transformations, scaling and rotations [32].

The domain of copy move forgery detection, or the generation of forged or tampered images is also common on the field of medical imaging [33]. The use of forging lead to extremely malicious practices which is usually performed for false insurance claims, to falsify the inherent medical condition, to defame public figures or famous entities, an example of which is increasing the present lesions by creating multiple copies. The paper [33] builds upon the existing algorithms for keypoint based forgery detection, and enhances the method by leveraging the minimum Eigen-value algorithm for distributing keypoints and obtaining reduced descriptors [33]. The method further uses the golden ball-based optimization technique which results in an overall rapid and robust forgery detection algorithm.

4. Conclusion

In law, media, and health, as well as in the business, digital picture authentication is becoming more necessary. The most prevalent counterfeit technique for digital photos is copy move forgery. Real-time solutions are becoming more critical for authenticating digital photographs in critical applications. The novel study as presented in this paper functions on offering a better-condensed description of the used methodologies and algorithms pertaining to copy move forgery detection. By understanding the current standards and the utility driven approaches as seen in real-time and other implementations are carefully assessed and a generalized and understandable description for the same is presented. The paper also analyzes the cutting edge and the extremely recent articles for a thorough review and perceiving process. For the future the authors aim to leverage the underlying techniques for image streams and for a computationally efficient experience to copy move forgery detection.

References

[1] Mohammed T M, Bunk J, Nataraj L, Bappy J H, Flenner A, Manjunath B S, Chandrasekaran S, Roy-Chowdhury A K and Peterson L 2018 (*Preprint* 1802.03154v2) URL `http://arxiv.org/abs/1802.03154v2`

[2] Muhammad G, Hussain M, Khawaji K and Bebis G 2011 Blind copy move image forgery detection using dyadic undecimated wavelet transform *2011 17th*

International Conference on Digital Signal Processing (DSP) (IEEE) URL `http://dx.doi.org/10.1109/icdsp.2011.6004974`

[3] Kumar S, Desai J V and Mukherjee S 2016 (*Preprint* `1612.03989v1`) URL `http://arxiv.org/abs/1612.03989v1`

[4] Manoharan J S 2021

[5] Vijayakumar T 2020 *Journal of Innovative Image Processing (JIIP)* **2** 121–127

[6] Lin S D and Wu T 2011 An integrated technique for splicing and copy-move forgery image detection *2011 4th International Congress on Image and Signal Processing* (IEEE) URL `http://dx.doi.org/10.1109/cisp.2011.6100366`

[7] Teerakanok S and Uehara T 2019 **7** 40550–40568 ISSN 2169-3536 URL `http://dx.doi.org/10.1109/access.2019.2907316`

[8] Thakur T, Singh K and Yadav A 2018 **179** 34–42 ISSN 0975-8887 URL `http://dx.doi.org/10.5120/ijca2018916108`

[9] Soni B, Das P K and Thounaojam D M 2018 **12** 167–178 ISSN 1751-9667 URL `http://dx.doi.org/10.1049/iet-ipr.2017.0441`

[10] Ali N H and Harun F 2019 **5** ISSN 2414-4592 URL `http://dx.doi.org/10.20474/jater-5.5.2`

[11] Lin H J, Wang C W, Kao Y T *et al.* 2009 **5**(5)

[12] Shivakumar B and Baboo L D S S 2010 *Global Journal of Computer Science and Technology*

[13] Christlein V, Riess C, Jordan J, Riess C and Angelopoulou E 2012 **7** 1841–1854 ISSN 1556-6013 URL `http://dx.doi.org/10.1109/tifs.2012.2218597`

[14] Gupta A, Saxena N and Vasistha S 2013 **3**

[15] Cozzolino D, Poggi G and Verdoliva L 2014 Copy-move forgery detection based on patchmatch *2014 IEEE International Conference on Image Processing (ICIP)* (IEEE) URL `http://dx.doi.org/10.1109/icip.2014.7026075`

[16] Uliyan D M, Jalab H A and Wahab A W A 2015 Copy move image forgery detection using hessian and center symmetric local binary pattern *2015 IEEE Conference on Open Systems (ICOS)* (IEEE) URL `http://dx.doi.org/10.1109/icos.2015.7377269`

[17] D'Amiano L, Cozzolino D, Poggi G and Verdoliva L 2015

[18] Li J, Li X, Yang B and Sun X 2014 **10**

[19] Bayram S, Sencar H T and Memon N 2008

[20] J S and Anver J 2016 *2016 Online International Conference on Green Engineering and Technologies (IC-GET)* 1–4

[21] Al-Hammadi M M and Emmanuel S 2016 Improving SURF based copy-move forgery detection using super resolution *2016 IEEE International Symposium on Multimedia (ISM)* (IEEE) URL `http://dx.doi.org/10.1109/ism.2016.0075`

[22] Dixit R, Naskar R and Sahoo A 2017 Copy-move forgery detection exploiting statistical image features *2017 International Conference on Wireless Communications, Signal Processing and Networking (WiSPNET)* (IEEE) URL `http://dx.doi.org/10.1109/wispnet.2017.8300165`

[23] Tarman and Saini H 2017 M-sift: A detection algorithm for copy move image forgery *2017 4th International Conference on Signal Processing, Computing and Control (ISPCC)* pp 425–430

[24] Zhang Z, Wang C and Zhou X 2018 **14**

[25] Kaur A, Walia S and Kumar K 2018 Comparative analysis of different keypoint based copy-move forgery detection methods *2018 Eleventh International Conference on Contemporary Computing (IC3)* (IEEE) URL http://dx.doi.org/10.1109/ic3.2018.8530489

[26] Khan U A, Kaloi M A, Shaikh Z A and Arain A A 2018 A hybrid technique for copy-move image forgery detection *2018 3rd International Conference on Computer and Communication Systems (ICCCS)* (IEEE) URL http://dx.doi.org/10.1109/ccoms.2018.8463337

[27] Chen B, Yu M, Su Q, Shim H J and Shi Y Q 2018 **6** 56637–56646 ISSN 2169-3536 URL http://dx.doi.org/10.1109/access.2018.2871952

[28] Kaur K 2018 Efficient and fast copy move image forgery detection technique *2018 Second International Conference on Intelligent Computing and Control Systems (ICICCS)* (IEEE) URL http://dx.doi.org/10.1109/iccons.2018.8663072

[29] Kaur N, Jindal N and Singh K 2022 ISSN 1380-7501 URL http://dx.doi.org/10.1007/s11042-022-13105-6

[30] Gan Y, Zhong J and Vong C 2022 **59** 102783 ISSN 0306-4573 URL http://dx.doi.org/10.1016/j.ipm.2021.102783

[31] Zhong J L, Gan Y F, Vong C M, Yang J X, Zhao J H and Luo J H 2022 **122** 108286 ISSN 0031-3203 URL http://dx.doi.org/10.1016/j.patcog.2021.108286

[32] Tahaoglu G, Ulutas G, Ustubioglu B, Ulutas M and Nabiyev V V 2022 ISSN 1380-7501 URL http://dx.doi.org/10.1007/s11042-021-11503-w

[33] Suganya D, Thirunadana Sikamani K and Sasikala J 2021 *International Journal of Computers and Applications* 1–9

Early heart disease prediction using ensemble learning techniques

K Rohit Chowdary[1], P Bhargav[1], N Nikhil[1], K Varun[1] and Dr D Jayanthi[1]

[1]Dept. of ECE, Gokaraju Rangaraju Institute of Engineering and Technology, Hyderabad, Telangana, India.

rohitchowdary1144@gmail.com, varunk0872@gmail.com, bhargav2907@gmail.com, nikhilnarla1527@gmail.com, jayanthivlsi@gmail.com

Abstract. Cardiovascular illnesses claim the lives of 18 million individuals each year (heart-related diseases). According to the WHO, heart disease is to blame for 31% of all deaths worldwide. In this study, a new machine learning model for predicting heart disease is provided. The proposed method was evaluated on Kaggle and the University of California, Irvine datasets. We used sample approaches and feature selection methods to identify the most useful characteristics in the dataset that was unbalanced. Eventually, classifier models were employed, and an ensemble classifier generated great accuracy. In two datasets, the proposed approach showed to be accurate in predicting heart disease. In all cases, Python was used.

Keywords: Disease Prediction; Ensemble Learning; Kaggle datasets; Python

1. Introduction

Cardiovascular disease is mostly caused by heart disease. There are numerous types of cardiovascular disease, such as coronary artery disease (CAD), peripheral artery disease (PAD), congenital heart disease (CHD), and stroke. There is a greater risk of heart attack or stroke in people with peripheral vascular disease. In general, the risk of developing PAD increases with age. Heart problems can begin as early as infancy for some people. Such an illness is congenital heart disease. This illness alters the flow of blood and causes heart-related issues. As a result of a cerebral vascular disease, blood vessels in the brain are harmed. There's a possibility of brain injury if the brain's blood supply is cut off. The most common causes of heart disease are poor eating habits, smoking, and excessive alcohol consumption [1]. Pain in the left shoulder or elbow and pain in the chest are two of the most typical symptoms of the condition. However, many diseases might go undetected for long periods of time if no symptoms are present. The United States is the country with the greatest number of deaths as a result of this disease. Analyzing the patient's prior medical records is a traditional strategy for diagnosing this disease, however this method is never sufficient. Alternatively, the patient's data can be collected and analysed for future precautionary measures. There are a variety of applications where machine learning and deep learning can be used. There is also a growing use of machine learning in healthcare. However, identifying high-quality datasets is critical to the machine learning model's performance. For the study of heart illness, we gathered two datasets from separate sources, namely Kaggle and UCI.

CVD (cardiovascular disease) is one of the leading causes of global mortality. In medical data analysis, this disease remains an extremely difficult problem to solve. Big Data, Artificial Intelligence, and the need for automated models have made it possible to construct a more accurate and efficient model for forecasting cardiac disease. Predicting cardiac disease has been the subject of numerous studies, but the focus on identifying the most essential traits is lacking. Therefore, it is critical to select the proper characteristics for the classification and diagnosis of cardiac disease. The primary goal of this research is to find and choose essential variables and machine learning approaches that can improve the classification models' prediction power for reliably predicting CVD. The results reveal that the hybrid ensemble model's proposed enhanced evolutionary feature selection surpasses the currently used methods as shown in figure 1.

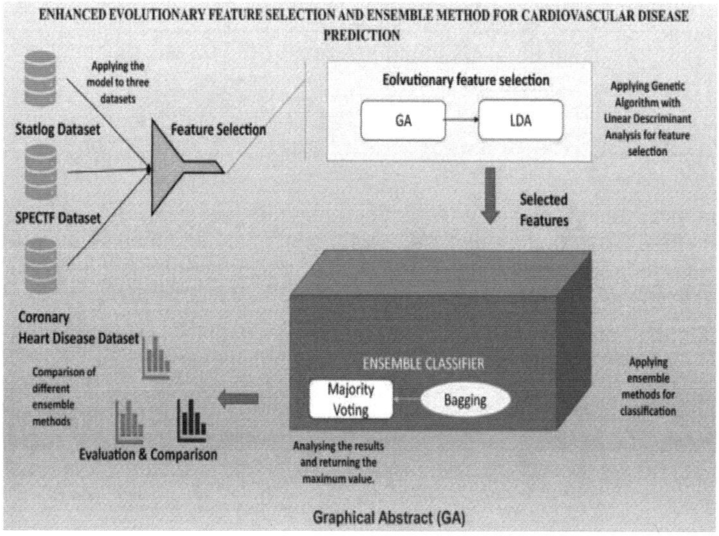

Figure 1. Cardiovascular Disease Prediction.

2. Literature Review

2.1. Robust Heart Disease Prediction: A novel approach based on significant feature and ensemble learning [2]

Heart disease is one of the most common causes of non-communicable and silent mortality in the world, according to the World Health Organization. Early detection of cardiac disease using clinical data is a difficulty. Machine learning's importance has risen steadily since it was first developed. Researchers in the field of medical sciences have been paying increasing attention to machine learning in the last few years. Machine learning methods and techniques are used to forecast diseases at an early stage. Researchers face a major hurdle when it comes to accurately predicting cardiac disease based on clinical data. Even though state-of-the-art outcomes have been achieved with diverse clinical data and different machine learning methods, there is still room for improvement. Using machine learning techniques, we propose to apply an unique method in this research for the detection of cardiac disease at an early stage. Essentially, the paper's goal is to identify correlations between features that can aid in predicting more accurate results. To do this, we used the UCI vascular heart disease dataset and ran our results against those of a recently published study.

2.2. Heart disease can be treated using the Extreme Learning Machine technique. [3]

The detection of heart disease, which affects millions of people, is one of the most important uses of machine learning systems. Heart disease patients share a wide range of characteristics that can be used to make an accurate diagnosis, such as age, gender, blood pressure, cholesterol levels, and a host of other variables. It is proposed that these variables be represented using an algorithm known as the Extreme Learning Machine (ELM). In place of expensive medical examinations, the suggested system provides a warning system for those who are at risk for heart disease. In order to test the system, the Cleveland Clinic Foundation has gathered data on roughly 300 patients. Heart disease may be predicted with roughly 80% accuracy with this design.

2.3. Cloud-based four-tier architecture for early detection of heart illness using machine learning techniques [4]

It's no secret that early identification and prognostication of heart disease are top priorities. The importance of early detection of heart disease cannot be overstated when it comes to health care (HCS). Expensive medicines and procedures that can be prohibitively expensive for underdeveloped countries are becoming more common in many health care systems. As a public health issue, heart disease is becoming more and more prevalent, particularly in the United States. Tobacco use, a sedentary lifestyle, a lack of exercise, and alcohol consumption are the primary causes of these disorders. Because of this, a cloud-based infrastructure is necessary to forecast and track health data properly. Machine learning has already been used to solve clinical issues and provide medical diagnosis. The prediction and monitoring of patient health information can be greatly enhanced by using the cloud-based, four-tier architecture described in this paper. To diagnose cardiac disease in its earliest stages, we used five basic machine learning techniques: As the primary objective of this inquiry, we are looking to see how well the chosen classification techniques work. A number of well-known metrics are also used to evaluate machine learning approaches. The five classifiers' performance was also evaluated using the ten-fold cross-validation approach. According to the results, the Artificial Neural Network (ANN) was the most effective model. Those in the medical profession who are trying to figure out which machine learning algorithms to apply can benefit from this study, on the other hand.

2.4. Machine learning algorithms can be used to forecast heart disease [6]

People nowadays are far too preoccupied with their careers and other commitments to take time to care for their own health and well-being. Due to their hectic schedules and disdain for their own health, the number of people falling ill is on the rise on a daily basis. Heart disease affects a large percentage of the population. Heart disease is the major cause of death for approximately a third of the global population, according to World Health Organization data (WHO). As a result, the ability to foretell the onset of heart disease is crucial in the medical industry. Massive amounts of data collected by the medical industry and hospitals may be difficult to evaluate. The application of machine learning techniques to anticipate outcomes and handle data can have a significant impact on medical professionals. As a result, we've talked about heart disease and the risk factors for it, as well as the basics of machine learning. A comparative evaluation of the machine learning algorithms used in the experiment yielded accurate predictions of cardiac illness. Predicting the likelihood of getting heart disease using machine learning is the primary goal of this investigation.

2.5. According to a recent study, machine learning can predict heart disease [7]

An accurate first-time diagnosis of cardiac disease can save a person's life; an incorrect diagnosis can be lethal. A wide range of machine learning and deep learning techniques are used to compare the dataset's results and analysis. The dataset comprises 14 key attributes for doing the analysis. An accuracy and confusion matrix has been used to verify a variety of possible outcomes. In order to increase the quality of the study, Isolation Forest is used to remove irrelevant features from the dataset. It is also mentioned how this research might be integrated with multimedia technologies, such as mobile devices. A 94.2 percent accuracy rate was achieved by using a deep learning approach as shown in figure 2.

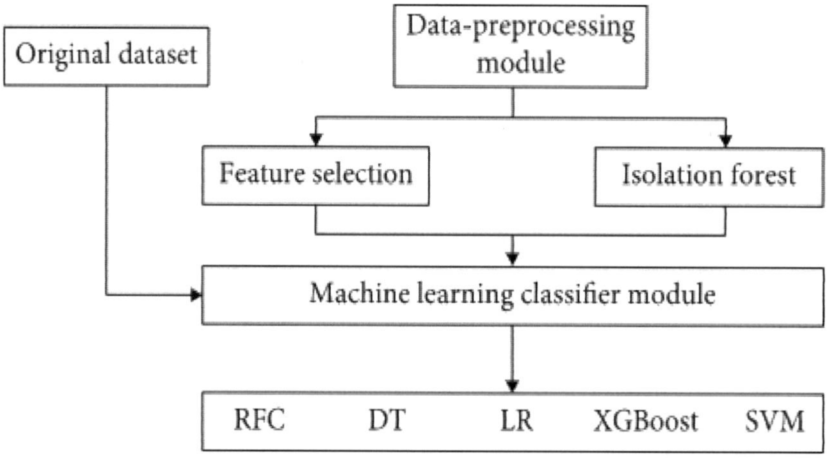

Figure 2. Machine learning classifiers.

2.6 Heart Disease Detection Using Machine Learning Majority Voting Ensemble Method [8]

Using the majority vote ensemble method proposed in this study, heart disease can be predicted. Any neighborhood clinic can do simple medical testing at a low cost. With real-life data from healthy and ill patients, the model is trained to enhance the project's goal of increasing patient trust and precision of medical diagnosis. The model uses the majority vote of several machine learning models to identify the patient and provide more accurate results. Finally, we achieved a 90% accuracy rate with our technique based on the hard voting ensemble model.

3. Implementation

Initial data sources were the UCI repository [10] and Kaggle [11]. We then tried out a variety of data preparation techniques on the datasets we had gathered. We next used feature selection techniques to identify the best features in the two datasets. We used three sampling strategies, namely Random Oversampling, Synthetic Minority Oversampling, and Adaptive Synthetic sampling, because one dataset contained unbalanced class labels. We used a variety of machine learning classifiers based on the selected features. Accuracy was high thanks to a team effort.

3.1. Feature selection

In order to reduce the complexity of the machine learning algorithms, the feature selection procedure is used. There are a variety of methods for deciding which features to include in a system. ANOVA F-value and Mutual Information were both used to narrow down the set of potential features. Using these two approaches, we were able to zero in on the best attributes. If the dataset is imbalanced, sampling strategies can be useful. 4238 samples are included in Dataset2 (Kaggle). Only 644 of these samples have the label 'TenYearCHD' as 1 and the rest have the label '0'. This data set is imbalanced, as evidenced by this dataset. Random oversampling, synthetic minority oversampling, and an adaptive synthetic sampling approach were all used in this study.

3.2. Sampling techniques

3.2.1. Irregular oversampling

Oversampling is a technique that uses many copies of a minority class to improve the organization of data. More than one instance of such an oversampling should be possible. This is one of the most basic approaches, and it has been shown to be energetic. Not all minority models need to be re-created; some may be replaced with a new model. It's as simple as inspecting current samples and then delivering new ones as shown in figure 3.

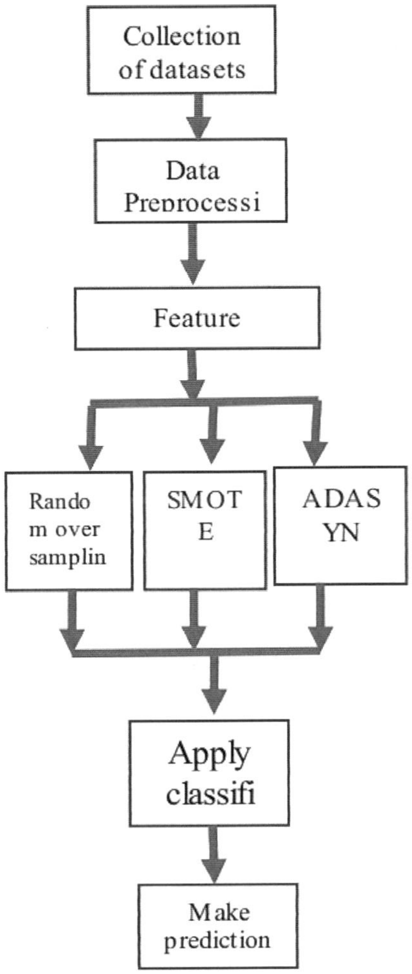

Figure 3. System architecture.

3.2.2. SMOTE (Synthetic Minority Oversampling)

Instead of repetition, this method relies on a heuristic approach. There are n tests and f highlights in the element space of some preparatory material. In light of these points, think about preparing for informational tests (with f-number features). k is the number of nearest neighbors in this technique, which takes one example from the given dataset. In order to make an engineered information point, it uses a vector between the current data point and one of the k neighbors, and expands the vector by the number x, which is within the range of 0 and 1. The addition of a half pint is meant to emphasize a different point.

3.2.3. ADASYN (Adaptive synthetic sampling approach)

The heuristic method is also used by ADASYN. Based on the SMOTE protocol. It shifts the order restriction from the majority to the minority, which is a pain. Tests for minority students are weighted according to their level of difficulty in learning. There is a significant difference between SMOTE and ADASYN when it comes to the kind of information tests that can be difficult to organize using the closest neighbors approach.

3.3. Table description

Two datasets from Kaggle and the University of California, Irvine (dataset- 1) were used (Dataset-2). There are 303 samples in Dataset1, and there are 14 characteristics. Age, gender, tptrestbps, chol, fbs, etc. are all independent traits, while 'target' is a dependent one. 4238 samples with 16 characteristics are included in dataset 2. Men, age, education, currentSmoker status, cigarettes-per-day, blood pressure medications, body mass index (BMI), and other variables are all examined irrespective of the 'TenYearCHD' feature. There are total samples, training rows and testing rows in the two datasets shown in Table 1. Dataset-1 has 1,242 training samples and 61 testing samples. dataset-2 is trained on 2,3390 samples and tested on 848 samples.

Table 1. Dataset description.

Dataset	Number of training samples	Number of testing samples	Total
Dataset-1(UCI)	242	61	303
Dataset-2(Kaggle)	3390	848	4238

4. Algorithms

4.1. Logistic regression classifier

To describe a binary dependent variable, the simplest form of a logistic function is utilized. However, there are other more complex extensions available. Logistic regression (or logit regression) can be used to estimate the parameters of a logistic model (a form of binary regression).

4.2. K-nearest neighbor classifier

The name KNN stands for "K-Nearest Neighbor," and that is exactly what it means. For training, it employs a machine learning algorithm. Classification and regression problems can be answered using this method. Predicting or categorizing a new unknown variable requires knowing how many of its nearest neighbors are known.

4.3. Support vector machine classifier

SVMs are a type of machine learning algorithm that can classify and predict data. Data are divided into two categories using the SVM supervised learning algorithm. An SVM outputs the margins between sorted data as far apart as possible.

4.4. Decision tree classifier

An attribute "test" is represented by each internal node, a branch indicates the result of the test, and a leaf node represents a decision tree class label (decision taken after computing all attributes).

4.5. Naive bayes classifier

When it comes to classification algorithms, the Nave Bayes Classifier is one of the simplest, most effective, and fastest options available. Predictions are made based on the probability that a given object will be encountered by the classifier.

4.6. Random forest classifier

Machine Learning that is supervised. Classification and regression applications commonly make use of algorithms like random forest. A decision tree is constructed from various samples, which are then combined and averaged for classification and regression purposes.

4.7. Ada-boost classifier
AdaBoost, or Adaptive Boosting, is a technique used in Machine Learning as part of an Ensemble Method. One-level decision trees, which have only one split, are the most widely employed algorithm with AdaBoost. These trees are referred to as Decision Stumps in the literature.

4.8. Stacking classifier
Using an ensemble learning technique known as level-one classifiers, the predictions of many classifiers are combined to form a single meta-classifier. Stacking in this manner is the most basic. The meta-classifier classifier can be anything you want it to be.

4.9. Voting classifier
A voting classifier can predict the outcome of a vote based on the findings of several base models. In order to construct the aggregation criterion, each estimator's output can be voted on separately.

5. Experimental results
Results are shown in figure 4-12.

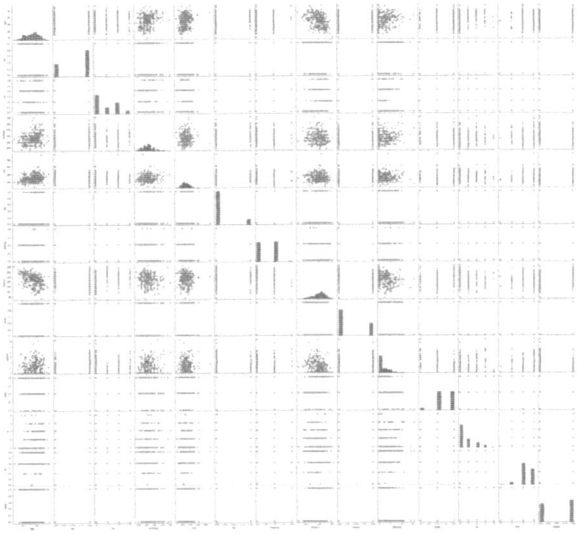

Figure 4. Data visualization

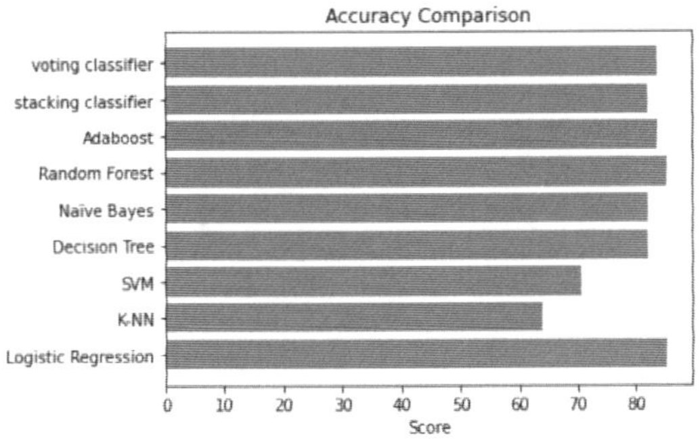

Figure 5. Accuracy comparison for dataset1.

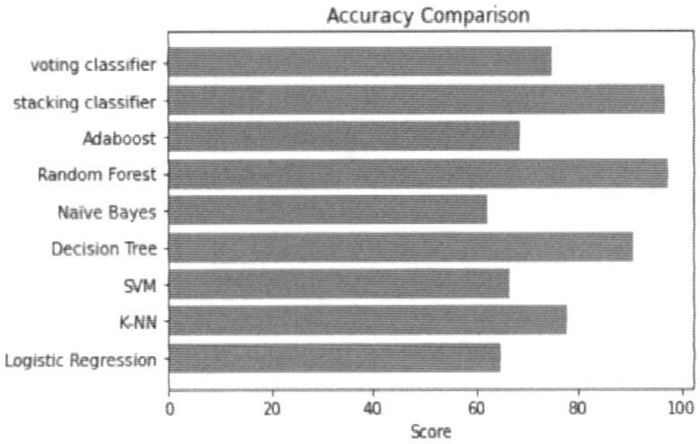

Figure 6. Accuracy comparison for dataset2 ROS.

Journal of Physics: Conference Series **2325** (2022) 012051 doi:10.1088/1742-6596/2325/1/012051

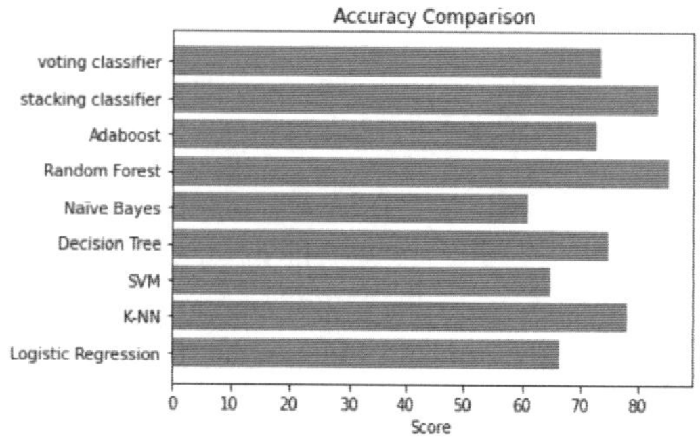

Figure 7. Accuracy comparison for dataset2 SMOTE.

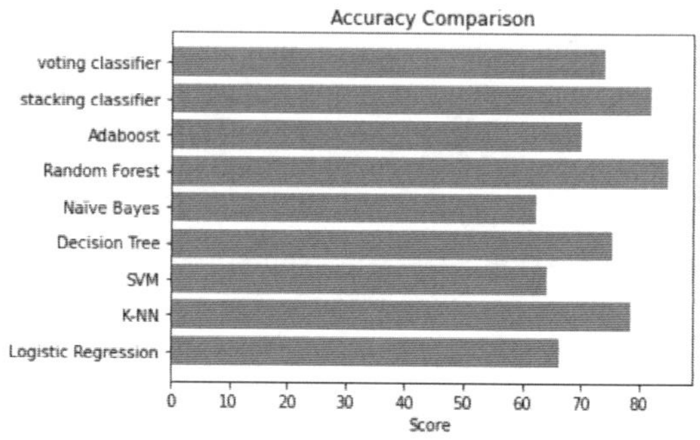

Figure 8. Accuracy comparison for dataset2 ADASYN.

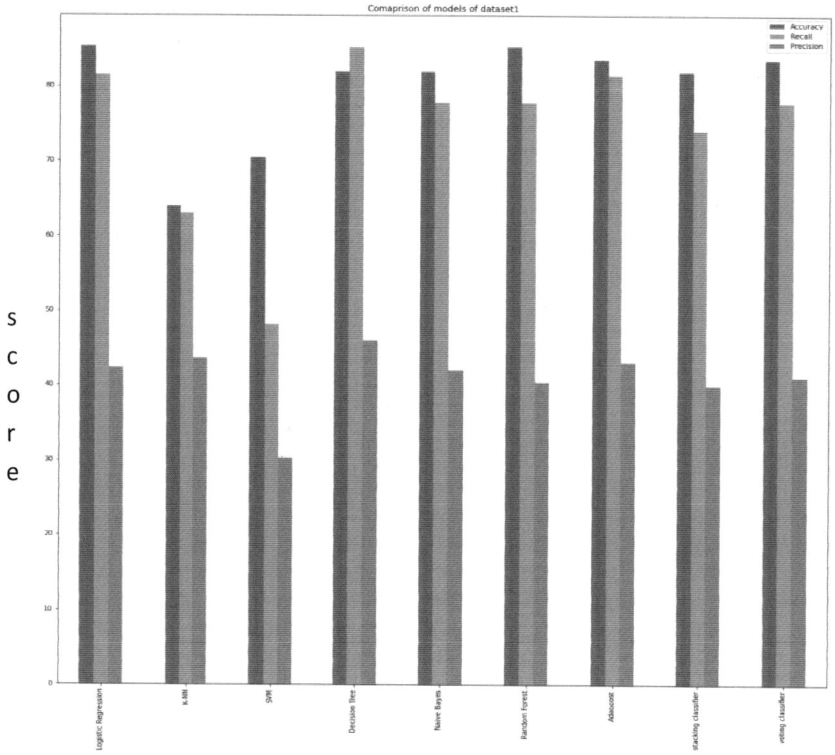

Figure 9. Comparison of models of dataset1

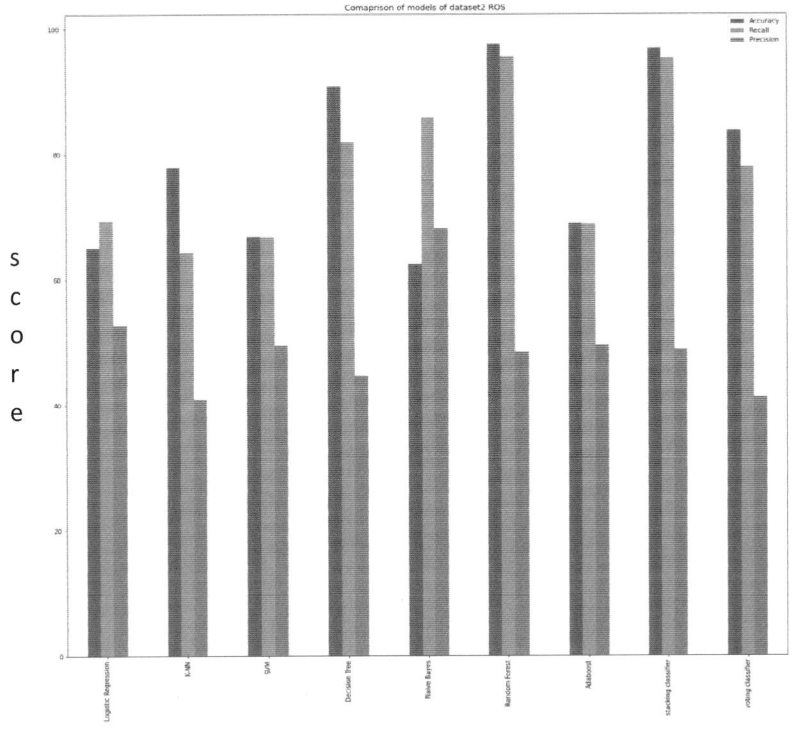

Figure 10. Comparison of models of dataset2 ROS

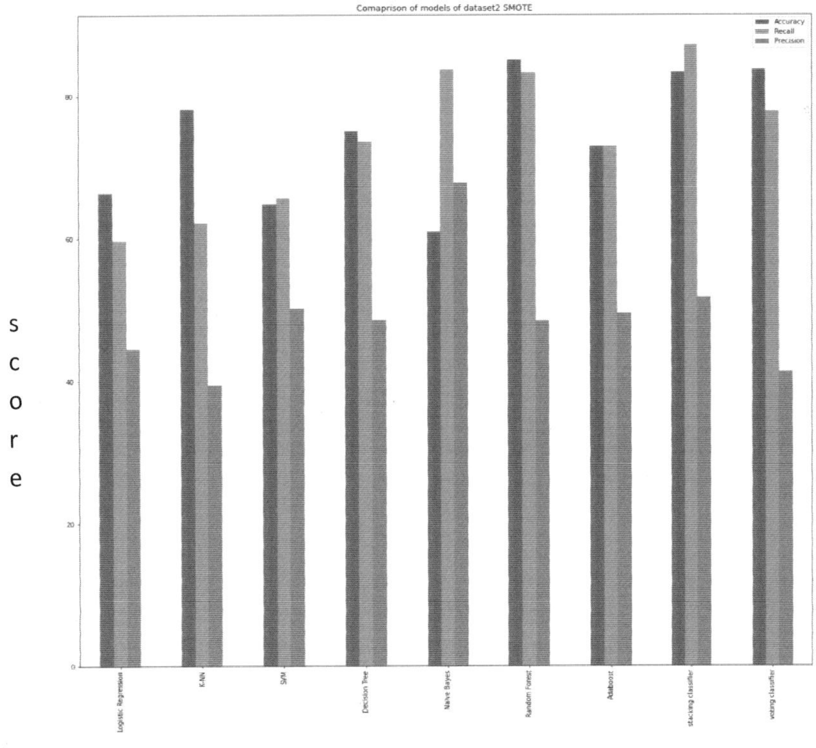

Figure 11. Comparison of models of dataset2 SMOTE

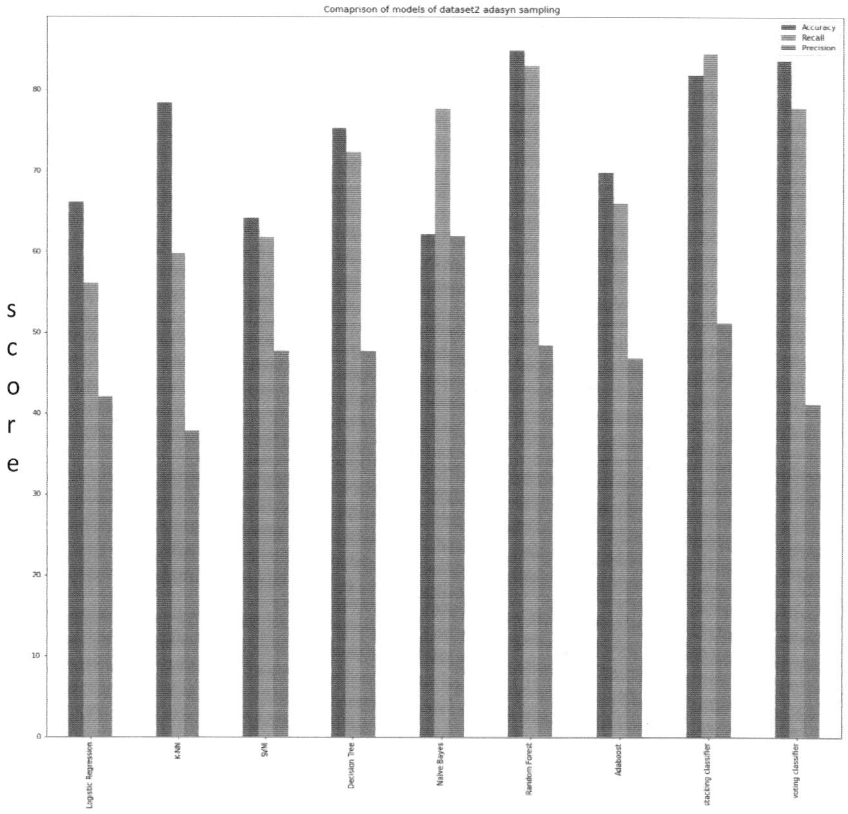

Figure 12. Comparison of models of dataset2 ADASYN

6. Conclusion

An ensemble classifier model for heart disease prediction was proposed in this research. For our tests, we employed two sets of data from UCI and kaggle. This was the first step in the process of identifying and selecting the greatest characteristics. With this classifier and sampling strategies, we were able to get a high detection rate. With stacking classifiers, random over sampling has 99 percent accuracy, SMOTE sampling has 93 percent accuracy, and the ADASYN sampling technique has 91 percent accuracy. Our suggested model outperforms other models in tests on two independent datasets.

7. Future scope

Despite the model's high accuracy, we were only able to use it on two datasets. In the future, we plan to use our ensemble method to predict heart disease in more real-world datasets.

8. References

[1] https://www.who.int/en/news-room/factsheets/detail/cardiovascular-diseases-(cvds)

[2] M. A. Alim, S. Habib, Y. Farooq and A. Rafay 2020 *Robust Heart Disease Prediction: A Novel Approach based on Significant Feature and Ensemble learning Model* 2020 3rd International Conference on Computing, Mathematics and Engineering Technologies (iCoMET), pp 1–5

[3] S. Ismaeel, A. Miri and D. Chourishi 2015 *Using the Extreme Learning Machine (ELM) technique for heart disease diagnosis* 2015 IEEE Canada International Humanitarian Technology Conference (IHTC2015), pp 1–3

[4] M. R. Ahmed, S. M. Hasan Mahmud, M. A. Hossin, H. Jahan and S. R. Haider Noori 2018 *A Cloud Based Four-Tier Architecture for Early Detection of Heart Disease with Machine Learning Algorithms* 2018 IEEE 4th International Conference on Computer and Communications (ICCC), pp 1951–1955

[5] A. U. HAQ et al., 2019 *Identifying the Predictive Capability of Machine Learning Classifiers for Designing Heart Disease Detection System* 2019 16th International Computer Conference on Wavelet Active Media Technology and Information Processing, pp 130–138

[6] A. Lakshmanarao,Y.Swathi and P.Sri Sai Sundareswar NOVEMBER 2019 *Machine Learning Techniques For Heart Disease Prediction* International Journal of Science & Technology Research Volume 8, Issue 11, pp 374–377

[7] A. Gavhane, G. Kokkula, I. Pandya and K. Devadkar 2018 *Prediction of Heart Disease Using Machine Learning* 2018 Second International Conference on Electronics, Communication and Aerospace Technology (ICECA), pp 1275–1278

[8] R. Atallah and A. Al-Mousa 2019 *Heart Disease Detection Using Machine Learning Majority Voting Ensemble Method* 2019 2nd International Conference on new Trends in Computing Sciences (ICTCS), pp 1–6

[9] Apurb Rajdhan, Avi Agarwal and Poonam Ghuli April 2020 *Heart Disease Prediction using Machine Learning* International Journal of Engineering Research & Technology (IJERT) Volume 09, Issue 04, pp 659–662

[10] https://archive.ics.uci.edu/ml/datasets/heart+disease

[11] https://www.kaggle.com/amanajmera1/framingham-heart-study-dataset

[12] Archana Singh and Rakesh Kumar 2020 *Heart Disease Prediction Using Machine Learning Algorithms* International Conference on Electrical and Electronics Engineering, IEEE-2020. pp 452–457

[13] S. Mohan, C. Thirumalai and G. Srivastava 2019 *Effective Heart Disease Prediction Using Hybrid Machine Learning Techniques* in IEEE Access, vol. 7, pp 81542–81554

[14] Shakya, Subarna, and P. P. Joby. 2021 *Heart Disease Prediction using Fog Computing based Wireless Body Sensor Networks (WSNs)*. IRO Journal on Sustainable Wireless Systems 3, no. 1 pp 49–58

[15] Valanarasu and Mr R. 2019 *Smart and secure IoT and AI integration framework for hospital environment* Journal of ISMAC 1, no. 03 pp 172–179

Dynamic Task Scheduling and Resource Allocation for Microservices in Cloud

Sanath Mugeraya, Kailas Devadkar

Department of Computer Engineering, Sardar Patel Institute of Technology, Mumbai, India.

sanath.mugeraya@spit.ac.in, kailas_devadkar@spit.ac.in

Abstract- With the emergence of new companies and the expansion of the information technology sector, the need for Cloud Computing becomes apparent. Currently, the enterprises are rapidly transitioning from monolithic architecture to microservice-driven architecture. This research study has discovered that all task scheduling algorithms were designed for a specific (set) number of virtual machines, which resulted in the bottleneck problem, where multiple tasks were assigned to the microservice scheduler and the execution time of processing the tasks was significantly increased. Therefore, to address this issue, a novel model was designed based on the number of tasks and accordingly the number of virtual machines were dynamically generated to send the tasks to the microservice scheduler one by one, and the difficulties with execution time were also addressed. The study also discovered that due to the multiple workloads on the microservices, resource allocation becomes extremely difficult. To address this issue, containerized microservices were discovered. Here, the microservices would be distributed in containers. To implement the dynamic work scheduling technique, a cloud microservice translator would be developed, where a user may upload a text file and quickly get it dynamically translated. The main aim of this research work is to improve the task scheduling and resource allocation in microservices.

Keywords- Task Scheduling, Microservices, Resource allocation, Cloud Computing

I. INTRODUCTION

With the increasing demand of data storage and web services, the new age technologies would run their applications on physical servers. Despite the hype, with the increase in the size of data and with increasing functionality, the traditional servers could not handle the large amount of data. This increases the need for companies and many other organizations to migrate all their data resources to cloud environment. The need for cloud microservices has increased in the recent times as it is more productive, flexible and cost-effective. Cloud microservices were introduced so as to tackle the problem of traditional cloud service-oriented architecture which puts the scalability and fault tolerance part in danger. But cloud microservices are a boon because they can execute independent instructions as compared to monoliths which are difficult to maintain and evaluate due to their complexity.

The main advantage of moving into cloud technology is that it is based on the concept of virtualization. Virtualization basically gives the most essentially important technology that powers cloud computing. So, in simple words it separates the software layer of the server from the hardware layer of the server.

In recent years the general trend that has been observed is that majority of multinational companies are quickly shifting from monolithic architecture to microservices due to the increasing complexity of monolithic type of architecture. So Microservice are an approach on how to develop an application as a set of small services running in its own process. Monolithic APIs (Application Programming Interface) are made up of small modules that are tightly coupled together and making changes to one part of the API requires changes to be made in the whole code which in turn results in the increased complexity.

The traditional style architecture is not made to run in cloud environment without making large modifications. So microservice is an attempt to arrange an application as a set of loosely coupled modules. In order to make the microservices scalable, the microservices are packaged as a container before deploying it any platform that helps in the containerized deployment.

II. RELATED WORK

[1] This paper speaks about how deploying microservices is a big task. It also speaks about the importance of bring a dynamic multi objective scheduling which would deal with the problem of poor resource allocation of microservices which would result in delayed response time and wastage of resource capacities. It also recommends bringing new technologies in orchestration frameworks which includes Kubernetes, which can result in proper scheduling of microservices and can improve the overall throughput of the system. For scheduling of the microservice, a knapsack problem was designed, where objects with values and their respective weights were the two variables. This algorithm was called Least Waste Fast First algorithm. So the proposed Least Waste Fast First algorithm was compared along First Come First Serve (FCFS). Spread algorithm and Binpack Algorithm. The proposed Least Waste Fast First algorithm showed great improvements with respect to resource utilization, CPU utilization, scheduling latency and execution time with respect to other algorithms.

[2] This paper proposes elastic scheduling algorithm which aims to better task scheduling and deals with the problem with respect to scaling of application. It also deals with the scheduling of tasks, which meets the deadline constraints. As a part of research for this paper it was found that all the existing algorithms ignore the importance of streaming workloads. The elastic scheduling also ensures that the cost of virtual machines is reduced. After identifying all the above problems an attempt was made so as to integrates scheduling of tasks and scaling configuration
In order to design a system which solves the problem of streaming workloads an optimization problem was taken into consideration.

The elastic scheduling which was proposed has three sections:

1. Configuration Solving
2. Urgency based Scheduling
3. Auto Scaling Configuration

So, in order to test this algorithm, a simulation platform was built, where the elastic scheduling was compared with many other algorithms such as ProCis, SCS and IC-OCPD2. So, it was found that the proposed elastic scheduling outperformed all the algorithms with respect to success ratio, auto scaling and better response time.

[3] In this particular paper an INTMA algorithm which is Interaction Aware Resource Allocation algorithm was proposed for deploying the microservices. In order to deploy the

microservices a new model was proposed which was called as Binary Quadratic Programming Problem. The INTMA algorithm was compared against INTRR algorithm which is Interaction aware Round Robin Algorithm. This paper also deals with problem with respect to microservice specific performance objective. What was also observed was that when it comes to designing microservice all the focus is on resource capacity, but the nature of the workload is always underestimated. This paper also stresses on the importance of microservice interaction. In order to compare the two algorithms and to evaluate the performance metrics of the two algorithms, Google Cloud platform was used. The microservices were tested for both INTMA and INTRR which has shown better response time, throughput and also improvement in the scheduling duration.

[4] This paper discusses the issues which are encountered when microservices are scheduled and highlights the problem with respect to resource management. Even though there are lot of advancements in technologies with respect to virtualization which includes container-based technology and hypervisor-based technology, but there isn't any framework or tech stack that manages a cluster of microservices. Also, while designing microservices, in order to have effective task scheduling we must take into consideration the number of requests it would handle.

III. SYSTEM DESIGN AND ANALYSIS

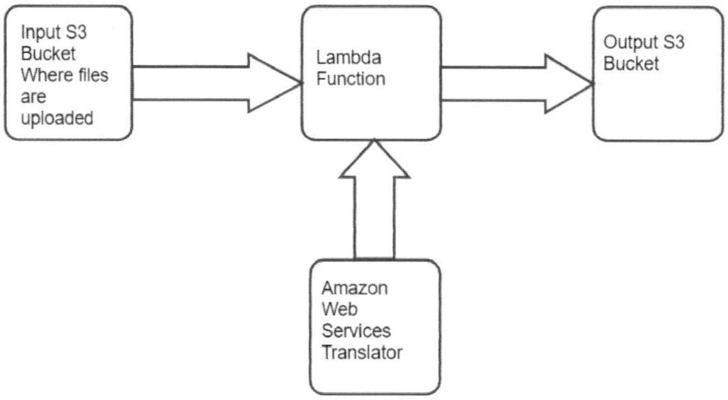

Figure 1. System Design of the designed Microservice Application

In order to build a lambda function, we first create an IAM (Identity and Access Management) in Amazon web services console, where we take access for S3 object trigger, AWS translate trigger and AWS cloud watch event.

So as to build a containerized microservice we make use of AWS ECR ie. Elastic Container Registry. We then take access for Elastic Container Registry from IAM role, where we get the credentials from AWS. After getting the access in ECR we create a repository inside which the docker file which was built and run is pushed.

After pushing the docker image to the ECR repository which was created, an image URI would be generated which would be used to create a lambda function using the option of creating lambda function with image.

In the AWS lambda console, we create an S3 trigger and a python code is written wherein any file uploaded on the input S3 bucket would in turn trigger lambda function, which would further trigger AWS translate resources.

The AWS translate would dynamically provide all the resources which are needed for the content translation. After which the file would be translated dynamically and the translated content would be uploaded on the output S3 bucket.

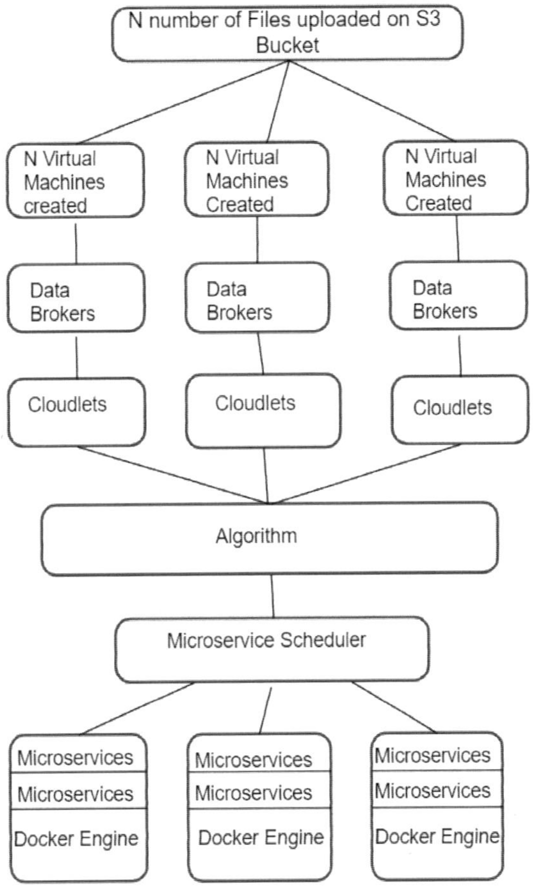

Fig. 2. Detailed Architecture of the Designed Microservice Application

In the proposed Dynamic Scheduling model, depending upon the number of files uploaded on the S3 bucket, that many number of virtual machines would dynamically be created as compared to the previous models where files would be given to static number of virtual machines. The file which is given to the virtual machine would further be given to data broker. Here the data broker would combine and integrate the multiple existing service into a new service. The file from the data broker is given to cloudlets. The file from the cloudlets would be further given to our algorithm, which in turn would give it to the microservice scheduler. The microservice scheduler would convert the file into a microservices. Since we created a containerized microservice using Docker, the docker which acts as an orchestrator which schedule and run the microservice in containers. The Amazon Web Services translator trigger which has been created would allocate the translate resources and would give the output in the output S3 bucket.

IV. PERFORMANCE EVALUATION

In this particular section, the experimental evaluation of the proposed dynamic scheduling and elastic scheduling is presented, which includes the experimental environment and the results.

1. EXPERIMENTAL ENVIRONMENT

In order to demonstrate the advantages of the proposed dynamic scheduling and elastic scheduling, a containerized microservice application was designed where any text file uploaded in the input S3 bucket would trigger the lambda function, which in turn would trigger the Amazon Web services (AWS) translator and would translate the text file and upload on the output S3 bucket. As soon as the files would be uploaded on the input S3 bucket many number of virtual machines would be dynamically created. The created virtual machines would further send the files to data broker. The data broker would further submit the files to the cloudlets. To show the differences in the two algorithms, at a time 5,10,15,20 files ranging from 5kb to 100kb were simultaneously uploaded on the input S3 bucket and the timestamps on the output S3 bucket were noted so as to demonstrate the advantages of the proposed dynamic scheduling model.

2. EXPERIMENTAL RESULTS

By noting the timestamps of the translated text files at the output S3 bucket, we get the time taken for parallel as well serial execution. So depending on the time taken for getting the output, we make a comparative study of the proposed dynamic scheduling and elastic scheduling. For the comparative study four parameters were considered for evaluating the performance. The parameters were Execution Time, Speedup, Efficiency and Throughput.

1. *Execution Time*

Number of Tasks	Dynamic Scheduling	Elastic Scheduling	Number of Virtual Machines	Improvement in both algorithms
5	14	17	5	21.42%
10	17	21	10	23.59%
15	20	25	15	25%
20	23	29	20	26.08%

Table 1. Execution Time of proposed Dynamic Scheduling and Elastic Scheduling where number of virtual machines would be same as the number of tasks given

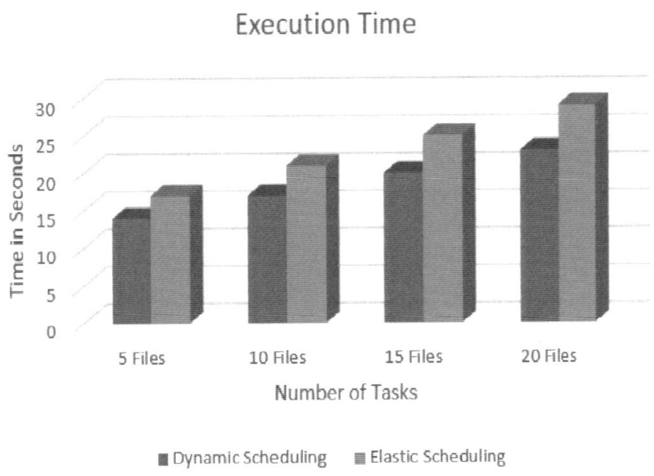

Figure 3. Comparison of execution time of proposed Dynamic Scheduling and Elastic Scheduling.

According to the results in Table 1, it is found that the execution time of proposed dynamic scheduling has been minimized by 24.0225% in comparison to elastic scheduling.

2. *Speedup*

Speedup is defined the ratio of time taken for serial execution to the time taken for parallel execution.

$$\text{Speedup}=\frac{\textit{Time taken for Serial Execution}}{\textit{Time taken for Parallel Execution}}$$

Number of Tasks	Elastic Scheduling	Dynamic Scheduling	Number of Virtual Machines	Improvement in both the algorithms
5	0.68	0.73	5	7.35%
10	1.809	2	10	10.55%
15	3.4	3.933	15	15.67%
20	4.79	5.91	20	23.382%

Table 2. Execution Time of proposed Dynamic Scheduling and Elastic Scheduling where number of virtual machines would be same as the number of tasks given

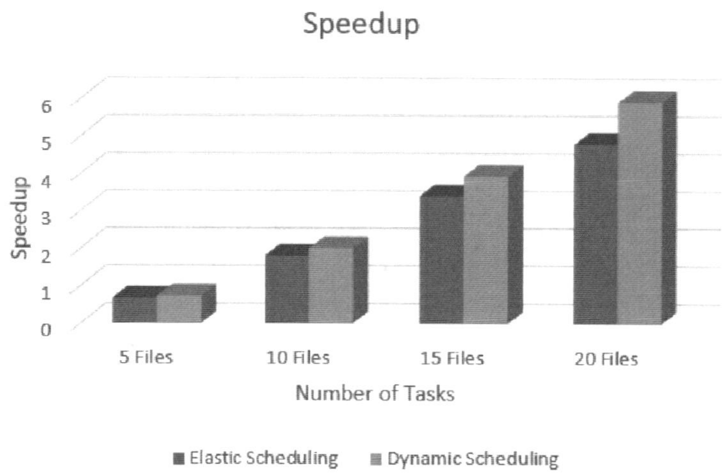

Figure 4. Comparison on the Speedup of Proposed Dynamic Scheduling and Elastic Scheduling.

According to the results in Table 2, it is found that the speedup of proposed dynamic scheduling has been increased by 14.238% in comparison to elastic scheduling.

3. *Efficiency*

$$\text{Efficiency}=\frac{\boldsymbol{Speedup}}{\boldsymbol{Number\ of\ Virtual\ Machines}}$$

Number of Tasks	Elastic Scheduling	Dynamic Scheduling	Number of Virtual Machines	Improvement in both the algorithms
5	0.136	0.146	5	7.37%
10	0.1809	0.2	10	10.558%
15	0.2266	0.2622	15	15.71%
20	0.2395	0.2955	20	23.382%

Table 3. Efficiency of proposed Dynamic Scheduling and Elastic Scheduling where number of virtual machines would be same as the number of tasks given

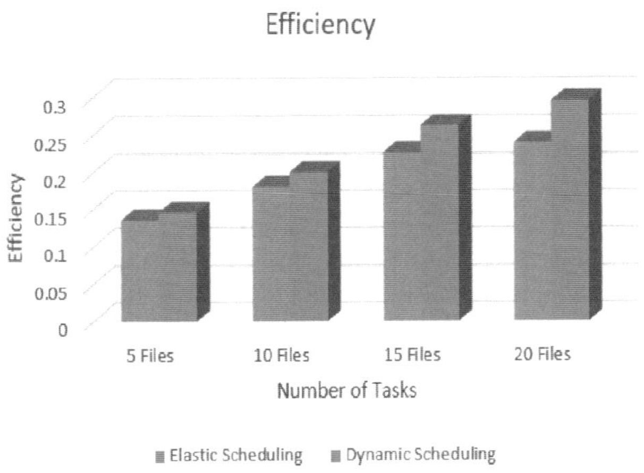

Figure 5. Comparison of Efficiency of proposed Dynamic Scheduling and Elastic Scheduling.

According to the results in Table 3, it is found that the efficiency of proposed dynamic scheduling has been increased by 14.255% in comparison to elastic scheduling.

4. Throughput

Here in this experiment throughput is measured by the number of tasks performed per second

Number of Tasks	Elastic Scheduling	Dynamic Scheduling	Number of Virtual Machines	Improvement in both the algorithms
5	0.294	0.3571	5	21.46%
10	0.476	0.588	10	23.52%
15	0.6	0.75	15	25%
20	0.69	0.869	20	25.94%

Table 4. Throughput of proposed Dynamic Scheduling and Elastic Scheduling where number of virtual machines would be same as the number of tasks given

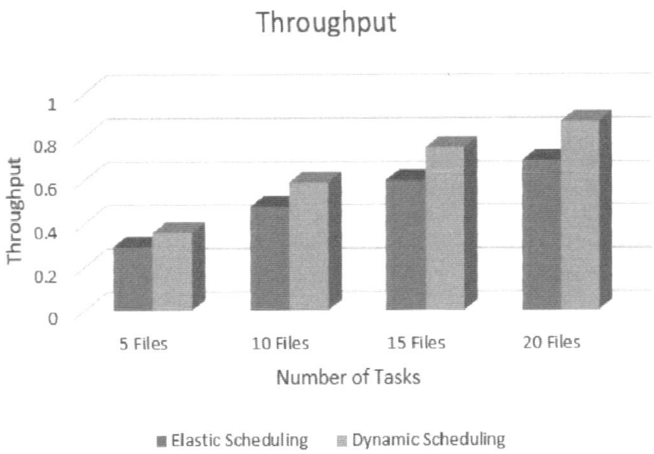

Figure 6. Comparison of Throughput of proposed Dynamic Scheduling and Elastic Scheduling.

According to the results in Table 4, it is found that the throughput of proposed dynamic scheduling has been increased by 23.98% in comparison to elastic scheduling.

V. CONCLUSION AND FUTURE WORK

This paper proposes Dynamic Task Scheduling algorithm for microservices in Cloud Computing Environment. The proposed algorithm aims to minimize the execution time, bring in improvement in speedup, betterment in efficiency and improved throughput of the designed application. The execution time of the proposed dynamic scheduling is reduced by 24.0255%. The speedup of the proposed dynamic scheduling is increased by 14.238%. The efficiency of the proposed dynamic scheduling is increased by 14.255%. The throughput of the proposed dynamic scheduling is increased by 23.98% respectively.

In near future, there is a scope for improvement since microservices is a diverse and huge topic. The task scheduling algorithm can be made even better and tested for various other parameters as well.

REFERENCES

[1] Hamid Mohammadi Fard "Dynamic Multi objective task scheduling of microservices in Cloud Technical University of Darmstadt" in 2020 IEEE/ACM 13th International Conference on Utility and Cloud Computing (UCC)

[2] Sheng Wang, Zhijun Ding "Elastic Scheduling for microservice applications in cloud Changjun Jiang" (2020)

[3] Christina Terese Joseph , Prof. K. Chandrasekaran Dynamic Interaction Aware Resource allocation for microservices in *Cloud Computing Lab, Department of Computer Science and Engineering, National Institute of Technology Karnataka, Surathkal* (2020) 101785

[4] Maria Fazio and Antonio Celesti "Issues in scheduling microservices in cloud" University of Messina 2325-6095/16/$33.00 © 2016 IEEE

[5] Aniello Castiglione " Challenges in delivering microservices in cloud University of Salerno" 2016 IEEE

[6] S. Soltesz , H. Pötzl , M.E. Fiuczynski , A. Bavier , L. Peterson , Container- based operating system virtualization: a scalable, high- performance alternative to hypervisors, in: ACM SIGOPS, vol. 41, ACM, 2019, pp. 275–287 .

[7] Y. Niu, F. Liu, and Z. Li, "Load balancing across microservices," in *Proc. IEEE INFOCOM*, 2018, pp. 199-207.

[8] C. Qu, R.N. Calheiros, and R. Buyya, "Auto-scaling web applications in
clouds: A taxonomy and survey," *ACM Compute. Surveys*, vol. 51, no. 6, pp. 173, Sep. 2019.

[9] M. Fazio , A. Celesti , R. Ranjan , C. Liu , L. Chen , M. Villari , Open issues in scheduling microservices in the cloud, IEEE Cloud Compute. 3 (5) (2018) 81–88

[10] E. Casalicchio, A study on performance measures for auto-scaling cpu- intensive containerized applications, Cluster Comput (2019

[11] D. Bernstein , Containers and cloud: from lxc to docker to kubernetes, IEEE Cloud Compute (2019) 81–84 .

[12] J.P. Martin , A. Kandasamy , K. Chandrasekaran , Exploring the support for high performance applications in the container runtime environment, Hum.-Centric Compute. Inf. Sci. 7 (1) (2019)

Detection and Classification of Arrhythmias by Deploying Deep Learning Models

T Padma[1], A Uday Kiran[2], C Jahnavi[3], S Rahul[4], Nehaa Raja[5], M Kamal Kumar[6]

[1] Professor, ECE, Gokaraju Rangaraju Institute of Engineering and Technology, Bachupally, Hyderabad, India.
[2,3,4,5,6] UG Students, ECE, Gokaraju Rangaraju Institute of Engineering and Technology, Bachupally, Hyderabad, India.

E-mail: profpadmat@gmail.com, adluriuday08@gmail.com, jahnavichsreenivas@gmail.com, siripuramrahul9@gmail.com, nehaaraja0804@gmail.com and mandrukamalkumar@gmail.com

Abstract. Arrhythmias can be detected using an ECG signal, which is an important tool in the healthcare industry. ECG overall variation trends, original variation features, and their relative positions are used to classify arrhythmias according to sphere knowledge and large-scale data analysis. They haven't been fully explored by being styles. CNN and hybrid CNN-LSTM models are used to address this problem. A LSTM and CNN are used to separate the ECG's overall variation trends and its unique features. In this project the implemented models are CNN and Hybrid LSTM models to check which model is better in identifying the arrythmias based on the ACC, SEN, and SPE scores. The Accuracy of the CNN model is 74.4 percent, respectively, while the Hybrid-CNN LSTM scores are 83.5 on the MIT-BIH arrhythmias dataset.

Keywords: Arrythmias, Convolutional Neural Network, Variation trends, Long Short-Term Memory, Electrocardiogram

1. Introduction

More than a third of all human deaths in 2016 were due to Cardio Vascular Disease (CVD) **[1]** and 86% of those deaths were due to a heart attack. The medical history and physical examination of each patient are the mainstays of the traditional approach to the diagnosis of CVD. The taxonomy of medical diseases is used to categorize patients in this study using quantitative medical parameters. As a result of the enormous amount of heterogeneous data that must be analysed and interpreted, the traditional rule-based diagnostic paradigm is often ineffective. The problem will be exacerbated if there aren't enough medical professionals and facilities, especially in developing countries. Thus, a monitoring and diagnosis system must be developed that is both accurate and affordable. using computer systems to assist in diagnosis Healthcare providers are increasingly requiring Computer-Aided Diagnosis (CAD) and appropriate medical assessments, which can be linked to this requirement CADS.

CADS are designed to monitor and evaluate an organ's performance by analyzing physiological signals. In order to educate people about their illnesses, CADs are convenient and easy to use. The ElectroCardioGram (ECG) is a non-stationary physiological signal that shows the heart's electrical activity. Other indications can be checked, such as the regularity of the pulse and the amount of psychological stress. A wide range of applications, including classification and prediction, have made use of deep neural networks (DNNs). For example, DNNs have had a significant impact on the accuracy of many medical tasks in recent years. Arrhythmias in ECG signals are now detected by modern CADS systems using DNN, which lowers the cost of continuous heart monitoring while also improving prediction accuracy. Automatic arrhythmia classification based on ECGs, on the other hand, typically faces a number of significant difficulties.

1.1. Arrhythmia Classification Challenges
CADS arrhythmia classification challenges can be summarized as:

- Even if arrhythmia symptoms are present, they may not be detected while the ECG signal is being recorded [2].
- A range of characteristics, such as age, gender, physical condition, and lifestyle, impact the amplitude and duration of people's ECGs. Finding a broad framework and corresponding standards that can be utilized by the general public is difficult.
- Physical activity, such as running, walking, and sleeping, can affect the ECG signal's morphology.
- ECG signal analysis requires a significant amount of data to be considered. As a result, it is more likely that an arrhythmia will be misdiagnosed.
- The recorded ECG signal may show morphological variations and discrepancies as a result of noise, artefacts, and interference.
- The model accuracy can be improved by eliminating noise by using high pass filter and notch filter.

2. Related work
Depending on the type of CA, each lead's ECG signal has unique characteristics [3]. The P wave or QRS shape (PAC/PVC) of early atrial and early ventricular contractions may be aberrant [4]. Both STE and STD must be greater than 0.1 mV in order to be considered abnormal [5].

To be able to correctly identify these complex ECG characteristics associated with CA, extensive training is necessary. Internal medicine doctors and cardiologists have been found to make diagnostic errors on occasion [6]. Physicians' workload is exacerbated by the rapid growth of ECG examinations. Computer algorithms that generate accurate and automated diagnoses could be beneficial to doctors. In spite of the wide range of ECG signal characteristics, recent years have seen significant progress in the area of ECG signal processing [7].

Using the first technique, ECG signals may be split into heartbeat, or cycles. For a small number of subjects, machine learning requires a huge amount of data to train predictive classification models using this beat-based approach. ECG signal delineation can be difficult because ECG morphological features are frequently extracted with high levels of imprecision [7]. This could be due to the fact that the same person creates both the training and test beats, resulting in a prediction accuracy of 99.9%. When test beats from patients who were not part of the training set were added in the cross-validation, the accuracy for six CA categories fell to 81.5 percent from 99.7%. [8].

Despite the fact that only ECG data from 48 and 452 subjects are publicly available, previous CA prediction studies have relied on these two databases regardless. A lack of subjects in neural network databases can lead to over-fitting issues in neural network algorithms [9]. For example, over-representation of certain CA types in the data can lead to over-fitting of models. UCIAD and MIT-BIH AD may be impacted as a result of these problems. Using UCIAD, it was possible to achieve a high level of CA classification accuracy (92%), but this accuracy dropped to 60% when training and testing were split 80/20. Muhammad and his co-workers (2018a) While UCIAD features (average width, amplitude, and so on) are only extracted for 12-leads in UCIAD's Only two leads of ECG data are included in the MIT-BIH AD ECG data (MIT-BIH AD ECG data only includes two leads) [10].

No downside to the second approach, which solves the beat-based problem completely. To put this technique into action, you'll need a lot of ECG data and a sophisticated artificial neural network with deep learning capabilities. Due to recent developments in both elements, the second strategy has grown more desirable. The 2017 Physio Net/Computing in Cardiology Challenge is now open for entries. **[11]** shared data from 8,528 people who had a variety of cardiac rhythms (AF, normal, other rhythms, and noise) to show how open-source research may benefit science. [12].

This article was written by [13]. Based on the complementing properties of ME and NCL, this research provides an ECG arrhythmia classification approach (NCL) [14]. ME includes a control parameter for NCL in its error function, allowing the training process to manage bias-variance-covariance trade-offs. [15] In addition to maintaining and alleviating the disadvantages of its basis approaches, the proposed classification method offers significantly better performance than the original methods.

3. System analysis

3.1. Existing system

In the Recent advances in deep learning, particularly LSTM networks and convolutional neural networks (CNNs), have demonstrated important information lines and befitting capabilities Disease detection and picture categorization, for example, are only a few of the uses. CNN is excellent at extracting the original features while LSTM is excellent at mining long-term time-series dependencies. Arrhythmias can be classified using LSTM or CNN, and researchers have had success. Despite this, there are a few things that could be done better.

3.2. Proposed system

This System Proposes CNN model which is an effective model for classification of irregular heartbeats and hybrid CNN-LSTM model which uses dropout layers to improve the model accuracy this paper involves a comparitive study between CNN and Hybrid CNN LSTM model.

4. Dataset

The MIT-BIH Arrhythmia Dataset and the PTB Diagnostic ECG Database are used to classify heartbeats. The heartbeat signals in this dataset are separated into two groups. To test deep learning models, researchers employed a dataset with a sampling frequency of 125 hertz and a total of 109446 ECG signals. The Dataset contains the various signals and separated into five sub classes these classes are a group of signals which contains various categories like Normal and Abnormal conditions of detected signals from ECG.

Journal of Physics: Conference Series **2325** (2022) 012053 doi:10.1088/1742-6596/2325/1/012053

5. Implementation

The following are the steps involved in putting the model into action.

- Data Collection
- Data Pre-processing
- Modelling

5.1. *Data Collection*

The information comes from the MIT-BIH database. The database is further divided into the two datasets called Training and Testing Datasets. In the figure.1 the heart beat for each target label of five sub-classes of arrythmias are plotted which are detected based on the number of occurrences in the database.

5.2. *Pre-processing*

The data is Pre-processed to eliminate unwanted noises by oversampling on training dataset to balance the dataset. In the figure.1 the heart beat for each target label of five sub-classes of arrythmias are plotted which are detected.

5.3. *Modelling*

Deep Learning models in Figure 1 is the future extraction and classification is done on the input and output which is generated by the using models such as CNN and Hybrid CNN LSTM.

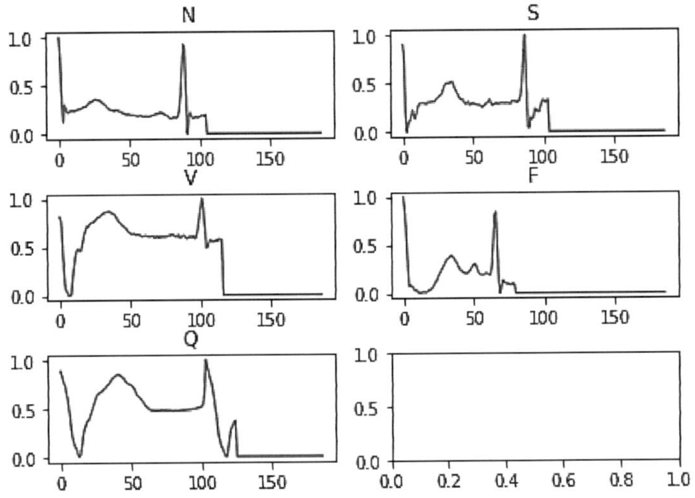

Figure 1. Sub-classes of Arrythmias

6. Block diagram

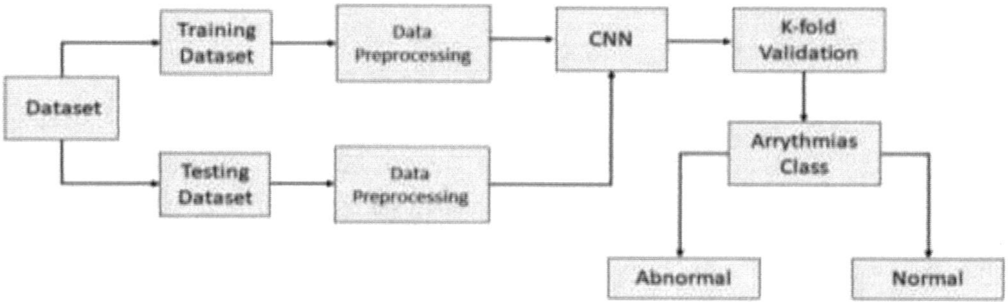

Figure 2. Block diagram of CNN

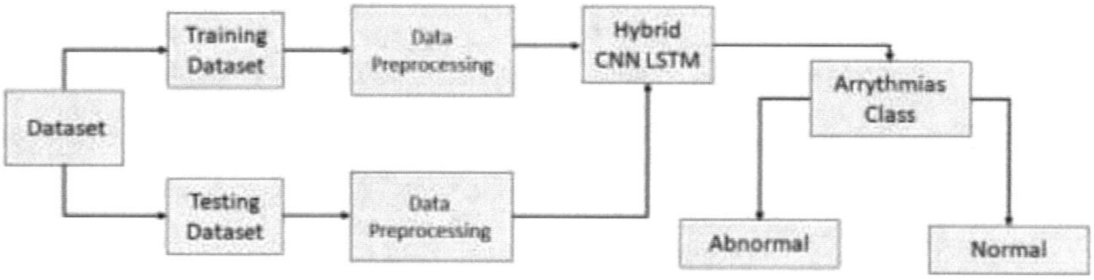

Figure 3. Block diagram of Hybrid CNN- LSTM

The Figure 2 and Figure 3 contains the functionality of each blocks used for classifying Arrythmias.

7. Deep learning models

7.1. Convolutional Neural Network

- CNNs are neurons that, like ordinary neural networks, have learnable weights and biases. Each neuron receives a large number of inputs, calculates a weighted sum of them, sends them via an activation function, and then outputs.
- As shown in Figure 4, CNN has three types of layers: convolution, pooling, and fully connected layers.
- Fully linked input layer (flatten): Takes the previous layer's output and "flattens" it into a single vector that may be used as an input for the following level.
- Unrolling (flattening): Converting a matrix to a vector.

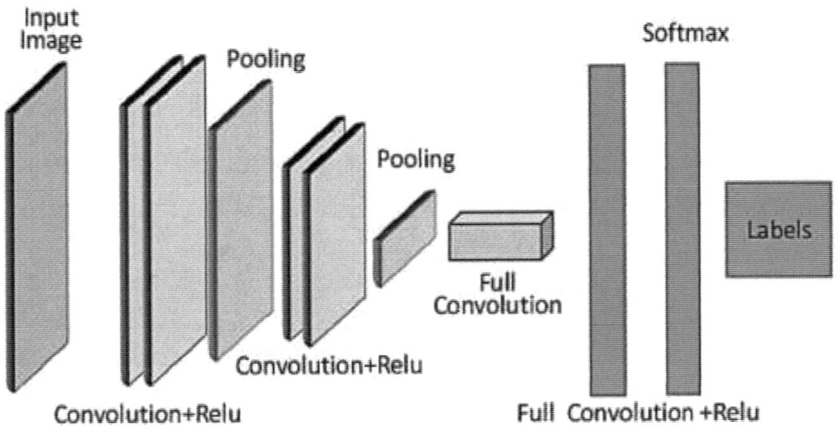

Figure 4. Basic CNN Architecture

7.2. Hybrid CNN LSTM

- CNN has only conv1D and max1D blocks. The result of the max1D block is transmitted to the input layer of the following system in hybrid system.
  yi (CNN xi) (1)
- xi is the CNN network's initial input vector, containing a class label for each type of data. The CNN network's output is yi. It's the feature vector created by CNN's max-pooling process. This information is integrated into the next deep learning architecture that is employed. In this study, CNN-LSTM, hybrid networks are used, and their performance is compared.

8. RESULTS AND DISCUSSION

The functional conditions or overall description papers comprise information on the product's viewpoint and features, the operating system and operational environment, graphics, design limits, and user evidence of the product's usefulness. Examining the project's circumstances and restrictions, which indicate how to address each, provides an overview of the project's strengths and shortcomings. This project was built on a PC, and the minimal hardware requirements for Enthought Python, Canopy, and VS Code users are different. As a result, processes that require a large number of calculations or tasks to be completed rapidly require more RAM and a quicker processor.
The proposed model's performance was trained with minimum of 50 epoch values and the following results are acquired which are shown in Table 1.

Table 1. Comparison of CNN And Hybrid CNN- LSTM

Parameters	**CNN**	**Hybrid LSTM**	**CNN**
Accuracy	74.4	83.5	

```
cnn_predictcv=np.argmax(cnnpredictionscv,axis=1)
cnn_actual_valuecv=np.argmax(Y_test,axis=1)
showResults(cnn_actual_valuecv, cnn_predictcv)

Accuracy  : 0.7441074365064864
Precision : 0.901016675905698
f1Score : 0.7937238346648685
[[13029  1856  1597  1153   483]
 [  107   405    28    14     2]
 [   76    41  1223    95    13]
 [    8     5     8   141     0]
 [   43     4    59    10  1492]]
```

Figure 5. Accuracy of CNN

```
predictcv=np.argmax(predictions,axis=1)
actual_valuecv=np.argmax(Y_test,axis=1)
showResults(actual_valuecv, predictcv)

Accuracy  : 0.8354193312625616
Precision : 0.9241890591689367
f1Score : 0.866545503401796
[[14914  1172   506  1004   522]
 [   83   449    10    10     4]
 [   76    39  1246    73    14]
 [    9     1     6   146     0]
 [   46    13    13     2  1534]]
```

Figure 6. Accuracy of Hybrid CNN- LSTM

9. Conclusion

To classify arrhythmias, this paper proposes two different models: a hybrid LSTM-CNN-CNN and a CNN. In this model structure appears to be the most effective based on the results of the tests. In terms of ACC which is shown in Figure 5 the CNN model achieves 74.4 percent accuracy. From the Figure 6 there is 83.5 percent accuracy in ACC for the hybrid CNN LSTM. An attention-based model should be developed to supplement the current model.

10. References

[1] Benjamin, Emelia J., et al. "Heart disease and stroke statistics—2019 update: a report from the American Heart Association." *Circulation* 139.10 (2019): e56-e528.

[2] Ceylan, R., & Özbay, Y. (2007). Comparison of FCM, PCA and WT techniques for classification ECG arrhythmias using artificial neural network. *Expert Systems with Applications*, *33*(2), 286-295.

[3] O'NEAL, WESLEY T., et al. "Atrial fibrillation and cardiovascular outcomes in the elderly." *Pacing and Clinical Electrophysiology* 39.9 (2016): 907-913.

[4] Senzaki, Hideaki, et al. "Ventricular afterload and ventricular work in Fontan circulation: comparison with normal two-ventricle circulation and single-ventricle circulation with Blalock-Taussig shunts." *circulation* 105.24 (2002): 2885-2892.

[5] Martin, Luis Cuadrado, et al. "Association between hypervolemia and ventricular hypertrophy in hemodialysis patients." *American journal of hypertension* 17.12 (2004): 1163-1169.

[6] Hannun, A. Y., Rajpurkar, P., Haghpanahi, M., Tison, G. H., Bourn, C., Turakhia, M. P., & Ng, A. Y. (2019). Cardiologist-level arrhythmia detection and classification in ambulatory electrocardiograms using a deep neural network. *Nature medicine*, *25*(1), 65-69.

[7] Gorter, Thomas M., et al. "Right heart dysfunction and failure in heart failure with preserved ejection fraction: mechanisms and management. Position statement on behalf of the Heart Failure Association of the European Society of Cardiology." *European journal of heart failure* 20.1 (2018): 16-37.

[8] Xue, Yong, et al. "Application of deep learning in automated analysis of molecular images in cancer: a survey." *Contrast media & molecular imaging* 2017 (2017).

[9] Zaroug, Abdelrahman, et al. "Lower limb kinematics trajectory prediction using long short-term memory neural networks." *Frontiers in Bioengineering and Biotechnology* 8 (2020): 362.

[10] Mustaqeem, Anam, Syed Muhammad Anwar, and Muahammad Majid. "Multiclass classification of cardiac arrhythmia using improved feature selection and SVM invariants." *Computational and mathematical methods in medicine* 2018 (2018).

[11] Yazdani, Sasan, et al. "Heart rhythm classification using short-term ECG atrial and ventricular activity analysis." *2017 Computing in Cardiology (CinC)*. IEEE, 2017.

[12] Clifford, Gari D., et al. "AF classification from a short single lead ECG recording: The PhysioNet/computing in cardiology challenge 2017." *2017 Computing in Cardiology (CinC)*. IEEE, 2017.

[13] Javadi, M., Arani, S. A. A. A., Sajedin, A., & Ebrahimpour, R. (2013). Classification of ECG arrhythmia by a modular neural network based on mixture of experts and negatively correlated learning. *Biomedical Signal Processing and Control*, *8*(3), 289-296.

[14] Vivekanandam, B. "*Speedy Image Crowd Counting by Light Weight Convolutional Neural Network.*" Journal of Innovative Image Processing 3, no. 3 (2021): 208-222.

[15] Vijayakumar, T. "*Neural network analysis for tumor investigation and cancerprediction.*" Journal of Electronics 1, no. 02 (2019): 89-98.

Analysis of object detection techniques for bird species identification

Saloni Saxena, Sneh Thorat, Prachi Jain, Rupal Mohanty, Trupti Baraskar

Dr. Vishwanath Karad's MIT World Peace University, S.No.124, Paud Road, Kothrud, Pune, Maharashtra 411038, India.

salonisaxena556@gmail.com, 1006sneh@gmail.com, prachijain200310@gmail.com, rupalmohanty2014@gmail.com, trupti.baraskar@mitwpu.edu.in

Abstract. The birdwatching community in India is extensive. Wildlife enthusiasts often face difficulty in identifying a particular bird. There exist, separately, object detection Machine Learning models as well as online directories for manual bird identification, but there is no approach combining the two for easy identification of birds in India. We present a technique that uses object detection algorithms such as Faster R-CNN and YOLOv5 to solve this challenge. Our dataset includes a total of 60 species of birds found in Maharashtra. Furthermore, these methods have been tested on datasets of various sizes to provide a thorough comparison of the two techniques and to better understand their behaviour as the dataset size and classes grow. The YOLOV5 model trained on 100 epochs for 3000 images achieved a mAP score of 0.78 whereas Detectron2 when trained on the same dataset achieved an IOU score of 0.001. These methods have not been tested on a dataset containing birds from peninsular India, particularly Maharashtra. As a result, we intend to extend the dataset and make bird identification easier for birdwatchers.

Keywords. Object Detection, Convolutional Neural Network (CNN), YOLOv5, Detectron2, mean Average Precision (mAP), Intersection Over Union (IOU)

1. Introduction

Birdwatching is an extremely significant activity. It is beneficiary not only for ecotourism but also for environmental wellbeing and conservation of local species [1]. In its essence, it is the observing of birds either as a hobby or for research purposes. There are certain professionals like ornithologists who approach birdwatching from a professional capacity. However, most birdwatchers are purely just hobbyists. In recent years, there has been an increase in the number of birdwatchers [2].

Birdwatching is also an arduous task, as birdwatchers are required to have extensive knowledge of various species to be able to locate their desired birds among the many local species. Professional birdwatchers also face difficulty in identifying birds and often have to go through comprehensive directories for identification [3]. Looking for a particular bird in a massive guide is a strenuous process as more than 500 bird species are observed just in Maharashtra itself [4]. Furthermore, numerous birds migrate to and from different states and countries, making the process of identification challenging.

With advancements in computer vision and machine learning, we have easier, more convenient and efficient ways of bird identification. Every bird has distinct features such as sound, habitat, region and appearance (colour, beak shape, beak length, wing span, etc). A study explains how birds can be identified using convolutional neural networks from their sounds [5]. However, it is far more reliable to identify birds by their appearance as multiple species of birds have analogous sounds which may be difficult to distinguish, and sound can be affected by surrounding noise. Various algorithms have been implemented on vast datasets to make bird identification easier, but no such solution has been implemented on a dataset of Indian bird species, specifically, of Maharashtra and Deccan Plateau region.

As a result, we propose using object detection techniques such as Faster R-CNN and YOLOv5 to identify birds in images from Peninsular India and Maharashtra. Faster RCNN is a faster and better form of Fast R-CNN, which is a variant of Regional-CNN. YOLOv5 is a brand-new, cutting-edge object identification method that builds on YOLO (You Only Look Once) [6]. This work[7] uses a 6-axis mechanical arm in conjunction with computer vision to investigate object tracking in plane examination using backprojection and the CAMSHIFT method.

Initially, we tested these algorithms on a dataset of 3000 images, with 50 images in each class. The findings were astounding, with YOLOv5 having the best mAP score of 0.78 and Faster R-CNN (using Detectron) having an AP score of 0.27 and not identifying most classes at all. Surprised by the poor results obtained from Faster RCNN, we decided to understand the impact of number of classes and size of dataset on these algorithms. To achieve a better comparative analysis, we implemented these algorithms on 250, 500, 1000, 1500, 2000, and 3000 images while maintaining the number of epochs constant at 100. Our initial prediction that Faster RCNN requires a substantially higher number of epochs than YOLOv5 and has difficulty training on big datasets with numerous classes was confirmed by the findings.

A comparison of the implemented algorithms in our study reveals that YOLOv5 is a more efficient method that produces better results in fewer epochs while also functioning well on bigger datasets. Thus, our approach provides a superior solution to the laborious bird identification process for a dataset that has not been covered before.

2. Dataset Collection

A total of 60 birds were selected for this study. 50 images of each bird were taken for training the model. To annotate and categorise our data, we used these 60 distinctive kinds of bird names as classes such as Brown Shrike, Glossy Ibis, Wooly Neck et al. 3000 images were collected in total. To train the model for a variety of circumstances, the photographs contain views from various perspectives and lightings as well as some images of birds flying. The image source is Oriental Bird Club Images Database [3].

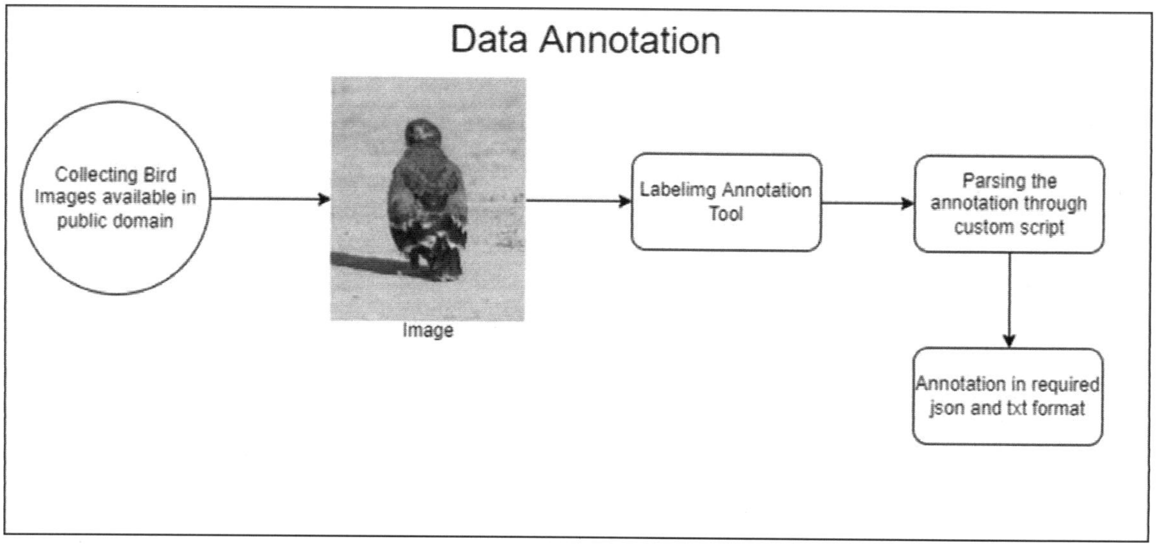

Figure 1. The initial step is to collect bird images, as shown in diagram. After gathering all of the images, each one is annotated with the LabelIMG annotator, in which we create bounding boxes around the bird to tell the model where it is in the image. These annotations are then parsed using a custom script, yielding annotations in both JSON and text format.

2.1 Data Preprocessing and Augmentation

This is a technique for transforming raw data into a usable and efficient format. Generally, in this stage the frequently used methods are orienting, resizing to the desired size, rescaling, changing the colour contrasts etc. Some object detection algorithms demand images to be of same size. The data preprocessing is also important for accuracy as data can be noisy and have some missing values. Data preprocessing helps to overcome this issue with feature selection and feature extraction. The data processing techniques used in this study are: Resize to 416px x 416px, Auto-orientation.

Apart from data preprocessing various data augmentation techniques were applied to the images. Data augmentation is the process of creating new augmented data by modifying existing data. Data augmentation increases the quantity of images in a data set while simultaneously reducing overfitting. Some data augmentation methods were used in this study which includes – flipping the images horizontally, adding shear of ±20° Horizontal, ±20° Vertical to the images, changing saturation between -50% and +50% and changing brightness between -34% and +34%. Using these methods helped us to increase the dataset substantially.

The images were divided into three sets: 70% in the training set, 20% in the validation set, and 10% in the test set.

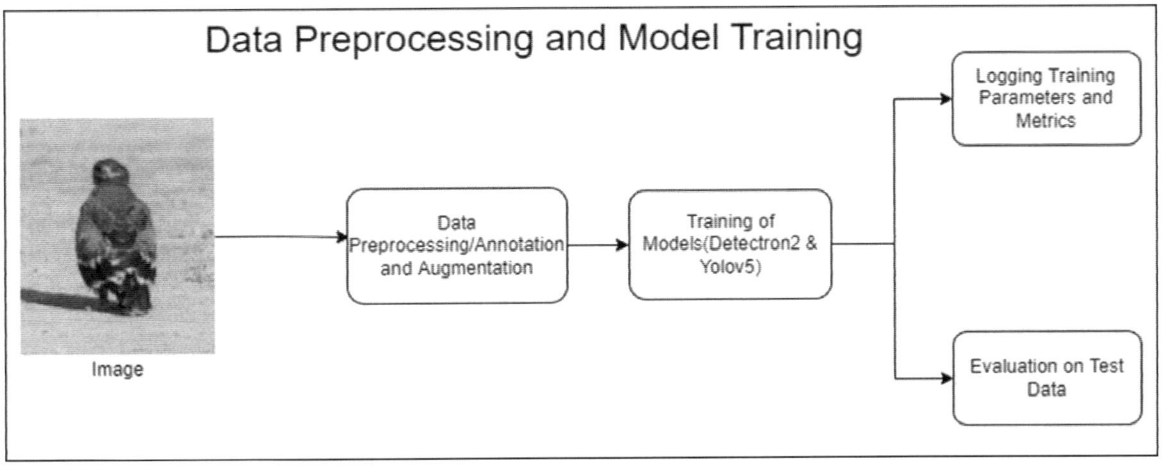

Figure 2. The images are preprocessed to convert the raw data into efficient data using the four steps that are data cleaning, data integration, data reduction, and data transformation and further trained using Detectron2 and Yolov5.

3. Models

3.1 YOLOv5

YOLO is an acronym for You Only Look Once. YOLOv5, the newest member of the family, is made up of three parts: the model backbone, neck, and head. Cross Stage Partial Networks, which extract features from the input image, serve as the system's backbone. PANet serves as the neck, obtaining feature pyramids and assisting in the identification of identical items when scaled differently. Finally, the head is assigned to carry out the final and most crucial detecting step. The SxS grid structure is used by YOLO algorithms to separate all of the input images. Each grid is responsible for object detection. The bounding boxes for the observed object have now been approximated by those grid cells. Each box includes 5 critical characteristics: x and y coordinates, object width and height in w and h, and a confidence score indicating the likelihood of the box containing that object.

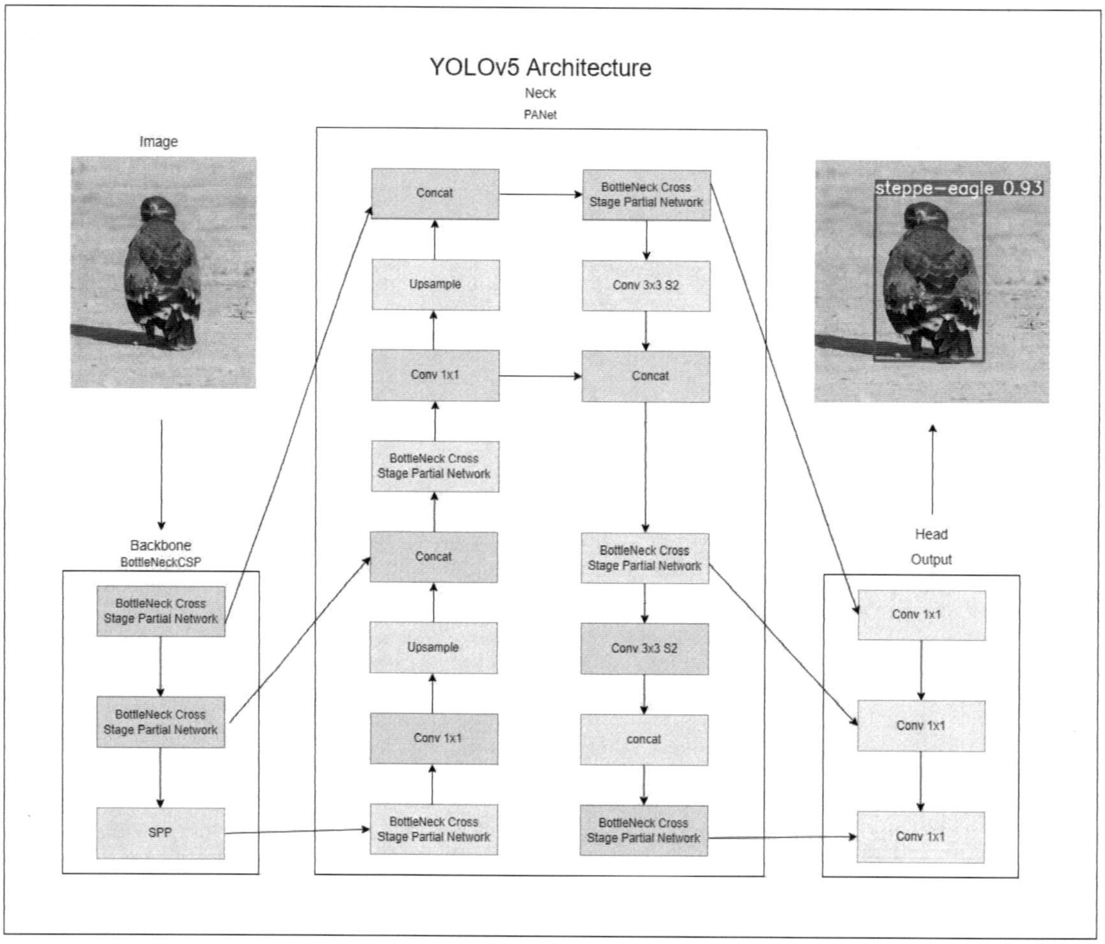

Figure 3. Backbone, Neck, and Head are the three primary sections of the YOLOv5 model's architecture. The backbone of the YOLOv5 network is a BottleNeck Cross Stage Partial Network. The ground truth photos are fed into the backbone, which extracts significant features. The feature pyramid is constructed at the model's neck. PANet is used by YOLOv5. The final detection element of the model is implemented in the model's head, which offers the images' final bounding boxes as well as other metrics, probabilities, and scores.

3.2 Detectron2

Facebook AI Research has released Detectron 2, a second-generation open-source object detecting technology. The three components of Detectron 2 are the convolution layers, the Region Proposal Network (RPN), and the Detection Network. The convolutional layers take an input picture, process it, and then pass the extracted features to the RPN. The second component of the Detectron 2 architecture is RPN. It's a basic network whose goal is to suggest potential regions for finding areas of interest. The Detection Network is the last component of the network, and it receives inputs from both the previous parts to build final bounding boxes around class objects depending on the computed confidence.

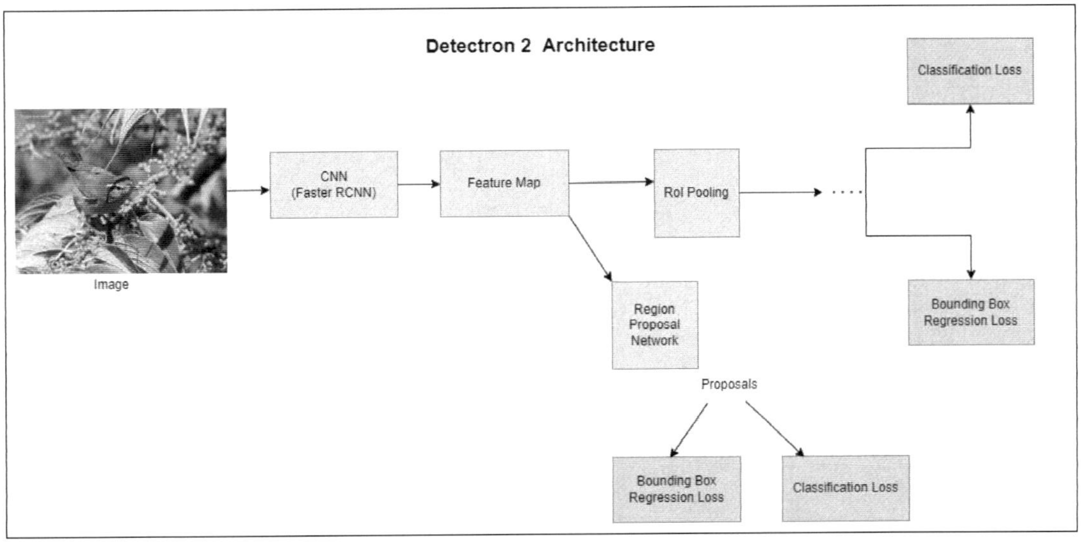

Figure 4. The architecture of Detectron 2 is made up of a convolution layer that captures features of the input images and feeds them to a regional proposal network that offers probable regions of interest for class objects and a detection network that creates bounding boxes around them.

4. Metrics and Losses

4.1 Precision

Precision is calculated by dividing the number of properly recognized Positive samples by the total number of Positive samples (either correctly or incorrectly). The accuracy of the model in classifying a sample as positive is assessed.

$$Precision = \frac{True\ Positive}{True\ Positive + False\ Positive}$$

4.2 Recall

The recall is obtained by dividing the total number of Positive samples by the number of true positive a. The recall of a model is a metric that measures how well it can understand Positive samples. The greater the recall, the greater the number of positive samples identified.

$$Recall = \frac{True\ Positive}{True\ Positive + False\ Negative}$$

4.3 Mean Average Precision

The mean average precision (mAP) is used to assess object detection models such as R-CNN and YOLO. The mAP calculates a score by comparing the ground-truth bounding box to the detected box. The higher the score, the better the model's detection accuracy.

4.4 Average Precision

The average precision (AP) is a way to summarize the precision-recall curve into a single value representing the average of all precisions. The AP is calculated according to the next equation. Using a loop that goes through all precisions/recalls, the difference between the c0urrent and next recalls is calculated and then multiplied by the current precision. In other words, the AP is the weighted sum of precisions at each threshold where the weight is the increase in recall.

4.5 Intersection over Union

The intersection over union (IOU), defined as the ratio between the union and intersection of the detected box and the ground truth box, is used as an indicator to determine whether an object is correctly detected, and the average precision (AP) value changes according to the IOU threshold.

The model predicts the bounding boxes of the detected objects. It is expected that the predicted box will not match exactly the ground-truth box.

$$Intersection\ over\ Union = \frac{Area\ of\ Intersection}{Area\ of\ Union}$$

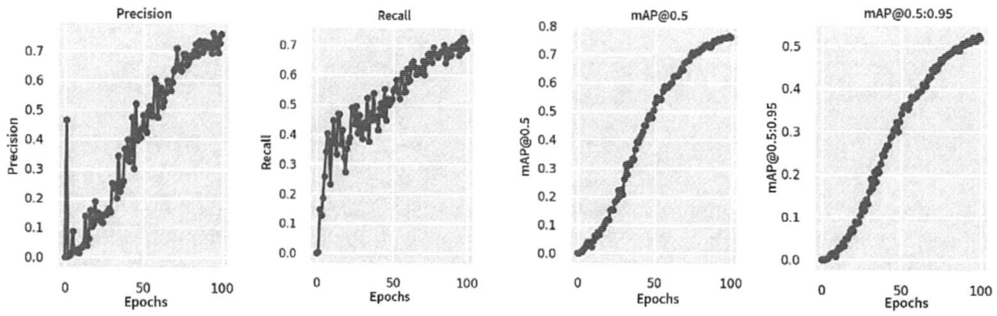

Figure 5. Precision, Recall and mAP graphs for YOLOv5.

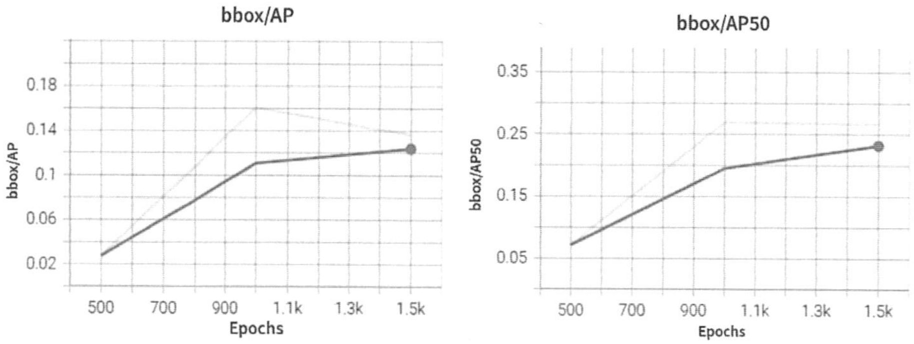

Figure 6. AP and AP50 graphs for YOLOv5.

5. Results

Initially, we used a dataset of 3000 images, i.e. 60 classes, to run both YOLOv5 and Faster R-CNN (Detectron2) on, for 100 epochs. The results were unexpected because Detectron2 produced poor results with an IoU score of 0.001 while YOLOv5 worked well in only 100 epochs with mAP of 78 percent.

The following metrics were used to evaluate YOLOv5: mAP, precision, and recall. The results can be seen in the graphs in Figure 5 and Figure 6.

Detectron2 was evaluated using the following metrics: IOU (0.5), IOU (0.5:0.95) and AP50. The result can be observed in Figure 7.

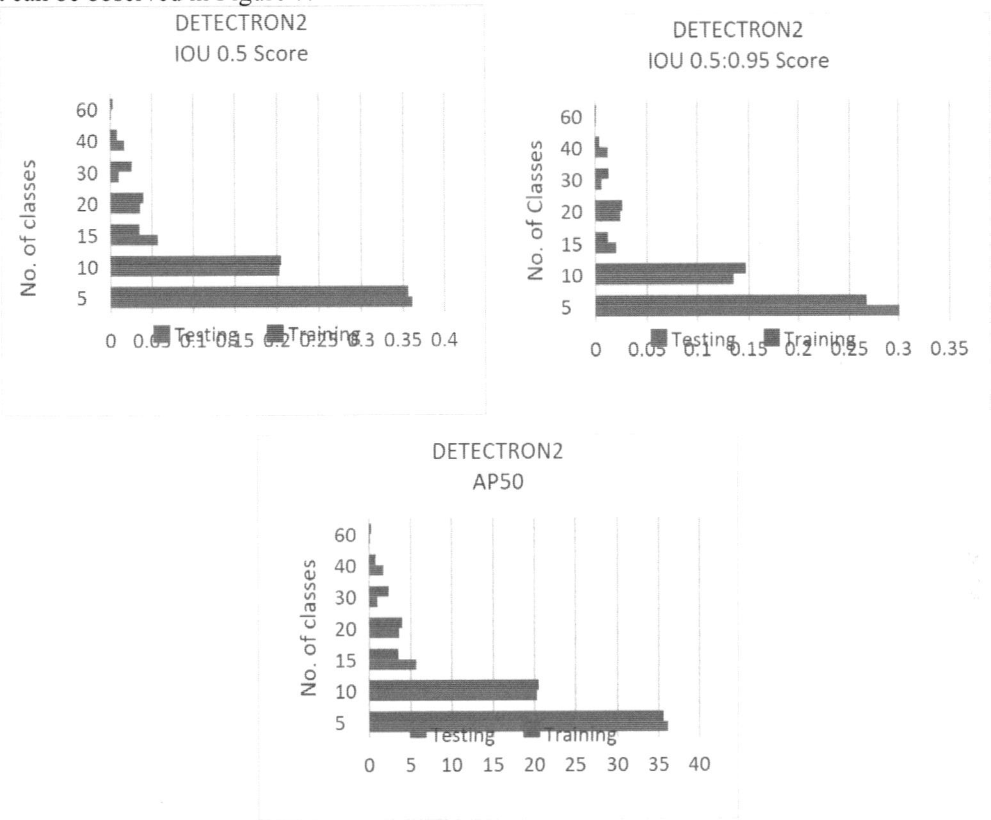

Figure 7. Visual representation of Detectron 2 metrics comparison of training and testing results.

Our initial hypothesis was that Detectron2 would require significantly more epochs to produce acceptable results, and that the number of classes would be inversely related to the results produced. Thus, we decided to broaden our research to a larger number of classes while maintaining the same number of epochs and images per class. As a result, our research now included 5, 10, 15, 20, 30, 40, 60 classes. It can be observed that although Detectron2 did not provide good results, the model is stable as the difference between the results of train and test is minimal. The findings are summarised in the table below. The train and test results can be seen in Figure 7.

Table 1. Results of Detectron 2.

No. of Classes	IOU 0.5	IOU 0.5:0.95	AP50
5	0.361	0.301	36.139
10	0.203	0.135	20.3
15	0.057	0.02	5.662
20	0.036	0.024	3.571
30	0.01	0.006	0.967
40	0.017	0.012	1.705
60	0.001	0.001	0.122

Table 2. Results of YOLOv5.

No. of classes	Best mAP	Last mAP
5	0.789	0.642
10	0.814	0.814
15	0.709	0.703
20	0.727	0.719
30	0.748	0.748
40	0.817	0.805
60	0.78	0.78

The YOLOv5 model could predict single birds with a confidence of 91 percent and 89 percent, as shown in the predictions in Figure 8. It could also recognise many birds in a single shot as well as birds from various perspectives.

The predictions of Detectron2 with 5 classes were as observed in Figure 9 and 10. As we can see in Figure 9, it correctly predicted the majority of birds. However, as the number of classes grew, the model was unable to correctly identify the majority of birds (Figure 10).

Figure 8. YOLOv5 Predictions

Figure 9. Detectron2 predictions with 5 classes

Figure 10. Detectron2 predictions with 60 classes

6. Conclusion

Initially, we used 3000 images to test YOLOv5 and Detectron2, however Detectron2 performed poorly. As a result, we experimented with a lesser number of classes, which yielded satisfactory results. As a corollary, we expanded the scope of our experiment to better analyse the behaviour of the two algorithms as the number of classes grew. The AP moved from A to B as the number of classes increased, indicating that Detectron2 grew increasingly inadequate with an increase in number of classes. We discovered that whereas YOLOv5 showed constant improvement with the adjustment, Detectron2 exhibited a drop. As a result of our research, we discovered some flaws in Detectron2(Architecture).

A dataset containing birds from peninsular India, particularly Maharashtra, has not been examined using similar approaches. As a result, we commenced with a dataset of the ARAI hills in Maharashtra, India. We were constrained by resources because we were constructing a custom dataset; however, our dataset can be expanded to a much larger one covering birds from all over Maharashtra and beyond. Birds' traits vary as they grow older, and male and female birds have extremely diverse features. By growing our dataset, we may be able to tackle these issues in the future.

7. References

[1] Sekercioglu C. Impacts of birdwatching on human and avian communities. Environmental Conservation. 2002;29(3):282-289.

[2] Raman S. [Internet]. Birdwatching is striking a chord with many amid pandemic. 2021 [cited 12 May 2022]. Available from: https://timesofindia.indiatimes.com/life-style/spotlight/birdwatching-is-striking-a-chord-with-many-amid-pandemic/articleshow/85351294.cms

[3] Oriental Bird Club Image Database : Welcome [Internet]. Orientalbirdimages.org. 2022 [cited 12 May 2022]. Available from: http://orientalbirdimages.org/

[4] eBird - Discover a new world of birding... [Internet]. Ebird.org. 2022 [cited 12 May 2022]. Available from: https://ebird.org/home

[5] Maharashtra, India - eBird [Internet]. Ebird.org. 2022 [cited 12 May 2022]. Available from: https://ebird.org/region/IN-MH
Incze A, Jancso H, Szilagyi Z, Farkas A, Sulyok C. Bird Sound Recognition Using a Convolutional Neural Network. 2018 IEEE 16th International Symposium on Intelligent Systems and Informatics (SISY). 2018

[6] Redmon J, Divvala S, Girshick R, Farhadi A. You Only Look Once: Unified, Real-Time Object Detection. 2016 IEEE Conference on Computer Vision and Pattern Recognition (CVPR). 2016;.

[7] Chen, Joy Iong-Zong, and Jen-Ting Chang. "Applying a 6-axis Mechanical Arm Combine with Computer Vision to the Research of Object Recognition in Plane Inspection." Journal of Artificial Intelligence 2, no. 02 (2020): 77-99

Real-Time Quality Assurance of Fruits and Vegetables with Artificial Intelligence

Jagannadha Swamy Tata, Naga Karthik Varma Kalidindi, Hitesh Katherapaka, Sharath Kumar Julakal, Mohan Banothu

Department of Electronics and Communication, Gokaraju Rangaraju Institute of Engineering and Technology, Hyderabad, Telangana, India.

jagan.tata@griet.ac.in, karthikvarma200@gmail.com, hiteshkatherapaka@gmail.com, sharathjs7@gmail.com, mohanbanothu786@gmail.com.

Abstract. Vegetables and fruits make up a major part of the human diet and finding a good grade of its quality is now a major issue in the market. To find a grade of the vegetable or fruit will be based on some parameters like size, shape, appearance, etc. The appearance is now a deciding factor for the market and affects the consumer's choice. So, we have designed an application that will classify fruits and grade them according to their quality with appearance as a parameter. This paper will describe the process which is involved in the application. This proposed system will use image processing to classify and grade the quality of fruits and vegetables by extracting features such as color, shape, and HOG (Histogram of Gradient) to classify the given fruit or vegetable. Image pre-processing techniques like data-augmentation and normalization along with Principle- Component Analysis (PCA), and also Deep learning (CNN) are used for getting good accuracy and for dimensional reduction. To speed up the identification and increase the usability, compared to current manual systems like a person checking every fruit and vegetable to grade which takes more time and energy, or by using embedded systems (sensors), we opted for a high-performance android application for quicker deployment.

Keywords --- Image Processing, Android Application, Fruits, Vegetables, Deep Learning, Machine Learning, Computer Vision.

1. Introduction

From the past, India was always predominantly an agricultural nation. The country produces a wide variety of fruits and vegetables throughout the seasons. In the world, India is one of the largest producers of fruits and vegetables. Due to the difficulty of classifying the quality of fruit using the traditional method in the agricultural or food industry, the image processing technique was developed to classify vegetables and fruits. Agriculture has always been a major sector in improving the Indian economy even in critical times due to its huge production and also consumption. Food is vital for human life and the safety of food has always been the concern of many healthcare treaties. Efficient vegetable and fruit quality checking are crucial in the present trend. In the field of agriculture, from the farmers' end or the consumer end performing the manual inspection is a difficult task and time taking process. These fruits and vegetables can be graded with two approaches such as non-visual and visual approaches. In the non-visual approach, the fruits or vegetables are graded using aroma, chemicals, etc., with the help of embedded systems using hardware and sensors to measure diseases and the quality of fruits and vegetables, however, these systems require many resources to operate, and they are usually useful only for the early stages [1][2][3][4][5]. In the visual approach, fruits and vegetables are graded based on their appearance which is found in the images of fruits and vegetables [6]. In terms of detecting objects and classifying them, artificial intelligence-based systems are popular [7]. In addition, these systems are used to detect the quality of fruits and vegetables and identify their diseases [8]. Artificial intelligence-based fruit or vegetable checking and grading systems are considered to potentially replace traditional manual inspection in the agricultural industry, automation increases the quality and production of the fruits and vegetables [9]. In terms of sensory characteristics, fruits and vegetables have an appearance that influences their market value, preference, and choice of the customers. In recent years, the food industry has begun to realize the benefits of computer machine vision and image processing techniques, especially when it comes to quality inspection and sorting [10].

The color and shape characteristics of fruits and vegetables are decisive for visual inspection. A capable autonomous system for sorting must be able to adequately identify both parameters. The outline of fruits or vegetables can simply be attained from a digital image using traditional techniques for image processing [11]. However, color identification includes many physical and psychological concepts, asking it is tough to properly model and process color in an image. There are wide varieties of color systems existing for the grading of fruits and vegetables based on colors. There are some techniques like Fuzzy logic, Neural networks; Based on Color histograms, Genetic algorithms, etc. [12]. There are also predefined models in Neural Networks for predicting [13]. To meet consumers' growing demands for higher-quality products, new computer vision inspection features have been developed. Currently, the use of computer vision in agriculture and food industries is being utilized in areas such as the inspection of fresh produce based on spectral images [14]. The development of automated tools in the agricultural and food industries has not taken full advantage of technologies such as image analysis and computer vision. As a general rule, the color of the fruit or vegetable denotes its maturity and also its defects. Fruits that have a good physical appearance will hold their value in the market, so we need to keep them in good condition after harvesting.

There are several primary motivations for the identification and classification of a quality detection system, such as the need for quality products from consumers, the consistency of machine behavior over that of humans, and labor shortages. As they are delicate materials, fruits and vegetables must be tested using non-destructive methods. Their size is the most important physical property, while their color is a measure of their visual properties [15]. Therefore, the classification of fruits and vegetables is necessary to evaluate agricultural products, meet quality standards, and increase the value of the market, as well as to effectively plan, package, and transport goods.

This paper presents an easy process for checking and grading the vegetables and fruits using software techniques which are also an economic and safe way to use and implement it by using on mobile. To grade the vegetables and fruits, the properties of these fruits and vegetables like size, shape, texture, etc., are required. We will be implementing the same process through machines for high accuracy and fewer faults. Classifying and grading samples will be time-consuming and error-prone if done through manual techniques. Using these quality measures as the basis of an automated system, the work will be quicker and error-free. The advance in Software development is significant in this color classification system. Using color and size information from the fruit surface, color-based classification classifies the fruits and vegetables based on their appearance. An artificial neural network (ANN) is used to detect the shape, size, and color of fruit samples. The designed model will require a widely used platform for using it by the users, so we have taken the android mobile as a platform for users to grade the fruits and vegetables. This android use can result in taking the proposed model to a huge number of users.

2. Literature survey

A variety of sensors are used to assess the quality of fruits and vegetables based on different parameters [1]. Estimating the freshness of the given fruit by finding edibility using sensors [2]. The inner quality of the watermelon is detected using acoustic impulse response [3]. The detection of quality and grading based on tissue using laser-induced fluorescence [4]. The fruits and vegetables are evaluated for freshness detection using ultrasonic technology [5]. Using image processing and deep learning strategy for detecting the mellowness of dragon fruit [6]. A method for identifying and classifying given mango fruits using image processing techniques [7]. With the use of image processing techniques to determine the nature of the given fruit or vegetable and the diseases associated with it [8]. Applying various image processing techniques for fruit quality assessment and these methods will be using different factors such as nutrition value, flavor, aroma, etc., in grading the fruit and this process will also reduce the human involvement in assessing the fruit [9]. Usage of image processing and computer vision in extracting the diverse types of features, and also explained different functional models for classifying the fruits [10]. Evaluating the quality of the given food image through computer machine vision by using image segmentation [11]. An Inspection system for grading the given tomatoes using image processing [12].

An application for food recommendations for users by evaluation received from the users using the AD-Net algorithm [13].

A computer vision method is used to evaluate the quality of a fruit or vegetable from a given image [14]. Methods for identifying the quality of fruit by employing non-destructive techniques and grading them [15]. The quality of the orange is detected using a machine learning algorithm-SVM [16]. Prediction of the quality of kiwifruit based on the firmness using ANN and linear regressions [17]. An image processing method for grading fruits and vegetables using various machine learning algorithms is presented and explained along with its effectiveness [18]. Android Application for finding whether the fruit is ripened naturally or artificially [19]. By utilizing image processing, designing an application that counts calories based on the features obtained from the given image [20]. The use of an ANN-based technique for measuring vegetable quality [21].

Some of the hardware and software techniques comparisons are shown in Table. 1 [3][4][5][7].

Testing Model	Reliability	Accuracy	Compatibility	Safety
Acoustic impulse	Not reliable	Nice	Thin-skinned	light damage
Laser-Induced Fluorescence	Highly reliability	High	Any	heavy damage
Ultra-Sonic	Low reliability	Bad	Thin-skinned	No damage
Computer Vision	Highly reliable	Very High	Any	No damage

Table 1. Fruit and Vegetable quality monitoring using hardware and software methods

3. Proposed system

In this proposed method, we have split the process into two parts: the first part is about the classification of fruit or vegetable and its quality grading using machine learning algorithms and image processing, and the second part is about developing an android application for carrying out the told process in the first part by converting the part one into a tensor-flow-lite model using Keras for deployment in android studio. For the machine learning model, we have verified several algorithms [14][16][17]. And the comparison is shown in Table. 2

Type	Accuracy	Execution Speed	Data-set Size
KNN	90.3%	Very Slow	Small
SVM	85%	Faster	Only for small
Linear Regression (LR)	92%	Slow	Moderate
Convolutional Neural Networks (CNN)	95%	Very Fast	Very Large

Table 2. Comparison of ML Algorithms

As we are dealing with a larger data set with various fruits and vegetables the Convolutional Neural Networks (CNN) model was found to be more efficient than other machine learning techniques.

3.1. *Quality Analysis*

In general, for detecting the quality of the fruits or vegetables five main modules or stages are used, described as images data-set collection as input, image pre-processing, feature extraction, feature
selection, classification, and detection. The general block diagram of detection of quality is shown in Fig. 1. This quality analysis will consist of two phases: Model Training Phase and Model Testing Phase.

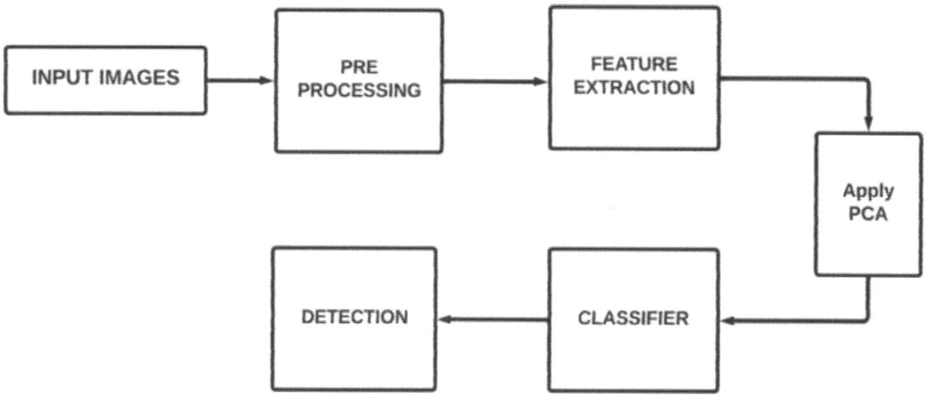

Figure 1. General block diagram of the quality detection system.

Dataset

The created dataset for the model to train is made using images from different sources like the internet, and real-life fruits and vegetable markets. The dataset created has around 2000 images for every type of fruit or vegetable i.e., we have taken three fruits and three vegetables as input for training the system. The dataset contains wanted and unwanted categories of fruits such as apples, bananas, oranges, and vegetables such as bitter gourd, capsicum, and tomatoes. The proposed system is implemented using Python and Jupiter notebook and the data-set of images is taken in .jpg format.

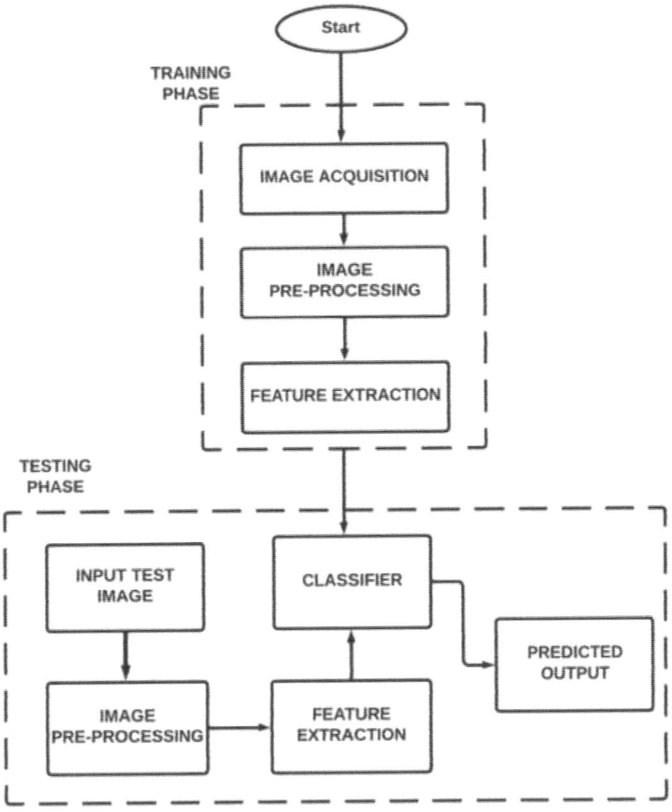

Figure 2. Workflow of the CNN Model.

Model Training Phase

In the training phase of the model, after creating the dataset for the model to train, the dataset images are pre-processed which is needed for eliminating the unwanted information from the image. Using image processing techniques, higher accuracy can be obtained from the given images [18]. In the pre-processing step, the images are first normalized through normalization for reducing the data loss of the images, and then the images are introduced to data augmentation for diversifying and increasing the dataset size when the classifier is trained. After the pre-processing, the features like color, shape, size, texture, etc., are extracted from the images. The extracted features are used to train the classifier. The proposed system uses Convolutional Neural Networks (CNN), a deep learning algorithm, which consists of a class of neural networks as a classifier for image recognition by a specialized way of processing on the grid of pixels. This process trains the system for classifying and grading [19]. For the training phase, the images of the dataset consist of fresh and stale classes of every fruit and vegetable.

Model Testing Phase

In the testing phase of the model, a set of testing images are given as input for the system. These images go through the image-processing and feature extraction processes. Based on the system classification the given input fruit or vegetable is graded and the result is displayed. The training and testing phases are shown in Fig.2.

CNN network has three additional hidden layers, namely an input layer, an output layer, and middle layers that are comprised of multiple convolution layers, pooling layers, as well as fully connected layers. The CNN usesconvolution and max pooling for extracting the features and flattens the values and then uses functions such as 'ReLu' and 'Softmax' for further reducing the matrices. The CNN model is shown in Fig.3.

This model is then checked to whether it is fixed for the given application and then the model is used for testing the inputs. With this CNN architecture, the system converts the given input image into an array of pixel values using the convolution and max pooling, after a certain number of times the convolution and max-pooling are done, then the data is flattened using 'ReLu' and 'Softmax'. The model runs through the data many times, these are called epochs. As the number of epochs increases, the more the model improves to a certain extent [20]. The Fig-4 will show the training, validation accuracy, and also training and validation loss of the model.

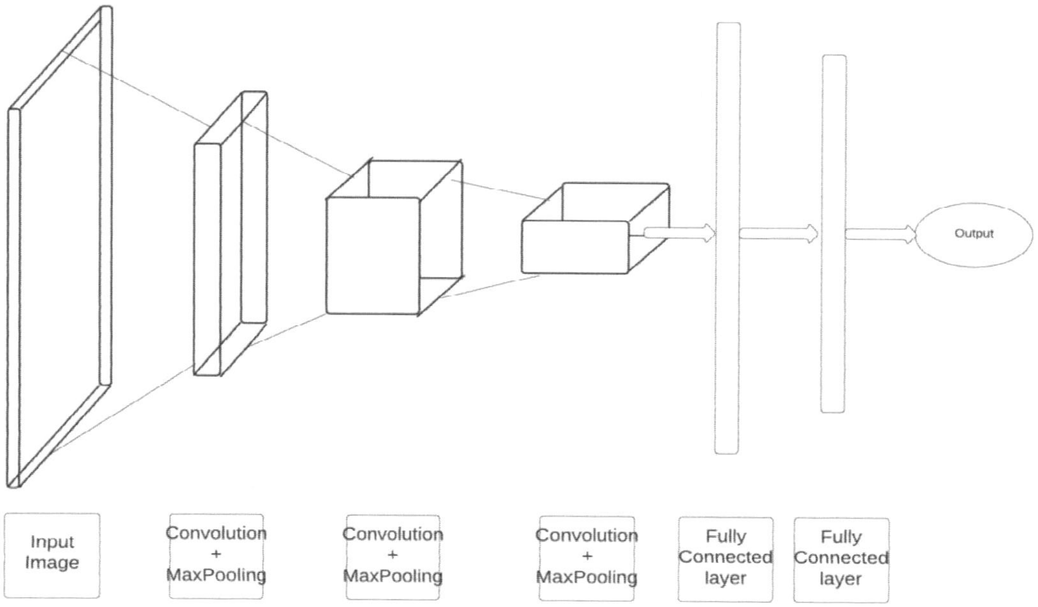

Figure 3. CNN Architecture

Tensor-Flow-Lite Model

After the testing of the model, the model is converted into a tensor-flow-lite model. The tensor-flow-lite model is created using the tensor-flow-lite-maker, an extension of Keras which simplifies the testing of a new image. This tensor-flow-lite model maker converts the CNN model into a 'tflite' model which is an optimized version of the CNN model and can be deployed on mobile or embedded devices.

3.2. *Android Application Development*

Android application is primarily software developed to run on Android devices or emulators. It provides a strong, flexible application framework that lets us develop mobile apps [13][21]. This android application development will consist of three steps: Configuration of the android studio, Development of the Application, and Deploying the android application on mobile. This android application can be used in fast processing the given process steps.

Configuration

In the first step, we used the Android Studio, in which we start a new android project and configure the system settings like which android versions of mobiles can access or use it, which helps in the smooth functioning of the application. This application is developed using the' Kotlin language. Using this 'Kotlin' language, we set up the functioning of the application.

Development

Using the 'Kotlin' language a program is executed for adding the functionalities of the application like access to the camera and gallery, also the User-Interface of the application. Then the tensor-flow-lite model is imported into the project.

Now, the work environment is done and the setup is started by making use of the tensor-flow-lite model. Then the program for using the tensor-flow-lite model is added for accessing the model. Now the written program is compiled for checking errors. If the code consists of errors, first the errors are rectified. The android system requires the application to be in'.apk' format. The'.apk' file is generated by making use of the function to generate'.apk' file in the android studio. After completing the procedure for generating the'.apk' file, it is stored in the same folder of the program file. The'.apk' file now is ready to be imported by the mobile and run as an android application.

Deployment

In the last step, we need to check or verify the execution of the application. After installing the application, it is checked with images of the fruit or vegetable which are captured. The application itself executes the process required and then the result is displayed in Fig.6. The entire android application setup and its execution are shown in Fig.4.

Working of the Application

After installation of the application, the application is opened. On the home page, an option can be chosen for selecting whether to select an image from the gallery or an image to be captured from the Mobile camera. After selecting the image, the model is used for classifying and grading the input image.

The application is designed in a way that the input image is given to the tensor-flow-lite model for classifying and grading and the output of the model is displayed in the result section of the interface of the Application with an option to read the result.

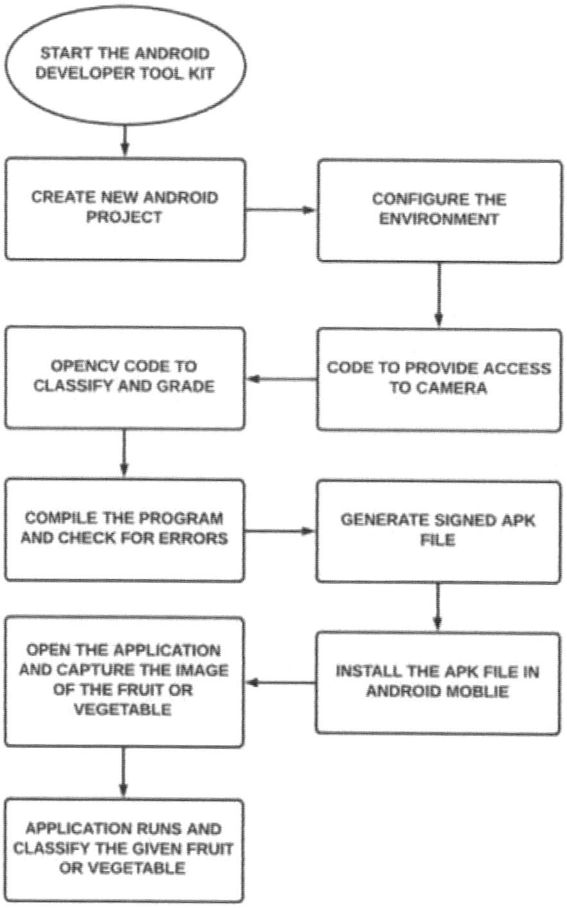

Figure 4. Flow chart of Android Application

4. Results and discussions

The result of this project is divided into two parts: one is the result obtained from the quality analysis process and the other is the result obtained from the Android application.

4.1. *Quality Analysis*

In the Quality Analysis process, the result can be defined as a classifier that is trained and can be used for testing an image. The training-validation accuracy graph and training-validation loss graph are shown in Fig.6. For the testing of the model, a dataset is given to the model and the obtained result is shown in Fig.5. Finally based on the model for classifying and grading the output is obtained in the form of message and voice.

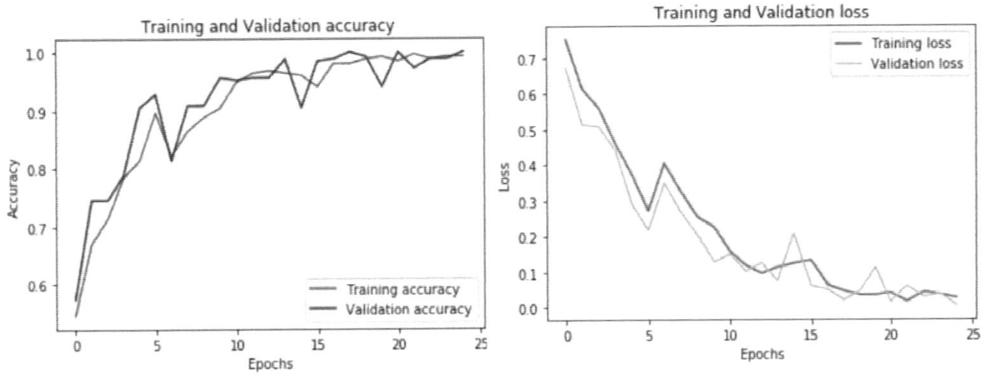

Figure 5. Training and Validation accuracy of model and Training and Validation loss of a model.

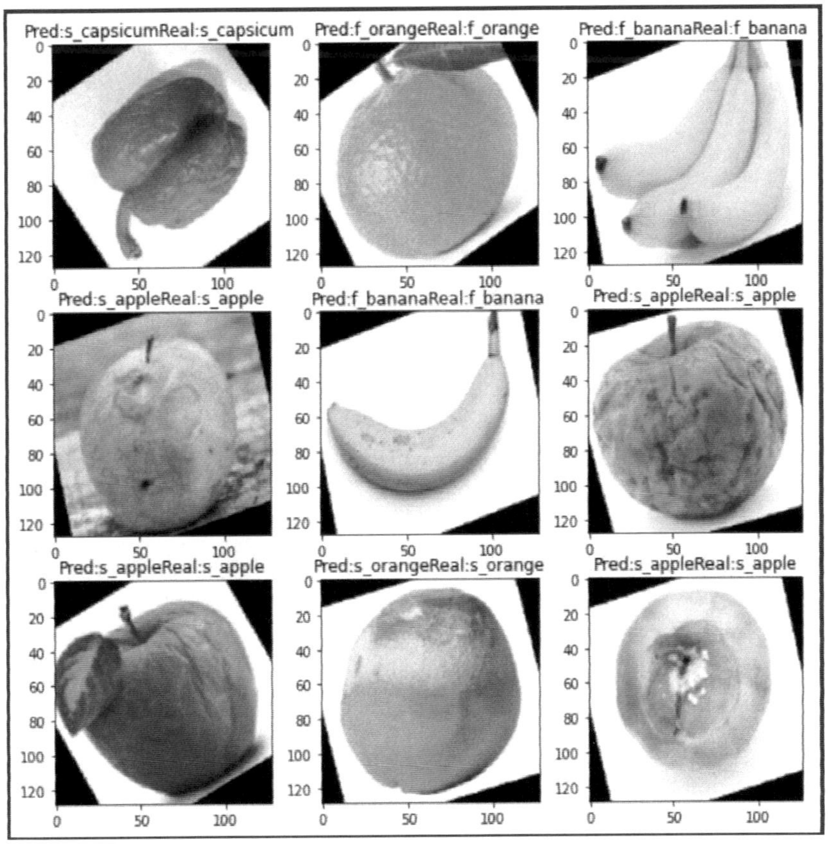

Figure 6. Test Result of the Model.

4.2. *Android Application*

As for the Android application, the result obtained is smooth installation on the Smartphone, running properly and also executing the code successfully, and displaying the output.

The output interface of the application is shown in Fig.7. Accordingly, the input image is evaluated based on the quality of the given fruit or vegetable. The result is displayed on the mobile interface and it can be viewed by the user with audio play it says the name of the fruit or vegetable and its quality.

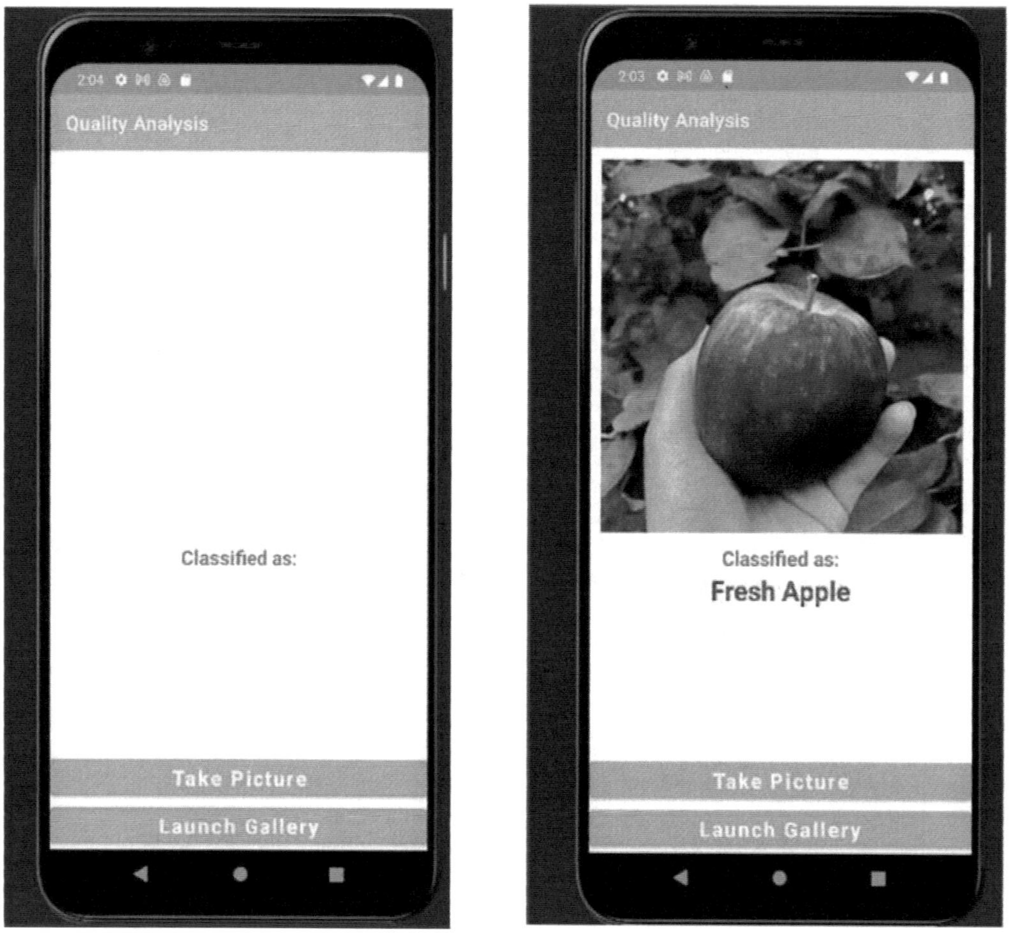

Figure 7. Mobile Interface of Android Application

5. Conclusion and future scope

In the traditional method of classifying and grading fruit or vegetable, there is a lot of human energy and time involved. By using the proposed system, we can reduce human efforts and time consumption. The accuracy and efficiency of the system are founded on two aspects which are feature extraction algorithms and the database used. With the help of the android application, the execution speed is further increased. The basic aim of this project is to grade the fruit or vegetable so that farmers can get added value for the products and also the customers can get high quality and healthy products in markets. The proposed system can be installed in a hardware device and run using cloud computation which will be helpful in fast processing and also in gaining higher accuracy. Even checking the products in large number. This system can be installed in a robot for checking the quality of the fruits and vegetables in the markets. This application system can be used as an information system between customers and shops such as markets, where the products can be graded and uploaded to the database from where the customer can get the required product within the shortest time.

References

[1] Khaled, D.E.; Novas, N.; Gazquez, J.A.; Garcia, R.M.; Manzano-Agugliaro, F. Fruit and Vegetable Quality Assessment via Dielectric Sensing. Sensors 2015, 15, 15363-15397. https://doi.org/10.3390/s150715363.

[2] S. Bala Naga Pranav, T. R. Kaushek Kumar, J. Hari Prakash, S. Sharan, and M. Ganesan, "Freshness Estimator for Fruits and Vegetables Using MQ Sensors," 2021 International Conference on Advances in Computing and Communications (ICACC), 2021, pp. 1-6, DOI: 10.1109/ICACC-202152719.2021.9708369.

[3] Diezma, Belén & Ruiz-Altisent, Margarita & Barreiro, Pilar. (2004). Detection of Internal Quality in Seedless Watermelon by Acoustic Impulse Response. Biosystems Engineering. 88. 221-230. 10.1016/j.biosystemseng.2004.03.007.

[4] Zude, M. (2003). DETECTION OF FRUIT TISSUE BROWNING USING LASER-INDUCED FLUORESCENCE SPECTROSCOPY. Acta Hortic. 628, 85-91 DOI: 10.17660/ActaHortic.2003.628.8

[5] A. Mizrach, Ultrasonic technology for quality evaluation of fresh fruit and vegetables in pre- and postharvest processes,Postharvest Biology and Technology,Volume 48, Issue 3,2008, Pages 315-330,ISSN 0925 5214,https://doi.org/10.1016/j.postharvbio.2007.10.018.(https://www.sciencedirect.com/science/article/pii/S0925521407003699)

[6] Vijayakumar, T., and Mr R. Vinothkanna. "Mellowness detection of dragon fruit using deep learning strategy." Journal of Innovative Image Processing (JIIP) 2, no. 01 (2020): 35-43

[7] D. Sahu and C. Dewangan, "Identification and Classification of Mango Fruits Using Image Processing," Int.J. Sci. Res. Comput. Sci. Eng. Inf. Technol., vol. 2, no. 2, pp. 203–210, 2017.]

[8] R Chandramma, Ramya C, S Vinitha, Shivani P R, Vidya M S, 2020, Automatic Fruit and Vegetable Detection and Disease Identification System, INTERNATIONAL JOURNAL OF ENGINEERING RESEARCH and TECHNOLOGY (IJERT) NCAIT – 2020 (Volume 8 – Issue 15).

[9] P. Pathmanaban, B. K. Gnanavel, and S. S. Anandan," Recent application of imaging techniques for fruit quality assessment," Trends in Food Science and Technology, vol. 94. pp. 32–42, 2019.

[10] Seema, A. Kumar, and G. S. Gill," Automatic Fruit Grading and Classification System Using Computer Vision: A Review," 2015 Second International Conference on Advances in Computing and Communication Engineering, 2015, pp. 598-603, DOI: 10.1109/ICACCE.2015.15.

[11] Nandhini. P, Dr. J. Jaya," Image Segmentation for Food Quality Evaluation Using Computer Vision System", Nandhini. P et al Int. Journal of Engineering Research www.ijera.com ISSN: 2248-9622, Vol. 4, Issue 2(Version 5), February 2014, pp.01-03.

[12] N. M. Z. Hashim, N. H. Mohamad, Z. Zakaria, H. Bakri, F. Sakaguchi," Development of Tomato Inspection and Grading System using Image Processing", International Journal of Engineering And Computer Science Volume 2 Issue 8 August 2013.

[13] Wang, Haoxiang. "Design of a Food Recommendation System using ADNet algorithm on a Hybrid Data Mining Process." Journal: Journal of Soft Computing Paradigm December 2022,

no. 4 (2021): 272-282.

[14] Bhargava, Anuja , Bansal, Atul. (2018). Fruits and Vegetables Quality Evaluation Using Computer Vision: A Review. Journal of King Saud University - Computer and Information Sciences. 33. 10.1016/j.jksuci.2018.06.002.

[15] Shital A. Lakare1, Prof: Kapale N.D2 "Automatic Fruit Quality Detection System".

[16] H. Patel, R. Prajapati and M. Patel, "Detection of Quality in Orange Fruit Image using SVM Classifier," 2019 3rd International Conference on Trends in Electronics and Informatics (ICOEI), 2019, pp. 74-78, doi: 10.1109/ICOEI.2019.8862758.

[17] Ali Mohammadi Torkashvand, Abbas Ahmadi, Niloofar Layegh Nikravesh, Prediction of kiwifruit firmness using fruit mineral nutrient concentration by artificial neural network (ANN) and multiple linear regressions (MLR), Journal of Integrative Agriculture,Volume 16, Issue 7,2017,Pages 1634-1644,ISSN 2095-3119, https://doi.org/10.1016/S2095-3119(16)61546-0.

[18] Prem Kumar, M.K., Parkavi, A. (2020). Quality Grading of the Fruits and Vegetables Using Image Processing Techniques and Machine Learning: A Review. In: Jayakumari, J., Karagiannidis, G., Ma, M., Hossain, S. (eds) Advances in Communication Systems and Networks. Lecture Notes in Electrical Engineering, vol 656. Springer, Singapore.

[19] Kaur, Mandeep, Sharma, Reecha. (2015). ANN-based Technique for Vegetable Quality Detection. 10. 62-70. 10.9790/2834-10516270.

[20] Balpande, Tejswini , Dhothkar, Nikita , Satpute, Heena , Durbude, Namrata , Chakole, Vijay , Mohammad, Sameer. (2020)." AUTOMATIC FRUITS DETECTION USING ARTIFICIAL INTELLIGENCE". SSRN Electronic Journal. 7. 86- 89

[21] S. Maheswaran, S. Sathesh, P. Priyadharshini, and B. Vivek," Identification of artificially ripened fruits using smart-phones," 2017 International Conference on Intelligent Computing and Control (I2C2), 2017, pp. 1-6, DOI: 10.1109/I2C2.2017.8321857.

A Deep Learning based Advanced Gait Detection

S. Sumathi[1], P. Bhuvaneshwari[2], Harikrishnaveni V[2], Sangami .S.A[2], Shivanee Bhuvaneshwari .N.S[2]

[1] Assistant Professor, Department of ECE, Velammal Engineering College, Chennai-600066, India.

[2] UG students, Department of ECE, Velammal Engineering College, Chennai-600066, India.

sumathi.s@velammal.edu.in, sangamianbus@gmail.com

Abstract: A gait is a biometric sign that people use to provide visual verification of viewing. Distinctive highlights need more than one movement cycle. In Gait Energy Image (GEI), for example, typical images of a frame with more than one cycle motion are analyzed. The unique stride highlights the required movement of an individual strolling more than one complete walking cycle. For instance, in Gait Energy Image (GEI), normal outline pictures of more than one entire walk cycle are processed. Be that as it may, there may be an incomplete step cycle in which information is accessible due to impediments. This research work has designed and developed a model for implementing stride arrangement and profound learning strategies. A separate walk dataset is used in our trial. The walk marks are removed from the Gait Energy Image (GEI) using the Convolution Neural Network (CNN). The classifier proposed here for grouping human gait postures is Mobile Net. Finally, the proposed methodology is evaluated on the OULP step dataset, affirming that the proposed design effectively reproduces total GEI even from the outrageously deficient walk cycles.

Keywords: Gait, Deep learning, Convolution Neural Network.

1. Introduction

GAIT acknowledgment is a biometric strategy used for perceiving the actions of different people by utilizing the highlights, which are separated from the walking style. Unlike numerous biometric modalities, step recognition has few more properties: a stride highlight include different characteristics for distinguishing people and it can be obtained from the oblivious walk of an agreeable subject at a reasonable separation from the camera, and from a somewhat low picture goal (e.g., an individual with a 30-pixel level in a picture). As a result, gait recognition applied to incompatible examination utilizing CCTV film. In any case, the step elements of uncooperative subjects might contain covariates that impact the actual stride presence of mobile individuals. The strength of such covariates is very significant for generating walk acknowledgment. However, there are no restrictions on the sea, strolling speeds, and clothing. Among the covariates, an adjustment of view happens again in the genuine circumstances to a great extent, and affects the presence of the strolling individual.

Matching a walk across various perspectives (cross-view coordinating) that come along these lines are considered as significant undertaking in the acknowledgment step. Two unique approaches are

Journal of Physics: Conference Series **2325** (2022) 012056 doi:10.1088/1742-6596/2325/1/012056

proposed for stride acknowledgment, they are appearance-based and model-based strategies. The appearance-based approach utilizes caught picture groupings to forward stride highlights, while the model-based criteria extricate model boundaries from the pictures. Specifically, 3D model-based approaches are ideal for cross-view matching attributable to their view-invariant nature. Despite trouble ascertaining 3D model boundaries with high precision from inferior quality pictures caught during an examination at a crime location.

•With the improvement of data and Internet innovation, the Advantages of a walk as a biometric

1. Non-prominent or Non-obtrusive:
2. Maintaining Secrecy:
3. Lesser Image Resolution Required:

Normal boundaries in walk examination

Gait Cycle:

A gait cycle addresses the principal fleeting element of handling in stride acknowledgment and compares it to an irregular cycle that travels from Rest to Right-Foot-Forward (RFF) to Rest to Left-Foot-Forward (LFF) to Rest position [1]. It covers the entire range of potential places where one goes through the usual stage of travel. Figure 1 shows the five trademark areas of the human movement cycle in the same place.

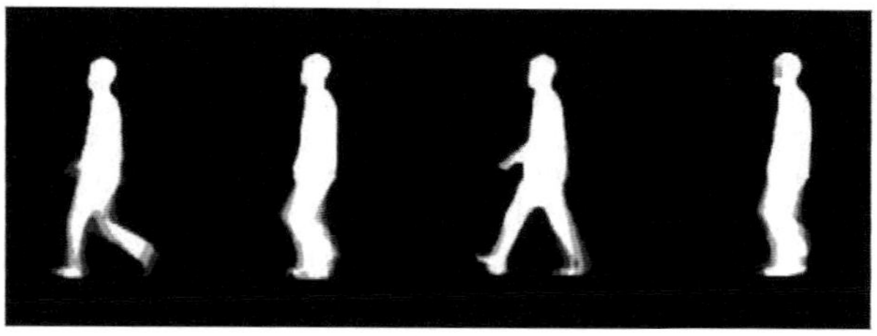

Figure 1 Trademark places run-of-the-mill human step cycle.

The step phase is the time expected for an individual to finish a single stride cycle. Yet, since the camera captures picture data as edges, and the casings are occasional (for example, 30 corners for each second or 30 fps), it is more helpful to get the step time of the individual as far as outlines. It can acquire by observing the number of edges slipped by between the beginning and finishing casings of a separated stride cycle. The step period works as an ID highlight in and of itself, but if the fps is low, crowded travel time falls into an easily accessible area. Accordingly, its prejudicial power diminishes, and it tends to be unique to pair with other, more unfair elements. The step period provides us with prompts about the momentum of the individual. For instance, assuming that the average step-time for a particular frame is 28 cases, and a person takes 45 edges per single travel cycle (45 step period), then, at that point, it tends to be reasoned that the speed of the individual is notably sluggish compared with the standard. It always works for frameworks that use steps to determine ages. Step length is the maximum length between individual appendages, which expect parts of the ID. The measurement is by setting a jump bar around the person's picture. The step length sets up in all the steps to produce hardy noise and errors. The outcomes are in the center value.

GAIT RECOGNITION

Gait recognition is an arising biometric innovation that affects individuals by simply scanning their walk. It has drawn interest as a strategy for proof since it is harmless and doesn't need the subject's collaboration. Walk recognition could use in helpful ways, such as identifying offenders in an inappropriate location. On the other hand, see healthcare implications for walk confirmation technology, which isn't just for security applications. For instance, perceiving changes in strolling designs from the beginning can assist in distinguishing conditions like Parkinson's sickness. Nonetheless, walk recognition innovation is still in its early stages. At this point, no model on both exact and appealing has developed.

There are two principal kinds of walk acknowledgment methods presently being developed. The first, walk acknowledgment because of the programmed examination of video symbolism, is the more broadly contemplated of the two. The directions of the joints and points over the long run are valued using video tests of the subject's walk. A numerical model of the movement is compared with other examples to decide their personality.

The following method uses a radar framework similar to that used by the police to distinguish high-speed vehicles. Radar records the cycle of action taken by humans as it moves. This data was compared with a few samples to separate.

An effort makes the action as accurate and practical as possible. For example, in special finger appearance or iris verification, it is assumed that the motion confirmation development will be carried over to the next five years and used in other biometrics as a visual strategy.

2. Related Work

Xin Chen et al. proposed making fake step tests under different viewpoints across different walk datasets by considering Multi-view Gait Generative Adversarial Network (MvGGAN). MvGGAN consolidates a lonely generator duplicate stride tests, a couple of strolling conditions, and a discriminator which sorts out not well-arranged jams and jams person information.

P. Sudeep SarkarJonathan Phillips investigated the identification of people's step plans. It has turned into an investigation issue. To give an approach to assessing progress and depicting the properties of stride acknowledgment, we present the HumanID Gait Challenge Problem.

John N. Carter and Mark S. Nixon devised another concept: free-stride biometrics. The structure is single-camera, does not require camera alignment, and operates with pictures. Accomplished through a process in which the step corrects itself.

Qiang Wu and Jian Zhang discovered that step is a significant biometric component of perceiving an individual from a distance, such as in video surveillance applications. Backsliding experiences in the growing experiences reliant on the adaptability embraced as backslide work, resulting in consistent relapse VTM model development.

The "OU-ISIR Gait Database, Enormous Population Dataset"—the exhibition's evaluation of a reliable measure of acceptance of the vision-based journey was by Haruyuki Iwama and Mayu Okumura. However, the existing step data has 185 topics considered, including a step-by-step knowledge base covering 4007 subjects and 1850 topics. The data set enables the selection of the most significant performance differences among the best images currently in use.

Yagi, Y. Makihara, A. Shiraishi, and D. Muramatsu inspected whether camera-based step acknowledgment is helpful for decently affirming a person or whether the objective of the acquired pictures isn't high enough. Notwithstanding, different viewpoints on the view of showing and testing reduce the affirmation of accuracy. In this situation, I propose a stage-based approval technique that uses a self-decisive view change scheme. The proposed system fabricates a changed framework related to the viewpoint on the game plan of the show and tests using a 3D step database made from non-center around various subjects' visual bodies.

Makihara and Y. Yagi proposed an iterative arrangement of spatial-fleeting close by concealing differences in establishment and chart cut division for frame extraction. The given allowance, spatial-transient establishment hide dealt with for fitting exhibited establishment tones to enter an establishment under another lighting condition. Later, nearer view tones are executed reliant upon the fit establishment, and spatial-worldly diagram slice calculation is applied to acquire a front-facing region/establishment division result.

Iwama and M. Okumura portrayed the world's data set, titled "OU-ISIR Gait Database, Large Population Dataset." an application to trustworthy person execution appraisal based on walk acknowledgment. While stride data sets consolidate everything considered for 185 subjects, as fabricate great walk data sets.

Y.-P. Tan and J. Lu investigated the step as a helpful biometric mark and have obtained creation revenue from PC vision researchers. In this paper, we propose another directed complex learning calculation, called UDSA, for view-invariant step signal figuring. The mark of UDSA is to search for an intention to project human step groupings accumulated to different viewpoints on the ground floor with lower layers so that internal group mathematical designs are protected and assisted between intermediate group planning stages.

Y.P. Tan, Liu, and J. Lu proposed a joint subspace learning (JSL), while a consistent-view approach to stride acknowledgment. The three-dimensional item can particularly tend to a weighted measure of several models' incomparable depiction coefficients are all-around for all dissimilar viewpoints, suggesting using these coefficients as consistent viewing factors in walk acknowledgment.

3. Proposed System

We foster a model for walking groups utilizing profound learning strategies. Single walking details are in our trial. The step energy picture (GEI) is a Convolutional neural network (CNN). The classifier we figure as data collection utilized for the order of human walk is Mobile Net, as displayed in Fig. These models are standard stride dataset, CASSIA, and the OUISIR information base. The exploratory will gather from both datasets.

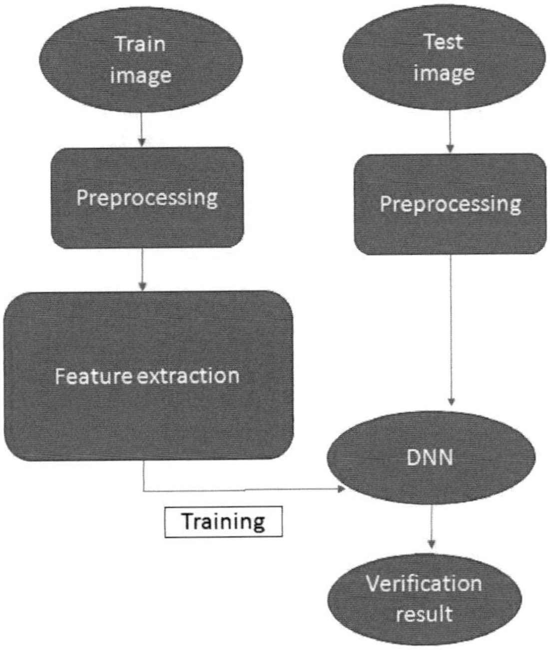

Figure 2 Methodology

Methodology

Examination outlined in Figure 1. The significant advances included: composing information assortment, information obtaining, classifier configuration, preparation, and testing. CNN uses MATLAB R2017b. Prepare on a PC with an NVIDIA GTX 960M Graphics Processing Unit (GPU) with 2 GB of memory.

Data Collection

Information acquired. It contains interactive coverings of three 23-layer template records deep into the Microsoft Kinect camera. Test performed from an indoor environment where every subject expected clothing. More than nine travel groups capture with a diagonal view from the top of the Kinect camera. More than nine sequences comprising 15–600 records focus on concentrations in every model.

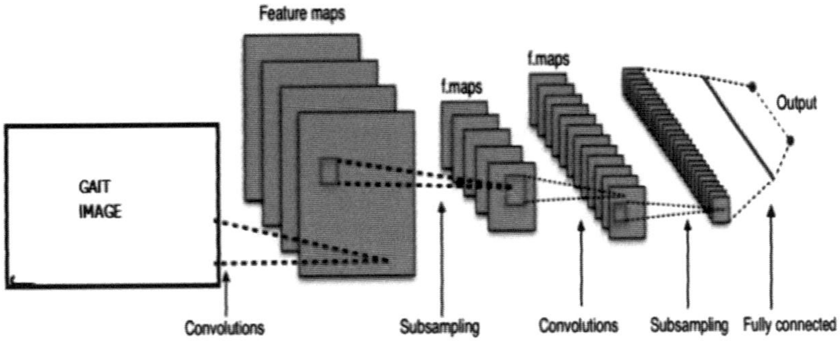

Figure 3 Data Collection

Data Acquisition

Step details of the entire travel collection were organized into related topics and translated into 60 by 26 three-layer exhibits to fit CNN's information needs.

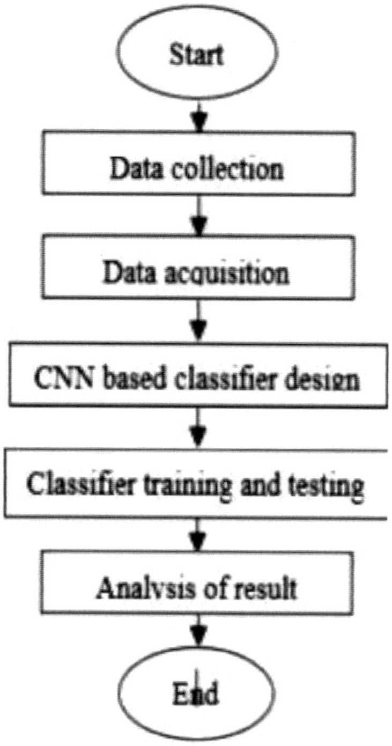

Figure 4 Data Acquisition

The section addresses every joint focus on a period, while the line addresses information collected at all times. Pieces of information were recreated, then travel information again separated and passed on to the preparation and testing of collections rate of fifty. Collections circulated haphazardly, furthering speculation.

Classifier Mobile Net Architecture

In this segment, we first describe the middle layers that the mobile network depends on, which are profundity-wise distinct channels. We then portray the Mobile Net construction and close depictions of the two models' contracting hyper-boundaries with multiplier and objective multiplier. Convolution visible profundity. The Mobile Net model relies on the most intelligent and divisive integration. The form of factorial convolution commonly incorporates factorial profundity-wise and point-wise convolution. For Mobile networks, the most intelligent conversions use a single channel for all data channels. Conviction of the point intelligence employs matrix unit modification to integrate the results of deep intellectual convolution. The cleverly divisive convolution divides this into two separable layers; one layer of filtering action and the other layer of integration. This category has the effect of effectively reducing statistical calculations and model sizes. Figure 2 shows standard deviation 2 (a) made of deep intellectual variable 2 (b) and 1×1 intelligent point 2 (c). The duplicate layer capture DF × DF × M data section F and generates a DF × DF × NF features that map the DF area width and square map1. M denotes the number of data channels. DG as an area width and square root result factor, and the N output term (yield value). The most common density of part K of DK × DK × M × N, where DK is part of the terminal of the structure, M is the number of data channels, and N is the number of output channels shown. The result includes a standard tolerance map for the first phase completion as follows:

$$G_{k,l,n} = \sum_{i,j,m} K_{i,j,m,n} \cdot F_{k+i-1,l+j-1,m}$$

Standard convolutions have the computational cost of:

$$D_K \cdot D_K \cdot M \cdot N \cdot D_F \cdot D_F$$

The calculation rate depends on the number of data M channels, the number of output N channels, the size of the episode Dk × Dk and the mapping component DF × DF.

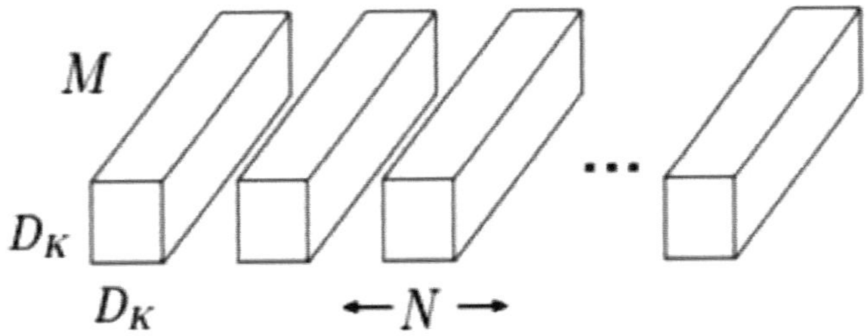

Figure 5 Standard Convolution Filters

The versatile Net model deals with these principles and their systematic efforts. In the beginning, it was deeply intelligent convolutions to break the correlation between the channel output number and the cycle size.

The Versatile Net models were ready in Tensor-stream [1] using RMS prop [33] with a non-simultaneous tendency to plunge like Origin V3 [31]. Regardless of planning a gigantic model, we use a less direct approach data augmentation method as smaller models experience fewer problems with over-equilibrium. When designing mobile nets, we do not use side heads or word slides and reduce the integrated image rotation by limiting the size of the small yields using large origin settings [31]. Furthermore, because channels have no limits, it is critical to have very little weight loss or absence (normal l2). In the Image Net benchmarks in the region corresponding to this topic, all models were ready with the same editing limits, paying less attention to the model size.

4. Results and Discussions

For the travel cycle analysis, images of human movement feature for classification and future reference. Following data collections were created in the form of a binary photo frame. Pieces of information of supervisory conditions are no prior knowledge of the item. In addition, the inner silhouettes from the referable video clips and the external silhouettes are in three different sets.

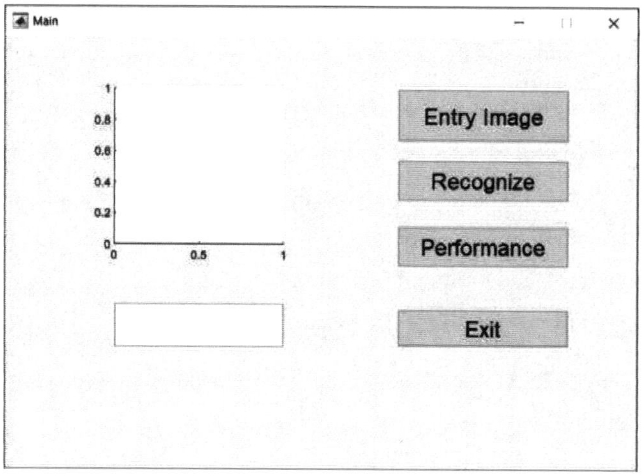

Figure 6 GUI Design

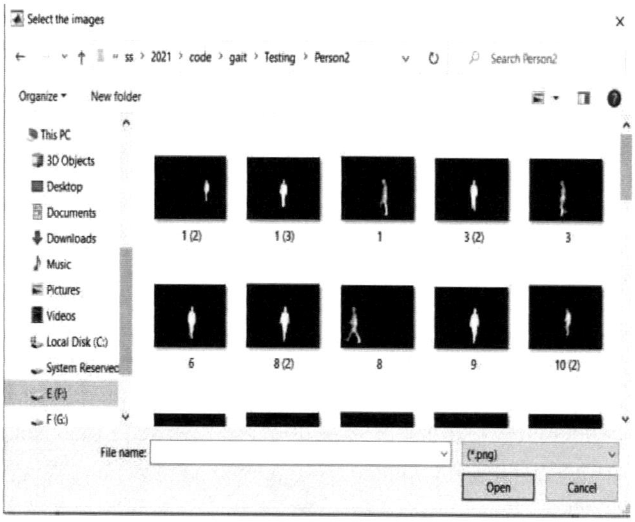

Figure 7 Trained images

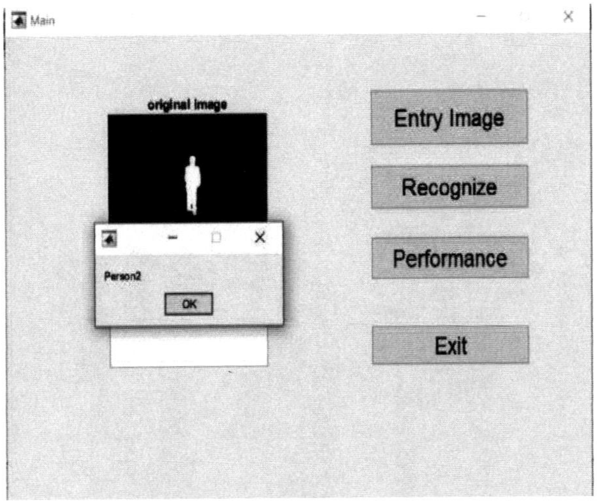

Figure 8 Recognition

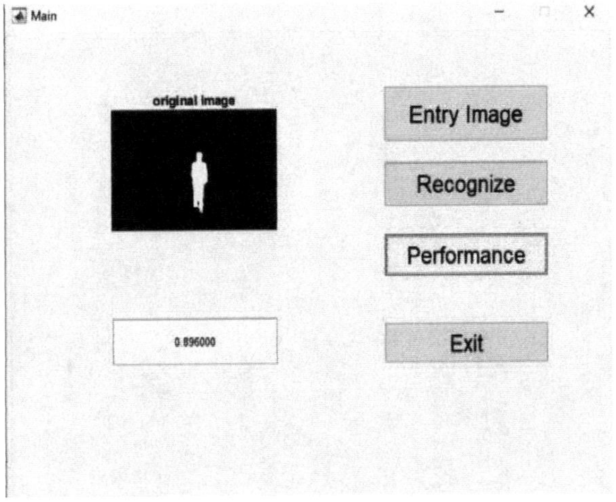

Figure 9 Performance

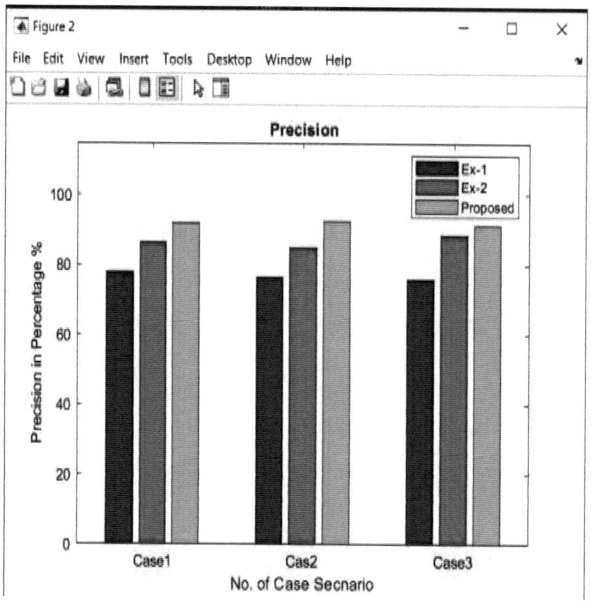

Figure 10 precision Analysis

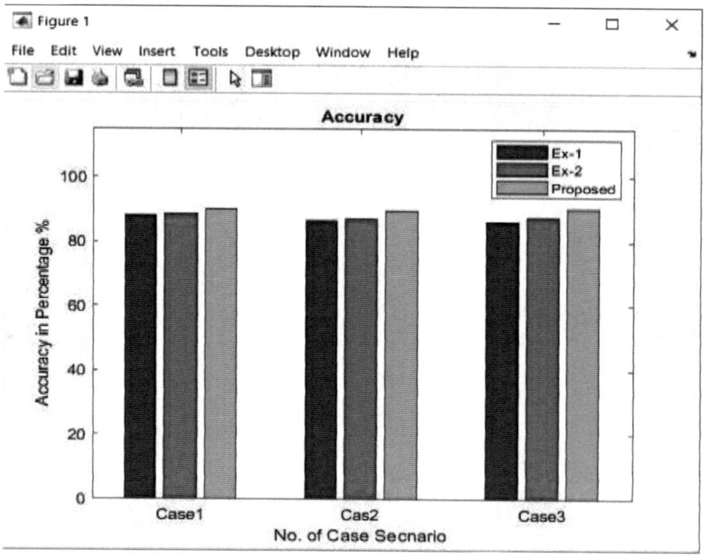

Figure 11 Accuracy Analysis

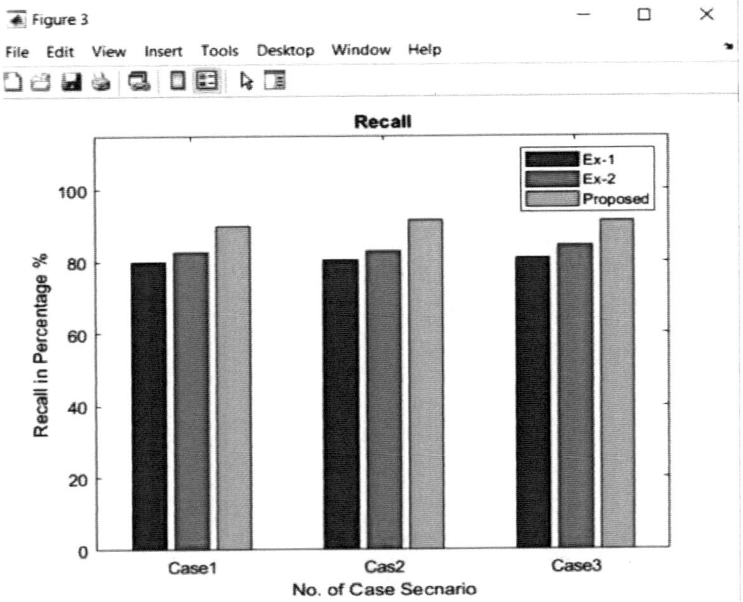

Figure12 Recall rate

Each clip has images at different viewing angles, and each clip splits into 25 frames per second using the CASIA37 website. All the participants' image frames (more than 16,821 images) separate into two disjoint sets: one for training and another for verification and testing. In the test, less than 80% of data was present in the network during training. A mobile network is modified based on input. Less than 20% of data uses network verification and individual assessment tests of Mobile networking. It assumes that the same participant is not in training and validates the test sets.
The proposed system starts the testing on a 64-bit PC with an Intel® 2.8GHzx-64 processor and 16GB of RAM using MATLAB R2020b. Smooth training curves and percentage of verification accuracy are shown in Fig.

5. Conclusion

The proposed venture affirmation approach achieves significant achievement in terms of planning and endorsement precision and means square bungles. The leading trial outcomes report executes different concerns of customer AI methodologies of previous significance step models, where images have fewer accuracy objectives and a massive extension set of data pictures. The well contemplated that the walking examination isn't only to extend the level of well-being and security locally, yet notwithstanding portrayed age and orientation. Later, steps taken by the individual affirmation related to various biometrics systems. This approach was highlighted from a multidisciplinary perspective and linked to another relevant information collection that would be proven.

6. References

[1] J. Fierrez-Aguilar, J. Ortega-Garcia, and J. Gonzalez-Rodriguez, "Target subordinate score procedures and their application to signature check," IEEE Trans. Syst., Man, Cyber C, Appl. Fire up., vol. 35, no. 3, pp. 418-425, Aug. 2005.

[2] S. Sarkar, P. J. Phillips, Z. Liu, I. R. Vega, P. Grother, and K. W. Bowyer, "The humanID stride challenge issue: Data sets, execution, and examination," IEEE Transactions on Design Analysis and Machine Intelligence, vol. 27, no. 2, pp. 162-177, Feb. 2005.

[3] M. Goffredo, I. Bouchrika, J. N. Carter, and M. S. Nixon, "Self-calibrating view-invariant stride biometrics," IEEE Transactions on Systems, Man, and Cybernetics, vol. 40, no. 4, pp. 997-1008, August 2010.

[4] W. Kusakunniran, Q. Wu, J. Zhang, and H. Li, "Stride acknowledgment under different review points in light of related movement relapse," IEEE Trans. Circuits System. Video Technol., vol. 22, no. 6, pp. 966-980, Jun. 2012.

[5] H. Iwama, M. Okumura, Y. Makihara, and Y. Yagi, "The OU-ISIR stride data set including the enormous populace dataset and execution assessment of walk acknowledgment," IEEE Trans. Inf. Crim. Security, vol. 7, no. 5, Oct. 2012.

[6] D. Muramatsu, A. Shiraishi, Y. Makihara, and Y. Yagi, "Erratic perspective on Change model for stride individual validation, "in Proc. IEEE Fifth Int. Conf. Biometrics, Theory, Appl. Syst., Sep. 2012, pp. 85-90.

[7] Y. Makihara and Y. Yagi, "Outline extraction because of iterative spatiotemporal neighborhood variety change and chart cut division," in Proc. of the nineteenth Int. Conf. Design Recognit., Tampa, FL, USA, Dec. 2008.

[8] H. Iwama, M. Okumura, Y. Makihara, and Y. Yagi, "The OU-ISIR stride information base contains the huge populace dataset and execution assessment of walk acknowledgment," IEEE Trans. Inf. Crim. Security, vol. 7, no. 5, pp. 1511–1521, Oct. 2012.

[9] J. Lu and Y.-P. Tan, "Uncorrelated discriminant simplex examination for view-invariant stride signal figuring," Pattern Recognit. Lett., vol. 31, no. 5, pp. 382–393, 2010.

[10] N. Liu, J. Lu, and Y.-P. Tan, "Joint subspace learning for view-invariant step acknowledgment," IEEE Signal Process. Lett., vol. 18, no. 7, pp. 431-434, Jul. 2011.

Driver Drowsiness Detection and MonitoringSystem using Machine Learning

Rohith Chinthalachervu, Immaneni Teja, M. Ajay Kumar, N. Sai Harshith, T. Santosh Kumar

Department of ECE, GRIET, Hyderabad, Telangana, India.

chinthalachervu18241a0407@grietcollege.com, narasimharao4864@gmail.com, methuku18241a0429@grietcollege.com, saiharshith126@gmail.com

Abstract: In the present world, lots of road accidents take place due to the lack of attention and alertness of driver. This is termed as driver drowsiness. This leads to a lot of unfortunate situations causing adverse damage to human lives. The main goal of this research is the detection of driver drowsiness and an appropriateresponse to the detection. There are many methods which are based on the motion of the vehicle or based on the driver's behavior. One of the methods is the physiological method which helps in distracting the driver from drowsiness and making him alert. And few methods require expensive sensors and deals with a lot of data. Therefore, this paper develops a system for detecting drowsiness in realtime with proper procedure and accuracy which is acceptable. In this system, the driver's facial expressions are captured andrecorded using a webcam. Every movement in each frame is detected using few techniques of image processing. The Eye Aspect Ratio,Mouth Opening Ratio, and Nose Length Ratio are calculated using the landmark points on the face. The calculated values are compared to the threshold values developed by the system and the difference in value leads to thedetection. At the same time, the machine learning algorithms are also implemented in offline manner. Based on the classification,the system has successfully achieved 95.58% of sensitivity and100% of specificity using Support Vector Machine. This modelsystem is compatible with all kinds of vehicles.

1. Introduction

Many road accidents which lead to death are because of drowsiness while driving. Drivers who drive long hours like truck drivers, bus drivers are likely to experience this problem. It is highly risky to drive with lack of sleep and driving for long hours will be more tiresome. Due to the drowsiness of thedriver causes very dangerous consequences, it is estimatedthat 70,000 to 80,000 injures & crashes happen worldwide in a year. Even deaths have reached 1000-2000every year. There are many unofficial deaths which are not confirmed by drivers that it was due to their drowsiness. This takes lives of many innocent people. It is a nightmare for alot of people who travel across world. It is very important to identify the driver drowsiness and alert the driverto prevent crash.

The goal of this research is the detection of the indication of this fatigue of the driver. The acquisition system, processing system and warning system are the three blocks that are present in the detection system. The video of the driver's front face is captured by the acquisition system and it is transferred to the next stage i.e., processing block. The detection is processed online and if drowsiness of driver is detected, then the warning systemgives a warning or alarm.

The methods to detect the drowsiness of the drive may be done using intrusive or nonintrusive method i.e., with and without the use of sensors connected to the driver. The cost of the system depends on the sensors used in the system. Addition of more parameters can increase the accuracy of the system to some extent. The motivation for the development of cost effective, and real-time driver drowsiness system with acceptable accuracy are the motivational factors of this work. Hence, the proposed system detects the fatigue of the driver from the facial images, and image processing technology and machine learning method are used to achieve a cost effective and portable system.

2. Methodology

The Machine Learning is an approach of getting the effect automatically. ML is the concept which is linked with many complex and advanced Algorithms, and it can be easily used with the Python programming language. The two librarieswhich contains the machine learning algorithms are:

1. scikit-learn Library
2. Pandas Library

The SVM algorithm used in this project is present in the library scikit- learn. It represents humans as machines, which is the Machine Learning. In order to gain high accuracy and high efficiency, it is necessary to implement Machine Learning algorithms.

2.1. Machine Learning with Python:

The Python properties are Platform Independence, Object Oriented and Robust Language. It has the machine learning algorithms in pandas and scikit-learn library and is an Open-source language.

2.2. Libraries:

1. NumPy: NumPy stands for Numerical Python. It usually includes all types of mathematical operations in the code. It performs linear algebra as well as matrices.

2. Pandas: Pandas are efficient libraries in Python Languages. These are imported and used for managing the datasets. In orderto train the machine using a dataset, dataset should be stored in a variable. This can be done using Pandas Library.

3. Seaborn: Seaborn is used for styling the pictorial representation such as bar graphs, curves etc. By using this library, the difference between the attributes is determined with the colors i.e., each attribute is distinguished with different color.

4. Sklearn: Sklearn is the most used library for implementing machine learning techniques. It is an open and free software

machine learning library. It is also used for performing Regression, Clustering, and Statistical Modeling. Everyone can access and use it, and sklearn is reusable in various contexts. Sklearn has many Machine Learning algorithms, and this library is also called as inbuilt Machine Learning technique. Sklearn has Logistic Regression, RF algorithm, feature extraction, classification report, and confusion matrix. These packages are inbuilt such that these can be accessed directly and can be used for the classification and prediction projects.

3. Existing Approach

There are three types of general methods to detect driver drowsiness; they are, vehicle-based, behavior-based and physiological-based methods. The steering wheel movement, the accelerator of vehicle or pattern of vehicle brakes, vehicle's speed, and deviation in position of lane are monitored continuously in the method which is based on vehicle [5].If there is any deviation in the values detected, it is considered as driver drowsiness. The sensors are not connected to the driver and this measurement is nonintrusive.

Visual behavior like blinking of eye, closing of eye, yawning, bending of head etc. are examined for drowsiness detection in behavioral based method [1]. A simple camera is used to take images to be sent as input to SVM algorithm to identify the above features and are called nonintrusive measurement. Monitoring the physiological signals like EOG, ECG, the heartbeat, EEG, pulse rate etc. [2,3] helps in detecting driver drowsiness based on physiological method and they are intrusive measurement due to the directconnection of sensors to the driver.

The current detection of the drowsiness methods are mainly based on the machine learning algorithms.

4. Proposed Approach

Almost all drivers have experienced this drowsiness problem while driving [7]. Youngsters and professional drivers are mostly affected by this drowsy driving because of continuous hours of driving without any rest. In many cities, auto drivers and cab drivers drive continuously overtime sometimes to complete their targets or at times to get bonus profit. Many of the poor workers in order to meet their daily expenses and for the sake of their loved families tend to work in night shifts for long time, this can be one of the main reasons for accidentstaking place because of drowsy driving.

Therefore, a driver drowsiness monitoring system has been developed in this paper. The block diagram of the proposed method for drowsiness monitoring is shown in figure 1.

A webcam has been used to record the video of the driver. The webcam is arranged in such a way that it captures the front facial image of the driver [8]. Once the video capturing is done, the recorded frames are then pulled out to get the 2-Dimensional images [9]. The object (Face) in the frames is detected by HOG and SVM algorithm.

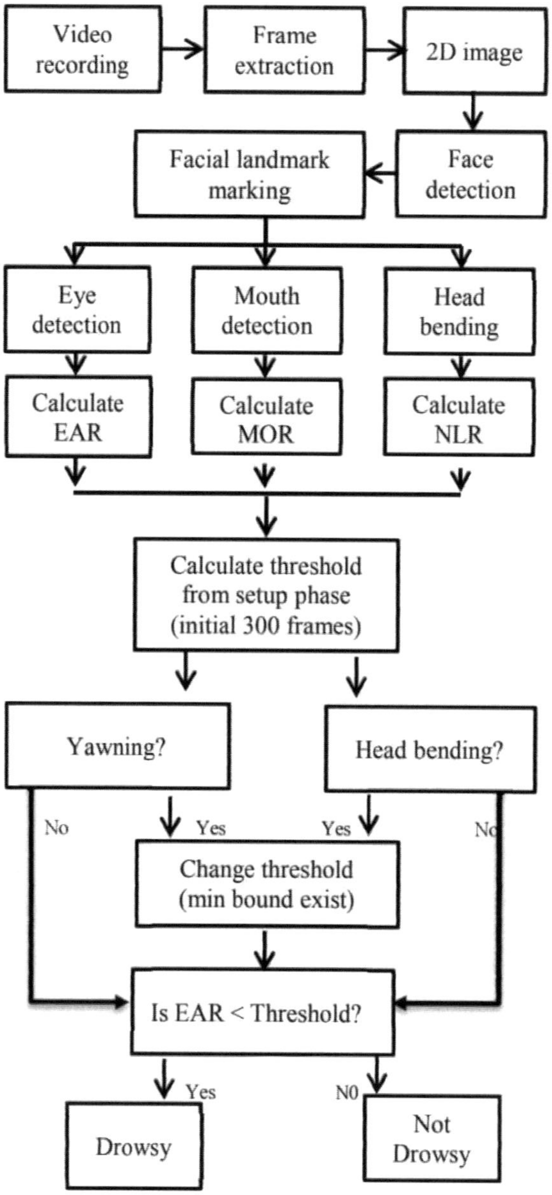

Fig.1 The block diagram of drowsiness detection system

5. Implementation

5.1 Acquiring Data:

At first, the video is captured and recorded using a laptop webcam and then the frames are released and proposed in a laptop. Then after the completion of extraction of frames, the picture implementation techniques are proposed on the 2-dimensional picture of the recorded video. Now, the required driver information is produced. The volunteers are asked to focus on the laptop webcam and perform activities like continual eye blinking& closing, mouth yawning, and head bending. And the webcam is adjusted to capture the video for 20 to 30 min.

5.2 Face Identification:

The driver's face is identified first after the frames have been extracted. In this paper, HOG & SVM algorithms [10] are used for face extraction. In this detection, only positive examples of the stable window size are taken for photographs and histogram of oriented gradients (HOG) descriptors are calculated on them. Following that, the negative examples of same size are considered for HOG descriptors and then results are evaluated. Typically, the count of negative examples is more than positive examples. Then after getting the characteristics of both the groups, a direct SVM algorithm is used for classifying the required task. To get more detection and better accuracy for SVM algorithm, strong negatives are used. In this detection, after the guidance, the SVM classifier is investigated on labeled data and the incorrect positive example characteristic values are reused for guidance purpose. To test the picture, the window used for positive examples are rendered over the image and then the required result or output will be classified for each window location. Then based on the results obtained and considering the various sample results, the higher value samples are considered for drowsiness identification and the boundary outline is drawn on the face. This minimum elimination steps will reduce the overlapping and redundant bounding in the bounding box.

5.3 Locating Face Points:

After the detection of faces, the further step is to locate the points on the human face like eyes, mouth, nose etc. Then the photograph that used for face detection has to be normalized to minimize the distance factor between camera and the driver. Therefore, the face photograph is resized with the width of 500 pixels and converted into gray scale pictures. Then normalization is done for regression trees [11]. And it will be approximate the positions of location points on the face from pixel intensities. In this process, the square error loss is reduced by using gradient boost learning. Different priors are used to discover various structures. By considering this procedure, the location of the boundaries of eyes, mouth and nose are noted and shown in Table I. Then the location points on face are marked and shown in fig.2. The red points on the figure are used for further identification.

Parts	**Landmark Points**
Mouth	[13-24]
Right eye	[1-6]
Left eye	[7-12]
Nose	[25-28]

Table I: The location points on face

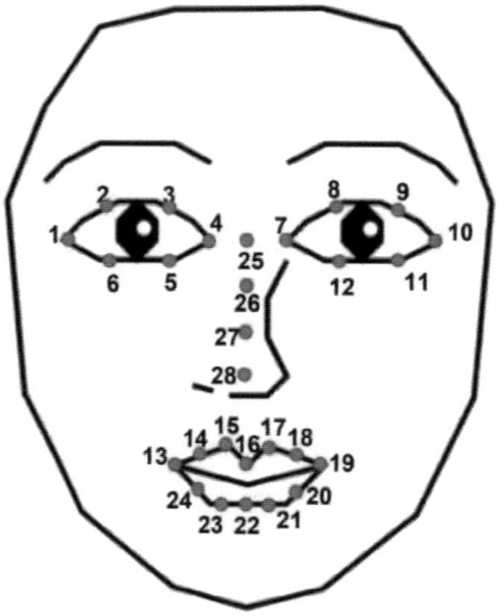

Fig.2 The landmark points on face

5.4 Feature Extraction:

After marking the points on face, the drowsiness features are calculated as given below.

Eye Aspect Ratio (EAR): From the boundary points on the eyes,the EAR is evaluated as inverse ratio of width of eyes to the heightof eyes. The mathematical formula for EAR is given as,

$$EAR = \frac{(p_2 - p_6) + (p_3 - p_5)}{2(p_4 - p_1)}$$

where i is the point marked as i in facial landmarking (i – j) gives thedistance between the points i and j. When the eyes are open EARis max and when the eyes are closed, the EAR is approximated to zero. So, the decreasing value of EAR indicates the closing of eyes andidentifies the drowsy behavior of the driver.

Mouth Opening Ratio (MOR): Mouth Opening Ratio is definedas the identification of yawning of the driver and ultimately drowsiness detection is done. The mathematical formula for calculating the MOR is given as,

$$MOR = \frac{(p_{15} - p_{23}) + (p_{16} - p_{22}) + (p_{17} - p_{21})}{3(p_{19} - p_{13})}$$

The MOR ratio is maximum when the mouth of driver is open and if it remains same for some time, then yawning condition is indicated. And if it decreases towards zero, then the condition is considered as normal, and not yawning.

Nose Length Ratio (NLR): When the head of the driver is down along withvertical axis, then by using the bending angle of head the drowsiness alert is identified. The focal plane of the webcam is directly proportional to head bending, and by using this head bending is calculated.

Normally, the nose makes an acute angle with the webcam. The acute angle increases when head moves upward and vice versa. Therefore, the nose length ratio is given as nose length to average length of nose, which also measures the head bending. The mathematical formula for NLR is given as,

$$NLR = \frac{nose\ length\ (p_{28} - p_{25})}{average\ nose\ length}$$

The facial points for nose length ratio and head bending are shown in figure.2.

5.5 Classification:

After obtaining the values of eyes, mouth and nose, the next step is to identify the drowsiness from the SVM frames. The flexible value of threshold is required for the drowsiness. Then, algorithms like SVM are used for the classification of the information.

For obtaining the threshold values for eyes, mouth and nose the initial condition of the driver is assumed to be in normal state. This step is known as setup phase. The EAR for nearly 100 to 300 frames will be recorded, and from this an average of150 to 200 higher values are recognized as strong threshold valuesfor EAR. The values which are high i.e., in which eyes are not closedis considered. If threshold is more than the tested value, then it is identified as eyes closed. For every person, the size of eye will be different; so, this makes the impact to decrease the setup for personto person. For computing MOR values, the threshold value of frames is calculated based on the condition that the mouth is not open. If the threshold is less than the test value i.e., when the mouth is open, the yawning is identified. Using the head bending feature, the angle between access and head can be determined in terms of nose length ratio. NLR values range from 0.8 to 1 in normal condition of headand varies with head bending up and down. The average of the NLR is measured as mean of lengths of nose in the setup assuming that head is not bend. After obtaining the threshold values, it is tested. Then, if at least one of the eyes, mouth and nose identifiers are not satisfied, then drowsiness alert is indicated. In practical, for example out of 75 frames at least 70 frames satisfy drowsiness conditions for one or more features, then the overall system is identified as drowsiness detection and the driver is alerted with the alarm sound.

To overcome the thresholding problem, a single value of thresholdis considered, and this threshold value depend on EAR. Toobtain the average value of EAR, 150 higher EAR values from 300 frames are considered. If the threshold value is more than EAR value, then the driver is at danger. By considering yawn andhead bending, this EAR threshold will be increased, and it isdispersed into more frames. The yawning and head bending frames are combined to get flexible value of threshold. If EAR value is more than obtained threshold value, thenthat condition is treated as drowsiness, and this indicates to alert. In head bending situation, if the head is down, then theframes are considered for a drowsiness alert. Table II shows the calculated parameters.

The machine learning algorithms and the threshold factors are used for the identification of driver's drowsiness, from the values obtained from EAR, MOR, and NLR. Earlier, these features were used for the analysis of classification from feature space to individual. However, here it is used for the principal component analysis [12].

Table II. Calculated parameter values of threshold

EAR from setup phase (average of 150 maximum values out of 300 frames)	0.34
Threshold = EAR - offset	0.34 - .045 = 0.295
At Yawning, (MOR > 0.6)	Threshold = Threshold +0.002 *Max bound exist
At Head Bending, (NLR < 0.7 OR NLR > 1.2)	Threshold = Threshold +0.001 *Max bound exist

After converting the values obtained from threshold, whether the features are significant for classes or not is tested. If three factors give five percent significance, this classification based on Bayesian Classifier and SVM algorithm is used [12].

6. Results And Discussion:

The webcam of the laptop is turned ON and the face of the driveris recorded and captured. The captured video is transferredto the processing block to identify the drowsiness detection. If mishap is detected, then alarm will buzz a sound to alertthe driver. The initial state of driver condition is given in Fig. 3. For this frame the values are given as,

EAR= 0.350, MOR= 0.341, NLR= 1.030

Fig.3 Initial condition of driver

Fig. 4(i) Drowsy identification based on Yawning

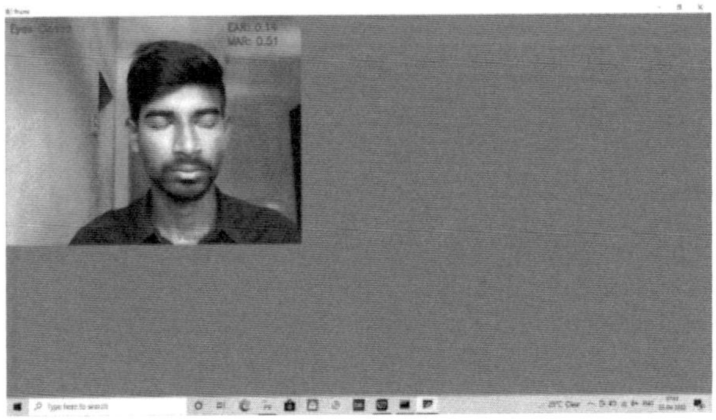

Fig. 4(ii) Drowsy identification based on closed eyes

Fig. 4(iii) Drowsy identification based on bent head

Various drowsiness situations are shown in Fig. 4. Figure 4(i) shows the example of alert for drowsiness due to yawning and Fig. 4(ii) shows the example of drowsiness alert due to eye closing. Fig. 4(iii) shows an example of drowsiness due to head bending. Table III shows the sample values of the parametersfor various states.

Table III: Sample values of distinct parameters & states

State	EAR	MOR	NLR
Normal	0.35	0.34	1.003
Yawning	0.22	0.77	0.76
Eye Closed	0.15	0.419	0.876
Head Bending	0.15	0.577	0.66

The system can also detect the driver's drowsiness along withspectacles as shown in Fig. 5.

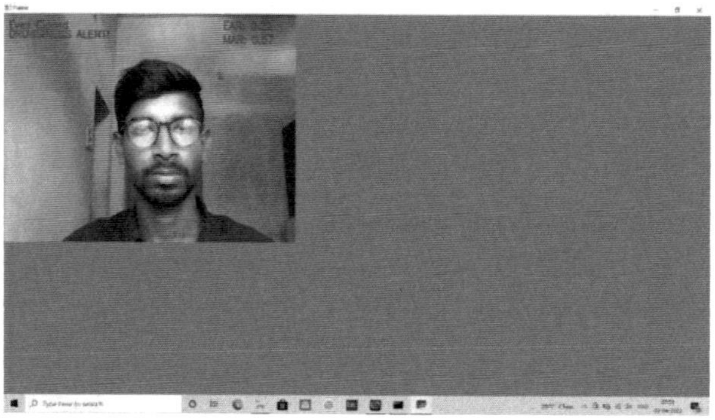

Fig. 5 Identification of eyes with glasses

The algorithms are used for the testing on the INVEDRIFAC dataset [13].

7. Conclusion

This paper has designed a system for the identification of drowsiness by implementing the machine learning algorithms andvisual behavior features like Eye Aspect Ratio, Mouth Opening Ratio and Nose Length Ratio. These features are calculated from the threshold values by recording the video using a webcam on laptop. This flexible technique is used for identifying the threshold value of drowsiness detection. The developed system works perfectly with the given data. Later, the values of threshold are stored and the algorithms of machine learning arecused. Here, algorithms like SVM and Bayesian classifier are used, that performs the perception of SVM which is 0.569. SVM algorithm gives more accuracy and the system developed using SVM algorithm gives ideal output. In future, the proposed model can be implemented in real life as hardware in car and bus to validate the developed system.

8. References

[1]. R. Kumaran, R. Deepak Kumar, M. Karthick Kumar, B. Rahul, 2020, Driver Drowsiness Monitoring System using Visual Behaviour and Machine Learning, INTERNATIONAL JOURNAL OF ENGINEERING RESEARCH & TECHNOLOGY (IJERT) RTICCT – 2020 (Volume 8 – Issue 12),

[2]. A. Sengupta, A. Dasgupta, A. Chaudhuri, A. George, A. Routray and R. Guha, "A Multimodal System for Assessing Alertness Levels Due to Cognitive Loading," in IEEE Transactions on Neural Systems and Rehabilitation Engineering, vol. 25, no. 7, pp. 1037-1046, July 2017, doi: 10.1109/TNSRE.2017.2672080.

[3]. K. T. Chui, K. F. Tsang, H. R. Chi, B. W. K. Ling and C. K. Wu, "An Accurate ECG-Based Transportation Safety Drowsiness Detection Scheme," in IEEE Transactions on Industrial Informatics, vol. 12, no. 4, pp. 1438-1452, Aug. 2016, doi: 10.1109/TII.2016.2573259.

[4]. S. Arefnezhad, A. Eichberger, M. Frühwirth, C. Kaufmann and M. Moser, "Driver Drowsiness Classification Using Data Fusion of Vehicle-based Measures and ECG Signals," 2020 IEEE International Conference on Systems, Man, and Cybernetics (SMC), 2020, pp. 451-456, doi: 10.1109/SMC42975.2020.9282867.

[5]. "IEEE Smart Grid Vision for Vehicular Technology: 2030 and Beyond," in IEEE Smart Grid Vision for Vehicular Technology: 2030 and beyond, vol., no., pp.1-65, 15 Jan. 2014, doi: 10.1109/IEEESTD.2014.6716939.

[6]. B. K. Savaş and Y. Becerikli, "Real Time Driver Fatigue Detection Based on SVM Algorithm," 2018 6th International Conference on Control Engineering & Information Technology (CEIT), 2018, pp. 1-4, doi: 10.1109/CEIT.2018.8751886.

[7]. K. Satish, A. Lalitesh, K. Bhargavi, M. S. Prem and T. Anjali.,"Driver Drowsiness Detection," 2020 International Conference on Communication and Signal Processing (ICCSP), 2020, pp. 0380- 0384, doi: 10.1109/ICCSP48568.2020.9182237.

[8]. S. Menon, S. J., A. S.K., A. P. Nair and S. S., "Driver Face Recognition and Sober Drunk Classification using Thermal Images," 2019 International Conference on Communication and Signal Processing (ICCSP), 2019, pp. 0400-0404, doi:10.1109/ICCSP.2019.8697908.

[9]. H. Kaufman, J. Woods, S. Dravida and A. Tekalp, "Estimation and identification of two-dimensional images," in IEEE Transactions on Automatic Control, vol. 28, no. 7, pp. 745-756, July 1983, doi: 10.1109/TAC.1983.1103311.

[10]. F. An, P. Xu, Z. Xiao and C. Wang, "FPGA-based object detection processor with HOG feature and SVM classifier," 201932nd IEEE International System-on-Chip Conference (SOCC), 2019, pp. 187-190, doi: 10.1109/SOCC46988.2019.1570558044.

[11]. J. Treboux, D. Genoud and R. Ingold, "Decision Tree Ensemble Vs. N.N. Deep Learning: Efficiency Comparison For A Small Image Dataset," 2018 International Workshop on Big Data and Information Security (IWBIS), 2018, pp. 25-30, doi:10.1109/IWBIS.2018.8471704.

[12]. K. J. Chabathula, C. D. Jaidhar and M. A. Ajay Kumara, "Comparative study of Principal Component Analysis based Intrusion Detection approach using machine learning algorithms," 2015 3rd International Conference on Signal Processing, Communication and Networking (ICSCN), 2015, pp. 1-6, doi: 10.1109/ICSCN.2015.7219853.

[13]. R. Kumaran, R. Deepak Kumar, M. Karthick Kumar, B. Rahul, 2020, Driver Drowsiness Monitoring System using Visual Behaviour and Machine Learning, INTERNATIONAL JOURNAL OF ENGINEERING RESEARCH & TECHNOLOGY (IJERT) RTICCT – 2020 (Volume 8 – Issue 12),

An Analysis of Machine Learning-Based Android Malware Detection Approaches

R. Srinivasan[1], Karpagam.S[2], M. Kavitha[3], R. Kavitha[4]

[1 3 4] Professor, Department of Computer Science and Engineering, Vel Tech University, Avadi, Chennai - 600062, Tamil Nadu, India.

[2] Associate Professor, Department of Mathematics, Vel Tech Multi Tech Dr Rangarajan Dr Sakunthala Engineering College, VelTech Rangarajan Dr Sagunthala R&D Institute of Science and Technology, Chennai, Tamil Nādu, India.

srinivasanrajkumar28@gmail.com, kavitha@veltech.edu.in, rkavitha@veltech.edu.in

Abstract: Despite the fact that Android apps are rapidly expanding throughout the mobile ecosystem, Android malware continues to emerge. Malware operations are on the rise, particularly on Android phones, it make up 72.2 percent of all smartphone sales. Credential theft, eavesdropping, and malicious advertising are just some of the ways used by hackers to attack cell phones. Many researchers have looked into Android malware detection from various perspectives and presented hypothesis and methodologies. Machine learning (ML)-based techniques have demonstrated to be effective in identifying these attacks because they can build a classifier from a set of training cases, eliminating the need for explicit signature definition in malware detection.

This paper provided a detailed examination of machine-learning-based Android malware detection approaches. According to present research, machine learning and genetic algorithms are in identifying Android malware, this is a powerful and promising solution. In this quick study of Android apps, we go through the Android system architecture, security mechanisms, and malware categorization.

Keywords: Malcode, Malware, Android Security, Machine learning, Feature extraction

1. Introduction

Due to the convenience and efficiency of many apps, as well as continuous improvements Smartphone usage and accompanying applications are rapidly increasing in today's hardware and software on smart devices. [7]. By 2023, 4.3 billion individuals are expected to own a smartphone. The Android operating system is the most popular among smartphone users (OS). In May 2021, it had a market share of 72.2 percent [8]. With a market share of 26.99 percent, Apple iOS is in second place, followed by Samsung, KaiOS, and other minor vendors with 0.81 percent[8].

The official app store for Android devices is Google Play. As of May 2021, there were around 2.9 million applications available on it. AppBrain classifies about 2.5 million of them as standard applications, whereas 0.4 million are labelled as low-quality apps[9]. Because of its global popularity, Android is a more appealing goal aimed at thieves then is extra vulnerable to malware and viruses. Many strategies for identifying these assaults have been proposed in studies, with one of the most prevalent being machine learning [10]. The taxonomical taxonomy of malware is depicted in Figure 1.

Android users and developers are notorious for making blunders that put them in danger and expose them to malware infection threats. As a result, in addition to virus detection, tactics for discovering these weaknesses are critical. The analysis of Android Application Packages (APKs) to construct an appropriate collection of features, the two basic components of malware detection utilising machine learning are training machine and deep learning (DL) algorithms on the produced characteristics to identify infected APKs. As a result, there is a discussion of the various methodologies for APK analysis, such as static, dynamic, and hybrid analysis.

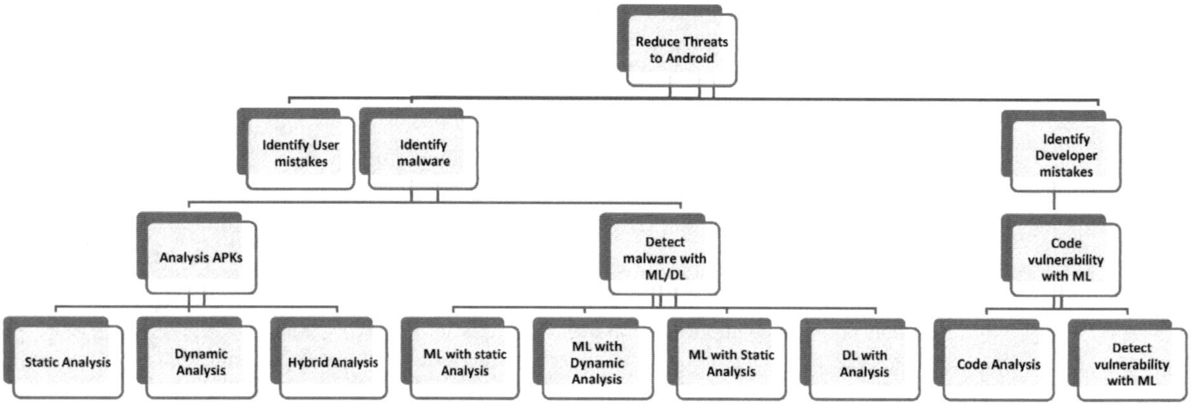

Figure 1. Taxonomy of the malware classification

Android applications are vulnerable to software theft on the open Smartphone market due to the ease with which they may be repackaged. Developers can modify or add to the original programme and then resell it as a new product on the open market. It's possible that the change is malicious. Repackaged malware accounts for 80.6 percent of all malware, illustrating its prevalence and severity, according to researchers. For detecting repackaged malware, there are two approaches: 1) similarity-based detection unique to repackaged malware, and 2) general purpose detection.

Various similarity tests are used by specific repackaged Android app solutions to find truly related apps. The similarity comparison is tied to how applications encode the user's navigation behaviours in View Droid, for example. DNA Droid examines the software dependency trees of users to assess code reuse. MassVet employs UI structures to compare the similarity of apps. Based on opcode sequence properties, Juxtapp and Droid MOSS compare code similarities.

The Android platform has become a key target for malware. Researchers have developed a variety of on-device detection ways to protect against Android malware. These on-device detection methods usually consist of two steps: I acquiring the app's behavioural characteristics from mobile devices, and (ii) sending the acquired features to remote servers for evaluation.

Methods can consistently detect malicious programmes by observing the activities of applications that operate on mobile devices (simply, software). On the other hand, in most cases, infrastructure is a barrier for mobile devices. Malware writers introduce malcode statically and change the control flow to ensure that it runs during repackaging. Repackaged malware is difficult to detect using traditional categorization methods because it behaves similarly to benign malware.

We provide a new approach for identifying repackaged malware on Android that focuses on the software's internal activities. Our method divides an app's code structure into several dependency-based structures, regions, and areas using code heterogeneity analysis (subsets of the code). Each area is defined by its own set of behavioural qualities.

We suggest a solution based on multiple reliance links that identifies security flaws, At the class and method levels, creates and builds graph partitioning code structures, and presents a solution based on numerous dependency linkages.

2. Background

This section covers the architecture and security aspects of Android, as well as potential Android attack vectors. It also contains an overview of the machine learning process and it will help non-ML readers grasp the content of this paper.

2.1 Architecture of Android

The Linux Kernel is used to create Android. Because Linux is open source, it checks the route evidence, offers networking drivers and protocols, virtual memory, device power, and security are all handled by this programme. [12]. The architecture of Android is layered [11].

The Android Runtime is used by Android apps and several system functions (ART). Before the ART, the runtime environment was Dalvik. Dalvik and ART were both created with Android applications in mind[13].

Because Android applications operate on with partial memory, memory and power management are particularly important challenges. As a result, the Android operating system is designed to effectively manage any resource[5]. If an application is not in use at the time, the Android OS, for example, will immediately suspend it in memory. The application life cycle's operating state is referred to as this. It can save power that can be used when the app reopens by doing so. Otherwise, until they are closed, the programmes are inactive [14].

2.2 Security Built-In

Security software is pre-installed on Android cellphones. It's a separate operating system for privileged users. [15]. The Android permission system, as well as the sandboxing method, assist to reduce the program's numerous risks and difficulties. Sandboxing on Android uses unique IDs based on the Linux environment to isolate executing programmes [16]. Apps can only access system resources with permissions granted by the user during installation or reconfiguration. The software will not function if specific permissions are not provided. Upgrades or enhancements to a system increase security and privacy in a variety of ways[16].

3. Related Work

Xu Jiang et al. propose a new method for detecting Android malware that is based on a fine-grained permission system that reflects the differences between malicious and benign applications as machine learning characteristics and, for the first time, provides information about the components' harmful permissions[1].

Tajuddin Manhar Mohammed et al, examine the permissions employed by malicious programmes in depth, as well as the efficacy of permission characteristics used to distinguish between harmful and benign apps[2].

FDP employs static techniques to capture all characteristics and analyse an application in an acceptable amount of time, according to Deqiang Li et al[3].

FDP is more successful at identifying more malware kinds and harmful programmes that declare a small number of damaging rights or permissions that are identical to those claimed by benign applications, according to Amiangshu Bosu et al tests[4].

4. Machine Learning to Detect Android Malware

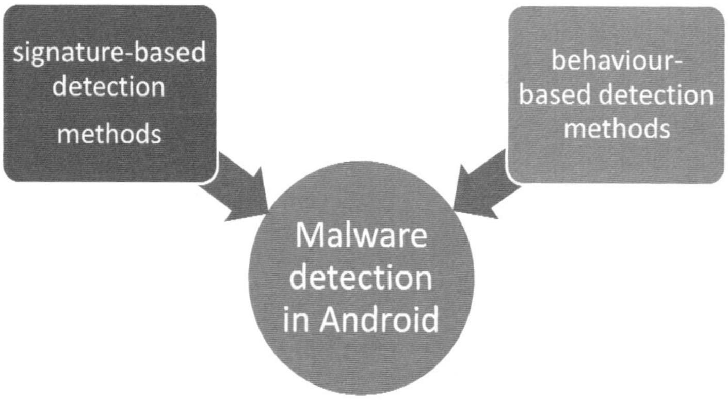

Figure 2. Types of detect Android Malware

Figure 2 shows types of malware to detected in android application and the signature-based detection method is simple, efficient, and produces minimal false positives. The binary code of the programme is compared to signatures in a known malware database. This technique, on the other hand, excludes the identification of unknown malware. As a result, behavior-based/anomaly-based detection is the most often used approach. In this case, machine learning and data science methodologies are widely applied [17-18].

4.1 Static, Dynamic, and Hybrid Analysis

As previously stated, some of the suggested ML algorithms in the literature involve analyzing APKs to extract characteristics. Static, dynamic, and hybrid analysis methods are characterized as three analysis strategies [19–21]. Instead of executing it on a mobile device, static analysis may be conducted by analyzing the bytecode and source code or re-engineered APK. Malware is detected via dynamic analysis, which analyses a programme while it is executing in a simulated or real-world context.

4.1.1 Static analysis with Machine Learning

Many research have been undertaken to identify Android malware using both standard ML-based techniques like Decision Trees (DT) and Support Vector Machines (SVM) as well as innovative DL-based models like Deep Convolutional Neural Network (Deep-CNN) [17] and Generative Adversarial Networks (GENAN) [18].

4.1.2 Dynamic Analysis with Machine Learning

The second sort of analysis is dynamic analysis. Malware may be discovered using machine learning after the application has been launched in a runtime environment using this approach. A network-based technique for detecting malware on Android was described in [23].

A detecting application was created as part of this effort. It is divided into three modules: network tracing, network feature extraction, and detection. Running programmers' network activity were tracked, and network traces were gathered on a regular basis in the traces collecting module. The features extraction module gathered information about the network that the apps used.

The detection module made use of the DT, LR, KNN, Bayes Network, and RF algorithms. The RF algorithm was the most accurate of the four (98.7 percent). This technique, on the other hand, was founded on network analysis. If dangerous programmes employed encrypted data, the malware detection accuracy would be lowered.

4.1.3 Hybrid Analysis with Machine Learning

The third approach that may be employed in ML-based Android malware detection is hybrid analysis. [93] outlined three ways to malware detection: signature-based, anomaly-based, and topic modeling-based tactics.

The behavioral-based technique outperformed the signature-based method. According to the findings of this investigation, the SVM classifier utilizing the hybrid analytic methodology beat the other ML approaches.

According to the research on ML-based techniques to malware detection, 65 percent of studies utilised static analysis, 15% used dynamic analysis, and the other 20% employed a mixed analytic strategy. This is seen in Figure 3. Because of the benefits it delivers, such as the ability to uncover more vulnerabilities, localise issues, and save money, static analysis may be more tempting than dynamic analysis. Table 3 shows the chart representation of ML algorithms commonly used for Android malware detection

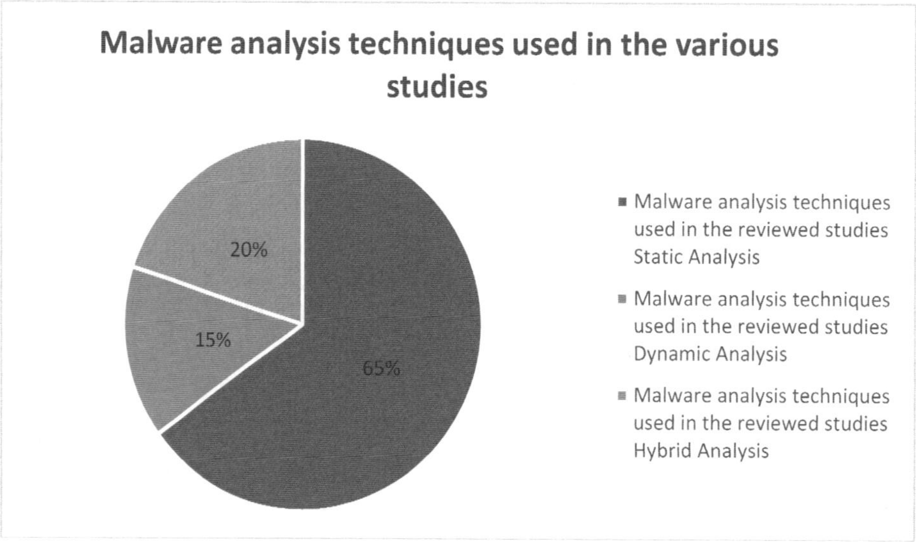

Figure 3. Malware analysis techniques used from various studies.

Table 1. ML algorithms commonly used for Android malware detection

Algorithm	Advantages	Disadvantages
Decision Trees(DT)	Possible with values Simple to comprehend	overfitting problem
k-nearest neighbours	Simple to implement Simple & quick	Sensitive to outliers
convolutional neural network(CNN)	Weight sharing and down sampling help to reduce non-essential factors.	High cost of calculation
Regression Models	Study with statistics	High-dimensional characteristics are difficult to deal with.
SVM	solve nonlinear high-dimensional small-scale issues.	Processing data has a high overhead.

5. Proposed System

The genetic algorithm (GA) is a search-based optimization method based on natural selection and genetic principles. It's a part of the method we recommend. It is widely used to find optimal or near-optimal solutions to tough problems that would otherwise take an eternity to solve. The suggested technique leverages an evolving Genetic Algorithm[5] to build the best-optimized feature subset that may be used to efficiently train machine learning algorithms.

If signature-based approaches fail to detect new types of malware that pose zero-day threats, machine learning-based algorithms are used to identify malware more successfully. Figure 4. Shows architecture Diagram of the proposed method

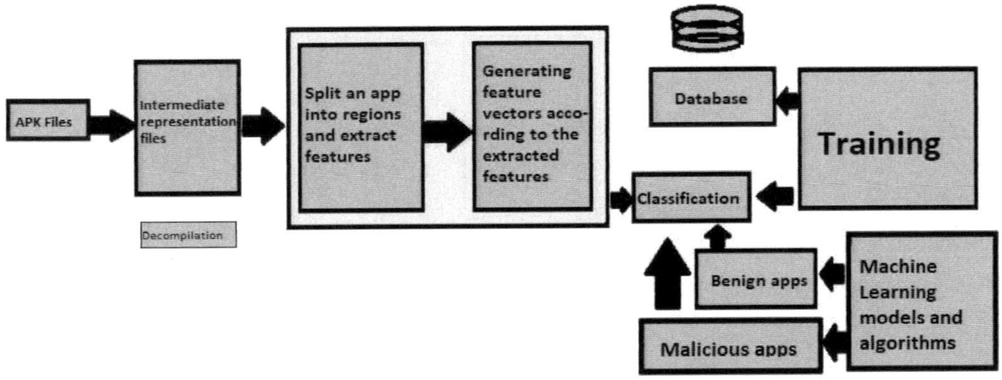

Figure 4. Architecture Diagram of the proposed method

As illustrated in the flowchart above, we'll start our tests at the beginning of each stage. The first section will concentrate on data and its preparation. Then, as a first step, we'll use Data Augmentation to try to enhance accuracy. To enhance capacity, the next stage is to make some modest changes to network infrastructures. After that, we'll look at a number of different loss functions, the majority of which are based on a single picture (and maybe with pair and triple of image). After that, armed with all of the knowledge we've gathered, we'll be able to select the best architecture for our requirements[6]. Figure 5. Show the flow of methodologies used in proposed work

6. Methodologies

Module 1: Data Set preparation

- Executable files (benign of malware file) are dissembled using a dissembler
- Assembly docs are parsed, and the data set is collected

Module 2: Using androguard tool.

To decode the AndroidManifest.xml or resources. arsc, use androguard axml and androguard arsc.

- APK or DEX files are analyzed and loaded
- Get a list of all the activities defined in the AndroidManifest.xml
- Update the data set

Module 3: Classification using Machine Leaning Algorithms

- Tree, SVM, K-Fold, KNN, Genetic Algorithm (GA)
- The steps involved in selecting features with GA:
- Step 1: Fill the method with binary-encoded feature subsets.
- Step 2: Begin the technique by creating a randomly generated starting population set.
- Step 3: Based on the supplied fitness function, assign a fitness score to the genetic algorithm.
- Step 4: Parental Selection: High-fitness chromosomes are given precedence over others in the next

generation of kids.

• Step 5: To generate offspring, use the specified probability of crossover and mutation to run crossover and mutation operations on the selected parents.

• Iteratively repeat Steps 3–5 until convergence is reached and the best chromosome from the population, i.e. the best feature subset, is acquired.

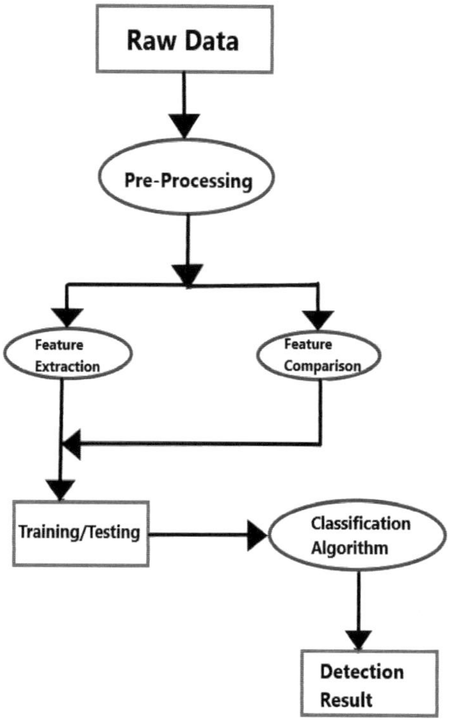

Figure 5. Flow of methodologies used in proposed work

6.1 Input and Output Model

When both the data and the correct answers for each object are provided, supervised learning is applied. There are two stages to supervised learning:

• Fitting a model to available training data and training a model.

• Obtaining predictions by applying the trained model to new samples.

The task:

• We're given a set of things to work with.

• Feature set X is used to represent each object.

• If an object relates to the right answer, it is marked with a Y. X might represent some file content or behaviour attributes, such as file statistics or a list of API calls executed, in the event of malware detection. Malware, innocuous, or a more precise categorization like virus, Trojan-Downloader, or adware might all be labelled Y.

The Android platform has the largest global market share because to its open-source nature and

Google's backing. It has drawn the attention of cyber thieves who exploit it to transmit dangerous malware since it is the most extensively used operating system on the globe. We use an evolutionary genetic strategy to detect Android malware, in which we use selected features to train machine learning classifiers, and then evaluate their ability to detect malware before and after feature selection.

According to the findings, the genetic algorithm produces the most efficient feature subset, reducing the feature dimension to less than half of the original feature set. Machine learning-based classifiers retain a classification accuracy of more than 90-91 percent after feature selection despite operating on a substantially reduced feature dimension, decreasing learning classifier compute complexity.

7. Conclusion:

It is vital to design a framework that can correctly identify malware as the number of threats posed to Android devices rises every day. These threats are transmitted mostly through malicious applications or malware. To produce the most optimised feature subset that may be utilised to train machine learning algorithms in the most efficient manner, the proposed technique employs an evolving Genetic Algorithm. Experiments show that employing, when working with lower dimension feature sets, use Support Vector Machine and Neural Network classifiers, a respectable classification accuracy of more than 90-91 percent may be maintained, decreasing the classifiers' training complexity. More work can be done to improve results by using larger data sets and evaluating the impact of other machine learning algorithms when used in conjunction with the Genetic Algorithm.

References:

[1] Xu Jiang, Baolei Mao, Jun Guan, Xingli Huang, 2020," Android Malware Detection Using Fine-Grained Features", Scientific Programming, vol. 2020, Article ID 5190138.

[2] Tajuddin Manhar Mohammed, Lakshmanan Nataraj, Satish Chikkagoudar, Shivkumar Chandrasekaran, B.S. Manjunath, Jan 2021" Malware Detection Using Frequency Domain-Based Image Visualization and Deep Learning", Hawaii International Conference on System Sciences (HICSS).

[3] Deqiang Li, Qianmu Li, June 2020" Adversarial Deep Ensemble: Evasion Attacks and Defenses for Malware Detection", in IEEE Transactions on Information Forensics and Security.

[4] Amiangshu Bosu, Fang Liu, Danfeng (Daphne) Yao, Gang Wang, April 2017," Collusive Data Leak and More: Large-scale Threat Analysis of Inter-app Communications", in ACM, Pages 71–85.

[5] Fang Liu, Haipeng Cai, Gang Wang, Danfeng Yao, Karim O. Elish, Barbara G. Ryder, December 2017," MR-Droid: A Scalable and Prioritized Analysis of Inter-App Communication Risks", in IEEE Security and Privacy Workshops (SPW).

[6] Muhammad Furqan Rafique, Muhammad Ali, Aqsa Saeed Qureshi, Asifullah Khan, Anwar Majid Mirza, October 2019," Malware Classification using Deep Learning based Feature Extraction and Wrapper based Feature Selection Technique", in arxiv.

[7]. Number of Mobile Phone UsersWorldwide from 2016 to 2023 (In Billions). Available online: https://www.statista.com/statistics/ 330695/number-of-smartphone-users-worldwide/ (accessed on 19 May 2021).

[8]. Mobile Operating System Market Share Worldwide. Available online: https://gs.statcounter.com/os-market-share/mobile/ worldwide/ (accessed on 19 May 2021).

[9]. Number of Android Applications on the Google Play Store. Available online: https://www.appbrain.com/stats/number-ofandroid- apps/ (accessed on 19 May 2021).

[10]. Gibert, D.; Mateu, C.; Planes, J. The rise of machine learning for detection and classification of

malware: Research developments, trends and challenges. J. Netw. Comput. Appl. 2020, 153, 102526. [CrossRef]

[11].PlatformArchitecture.Availableonline:https://developer.android.com/guide/platform(accessed on 19 May 2021).

[12]. Khan, J.; Shahzad, S. Android Architecture and Related Security Risks. Asian J. Technol. Manag. Res. [ISSN: 2249–0892] 2015, 5, 14–18. Available online: http://www.ajtmr.com/papers/Vol5Issue2/Vol5Iss2_P4.pdf(accessed on 19 May 2021).

[13].Cai, H.; Ryder, B.G. Understanding Android application programming and security: A dynamic study. In Proceedings of the 2017 IEEE International Conference on Software Maintenance and Evolution (ICSME), Shanghai, China, 17–22 September 2017; pp. 364–375. [CrossRef]

[14]. Liu, K.; Xu, S.; Xu, G.; Zhang, M.; Sun, D.; Liu, H. A Review of Android Malware Detection Approaches Based on Machine Learning. IEEE Access 2020, 8, 124579–124607. [CrossRef]

[15].Gilski, P.; Stefanski, J. Android os: A review. Tem J. 2015, 4, 116. Available online:https://www.temjournal.com/content/41/14/ temjournal4114.pdf(accessed on 19 May 2021).

[16]. Privacy in Android 11 | Android Developers. Available online: https://developer.android.com/about/versions/11/privacy (accessed on 19 May 2021).

[17]. Chen, T.; Mao, Q.; Yang, Y.; Lv, M.; Zhu, J. TinyDroid: A lightweight and efficient model for Android malware detection and classification. Mob. Inf. Syst. 2018, 2018. [CrossRef]

[18]. Nisa, M.; Shah, J.H.; Kanwal, S.; Raza, M.; Khan, M.A.; Damaševi˘cius, R.; Blažauskas, T. Hybrid malware classification method using segmentation-based fractal texture analysis and deep convolution neural network features. Appl. Sci. 2020, 10, 4966. [CrossRef]

[19]. Alqahtani, E.J.; Zagrouba, R.; Almuhaideb, A. A Survey on Android Malware Detection Techniques Using Machine Learning Algorithms. In Proceedings of the 2019 Sixth International Conference on Software Defined Systems (SDS), Rome, Italy, 10–13 June 2019; pp. 110–117. [CrossRef]

[20]. Lopes, J.; Serrão, C.; Nunes, L.; Almeida, A.; Oliveira, J. Overview of machine learning methods for Android malware identification. In Proceedings of the 2019 7th International Symposium on Digital Forensics and Security (ISDFS), Barcelos, Portugal, 10–12 June 2019; pp. 1–6. [CrossRef]

[21]. Choudhary, M.; Kishore, B. HAAMD: Hybrid analysis for Android malware detection. In Proceedings of the 2018 International Conference on Computer Communication and Informatics (ICCCI), Coimbatore, India, 4–6 January 2018; pp. 1–4. [CrossRef]

[22]. Garg, S.; Baliyan, N. Android Security Assessment: A Review, Taxonomy and Research Gap Study. Comput. Secur. 2020, 100, 102087. [CrossRef]

[23]. Garg, S.; Peddoju, S.K.; Sarje, A.K. Network-based detection of Android malicious apps. Int. J. Inf. Secur. 2017, 16, 385–400. [CrossRef]

[24]. Janaka Senanayake 1,* , Harsha Kalutarage 1 and Mhd Omar Al-Kadri, "Android Mobile Malware Detection Using Machine Learning: A Systematic Review", 10, Electronics 2021.

[25]. Xu, K.; Li, Y.; Deng, R.H.; Chen, K. Deeprefiner: Multi-layer android malware detection system applying deep neural networks. In Proceedings of the 2018 IEEE European Symposium on Security and Privacy (EuroS&P), London, UK, 24–26 April 2018; pp. 473–487. [CrossRef]

AUTHOR INDEX